Advances in

QUANTUM CHEMISTRY

VOLUME **55**

Advances in
QUANTUM CHEMISTRY

APPLICATIONS OF THEORETICAL METHODS TO ATMOSPHERIC SCIENCE

VOLUME 55

Editors

JOHN R. SABIN
Quantum Theory Project
University of Florida
Gainesville, Florida

ERKKI BRÄNDAS
Department of Quantum Chemistry
Uppsala University
Uppsala, Sweden

Founding Editor

PER-OLOV LÖWDIN
1916–2000

Guest Editors

MICHAEL E. GOODSITE
Department of Physics and Chemistry
University of Southern Denmark
Odense, Denmark

MATTHEW S. JOHNSON
Copenhagen Centre for Atmospheric Research
Department of Chemistry, University of Copenhagen
Denmark

ELSEVIER

Amsterdam • Boston • Heidelberg • London
New York • Oxford • Paris • San Diego
San Francisco • Singapore • Sydney • Tokyo
Academic Press is an imprint of Elsevier

Academic Press is an imprint of Elsevier
84 Theobald's Road, London WC1X 8RR, UK
Radarweg 29, PO Box 211, 1000 AE Amsterdam, The Netherlands
Linacre House, Jordan Hill, Oxford OX2 8DP, UK
30 Corporate Drive, Suite 400, Burlington, MA 01803, USA
525 B Street, Suite 1900, San Diego, CA 92101-4495, USA

First edition 2008

ISBN: 978-0-12-374335-0
ISSN: 0065-3276

For information on all Academic Press publications
visit our website at books.elsevier.com

Printed and bound in USA

08 09 10 11 12 10 9 8 7 6 5 4 3 2 1

CONTENTS

Contributors xi

1. Applications of Theoretical Methods to Atmospheric Science **1**
Matthew S. Johnson and Michael E. Goodsite

Acknowledgements 3
References 4

2. Mass-Independent Oxygen Isotope Fractionation in Selected Systems. Mechanistic Considerations **5**
R.A. Marcus

1. Introduction 6
2. The MIF in Ozone Formation 10
3. Quantum Dynamical Computations 12
4. Individually Studied Ratios of Isotopomeric Reaction Rate Constants 13
5. Rate Constant Ratios and Enrichments for Other Reactions 15
6. Oxygen Isotopic Fractionation for $CO + OH \rightarrow CO_2 + H$ 16

Acknowledgements 18
References 18

3. An Important Well Studied Atmospheric Reaction, $O(^1D) + H_2$ **21**
João Brandão, Carolina M.A. Rio and Wenli Wang

1. Introduction 22
2. Characterization of the $O(^1D) + H_2$ Reaction 23
3. Potential Energy Surfaces 25
4. Dynamical Studies 26
5. Results and Comparison with Experiment 28
6. Final Remarks and Conclusions 39

Acknowledgements 39
References 39

4. Gaseous Elemental Mercury in the Ambient Atmosphere: Review of the Application of Theoretical Calculations and Experimental Studies for Determination of Reaction Coefficients and Mechanisms with Halogens and Other Reactants **43**
Parisa A. Ariya, Henrik Skov, Mette M.-L. Grage and Michael Evan Goodsite

1. Introduction 44
2. Kinetic and Product Experiments 45

3. Theoretical Evaluation of Kinetic Data 50
4. Perspectives 53
Acknowledgements 53
References 54

5. Photolysis of Long-Lived Predissociative Molecules as a Source of Mass-Independent Isotope Fractionation: The Example of SO_2 57
James R. Lyons

1. Introduction 58
2. Absorption Spectra for SO_2 Isotopologues 59
3. Photolysis of SO_2 Isotopologues in a Low O_2 Atmosphere 64
4. Photolysis of SO_2 in the Modern Atmosphere 68
5. Improvements to Spectra and Additional Sources of S-MIF 70
6. Conclusions and Broader Implications 73
Acknowledgements 73
References 74

6. A New Model of Low Resolution Absorption Cross Section 75
R. Jost

1. Introduction and Motivations 75
2. Improved Model of Low Resolution Absorption Cross Section (XS) for Diatomic Molecules 78
3. Quantum Correction to the Low Resolution Absorption Cross Section of Diatomic Molecules 80
4. The Absorption Cross Section of Cl_2 Molecule 84
5. A 3D Version of the Model and Its Application to Triatomic Molecules 89
6. Conclusions and Perspectives 96
Acknowledgements 97
References 100

7. Isotope Effects in Photodissociation: Chemical Reaction Dynamics and Implications for Atmospheres 101
Solvejg Jørgensen, Mette M.-L. Grage, Gunnar Nyman and Matthew S. Johnson

1. Introduction 102
2. Electronic Structure Calculations 103
3. Construction of the Time-Independent Hamiltonian Operator 108
4. Time-Independent Methods 109
5. Time-Dependent Methods 111
6. Examples of Photodissociation 115
7. Perspective 128
Acknowledgements 128
References 129

8. Atmospheric Photolysis of Sulfuric Acid 137
Henrik G. Kjaergaard, Joseph R. Lane, Anna L. Garden, Daniel P. Schofield, Timothy W. Robinson and Michael J. Mills

1. Introduction 138
2. Vibrational Transitions 141

3. Electronic Transitions 149
4. Atmospheric Simulations 153
5. Conclusion 155
Acknowledgements 156
References 156

9. Computational Studies of the Thermochemistry of the Atmospheric Iodine Reservoirs HOI and $IONO_2$ 159
Paul Marshall

1. Introduction 160
2. Methodology and Results 161
3. Discussion 165
4. Conclusions 173
Acknowledgements 173
References 174

10. Theoretical Investigation of Atmospheric Oxidation of Biogenic Hydrocarbons: A Critical Review 177
Jun Zhao and Renyi Zhang

1. Introduction 177
2. Theoretical Approaches in Atmospheric Hydrocarbon Oxidation Research 178
3. Theoretical Investigation of Biogenic Hydrocarbon Oxidation 183
4. Conclusions and Future Research 207
Acknowledgements 209
References 209

11. Computational Study of the Reaction of *n*-Bromopropane with OH Radicals and Cl Atoms 215
Claudette M. Rosado-Reyes, Mónica Martínez-Avilés and Joseph S. Francisco

1. Introduction 216
2. Computational Methods 219
3. Results and Discussion 220
4. Atmospheric Implications 241
5. Conclusion 242
Acknowledgements 243
References 243

12. Atmospheric Reactions of Oxygenated Volatile Organic Compounds + OH Radicals: Role of Hydrogen-Bonded Intermediates and Transition States 245
Annia Galano and J. Raúl Alvarez-Idaboy

1. Introduction 246
2. Kinetics 247
3. Energies 251
4. Aliphatic Alcohols 252
5. Aldehydes 256
6. Ketones 258

7. Carboxylic Acids 264
8. Multifunctional Oxygenated Volatile Organic Compounds 266
9. Concluding Remarks 268
References 270

13. Theoretical and Experimental Studies of the Gas-Phase Cl-Atom Initiated Reactions of Benzene and Toluene **275**
A. Ryzhkov, P.A. Ariya, F. Raofie, H. Niki and G.W. Harris

1. Introduction 276
2. Methods 277
3. Results and Discussions 279
4. Conclusions 292
References 294

14. Tropospheric Chemistry of Aromatic Compounds Emitted from Anthropogenic Sources **297**
Jean M. Andino and Annik Vivier-Bunge

1. Introduction 298
2. Reactions 298
3. Summary: Areas for Future Work 309
References 309

15. Elementary Processes in Atmospheric Chemistry: Quantum Studies of Intermolecular Dimer Formation and Intramolecular Dynamics **311**
Glauciete S. Maciel, David Cappelletti, Gaia Grossi, Fernando Pirani and Vincenzo Aquilanti

1. Introduction 312
2. Major Atmospheric Components and Their Dimers 313
3. Aspects of Interactions Involving Minor Atmospheric Components: H_2O and H_2S 319
4. Peroxides and Persulfides 324
5. Concluding Remarks and Perspectives 328
Acknowledgements 328
References 328

16. The Study of Dynamically Averaged Vibrational Spectroscopy of Atmospherically Relevant Clusters Using *Ab Initio* Molecular Dynamics in Conjunction with Quantum Wavepackets **333**
Srinivasan S. Iyengar, Xiaohu Li and Isaiah Sumner

1. Introduction 334
2. Computational Methods for Soft Vibrational Mode Clusters 335
3. Cluster Dynamics Simulations using ADMP and QWAIMD 342
4. Conclusions 348
Acknowledgements 348
References 348

17. From Molecules to Droplets **355**
Allan Gross, Ole John Nielsen and Kurt V. Mikkelsen

1. Introduction 356
2. Energy Functional and Hamiltonians 358
3. The Introduction of the Multiconfigurational Self-Consistent Wave Function 366
4. Derivation of Response Equations for Quantum–Classical Systems 369
5. Brief Overview of Results 381
6. Conclusion 382
Acknowledgement 383
References 383

18. Theoretical Studies of the Dissociation of Sulfuric Acid and Nitric Acid at Model Aqueous Surfaces **387**
Roberto Bianco, Shuzhi Wang and James T. Hynes

1. Introduction 388
2. Methodology 389
3. First Acid Dissociation of Sulfuric Acid 393
4. Acid Dissociation of Nitric Acid 397
5. Concluding Remarks 401
Acknowledgements 402
References 402

19. Investigating Atmospheric Sulfuric Acid–Water–Ammonia Particle Formation Using Quantum Chemistry **407**
Theo Kurtén and Hanna Vehkamäki

1. Introduction 407
2. Theoretical Methods for Free Energy Calculations 411
3. Applications of Quantum Chemistry to Atmospheric Nucleation Phenomena 415
4. Challenges 423
5. Conclusions 424
Acknowledgements 425
References 425

20. The Impact of Molecular Interactions on Atmospheric Aerosol Radiative Forcing **429**
Shawn M. Kathmann, Gregory K. Schenter and Bruce C. Garrett

1. Introduction 429
2. Atmospheric Physics 434
3. Background on Nucleation Theories 438
4. Dynamical Nucleation Theory 440
5. Why Accurate Chemical Physics is Important to Nucleation 444
6. Summary and Future Directions 445
Acknowledgements 446
References 446

21. Computational Quantum Chemistry: A New Approach to Atmospheric Nucleation **449**
Alexey B. Nadykto, Anas Al Natsheh, Fangqun Yu, Kurt V. Mikkelsen and Jason Herb

1. Introduction 450
2. Nucleation Theory 454
3. Why Should We Apply the Quantum Theory to Atmospheric Problems? 455
4. Quantum Methods 456
5. Application of Quantum Methods to Atmospheric Species 457
6. Concluding Remarks 475

Acknowledgements 475
References 476

Subject Index **479**

CONTRIBUTORS

Numbers in parentheses indicate the pages where the authors' contributions can be found.

Anas Al Natsheh (449)
Kajaani University of Applied Sciences, Kuntokatu 5, 87101 KAJAANI, Finland

J. Raúl Alvarez-Idaboy (245)
Facultad de Química, Universidad Nacional Autónoma de México, Ciudad Universitaria, 04510, México D.F., México

Jean M. Andino (297)
Arizona State University, Department of Civil & Environmental Engineering and Department of Chemical Engineering, PO Box 875306, Tempe, AZ 85287-5306, USA

Vincenzo Aquilanti (311)
Dipartimento di Chimica, Università di Perugia, 06123 Perugia, Italy

Parisa A. Ariya (43)
Departments of Chemistry and Atmospheric and Oceanic Sciences, McGill University, 801 Sherbrooke St. W., Montreal, PQ, Canada, H3A 2K6

P.A. Ariya (275)
Department of Chemistry and Department Atmospheric and Oceanic Sciences, McGill University, 801 Sherbrooke St. W., Montreal, Quebec, Canada, H3A 2K6

Roberto Bianco (387)
Department of Chemistry and Biochemistry, University of Colorado, Boulder, CO 80309-0215, USA

João Brandão (21)
Dept. Química, Bioquímica e Farmácia, Universidade do Algarve, 8005-139 Faro, Portugal

David Cappelletti (311)
Dipartimento di Ingegneria Civile ed Ambientale, Universitá di Perugia, 06100 Perugia, Italy

Joseph S. Francisco (215)
Department of Chemistry and Department of Earth and Atmospheric Sciences, Purdue University, West Lafayette, IN 47907, USA

Annia Galano (245)
Instituto Mexicano del Petróleo, Eje Central Lázaro Cárdenas 152, 07730, México D.F., México and Departamento de Química, Universidad Autónoma Metropolitana-Iztapalapa, San Rafael Atlixco 186, Col. Vicentina, Iztapalapa, C.P. 09340, México D.F., México

Anna L. Garden (137)
Department of Chemistry, University of Otago, P.O. Box 56, Dunedin, New Zealand

Bruce C. Garrett (429)
Chemical & Materials Sciences Division, Pacific Northwest National Laboratory, Richland, WA 99352, USA

Michael E. Goodsite (4)
Department of Physics and Chemistry, University of Southern Denmark, Campusvej 55, DK-5230 Odense, Denmark and Department of Atmospheric Environment, National Environmental Research Institute, Aarhus University, Frederiksborgvej 399, P.O. Box 358, DK-4000 Roskilde, Denmark

Michael Evan Goodsite (43)
National Environmental Research Institute, Aarhus University, Frederiksborgvej 99, Roskilde, Denmark and Department of Physics and Chemistry, University of Southern Denmark, Campusvej 55, DK 5230, Odense M, Denmark

Mette M.-L. Grage (43)
Department of Chemistry, University of Gothenburg, S-412 96 Göteborg, Sweden and University of Copenhagen, Universitetsparken 5, DK 2100, Copenhagen Ø, Denmark

Mette M.-L. Grage (101)
Copenhagen Center of Atmospheric Research, Department of Chemistry, University of Copenhagen, Universitetsparken 5, DK-2100 Copenhagen Ø, Denmark and Department of Chemistry, Göteborg University, S-412 96 Göteborg, Sweden

Allan Gross (355)
Research Department, Danish Meteorological Institute, Lyngbyvej 100, DK-2100 Copenhagen Ø, Denmark

Gaia Grossi (311)
Dipartimento di Chimica, Università di Perugia, 06123 Perugia, Italy

G.W. Harris (275)
Centre for Atmospheric Chemistry, York University, 4700 Keele Street, Toronto, Ontario, Canada, M3J 1P3

Jason Herb (449)
Atmospheric Sciences Research Center, State University of New York at Albany, 251 Fuller Rd., Albany 12203, NY, USA

James T. Hynes (387)
Department of Chemistry and Biochemistry, University of Colorado, Boulder, CO 80309-0215, USA and Département de Chimie, CNRS UMR 8640 PASTEUR, Ecole Normale Supérieure, 24 rue Lhomond, Paris 75231, France

Srinivasan S. Iyengar (333)
Department of Chemistry and Department of Physics, Indiana University, 800 E. Kirkwood Ave, Bloomington, IN 47405, USA

Fangqun Yu (449)
Atmospheric Sciences Research Center, State University of New York at Albany, 251 Fuller Rd., Albany 12203, NY, USA

Matthew S. Johnson (4, 135)
Copenhagen Center for Atmospheric Research, Department of Chemistry, University of Copenhagen, Universitetsparken 5, DK-2100 Copenhagen Ø, Denmark

Solvejg Jørgensen (101)
Copenhagen Center of Atmospheric Research, Department of Chemistry, University of Copenhagen, Universitetsparken 5, DK-2100 Copenhagen Ø, Denmark

R. Jost (75)
Laboratoire de Spectrométrie Physique, Université Joseph Fourier de Grenoble, B.P. 87, F-38042 Saint Martin d'Hères Cedex, France

Shawn M. Kathmann (429)
Chemical & Materials Sciences Division, Pacific Northwest National Laboratory, Richland, WA 99352, USA

Henrik G. Kjaergaard (137)
Department of Chemistry, University of Otago, P.O. Box 56, Dunedin, New Zealand and Lundbeck Foundation Center for Theoretical Chemistry, Department of Chemistry, Aarhus University, DK-8000 Aarhus, Denmark

Theo Kurtén (407)
Department of Physical Sciences, P.O. Box 64, 00014 University of Helsinki, Finland

Joseph R. Lane (137)
Department of Chemistry, University of Otago, P.O. Box 56, Dunedin, New Zealand

Xiaohu Li (333)
Department of Chemistry and Department of Physics, Indiana University, 800 E. Kirkwood Ave, Bloomington, IN 47405, USA

James R. Lyons (57)
Institute of Geophysics and Planetary Physics, and Department of Earth and Space Sciences, University of California, Los Angeles, CA 90095, USA

Glauciete S. Maciel (311)
Dipartimento di Chimica, Università di Perugia, 06123 Perugia, Italy

R.A. Marcus (5)
Noyes Laboratory of Chemical Physics, MC 127-72, California Institute of Technology, Pasadena, CA 91125-0072, USA

Paul Marshall (159)
Center for Advanced Scientific Computing and Modeling, Department of Chemistry, University of North Texas, P.O. Box 305070, Denton, Texas 76203-5070, USA

Mónica Martínez-Avilés (215)
Department of Chemistry and Department of Earth and Atmospheric Sciences, Purdue University, West Lafayette, IN 47907, USA

Kurt V. Mikkelsen (355, 449)
Department of Chemistry, University of Copenhagen, Universitetsparken 5, DK-2100 Copenhagen Ø, Denmark

Michael J. Mills (137)
LASP/PAOS, University of Colorado, Boulder, Colorado, 80309, USA

Alexey B. Nadykto (449)
Atmospheric Sciences Research Center, State University of New York at Albany, 251 Fuller Rd., Albany 12203, NY, USA

Ole John Nielsen (355)
Department of Chemistry, University of Copenhagen, Universitetsparken 5, DK-2100 Copenhagen Ø, Denmark

H. Niki (275)
Centre for Atmospheric Chemistry, York University, 4700 Keele Street, Toronto, Ontario, Canada, M3J 1P3

Gunnar Nyman (101)
Department of Chemistry, University of Gothenburg, S-412 96 Göteborg, Sweden

Fernando Pirani (311)
Dipartimento di Chimica, Università di Perugia, 06123 Perugia, Italy

F. Raofie (275)
Department of Chemistry and Department Atmospheric and Oceanic Sciences, McGill University, 801 Sherbrooke St. W., Montreal, Quebec, Canada, H3A 2K6

Carolina M.A. Rio (21)
Dept. Química, Bioquímica e Farmácia, Universidade do Algarve, 8005-139 Faro, Portugal

A. Ryzhkov (275)
Department of Chemistry and Department Atmospheric and Oceanic Sciences, McGill University, 801 Sherbrooke St. W., Montreal, Quebec, Canada, H3A 2K6

Timothy W. Robinson (137)
Department of Chemistry, University of Otago, P.O. Box 56, Dunedin, New Zealand

Claudette M. Rosado-Reyes (215)
Department of Chemistry and Department of Earth and Atmospheric Sciences, Purdue University, West Lafayette, IN 47907, USA

Gregory K. Schenter (429)
Chemical & Materials Sciences Division, Pacific Northwest National Laboratory, Richland, WA 99352, USA

Daniel P. Schofield (137)
Department of Chemistry, University of Otago, P.O. Box 56, Dunedin, New Zealand

Henrik Skov (43)
National Environmental Research Institute, Aarhus University, Frederiksborgvej 99, Roskilde, Denmark

Isaiah Sumner (333)
Department of Chemistry and Department of Physics, Indiana University, 800 E. Kirkwood Ave, Bloomington, IN 47405, USA

Hanna Vehkamäki (407)
Department of Physical Sciences, P.O. Box 64, 00014 University of Helsinki, Finland

Annik Vivier-Bunge (297)
Área de Química Cuántica, Departamento de Química, Universidad Autónoma Metropolitana, Ixtapalapa, C.P. 09340, Mexico D.F., Mexico

Shuzhi Wang (387)
Department of Chemistry and Biochemistry, University of Colorado, Boulder, CO 80309-0215, USA

Wenli Wang (21)
Dept. Química, Bioquímica e Farmácia, Universidade do Algarve, 8005-139 Faro, Portugal

Renyi Zhang (177)
Department of Atmospheric Sciences and Department of Chemistry, Texas A&M University, College Station, TX 77843, USA

Jun Zhao (177)
Department of Atmospheric Sciences and Department of Chemistry, Texas A&M University, College Station, TX 77843, USA

CHAPTER 1

Applications of Theoretical Methods to Atmospheric Science

Matthew S. Johnson* and **Michael E. Goodsite**,*****

Theoretical chemistry involves explaining chemical phenomenon using natural laws. The primary tool of theoretical chemistry is quantum chemistry, and the field may be divided into electronic structure calculations, reaction dynamics and statistical mechanics. These three play a role in addressing an issue of primary concern: understanding photochemical reaction rates at the various conditions found in the atmosphere. Atmospheric science includes both atmospheric chemistry and atmospheric physics, meteorology, climatology and the study of extraterrestrial atmospheres.

The chemical side of atmospheric science has grown considerably in the past generation because it is now recognized that chemistry is at the heart of a wide range of atmospheric phenomenon including acid rain, ozone depletion, air pollution, the atmospheric transport, conversion and deposition of pollutants such as mercury and perfluorinated species, greenhouse gas budgets, determining the impact of biofuels and biogenic emissions, and cloud formation. This list is incomplete, but the point is that atmospheric chemistry provides insight into issues that are of central importance to modern society. At the same time theoretical techniques have revolutionized the study of chemistry. Due to the parallel development of theoretical methods and computing power, quantum chemistry has become fast and flexible, often yielding detailed insight into chemical processes that cannot be obtained in the laboratory.

One may debate the relative importance of theory and experiment. The late Gert Due Billing [1] said that if the theory and the experiment disagree, check

* Copenhagen Center for Atmospheric Research, Department of Chemistry, University of Copenhagen, Universitetsparken 5, DK-2100 Copenhagen Ø, Denmark
** Department of Physics and Chemistry, University of Southern Denmark, Campusvej 55, DK-5230 Odense, Denmark
*** Department of Atmospheric Environment, National Environmental Research Institute, Aarhus University, Frederiksborgvej 399, P.O. Box 358, DK-4000 Roskilde, Denmark

Advances in Quantum Chemistry, 55
ISSN 0065-3276, DOI: 10.1016/S0065-3276(07)00201-8

the experiment, and if they agree, check the theory. Atmospheric science offers a third alternative: check the atmosphere. The atmosphere is rich hunting ground for theorists looking for interesting systems to study. The atmosphere is an engine for generating problems that matter, examples including ozone depletion, air pollution and the nucleation of new particles. In addition the proliferation of ground-based monitoring stations, balloons and satellites has provided a large database of measurements that can be used to inspire and anchor theoretical results.

The earliest report we know of combining quantum chemistry with the atmosphere deals with the excitation of nitrogen and oxygen molecules as the primary event in the photo-excitation of the atmosphere. Krauss et al. reported theoretical calculations of the electronic properties of the molecules required for the prediction of optical emissions from normal and disturbed atmospheres, including the analysis of the electronic structure of excited states of nitrogen [2].

In 1984 Krauss and Stevens described tests and applications of the effective potential method used to gain knowledge of the electronic structure of the molecules in order to analyze the accuracy of the experimentally deduced dissociation energies of refractory metal salts [3]. They used the development of *ab initio* theoretical methods for the calculation of potential energy surfaces, which further allowed the direct computation of certain rate constants. Transition state theory was also utilized for this computation of some rate constants. However, as discussed by Krauss and Stevens, as of the mid 1980's computational techniques were not yet readily applied to atmospheric science. Computing power and theoretical methods since these seminal reports have been greatly advanced.

The anomalous enrichment of ^{17}O and ^{18}O in stratospheric ozone provides a good example of the interplay of laboratory experiments, field studies and theory. The effect was first observed in the Thiemens laboratory in 1983 [4] and then in the stratosphere by Mauersberger [5]. The first truly successful explanation of the underlying mechanism did not appear for nearly two decades in a series of articles by Rudy Marcus and coworkers [6–9], a discussion which is extended in this issue. The unique distribution of oxygen isotopes in ozone has proved to be a useful tracer for diverse atmospheric phenomena, including the exposure of CO_2 to excited oxygen atoms in the stratosphere [10], the productivity of the biosphere [11] and the origin of nitrate found in polar ice [12].

Models of atmospheric chemistry have grown with the abilities of computers and the demands of simulating climate change. One example is the simulation of isotopic fractionation in the photolysis of nitrous oxide isotopologues using time-dependent wavepacket propagation [13]. The isotope-dependent absorption cross sections were used in a three dimensional global circulation model to calculate the isotopic fractionation occurring in stratospheric photolysis. The back flux of isotopically enriched N_2O from the stratosphere to the troposphere could explain the observed distribution of N_2O in the troposphere, reducing uncertainties in the budget of this key greenhouse gas [14]. The study was unique in proceeding directly from wavepacket propagation simulations to atmospheric modeling, the results of which could be directly and favorably compared with data from field studies of the distributions of the isotopomers and isotopologues of N_2O. The theory and the atmosphere agreed, necessitating laboratory experiments which largely confirmed the theoretical studies [15]. Time-independent models have also been applied to the problem [16].

As writing was completed for this Special Issue in 2007, the articles testify to the growth and vitality of the field due to refinements of modeling and theories, advancements in computational science, and collaboration across scientific disciplines.

The roles and opportunities for the theoretical chemist as part of an atmospheric science investigative team have become both more defined and diverse. Since Krauss and Stevens [3], many of the topics they raised have been used, and continue to need to be used to advance our understanding of the chemistry of the atmosphere. For example, theoretical methods are used to predict and verify theories of the mechanistic pathways for the photooxidation of mercury (Ariya et al., this edition). The paper shows how heats of reaction are calculated by various methods, results which are then used to rule out reaction schemes.

Quantum chemistry provides data that improves understanding of chemical kinetics. The data is further used as input for parameterizing transport and deposition models or chemical reaction schemes in models of various other atmospheric processes. As documented in many of the articles in this special edition, theoretical techniques are tested through comparison to laboratory measurements and atmospheric observations, and then further applied towards predicting mechanisms and reaction rates which are currently unknown.

One of the key uncertainties in climate models is determining the radiative impact of clouds, and there is particular interest in understanding the chemistry of clouds: particle nucleation, particle growth and heterogeneous reactivity. We have organized the issue, roughly, according to the size of the systems considered, starting with three-atom systems such as ozone and $O\ (^1D) + H_2$ and proceeding to particle nucleation and the consideration of molecules within droplets.

It is our intention that this special edition serve as a useful catalyst and inspiration, to give the readers an idea of the state of the art and its utility in applications of theoretical methods to atmospheric science and we encourage further work in the areas specified by the authors in their contributions. All promise to have a global impact on science and the environment, and perhaps peace [17].

ACKNOWLEDGEMENTS

We are overwhelmed by the variety and quality of the articles that have come forward for this issue and are grateful to the scientific community for their efforts. This special edition was the result of the entire community coming together to write and/or review work covering a broad range of current topics. Through several open calls to the community as well as direct invitations, the editors hope that everyone who wanted to contribute was afforded the opportunity to do so. The editors thank the authors and reviewers for their hard work, as well as the editorial staff and our families and work places in supporting the production of this Special Issue.

REFERENCES

[1] M. Baer, C. Coletti, G.C. Schatz, S. Toxvaerd, L. Wang, *Gert D. Billing Memorial Issue, J. Phys. Chem. A* **108** (41) (2004).

[2] M. Krauss, F.H. Mies, D. Neumann, P.S. Julienne, *Application of Quantum Chemistry to Atmospheric Chemistry, National Bureau of Standards*, Physical Chemistry Division, Washington, DC, 1976.

[3] M. Krauss, W.J. Stevens, *Application of Quantum Chemistry to Atmospheric Chemistry. National Bureau of Standards*, Quantum Chemistry Group, Gaithersburg, MD, 1984.

[4] M.H. Thiemens, J.E. Heidenreich, The mass-independent fractionation of oxygen—A novel isotope effect and its possible cosmochemical implications, *Science* **219** (1983) 1073–1075.

[5] K. Mauersberger, Ozone isotope measurements in the stratosphere, *Geophys. Res. Lett.* **14** (1987) 80–83.

[6] R.A. Marcus, An intramolecular theory of the mass-independent isotope effect for ozone. II. Numerical implementation at low pressures using a loose transition state, *J. Chem. Phys.* **113** (2000) 9497;
R.A. Marcus, *J. Chem. Phys.* **124** (2006) 034712.

[7] Y.Q. Gao, R.A. Marcus, On the theory of the strange and unconventional isotopic effects in ozone formation, *J. Chem. Phys.* **116** (2002) 137–154.

[8] Y.Q. Gao, R.A. Marcus, Strange and unconventional isotope effects in ozone formation, *Science* **293** (2001) 259–263.

[9] B.C. Hathorn, R.A. Marcus, An intramolecular theory of the mass-independent isotope effect for ozone. I, *J. Chem. Phys.* **111** (1999) 4087–4100.

[10] Y.L. Yung, W.B. DeMore, J.P. Pinto, Isotopic exchange between carbon-dioxide and ozone via $O(^1D)$ in the stratosphere, *Geophys. Res. Lett.* **18** (1991) 13–16.

[11] B. Luz, E. Barkan, M.L. Bender, M.H. Thiemens, K.A. Boering, Triple-isotope composition of atmospheric oxygen as a tracer of biosphere productivity, *Nature* **400** (1999) 547–550.

[12] J. Savarino, J. Kaiser, S. Morin, D.M. Sigman, M.H. Thiemens, Nitrogen and oxygen isotopic constraints on the origin of atmospheric nitrate in coastal Antarctica, *Atmos. Chem. Phys.* **7** (2007) 1925–1945.

[13] M.S. Johnson, G.D. Billing, A. Gruodis, M.H.M. Janssen, Photolysis of nitrous oxide isotopomers studied by time-dependent hermite propagation, *J. Phys. Chem. A* **105** (2001) 8672–8680.

[14] C.A. McLinden, M.J. Prather, M.S. Johnson, Global modeling of the isotopic analogues of N_2O: Stratospheric distributions, budgets, and the O^{17}–O^{18} mass-independent anomaly, *J. Geophys. Res.-Atmos.* **108** (D8) (2003), Art. No. 4233.

[15] P. von Hessberg, J. Kaiser, M.B. Enghoff, C.A. McLinden, S.L. Sorensen, T. Rockmann, M.S. Johnson, Ultra-violet absorption cross sections of isotopically substituted nitrous oxide species: (NNO)–N^{14}–N^{14}, (NNO)–N^{15}–N^{14}, (NNO)–N^{14}–N^{15} and (NNO)–N^{15}–N^{15}, *Atmos. Chem. Phys.* **4** (2004) 1237–1253.

[16] M.K. Prakash, J.D. Weibel, R.A. Marcus, Isotopomer fractionation in the UV photolysis of N_2O: Comparison of theory and experiment, *J. Geophys. Res.-Atmos.* **110** (2005) D21315.

[17] The Nobel Peace Prize awarded in equal parts to Al Gore and the IPCC panel, http://nobelpeaceprize.org/eng_lau_announce2007.html, accessed October 14th, 2007.

CHAPTER 2

Mass-Independent Oxygen Isotope Fractionation in Selected Systems. Mechanistic Considerations

R.A. Marcus*

Contents

1. Introduction 6
2. The MIF in Ozone Formation 10
3. Quantum Dynamical Computations 12
4. Individually Studied Ratios of Isotopomeric Reaction Rate Constants 13
5. Rate Constant Ratios and Enrichments for Other Reactions 15
6. Oxygen Isotopic Fractionation for $CO + OH \rightarrow CO_2 + H$ 16
Acknowledgements 18
References 18

Abstract

Studies of the mass-independent effect on oxygen isotope fractionation (MIF) are summarized, focusing on the MIF in ozone formation in the laboratory, a similar effect being found in the atmosphere. (As used here the MIF for O isotopomers means that the usual three-isotope plot has a slope of about unity instead of the conventional 0.52.) The marked difference of results obtained with systems when there is extensive isotopic exchange ("scrambled") and experimental conditions where isotopic exchange is minimized ("unscrambled") is also discussed. Ratios of rate constants can be measured only in the latter and show large isotope effects. These large isotope effects were shown to cancel exactly when there is extensive isotopic exchange, and one then obtains instead an MIF for ozone. We also consider a possible role of the chaperon versus energy transfer mechanisms for ozone formation in influencing the temperature dependence of the MIF. MIF in other systems are noted. Separately we consider the $CO + OH \rightarrow CO_2 + H$ reaction, an isotopically anomalous reaction. It

* Noyes Laboratory of Chemical Physics, MC 127-72, California Institute of Technology, Pasadena, CA 91125-0072, USA

Advances in Quantum Chemistry, Vol. 55
ISSN 0065-3276, DOI: 10.1016/S0065-3276(07)00202-X

has sometimes been called "mass-independent" but has no O atom symmetry for the HOCO* intermediate and so no MIF in the above sense (*ca.* unit slope) would be expected on theoretical grounds.

1. INTRODUCTION

The principal focus of the present article is on the "mass-independent isotope fractional effect" (MIF) found in atmospheric and laboratory produced ozone. When this MIF occurs, a plot of the positive or negative "enrichment" $\delta^{17}O$ in samples versus that of $\delta^{18}O$ in those same samples has a slope of approximately unity, rather than its typical value of about 0.52. The 0.52 is the value expected using conventional transition state theory when nuclear tunneling effects are absent. For an isotope Q, δQ is defined in per mil as $1000\,[(Q/O)/(Q/O)_{std} - 1]$, where Q/O is the ratio of Q to ^{16}O in the sample and std refers to its value in some standard sample, standard mean ocean water. An example of a three-isotope plot showing a slope of 0.52 is given in Figure 2.1.

The MIF phenomenon was first observed by Clayton in 1973 for the isotopic oxygen content in the earliest solids in the solar system, the so-called calcium–aluminum-rich inclusions (CAIs) in carbonaceous chondritic meteorites [1]. The slope of $\delta^{17}O$ versus $\delta^{18}O$ plot for the CAIs was close to unity, the CAIs being equally deficient in the heavy O isotopes, deficient in the δ notation sense, while the ozone is equally enriched in those isotopes in that sense, as in Figure 2.2. Both are examples of an MIF. Interest in this striking phenomenon for the CAIs is motivated by what it may reveal about the formation of the early solar system. Standard reaction rate transition state theory [3], and behavior of oxygen an other isotope fractionation in many other systems, would have led, instead, to the slope

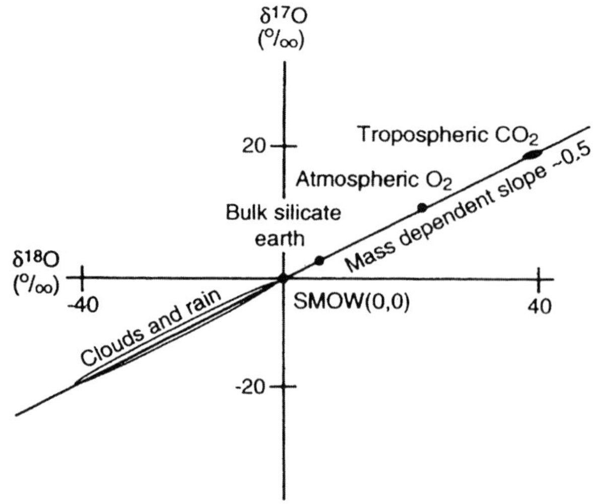

FIGURE 2.1 Three-isotope plot with a conventional slope [59].

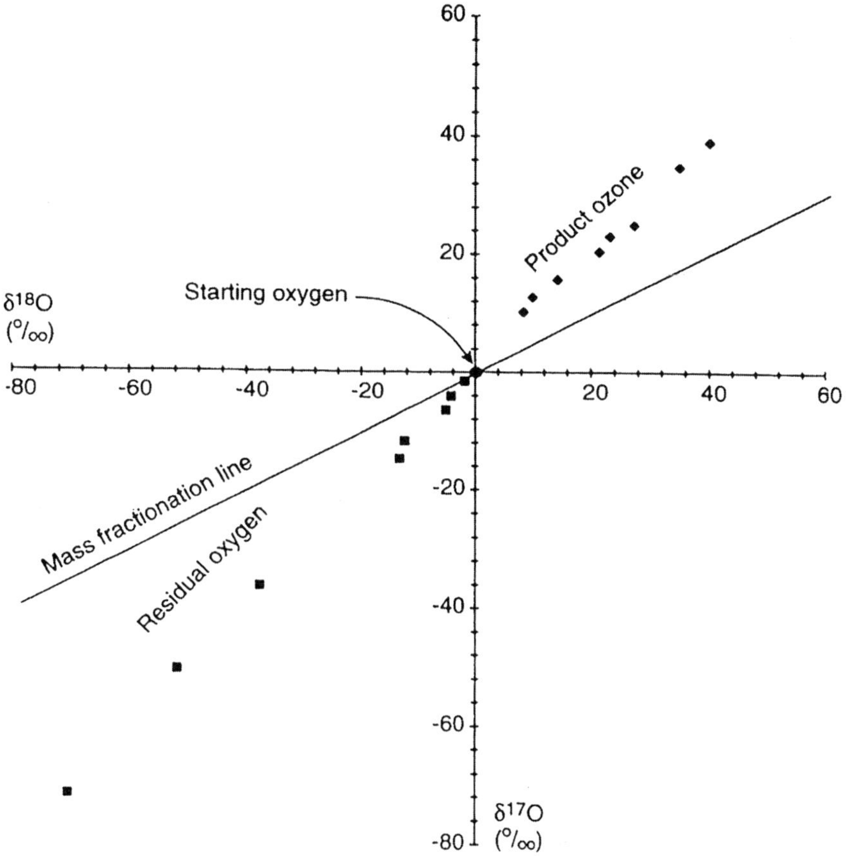

FIGURE 2.2 Three-isotope plot for ozone [5].

being about 0.52 [4]. This present chapter draws, in part, upon previous work in my group and on a recent contribution to a multi-authored chapter written for a quite different audience, one focused on the CAIs [2].

The MIF oxygen isotope effect in the gas phase was first discovered by Thiemens and Heidenreich in 1983 for ozone formation in the laboratory, the ozone being equally enriched in ^{17}O and ^{18}O relative to some standard [5], as in Figure 2.2. In these experiments the O_3 was produced by an electric discharge of O_2, followed by a reaction $O + O_2 \rightarrow O_3$, which proceeds via a vibrationally excited intermediate O_3^*, as in Eq. (1) below. Later, this MIF was also found by Yang and Epstein [6]. Anomalous oxygen isotope fractionation was found in the stratosphere by Mauersberger [7] and the anomaly has proven to be an important tracer in Earth's atmosphere (e.g., [8]). By studying both ^{17}O and ^{18}O the MIF was later found for O_3 in the stratosphere in 1990 by Morton *et al.* [9]. Studies have shown MIF effects in reactions involving CO and CO_2 [10,11].

Understanding the mass-independent isotope fractionation effect for ozone in the laboratory [5] and stratosphere [9] poses interesting challenges. These chal-

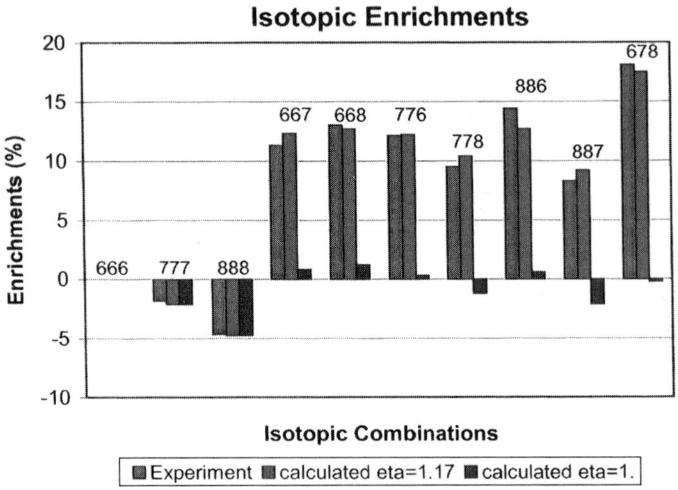

FIGURE 2.3 Heavy oxygen isotope enriched mass independent fractionation, experiment [59] and theory [13].

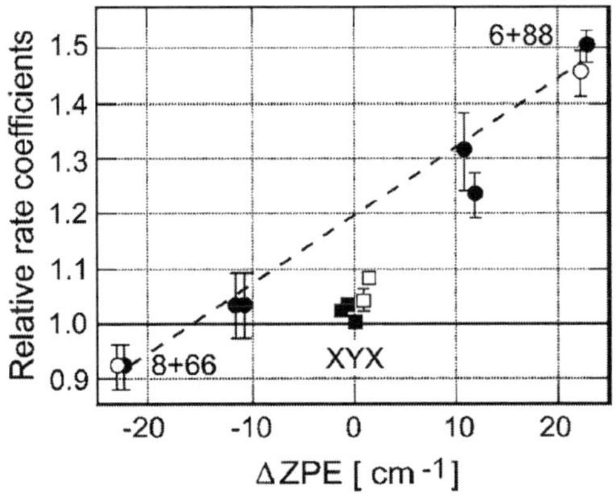

FIGURE 2.4 Ratios of rate constants ($k^r_{X+ZZ \to XYZ}/k^r_{6+66 \to 666}$) vs. zero-point energy difference of exit channels [45]. The clustering of points for the symmetric isotopomers at $\Delta ZPE = 0$ is due to the points having been slightly displaced horizontally for clarity [45].

lenges are both in the stratospheric/atmospheric interest in the phenomenon itself and in understanding the new chemical physics responsible for the phenomenon. Much is now understood about the MIF in ozone formation in the gas phase [12–16], but there remain interesting questions, posed below, to be addressed. The experimental data themselves involve the effect of pressure [9,17] and temperature [9] on the MIF, and the observation of the MIF for all possible isotopomers in systems enriched in the heavy isotopes ^{17}O and ^{18}O, the latter depicted in Figure 2.3. There are also data that are interesting in their own right but are not relevant to the MIF, namely a large mass-dependent but anomalous isotope effect

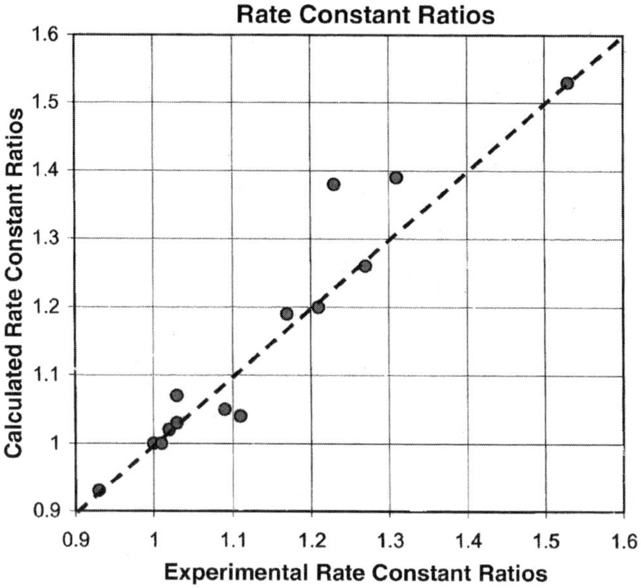

FIGURE 2.5 Experimental vs. calculated ratios of rate constants [13].

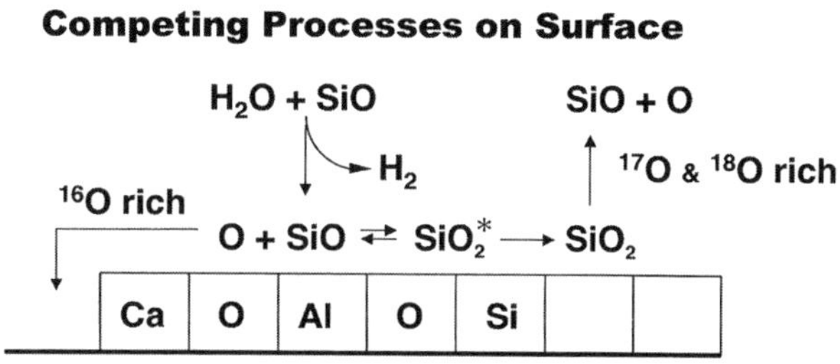

Schematic diagram of competing processes on the surface of a growing CAI mineral.

FIGURE 2.6 Theory for forming CAIs on dust surfaces at 1500–2000 K and separating oxygen isotopes in a mass-independent way [27].

for ozone formation observed under very special experimental conditions [18], as in Figures 2.4 and 2.5. We discuss it later. Recent review articles on MIF and related behavior in the atmosphere include [19–26].

While the main focus of the present article is on ozone, we note briefly in passing that several mechanisms have been postulated for the MIF in the CAIs, in particular an on-dust chemical mechanism [27] that is a surface reaction analog of the ozone formation mechanism discussed here, depicted schematically in Figure 2.6, and spectral self-shielding mechanisms [28–30]. In the latter, CO is dissociated photochemically in the early solar system, so producing a ^{16}O domain that is largely physically separated from that of the less abundant isotopes ^{17}O

and ^{18}O, because of spectral shielding (saturation). Subsequent mixing of the two domains in different proportions would yield gases that have a slope near unity for a plot of $\delta^{17}O$ versus $\delta^{18}O$. The details of how the separated O atom isotopes in the domains are transformed chemically into the MIF in CAIs remain to be described. We also note that unusual isotope effects have been reported for ionic systems [31–33].

2. THE MIF IN OZONE FORMATION

We first summarize a theory applied to the MIF effect in ozone formation [12–15]. The theory of the MIF in ozone formation involves several concepts, some of which are well established in the literature by many studies on the rates of unimolecular dissociation and bimolecular recombination reactions [34–39]. The main features of this analysis of the MIF in ozone formation are summarized below.

Oxygen atoms are typically formed in the stratosphere or laboratory by photodissociation of O_2 and in the laboratory also by electric discharge in O_2. The O and O_2 in the gas phase recombine on collision to form a vibrationally excited ozone molecule O_3^*:

$$O + O_2 \rightleftarrows O_3^*. \tag{1}$$

The O_3^* subsequently redissociates, as in the reverse of reaction (1), or is stabilized by collisions involving loss of vibrational energy to any molecule M:

$$O_3^* + M \rightarrow O_3 + M. \tag{2}$$

After formation of O_3^* an energy redistribution ("intramolecular energy randomization") occurs among the vibrations and rotations of the O_3^* molecule. This redistribution is typically assumed to provide on the average an equipartitioning of the excess energy among all the coordinates of this vibrationally hot O_3, subject to the constraint of the fixed total energy E and total angular momentum J of the O_3^* that exists prior to the next collision. The energy redistribution among the coordinates is due to anharmonic couplings of the molecular vibrations of O_3^* and to coriolis, centrifugal distortion, and other couplings of the vibrations and rotations. Similar remarks apply to other dissociating molecules.

The most common theory used to treat bimolecular recombination and unimolecular dissociation reactions in the literature is a statistical theory, "RRKM" theory (Rice, Ramsperger, Kassel, Marcus) [34–39]. However, symmetrical isotopomers such as $^{16}O^{16}O^{16}O$, $^{16}O^{17}O^{16}O$ and $^{16}O^{18}O^{16}O$ have fewer intramolecular dynamical couplings for an intramolecular energy redistribution (fewer "quantum mechanical coupling matrix elements"), because of symmetry restrictions, as compared with asymmetric O_3 isotopomers, such as $^{16}O^{16}O^{17}O$ and $^{16}O^{16}O^{18}O$ [15]. Because of the reduced number of coupling elements in the symmetric isotopomers, we have assumed that the symmetric isotopomers have less energy redistribution of the energy of the newly formed chemical bond among the other coordinates than do the asymmetric isotopomers ("less statistical") [15]. Thereby, this O_3^* occupies less "phase space" (as depicted schematically in Figure 2.7) and,

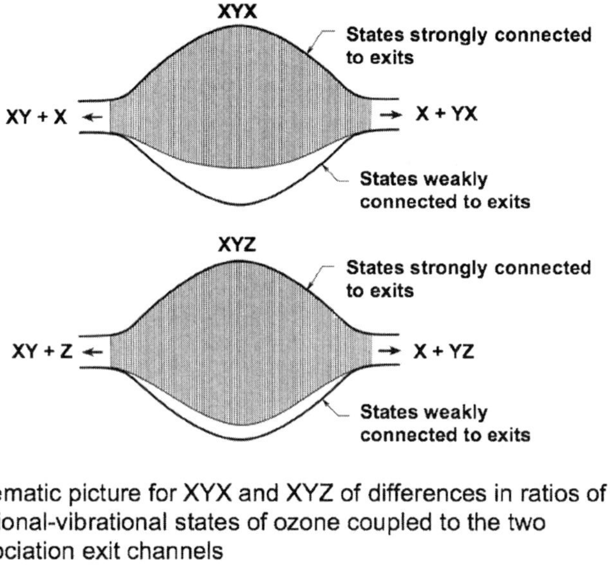

FIGURE 2.7 Schematic diagram illustrating the reduced effective phase space for a symmetric ozone isotopomer, so leading to an enhanced dissociation rate [12].

consequently, in terms of unimolecular reaction theory it has a shorter lifetime for redissociation. The shorter lifetime of an O_3^* isotopomer means that there is less chance of its being stabilized by loss of energy in a collision in reaction (2) and so the rate of formation of the stabilized O_3 molecule is less. For symmetric molecules the conventional expression for the density of states of the vibrationally excited molecule is divided by a factor η for symmetric isotopomers and by unity for asymmetric ones. In the calculations, to fit the experimental result on the MIF, η was chosen to be about 1.18 [13,14].

This assumption of less energy redistribution in the symmetric isotopomers, compared with asymmetric isotopomers, remains to be tested by *ab initio* quantum mechanical calculations, and as well as by a more direct experiment, as discussed later. Such fundamental quantum mechanical calculations would yield an *a priori* value of η. In principle, η can also be inferred from the more direct but more difficult experiment.

This property of a reduced number of coupling elements for symmetric systems is the same for all symmetrical isotopomers regardless of isotopic masses, since all have the same common symmetry property. The formation of symmetric heavy atom isotopomers, $^{16}O^{17}O^{16}O$ and $^{16}O^{18}O^{16}O$ in the isotopically scrambled system dilutes the magnitude of the MIF by about 1/3 but doesn't eliminate it.

In summary, because of a dynamical consequence of symmetry in this model, the vibrationally excited asymmetric isotopomers, such as QOO*, have approximately equal lifetimes that are longer than that of the symmetric isotopomers, such as OOO* and OQO*, Q being ^{17}O or ^{18}O. At low pressures they have thereby an improved chance of being deactivated by a collision and so of forming a stable ozone molecule, so leading to an equal (mass-independent) fractionation of the heavy isotopes in the ozone, relative to some standard.

This symmetry/asymmetry behavior is not restricted to O_3 but would apply to all triatomic or larger molecules that have the possibility of forming symmetric and asymmetric isotopomers, although its extent will depend on the molecule and the temperature, e.g., it can differ in magnitude for O_3^*, CO_2^*, SiO_2^*, and O_4^{+*}. Of course, when a highly vibrationally excited molecule such as O_3^* undergoes large amplitude vibrations, it will for short instants deviate from its simple symmetry (C_{2V} in the case of the symmetric isotopomer O_3 or OQO). Such effects could reduce the magnitude of the MIF, though not eliminate it.

Experimental studies permit some tests of these ideas. For example, the universality of the effect among all types of ozone isotopomers is seen in oxygen mixtures heavily enriched in ^{17}O and ^{18}O [18] and compared with theory in Figure 2.3 [12,13]. It is also seen in Figure 2.4, the points at $\Delta ZPE = 0$ being about 18% lower than those for the asymmetric cases. The effect of pressure on the MIF from 10^{-2} to 10^2 bar [16,17] has been measured and the theory tested by comparison with the data [12,13]. We note that at higher pressures all of the O_3^* formed is collisionally deactivated to form O_3 and so the anomaly reflected in the differences in lifetimes of the symmetric and asymmetric isotopomers of O_3^* disappears and so the MIF disappears, experimentally and theoretically.

The effect of temperature on the MIF is found in experiments [16] to increase with increasing temperature. Thus, the observed MIF is not a "threshold energy effect". The origin of this temperature effect poses an interesting question: is it because the higher the temperature the higher the energy of the typical O_3^*, the shorter its lifetime, the less time there is for energy redistribution in the excited ozone, and so the greater the non-statistical effect and thereby a larger MIF? Or does it indicate another phenomenon? We return to this point in a discussion of a 'chaperon' versus energy transfer mechanism for ozone formation.

Missing is a very direct experiment on energy redistribution in ozone formation: The dissociative lifetime behavior of the vibrationally excited O_3^* has not been studied under well-defined collision-free conditions. Under suitable conditions O_3^* could be prepared with a known vibrational energy and its time-evolution could be studied using a two laser "pump-dump" method in a molecular beam. A single-exponential decay of the O_3^* would indicate full statistical intramolecular mixing of the energy, while a more complex time decay would indicate incomplete mixing ("non-RRKM" behavior), e.g., [40]. Different isotopomers of O_3^* could be similarly studied, together with the effect of increased energy on the distribution of lifetimes. An increased energy is expected to increase the difference in the lifetimes of the symmetric and asymmetric isotopomers, based on one interpretation of the observed effect of temperature one the MIF noted above. An alternative possible explanation is considered in Section 4 in terms of a competition of two mechanisms.

3. QUANTUM DYNAMICAL COMPUTATIONS

There have been several preliminary quantum mechanical computations of some elementary steps in ozone formation [41–43]. In some studies the zero-point energy exit channel aspect has been addressed, and there has also been some study of Feshbach resonances for the collision, $O+O_2 \rightleftarrows O_3^*$. The fully quantum mechanical calculations are typically at $J = 0$. The complete scheme involves, in addition to these scattering resonances of $O + O_2 \rightleftarrows O_3^*$ at all J, the role of collisions with a third-body M in reaction (2) in forming the collisionally stabilized O_3 molecules. At present quantum mechanical calculations of these processes are still in a very early stage.

4. INDIVIDUALLY STUDIED RATIOS OF ISOTOPOMERIC REACTION RATE CONSTANTS

In the standard experiments on the MIF there is extensive isotopic exchange ("isotopic scrambling"), such as $Q+OO \rightleftarrows QOO^* \rightleftarrows QO+O$. However, there is also a body of experimental data on ozone formation obtained under very special experimental conditions [18] in which isotopic exchange is minor. In these experiments ratios of recombination rate constants, such as the $[^{16}O + {}^{18}O^{18}O]/[^{16}O + {}^{16}O^{16}O]$ ratio of ks are measured directly. (The unmeasured ^{16}O atom concentration cancels in this ratio of rates.) The results show very large isotope-specific effects, very different from the MIF. The new data are now well understood in terms of differences of zero-point energies of the transition states for the two competing modes of dissociation of an asymmetric ozone molecule, e.g., $OOQ \rightarrow O+OQ$ and $\rightarrow OO+Q$, e.g., [12–14]. Earlier, excellent correlations of these rates with a variety of molecular properties had been noted [44,45] but the argument showing that of these properties it was the zero-point energy difference of the two exit channels of the dissociating asymmetric isotopomers that was the origin was given in a detailed theoretical analysis [14].

As interesting as these special "exit-channel" effects are in their own right, it has been shown that because of a cancellation, they have no bearing on the MIF phenomenon [15]. We stress this point, since occasionally it is assumed in the literature that the special exit channel effect in the ratios is a key to understanding the MIF. Instead, the mass-independent effect of "scrambled" systems and the anomalously large mass-dependent effect for reactions of the type $Q + OO \rightleftarrows QOO^* \rightarrow QOO$ and $QO + O$, have very different origins and are unrelated. Perhaps these remarks may seem paradoxical. The various rate constants for these "isotopically unscrambled" reactions can be used to compute the observables for the isotopically scrambled system, and so compute $\delta^{17}O$ and $\delta^{18}O$. However, the detailed analysis [15] showed that there is much cancellation, summarized below, and that the theoretical expression for the MIF conditions is now simpler than would appear from the expression for the MIF in terms of the individual rate constants [15]. In particular, the zero-point energy effect, important for the individual isotope rate constants, disappears when the combination of them that determines the MIF is calculated.

Some physical insight into how this cancellation arises is considered next. In particular, we comment on why the dramatic and large exit channel mass-dependent isotope effect observed [18] under "unscrambled" experimental conditions is not relevant when the experimental conditions are, instead, the usual ones typical for MIF observations, "scrambled". We first note that under scrambled conditions an intermediate such as OQQ* formed at *low enough* energies from O + QQ can only dissociate back into O + QQ and not into the other channel, OQ + Q, because OQ has a higher zero-point energy than QQ. In contrast, in the comparison reaction in the ratio, O + OO → O_3^*, there are always two energetically accessible dissociation channels at all energies. So this OQQ* has a longer lifetime on the average than the O_3^* aside from any conventional isotope effects, and is therefore more likely to be stabilized by a deactivating collision. Thereby, the ratio of rates of formation of the stabilized OQQ compared with that of O_3 is larger than unity. The opposite is true for a reaction such as Q + OO → QOO, in which the reactant, O_2, has a higher zero-point energy than does the product of the dissociation into QO + O. In this case the dissociation via QOO* is faster than if it didn't have this energy excess, and so the ratio of the product QOO to that of the product of the comparison reaction, Q + QQ → QQQ can be lower than unity.

The cancellation noted above occurs via the partitioning factors in the expression for each rate constant: These factors describe the relative numbers of quantum states in the transition states of the two dissociating channels of the vibrationally hot dissociating asymmetric isotopomer [15]. The difference is magnified by difference in the zero-point energy of the two competing dissociating channels and leads to the large differences in specific rate constants. For systems where isotopic exchange is extensive, these two partitioning factors sum to unity in the theoretical expression for the enrichment [15], and so the ΔZPE effect is now absent.

The situation when the gas is isotopically scrambled, however, is very different and indeed the experimentally observed measured quantity is also very different. When the gas is isotopically scrambled, one does not measure these specific ratios of rate constants. Instead, a statistical steady-state, such as Q + OO ⇄ QOO* ⇄ QO + O and in the above example O + QQ ⇄ OQQ* ⇄ OQ + Q, exists at all energies, and now the energy distribution of the vibrationally excited intermediates is that which is dictated by the steady-state equations for the above reactions, and not by that of a vibrationally hot intermediate formed solely via one channel. Under such conditions all energies of the intermediate are statistically accessible, if not from one side of the reaction intermediate then from the other. Phrased differently, the isotopic composition of the collisionally stabilized product O_3 or QO_2 or will typically differ from that of the vibrationally excited species O_3^* or QO_2^*, since the intrinsic lifetime of the latter is isotope-dependent, as discussed in [15]. The usual RRKM-type pressure-dependent rate expression and conventional isotope effect results, modified by the nonstatistical effect discussed earlier [15].

Another interesting result obtained in experiments on these ratios of rate constants is the lack of the dependence of the ratio on the nature of the gas collisionally deactivating the vibrationally excited ozone isotopomer [46]. Different mechanisms have been postulated for the collisional deactivation, including the energy transfer mechanism used here and most commonly used elsewhere, and a 'chaperon' mechanism in which the third-body collision partner forms collision complex with the O or with the O_2 prior to the recombination step. Recently the chaperon mechanism was revisited for ozone formation, analyzing pressure and temperature dependent data on the recombination rate [47]. Since the ratios of rate constants of formation of the different isotopomers are independent of the nature of the third body [46] for the ratios studied, any chaperon mechanism and indeed any mechanism needs to be consistent with that observation. These different gases do affect the rate of ozone formation [48].

In an interpretation of the data [46] and in a calculation [49] it has been proposed that the energy transfer mechanism (Eqs. (1) and (2)) is dominant at higher temperatures and the chaperon mechanism at lower temperatures. In that case we can explore the consequences for MIF: From a theoretical point of view the chaperon mechanism does not have the long-lived O_3^* and so does not have the isotopic symmetry effect present in the energy transfer mechanism of Eqs. (1) and (2), and so would not be expected to have an MIF. In this case, therefore, we have an alternative explanation of why the MIF is larger at higher temperatures. At higher temperatures the energy transfer mechanism dominates and so there would be an MIF for ozone formation. At lower temperatures the chaperon mechanism dominates and so the MIF for ozone would tend to disappear. If the relative importance of the two mechanisms depends on the nature of the colliding gas, then the MIF would also, a result that can be tested experimentally. The test would be least ambiguous at lower pressures, before the decrease of the MIF with increasing pressure becomes important.

5. RATE CONSTANT RATIOS AND ENRICHMENTS FOR OTHER REACTIONS

The ratio of rate constants and the pressure dependence of the MIF for other reactions have also been treated theoretically, reactions such as $CO + O \rightarrow CO_2$, $SO + O \rightarrow SO_2$, $O + NO \rightarrow NO_2$ [49]. In a calculation of the ratios of rates and of the MIF for these other reactions, some of which may be "spin-forbidden" we should consider whether the odd spin of ^{17}O would affect the reaction rate and hence affect other properties such as an MIF. At first glance a potential objection to observing an MIF in a reaction such as $CO + O \rightarrow CO_2^*$, an objection that is easily removed however, is that the desired reaction is electronically spin-forbidden (singlet CO + triplet O $\rightarrow$ singlet CO_2). The presence of an odd numbered nucleus ^{17}O in a reactant reduces the spin-forbidden impediment for this reaction, due to electron spin-nuclear spin coupling, and so catalyzes the recombination and destroys any mass-independence. However, this effect would occur both in the formation, $CO + O \rightarrow CO_2^*$, and in the redissociation, $CO_2^* \rightarrow CO + O$, of the CO_2^*. Since the redissociation of the CO_2^* dominates over collisional stabilization of the CO_2^* at low pressures this spin–spin effect favoring formation of CO_2^* containing ^{17}O also favors this reverse process and so the effect cancels at low pressures. At sufficiently high pressures, however, pressures not relevant here, there would be little redissociation of the CO_2^* because of collisional deactivation, and then the spin–spin coupling would indeed favor ^{17}O enrichment in the CO_2 or in any of the other reactions. An approximate formula for the ratios of rate constants and for the pressure dependence for such systems is given elsewhere [50].

There are also well known isotope effects associated with odd nuclear spin, spin-electron spin coupling, e.g., [10]. Such an effect that would favor ^{17}O formation over even spin nuclei absent in ozone formation, presumably because the ozone formation occurs on a singlet electronic state surface: While there are some 27 electronic states of an O/O_2 pair that in principle could be involved in ozone formation, only the lowest singlet state of O_3 is deep enough to be sufficiently long-lived to be collisionally stabilized to lead to O_3 formation at atmospheric pressure. Only it was assumed [12–15] to play a significant route to O_3 formation under those conditions. The "no reaction" posed by the other entrance channels reduced the calculated rate of O_3 formation accordingly [15].

6. OXYGEN ISOTOPIC FRACTIONATION FOR $CO + OH \rightarrow CO_2 + H$

The $CO + OH \rightarrow CO_2 + H$ reaction has been reported as showing mass independent fractionation [11]. This term, in the strict sense of "mass-independence" used earlier means that the slope of a $^{17}O/^{16}O$ fractionation plotted versus the corresponding $^{18}O/^{16}O$ fractionation would be about unity, rather than having the conventional mass-dependent value of 0.52. However, what is frequently meant by MIF of oxygen instead is that the quantity $\Delta^{17}O = \delta^{17}O - 0.52\delta^{18}O$ is different from zero. The number of experiments needed to determine $\Delta^{17}O$ is only one, instead of the number needed to determine the slope of a 3-isotope and is commonly used in the literature. Indeed, sometimes the data are such that a three-isotope plot cannot be obtained from the available data. To avoid possible confusion one might term reactions that have a non zero value for $\Delta^{17}O$ but for which the $\delta^{17}O$ vs. $\delta^{18}O$ slope has not been measured, as being mass-anomalous fractionated, MAF. However, the term MIF is in widespread usage, and includes both the strictly MIF and any mass anomalous fractionation (MAF) reactions.

The reaction $CO + OH \rightarrow CO_2 + H$ is important for controlling the CO and OH concentration in the atmosphere and has an anomalous $\Delta^{17}O$ [11,51] and an ^{18}O isotope effect has also been measured in [52]. In theoretical studies on this reaction results have been obtained for C [53], H [53] and O [54] isotope effects. Inasmuch as the intermediate in this reaction, HOCO*, has no symmetry with respect to isotopes, the symmetry effect that appears in the ozone and the CAI problems is now absent. Any anomalies that appear in $\Delta^{17}O$ are then due to other reasons. Available three-isotope plots for this reaction are too sparse to define its slope reliably. The theory [53] provides an understanding of the H/D and $^{12}C/^{13}C$ data, and its pressure dependence, even though the C isotope effect is very small (−5 or so mil [11,55] at the lowest pressures). In calculations both nuclear tunneling of the H and complexity of mechanism can cause some deviation from the slope of a 3-isotope plot of 0.52 [54]. So they too can be a source of non-zero $\Delta^{17}O$ effect. Thus far, both the calculated [54] and the experimental O isotope effect are small, but unlike the C isotope data they do not agree. The problem in understanding the O-isotope behavior lies either in a sensitivity of the theoretical calculations of a small effect, though the calculation of the equally small ^{13}C effect agreed with the data, or it lies in the experimental complexity of the reactions. In any case the effect is considerably smaller by more than an order of magnitude than the isotopic anomaly in ozone.

In one study of the HO + CO reaction there were many reaction steps, including the presence of ozone among the reactants [51]. In others the presence of vibrationally excited OH reactants is possible when the source of the OH is photochemical [11,51,56,57]. Isotopic exchange would also be one source of discrepancy between theory and experiment for the ^{18}O isotope effect [54], if it occurred: It was estimated [54] that an isotopic exchange rate constant of about 2 to 3×10^{-15} cm^3 molecule^{-1} s^{-1} could explain the difference between the calculated and experimentally observed O isotope effect. In one case an O isotopic exchange was reported [56] that disappeared at higher pressure. In another experiment, consistently no O isotopic exchange was reported at that higher pressure [57]. In another experiment showing no isotopic exchange the upper limit set for O isotopic exchange [58] was improved, being a factor of two or three smaller than the rough value given above needed to explain the observed discrepancy in the ^{18}O isotope effect. There are also results on $\Delta^{17}O$ for this reaction [11], and a nonzero $\Delta^{17}O$ was obtained in the theory [54].

An experiment that may be simpler in its reactions, more sensitive than the detection limit of O isotope exchange need to explain the discrepancy, and as free as possible from OH vibrational excitation, would be helpful to explore the issue further. On the other hand the isotope effect is very small and the system sufficiently complicated that the theoretical calculations themselves may be the source of the discrepancy. In either case we infer that the reaction itself is mass anomalous rather than mass-independent in the strict sense of the term, both experimentally and in terms of current theoretical insight. In concluding this section we comment in physical terms on the small isotope effects, about -5 to -10 per mil, observed for ^{13}C and ^{18}O in this system at low pressures.

At low pressures the collisional deactivation of the HOCO* can be neglected, so simplifying considerably the analysis since the reaction rate will no longer depend on the density of states of the intermediate HOCO*. There are two transition states for the reaction, one for the formation of HOCO* from HO + CO and the other for the formation of H + OCO from HOCO*. Judging from the significant H isotope effect, the second transition state dominates (is rate controlling) at low pressures. In any case we will assume it for simplicity in the present discussion. Increasing the mass of the C or O has two effects: from its small contribution to the reaction tunneling coordinate (dominantly an H coordinate) the effective mass of the H is also increased, resulting in a reduced tunneling rate, so making the reaction rate smaller for the heavier isotope. On the other hand the extra mass increases the number of states in the transition state for the dissociation of HOCO* to OCO + H, relative to increasing the statistically average number for the reactants [54], so increasing the reaction rate. Thus an increase of the C or O mass has two conflicting consequences for the reaction rate.

The C in HOCO* is coupled to several vibrations in the HOCO* and so this "number of states effect" can be dominant for C and cause $^{12}k < ^{13}k$, in agreement with experiment. For the O in CO leading to HOCO* it is coupled to fewer vibrations in the HOCO*, compared with the number coupled to the C and, in the theoretical calculation, did not dominate the tunneling contribution and so in the theoretical calculations we had $^{16}k > {}^{18}k$, in disagreement with experiment. In this way one can understand the ^{13}C isotope effect, but not the current discrepancy for the ^{18}O effect. Nevertheless the effect is small, the result of opposing tendencies, and further experiments and theory may be helpful.

ACKNOWLEDGEMENTS

It is a pleasure to acknowledge the support of this research by the National Science Foundation and the very helpful comments of anonymous reviewers.

REFERENCES

[1] R.N. Clayton, L. Grossman, T.K. Mayeda, *Science* **182** (1973) 458.
[2] E.D. Young, K. Kuramoto, R.A. Marcus, H. Yurimoto, S.B. Jacobsen, Mass-independent oxygen isotope variation in the Solar nebula, *Rev. Mineral. Geochem.* **68** (2008) 187.
[3] E.g., J. Bigeleisen, M.G. Mayer, *J. Chem. Phys.* **15** (1947) 261.
[4] R.E. Weston Jr., *Chem. Rev.* **99** (1999) 2115.
[5] M.H. Thiemens, H.E. Heidenreich, *Science* **219** (1983) 1073.
[6] J. Yang, S. Epstein, *Geochim. Cosmochim. Acta* **51** (1987) 2011.
[7] K. Mauersberger, *Geophys. Res. Lett.* **14** (1987) 80.
[8] Y.L. Yung, W.B. De Mine, J.P. Pinto, *Geophys. Res. Lett.* **18** (1993) 13.
[9] J. Morton, J. Barnes, B. Schueler, K. Mauersberger, *J. Geophys. Res.* **95** (1990) 901.
[10] S.K. Bhattacharya, J. Savarino, M.H. Thiemens, *Geophys. Res. Lett.* **27** (2000) 1459.
[11] T. Röckmann, C.A.M. Brenninkmeijer, G. Saueressig, P. Bergamaschi, J.N. Crowley, H. Fischer, P.J. Crutzen, *Science* **281** (1998) 544.
[12] Y.Q. Gao, R.A. Marcus, *Science* **293** (2001) 259.
[13] Y.Q. Gao, R.A. Marcus, *J. Chem. Phys.* **116** (2002) 137.
[14] Y.Q. Gao, W.-C. Chen, R.A. Marcus, *J. Chem. Phys.* **117** (2002) 1536.
[15] B.C. Hathorn, R.A. Marcus, *J. Chem. Phys.* **111** (1999) 4087.
[16] D. Krankowsky, P. Lammerzahl, K. Mauersberger, *Geophys. Res. Lett.* **27** (2000) 2593.
[17] M.H. Thiemens, T. Jackson, *Geophys. Res. Lett.* **17** (1990) 717.
[18] K. Mauersberger, B. Erbacher, D. Krankowsky, J. Gunther, R. Nickel, *Science* **283** (1999) 370.
[19] G. Michalski, Z. Scott, M. Kabiling, M.H. Thiemens, *Geophys. Res. Lett.* **30** (2003) 1870.
[20] S. Morin, J. Savarino, S. Bekki, S. Gong, J.W. Bottenheim, *Atmos. Chem. Phys.* **7** (2007) 1451.
[21] M.H. Thiemens, *Annu. Rev. Earth Planet. Sci.* **34** (2006) 217.
[22] K. Mauersberger, D. Krankowsky, C. Janssen, R. Schinke, *Adv. At. Mol. Opt. Phys.* **50** (2005) 1.
[23] C.A.M. Brenninkmeijer, C. Janssen, J. Kaiser, T. Rockmann, T.S. Rhee, S.S. Assonov, *Chem. Rev.* **103** (2003) 5125.
[24] M.S. Johnson, K.L. Feilberg, P. von Hessberg, O.J. Nielsen, *Chem. Soc. Rev.* **31** (2002) 313.
[25] D. Krankowsky, F. Bartecki, G.G. Klees, K. Mauersberger, K. Schellenbach, J. Stehr, *Geophys. Res. Lett.* **22** (1995) 1713.
[26] J.C. Johnson, M.H. Thiemens, *J. Geophys. Res.* **102** (1997) 25395.
[27] R.A. Marcus, *J. Chem. Phys.* **121** (2004) 8201.
[28] R.N. Clayton, *Nature* **415** (2002) 860.
[29] J. Lyons, E. Young, *Nature* **435** (2005) 317.
[30] H. Yurimoto, K. Kuramoto, *Science* **317** (2004) 231.
[31] G. Gellene, *J. Chem. Phys.* **96** (1992) 4387.
[32] J. Xie, B. Poirier, G. Gellene, *J. Am. Chem. Soc.* **127** (2005) 1699.
[33] K. Griffith, G. Gellene, *ACS Symp. Ser.* **502** (1992) 210.

[34] R.G. Gilbert, S.C. Smith, *Theory of Unimolecular and Recombination Reactions*, Blackwell Scientific Publications, Oxford, UK, 1990.
[35] K.A. Holbrook, M.J. Pilling, S.H. Robertson, *Unimolecular Reactions*, second ed., Wiley, New York, 1996.
[36] W. Forst, *Unimolecular Reactions: A Concise Introduction*, Cambridge University Press, New York, 2003.
[37] T. Baer, W.L. Hase, *Unimolecular Reaction Dynamics, Theory and Experiment*, Oxford University Press, New York, 1996.
[38] D.M. Wardlaw, R.A. Marcus, *Adv. Chem. Phys.* **70** (1988) 231.
[39] R.A. Marcus, *J. Chem. Phys.* **20** (1952) 359.
[40] R.A. Marcus, W.L. Hase, K.N. Swamy, *J. Chem. Phys.* **88** (1984) 6717.
[41] D. Charlo, D.C. Clary, *J. Chem. Phys.* **120** (2004) 2700.
[42] R. Schinke, S.Y. Grebenshchikov, M.V. Ivanov, P. Fleurat-Lessard, *Annu. Rev. Phys. Chem.* **57** (2006) 625.
[43] D. Babikov, B.K. Kendrick, R.B. Walker, P. Fleurat-Lessard, R. Schinke, *J. Chem. Phys.* **119** (2003) 2577.
[44] B.C. Hathorn, R.A. Marcus, *J. Chem. Phys.* **114** (2000) 9497.
[45] C. Janssen, J. Guenther, K. Mauersberger, D. Krankowsky, *Phys. Chem. Chem. Phys.* **3** (2001) 4718.
[46] J. Guenther, D. Krankowsky, K. Mauersberger, *Chem. Phys. Lett.* **324** (2000) 31.
[47] K. Luther, K. Lum, J. Troe, *Phys. Chem. Chem. Phys.* **7** (2005) 2764.
[48] H. Hippler, R. Rahn, J. Troe, *J. Chem. Phys.* **93** (1990) 6560.
[49] M.V. Ivanov, R. Schinke, *J. Chem. Phys.* **124** (2006) 104303.
[50] Y.Q. Gao, R.A. Marcus, *J. Chem. Phys.* **127** (2007), Art. No. 244316.
[51] K.L. Feilberg, M.S. Johnson, C.J. Nielsen, *Phys. Chem. Chem. Phys.* **7** (2005) 2318.
[52] C.M. Stevens, L. Kaplan, R. Gorse, S. Durkee, M. Compton, S. Cohen, K. Bielling, *Int. J. Chem. Kinet.* **12** (1980) 935.
[53] W.-C. Chen, R.A. Marcus, *J. Chem. Phys.* **123** (2005) 094307.
[54] W.-C. Chen, R.A. Marcus, in Ph.D. thesis Caltech, 2008, to appear.
[55] H.G.J. Smit, A. Volz, D.R. Erhalt, H. Knappe, in: H.-L. Schmidt, H. Foerster, K. Heinzinger (Eds.), *Stable Isotopes*, Elsevier Publishing Co., Amsterdam, 1982, p. 147.
[56] M.J. Kurylo, A.H. Laufer, *J. Chem. Phys.* **70** (1979) 2032.
[57] H. Niki, P.D. Maker, C.M. Savage, L.P. Breitenbach, *J. Phys. Chem.* **88** (1984) 2116.
[58] G.D. Greenblatt, C.J. Howard, *J. Phys. Chem.* **93** (1989) 1035.
[59] K. Mauersberger, J. Morton, B. Schueler, J. Stehr, S.M. Anderson, *Geophys. Res. Lett.* **20** (1993) 1031.

CHAPTER 3

An Important Well Studied Atmospheric Reaction, O (^1D) + H_2

João Brandão*, **Carolina M.A. Rio*** and **Wenli Wang***

Contents

1. Introduction 22
2. Characterization of the O (^1D) + H_2 Reaction 23
3. Potential Energy Surfaces 25
4. Dynamical Studies 26
   4.1 Quantum calculations 26
   4.2 Quasiclassical calculations 27
5. Results and Comparison with Experiment 28
   5.1 Differential cross sections 28
   5.2 Product energy distributions 30
   5.3 Isotopic effects 31
   5.4 Total reactive cross sections 35
   5.5 Thermal rate constants 37
6. Final Remarks and Conclusions 39

Acknowledgements 39

References 39

Abstract

Among the chemical reactions in atmosphere, the reaction of an excited oxygen atom, O (^1D), with ground state molecular hydrogen, H_2 ($X\,^1\Sigma_g^+$), has been one of the most studied both experimentally and theoretically. To describe this reaction, various potential energy surfaces have been calibrated and their dynamics has been studied using quantum mechanical and quasiclassical trajectory methods. The theoretical results have shown to be in good agreement with experiment. The main uncertainties arise in the low temperature rate constants and in the isotopic branching ratio when reacting with HD.

* Dept. Química, Bioquímica e Farmácia, Universidade do Algarve, 8005-139 Faro, Portugal

Advances in Quantum Chemistry, Vol. 55
ISSN 0065-3276, DOI: 10.1016/S0065-3276(07)00203-1

1. INTRODUCTION

The absorption of radiation of wave lengths between 200 and 300 nm in the ozone stratospheric layer is a process of key importance in the reduction of ultra-violet radiation, which is known to cause biological mutations, solar burnings and other physiological effects [1]. The resultant oxygen atom in its lowest electronically excited state, O (^1D), see Eq. (1), is a highly reactive species that plays a significant role in initiating much stratospheric chemistry [2,3].

$$O_3 + h\nu\ (< 300\ \text{nm}) \rightarrow O\,(^1D) + O_2. \tag{1}$$

In particular, the excited oxygen atom quickly attacks methane, water vapor or molecular hydrogen present in stratosphere, see Figure 3.1, producing hydroxyl radicals according to reactions [4]

$$O\,(^1D) + H_2O \rightarrow OH + OH, \tag{2}$$

$$O\,(^1D) + CH_4 \rightarrow CH_3 + OH, \tag{3}$$

$$O\,(^1D) + H_2 \rightarrow H + OH. \tag{4}$$

These reactions constitute a source of OH in upper stratosphere, which ultimately controls the upper boundary of the ozone layer through the OH/HO_2 catalytic destruction cycles [5] getting rid of about 10% of the existing ozone [6]. The decrease of the ozone layer is a well known problem, see *e.g.* [3,7–9].

Due to its important role in atmospheric chemistry, the reaction

$$O\,(^1D) + H_2 \rightarrow H + OH, \qquad \Delta H_0^\circ = -181.4\ \text{kJ mol}^{-1}, \tag{5}$$

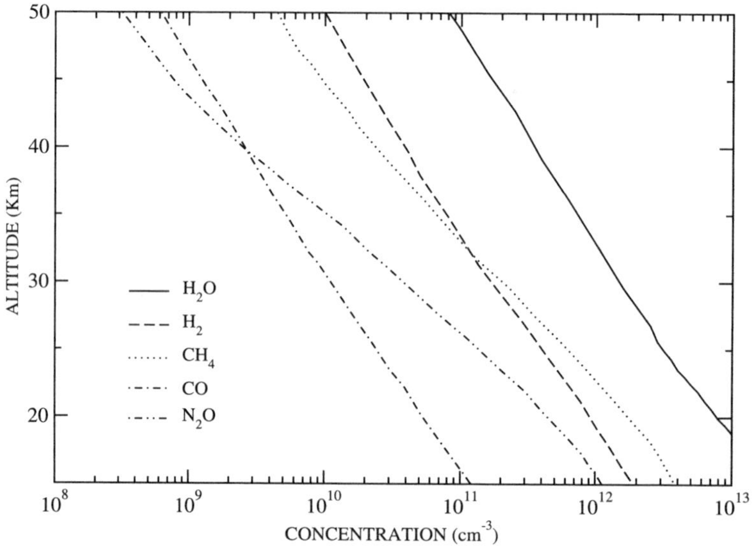

FIGURE 3.1 Gases present in stratosphere, adapted from reference [4].

and its isotopic variants

$$O\,(^1D) + D_2 \rightarrow D + OD, \qquad \Delta H_0^\circ = -179.7\ \mathrm{kJ\,mol^{-1}}, \qquad (6)$$

$$O\,(^1D) + HD \rightarrow D + OH, \qquad \Delta H_0^\circ = -177.9\ \mathrm{kJ\,mol^{-1}}, \qquad (7)$$

$$O\,(^1D) + HD \rightarrow H + OD, \qquad \Delta H_0^\circ = -183.8\ \mathrm{kJ\,mol^{-1}}, \qquad (8)$$

have been the subject of several experimental studies such as those observed in references [10–17]. Numerous theoretical studies of its dynamics have also been carried out, for example see [11,13,16–31], using several potential energy surfaces (PESs) published for this system [29,32–43]. These works have been the subject of partial reviews by Liu [44], Althorpe and Clary [45], Balucani *et al.* [46] and Aoiz *et al.* [47].

The title reaction is also of considerable interest to combustion [48] and laser chemistries [49]. Theoretically it is one of the best known complex-forming reactions from a fundamental point of view [50].

2. CHARACTERIZATION OF THE O (^1D) + H_2 REACTION

A general overview of the different potential energy surfaces relevant to the reactions involving an oxygen atom and the hydrogen molecule has been sketched in the work of Durand and Chapuisat [51].

As it can be seen in Figure 3.2, five potential energy surfaces are accessible to reactants but the mainly contribution comes from the lowest $\tilde{X}\,^1A'$ PES, which correlates with the $\tilde{X}\,^1A_1$ ground state of the H_2O molecule. In this PES the title reaction proceeds without energy barrier from reactants to products through a highly excited water molecule.

The first excited state, $\tilde{A}\,^1A''$, correlates with the ground state products OH ($X\,^2\Pi$) + H and must also be considered for energies higher than 8.4 kJ mol^{-1} [22], or temperatures above 500 K. The second excited state, $\tilde{B}\,^1A'$, adiabatically correlates with excited state products, OH ($A\,^2\Sigma^+$) + H. At collinear geometries these two surfaces correspond to a double degenerate Π state.

The other two upper PESs ($\tilde{B}\,^1A''$ and $\tilde{C}\,^1A'$ states) are repulsive and correspond to a double degenerate Δ state at collinear geometries and correlate with excited state products. Their contribution to the title reaction is negligible [22].

As it can be seen in Figure 3.2, at collinear geometries the ground $\tilde{X}\,^1A'$ state has Σ symmetry and adiabatically correlates with an excited $A\,^2\Sigma^+$ state of the hydroxyl radical. As a result there is a crossing between the Σ and Π states. Schatz and co-workers [12,22,23,52] estimated a constant contribution at collinear geometries from the $^1\Pi$ state of approximately 10%, due to the non adiabatic electrostatic coupling between the $^1\Sigma^+$ ($\tilde{X}\,^1A'$) and $^1\Pi$ ($\tilde{B}\,^1A'$ and $\tilde{A}\,^1A''$) states.

As a conclusion, in addition to the role of the lowest ($\tilde{X}\,^1A'$) PES the contribution of the two first excited states should be considered.

The reactions of ground state oxygen atom, O (^3P), with molecular hydrogen proceed in PESs with a high energy barrier, being the room-temperature rate con-

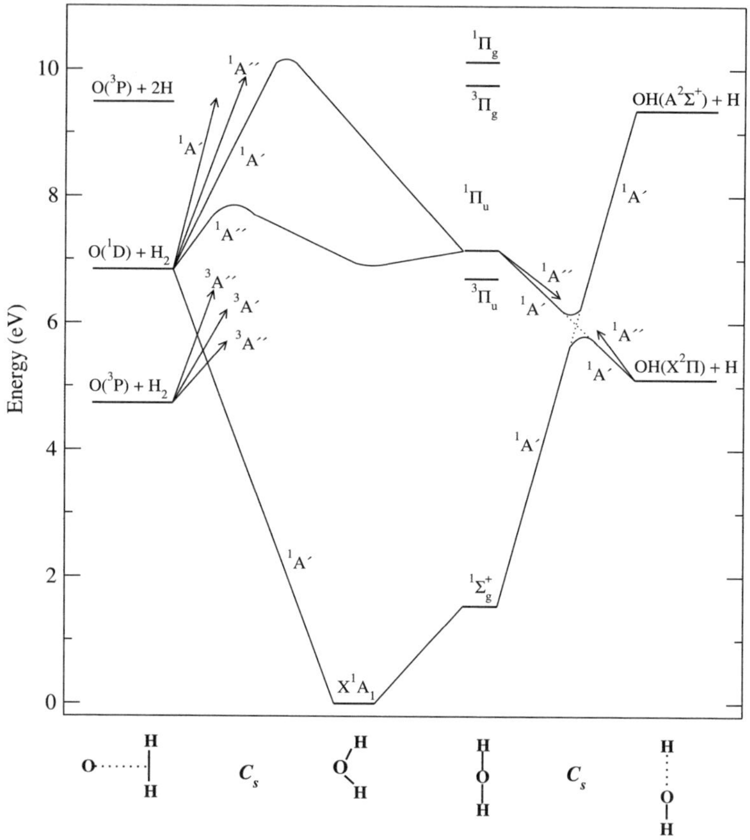

FIGURE 3.2 Correlation diagram of the potential energy surfaces involved in the O (^1D) + H_2 reaction, adapted from reference [51].

stant 10^{-7} times lower than that of O (^1D) + H_2 [53]. Most recently, Maiti and Schatz [54] and Chu *et al.* [55] have studied the intersystem crossing effects between the triplet and singlet states. Mainly focused on the O (^3P) + H_2 reaction, they found a negligible contribution to this reaction. In respect to the O (^1D) + H_2 reaction, Maiti and Schatz, using quasiclassical variant of the Trajectory Surface Hopping (TSH) approach [56], found that the intersystem cross effects on the O (^1D) + H_2 may decrease the adiabatic reactivity up to 10% through non reactive transitions to the triplet PESs. Chu *et al.* found a small reactivity on these PESs from a wave packet initiated on the $\tilde{X}\,^1A'$ PES. The net effect of the intersystem cross between triplet and singlet PESs is still an open question.

Although not shown in Figure 3.2, the H_2O ($\tilde{X}\,^1A'$) PES also correlates with the ground state triplet oxygen atom, O (^3P), and the triplet state hydrogen molecule, H_2 ($a\,^3\Sigma_u^+$). This is the lowest channel at large H–H distances [57] and should be also considered on a global PES for this system. Due to the very high energy of this region, the effects of this crossing on the dynamics of the title reaction should be negligible. Nevertheless, this channel should be taken into account on a global PES.

3. POTENTIAL ENERGY SURFACES

To our knowledge, the first PES suitable for dynamical studies of the title reaction was proposed in 1976 by Murrell and Sorbie [58] from spectroscopic data, but the first dynamical studies have been carried out in 1980 by Schinke and Lester in a PES fitted from *ab initio* data [32]. To better reproduce the experimental rate constant, those authors propose two different potential energy surfaces named SL1 and SL3. Subsequently, Murrell, Carter, Mills and Guest have published a double-valued PES [57] to correctly account for the different crossings between the diabatic surfaces and Murrell and Carter proposed a simpler single-valued PES that approximates it [59]; this last one is known as MC PES.

The SL3 and MC PESs have shown to give quite different dynamical results for the O (^1D) + H_2 reaction. In a comparative study, Fitzgerald and Schatz [19] have shown that the MC PES favours a collinear approach of the O atom to the H_2 molecule, while the SL3 PES favours an insertion mechanism preceded by a perpendicular approach.

Kuntz *et al.* [60] using the diatomics-in-molecule (DIM) method constructed three slightly dfferent DIM PESs for the two lowest states. All these DIM PESs exhibit no barrier to reaction (insertion or abstraction) on the lowest surface.

Ten years ago Schatz and co-workers [61] proposed a new potential energy surface for the ground state $\tilde{X}\,^1A'$ H_2O PES fitted from *ab initio* MR-CI calculations using a triple zeta basis set and another for the first excited state $\tilde{A}\,^1A''$ H_2O PES [52]. Those PESs are known as K PESs. Lin and Guo [62] have observed that the $\tilde{X}\,^1A'$ H_2O K PES has an unphysical minimum of 50 cm^{-1} in the collinear H + OH asymptote, which caused convergence problems in calculating vibrational levels near the dissociation threshold [63].

One year later Dobbyn and Knowles presented three potential energy surfaces [64] that reproduce both $\tilde{X}\,^1A'$, $\tilde{A}\,^1A''$ and $\tilde{B}\,^1A'$ PESs and the non-adiabatic coupling between the $\tilde{X}\,^1A'$ and $\tilde{B}\,^1A'$ states [40]. Although extensively used in many calculations, as far as we know, there is not any publication describing in detail these DK PESs, only a brief presentation can be found in the work of Aoiz *et al.* [15].

In addition to the studies of the title reaction, the potential energy surface for the ground state water molecule is of great importance for roto-vibrational spectroscopic studies of this system. Therefore some PESs that accurately describe the bottom well have been published [35,37,65–67]. On the other hand those PES do not dissociate correctly and are not suitable for dynamical studies. In order to correct this behaviour, Varandas used an energy switching approach, joining the spectroscopic description of the bottom well with a semiempirical description of the van der Waals interactions between reactants or products, to build a global single-valued PES for the ground state water molecule [39], ES PES. This approach has been complemented with double and triple-valued potential energy surfaces (ES-2v and ES-3v PESs) for this system [41,42].

Recently, mainly based on the very accurate *ab initio* results of Partridge and Schwenke [66] complemented with other data [68,69] and on a careful description of the long range interactions between the different dissociation channels [70], Brandão and Rio presented another double-valued PES for this system [43]. This BR PES displays a small van der Waals minimum and a small saddle point under the dissociation limit for the C_{2v} approach of the O (^1D) atom to the H_2 molecule and a very small barrier (< 0.4 kJ mol^{-1}) to collinear addition ($^1\Sigma^+$ surface) in agreement with the findings of Walch and Harding [68].

4. DYNAMICAL STUDIES

The different potential energy surfaces have been used in several calculations of the dynamics of the title reaction using quasiclassical trajectory methods and quantum calculations, see references [71,72] for a general description of these methods. Some capture and statistical calculations have also been reported for this reaction.

4.1 Quantum calculations

Accurate quantum dynamical calculations for reactions with deep wells have been a major challenge to theoreticians. In this regard, due to the potential well depth of 7 eV corresponding to the stable water molecule, the reaction O (^1D) + H_2 $\rightarrow$ OH + H poses a formidable obstacle to exact quantum dynamical treatment. As a result, the earliest quantum calculation on this reaction was carried out by Badenhoop *et al.* [73] using a 2D model where the bending motion was treated by a sudden approximation.

With the increase of modern computer capacity and the development of quantum reactive scattering theories, such as the increasing implementation of Time Dependent Quantum Mechanical (TDQM) wave packet methods, full dimensional quantum calculations on the title reaction have been reported since 1996, first by Peng *et al.* [74] followed by Dai [75], by Balint-Kurti *et al.* [76] and by Gray *et al.* [77]. All these earlier quantum dynamical studies employed TDQM wave packet methods on the adiabatic ground electronic $\tilde{X}\,^1A'$ state having in mind the reaction mechanism (or the role of direct abstraction vs. insertion in the reaction mechanism). Among these works, Peng [74] and Dai [75] used the SL1 PES, while the works of Balint-Kurti [76] and Gray [77] were based on the K surface. Subsequent TDQM wave packet calculations [23,24,78–80] have been carried out mainly on the ground DK PES. The most recent wave packet calculations by Lin and Guo [62] used the BR PES to compute total cross sections and thermal rate constants.

Among those TDQM studies, exact quantum dynamical calculations were usually limited to the total angular momentum $J = 0$. For $J > 0$, most of the authors used a capture model (or L-shift model) [77] to estimate the reaction probability from the $J = 0$ results. Even the direct calculations of reaction probabilities for $J > 0$ were performed using the centrifugal sudden (CS) approximation. Carroll and Goldfield [79] have reported wave packet calculations for $J > 0$ with inclusion of the Coriolis coupling terms. In comparison with the CS approximation results, these authors concluded that the CS approximation should yield accurate estimates of the reaction cross sections and rate constants for the title reaction.

The first exact quantum calculations of integral and differential cross sections on the adiabatic $\tilde{X}\,^1A'$ state were reported in 2001 by Honvault and Launay [15,81]. They have carried out quantum reactive scattering calculations of the title reaction on the DK PES within the Time Independent Quantum Mechanical (TIQM) framework using the hyperspherical close-coupling method.

The presence of five PESs correlating with the O (^1D) + H_2 reactants constitutes a further difficulty for the theoretical studies of the title reaction. Drukker and Schatz [22] have studied the effect of electronic Coriolis coupling in the entrance region for O (^1D) + H_2 using all five potential surfaces. In this work, the DK surfaces were used to describe the lowest three states while two DIM surfaces were employed for the upper two states. Using the vibrationally adiabatic coupled-channel approximation, they found that the upper two PESs ($\tilde{B}\,^1A''$ and $\tilde{C}\,^1A'$) are important only when considering the electronic fine structure of the reagents and the overall reactivity is dominated by the ground $\tilde{X}\,^1A'$ state with some influence of the first two excited $\tilde{A}\,^1A''$ and $\tilde{B}\,^1A'$ states. The participation of the $\tilde{A}\,^1A''$ and $\tilde{B}\,^1A'$ states in the title reaction has then been further investigated by a series of nonadiabatic quantum reactive scattering calculations using TDQM [23,24] or TIQM [26] as well as by some adiabatic TDQM or TIQM calculations on the $\tilde{A}\,^1A''$ surface [15,16,24,27,80,82].

Recently, Alexander *et al.* [29] carried out an *ab initio* study of the four states ($\tilde{X}\,^1A', \tilde{A}\,^1A'', \tilde{a}\,^3A''$ and $\tilde{a}\,^3A'$) that correlate to OH–H in the product region and the electronic and spin-orbit couplings between them. These PESs were used to study the nonadiabatic effects on the branching between the product OH ($^2\Pi$) multiplet levels, using a statistical model of atom–diatom insertion reactions combined with coupled-states capture theory.

More recently and related to the title reaction and to the nonadiabatic effects, Chu *et al.* [55] reported an exact quantum wave-packet study of intersystem crossing effects for the O ($^3P_{2,1,0}$, $^1D_2$) + H_2 reaction, in which three triplets, $\tilde{a}\,^3A''$, $\tilde{b}\,^3A''$ and $\tilde{a}\,^3A'$, and one singlet, $\tilde{X}\,^1A'$, electronic states were employed.

4.2 Quasiclassical calculations

Almost all the published PESs have been used to perform quasiclassical trajectory (QCT) [71,83] studies of the title reaction. Those calculations have been mainly focused in adiabatic studies in each PES. In addition, the trajectory surface hopping method [56], has been used to study the nonadiabatic effects between the $\tilde{X}\,^1A'$ and $\tilde{B}\,^1A'$ PESs.

Excluding some details on the rotational energy distribution of the products, a general agreement has been found between the classical and quantum results when performed on the same PES. It's important to notice that the final attribution of vibrational and rotational quantum levels obtained from quasiclassical calculations is a crude estimation. We also note that, due to the orbital and spin angular moments of the product OH $^2\Pi$ state, the classical angular moment of the OH diatomic, as computed in classical trajectories, differs from the nuclear rotational quantum number, N'.

At low temperatures, where quantum effects such as resonances can play an important role, the classical and quantum results for the thermal rate constant also diverge, as shown in Figure 3.11 of Section 5.5.

5. RESULTS AND COMPARISON WITH EXPERIMENT

Due to its importance, the reaction of O (^1D) + H_2 ($X\,^1\Sigma_g^+$) has been the subject of many experimental studies that can be compared with the theoretical predictions. For a recent review of the molecular beam experiments on the title reaction see the work of Balucani *et al.* [46].

5.1 Differential cross sections

The differential cross section is one of the most important results from molecular beam experiments. There we can see the differences between the insertion mech-

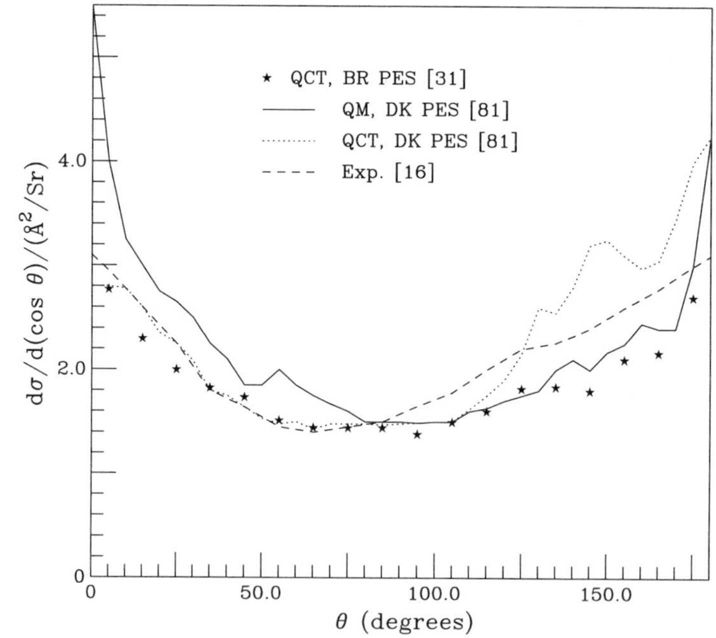

FIGURE 3.3 Angular distribution of the products for the reaction O (^1D) + H_2, at the collision energy of 5.4 kJ mol^{-1}.

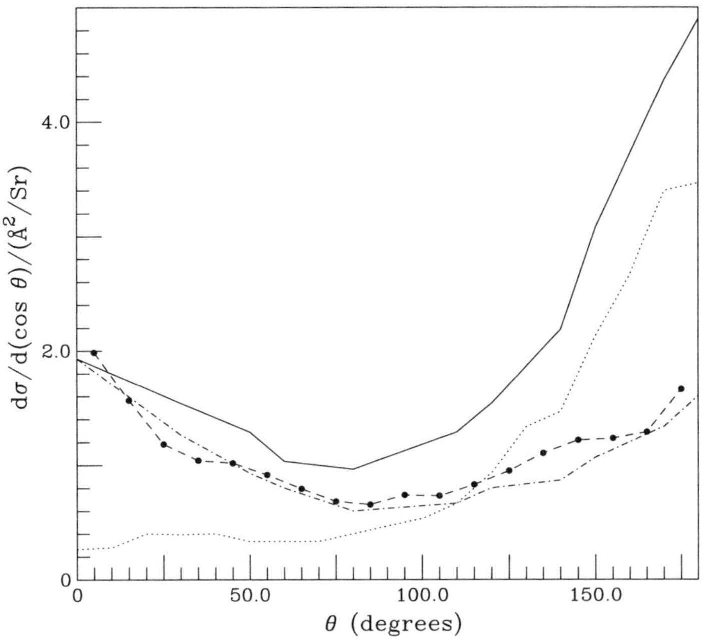

FIGURE 3.4 Angular distribution of the products for the reaction O (^1D) + HD → OD + H, at the collision energy of 19.0 kJ mol^{-1}. (—, · · ·, - - - ·) represent the total, abstraction and insertion components obtained from the work of Hsu, Pederson and Schatz [12]. (- - -) displays the QCT results on the ground state BR PES [30].

anism of the reaction on the ground state PES that leads to forward-backward symmetry and the abstraction mechanism when the title reaction proceeds on the first excited state.

For this reason this property has been extensively computed using both quantum and classical methods, being the agreement between the experimental and the theoretical results using the most recent PESs generally good.

Figure 3.3 displays a comparison of the experimental differential cross section results at the collision energy of 5.4 kJ mol^{-1} with quantum and classical results on the $\tilde{X}\,^1A'$ DK and BR PESs. It is expected that only the ground state PES should contribute to reaction at this collision energy. Peculiarly, the DK PES gives a small anisotropy in opposite directions when comparing quantum and quasiclassical results. This discrepancy has been explained by Balucani *et al.* [17] as a result of quantum tunneling of the collisions with large angular moment through the centrifugal barrier, which is not accounted for in classical trajectory methods.

A different pattern is observed at 19.0 kJ mol^{-1} collision energy for the O + HD → OD + H reaction. At this higher energy we can see an additional contribution from the first excited state as plotted in Figure 3.4.

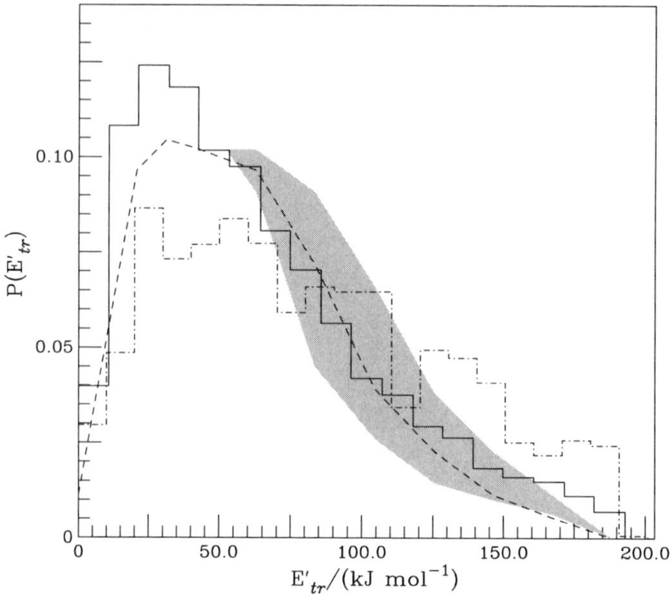

FIGURE 3.5 Translational energy distribution of the products of the reaction O (^1D) + H_2 at collision energy of 8.0 kJ mol^{-1} obtained by the QCT method (—) on BR PES and using statistical phase space calculations (- · · -) [31]. The line (- - -) and the shaded area represent the experimental results and respective error bars [10].

5.2 Product energy distributions

The comparison between phase space [84] statistical calculations [31], experimental data and dynamical studies of the title reaction have shown that the dynamics of the OH_2 complex plays an important role on the energy distribution of the products. The higher rotational energy of the products has already been noted by Buss *et al.* [18] and has been explained as a result of the excitation of the vibrational bending mode due to the insertion mechanism. Despite the large exothermicity of this reaction, the long-range interactions have shown to play an important role on vibrational and rotational anisotropies of the OH product [85]. When using the DK PES and the coupled-channel statistical theory, Rackham *et al.* [86] have found a rotational distribution in close agreement with accurate quantum calculations on the same PES.

In Figure 3.5 we compare the experimental [10] translational energy distribution of the products of the reaction O (^1D) + H_2 at the collision energy of 8.0 kJ mol^{-1} with QCT results on the BR PES and statistical phase space results [31]. The agreement between the theoretical and experimental results seems reasonable.

The good agreement between theory and experiment is shown in Table 3.1 where we compare the experimental results of Aoiz *et al.* [14,15] at an average collision energy of 11.7 kJ mol^{-1} with QM and QCT calculations on the BR and DK PESs. In that work [14,15] the authors performed QCT and QM calculations

TABLE 3.1 Relation $P(\upsilon = 4)/P(\upsilon = 3)$

	BR PES QCT [31]	DK PES		Exp. [14,15]
		QCT [14,15]	QM [14,15]	
$\tilde{X}\,^1A'$ (only)	0.507	0.47 ± 0.01	0.53	–
$\tilde{X}\,^1A' + \tilde{A}\,^1A''$	–	0.60 ± 0.01	0.61	0.59 ± 0.05

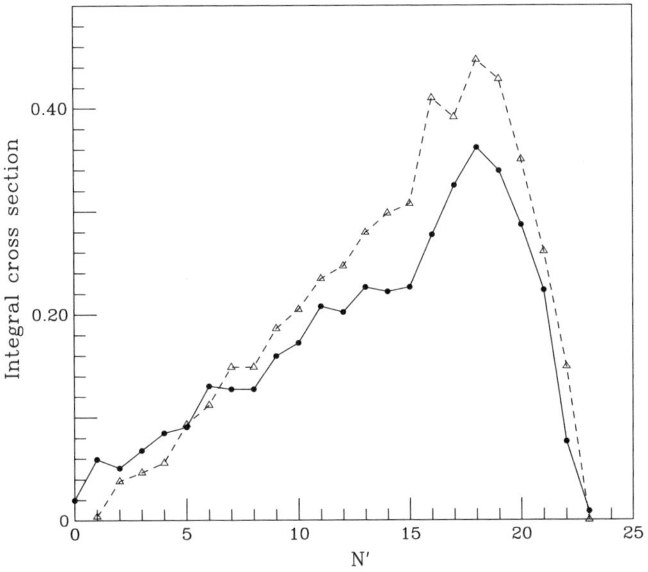

FIGURE 3.6 Experimental vs QCT rotationally integral cross section (in Å^2) for the O (^1D) + H_2 reaction, at 5.4 kJ mol^{-1} collision energy for $\upsilon' = 2$. Line (- - -, △), experimental data [16]; (—, ●) correspond to QCT results on BR PES [31].

both on the K and DK PESs and concluded that the DK surfaces present a better description for the dynamics of the reaction O (^1D) + H_2.

The inverted distribution of the rotational energy of the products is a characteristic property of this insertion reaction. Having in mind the above referred trouble in assigning the nuclear rotational quantum number, N', from classical trajectories (see Section 4.2), we show in Figure 3.6 a comparison between experimental [16] and QCT results on the BR PES [31] for the rotational integral cross section of the O (^1D) + H_2 reaction, at 5.4 kJ mol^{-1} collision energy for $\upsilon' = 2$. Here, the agreement is reasonably good.

5.3 Isotopic effects

The isotopic branching ratio, $\Gamma_{OD/OH} = \sigma_r(OD{+}H)/\sigma_r(OH{+}D)$, of the products of the O (^1D) + HD reaction is the major divergence between theory and experiment

TABLE 3.2 Experimental isotopic branching ratio for O (^1D) + HD

E_{col} (kJ mol^{-1})	$\Gamma_{OD/OH}$	Ref.
2.1–6.7	1.03	[50]
8.6	1.17 ± 0.10	[50,88]
10.0	1.5 ± 0.2	[89]
10.0	1.3 ± 0.1	[89]
13.5	1.35 ± 0.20	[90]
14.2	1.4 ± 0.2	[89]
15.5	1.34	[50]
18.8	1.49	[50]
18.9	1.5	[91]
Room temp.	1.13 ± 0.08	[92]
$T = 298$ K	1.33 ± 0.07	[93]

TABLE 3.3 Reactive cross sections for O (^1D) + HD and isotopic branching ratio, $\Gamma_{OD/OH}$, using different PESs. All results but the last one refer to QCT calculations

E_{col} (kJ mol^{-1})	PES	$\sigma_{r(OH+D)}$ (Å^2)	$\sigma_{r(OD+H)}$ (Å^2)	$\Gamma_{OD/OH}$	Ref.
2.1	SL1	–	–	1.85 ± 0.24	[19]
	MC	–	–	1.08 ± 0.06	[19]
8.6	$^1A'$ DK	7.24	14.03	1.94	[87]
	$^1A'$ BR	6.19 ± 0.11	13.29 ± 0.12	2.15 ± 0.09	[30]
14.5[a]	$^1A'$ BR	5.50 ± 0.11	13.15 ± 0.16	2.38 ± 0.10	[30]
15.5	$^1A'$ BR	5.56 ± 0.10	11.76 ± 0.12	2.12 ± 0.10	[30]
19.0	$^1A'$ DK	6.28	11.46	1.82	[87]
	$^1A'$ K	6.94	10.59	1.53	[87]
	SL1	6.144	9.482	1.56 ± 0.02	[94]
	$^1A'$ BR	5.49 ± 0.10	11.27 ± 0.12	2.05 ± 0.10	[30]
20.9	SL1	–	–	1.4	[19]
	MC	–	–	1.09	[19]
Room temp.	WMF	–	–	4.5	[92]
	MCMG	–	–	2.0 ± 0.3	[95]
21.0–51.0	DK PESs	–	–	≈ 2	[26]

a Average value.

in this system. We collect in Table 3.2 the experimental data found in the literature for this branching ratio and, for comparison, we display in Table 3.3 the theoretical estimates using several potential energy surfaces for the lower adiabatic state. Comparing those tables we can see that the experimental value range is consistently lower than the theoretical ones, being these last results strongly dependent on the PES used in the calculations. From the most recent PESs, only the theoreti-

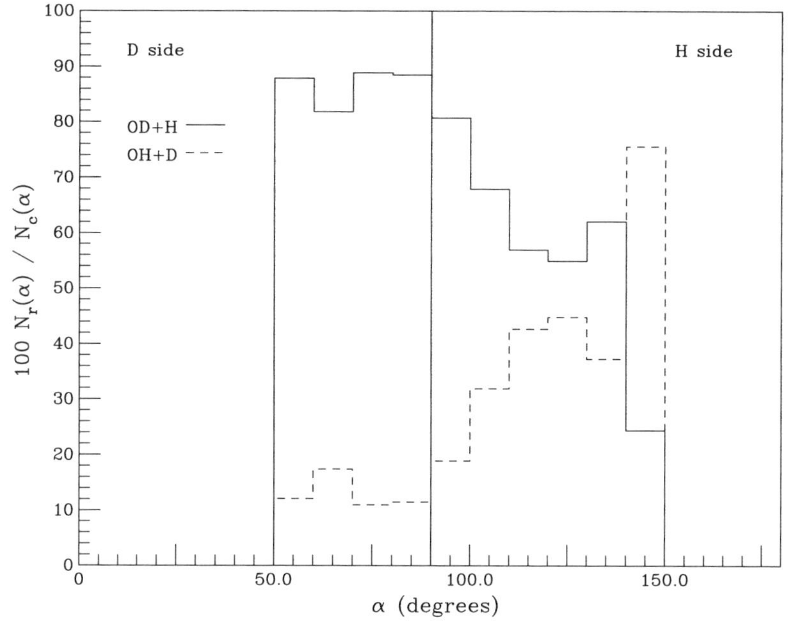

FIGURE 3.7 Reaction channel probability, in %, for each approach angle, α, for O (^1D) + HD at 2.05 kcal mol^{-1}.

cal results on the ground K PES agree with the higher experimental estimates. In addition, the role of the excited PES is expected to be small and contribute to a higher isotopic ratio [87].

To clarify the theoretical results, Fitzcharles and Schatz [19] discussed the angle of approach in the MC and SL1 PESs. They conclude that the MC PES favours a collinear collision where the first bond formed becomes the product diatomic, which explains an isotopic branching ratio close to one; on the other hand the SL1 PES leads to the formation of a tight complex, which dissociates into OD + H predominantly. Rio and Brandão [30] performed QCT results on the BR PES and found that, although the reaction proceeds through an insertion mechanism, those trajectories that reach the D atom side will dissociate in OD + H; but only those trajectories reaching the H side at a very high angle will dissociate as OH + D predominantly, as shown in Figure 3.7. In addition they found that during the complex lifetime there is an energy transfer process that favours the OD + H channel.

This divergence between theory and experiment is not so clear when we look at the distribution of the energy of the different dissociation channels. In Table 3.4 we compare the energy distribution of the products at the collision energies of 8.6 kJ mol^{-1} and 19.0 kJ mol^{-1}. We see a good agreement between the calculated and experimental results at the lower energy but at 19.0 kJ mol^{-1} the experimental results are too disperse to allow any conclusion. As far as we know, there is no experimental information on the partition of the internal energy of the diatomic.

TABLE 3.4 Distribution of translational, vibrational and rotational energy of the products of O (^1D) + HD

E_{col} ($kJ\,mol^{-1}$)	Source	Channel	E_{tr} (%)	E_{vib} (%)	E_{rot} (%)	Ref.
8.6	BR PES	OD + H	32.44	41.50	26.06	[30]
		OH + D	26.50	37.53	35.97	[30]
	EXP.	OD + H	32	–	–	[11]
		OH + D	25	–	–	[11]
19.0	BR PES	OD + H	32.42	40.46	27.12	[30]
		OH + D	28.40	36.55	35.05	[30]
	SL1 PES	OD + H	33	40	28	[94]
		OH + D	28	39	32	[94]
	EXP.	OD + H	41 ± 9	–	–	[89]
		OH + D	19 ± 4	–	–	[89]
		OD + H	41 ± 7	–	–	[90]
		OH + D	32 ± 5	–	–	[90]
		OD + H	30	–	–	[91]
		OH + D	22	–	–	[91]

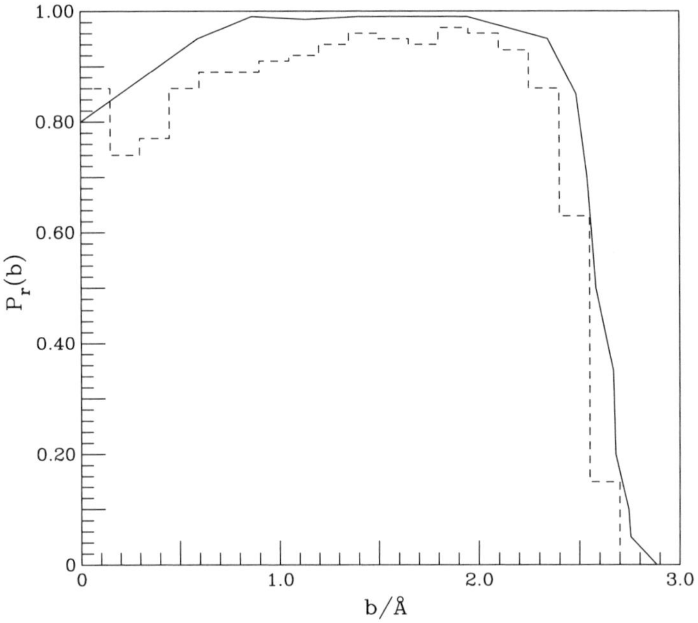

FIGURE 3.8 Opacity functions for the reaction O (^1D) + H_2 on the ground state PES. (—) corresponds to QM results on the DK PES at the collision energy of 9.6 $kJ\,mol^{-1}$ [85] and (- - -) correspond to QCT results on BR PES at 8.0 $kJ\,mol^{-1}$ collision energy [31].

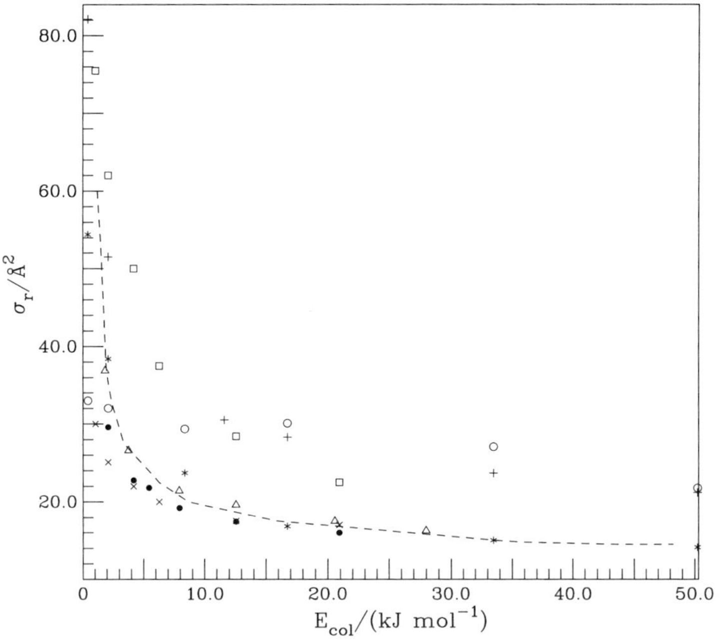

FIGURE 3.9 Comparison of QCT results of the total reactive cross section as a function of the collision energy for the reaction O (^1D) + H_2 using different published PESs. (●) BR PES [31]; (- - -) K PES [96]; (△) DK PES [24]; (+) ES PES [21]; (∗) ES-2v II PES [25]; (○) ES-2v III PES [25]; (×) SL1 PES [32] and (□) SL3 PES [32].

5.4 Total reactive cross sections

Due to limitations of the experiment, the molecular beam experiments are unable to measure the total cross section of the title reaction. As a consequence, here we can only compare the computed total cross sections using the different published PESs.

Closely related to the total cross section, the opacity function gives us the probability, $P_r(b)$, of a collision to be reactive at a given impact parameter, b. On the ground state PES, the title reaction proceeds without a barrier, being the reaction defined by the capture probability. As a consequence, the opacity function computed using the QCT method on the BR PES at different collision energies presents a similar form. As shown in Figure 3.8, the reaction probability is close to 0.9 for all collision energies at impact parameters less than 2.2 Å, vanishing quickly for larger impact parameters [31]. Rackham *et al.* [85] plotted the opacity function from accurate QM calculations [81] using the ground DK PES at the collision energy of 9.6 kJ mol^{-1}. These results, also plotted in Figure 3.8, display a similar trend with larger probability than on the BR PES at 8.0 kJ mol^{-1}, which is coherent with the larger reactivity found in the DK PES.

In Figure 3.9 we compare the QCT results for the total reactive cross section as a function of the collision energy computed by using different PESs. Varandas

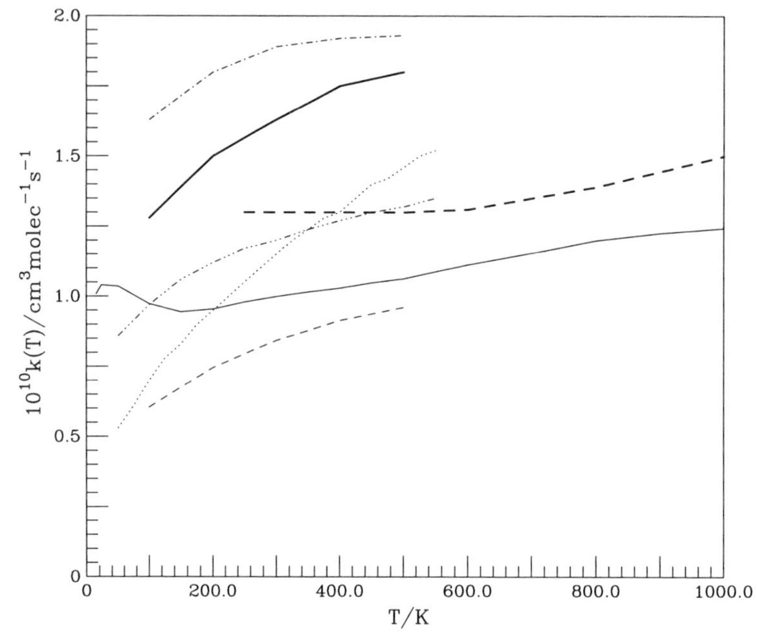

FIGURE 3.10 QCT thermal rate constants computed on the ground H_2O ($\tilde{X}\,^1A'$) PES. (—) BR PES [28]; (- - -) SL1 PES [32]; (- · · -·) SL3 PES [32]; (—) ES [21]; (- · · · · · · ·) ES-2v II [25]; (· · ·) ES-2v III [25] and (– – –) K PES [52].

et al. [25] have shown that the reactivity on the ES-2v PES strongly depends on the shape of the small bump for the C_{2v} insertion. Adding a Gaussian term, they have been able to propose two modifications of this PES, namely ES-2v II and ES-2v III, with different reactivities as shown in this figure.

We can see that the ES, ES-2v III and SL3 PES have a very different pattern with larger reactive total cross sections. The DK and K PESs have a very similar behaviour. Different is the ES-2v II PES which is similar to these PESs for energies above 17 kJ mol^{-1}, but has larger reactive cross section at 8 kJ mol^{-1} becoming smaller at very low collision energies. The BR and SL1 PESs are very similar being the BR more reactive at low collision energies.

These reactive total cross sections can be considered as the sum of a capture term, which decreases exponentially with energy, and a rigid sphere term, which is constant and should dominate at high collision energies [31].

$$\sigma(E) = \sigma_{cap}(E) + \sigma_{rs} = AE^{-m} + B. \tag{9}$$

The QM integral cross-section results of Lin and Guo [62] on the BR PES are not plotted in this figure. As shown in [31] these results are in good agreement with the QCT calculations in the same PES and somehow validates the use of classical mechanics in these studies.

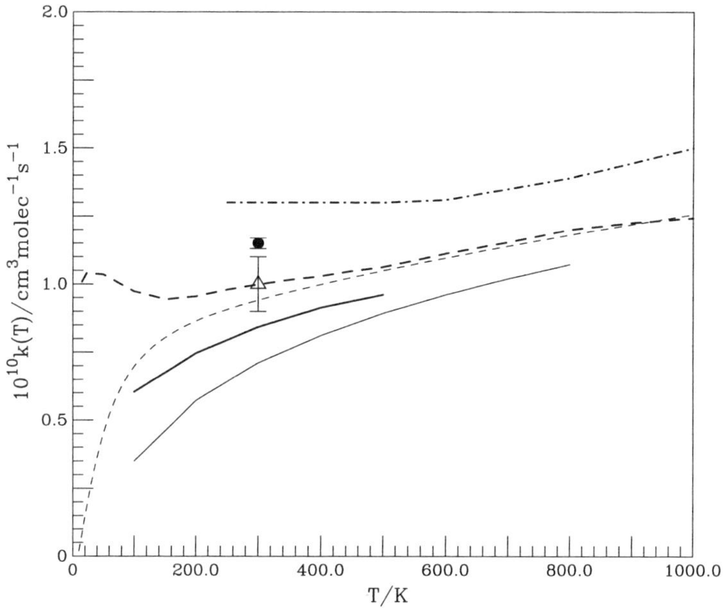

FIGURE 3.11 Comparison of QM and QCT thermal rate constants for the reaction O (^1D) + H_2, at different temperatures. (– – –) QCT [28] and (- - -) QM [62] results using the BR PES; (—) QCT [74] and (—) QM [74] results from the SL1 PES; line (- · · ·) QCT [52] and (●) QM results from the K PES [77] and (△) QM results on DK PES [77].

5.5 Thermal rate constants

A direct consequence of the different results for the total reactive cross section shown in Figure 3.9 is the diversity of the thermal rate constant computed using only the contribution of the ground H_2O ($\tilde{X}\,^1A'$) PES shown in Figure 3.10. In this figure we can see the higher reactivity of the SL3 and ES PESs and the high activation energy of the ES-2v III PES, which is a result of an almost constant reactive cross section. As expected the K and BR PESs display a parallel trend being the K PES more reactive. Also predictable from the low energy total cross sections is the behaviour of the SL1 PES, which has lower reactivity than the BR PES. The ES-2v II PES has lower reactivity at low temperatures but approaches (and even surpasses) the K PES as the temperature rises. So far as we know, there are not any published QCT results for the thermal constant computed using the DK PESs.

It is important to assess the role of quantum effects on the thermal rate constant. Unfortunately we only found QM thermal rate constant results for the SL1 [74] and BR PESs [62] at a large range of temperatures and two estimates for the K and DK PESs [77] at 300 K with a large uncertainty in the last one. In Figure 3.11 we compare these QM results with the correspondent QCT estimates. In all cases the QCT results lie above the QM predictions, being this difference larger in the SL1 PES. At temperatures above 400 K the QM and QCT results on the BR PES agree with each other, but at lower temperatures the differences become consid-

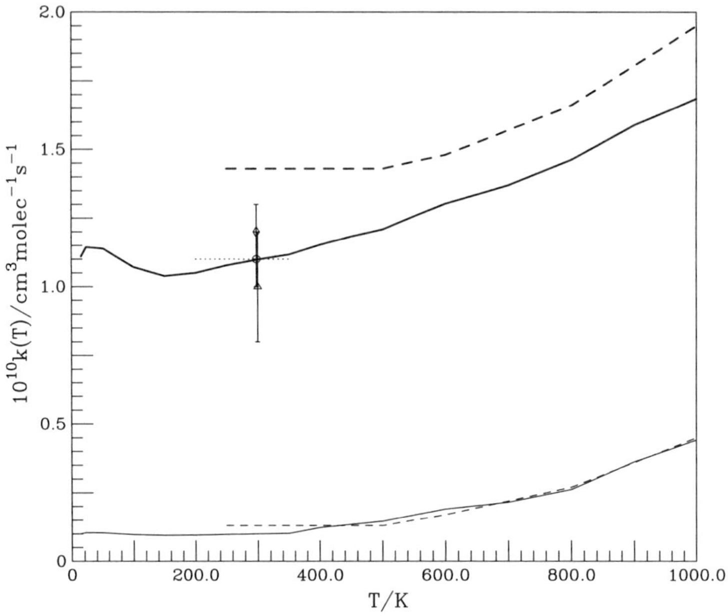

FIGURE 3.12 Thermal rate constants for the reaction O (^1D) + H_2, including the contribution from the excited PESs. (—) correction term and (—) estimated rate constant using the BR PES [28]. Line (- - -) correction term and (– – –) estimated rate constant using the K PES [52]. Most recent experimental data: (△) NASA report [3]; (· · ·) and (⊙) Atkinson [97] and (◇) Talukdar [93].

erable. Although some differences could reflect the restrictions of the quantum calculations, quantum effects should play an important role at low temperatures. This difference may be an indication of quantum effects due to resonances near the threshold. Another possible explanation is that, despite the agreement in the low energy cross sections, the QCT calculations are unable to account for the quantification of the angular moment, important at low energies. We also note that the QM results are not averaged over the possible internal states of the reactant diatomic.

As stated above when characterizing this reaction (see Section 2), two additional contributions should be taken into account before comparison with experiment. One is the abstraction reaction occurring in the first excited PES, H_2O ($\tilde{A}\,^1A''$), for temperatures higher than 500 K [22], being this contribution for the temperature of 1000 K approximately 0.3×10^{-10} cm^3 molec^{-1} s^{-1}. The other is the constant contribution from the $^1\Pi$ state of approximately 10% proposed by Schatz and co-workers [12,22,23,52], due to the non-adiabatic electrostatic coupling between the $^1\Sigma^+$ and $^1\Pi$ states at collinear geometries.

In Figure 3.12 we plot the estimated corrections from the upper PESs and a corrected estimate for the total thermal rate constant using both the BR PES [28] and K PES [52]. For comparison we also plot in this figure the experimental estimates for this thermal rate constant. There is a close agreement between the most recent experimental data and the QCT+corrections estimates of the thermal rate constant from the BR PES. The higher reactivity found for the K PES is in accordance with the greater cross sections exhibit by this PES, as shown in Figure 3.9.

6. FINAL REMARKS AND CONCLUSIONS

We can conclude that there are accurate potential energy surfaces to describe the reaction O (^1D) + H_2, which plays an important role in the ozone depletion cycle. The most recent PESs correctly reproduce the molecular beam experimental results, namely, the differential cross sections and energy distribution of the products, including the contribution of the abstraction mechanism in the first excited PES, within the present experimental resolution.

In spite of these results, some subjects still need to be clarified. One is the isotopic branching ration in the reaction O (^1D) + HD, which shows a disagreement between the predictions of these PESs and the experimental results. This should be related with the details of the PES and the mechanism of energy transfer in the H_2O complex favouring one or other bond breaking. Another is the low temperature, below 200 K, thermal rate constant, where we found a clear divergence between classical and quantum results. This should clarify the role of long range van der Waals interactions in the dynamics of this reaction. A third still open question is the magnitude of the intersystem crossing effects between the singlet and triplet PESs on the title reaction.

We have entitled this work, "a well studied reaction", but the more we study this system the more questions need to be answered.

ACKNOWLEDGEMENTS

This work was supported by the FCT under the POCTI/CTA/41252/2001 Research Project, co-financed by the European Community Fund, FEDER.

REFERENCES

[1] T.E. Graedel, P.J. Crutzen, *Atmospheric Change—An Earth System Perspective*, W.H. Freeman and Company, New York, 1993.
[2] J.G. Anderson, *Ann. Rev. Phys. Chem.* **38** (1987) 489.
[3] S.P. Sander, *et al.*, Chemical Kinetics and Photochemical Data for Use in Atmospheric Studies, JPL Publication 06-2, NASA-Jet Propulsion Laboratory, Pasadena, California, 2006.
[4] M. Nicolet, *Adv. Chem. Phys.* **55** (1985) 63.
[5] S. Koppe, *et al.*, *Chem. Phys. Lett.* **214** (1993) 546.
[6] H.S. Johnston, *Annu. Rev. Phys. Chem.* **26** (1975) 315.
[7] J.W. Chamberlain, D.M. Hunten, *Theory of Planetary Atmospheres—An Introduction to Their Physics and Chemistry*, Academic Press, San Diego, 1987.
[8] J.R. Barker, A brief introduction to atmospheric chemistry, in: J.R. Barker (Ed.), *Progress and Problems in Atmospheric Chemistry*, World Scientific, Singapore, 1995, pp. 1–33.
[9] J.H. Seinfeld, Chemistry of ozone in the urban and regional atmosphere, in: J.R. Barker (Ed.), *Progress and Problems in Atmospheric Chemistry*, World Scientific, Singapore, 1995, pp. 34–57.
[10] M. Alagia, *et al.*, *J. Chem. Phys.* **108** (1998) 6698.

[11] Y.-T. Hsu, K. Liu, L.A. Pederson, G.C. Schatz, *J. Chem. Phys.* **111** (1999) 7921.
[12] Y.-T. Hsu, K. Liu, L.A. Pederson, G.C. Schatz, *J. Chem. Phys.* **111** (1999) 7931.
[13] M. Ahmed, D.S. Peterka, A.G. Suits, *Chem. Phys. Lett.* **301** (1999) 372.
[14] F.J. Aoiz, *et al.*, *Dynamics of the O* (^1D) + H_2 $\rightarrow$ OH ($v = 3,4;N$) + H, book of 13th European Conference on Dynamics of Molecular Collisions (MOLEC 2000), A.26, Jerusalém, Israel, 17–22 September, 2000, p. 78.
[15] F.J. Aoiz, *et al.*, *Phys. Rev. Lett.* **86** (2001) 1729.
[16] F.J. Aoiz, *et al.*, *J. Chem. Phys.* **116** (2002) 10692.
[17] N. Balucani, *et al.*, *Mol. Phys.* **103** (2005) 1703.
[18] R.J. Buss, P. Casavecchia, S.J. Sibener, Y.T. Lee, *Chem. Phys. Lett.* **82** (1981) 386.
[19] M.S. Fitzcharles, G.C. Schatz, *J. Phys. Chem.* **90** (1986) 3634.
[20] T.-S. Ho, T. Hollebeek, H. Rabitz, L.B. Harding, G.C. Schatz, *J. Chem. Phys.* **105** (1996) 10472.
[21] A.J.C. Varandas, A.I. Voronin, A. Riganelli, P.J.S.B. Caridade, *Chem. Phys. Lett.* **278** (1997) 325.
[22] K. Drukker, G.C. Schatz, *J. Chem. Phys.* **111** (1999) 2451.
[23] S. Gray, C. Petrongolo, K. Drukker, G. Schatz, *J. Phys. Chem. A* **103** (1999) 9448.
[24] S.K. Gray, *et al.*, *J. Chem. Phys.* **113** (2000) 7330.
[25] A.J.C. Varandas, A.I. Voronin, P.J.S.B. Caridade, A. Riganelli, *Chem. Phys. Lett.* **331** (2000) 331.
[26] T. Takayanagi, *J. Chem. Phys.* **116** (2002) 2439.
[27] F.J. Aoiz, L. Banares, J.F. Castillo, V.J. Herrero, B. Martinez-Haya, *Phys. Chem. Chem. Phys.* **4** (2002) 4379.
[28] J. Brandão, C.M.A. Rio, *Chem. Phys. Lett.* **377** (2003) 523.
[29] M.H. Alexander, E.J. Rackham, D.E. Manolopoulos, *J. Chem. Phys.* **121** (2004) 5221.
[30] C. Rio, J. Brandão, *Chem. Phys. Lett.* **433** (2007) 268.
[31] C. Rio, J. Brandão, *Mol. Phys.* **105** (2007) 359.
[32] R. Schinke, W.A. Lester Jr., *J. Chem. Phys.* **72** (1980) 3754.
[33] J.N. Murrell, S. Carter, S.C. Farantos, P. Huxley, A.J.C. Varandas, *Molecular Potential Energy Functions*, Wiley, Chichester, 1984.
[34] L. Halonen, T. Carrington, *J. Chem. Phys.* **88** (1988) 4171.
[35] P. Jensen, *J. Mol. Spectrosc.* **133** (1989) 438.
[36] E. Kauppi, L. Halonen, *J. Phys. Chem.* **94** (1990) 5779.
[37] O. Polyansky, P. Jensen, J. Tennyson, *J. Chem. Phys.* **101** (1994) 7651.
[38] A.J.C. Varandas, A.I. Voronin, *Mol. Phys.* **85** (1995) 497.
[39] A.J.C. Varandas, *J. Chem. Phys.* **105** (1996) 3524.
[40] A.J. Dobbyn, P.J. Knowles, *Mol. Phys.* **91** (1997) 1107.
[41] A.J.C. Varandas, *J. Chem. Phys.* **107** (1997) 867.
[42] A.J.C. Varandas, A.I. Voronin, P.J.S.B. Caridade, *J. Chem. Phys.* **108** (1998) 7623.
[43] J. Brandão, C.M.A. Rio, *J. Chem. Phys.* **119** (2003) 3148.
[44] K. Liu, *Annu. Rev. Phys. Chem.* **52** (2001) 139.
[45] S.C. Althorpe, D.C. Clary, *Annu. Rev. Phys. Chem.* **54** (2003) 493.
[46] N. Balucani, G. Capozza, F. Leonori, E. Segoloni, P. Casavecchia, *Int. Rev. Phys. Chem.* **25** (2006) 109.
[47] F. Aoiz, L. Banares, V. Herrero, *J. Phys. Chem. A* **110** (2006) 12546.
[48] G. Dixon-Lewis, D.J. Williams, The oxidation of hydrogen and carbon monoxide, in: C.H. Bamford, C.F.H. Tipper (Eds.), in: *Comprehensive Chemical Kinetics*, vol. 17, Elsevier, Amsterdam, 1977, pp. 1–248.
[49] A.B. Callear, H.E. Van den Bergh, *Chem. Phys. Lett.* **8** (1971) 17.
[50] Y.-T. Hsu, J.-H. Wang, K. Liu, *J. Chem. Phys.* **107** (1997) 2351.
[51] G. Durand, X. Chapuisat, *Chem. Phys.* **96** (1985) 381.
[52] G.C. Schatz, *et al.*, *J. Chem. Phys.* **107** (1997) 2340.
[53] R.N. Dubinsky, D.J. McKenney, *Can. J. Chem.* **53** (1975) 3531.
[54] B. Maiti, G.C. Schatz, *J. Chem. Phys.* **119** (2003) 12360.
[55] T.-S. Chu, X. Zhang, K.-L. Han, *J. Chem. Phys.* **122** (2005) 214301.
[56] J.C. Tully, *J. Chem. Phys.* **93** (1990) 1061.
[57] J.N. Murrell, S. Carter, I.M. Mills, M.F. Guest, *Mol. Phys.* **42** (1981) 605.
[58] K.S. Sorbie, J.N. Murrell, *Mol. Phys.* **29** (1975) 1387.
[59] J.N. Murrell, S. Carter, *J. Phys. Chem.* **88** (1984) 4887.

[60] P.J. Kuntz, B.I. Niefer, J.J. Sloan, *J. Chem. Phys.* **88** (1988) 3629.
[61] G.C. Schatz, *et al.*, *J. Chem. Phys.* **105** (1996) 10472.
[62] S.Y. Lin, H. Guo, *Chem. Phys. Lett.* **385** (2004) 193.
[63] S. Gray, E. Goldfield, *J. Phys. Chem. A* **105** (2001) 2634.
[64] A.J. Dobbyn, P.J. Knowles, *Faraday Discuss.* **110** (1998) 247.
[65] P. Jensen, S.A. Tashkun, V.G. Tyuterev, *J. Mol. Spectrosc.* **168** (1994) 271.
[66] H. Partridge, D.W. Schwenke, *J. Chem. Phys.* **106** (1997) 4618.
[67] O.L. Polyansky, P. Jensen, J. Tennyson, *J. Chem. Phys.* **105** (1996) 6490.
[68] S.P. Walch, L.B. Harding, *J. Chem. Phys.* **88** (1988) 7653.
[69] F. Schneider, F.D. Giacomo, F.A. Gianturco, *J. Chem. Phys.* **104** (1996) 5153.
[70] J. Brandão, C.M.A. Rio, *Chem. Phys. Lett.* **372** (2003) 866.
[71] D.G. Truhlar, J.T. Muckerman, Reactive scattering cross sections iii. Quasiclassical and semiclassical methods, in: R.B. Bernstein (Ed.), *Atom—Molecule Collision Theory*, Plenum Press, New York, 1979, pp. 505–566.
[72] J.Z.H. Zhang, *Theory and Application of Quantum Molecular Dynamics*, World Scientific, Singapore, 1999.
[73] K. Badenhoop, H. Koizumi, G.C. Schatz, *J. Chem. Phys.* **91** (1989) 142.
[74] T. Peng, D.H. Zhang, J.Z.H. Zhang, R. Schinke, *Chem. Phys. Lett.* **248** (1996) 37.
[75] J. Dai, *J. Chem. Phys.* **107** (1997) 4934.
[76] G.G. Balint-Kurti, A.I. Gonzalez, E.M. Goldfield, S.K. Gray, *Faraday Discuss.* **110** (1998) 169.
[77] S.K. Gray, E.M. Goldfield, G.C. Schatz, G.G. Balint-Kurti, *Phys. Chem. Chem. Phys.* **1** (1999) 1141.
[78] M. Hankel, G.G. Balint-Kurti, S.K. Gray, *J. Chem. Phys.* **113** (2000) 9658.
[79] T. Carroll, E. Goldfield, *J. Phys. Chem. A* **105** (2001) 2251.
[80] M. Hankel, G. Balint-Kurti, S. Gray, *J. Phys. Chem. A* **105** (2001) 2330.
[81] P. Honvault, J.-M. Launay, *J. Chem. Phys.* **114** (2001) 1057.
[82] F.J. Aoiz, L. Banares, J.F. Castillo, B. Martinez-Haya, M.P. de Miranda, *J. Chem. Phys.* **114** (2001) 8328.
[83] J.T. Muckerman, *J. Chem. Phys.* **54** (1971) 1155.
[84] P. Pechukas, J.C. Light, C. Rankin, *J. Chem. Phys.* **44** (1966) 794.
[85] E.J. Rackham, T. Gonzalez-Lezana, D.E. Manolopoulos, *J. Chem. Phys.* **119** (2003) 12895.
[86] E.J. Rackham, F. Huarte-Larranaga, D.E. Manolopoulos, *Chem. Phys. Lett.* **343** (2001) 356.
[87] F.J. Aoiz, L. Bañares, M. Brouard, J.F. Castillo, V.J. Herrero, *J. Chem. Phys.* **113** (2000) 5339.
[88] S.-H. Lee, K. Liu, *Chem. Phys. Lett.* **290** (1998) 323.
[89] Y. Matsumi, K. Tonokura, M. Kawasaki, H.L. Kim, *J. Phys. Chem.* **96** (1992) 10622.
[90] T. Laurent, *et al.*, *Chem. Phys. Lett.* **236** (1995) 343.
[91] D.C. Che, K. Liu, *J. Chem. Phys.* **103** (1995) 5164.
[92] K. Tsukiyama, B. Katz, R. Bersohn, *J. Chem. Phys.* **83** (1985) 2889.
[93] R.K. Talukdar, A.R. Ravishankara, *Chem. Phys. Lett.* **253** (1996) 177.
[94] A.J. Alexander, F.J. Aoiz, M. Brouard, J.P. Simons, *Chem. Phys. Lett.* **256** (1996) 561.
[95] L.J. Dunne, *Chem. Phys. Lett.* **158** (1989) 535.
[96] A.J. Alexander, *et al.*, *Chem. Phys. Lett.* **278** (1997) 313.
[97] R. Altkinson, *et al.*, *J. Phys. Chem. Ref. Data* **21** (1992) 1125.

CHAPTER 4

Gaseous Elemental Mercury in the Ambient Atmosphere: Review of the Application of Theoretical Calculations and Experimental Studies for Determination of Reaction Coefficients and Mechanisms with Halogens and Other Reactants

Parisa A. Ariya[*], **Henrik Skov**[**], **Mette M.-L. Grage**[***,****] and **Michael Evan Goodsite**[**,*****]

Contents

1. Introduction	44
2. Kinetic and Product Experiments	45
3. Theoretical Evaluation of Kinetic Data	50
4. Perspectives	53
Acknowledgements	53
References	54

[*] Departments of Chemistry and Atmospheric and Oceanic Sciences, McGill University, 801 Sherbrooke St. W., Montreal, PQ, Canada, H3A 2K6
E-mail: parisa.ariya@mcgill.ca

[**] National Environmental Research Institute, Aarhus University, Frederiksborgvej 99, Roskilde, Denmark
E-mail: hsk@dmu.dk

[***] Department of Chemistry, Göteborg University, S-412 96 Göteborg, Sweden

[****] University of Copenhagen, Universitetsparken 5, DK 2100, Copenhagen Ø, Denmark
E-mail: mgrage@kl5.ki.ku.dk

[*****] Department of Physics and Chemistry, University of Southern Denmark, Campusvej 55, DK 5230, Odense M, Denmark
E-mail: meg@ifk.sdu.dk

Advances in Quantum Chemistry, Vol. 55
ISSN 0065-3276, DOI: 10.1016/S0065-3276(07)00204-3

Abstract

Understanding the kinetics and mechanisms associated with the atmospheric chemistry of mercury is of great importance to protecting the environment. This review will focus on theoretical calculations to advance understanding of gas phase oxidation of gaseous elemental mercury (GEM) by halogen species. Understanding the gas phase oxidation process between atmospheric mercury and halogen compounds is particularly important as all studies indicate that this interaction is the primary conversion mechanism in the troposphere leading to deposition of mercury. Theoretically predicting the thermochemistry of mercury containing species in the atmosphere is important because of the lack of experimental results. In this article a review of theoretical calculations of rate constants and reaction products is presented. Available laboratory data are listed and discussed as well in order to highlight the subjects where theoretical calculations in particular can be of value in the future.

1. INTRODUCTION

This review will focus on theoretical calculations to advance understanding of gas phase oxidation of gaseous elemental mercury (GEM) by halogen species. Computational and experimental studies to help parameterize models have been performed to make a more reliable description of the dynamics of mercury in the atmosphere so that the consequences of abatement strategies can be assessed. Quantum chemical calculations are the only way to viably investigate the mechanisms and advance what is observed in field and laboratory studies.

As the atmosphere plays a significant role as a medium for chemical and physical transformation, it is imperative to understand the fundamental of the kinetics and thermodynamics of the elementary and complex reactions that remove and generate GEM in the atmosphere. In Arctic and Antarctic regions after polar sunrise, field studies have demonstrated that GEM is rapidly oxidized to Hg(Π) compounds, known operationally as reactive gaseous mercury (RGM), which is quickly deposited to the surface, a phenomenon known as Arctic Mercury Depletion Events (AMDEs). After deposition, RGM may be reduced back to GEM and reemitted to the atmosphere. The true constituents of RGM are not known and the composition is estimated based on observed correlation of GEM with a competing depletion of surface ozone concentration, in accordance with what is known about the depletion of ozone in the Arctic atmosphere (ozone depletion events, ODEs). Observations of AMDEs and ODEs in the polar boundary layer have provoked several theories on reactions of elemental mercury with various atmospheric oxidants. Atmospheric mercury has thus been theorized to be oxidized by photochemically initiated catalytic reactions involving halogen species, particularly Br and BrO, though other halogens species such as IO may play a role in the marine boundary layer, and the Antarctic as well. The reaction of GEM and BrO to produce HgO and Br was first thought to be the dominant reaction, but recent theoretical studies have decisively shown that this reaction is highly endothermic. However, this result is in conflict with experimental data on the energetics of the species existing in the vapour over heated HgO(s) [1]. The oxidized mercury can further deposit on the Earth's surface and may bioaccumulate in the aquatic food chain, through complex, but not well understood mechanisms. A significant part of the deposited mercury is photo-reduced and re-emitted as GEM [2–7]. Recent field studies indicate that despite the efficient photoreduction, a net sea ice/snow interface may constitute a site for Hg accumulation [8]. Hence there is a net deposition of mercury is occurring in the Arctic, particularly in locations where bioaccumulation are expected [8]. Recent regional and global modeling studies have considered the oxidation as well as photoreduction and re-emission, and they report a net deposition of atmospheric mercury overall in the Arctic region [9,10] in accordance with previous field studies (cf. [11]). There have been several excellent review articles on mercury transformation in the atmosphere [12–16], particularly on its properties, sources, sinks, and fluxes. We review the current state of knowledge of the kinetics, product distribution, and thermo-chemical studies of elemental mercury with known atmospheric oxidants, in this case with halogen compounds. We focus on a comprehensive review of the experimental and theoretical kinetic evaluation of gaseous elemental mercury reactions with atmospheric halogen oxidants. We do not consider the body of research of mercury halogen interaction at higher than ambient temperatures, such as those found in industrial processes. We will outline major gaps and some future research directions.

Unlike the reactions of GEM in solution, experimental data on the gas-phase reactions of elemental mercury with some atmospheric oxidants are limited due to challenges including complexity of reactions, the low concentrations of species at atmospheric conditions, the low volatility of products, sensitivity to temperature and pressure, and the strong effects of water vapour and surface on kinetics. The possible effects and distribution of mercury isotope fractionation have not been analysed in any of the studies. The isotopes dilute the signal and mean that with current mass spectrometry techniques, ambient RGM compounds can not be identified. The possibility of theoretically predicting the thermochemistry of mercury-containing species of atmospheric interest is important and is complementary to laboratory and field studies.

Observations by Schroeder *et al.* [17] on concomitant rapid depletion of elemental mercury and ozone in the boundary layer indeed provoked several theoretical, laboratory and field studies on reactions of elemental mercury with various atmospheric oxidants see Table 4.1 and references therein.

2. KINETIC AND PRODUCT EXPERIMENTS

The rate of the atmospheric chemical transformation of elemental mercury with a given oxidant is dependent on two factors. The first factor is the reactivity of mercury towards a given oxidant at environmentally relevant conditions, such as temperature, pressure, oxygen concentration, and relative humidity. The second factor is the concentration (or mixing ratio) of the oxidant. The existing laboratory studies of mercury kinetic reactions have been obtained using steady state reaction

TABLE 4.1 Available rate constants for selected atmospheric reactions from the literature at room temperature (296 ± 2 K) or other temperatures (modified after Steffen *et al.* [16])[a]

Reaction 1	Rate constants (molecule cm^{-3} s^{-1})	Reference	Comments
$Hg^0 + O_3$ → products	$(3 \pm 2) \times 10^{-20}$	Hall [18]	
	$(7.5 \pm 0.9) \times 10^{-19}$	Pal and Ariya [19]	Temperature dependence is evaluated
	$(6.4 \pm 2.3) \times 10^{-19}$	Sumner *et al.* [20]	– given reaction rate is extrapolated, – at zero relative humidity.
$Hg^0 + HO$ → products	$(8.7 \pm 2.8) \times 10^{-14}$	Sommar *et al.* [21]	
	$(1.6 \pm 0.2) \times 10^{-12}$	Miller *et al.* [22]	70 °C
	$(9.3 \pm 1.3) \times 10^{-14}$	Pal and Ariya [19]	Temperature dependence evaluated at 100 and 400 Torr He and air
	$<10^{-13}$	Bauer *et al.* [23]	
$Hg^0 + Cl$ → products	$(1.0 \pm 0.2) \times 10^{-11}$	Ariya *et al.* [24]	
	$(1.5) \times 10^{-11}$	Horne *et al.* [25]	383–443 K
	2.8×10^{-11}	Khalizov *et al.* [26]	Theo.—calc. at 298 K, 760 Torr
	7.6×10^{-13}	Goodsite *et al.* [27]	Theo.—calc. at 298 K, 760 Torr
		Donohoue *et al.* [28]	Second-order rate was calculated at 260 K and 760 Torr
$Hg^0 + Cl_2$ → products	$(2.7 \pm 0.2) \times 10^{-18}$	Ariya *et al.* [24]	
	$(2.5 \pm 0.9) \times 10^{-18}$	Sumner *et al.* [20]	
$Hg^0 + Br$ → products	$(3.2 \pm 0.3) \times 10^{-12}$	Ariya *et al.* [24]	1 atm, 298 K
	10^{-13}	Grieg *et al.* [29]	383–443 K
See also Holmes *et al.* [30] Table 1 for k_2, k_3 and k_4 constants	1.0×10^{-12}	Khalizov *et al.* [26]	Theo.—calc. at 298 K, 760 Torr
	1.1×10^{-12}	Goodsite *et al.* [27]	Theo.—calc. at 298 K, 760 Torr
	3.6×10^{-13}	Donohoue *et al.* [31]	243–298 K, 200–600 Torr
	0.8	Skov *et al.* [32]	Interpretation of field study assuming[a] −40 °C
	1.2	Skov *et al.* [32]	Interpretation of field study assuming[a] −10 °C
$Hg^0 + BrO$ → products	$1 \times 10^{-15} < k < 1 \times 10^{-13}$	Raofie and Ariya [33]	
	1×10^{-14}	Sumner *et al.* [20]	
$Hg^0 + Br_2$ → products	$\leqslant (9 \pm 2) \times 10^{-17}$	Ariya *et al.* [24]	No reaction was observed under experimental conditions employed.
	No reaction	Sumner *et al.* [20]	
	2.7×10^{-31} and 3.4×10^{-31}	Balabanov *et al.* [34]	Theo.—calc.
$Hg^0Br + Br$ → products	2.5×10^{-10}	Goodsite *et al.* [27]	Theo.—calc. at 298 K, 760 Torr
	$1.05 \pm 0.14 \times 10^{-10}$	Balabanov *et al.* [34]	Theo.—calc.
$Hg^0 + F_2$ → products	$(1.8 \pm 0.4) \times 10^{-15}$	Sumner *et al.* [20]	
$Hg^0 + NO_3$ → products	$\leqslant 4 \times 10^{-15}$	Sommar *et al.* [35]	
	$\leqslant 7 \times 10^{-15}$	Sumner *et al.* [20]	
$Hg^0 + H_2O_2$ → products	$\leqslant 8 \times 10^{-19}$	Tokos *et al.* [36]	
$Hg(CH_3)_2 + HO$ → products	$(1.97 \pm 0.23) \times 10^{-11}$	Niki *et al.* [37]	
$Hg(CH_3)_2 + Cl$ → products	$(2.75 \pm 0.3) \times 10^{-10}$	Niki *et al.* [38]	
$Hg(CH_3)_2 + NO_3$ → products	$(7.4 \pm 2.6) \times 10^{-14}$	Sommar *et al.* [39]	

a The difference in results at −10 and −40 °C is due to the temperature dependence of the competing reaction between ozone and Br.

chamber or fast flow tubes. A single study has been carried out on the analysis on field data. Both relative and absolute techniques were used in these studies (cf. [40]).

Both absolute and relative techniques have advantages as well as disadvantages. The disadvantage of the relative rate is that the calculated reaction rate constant is only as good as the original value of the reaction rate constant for the reference molecule used, and this is why most detailed relative rate studies include several reference molecules to overcome this challenge. Another disadvantage is the complexity of the reactants and enhanced potential for side reactions. This challenge can be overcome with careful experimental setup and additional targeted experiments to minimize and characterize the extent of undesired reactions. An advantage of a detailed relative study is that one can readily perform the experiments under simulated tropospheric conditions, and also that the reaction chambers can be coupled with several state-of-the-art instruments for simultaneous analysis, which allows detailed product analysis as well as kinetic determinations.

The advantage of the absolute method is clearly that there is no need for incorporation of errors due to the reference molecules. However, in many absolute studies, one can follow merely one or two reactants, and considering the complexity of mercury reactions, and the extent of secondary reactions, the calculated values may be affected. Another challenge is that absolute rate studies often are performed at lower pressure than tropospheric boundary layer pressure ($\sim$740 Torr) and at concentrations orders of magnitude higher than tropospheric levels. Hence the data obtained under such conditions must be properly corrected for the ambient tropospheric situation, particularly in the case of complex mercury adduct reactions, and given the lack of detailed product analysis, and different carrier gases, this is not trivial. However, as shown in Pal and Ariya [19], both relative and absolute studies of the same reaction can yield the same values of rate constants within the experimental uncertainties, and thus increase the confidence in the overall result.

There are limited temperature dependence studies of reactions of elemental mercury with atmospheric oxidants (e.g., with O_3, HO, Br and Cl). Some reactions are expected to have slight temperature dependence and hence the data can be directly used for wide range of atmospheric temperature. However, some others can exhibit stronger temperature dependence. We hence require temperature dependence kinetic data that can reflect the general conditions in the troposphere. We thus recommend strongly to have the available data over a wide range of temperatures.

Mercury and halogen interaction has been experimentally studied under the conditions as summarized below.

Methyl iodide was shown to be non-reactive toward GEM under atmospheric conditions ($k < 1 \times 10^{-21}$ cm^3 molecule^{-1} s^{-1}) [36].

Ultraviolet bands of the HgCl, HgBr and HgI molecules have been investigated [41–43]. The authors coupled their measurements with computer analysis to determine precise vibrational constants for the molecules. For HgI a vibrational analysis was proposed for the E–X system. Approximate rotational constants were also determined for the C state.

The first study of Hg^0 and chlorine atoms was published in 1968 [25] HgCl was measured by time resolved absorption spectroscopy in the temperature range 383–443 K significantly higher than ambient temperatures:

$$Hg + Cl \rightarrow HgCl. \tag{1}$$

The rate constant, k_1 for the reaction of mercury with chlorine atoms was then derived to be 5.0×10^{-11} $cm^3\,molecule^{-1}\,s^{-1}$ in 720 Torr CF_3Cl and 1.5×10^{-11} $cm^3\,molecule^{-1}\,s^{-1}$ in 10 Torr CF_3Cl + 710 Torr Ar. The authors [25] mentioned that k_1 has an uncertainty of a factor of three because of the accumulation of experimental errors in evaluating the separate terms, and the rate constant can be considered to be more accurate than the order of magnitude when the results are transferred to atmospheric conditions.

Molecular chlorine was suggested to have a relatively modest reaction rate, 4×10^{-16} $cm^3\,molecule^{-1}\,s^{-1}$ [13,44–47] though the reaction was found to be strongly surface catalysed [45,46], and the experimental value should be considered as an upper limit.

In 2002 extensive kinetic and product studies on the reactions of gaseous Hg^0 with molecular and atomic halogens (X/X_2 where X = Cl, Br) were performed at atmospheric pressure (750 ± 1 Torr) and room temperature (298 ± 1 K) in air and N_2 [24]. Kinetics of the reactions with X/X_2 were studied using both relative and absolute techniques. Cold vapour atomic absorption spectroscopy (CVAAS) and gas chromatography with mass spectroscopic detection (GC-MS) were the analytical methods applied. The measured rate constants for the reactions of Hg^0 with Cl_2, Cl, Br_2, and Br were $(2.6 \pm 0.2) \times 10^{-18}$, $(1.0 \pm 0.2) \times 10^{-11}$, $< (0.9 \pm 0.2) \times 10^{-16}$, and $(3.2 \pm 0.3) \times 10^{-12}$ $cm^3\,molecule^{-1}\,s^{-1}$, respectively. Thus Cl_2 and Br_2 are not important reactants in the troposphere for the Cl_2 and Br_2 concentrations reported in literature [24].

Chlorine and bromine atoms were generated using UV and visible photolysis of molecular chlorine and bromine, respectively, in addition to UV ($300 \leqslant \lambda \leqslant 400$ nm) photolysis of chloroacetyl chloride and dibromomethane. The reaction products were analyzed in the gas-phase, in suspended aerosols and on the wall of the reactor using MS, GC-MS and inductively coupled plasma mass spectrometry (ICP-MS). The major products identified were $HgCl_2$ and $HgBr_2$ adsorbed on the wall. Suspended aerosols, collected on the micron filters, contributed to less than 0.5% of the reaction products under the experimental conditions.

Donohoue *et al.* [31] has reported two other kinetic data sets for Cl and Br reactions using a pulsed laser photolysis-pulsed laser induced fluorescence spectroscopy. These data sets are obtained using pseudo-first order conditions with respect to halogens or mercury and experiments were performed at a broad range of temperatures. The authors of these studies indicate an uncertainty estimation of ±50% in the rate coefficients due to the determination of absolute concentrations of chlorine and bromine atoms [31]. Sumner *et al.* [20] reinvestigated both reactions using a 17.3 m^3 environmental chambers equipped with fluorescent lamps and sun lamps to mimic environmental reactions, and evaluated the rate constants to be in the order of 10^{-12} cm^3 $molecule^{-1}$ s^{-1} and 10^{-11} cm^3 $molecule^{-1}$ s^{-1} for reactions of Br and Cl, respectively.

Reactions of mercury with Br-containing radicals, either Br or BrO were necessary to investigate once satellite 'BrO' total surface column measurements showed correlation with the geographical and temporal extension, where models predict AMDEs [32,48–51]. Experimental studies of XO reactions are very scarce. To our knowledge there is only one published laboratory kinetic study on the reaction of BrO with elemental mercury [33] during which, using the relative rate methods, the room temperature bimolecular rate constant for $BrO + Hg^0$ was estimated to lie within the range $10^{-15} < k < 10^{-13}$ cm^3 $molecule^{-1}$ s^{-1}, and in the later product study this reaction was estimated to be $<5 \times 10^{-14}$ cm^3 $molecule^{-1}$ s^{-1} (Raofie and Ariya, personal communication). The faster end of this range makes BrO a significant potential contributor to mercury depletion events in the Arctic, however the lower range renders this radical less effective than Br radicals using both lower and upper limit of existing experimental literature data. This is somehow in contradiction with theoretical calculations, see next section. A report was published on the first experimental product study of BrO-initiated oxidation of elemental mercury at atmospheric pressure of ~740 Torr and $T = 296 \pm 2$ K [52]. The authors used chemical ionization and electron impact mass spectrometry, gas chromatography coupled to a mass spectrometer, a MALDI-TOF mass spectrometer, a cold vapour atomic fluorescence spectrometer, and high-resolution transmission electron microscopy coupled to energy dispersive spectrometry. BrO radicals were formed using visible and UV photolysis of Br_2 and CH_2Br_2 in the presence of ozone. They analyzed the products in the gas phase, on suspended aerosols and on wall deposits, and identified HgBr, HgOBr or HgBrO, and HgO as reaction products. Experimentally, they were unable to distinguish between HgBrO and HgOBr. The existence of stable Hg^+ in form of HgBr, along with Hg^{2+} upon BrO-initiated oxidation of Hg^0, suggests that in field studies, it is fundamental to selectively quantify various mercury species in mercury aerosols and deposits both in oxidation state I and II. The majority of mercury containing products were identified as deposits, however, aerosols accounted for a substantial portion of products. Noticeably the authors pointed out that although the extent of heterogeneous reactions in their experiments were reduced significantly, the existence of these reactions should have not been totally neglected under their experimental conditions.

No definite conclusions on the potential primary or secondary reactions of BrO, can be made at this stage. Even considering one order of magnitude uncertainties in the existing kinetic data, Br reactions make it the likely radical to explain elemental mercury depletion in the Arctic. Two independent studies [10,32] support this conclusion. The existing kinetic results indicate that the direct BrO impact is less important than Br, but further studies are required to examine this conclusion. For example, A. Saiz-Lopez *et al.* [53] have unexpectedly, recently discovered via long path DOAS measurements, significant amounts of iodine oxide (IO) above the Antarctic ice, and that bromine persists there for several months, throughout the summer, thus giving rise to a greater oxidizing effect than formerly thought possible, given observations in the Arctic.

Interesting models have been developed on the importance of iodine chemistry and its potential impact on mercury depletion events [54,55]. Since there is no existing laboratory study on kinetics and products of I_2, I and IO with elemental mercury, we encourage additional studies in this domain to evaluate further the implication of iodine chemistry in the troposphere.

Significant concentrations of halogens are observed predominantly over the polar regions and over the marine boundary layer and not generally over the continents, with the exception of coastal regions, high halide source regions such as salt lakes, and some industrial regions where halogens are widely used [56].

3. THEORETICAL EVALUATION OF KINETIC DATA

The possibility of theoretically predicting the thermochemistry of mercury-containing species of atmospheric interest is of strong importance due to the relative lack of accurate experimental information at all temperatures relevant to ambient air, especially to temperatures different than room temperature. They also serve fundamentally to further comprehend the complex reaction mechanisms. Accurate *ab initio* studies for measurements such as heats of formation, reaction enthalpies, and activation energies are particularly challenging, particularly in light of the large nuclear charge (80) and large number of electrons intrinsic to mercury. There is a detailed review on *ab initio* thermochemical and kinetic studies on mercury reactions [15] and hence we discuss previous studies mainly in relation to experimental results. The existing theoretical kinetic data are also shown in Table 4.1. *Ab initio* calculations rely on careful choice of electron correlation method, treatment of relativistic effects, basis set truncation errors, etc., in order to obtain accurate kinetic data. The latter depends intimately on the underlying potential energy surface. A rigorous calculation of the rate coefficient for a given reaction generally involves either quantum scattering or classical trajectory calculations, which in turn require a global or semi-global potential energy surface (PES). While these treatments are feasible for relatively small systems depending on the required accuracy of the underlying PES, most studies employ more approximate treatments of the reaction dynamics, e.g., transition state theory (TST) or RRKM theory (Rice–Ramsberger–Kassel–Marcus theory).

For bimolecular reactions involving a barrier, transition state theory is often used. The basic tenet is that there exists a critical configuration lying between reactants and products where all trajectories arising from reactants are assumed to irreversibly lead to products [57,58]. For reactions that proceed without a barrier, e.g., unimolecular dissociation or recombination reactions, RRKM theory is often employed. The use of RRKM involves two central approximations (cf. Steinfeld *et al.* and Gilbert and Smith [57,59] and references therein): (i) as with transition state theory, RRKM assumes the existence of a critical configuration between reactants and products which is not recrossed and (ii) the energy of the excited reactant is distributed randomly throughout all the available molecular states. To satisfy the first approximation, it is generally very important to employ the variational version, which is equivalent to a microcanonical VTST calculation. So as above, one needs to calculate structures and vibrational frequencies along the reaction path. In order to satisfy the second criterion, the reactant must be a molecule large enough to provide efficient intramolecular vibrational energy redistribution. Hence, the use of RRKM for atom-atom recombination reactions should probably not be used as the main approach for theoretical kinetic evaluation. In these cases, quasiclassical trajectory calculations would seem to be the most reliable method. For reaction rate constant calculations for barrierless reactions using RRKM and VTST-like methods, the Variflex program [60] is a convenient choice for polyatomic systems, since it also allows several options for the calculation of pressure effects on the rate constant (standard VTST yields only a high pressure limit rate constant).

Since 2003 several studies have contributed to the understanding of the reaction system of reactions (2a), (2b), (3a), (3b) and (4) [26,27,34,61,62,65]

$$\mathrm{Hg} + \mathrm{X} \rightarrow \mathrm{HgX}, \tag{2a}$$

$$\mathrm{HgX} \rightarrow \mathrm{Hg} + \mathrm{X}, \tag{2b}$$

$$\mathrm{HgX} + \mathrm{X} \rightarrow \mathrm{HgX_2}, \tag{3a}$$

$$\mathrm{XHg} + \mathrm{X} \rightarrow \mathrm{X} + \mathrm{HgX}, \tag{3b}$$

$$\mathrm{Hg} + \mathrm{XO} \rightarrow \mathrm{HgO} + \mathrm{X}, \tag{4}$$

where X is either Cl, Br or I.

The contributions will be briefly sketched here following the chronological order of appearance in the literature. Shepler and Peterson [61] calculated the potential energy curves of HgO using multi reference configuration interaction (MRCI) as well as coupled cluster theory (CCSD(T)). Their result showed that the reaction $\mathrm{Hg} + \mathrm{BrO} \rightarrow \mathrm{HgO} + \mathrm{Br}$ is strongly endothermic. Notably they also found that HgO was stable with respect to Hg^0 by just 4 kcal mol^{-1}, suggesting that the previous experimental results from the nineteeneighties were erroneous [63,64].

Khalizov *et al.* [26] used DFT and *ab intio* calculations at B3LYP and QCISD level of theory to determine geometry optimisations and frequencies for various molecules, HgX and $\mathrm{HgX_2}$, where X = F, Cl or Br. Furthermore reaction enthalpies were calculated for the nine possible reactions (2a), (3a) and $\mathrm{Hg} + \mathrm{X_2} \rightarrow \mathrm{HgX_2}$. The back reaction rate constants for the reactions (2b) were calculated in the high pressure limit using collision theory comparing to calculations using canonical VTST. These rate constants are tabulated in Table 4.1.

Tossell [62] calculated energetics for oxidation of GEM for various reactions including (2a), (3a) and (4), with X = Br and Cl. The methods and levels of theory included Hartree–Fock (HF), Moller–Plesset to the second order (MP2), quadratic CI (QCISD) and CCSD(T). Novel results included findings of the optical transitions of HgO, HgX and $\mathrm{HgX_2}$. These indicated that HgO as well as HgX would be unstable towards sunlight in the troposphere, whereas $\mathrm{HgX_2}$, would be quite stable for energies in the visible region.

Goodsite *et al.* [27] calculated optimised geometries and molecular parameters at B3LYP level of theory for HgX and HgXY where X = Br (and I and OH) and Y = Br, I, OH and $\mathrm{O_2}$. This was followed by RRKM theory to yield the reaction rate constants for reactions (2a) and (2b) for X = Br and I, and (3a) for X = Br. The reaction rates for X = Br are tabulated in Table 4.1.

Balabanov *et al.* [34] computed the potential energy surfaces for $HgBr_2$ using internally contracted MRCI. The potential surfaces were then used for quasi classical trajectory (QST) and VTST calculations to evaluate the rate constants for the seven reactions including (3a) and (3b), $Hg + Br_2 \rightarrow HgBr_2$ and $HgBr_2 \rightarrow$ products. The calculations yielded some very interesting results, i.e. they found the rates for the reactions $Hg + Br_2 \rightarrow HgBr_2$ and $Hg + Br_2 \rightarrow HgBr + Br$ to be as small as of $k = 2.7 \times 10^{-31}\ cm^3\ mol^{-1}\ s^{-1}$ and $k = 3.4 \times 10^{-31}\ cm^3\ mol^{-1}\ s^{-1}$, respectively. They discuss the likeness of these reaction paths to the ones of the reactions of Hg with BrO, and suggest that a more complex mechanism like the one introduced in the modeling work of Calvert and Lindberg [54,55] is considered for explaining the larger reaction rates found in experiments [24,33]:

$$HgBr + Br_2 \rightarrow HgBr_2 + Br. \tag{5}$$

As can be seen in Table 4.1 they also found the reaction rate for reaction (3a) to be lower than Goodsite *et al.* [27], and explained this by the higher level of theory used in the Balabanov *et al.* [34] calculations.

Additionally we want to mention two investigations by Shepler *et al.* [65,66]. The first [65] is of the reactions Hg + IX for X = I, Br, Cl and O with respect to enthalpies; for the stable triatomics also heats of formation, bond lengths and harmonic vibrational frequencies, together with dissociation energies, equilibrium bond lengths, and harmonic vibrational frequencies for the diatomics involved. The calculations were carried out employing the CCSD(T) method. There are a wealth of results that are compared to experimental outcomes and that can be used in atmospheric modeling. We repeat here our call for laboratory studies on the reaction chemistry of mercury and the iodine compounds I_2, I and IO, as these are missing in this context.

The other study [66] is a very recent one on aqueous micro solvation of mercury halide species. The methods and level of theory employed were MP2 and DFT/B3LYP. The general trend was that oxidation of mercury halide species was seen to be favoured by the presence of water molecules. Notably the reactions (2a) and (3a)-like channel $HgX + Y \rightarrow HgXY$ together with the reaction $Hg + XY \rightarrow HgXY$ became more exothermic in the presence of water, whereas the abstraction channel $HgX + Y \rightarrow Hg + XY$ became less exothermic.

To summarise the outcome of these papers with respect to the reaction system (2)–(4) Reaction (4) was investigated for X = Br [61,62,67] and X = I [65] to be endothermic and most probably without any importance in the atmosphere, in the absence of water, whereas reaction (2) is exothermic [26,27,34,62]. Khalizov concluded that Hg + Br might be the dominant process for atmospheric mercury depletion episodes (AMDE) occurring during Arctic Spring. This conclusion is in line with previous laboratory kinetic studies of the same reaction system [24]. This conclusion is further supported by Goodsite *et al.* [27] that studied the temperature dependence of the reaction and showed that the HgBr intermediate is stabilised towards uni-molecular degradation at low temperatures which permits the addition of the Br reaction (3) and thus $HgBr_2$ is a possible candidate for the formation of the otherwise unknown RGM. Balabanov *et al.* [34] suggest a more complex reaction path (5), also including HgBr.

4. PERSPECTIVES

Despite the novel positive acquisitions of knowledge from experimental and theoretical studies of gas-phase elemental mercury chemistry there are still large gaps before a complete understanding of the fate of mercury in the atmosphere is obtained. It is essential to provide kinetic data and information about formed products. There are some limited studies on the kinetics of gas-phase elemental mercury oxidation on surfaces [68–70]. However, experimental studies on uptake or kinetics of heterogeneous reactions of mercury on various environmentally relevant surfaces such as ice, snow, and aerosols and biomaterials, are needed.

The present paper describes the most important progress that has been made within the understanding of the atmospheric chemistry of mercury within the application of theoretical calculations and experimental studies for determination of reaction coefficients and mechanisms with halogens and other reactants. There are still large uncertainties to cope with before a reliable description of dynamics and fate of mercury can be established. Theoretical calculations represent a very cost effective method to get the first information about rate constants, reaction products and as to what experimentalists should examine. Finally, theoretical calculations can document that we actually have a full understanding of the fundamental processes of atmospheric mercury. The study of IO [53] in the Antarctic opens the possibility that I and IO plays an important role in the oxidation of Hg^0. These reaction mechanisms should continue to be studied in the field and with theoretical methods. As most laboratory studies of the oxidation mercury in the atmosphere are carried out at room temperature it is very important that theoretical calculations state the temperature dependence of the various reaction steps and the thermally stability of the reaction intermediates and end products.

A particular challenge is the reaction between Hg^0 and ozone to form HgO and O_2. There are large discrepancies in the existing laboratory study and theoretical studies are difficult because the spin of reactants are different from those of the reaction intermediates, such as HgO_3, and the end product.

ACKNOWLEDGEMENTS

We acknowledge National Science and Engineering Research Council of Canada (NSERC), and Canadian Foundation for innovation for financial support. The Danish Environmental Protection Agency financially supported this work with means from the MIKA/DANCEA funds for Environmental Support to the Arctic Region. The findings and conclusions presented here do not necessarily reflect the views of the Agency. Michael E. Goodsite was financially supported by NERI and the University of Southern Denmark, Faculty of Science. Mette M.-L. Grage gratefully acknowledges The Swedish Research Council. The authors thank Karen

Cauthery for her editorial expertise and assistance as well as the referees and A.J. Hynes for comments which helped us greatly improve this paper.

REFERENCES

[1] J.A. Tossell, *J. Phys. Chem. A* **110** (2006) 2571.
[2] J.D. Lalonde, A. Poulain, M. Amyot, *Environ. Sci. Technol.* **36** (2002) 174.
[3] A. Dommergue, C.P. Ferrari, L. Poissant, *Environ. Sci. Technol.* **37** (2003) 3289.
[4] C.P. Ferrari, A. Dommergue, C.F. Boutron, *Atmos. Environ.* **38** (2004) 2727.
[5] S.B. Brooks, A. Saiz-Lopez, A.H. Skov, *et al.*, *Geophys. Res. Lett.* **33** (2006), Art. No. L13812 JUL 13.
[6] A. Ryaboshapko Jr., O. Russell Bullock, J. Christensen, M. Cohen, A. Dastoor, I. Ilyin, G. Petersen, D. Syrakov, R.S. Artz, D. Davignon, *Sci. Total Environ.* **376** (2007) 228.
[7] K. Aspmo, C. Temme, T. Berg, C. Ferrari, P.A. Gauchard, X. Fain, G. Wibetoe, *Environ. Sci. Technol.* **40** (2006) 4083.
[8] A.J. Poulain, E. Garcia, M. Amyot, P.G.C. Campbell, P.A. Ariya, *Geochim. Cosmochim. Acta* **71** (2007) 3419.
[9] A.P. Dastoor, D. Davignon, N. Theys, M. Van Roozendael, A. Steffen, P.A. Ariya, *Environ. Sci. Technol.* (2007), in press.
[10] P.A. Ariya, A.P. Dastoor, M. Amyot, W.H. Schroeder, L. Barrie, K. Anlauf, F. Raofie, A. Ryzhkov, D. Davignon, J. Lalonde, A. Steffen, *Tellus B* **56** (2004) 397.
[11] S.E. Lindberg, S.B. Brooks, C.-J. Lin, K.J. Scott, M.S. Landis, R.K. Stevens, M. Goodsite, A. Richter, *Envirn. Sci. Technol.* **36** (2002) 1245.
[12] J. Sommar, X.B. Feng, K. Gardfeldt, *J. Environ. Monit.* **1** (1999) 435.
[13] W.H. Schroeder, G. Yarwood, H. Niki, *Water, Air, Soil Pollut.* **56** (1991) 653.
[14] C.J. Lin, S.O. Pehkonen, *Atmos. Environ.* **33** (1999) 2067.
[15] P.A. Ariya, K. Peterson, in: N. Pirrone (Ed.), *Atmospheric Chemical Transformation of Elemental Mercury, Mercury in Environment*, Kluwer, Dordrecht, 2005.
[16] A. Steffen, T. Douglas, M. Amyot, P. Ariya, K. Aspmo, T. Berg, J. Bottenheim, A. Dastoor, R. Ebinghaus, C. Ferrari, K. Gårdfeldt, M. Goodsite, D. Lend, A. Poulain, C. Scherz, H. Skov, J. Sommar, C. Temme, *Atmos. Chem. Phys. Discuss.* **7** (4) (2007) 9283.
[17] W.H. Schroeder, K.G. Anlauf, L.A. Barrie, J.Y. Lu, A. Steffen, D.R. Schneeberger, T. Berg, *Nature* **394** (1998) 331.
[18] B. Hall, *Water, Air, Soil Pollut.* **80** (1995) 301.
[19] B. Pal, P.A. Ariya, *J. Phys. Chem.-Chem. Phys.* **6** (2004) 752.
[20] A.-L. Sumner, C. Spicer, in: N. Pirrone, K.R. Mahaffey (Eds.), *Dynamics of Mercury Pollution on Regional and Global Scales: Atmospheric Processes and Human Exposures Around the World*, Kluwer, Dordrecht, 2005.
[21] J. Sommar, K. Gårdfeldt, D. Stromberg, X. Feng, *Atmos. Environ.* **35** (2001) 3049.
[22] G.C. Miller, J. Quashnick, V. Hebert, *Abstr. Pap. Am. Chem. Soc.* (2001), 221st:AGRO-016.
[23] D. Bauer, L. D'Ottone, P. Campuzaon-Jos, A.J. Hynes, *J. Photochem. Photobiol.* **157** (2003) 247.
[24] P.A. Ariya, A. Khalizov, A. Gidas, *J. Phys. Chem. A* **106** (2002) 7310.
[25] D.G. Horne, R. Gosavi, O.P. Strausz, *J. Chem. Phys.* **48** (1968) 4758.
[26] A.F. Khalizov, B. Viswanathan, P. Larregaray, P.A. Ariya, *J. Phys. Chem. A* **107** (2003) 6360.
[27] M.E. Goodsite, J.M.C. Plane, H. Skov, *Environ. Sci. Technol.* **38** (2004) 1772.
[28] D.L. Donohoue, D. Bauer, A.J. Hynes, *J. Phys. Chem. A* **109** (2005) 7732.
[29] G. Grieg, H.E. Gunning, O.P. Strausz, *J. Chem. Phys. Lett.* **52** (1970) 3684.
[30] C.D. Holmes, D.J. Jacob, X. Yang, *Geophys. Res. Lett.* **33** (2006) L20808, doi:10.1029/2006GL027176.
[31] D.L. Donohoue, D. Bauer, B. Cossairt, A.J. Hynes, *J. Phys. Chem. A* **110** (2006) 6623.
[32] H. Skov, J.H. Christensen, M.E. Goodsite, N.Z. Heidam, B. Jensen, P. Wåhlin, G. Geernaert, *Environ. Sci. Technol.* **38** (2004) 2373.
[33] F. Raofie, P.A. Ariya, *J. Phys. IV* **107** (2003) 1119.
[34] N.B. Balabanov, B.C. Shepler, K.A. Peterson, *J. Phys. Chem. A* **109** (2005) 8765.
[35] J. Sommar, M. Hallquist, E. Ljungström, O. Lindqvist, *J. Atmos. Chem.* **27** (1997) 233.

[36] J.J.S. Tokos, B. Hall, J.A. Calhoun, E.M. Prestbo, *Atmos. Environ.* **32** (1998) 823.
[37] H. Niki, P.S. Maker, C.M. Savage, L.P. Breitenbach, *J. Phys. Chem.* **87** (1983) 3722.
[38] H. Niki, P.S. Maker, C.M. Savage, L.P. Breitenbach, *J. Phys. Chem.* **87** (1983) 4978.
[39] J. Sommar, M. Hallquist, E. Ljungstrom, *Chem. Phys. Lett.* **257** (1996) 434.
[40] B.J. Finlayson-Pitts, J.N. Pitts Jr., *Chemistry of the Upper and Lower Atmosphere*, Academic Press, San Diego, 2000.
[41] A.K. Rai, S.B. Rai, D.K. Rai, *J. Phys. B: At. Mol. Phys.* **15** (1982) 3239.
[42] A.K. Rai, S.B. Rai, D.K. Rai, *J. Phys. B: At. Mol. Phys.* **16** (1983) 1907.
[43] A.K. Rai, S.B. Rai, D.K. Rai, *J. Phys. B: At. Mol. Phys.* **17** (1984) 1817.
[44] R. Menke, G. Wallis, *Am. Ind. Hyg. Assoc. J.* **41** (1980) 120.
[45] A.K. Medhekar, M. Rokni, D.W. Trainor, J.H. Jacob, *Chem. Phys. Lett.* **650** (1979) 600.
[46] I. Skare, R. Johansson, *Chemosphere* **24** (1992) 1633.
[47] C. Seigneur, J. Wrobel, E. Constantinou, *Environ. Sci. Technol.* **28** (1994) 1589.
[48] A. Richter, F. Wittrock, M. Eisinger, J.P. Burrows, *Geophys. Res. Lett.* **25** (1998) 2683.
[49] A. Richter, F. Wittrock, A. Ladstätter-Weissenmayer, J.P. Burrows, *Adv. Space Res.* **29** (2002) 1667.
[50] R.W. Müller, H. Bovensmann, J.W. Kaiser, A. Richter, A. Rozanov, F. Wittrock, J.P. Burrows, *Adv. Space Res.* **29** (2002) 1655.
[51] M. van Roozendael, T. Wagner, A. Richter, I. Pundt, D.W. Arlander, J.P. Burrows, M. Chipperfield, C. Fayt, P.V. Johnston, J.-C. Lambert, K. Kreher, K. Pfeilsticker, U. Platt, J.-P. Pommereau, B.-M. Sinnhuber, K.K. Toernkvist, F. Wittrock, *Adv. Space Res.* **29** (2002) 1661.
[52] F. Raofie, P.A. Ariya, *Environ. Sci. Technol.* **38** (2004) 4319.
[53] A. Saiz-Lopez, *et al.*, *Science* **317** (2007), doi:10.1126/science.1141408 348.
[54] J.G. Calvert, S.E. Lindberg, *Atmos. Environ.* **38** (2004) 5087.
[55] J.G. Calvert, S.E. Lindberg, *Atmos. Environ.* **38** (2004) 5105.
[56] W.R. Simpson, R. von Glasow, K. Riedel, P. Anderson, P. Ariya, J. Bottenheim, J. Burrows, L. Carpenter, U. Friess, M.E. Goodsite, D. Heard, M. Hutterli, H.-W. Jacobi, L. Kaleschke, B. Neff, J. Plane, U. Platt, A. Richter, H. Roscoe, R. Sander, P. Shepson, J. Sodeau, A. Steffen, T. Wagner, E. Wolff, *Atmos. Chem. Phys. Discuss.* **7** (2007) 4285, www.atmos-chem-phys-discuss.net/7/4285/2007/.
[57] J.I. Steinfeld, J.S. Francisco, W.L. Hase, *Chemical Kinetics and Dynamics*, Prentice-Hall, Englewood Cliffs, NJ, 1989.
[58] D.G. Truhlar, B.C. Garrett, S.J. Klippenstein, *J. Phys. Chem.* **100** (1996) 12771.
[59] R.G. Gilbert, S.C. Smith, *Theory Unimolecular and Recombination Reactions*, Blackwell Scientific, Oxford, 1990.
[60] S.J. Klippenstein, A.F. Wagner, S.H. Robertson, R.C. Dunbar, D.M. Wardlaw, http://chemistry.anl.gov/variflex (1999).
[61] B.C. Shepler, K.A. Peterson, *J. Phys. Chem. A* **107** (2003) 1783.
[62] J.A. Tossell, *J. Phys. Chem. A* **107** (2003) 7804.
[63] M. Grade, H. Hirschwald, *Ber. Bunsen-Ges. Phys. Chem.* **86** (1982) 899.
[64] M.W. Chase Jr., C.A. Davies, J.R. Downey Jr., D.J. Frurip, R.A. McDonal, A.N.J. Syverud, *Phys. Chem. Ref. Data* **14** (Suppl. No. 1) (1985).
[65] B.C. Shepler, N.B. Balabanov, K.A. Peterson, *J. Phys. Chem. A* **109** (2005) 10363.
[66] B.C. Shepler, A.D. Wright, N.B. Balabanov, K.A. Peterson, *J. Phys. Chem. A* **111** (2007) 11342.
[67] A.D. Isaacson, D.G.J. Truhlar, *Chem. Phys.* **76** (1982) 1380.
[68] T.G. Lee, P. Biswas, E. Hedrick, *Ind. Eng. Chem. Res.* **43** (2004) 1411.
[69] J.R.V. Flora, R.D. Vidic, W. Liu, R.C. Thurnau, *J. Air Waste Manage. Ass.* **48** (1998) 1051.
[70] R.D. Vidic, M.-T. Chang, R.C. Thurnau, *J. Air Waste Manage. Ass.* **48** (1998) 247.

CHAPTER 5

Photolysis of Long-Lived Predissociative Molecules as a Source of Mass-Independent Isotope Fractionation: The Example of SO_2

James R. Lyons*

Contents

1. Introduction 58
2. Absorption Spectra for SO_2 Isotopologues 59
3. Photolysis of SO_2 Isotopologues in a Low O_2 Atmosphere 64
4. Photolysis of SO_2 in the Modern Atmosphere 68
5. Improvements to Spectra and Additional Sources of S-MIF 70
6. Conclusions and Broader Implications 73

Acknowledgements 73

References 74

Abstract

Laboratory experiments have demonstrated that a mass-independent fractionation (MIF) signature is present in elemental sulfur produced during SO_2 photolysis, but the underlying mechanism remains unknown. I report here the results of chemical kinetics modeling of self-shielding during photodissociation of SO_2 in the $\tilde{C}^1B_2 - \tilde{X}^1A_1$ bands from 190 to 220 nm. This band system is dominated by a bending mode progression that produces shifts in the absorption spectrum upon sulfur isotope substitution. Self-shielding in the rotationally-resolved lines of $^{32}SO_2$ produces MIF signatures in SO and residual SO_2. Using approximate synthetic spectra for the sulfur isotopologues of SO_2, I show that SO_2 photolysis yields a sulfur MIF signature that can account for much of the laboratory MIF measured, and is in qualitative agreement with $\Delta^{33}S$ and $\Delta^{36}S$ values observed in Archean rocks.

* Institute of Geophysics and Planetary Physics, and Department of Earth and Space Sciences, University of California, Los Angeles, CA 90095, USA
E-mail: jrl@ess.ucla.edu

Advances in Quantum Chemistry, Vol. 55
ISSN 0065-3276, DOI: 10.1016/S0065-3276(07)00205-5

1. INTRODUCTION

The origin of mass-dependent fractionation (MDF) in isotope systems lies in the mass dependence of the molecular properties (e.g., zero-point energy) and physical processes (e.g., evaporation) affecting the compound. If a compound comprised of atoms with 3 or more stable isotopes, such as oxygen or sulfur, deviates from a mass-dependent relationship, the compound is said to exhibit mass-independent fractionation (MIF). MIF signatures are not affected by mass-dependent processes, and so are excellent tracers of the small number of mass-independent processes that exist in nature.

MIF in oxygen isotopes (O-MIF) is well known in primitive meteorites [1] and in atmospheric O_3 [2,3], and MIF in sulfur isotopes (S-MIF) has been discovered in Archean rocks [4] and modern ice core sulfates [5]. In terrestrial environments O-MIF is present in many atmospheric molecules and aeolian sediments, and is nearly always a result of interactions with atmospheric ozone [6]. It is believed that MIF in O_3 results from the non-statistical randomization of energy in vibrationally excited O_3 during the O_3 formation reaction, $O + O_2 \rightarrow O_3$, in a manner that depends on the symmetry of the O_3 isotopomer [7]. The source of O-MIF in primitive meteorites is unknown but has been attributed to self-shielding during photodissociation of CO in the solar nebula [3,8–10], and also to ozone-like non-statistical reactions on mineral grain surfaces [11], a hypothesis not yet verified in the laboratory.

The formation of MIF during self-shielding photolysis reactions is the focus of this article, and is well exemplified by CO. Self-shielding during CO photodissociation is possible because long-lived predissociative states excited by 90–110 nm photons result in numerous rotationally resolved absorption bands. Line positions in these bands are a function of isotope masses for O (and C), and saturation of the lines of a given isotopologue are dependent on the abundance of the isotopologue. The dependence of photolysis rates on isotopologue abundance rather than mass means that photodissociation in a natural mixture of isotopologues yields products that have a strong MIF signature, with the magnitude of MIF dependent on the difference between the line peaks and the absorption continuum. Many molecules that undergo indirect photodissociation exhibit line-type spectra (e.g., CO, N_2, SO, O_2, SO_2), but others, especially those that undergo direct photodissociation, have very little structure or only broad vibronic structure (e.g., N_2O, O_3). Photolysis of molecules of the latter type do yield mass-dependent isotope effects [12], but MIF due to self-shielding in these systems is small to negligible because of the lack of rotational features. Other possible sources of MIF during photolysis include cross section amplitude effects resulting from changes in the shape of the wavefunction upon isotope substitution (e.g., [12,13]), and perhaps dynamical effects associated with resonances.

The discovery of S-MIF in Archean and Paleoproterozoic sedimentary sulfides and sulfates [4] promises to yield both qualitative and quantitative insights into the composition of the paleoatmosphere. To fully realize the potential of this geochemical record, the mechanism causing the S-MIF must be quantitatively understood. Closed-system photolysis experiments on SO_2, the most abundant sulfur gas emitted during volcanism, have shown that SO_2 photolysis at a variety of wavelengths produces MIF in elemental sulfur residue [14,15]. However, extracting a mechanism from such experiments is difficult, especially for sulfur species. Here, using approximate SO_2 isotopologue spectra, I show that self-shielding during SO_2 photodissociation can produce a S-MIF signature in the product SO and undissociated SO_2. In a low O_2 atmosphere, S-MIF signatures were preserved and eventually recorded in sedimentary rocks.

For relatively small isotopic fractionations, the magnitudes of sulfur MIF for the four stable isotopes of sulfur are given by the linear expressions

$$\Delta^{33}S = \delta^{33}S - 0.515\delta^{34}S, \tag{1a}$$

$$\Delta^{36}S = \delta^{36}S - 1.90\delta^{34}S, \tag{1b}$$

where and $\delta^{x}S = 10^3((^{x}S/^{32}S)_{\text{sample}}/(^{x}S/^{32}S)_{\text{ref}} - 1)$ for $x = 33$, 34 or 36 and for a Vienna Canyon Diablo Troilite (VCDT) reference. The values 0.515 and 1.90 are determined from the ratios of equilibrium partition functions for the appropriate isotope exchange reactions. The ratio $\Delta^{36}S/\Delta^{33}S$ is useful in identifying MIF mechanisms, and in comparing observations from the rock record to experiments and theory [4,14,16]. Archean sulfides and sulfates have been found to have $\Delta^{36}S/\Delta^{33}S \sim -0.9$ [4]. Sulfate produced in the plumes of recent Plinian eruptions and collected in Antarctic ice cores [5,17] have fractionation signatures distinct from Archean sedimentary sulfides and sulfates (e.g., $\Delta^{36}S/\Delta^{33}S \sim -3.3$ for the Agung eruption), and are believed to result from either SO_2 photooxidation [5,17] or SO_3 photolysis [18].

2. ABSORPTION SPECTRA FOR SO_2 ISOTOPOLOGUES

In laboratory experiments S-MIF has been observed in photolysis of SO_2 [14,15] and CS_2 [19], but negligible to only very slight MIF is seen in H_2S photolysis [20]. Elemental sulfur (S_{el}), or a CS_x polymer in the case of CS_2, is produced in all three photolysis experiments by a complex sequence of photolysis, bimolecular and trimolecular gas phase reactions, and wall reactions. Comparison of absorption spectra for SO_2, H_2S and CS_2 reveals vibronic structure in SO_2 and CS_2 spectra but very little structure in the H_2S spectrum (Figure 5.1), and suggests that the act of photodissociation of the parent gas could be the source of S-MIF for SO_2 and CS_2.

In a low-O_2 atmosphere with negligible O_3 absorption SO_2 photodissociation occurs by

$$SO_2(\tilde{X}^1A_1) + h\nu(165\text{–}220\text{ nm}) \rightarrow SO_2(\tilde{C}^1B_2) \rightarrow SO(^3\Sigma^-) + O(^3P). \tag{R1}$$

Absorption by CO_2 and H_2O in the ancient atmosphere is expected to limit the short wavelength photons to wavelengths greater than about 190 nm. Figure 5.2a (black curve) shows a high-resolution spectrum for $^{32}SO_2$ in the region 190–220 nm at 213 K [21]. The absorption continuum is about a factor of 3 below the line peaks. The vibronic structure is primarily due to a (1, v_2, 2) bending mode progression

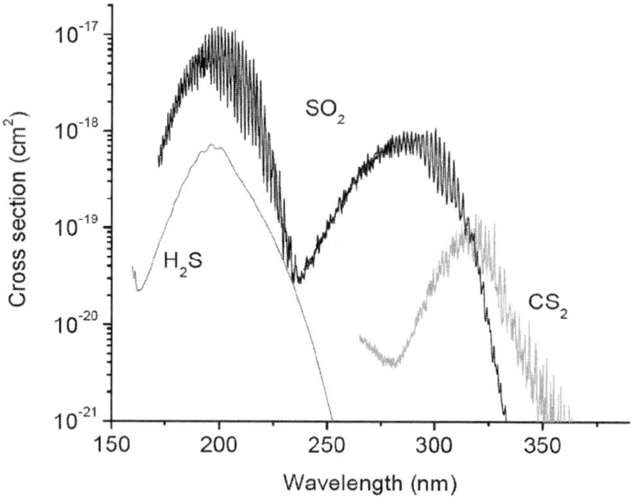

FIGURE 5.1 Absorption cross section spectra of SO_2, H_2S and CS_2. Vibronic bands are evident in SO_2 and CS_2 but are not present in H_2S. The spectra resolutions are 0.05–0.1 nm for SO_2 [44,45], 0.05 nm for CS_2 [46], and 0.005 nm for H_2S [47]. The H_2S cross sections have been reduced by a factor of 10.

with an additional weaker $(3, v_2, 0)$ bending mode progression [22]. The spectrum from 190 to 220 nm includes bands with an upper state vibrational quantum number from $v_2' = 4$ to 22. Although the band assignments become progressively more ambiguous because of anharmonic coupling [22], I will designate bands by v_2 in the $(1, v_2, 2)$ progression down to 190 nm. The rotational line structure (Figure 5.2b) is resolved but very congested due to perturbation from other states. Line widths are narrow or even Doppler-limited ($\sim$0.04 cm^{-1} at 295 K) for wavelengths less than 220 nm [23,24], indicating predissociation lifetimes longer than 100 ps.

For a bending mode progression in a harmonic oscillator, the shifts in the band head locations upon sulfur isotope substitution are given by [25]

$$\nu^*(v_2) - \nu(v_2) = \sum_i (\omega_i^{*\prime} - \omega_i^{\prime})\left(v_i^{\prime} + \frac{1}{2}\right) - \sum_i (\omega_i^{*\prime\prime} - \omega_i^{\prime\prime})\left(v_i^{\prime\prime} + \frac{1}{2}\right), \tag{2}$$

where ω_i are the three harmonic vibrational frequencies for SO_2, and v_i are the corresponding vibrational quantum numbers, primed (double primed) quantities indicate the excited (ground) electronic state, and the asterisk superscript indicates a rare sulfur isotope. The second term on the right hand side of Eq. (2) is the change in the ground state zero-point energy upon isotope substitution. The quantities $\omega_i^{*\prime} - \omega_i^{\prime}$ and $\omega_i^{*\prime\prime} - \omega_i^{\prime\prime}$ can be computed from *ab initio* modeling of SO_2 and from previously published excited state and ground state harmonic frequencies (e.g., [26,27]).

Alternatively, band peak shifts can be computed from *ab initio* absorption spectra for SO_2 isotopologues [28], which is the method I've used here. The *ab initio* spectra were computed by time domain methods (Lanczos recursion and Chebyshev propagation) using a new potential energy surface for SO_2. The *ab initio*

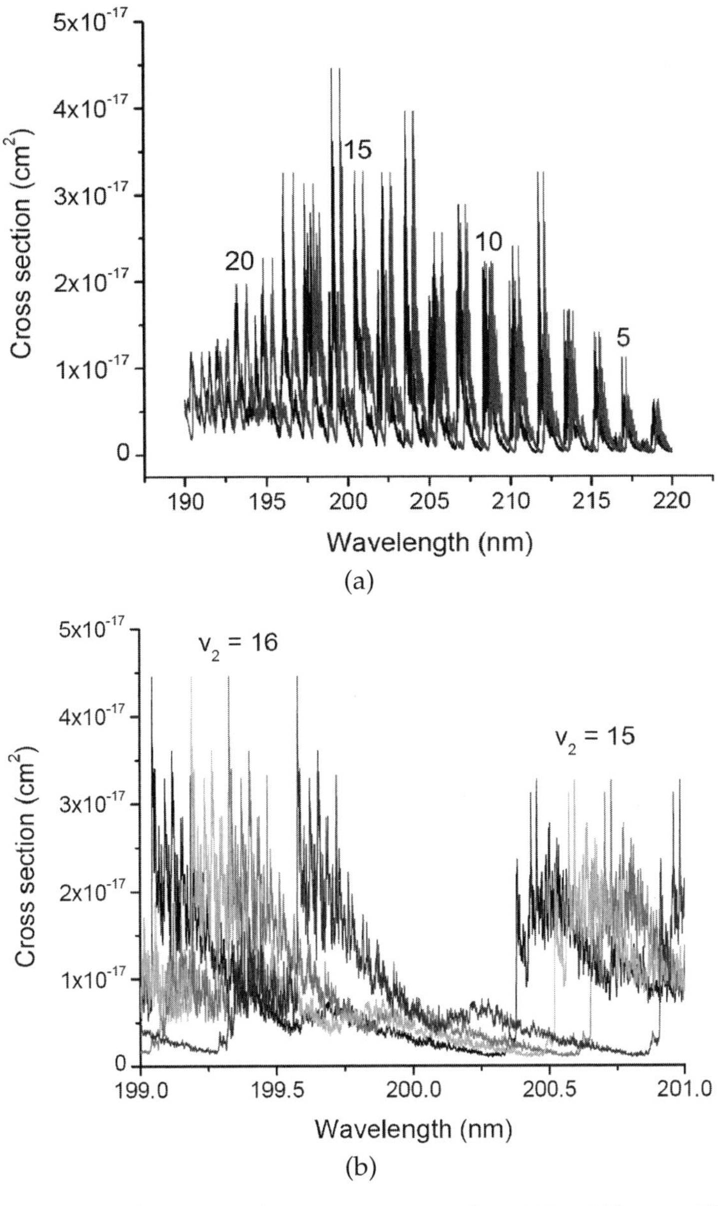

FIGURE 5.2 (a) High resolution SO_2 absorption spectrum from 190 to 220 nm at 213 K [21]. Approximate spectrum for $^{36}SO_2$ (blue) is also shown as a red-shifted version of the $^{32}SO_2$ spectrum (black). Band vibrational assignments are $(v_1, v_2, v_3) = (1, v_2, 2)$, with v_2 as labeled. At shorter wavelengths anharmonic coupling renders v_2 a less meaningful quantum number [22], but it remains a convenient band label. (b) Two vibronic bands ($v_2 = 16$ and $v_2 = 15$) in the model SO_2 absorption spectrum with SO_2 isotopologue spectra offset according to their bending mode quantum number (black—$^{32}SO_2$, green—$^{33}SO_2$, red—$^{34}SO_2$, blue—$^{36}SO_2$). Based on $^{32}SO_2$ cross section data [21]. (For interpretation of the references to color in this figure legend, the reader is referred to the web version of this book.)

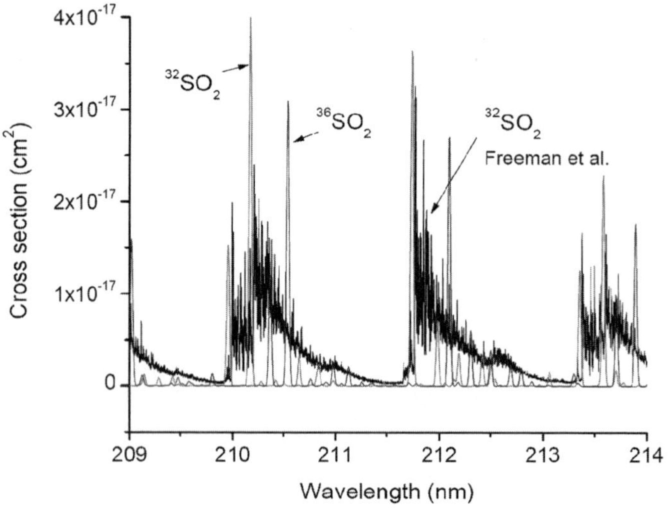

FIGURE 5.3 Normalized *ab initio* spectra for $^{32}SO_2$ (red) and $^{36}SO_2$ (blue) [28], and laboratory absorption spectra for $^{32}SO_2$ at 213 K (black) [21]. Note that the peaks of the vibrational progressions agree, but the *ab initio* spectra do not match well with the measured absorption continuum. The agreement between the *ab initio* and measured vibronic band locations deteriorates at wavelengths shortward of ~200 nm. (For interpretation of the references to color in this figure legend, the reader is referred to the web version of this book.)

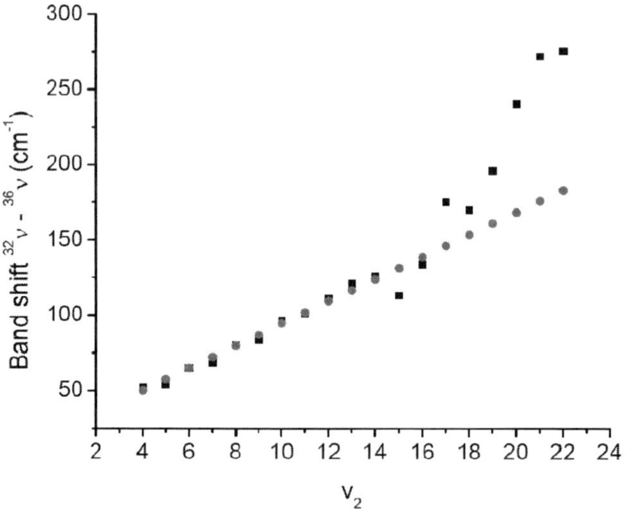

FIGURE 5.4 Band shifts between $^{32}SO_2$ and $^{36}SO_2$ as determined from Ran et al. [28] *ab initio* results versus the bending mode quantum number. The difficulty of correlating *ab initio* and measured [21] band features at wavelengths shortward of ~200 nm yields several suspect shifts for $v_2 > 15$. A harmonic (linear in v_2) fit to the *ab initio* results from 200 to 220 nm allows an estimate of the expected shifts at $v_2 = 15, 17, 19, 20–22$. Shifts determined directly from the *ab initio* data are shown as black squares, and the harmonic fit is shown as red squares.

spectra could, in principle, be used directly as SO_2 synthetic spectra, and would yield very large, and most likely unrealistic, S-MIF effects during SO_2 photolysis (results not shown). However, as shown in Figure 5.3, the agreement between the measured absorption spectrum for $^{32}SO_2$ at 213 K [21] and the *ab initio* spectrum [28] is poor. The *ab initio* spectrum lacks the rotational congestion of the observed spectrum, and is therefore a poor match to the measured absorption continuum at 213 K. On other hand, the locations of the band peaks are fairly accurate at wavelengths longward of 200 nm (Figure 5.3), and I used the computed band peak positions for each SO_2 isotopologue to compute the shifted spectra used for sulfur isotope calculations below. Because band positions in the *ab initio* spectra [28] become progressively less accurate at wavelengths <200 nm, and because band features become more difficult to identify, I computed a linear fit to $\nu^* - \nu$ in v_2 from 200 to 220 nm and extrapolated the fit to 190 to 200 nm (Figure 5.4). The computed isotopologue shifts from the Ran et al. [28] results are shown in Table 5.1.

TABLE 5.1 Isotopologue band shifts determined from *ab initio* results of Ran et al. [28] and bandhead locations measured by Freeman et al. [21] for $^{32}SO_2$

Bending mode (v_2)	$\nu^{32} - \nu^{33}$ (cm^{-1})	$\nu^{32} - \nu^{34}$ (cm^{-1})	$\nu^{32} - \nu^{36}$ (cm^{-1})	$^{32}SO_2$ band locations [21] (nm)
4	14.00	27.36	51.97	218.70
5	14.85	28.85	54.06	216.85
6	17.70	34.09	64.90	215.15
7	18.85	36.14	68.52	213.40
8	21.85	42.12	80.16	211.65
9	22.63	44.56	83.62	209.95
10	25.99	50.58	96.01	208.25
11	26.88	53.27	101.06	206.65
12	30.14	58.35	111.12	204.96
13	32.84	64.18	120.99	203.45
14	31.83	64.10	126.09	201.78
15[a]	35.40	69.82	131.65	200.35
16	36.86	71.15	134.06	198.77
17[a]	39.41	77.88	146.47	197.26
18	42.13	92.23	170.17	195.86
19[a]	43.44	85.94	161.29	194.32
20[a]	45.44	89.97	168.70	193.03
21[a]	47.46	94.00	176.11	191.84
22[a]	49.47	98.03	183.52	190.0[b]
α (cm^{-1})	5.25	9.37	20.50	
β (cm^{-1})	2.01	4.03	7.41	

a Band shifts for these values of v_2 determined by harmonic fit, $\nu^{32} - \nu^x = \alpha_x + \beta_x v_2$, to Ran et al. [28] *ab initio* data from 200 to 220 nm.
b Lowest wavelength included in photochemical model, rather than bandhead location.

The shifts were applied to a measured $^{32}SO_2$ spectrum [21], with band positions shown in the last column of Table 5.1. The coefficients of the linear fit to $\nu^* - \nu$ are also given in Table 5.1. The model S-MIF signatures computed during SO_2 photolysis (see below) exhibit little sensitivity to whether the Ran et al. shifts (black squares in Figure 5.4) are used versus the Ran et al. shifts from 200 to 220 nm and the linear fit from 190 to 200 nm. (All results presented below use the latter scheme). However, the S-MIF results were different (e.g., smaller $\Delta^{36}S/\Delta^{33}S$ ratio) if only the linear fit was used from 190 to 220 nm, possible indicating that fortuitous line overlap was more likely with the purely linear fit, an effect not expected with actual spectra.

Figure 5.2a shows the resulting shifted spectra for the $^{32}SO_2$ and $^{36}SO_2$ isotopologues from 190–220 nm. Note that the rotational features (Figure 5.2b) for the rare isotopologues are simply shifted versions of those of $^{32}SO_2$. In fact, significant variations in line positions and amplitudes and smaller variations in the level of the absorption continuum between the various isotopologues will occur. Amplitude variations in lines are evident in the *ab initio* spectra [28]. For this reason the isotope results presented here, which are a result only of the shift in band peak positions in the SO_2 isotopologue spectra, must be regarded as approximate.

3. PHOTOLYSIS OF SO_2 ISOTOPOLOGUES IN A LOW O_2 ATMOSPHERE

Radiative transfer calculations were performed in the context of modeling the photochemistry of sulfur in the early Earth atmosphere. Rather than present an in-depth discussion of the many photochemical possibilities [29], my emphasis here is on the dissociation rates of SO_2 isotopologues. However, because absorption by CO_2, H_2O and other possible compounds will alter the dissociation rates of SO_2, particularly at short wavelengths (<200 nm), the radiative transfer and photochemistry are necessarily linked. The photodissociation rate coefficient for the $^xSO_2$ isotopologue ($x = 32, 33, 34$ or 36) from 190 to 220 nm is given by

$$J_x(z) = \int_{190}^{220} \phi_x(\lambda)\sigma_x(\lambda)F_0(\lambda)e^{-\tau(\lambda,z)}d\lambda, \qquad (3)$$

where ϕ_x is the photodissociation quantum yield (i.e., fraction of photoexcited SO_2 that dissociates), σ_x is the absorption cross section for $^xSO_2$, and F_0 is the photon flux at the top of the model. The opacity is $\tau(\lambda, z) = \sum_i N_i(z)\sigma_i(\lambda)$ where N_i is the column density of absorber or scatterer i, and the sum is over all absorbers and scatterers, including all SO_2 isotopologues and CO_2. In Eq. (3) σ_x is as shown in Figure 5.2 for each isotopologue $^xSO_2$, and F_0 and N_i are determined in the atmospheric model. For the (1, v_2, 2) progression the fluorescence quantum yield is <0.2 in the dissociation region [22]. In Eq. (3) $\phi_x = 1 - \phi_f$, where ϕ_f is the fluorescence quantum yield [22], which I have assumed to be identical for all isotopologues. I have used a polynomial fit to the fluorescence data to describe ϕ_x in Eq. (3). I have also assumed that the low values of ϕ_f (~0.003) determined for

206 to 210 nm [22] extends to 190 nm. Nearly identical results are obtained if ϕ_x is assumed to be unity from 190 to 220 nm.

The photodissociation rate coefficients are included as source and sink terms in a system of time-dependent continuity equations for the atmosphere. Modern values for vertical (eddy) diffusion and solar photon flux are utilized. The system of 2nd-order ordinary differential equations is solved by integration, and yields chemical species abundances as a function of time and altitude. The isotope atmospheric chemistry includes only SO_2 isotopologue photodissociation reactions and production of SO isotopologues. Additional isotopic reactions such as SO_2 oxidation by OH, SO photolysis, SO disproportionation during self-reaction, and SO dimmer formation, have been neglected. My objective here is to focus only on SO_2 photolysis as a S-MIF mechanism.

Figure 5.5 shows the photodissociation rate coefficients and delta values of the rate coefficients as a function of altitude for 10 ppb initial SO_2 and 0.02 CO_2. The calculations are at a time corresponding to a 10% reduction in column SO_2. The photodissociation rate coefficients (Figure 5.5a) undergo a monotonic increase in value with increase in S isotope mass, illustrating the importance of the magnitude of the shift in vibronic bands. Delta-values (Figure 5.5b) of the photodissociation rate coefficients are computed as $\delta^x J_{SO_2} = 10^3(J_x/J_{32} - 1)$, and are largest in magnitude for $\delta^{36}S$. The mass-independent signatures, $\Delta^{33}S$ and $\Delta^{36}S$ (Figure 5.5c), are

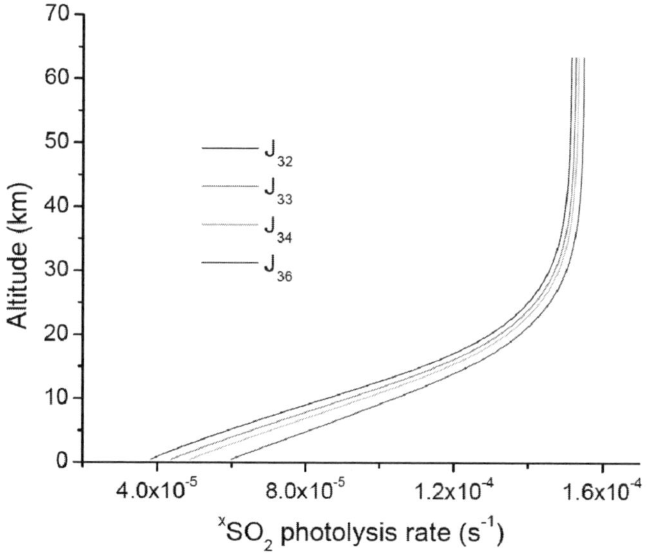

(a)

FIGURE 5.5 The variation with altitude of SO_2 isotopologue photolysis rates and rate coefficient delta values in the model atmosphere. Results are shown at a time corresponding to a 10% reduction in the initial SO_2 column density. (a) Photolysis rates for each isotopologue, illustrating that the rate increases with the magnitude of the band shift. (b) Rate coefficient δ-values computed as $\delta^x J_{SO_2} = 10^3(J_x/J_{32} - 1)$. (c) Rate coefficient isotope anomalies, $\Delta^{33}J_{SO_2}$ and $\Delta^{36}J_{SO_2}$, and their ratio.

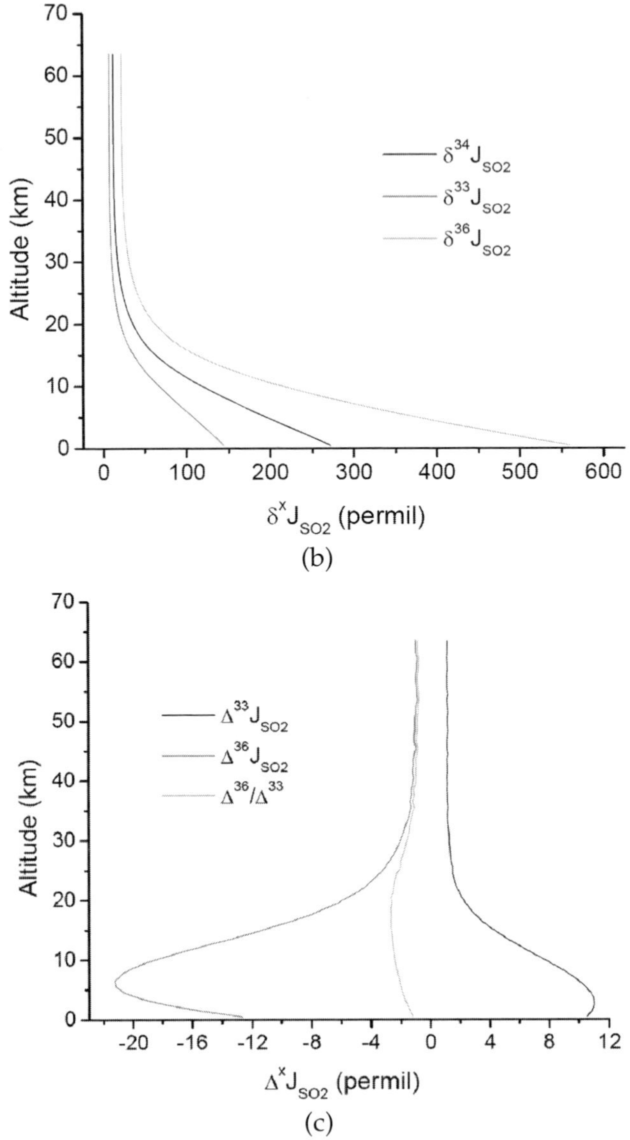

FIGURE 5.5 *(Continued.)*

positive for $\Delta^{33}S$ and negative for $\Delta^{36}S$, and have a ratio $(\Delta^{36}S/\Delta^{33}S) = -1.3$ at the surface. Figure 5.6a shows the mixing ratios of ^{32}SO and $^{32}SO_2$ as a function of altitude for a 10% reduction in initial SO_2. The corresponding delta-value plots (Figures 5.6b and 5.6c) for SO is similar to the delta-values for the photodissociation rate coefficients (Figures 5.5b and 5.5c).

Atmospheric model results are shown in Figure 5.7 for the column densities of residual SO_2 and product SO for $pSO_2 = 10$ ppb and $pCO_2 = 0.02$. Large mass-dependent (Figure 5.7a) and mass-independent (Figure 5.7b) effects are predicted, with $\delta^{34}S(SO) \sim 130‰$ and $\Delta^{33}S(SO) \sim 7‰$ with 90% of SO_2 remaining. The model also predicts $\Delta^{36}S/\Delta^{33}S \sim -2$, which lies between the Antarctic ice core sulfates data (ratio ~ -3.3) and the Archean rock data (ratio ~ -0.9). The S-MIF signature in photoproduct SO is a function of initial pSO_2 and a weak function of pCO_2 (Figure 5.8). The model $\Delta^{36}S/\Delta^{33}S$ ratio varies moderately with pSO_2 and pCO_2 for $pSO_2 \leqslant 10$ ppb with most values between -1 and -3. A much larger range in model $\Delta^{36}S/\Delta^{33}S$ occurs for $pSO_2 = 100$ ppb, with val-

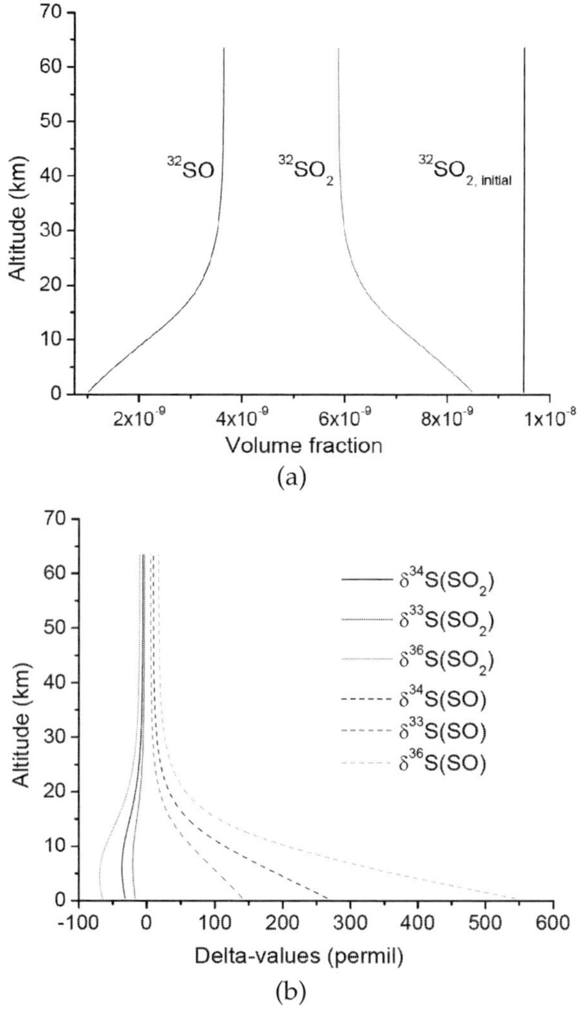

FIGURE 5.6 The variation with altitude of SO and SO_2 abundance and delta values for the same conditions as in Figure 5.5. (a) Volume fraction (or mixing ratio) of ^{32}SO, $^{32}SO_2$ and the initial $^{32}SO_2$ versus altitude. (b) Delta-values relative to initial SO_2 sulfur isotope ratios. (c) Mass-independent signatures ($\Delta^{33}S$ and $\Delta^{36}S$) for SO and SO_2.

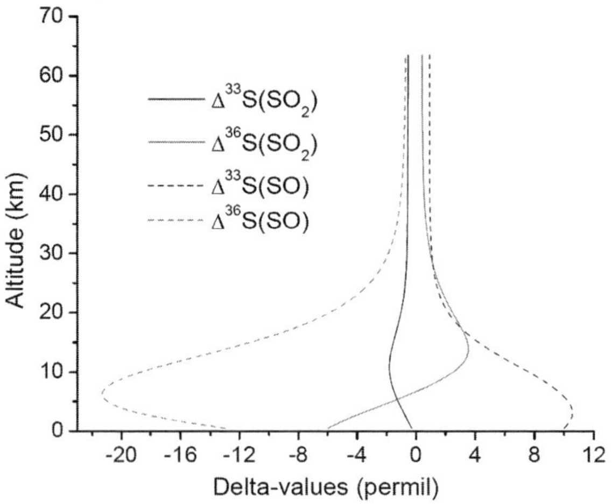

(c)

FIGURE 5.6 (*Continued.*)

ues ranging from ∼−4 to +4. The positive $\Delta^{36}S/\Delta^{33}S$ values are a consequence of self-shielding in ^{34}S. The $\Delta^{33}S$(SO) values in Figure 5.8 correspond to a 10% reduction in initial pSO_2. Comparing these values to the maximum $\Delta^{33}S$ measured in pyrites, which are ∼8–10‰ [30,36], suggests $pSO_2 \sim 10$ ppb during generation of the largest S-MIF signatures, or about 100 times the present-day global SO_2 concentration. Deposition of the elemental S derived from SO into a large reservoir of sulfur (e.g., ocean) may dilute the MIF signatures shown in Figure 5.8. The large $\delta^{34}S$(SO) values computed with the model are much larger than the $\delta^{34}S$ values observed values in rocks, which are typically in the range of −10 to +20 ‰ [30,36]. This discrepancy may indicate that mixing and additional mass-dependent fractionation occurred during the formation of elemental sulfur and sulfur-bearing minerals. The MIF signatures computed here could also be derived from localized enhancements in SO_2 concentration lasting for weeks following a volcanic eruption, although additional UV absorbers such as volcanic ash could create a more complex UV transmission environment than has been considered here.

4. PHOTOLYSIS OF SO_2 IN THE MODERN ATMOSPHERE

The model results (Figures 5.6–5.8) presented here are for a low-O_2 atmosphere in which SO does not rapidly reform SO_2 by reaction with O_2. For the ice core sulfate data it is unlikely that reaction (R1) in the modern atmosphere is the source of the sulfur MIF. Instead, SO_2 photoexcitation [31],

$$SO_2(\tilde{X}^1A_1) + h\nu(260\text{–}340\text{ nm}) \rightarrow SO_2(\tilde{B}^1B_1, \tilde{A}^1A_2) \qquad \text{(R2)}$$

followed by disproportionation during self-reaction [32],

$$SO_2(\tilde{B}^1B_1) + SO_2(\tilde{X}^1A_1) \rightarrow SO_3(\tilde{X}^1A_1) + SO(a^1\Delta) \qquad \text{(R3)}$$

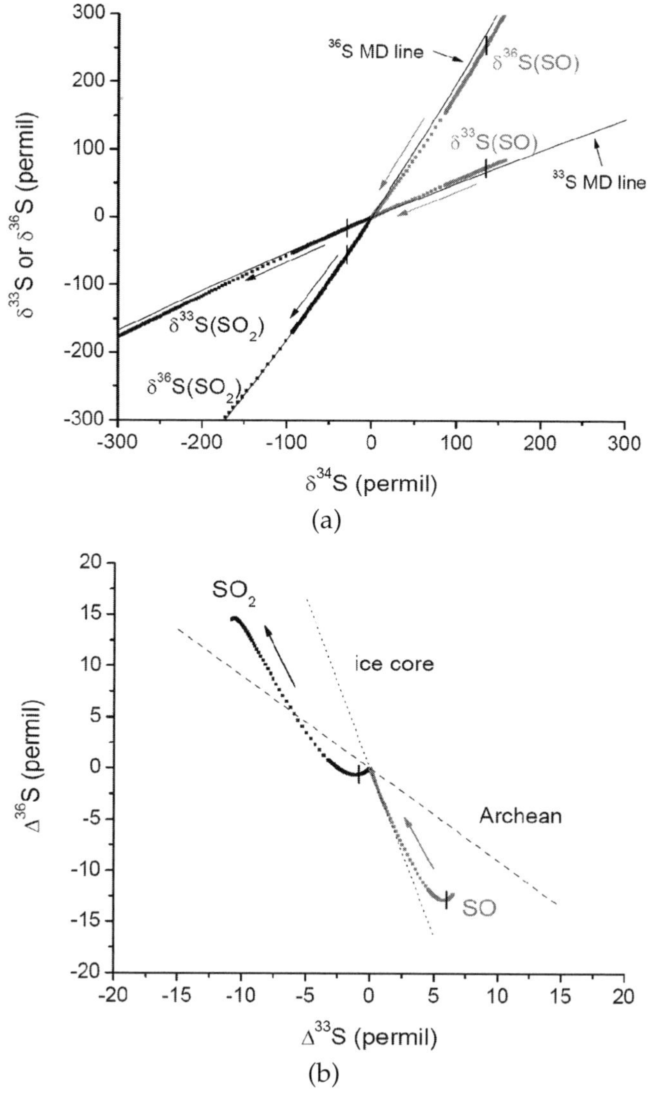

FIGURE 5.7 Model delta-values and S-MIF signatures for atmospheric column densities of SO_2 and SO relative to initial SO_2. Isotope fractionation occurs during SO_2 dissociation from 190 to 220 nm using a present-day solar flux. The model assumes $pCO_2 = 0.02$ and initial $pSO_2 =$ 10 ppb. The arrows show the direction of the time evolution of the delta values and S-MIF signature. As a greater fraction of SO_2 is photolyzed the magnitude of the delta values of residual SO_2 (labeled as SO_2) increase and those of SO decrease. The locations at which initial SO_2 has been reduced by 10% is indicated by vertical bars. (a) Model $\delta^{33}S$ and $\delta^{36}S$ values for SO and residual SO_2, and the mass-dependent curves (MD). (b) Model $\Delta^{33}S$ and $\Delta^{36}S$ values for SO and residual SO_2. The line labeled 'ice core' has a slope of -3.3 and was determined from measurement of Antarctic ice core sulfates derived from the Agung eruption [17]. The 'Archean' line has a slope of -0.9 and was determined from measurement of Archean and Paleoproterozoic sulfides and sulfates [4].

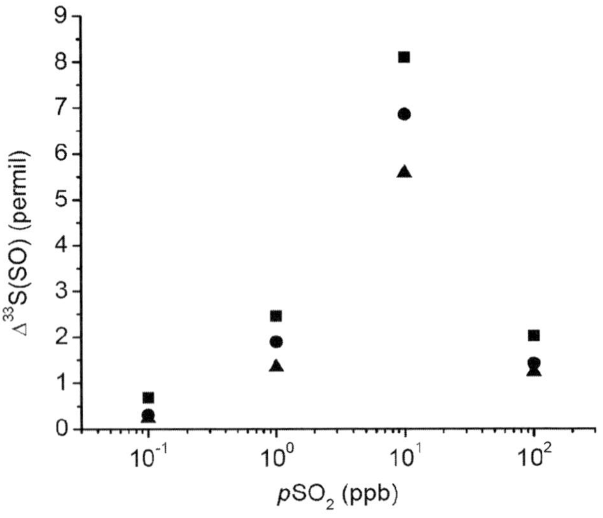

FIGURE 5.8 $\Delta^{33}S$ of the column density of photolysis product SO that corresponds to a 10% reduction in the initial column density of SO_2. Results are shown for $pCO_2 = 0.002$ (solid square), $pCO_2 = 0.02$ (solid circle), and $pCO_2 = 0.2$ (solid triangle). The magnitude of the MIF signature due to the SO_2 self-shielding mechanism scales with the concentration of SO_2. The peak in model $\Delta^{33}S$ values shown occurs at 10 ppb SO_2 and is comparable to the maximum $\Delta^{33}S$ measured in Archean pyrites [30].

is a more likely pathway for creating an additional sulfur reservoir (i.e., SO_3) with a sulfur MI signature [5]. The 280–320 nm region of the SO_2 spectrum (Figure 5.1) has a rotationally-resolved band structure similar to that near 200 nm. The band structure is again due to a progression of vibrational modes [33]. By the same arguments I have made above for SO_2 dissociation in the $\tilde{C}^1B_2$ state, isotope-selective photoexcitation of SO_2 will occur in a manner qualitatively similar to that for production of SO by photodissociation in Figures 5.7a and 5.7b. That is, $\Delta^{33}S(^1SO_2) > 0$, where $^1SO_2$ denotes the product in reaction (R2). If reaction (R3) proceeds by abstraction of the oxygen atom from ground state SO_2, then $\Delta^{33}S(SO_3) > 0$. The isotopic anomalies observed in recent Antarctic sulfates would then first exhibit $\Delta^{33}S > 0$ because of the rapid conversion of SO_3 to sulfate aerosol, followed by $\Delta^{33}S < 0$ as the remaining SO_2 with $\Delta^{33}S < 0$ is more slowly converted to sulfate, in agreement with recent observations and analysis [17]. The suggestion that SO_3 photodissociation [18] is the source of the MI signature in ice core sulfates cannot be evaluated by the self-shielding mechanism proposed here due to a lack of sufficiently high resolution cross section data for SO_3.

5. IMPROVEMENTS TO SPECTRA AND ADDITIONAL SOURCES OF S-MIF

The approximate SO_2 isotopologue spectra constructed here can be improved in several ways. First, less prominent vibrational mode progressions should be included. For example, I have neglected the (3, v_2, 0) progression, which is clearly present in LIF spectra, and is weakly present in absorption [22]. Second, I have used $^{32}SO_2$ absorption data [21] acquired at 213 K, which reduces the significance of SO_2 hot bands. The temperature of the Archean atmosphere is unknown. If it were considerably warmer than today, or even comparable to the present day, hot bands of SO_2 would become more important. Third, pressure broadening has not been accounted for here, and will cause line broadening in the lower stratosphere and troposphere. Such line broadening may reduce the magnitude of the resulting S-MIF (Figure 5.8). Fourth, the harmonic approximation used in extrapolating isotopologue band shifts at high v_2 (Figure 5.4) neglects potentially significant anharmonicity. Improved *ab initio* calculations and high-resolution laboratory spectra for the rare isotopologues of SO_2 will be essential to finally arriving at accurate isotopologue spectra that span the full wavelength range of interest for atmospheric chemistry (~180 to 400 nm). The work of Ran et al. [28] is a significant step in that direction.

In addition to SO_2 self-shielding many other possible sources of S-MIF can be identified. The model $\Delta^{36}S/\Delta^{33}S$ results for the case of a low-O_2 atmosphere (e.g., Figure 5.7b) are in qualitative but not quantitative agreement with the ancient rock record. Elemental S, derived from SO, is predicted here to have $\Delta^{33}S > 0$ and $\Delta^{36}S < 0$, which is consistent with observations of most pyrites [4], but the magnitude of the $\Delta^{36}S/\Delta^{33}S$ ratio is about a factor of 2 to 3 too high (~-2.5 vs. ~-0.9). This is a significant discrepancy, and may indicate that MI processes in addition to SO_2 photodissociation are at work. One such MI process almost certain to be important in a low-O_2 atmosphere is SO photodissociation. Isotope-selective photolysis will occur in SO at wavelengths ~190–230 nm, but rotationally-resolved spectra, either laboratory or synthetic, are needed to estimate the MI effect. In addition to S-MIF due to SO photolysis, SO_2 photoexcitation (~280–330 nm) and SO_3 photolysis [18] must also be considered as possible contributors to S-MIF in the ancient atmosphere. S-MIF due to these photo-processes will be considered in future work.

Another possible source of S-MIF is the variation in cross section amplitude of an absorption band or the underlying absorption continuum upon sulfur isotope substitution. The difference in zero-point energies for the different sulfur (and oxygen) isotopologues of SO_2 implies small changes in the shape of the wavefunction. For the case of direct photodissociation, the wavefunction becomes more peaked at the center of the band and drops off near the edges for substitution of heavier isotopes. I have utilized the method of Liang et al. [12] to compute the new cross sections for SO_2 isotopologues and evaluate the expected S-MIF, assuming that SO_2 undergoes direct photodissociation. Clearly, SO_2 does not undergo direct dissociation, so this is only a test to see if cross section amplitude variations can produce S-MIF in the simplest case. Figure 5.9 shows the expected direct dissociation wavefunction for each isotopologue, normalized to the $^{32}SO_2$ wavefunction, using computed zero-point energy shifts [37]. Isotopologue cross sections were computed from $^{32}SO_2$ data [21] by scaling by the normalized wavefunctions. The band shifts (Table 5.1) were set to zero to isolate any amplitude-related S-MIF. The largest MIF computed was $\Delta^{36}S(SO) \sim 0.2‰$, far below the self-shielding MIF sig-

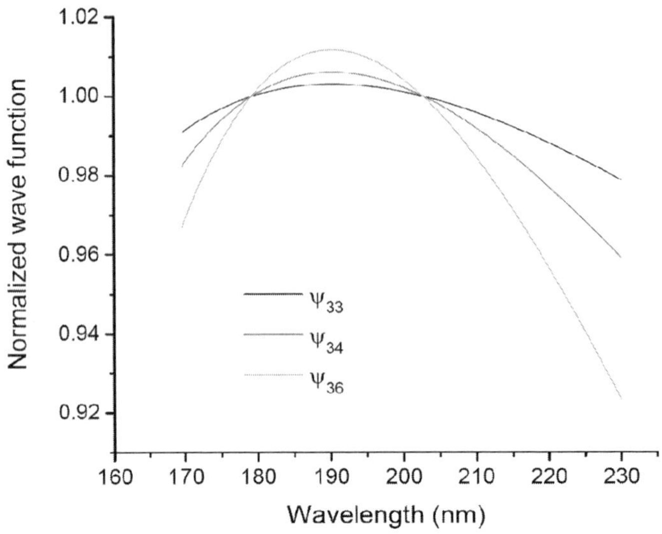

FIGURE 5.9 Hypothetical wave functions for $^{33}SO_2$, $^{34}SO_2$, and $^{36}SO_2$ normalized to $^{32}SO_2$ showing the effect of zero-point energy change on the shape of the wavefunction. The wavefunctions were computed by the method of Liang et al. [12] for a molecule that undergoes direct photodissociation. Because SO_2 undergoes indirect dissociation in the 190–220 nm range, the wavefunctions are meant to be illustrative only.

natures. Indirect photodissociation and high-order excited state vibrational levels, both of which occur for SO_2, may produce large MIF signatures, but the direct dissociation case does not seem to yield significant MIF.

Some recent meteorite and geochemical measurements [38,39] show evidence for non-zero MIF in ^{33}S but very small or negligible MIF in ^{36}S, i.e., $|\Delta^{33}S| \gg |\Delta^{36}S|$. MIF signatures of this type might fortuitously arise during SO_2 photodissociation for very high SO_2 abundance (~100 ppb), but more likely are the result of hyperfine effects associated with the non-zero nuclear spin of ^{33}S. A photochemical hyperfine effect may occur, analogous to spin-forbidden CO_2 photolysis [40], at near-ultraviolet wavelengths,

$$SO_2(\tilde{X}^1A_1) + h\nu(370\text{–}400\text{ nm}) \rightarrow SO_2(\tilde{a}^3B_1) \quad \text{(R4)}$$

followed again by disproportionation during self-reaction

$$SO_2(\tilde{a}^3B_1) + SO_2(\tilde{X}^1A_1) \rightarrow SO_3(\tilde{X}^1A_1) + SO(X^3\Sigma^-). \quad \text{(R5)}$$

The $^3SO_2$ product in reaction (R4) would be highly enriched in ^{33}S, which would be passed to SO_3 and/or SO.

Chemical, as opposed to photolytic, sources of S-MIF are also possible in the Archean atmosphere. By analogy with oxygen MIF produced during O_3 formation [2], which is attributed to a symmetry-dependent non-statistical redistribution of internal energy in the vibrationally excited O_3 [7], it is likely that reactions such as $S + S_2 \overset{M}{\rightarrow} S_3$ and $S_2 + S_2 \overset{M}{\rightarrow} S_4$ would also produce MIF signatures. However, the partial pressures of S_2, S_3 and S_4 in a sulfur vapor are extremely low at temperatures ~30 °C (10^{-16}, 10^{-17} and 10^{-17} atmospheres, respectively; $S_{el} \sim 10^{-9}$ atmospheres [34]), making it unlikely that S_3 and S_4 (and S_{el}) formed in the gas phase in the Archean atmosphere. It is known, albeit from sparse laboratory data, that O_3 formed in wall reactions does not possess an O-MIF signature [35]. Similarly, I suggest that S_3 and S_4 produced in wall reactions or on the surfaces of atmospheric/volcanic particulates will not exhibit a S-MIF signature unless the reactants (S and S_2) already have such a signature.

Another possible source of chemical MIF is self-reaction of the SO radical, which either yields S and SO_2 as disproportionation products [41], or the SO dimer as an association product [42]. Two of the four lowest energy singlet states of the SO dimer [43] have a reduced point group symmetry upon sulfur isotope substitution. If sulfur atom exchange is another possible outcome in the self reaction of SO, then a symmetry-derived mass-independent effect analogous to O_3 formation may be possible. A similar conclusion may be drawn for the triplet states of the SO dimer.

6. CONCLUSIONS AND BROADER IMPLICATIONS

The calculations I've presented here support the idea that SO_2 self-shielding is at least in part responsible for the S-MIF observed in Archean rocks. Because of the approximations made in the SO_2 isotopologue synthetic spectra, it cannot be claimed that SO_2 photolysis is the only source of MIF. Experimental data needed to improve the isotopologue spectra include measurements of vibronic band shifts for all four sulfur isotopologues and for the $^{32}S^{16}O^{18}O$ and $^{32}S^{16}O^{17}O$ isotopologues, high resolution absorption measurements showing the shifts in rotational features, and accurate measurements of the amplitude changes of both vibronic bands and rovibronic lines.

The qualitative agreement of the model S-MIF signatures with Archean pyrites suggests that self-shielding is a plausible source of MIF. If indeed self-shielding in sulfur gases is the mechanism that creates S-MIF in terrestrial rocks, it demonstrates that large MIF signatures in terrestrial rocks and meteorites can be derived from purely photochemical effects independent of molecular symmetry, and provides conceptual support to the idea that O-MIF in meteorites is a result of self-shielding in CO [2,8–10].

ACKNOWLEDGEMENTS

The author thanks Y. Ueno and an anonymous referee for many helpful comments and suggestions. The author also thanks E. Schauble at UCLA for use of and assistance with his cluster for preliminary SO_2 *ab initio* calculations, H. Guo for theoretical SO_2 absorption spectra, and J. Kasting for providing a copy of his photochemical code. Support from the UCLA IGPP Center for Astrobiology and the NASA Astrobiology Institute, and from the NASA Exobiology and Evolutionary Biology program (grant # NNX07AK63G) is gratefully acknowledged.

REFERENCES

[1] R.N. Clayton, L. Grossman, T.K. Mayeda, *Science* **182** (1973) 483.
[2] M.H. Thiemens, H.E. Heidenreich III, *Science* **219** (1983) 1073.
[3] K. Mauersberger, *Geophys. Res. Lett.* **14** (1987) 80.
[4] J. Farquhar, H. Bao, M. Thiemens, *Science* **289** (2000) 756.
[5] J. Savarino, A. Romero, J. Cole-Dai, S. Bekki, M.H. Thiemens, *Geophys. Res. Lett.* **30** (2003) 2131.
[6] M.H. Thiemens, *Annu. Rev. Earth Planet. Sci.* **34** (2006) 217.
[7] Y.Q. Gao, R.A. Marcus, *Science* **293** (2001) 259.
[8] O. Navon, G.J. Wasserburg, *Earth Planet. Sci. Lett.* **73** (1985) 1.
[9] R.N. Clayton, *Nature* **415** (2002) 860.
[10] J.R. Lyons, E.D. Young, *Nature* **435** (2005) 317.
[11] R.A. Marcus, *J. Chem. Phys.* **121** (2004) 8201.
[12] M.C. Liang, G.A. Blake, Y.L. Yung, *J. Geophys. Res.* **109** (2004), Art. # D10308.
[13] C.E. Miller, R.M. Onorato, M.C. Liang, Y.L. Yung, *Geophys. Res. Lett.* **32** (2005), Art. # L14814.
[14] J. Farquhar, J. Savarino, S. Airleau, M.H. Thiemens, *J. Geophys. Res.* **106** (2001) 32829.
[15] B.A. Wing, J.R. Lyons, J. Farquhar, paper presented at the *228th Meeting of the American Chemical Society*, Philadelphia, PA, 22 August 2004.
[16] J. Farquhar, B.A. Wing, *Earth Planet. Sci. Lett.* **6707** (2003) 1.
[17] M. Baroni, M.H. Thiemens, R.J. Delmas, J. Savarino, *Science* **315** (2007) 84.
[18] A.A. Pavlov, M.J. Mills, O.B. Toon, *Geophys. Res. Lett.* **32** (2005) L12816.
[19] P. Zmolek, X. Xu, T. Jackson, M.H. Thiemens, W.C. Trogler, *J. Phys. Chem. A* **103** (1999) 2477.
[20] J. Farquhar, J. Savarino, T.L. Jackson, M.H. Thiemens, *Nature* **404** (2000) 50.
[21] D.E. Freeman, K. Yoshino, J.R. Esmond, W.H. Parkinson, *Planet. Space Sci.* **32** (1984) 1125.
[22] A. Okazaki, T. Ebata, N. Mikami, *J. Chem. Phys.* **107** (1997) 8752.
[23] J.P. Koplow, D.A.V. Kliner, L. Goldberg, *Appl. Opt.* **37** (1998) 3954.
[24] G. Stark, P.L. Smith, J. Rufus, A.P. Thorne, J.C. Pickering, G. Cox, *J. Geophys. Res.* **104** (1999) 16585.
[25] G. Herzberg, *Molecular Spectra and Molecular Structure III. Electronic Spectra and Electronic Structure of Polyatomic Molecules*, Van Nostrand Reinhold, Princeton, 1967.
[26] I. Dubois, *J. Mol. Struct.* **3** (1969) 269.
[27] J.P. Flament, N. Rougeau, M. Tadjeddine, *Chem. Phys.* **167** (1992) 53.
[28] H. Ran, D. Xie, H. Guo, *Chem. Phys. Lett.* (2007), doi:10.1016/j.cplett.2007.03.103.
[29] K. Zahnle, M. Claire, D. Catling, *Geobiology* **4** (2006) 271.
[30] S. Ono, J.L. Eigenbrode, A.A. Pavlov, P. Kharecha, D. Rumble, J.F. Kasting, K.H. Freeman, *Earth Planet. Sci. Lett.* **213** (2003) 15.
[31] H. Okabe, *Photochemistry of Small Molecules*, Wiley, New York, 1978.
[32] R.P. Turco, R.C. Whitten, O.B. Toon, *Rev. Geophys. Space Phys.* **20** (1982) 233.
[33] K. Yamanouchi, K. Okunishi, M. Endo, Y. Tsuchiya, *J. Mol. Struct.* **352** (1995) 541.
[34] H. Rau, T.R.N. Kutty, J.R.F. Guedes de Carvalho, *J. Chem. Thermodyn.* **5** (1973) 833.
[35] K. Mauersberger, D. Krankowsky, C. Janssen, R. Schinke, *Adv. At. Mol. Opt. Phys.* **50** (2005) 1.
[36] B.S. Kamber, M.J. Whitehouse, *Geobiology* **5** (2007) 5.
[37] G. Michalski, R. Jost, D. Sugny, M. Joyeux, M. Thiemens, *J. Chem. Phys.* **121** (2004) 7153.
[38] V. Rai, T.L. Jackson, M.H. Thiemens, *Science* **309** (2005) 1062.
[39] J. Farquhar, S.-T. Kim, A. Masterson, *Earth Planet. Sci. Lett.* **264** (2007) 1.
[40] S.K. Bhattacharya, J. Savarino, M.H. Thiemens, *Geophys. Res. Lett.* **27** (2000) 1459.
[41] R.I. Martinez, J.T. Herron, *Int. J. Chem. Kinet.* **15** (1983) 1127.
[42] J.T. Herron, R.E. Huie, *Chem. Phys. Lett.* **76** (1980) 322.
[43] C.J. Marsden, B.J. Smith, *Chem. Phys.* **141** (1990) 335.
[44] C.Y.R. Wu, B.W. Yang, F.Z. Chen, D.L. Judge, J. Caldwell, L.M. Trafton, *Icarus* **145** (2000) 289.
[45] K. Bogumil, J. Orphal, T. Homann, S. Voigt, P. Spietz, O.C. Fleischmann, A. Vogel, M. Hartmann, H. Bovensmann, J. Frerick, J.P. Burrows, *J. Photochem. Photobiol. A Photochem.* **157** (2003) 167.
[46] E.-P. Röth, R. Ruhnke, G. Moortgat, R. Meller, W. Schneider, UV/VIS absorption cross sections and quantum yields for use in photochemistry and atmospheric modeling. Part 2. Organic substances, Berichte des Forschungszentrums Jülich, jül-3341, Jülich, Germany, 1997.
[47] C.Y.R. Wu, F.Z. Chen, *J. Quant. Spectrosc. Radiat. Transfer* **60** (1998) 17.

CHAPTER 6

A New Model of Low Resolution Absorption Cross Section

R. Jost*

Contents

1. Introduction and Motivations 75
2. Improved Model of Low Resolution Absorption Cross Section (XS) for Diatomic Molecules 78
3. Quantum Correction to the Low Resolution Absorption Cross Section of Diatomic Molecules 80
4. The Absorption Cross Section of Cl_2 Molecule 84
5. A 3D Version of the Model and Its Application to Triatomic Molecules 89
   5.1 The Abs. Cross Section of the Hartley band of O_3 91
   5.2 The SO_2 Abs. Cross Section 93
   5.3 The CO_2 Abs. Cross Section 93
6. Conclusions and Perspectives 96
Acknowledgements 97
Appendix A. A Polynomial Version of Formula (12′) 97
Appendix B. A Polynomial Version of Formula (27′) 98
Appendix C. The Temperature Dependence of the Absorption Cross Section 99
References 100

1. INTRODUCTION AND MOTIVATIONS

The Absorption Cross Section (Abs. XS or XS or $\sigma(E)$) of molecules in the visible and or UV range is due to (at least one) electronic transition. At low resolution, each Abs.XS has a bell shape which can be described with very few (3 or 4, see below) parameters. This bell shape can be understood as the "reflection" of the ground state vibrational wavefunction (which can be approximated as a one-dimensional or multidimensional Gaussian) on the potential energy surface (PES)

* Laboratoire de Spectrométrie Physique, Université Joseph Fourier de Grenoble, B.P. 87, F-38042 Saint Martin d'Hères Cedex, France

Advances in Quantum Chemistry, Vol. 55
ISSN 0065-3276, DOI: 10.1016/S0065-3276(07)00206-7

of an upper electronic state. The "reflection principle" was introduced by Condon [1] and applied first to diatomic molecules [2–4] and then extended to polyatomic molecules [5,6].

When applied to diatomic molecules, the simplest form of the reflection method requires only three parameters: $-V_e$, the vertical excitation energy at the equilibrium geometry of the ground state, $-\beta$, the inverse of the width of the initial wavefunction, and $-V'_e$, the slope of the upper PES at the ground state equilibrium geometry. Surprisingly, the low resolution Abs. XSs of a large number of polyatomic molecules also have a bell shape, which can be described by an analytic function involving only 3 or 4 parameters. Strictly speaking, polyatomic molecules require $(3N-6)\beta$ parameters and $(3N-6)$ components of the upper PES slope but we will see that a single effective V'_e can still be defined. The model described below is very useful to compare the Abs. XS of various isotopologues of a given molecule because the isotope effect is easy to calculate with this analytic model. The ratio of Absorption XSs for various isotopologues plays an important role when isotope budget (or isotope ratios) are considered, because the photochemistry (and then the Abs. XS) of many molecules like O_3, NO_2, SO_2, N_2O, H_2CO involves many cycles of photodissociation and formation (recombination). In other words, differences in the formation rate and/or photodissociation rate are able to induce "anomalies" (e.g. Mass Independent Fractionation or MIF) in isotope ratios. The best known example of selectivity in the isotopologue formation rate is that of O_3: the formation rate of the $^{18}O^{16}O^{16}O$ asymmetric isotopologue is larger by a factor of 1.42 than that of the symmetric isotopologues, $^{16}O_3$ and $^{16}O^{18}O^{16}O$, which have almost the same rate [7]. Analogously, the ratios of photodissociation rates of N_2O isotopologues have been studied both experimentally [8,12] and theoretically [9,10]. For symmetric triatomic molecules with two identical oxygen atoms such as SO_2, CO_2, NO_2 (or the edge atoms for O_3), any single oxygen isotope substitution of ^{16}O by ^{17}O or ^{18}O leads to a symmetry breaking effect (from the C_{2v} to the C_s symmetry) which may be significant. One important aspect of this C_{2v} to C_s symmetry breaking is energy splitting between the two dissociation channels: C_s isotopologue of $^{18}OX^{16}O$ composition (X = S, N, C or O) can dissociate either into $^{18}OX + {}^{16}O$ (channel A) or into $^{16}OX + {}^{18}O$ (channel B) and these two channels are not energetically degenerate because of the difference between the zero point energies (ZPE) of ^{18}OX and ^{16}OX. Consequently, the branching ratio between these two dissociation channels is not unity because there is a propensity in favour of channel A. This propensity should vanish at photon energies well above the threshold (e.g. for ozone) but should be significant near the threshold if the corresponding Abs. XS is large (e.g. for NO_2). When we consider a mixture of isotopologues (e.g. in the atmosphere) illuminated by the actinic flux, we must consider the overall probability of dissociation of each isotopologue which is proportional to its Abs. XS, but we must also consider the branching ratios between the two dissociation channels of the asymmetric isotopologue(s). For most molecules, only the Abs. XS of the main isotopologue is known, the main exceptions being those of O_3 (only $^{18}O_3$) [11], N_2O [12] OCS [13] and NO_2 [14]. The models presented below allow prediction of the XS dependence on isotopic substitution. For all the molecules, the Abs. XS of the various isotopologues are expected to be very similar and their tiny differences are (and will remain) difficult to determine experimentally. Moreover, some of the Abs. cross sections of polyatomic molecules display a vibronic structure superimposed on the envelope of each Abs. XS. Each vibronic structure is specific to an isotopologue and it is rather difficult to derive the vibronic structure from one (main) isotopologue relative to another, except near the bottom of the PES where the normal modes are relevant. In contrast, the Abs. XS envelopes (i.e. the Abs. XS at very low resolution) of various isotopologues can be compared and related using the reflection method [15].

The two main requirements of the reflection method are knowledge of the initial state wavefunction (usually the $(0,0,0)$ state of the ground electronic surface) and of the shape of the PES of the upper electronic state. The key feature of isotopic substitution is the fact that this upper PES is independent of the isotopic composition. In contrast, the initial wavefunction is isotopically dependent but the effect can be easily calculated for the $(0,0,0)$ ground state. The ZPE, which is isotope dependent, plays an important role in shifting the maximum of the Abs. XS envelope.

Some of these ideas were recently used to calculate the isotope effect for the photodissociation of N_2O and O_3 by Liang *et al.* [15]. They predict the isotope effect based on a numerical analysis in which they describe the Abs. XS divided by E (where E is the photon energy, see below) as the product of a Gaussian and a numerical γ function assumed to be the same for all isotopologues. The isotope effect is taken into account in the 3 parameters of the Gaussian wavefunction and then reported on the Abs. XS of each isotopologue. From this the related fractionation constant can be determined.

In this paper, we introduce a new analytic form of the Abs. XS which is an improved version of the well-known simple Gaussian introduced by several authors [2–6] and adapted to triatomic molecules by Schinke in his book "Photodissociation dynamics" [6].

When applied to isotope effects, the main weakness of the reflection method is the assumption that the transition dipole moment is constant for all isotopologues. This weakness remains in the improved model presented below. Only *ab initio* calculations are able to go beyond this approximation. However, the dependence of the transition dipole moment along the nuclear coordinates can be introduced (numerically or analytically) in the model below, even if a less compact analytic form is expected. This paper is organized as follows: in Section 2 the "standard" reflection model is improved by taking into account the curvature of the upper state potential (in addition to its slope). In Section 3, the quantum character of the final state is taken into account by replacing the Dirac function by an Airy function. In Section 4 the model is applied to the Cl_2 molecule. In Section 5 the model is adapted and applied to the O_3, SO_2 and CO_2 triatomic molecules. Conclusions and perspectives are presented in Section 6.

2. IMPROVED MODEL OF LOW RESOLUTION ABSORPTION CROSS SECTION (XS) FOR DIATOMIC MOLECULES

The model is first presented for a diatomic molecule (1D PES) for the sake of simplicity. An adaptation to triatomic molecules, with a special attention to $C_{2v} \leftarrow C_{2v}$ electronic transitions, is presented in Section 4.

We use the treatment presented by Schinke [6], pages 110 to 113, and some hints of Heller [5] and Pack [16]. The reader may refer to Figure 1 of Ref. [15] or to Figure 6.1 of Ref. [6] for a qualitative description of the reflection method.

In the simple version of the reflection method, the potential of the upper electronic state is linearly approximated to:

$$V(R) = V_e - V'_e \cdot (R - R_e), \tag{1}$$

where V_e is the vertical excitation energy (below we use $V_0 = V_e - \text{ZPE}$, where ZPE stands for zero point energy of the ground state); R_e is the equilibrium nuclear distance of the ground state allowing us to define the vertical excitation energy V_e. V'_e is the absolute value of the derivative of the upper potential at R_e. In (1), we follow the convention of previous authors by using the minus sign in front of V'_e, because the first derivative of the excited potential is almost always negative.

Let's consider the wavefunction of the initial state, before absorption, which will be approximated using a Gaussian [17]:

$$\psi_0(R) = \beta^{1/2}\pi^{-1/4} \exp\left[-1/2 \cdot \beta^2 (R - R_e)^2\right], \tag{2}$$

where $\beta = (\mu\omega/\hbar)^{1/2}$ is the inverse of a length. Alternatively, we could use $\beta = ((\mu k)^{1/2}/\hbar)^{1/2}$ if we wish to characterize an isotopic substitution effect (i.e. change in the reduce mass μ) for which k is a constant (but not ω!).

Considering $|\psi_0(R)|^2$ as the initial probability density, using the reflection method and setting $V_0 = V_e - \text{ZPE}$, the Absorption Cross Section (XS) can be written:

$$\sigma(E) = AE(1/V'_e) \exp\left(-\left((E - V_0)/\left(V'_e/\beta\right)\right)^2\right), \tag{3}$$

where A includes the square of the transition dipole moment and allows characterizing the maximum amplitude of the cross section. This expression is obtained using the Condon approximation which assumes that the final state of the optical transition is a Dirac function located at $R = R_t$ such that $V(R_t) = E + \text{ZPE}$ where E is the photon energy.

Instead of $\sigma(E)$, we will consider the ratio $\sigma(E)/E$:

$$\sigma(E)/E = A(1/V'_e) \exp\left(-\left((E - V_0)/\left(V'_e/\beta\right)\right)^2\right), \tag{3'}$$

which is a Gaussian that depends on three parameters; the amplitude A (below we will use $A' = A/V'_e$), the centre V_0 and the half width at $1/e$, defined by the ratio V'_e/β. At this lowest level of the theory, $\sigma(E)/E$ is a Gaussian and is symmetric. The main advantage of using $\sigma(E)/E$ is that any asymmetry in the experimental $\sigma(E)/E$ will reveal a deviation from the simplest reflection model which predicts a symmetric $\sigma(E)/E$.

We get a compact expression:

$$\sigma(E)/E = A' \exp[-X^2], \tag{3''}$$

if we set $A' = A(1/V'_e)$ and:

$$X = (E - V_0)/(V'_e/\beta). \tag{4}$$

As already mentioned, one of the main weaknesses of the simple reflection method is the fact that the electronic transition dipole moment, (or the transition dipole moment surface, TDMS for polyatomic molecules in Section 4) is assumed to be constant. This weakness will remain in the Formulae (12), (27) and (29) derived below. The average value of the square of the TDM (or TDMS) is then included in amplitude A and $A' = A|1/V'_e|$. In Formulae (3), (3′) and (3″) the mass (or isotopologue) dependent parameters are β and the ZPE. In contrast, V_e and V'_e, which define the upper potential, are mass independent. This Formula (3) is already known even if different notations have been used by various authors. As an example, Schinke has derived the same formula in his book [6], pages 81, 102 and 111. Now, the model will be improved by including the contribution of the second derivative of the upper potential at R_e. The polynomial expansion of the upper potential up to *second order* in $(R - R_e)$ can be expressed as:

$$V(R) = V_e - V'_e(R - R_e) + 1/2V''_e(R - R_e)^2, \tag{5}$$

where V_e, V'_e and R_e are defined above and V''_e is the second derivative of $V(R)$ at R_e.

The second derivative, V''_e, is expected to be positive for most of the molecules, in contrast with the first derivative, which is almost always negative (see above). The absolute value of the first derivative of $V(R)$ around R_e becomes:

$$|V'_e(R)| = V'_e - V''_e(R - R_e) \tag{6}$$

assuming $|V'_e| > |V''_e(R - R_e)|$ or $|(R - R_e)| < V'_e/V''_e$.

Replacing V'_e by $V'_e(R)$ in Formula (3′), we get:

$$\sigma(E)/E = A(V'_e - V''_e(R - R_e))^{-1} \exp[-\{\beta(E - V_0)/(V'_e - V''_e(R - R_e))\}^2]. \tag{7}$$

The "trick", which is the essence of the reflection method, already used to derive (3) and (3′) [6], is to replace the $R - R_e$ dependence by the $E - V_0$ dependence. To *first* order in $(R - R_e)$, we get:

$$E = V(R) - \text{ZPE} \cong V_e - V'_e(R - R_e) - \text{ZPE} = V_0 - V'_e(R - R_e) \tag{8}$$

or

$$R - R_e = -(E - V_0)/V'_e. \tag{9}$$

Using (9) in (6), Formula (7) becomes:

$$\sigma(E)/E = A(V'_e + V''_e(E - V_0)/V'_e)^{-1}$$
$$\times \exp[-\{\beta(E - V_0)/(V'_e + V''_e(E - V_0)/V'_e)\}^2]. \tag{10}$$

Setting

$$r = V''_e / \beta V'_e \tag{11}$$

as a new (dimensionless) parameter and using (4) as before, we get, for $rX < 1$, corresponding to $|(R - R_e)| < V'_e / V''_e$:

$$\sigma(E)/E = A\big(V'_e(1 + rX)\big)^{-1} \exp\big[-(X^2/(1 + rX)^2)\big]. \tag{12}$$

We may approximate (12), assuming $rX \ll 1$ and setting $A' = A/V'_e$, by:

$$\sigma(E)/E \cong A'(1 - rX + r^2X^2) \exp\big[-(X^2) \cdot \{1 - 2rX + 3r^2X^2\}\big]. \tag{12'}$$

Note that A' is not exactly the maxima of $\sigma(E)/E$ and V_0 is not exactly the energy location of this maximum. The form (12′) was chosen to allow comparison (at the end of Section 3) to a similar expression for $\sigma(E)/E$ derived from another approach in Section 3.

Formula (12), (or (12′)) which is a new and much better version of Formula (3″), depends on four parameters, three of them being the same as in (3″) and the fourth parameter, r, proportional to V''_e, describing the asymmetry of $\sigma(E)/E$. Note that only β and ZPE (hidden in V_0) are isotopologue dependent (as for (3″)) and that V''_e is mass independent, like V_e and V'_e. It may be convenient to approximate this formula as a product of a Gaussian function (corresponding to the linear potential) with a correction function $\gamma(E)$ (here we use Liang's notation [15] for the sake of comparison). The function $\gamma(E)$ is given in Appendix A as a polynomial, up to the second order in r and to the fourth order in X (or $E - V_0$). The function $\gamma(E)$ was determined numerically for the O_3 and N_2O Abs. XSs by Liang *et al.* [15], but their analytic expression of $\gamma(E)$ disagree with the ones given in Appendices A and B.

3. QUANTUM CORRECTION TO THE LOW RESOLUTION ABSORPTION CROSS SECTION OF DIATOMIC MOLECULES

The reflection method used in Section 2 is "semi-classical", in the sense that the final state of the optical transition is assumed to be a Dirac function, even if the initial wavefunction and its ZPE are taken into account. In this section we derive the Abs. XS using a pure quantum calculation describing the final state as an Airy function. This approach was discussed and used by Child [18], Tellinghuisen [19] and Schinke [6]. In short, the Abs. XS for photons of energy E_{ph} is proportional to the square of the overlap between the initial wavefunction and the final wavefunction at energy $E = E_{ph}$+ZPE (the energy is taken from the bottom of the G.S. PES). As before (see Section 2), the initial wavefunction is a Gaussian defined in (3), but the final wavefunction is an Airy function [20] instead of a Dirac function. The Airy function is an exact form for the final wavefunction when the upper potential is the *linear* function of R given in (1). We define the classical turning point, R_t, from $V(R_t) = E_{ph}$+ZPE. Using (1), we derive a relation between R_t and the photon energy:

$$R_t - R_e = (E + \text{ZPE} - V_e)/V'_0 = (E - V_0)/V'_0. \tag{13}$$

The Airy wavefunction [20], describing the final state after photon absorption, is [6,21],

$$\psi_1(R;R_t) = \mathrm{Ai}\{-(2\mu V_0'/\hbar^2)^{1/3}(R-R_t)\}. \tag{14}$$

In order to calculate the integral between the initial and final state, the R variable should be converted into the reduced x variable, defined as $x=\beta(R-R_e)$, and used to define the initial ground state wavefunction (this expression is equivalent to (2)):

$$\psi_0(x) = \pi^{-1/4}\exp(-x^2/2). \tag{15}$$

The Airy wavefunction, expressed as a function of the same reduced variable x becomes:

$$\psi_1(X;X_0) = \mathrm{Ai}\{-\alpha(x-x_0)\}. \tag{16}$$

In this formula, the dimensionless parameter α is

$$\alpha = \left[2\mu^{1/4}V_e'/\hbar^{1/2}k^{3/4}\right]^{1/3} \tag{17}$$

and x_0 is related to the photon energy, E, via R_t, the classical turning point as follows: the relation $(R-R_t)=(R-R_e)-(R_t-R_e)$ allows for the introduction of the same variable x used in Ref. [6]. From this, we define x_0 according to:

$$x_0 = \beta(R_t - R_e). \tag{18}$$

If, as before, the transition dipole moment is assumed to be constant (i.e. not a function of x), the Abs. XS is proportional to the square of the overlap between $\psi_0(x)$ the Gaussian (describing the initial wavefunction) and $\psi_1(x;x_0)$, the Airy function describing the final state at energy E + ZPE after absorption of a photon of energy E. Note that, below, the energy dependence of the XS is hidden in x_0 in Formulae (19), (22), (23), (24) and (26). Integrating along the reduced variable x, the integral overlap between $\psi_0(x)$ and $\psi_1(x;x_0)$ is:

$$S_{01}(x_0;\alpha) = 2^{1/2}\pi^{1/4}\exp\left[(\alpha^6/12)+\alpha^3x_0/2\right]\cdot\mathrm{Ai}\left[(\alpha^4/4)+\alpha x_0\right]. \tag{19}$$

This analytic result is given by Child, page 123 of his book [18].

If

$$\alpha^4 \gg \alpha x_0 \gg 4 \tag{20}$$

an asymptotic form of the Airy function appearing in (19) can be used [20]:

$$\mathrm{Ai}(-z) \cong \frac{1}{2}\pi^{-1/2}|z|^{-1/4}\exp\left[-2/3(-z)^{3/2}\right]. \tag{21}$$

Note that the upper limit ($\alpha \to \infty$) of the condition (20), corresponds to the Condon approximation for which the Airy function can be approximated by a Dirac function.

From (19) and (21), setting $z=-[(\alpha^4/4)+\alpha x_0]$, we get:

$$S_{01}(x_0;\alpha) \cong \pi^{-1/4}\exp\left[(\alpha^6/12)+\alpha^3x_0/2\right]1/\alpha\left(1+4x_0/\alpha^3\right)^{-1/4} \times \exp\left[-\alpha^6/12\left(1+4x_0/\alpha^3\right)^{3/2}\right]. \tag{22}$$

As $\alpha^4 \gg \alpha x_0$ (see (20)) or $4x_0/\alpha^3 \ll 1$, the second exponential of (22) can be expanded in a power series of $4x_0/\alpha^3$. Then, the two leading terms of the expansion cancel with the first exponential of (22), and we get:

$$S_{01}(x_0;\alpha) \cong \pi^{-1/4}\alpha^{-1}\left(1+4x_0/\alpha^3\right)^{-1/4} \times \exp\left[-\left(\left(x_0^2\right)/2\right)\left\{1-2/3\cdot\left(x_0/\alpha^3\right)+\left(x_0/\alpha^3\right)^2\right\}\right]. \tag{23}$$

This formula is given here to the fourth order in x_0 but can be calculated to any order.

From (23) we can derive $\sigma(E)/E$ which is proportional to $|S_{01}|^2$:

$$\sigma(E)/E \cong A\alpha^{-2}\left(1+4x_0/\alpha^3\right)^{-1/2} \times \exp\left[-\left(x_0^2\right)\left\{1-2/3\cdot\left(x_0/\alpha^3\right)+\left(x_0/\alpha^3\right)^2\right\}\right], \tag{24}$$

where A is a constant including the square of the transition dipole moment. Using (13) and (18) we can express x_0 as a linear function of E and then as a linear function of $X = (E - V_0)/(V_e'/\beta)$ and we get $x_0 = X$. Doing so for $4x_0/\alpha^3 \ll 1$, (24) is reduced to Formula (3″) which is the simplest form of the reflection method. According to (4), and using (17) and (18), we get the dimensionless ratio:

$$x_0/\alpha^3 = tX, \tag{25}$$

where

$$t = \frac{1}{2}\cdot\hbar\omega/(V_e'/\beta) = \frac{1}{2}\cdot\hbar^{1/2}k^{3/4}\mu^{-1/4}/V_e'. \tag{26}$$

The parameter t (describing the quantum effect) decreases when μ increases (or when $\hbar$ decreases) as expected for the classical limit. From (24) and using (25), the $\sigma(E)/E$ can be approximated, to the second order in t (for $4tX \ll 1$), as:

$$\sigma(E)/E \cong A'(1+4tX)^{-1/2}\exp\left[-X^2\left\{1-2/3\cdot tX+t^2X^2\right\}\right] \tag{27}$$

or

$$\sigma(E)/E \cong A'(1-2tX+6t^2X^2)\exp\left[-X^2\left\{1-(2/3)tX+t^2X^2\right\}\right], \tag{27'}$$

where A' stands for the amplitude of $\sigma(E)/E$ at $X = 0$. Note that A' is not exactly the maximum of $\sigma(E)/E$ and V_0 is not exactly the energy location of this maximum.

Formula (27′) is similar to Formula (12′). The differences between the two are discussed below and numerically in Section 4. To simplify the comparison, both formulae have been expanded to the fourth order in X, and to second order in r or t in Appendices A and B. It should be noted first that, numerically (see Section 4), the two prefactors, $(1 + 4tX)^{-1/2} \approx (1 - 2tX + 6t^2X^2)$ of (27′) or $(1 + rX)^{-1} \approx (1 - rX + r^2X^2)$ in (12′) play a minor role. Thus, comparing the arguments of the exponentials in (12′) and (27′), the leading term, $-2/3\cdot tX$, of $\{1-(2/3)tX+t^2X^2\}$ in Formula (27′), can be considered as equivalent to the leading term, $-2rX$, of $\{1 - 2rX + 3r^2X^2\}$ in Formula (12′) if we set $t = 3r$. These two leading terms are both able to characterize numerically the asymmetry of a $\sigma(E)/E$, but they correspond to two different physical interpretations of a given experimental asymmetry of $\sigma(E)/E$: the difference between (12) (or (12′)) and (27) (or (27′)) is the fact that Formula (27) does not involve a curvature of the upper potential and depends on only *three* parameters, the amplitude A', the centre V_0 and the slope V'_e because the fourth parameter t (see Formula (26)) is proportional to the inverse of V'_e (β and μ are assumed to be known). In contrast Formula (12) (or (12′)) depends on *four* parameters, the fourth parameter, r or V''_e, describing the curvature of the upper potential, is neglected in (27′). Note that the two quadratic terms, t^2X^2 and $3r^2X^2$ (see above) are not equivalent if we set $t = 3r$, but they do not contribute to the asymmetry of $\sigma(E)/E$. They do however contribute to the width. So, the two widths, V'_e/β, fitted respectively with (12′) and (27′) are slightly different, as expected (see Table 6.2 below).

The observed asymmetry of an experimental $\sigma(E)/E$ reflects the *combined effect* of the upper potential curvature (described by $r = V''_e/(V'_e/\beta)$) *and* the quantum effect described by $t = \frac{1}{2}\hbar\omega/(V'_e/\beta)$. We tried without success to find an *exact* analytic expression of the integral, similar to (19), which would take into account the effects of the curvature described in Section 2 *and* of the quantum effect described in this section. In fact, the two effects are rather similar (both induce an asymmetry of $\sigma(E)/E$) and cannot be discriminated numerically. Instead, we will use a single parameter, s, which s is an *effective asymmetry parameter* which reflect the above mentioned combined effect. Tentatively, we define the s parameter as:

$$s = \left[(2r)^2 + (2t/3)^2\right]^{1/2} \tag{28}$$

in order to have the correct behaviour when $r = 0$ (then $s = 2t/3$) as expected from (27′) or $t = 0$ (then $s = 2r$ as expected from (12′)).

Moreover, we prefer to use a simpler formula in which we neglect the prefactor (which is either $(1 - rX + r^2X^2)$ in (12′) or $(1 - 2tX + 6t^2X^2)$ in (27′)). These two simplifications lead to a rather simple formula which will be used below:

$$\sigma(E)/E \cong A' \exp\left[-(X^2)\cdot\left\{1 - sX + \left(\frac{3}{4}\right)s^2X^2\right\}\right]. \tag{29}$$

Note that the coefficient $3/4$ in front of s^2X^2 in (29) match the coefficient $3r^2$ of X^2 in (12′) but do not match the coefficient t^2 of X^2 in (27′). However, this choice is correct when $r \gg t$, i.e. when the curvature effect is dominant compared to the quantum effect. The quantitative interpretation of the effective parameter s, obtained numerically is uncertain because we do not know if the effect of the curvature (characterized by r) and the quantum effect (characterized by t) follows the tentative Formula (28). In Section 4 we compare numerically the above-presented models by fitting the experimental $\sigma(E)/E$ of the Cl_2 molecule with Formulae (12′), (27′) and (29). We compare the fitted values of r and t with their expected values given in Formulae (11) and (26) and we interpret the fitted value of s when using (29).

4. THE ABSORPTION CROSS SECTION OF Cl_2 MOLECULE

The Abs. XS of Cl_2 molecule was chosen because it is very smooth and because the Abs. XS published by Maric *et al.* [22] in 1993 seems to be accurate. The smoothed $\sigma(E)/E$, shown on Figure 6.1, goes from 18000 to 42000 cm^{-1}. This experimental XS is a mixture of two overlapping electronic transitions: a strong singlet–singlet transition ($^1\Pi_{1u} \leftarrow {}^1\Sigma_g^+$) centred around 30000 cm^{-1} and a twenty times weaker singlet–triplet transition ($^3\Pi_{0u} \leftarrow {}^1\Sigma_g^+$) centred around 25000 cm^{-1} as displayed on Figure 6.1. Consequently, due to the strong overlap between these two transitions, the various fits discussed below involve a simultaneous fit of both transitions. The weak singlet–triplet XS was fitted with a Gaussian (3 parameters) in order to limit the total number of parameters to 6 or 7 (see below). We must consider the Cl_2 gas at the natural abundance, consisting of 65% of ^{35}Cl and 35% of ^{37}Cl. Therefore, we expect a statistical mixture of the three Cl_2 isotopologues: $^{35}Cl_2$ (42%) $^{35}Cl^{37}Cl$ (46%) and $^{37}Cl_2$ (12%). Given the vibrational frequencies of these three isotopologues, 559.72, 552.13 and 544.59 cm^{-1}, the corresponding weighted average value to be used is $\hbar\omega_e/hc = 554.4$ cm^{-1}. Similarly, the average ZPE of the mixture is close to 275 cm^{-1}. The tiny differences between the ZPE of these three isotopologues (± 7 cm^{-1} from the average) is negligible compared to $V_0 \approx 30400$ cm^{-1}. In the following, we will interpret the experimental $\sigma(E)/E$ of the Cl_2 gas mixture as the $\sigma(E)/E$ of a single molecule with a single set of parameters, A', V_0, V'_e/β and t or r or s (see below).

Three published XS of Cl_2, by Burkholder [4], Maric [22] and JPL [23][1], were compared in order to check the consistency of these experimental XSs. This comparison is done in Table 6.1, using the fitted parameters obtained from Formula (29). These comparisons show that the parameters of the main S–S transition are in good agreement and those of the weak S–T transition are poorly determined for the JPL and Burkholder XS.

It should be noted that two of these three XSs have very few data points and we cannot expect a good accuracy for the r parameter describing the asymmetry, even if the *numerical* uncertainty (Chi^2/DoF) of the fit is small. Note that the JPL data points are not the raw data but are determined from a fit of experimental data with a different analytic formula [23]. Below, the experimental XS of Maric [22] is used to test and compare the various approximations introduced in Sections 2 and 3 of this paper.

Various sets of fitted parameters corresponding to different versions of the above-presented models (Formulae (12′), (27′) and (29)) are compared in Table 6.2. The strong S–S transition (see Figure 6.1) was fitted with six different functions

[1] The JPL-2006 recommendation for the Absorption Cross Section is calculated at 5-nm intervals using the semiempirical function given below which is the Eq. (12) of [23]

$$\sigma(\lambda, T) = 2.73 \times 10^{-19}\ \mathrm{cm}^2 \cdot \tanh^{0.5} \cdot \exp\{-99.0 \times \tanh \times [\ln(329.5\ \mathrm{nm}/\lambda)]^2\}\ \text{(singlet–singlet)}$$

$$+\, 9.32 \times 10^{-21}\ \mathrm{cm}^2 \cdot \tanh^{0.5} \cdot \exp\{-91.5 \times \tanh \times [\ln(406.5\ \mathrm{nm}/\lambda)]^2\}\ \text{(singlet–triplet)},$$

with $\tanh = \tanh(hcX559.751\ \mathrm{cm}^{-1}/2kT) = \tanh(402.7/T)$; λ in nm, $250 < \lambda < 550$ nm, and T in K; $300\ \mathrm{K} > T > 195$ K. The first component at 329.5 nm (25316 cm^{-1}) corresponds to the dominant singlet–singlet transition and the second component at 406.5 nm (24600 cm^{-1}) corresponds to the weaker singlet–triplet electronic transition.

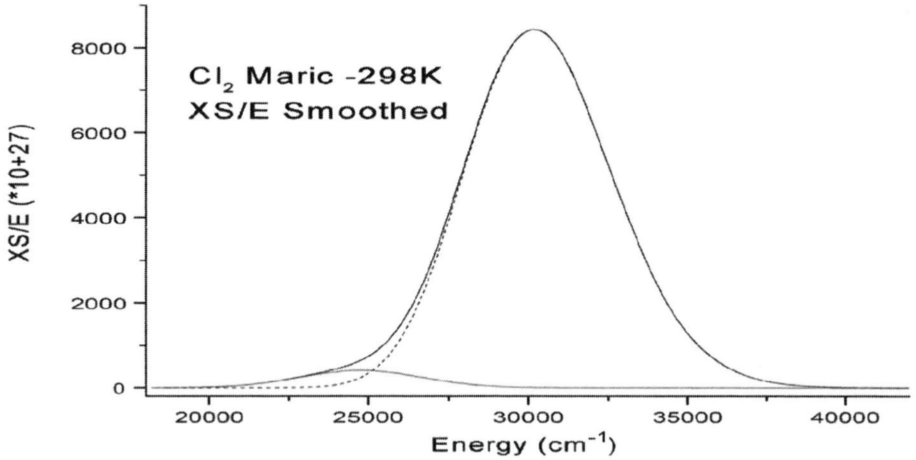

FIGURE 6.1 Plot of the smoothed experimental Cl_2 Abs. Cross Section (XS/E) (full line), and the two contributions of the fitted XS/E: the strong singlet–singlet (S–S) transition is dashed line, and the weak singlet–triplet (S–T) transition is a full gray line. The analytic forms and the parameters of the S–S and S–T components are given in Table 6.1, Fit-3.

TABLE 6.1 Comparison of the fitted parameters for the three experimental XSs of Maric (1993), JPL (2002) and Burkholder (1983). The three fits have been performed with the same analytic expression which is the sum of two components: the main singlet–singlet transition is described by Formula (29) and the weak singlet–triplet transition by a Gaussian. The low Chi^2/DoF for the JPL XS of 17 is mainly due to the very low number (30 pts) of data points

	Maric (1993) (298 K) (4644 points)	JPL (2002) (298 K) (30 points)	Burkholder (301 K) (35 points)
	Parameters of the singlet–singlet electronic transition		
Amplitude ($\times 10^{+27}$)	8430.5 ± 0.34	8436.8 ± 5	8586.3 ± 18
V_0 (cm^{-1})	30158.78 ± 0.2	30166.2 ± 2.6	30131 ± 12
$\beta V'_e$ (cm^{-1})	3230.11 ± 0.5	3252.2 ± 5.3	3291 ± 13
$2t/3$ or $2r$ or s	0.122 ± 0.001	0.102 ± 0.003	0.081 ± 0.005
	Parameters of the singlet–triplet electronic transition		
Amplitude ($\times 10^{+27}$)	384.8 ± 0.8	336 ± 7	322 ± 34
V_0 (cm^{-1})	24612 ± 8	24253 ± 78	24040 ± 380
$\beta V'_e$ (cm^{-1})	2614.7 ± 6	2446 ± 74	1509 ± 552
	Fit residual		
Chi^2/DoF	12	17	778

based on (3′), (12′) (with or without the prefactor) (27′) (with t fixed or free and with or without the prefactor), in order to check and compare numerically various approximations. Note that (12′) without the prefactor is equivalent to For-

TABLE 6.2 Various sets of parameters fitted on the Cl_2 Abs. XS of Cl_2 of Maric (1993). Six sets of fitted parameters are compared. Each set corresponds to slightly different fitted functions. For the six columns, the 3 fitted parameters of the Gaussian describing the weak singlet–triplet transition are given in the lower part of this table. In the upper part of the table we give the (3 or 4) parameters describing the strong singlet–singlet transition for six slightly different fits. The first column gives the 3 parameters of a (second) Gaussian. The second column gives the four fitted parameters of Formula (12′) including the prefactor $(1 - rX + r^2X^2)$. The third column gives the four parameters of Formula (29) which is equivalent to Formula (12′) without its prefactor. The fourth column gives the three fitted parameters using Formula (27′) in which t is constrained to $t = \frac{1}{2}\hbar\omega/\beta V'_e = 0.0565$. The fifth column gives the four fitted parameters of model (27′) when t is not constraint. The sixth column gives the four fitted parameters of (27′) when the prefactor is omitted

	FIT 1	FIT 2	FIT 3	FIT 4	FIT 5	FIT 6
	Fit with 2 Gaussians (6 parameters)	Fit Formula (12′) with prefactor (7 parameters)	Fit Formula (29) with prefactor (7 parameters)	Fit Formula (27′) with prefactor and t fixed (6 parameters)	Fit Formula (27′) with prefactor and t free (7 parameters)	Fit Formula (27′) without prefactor (7 parameters)
Parameters of the singlet–singlet electronic transition $(^1\Pi_{1u} \leftarrow {}^1\Sigma_g^+)$						
Amplitude $(\times 10^{+27})$	8400.4 ± 8	8422.57 ± 0.34	8430.5 ± 0.34	8344.1 ± 3.4	8225.6 ± 2.2	8426 ± 2.4
Centre V_0 (cm^{-1})	30271.0 ± 2.5	30259.47 ± 0.26	30158.78 ± 0.2	30483.7 ± 1.2	30652.8 ± 2	30165.7 ± 1.5
Width V'_e/β (cm^{-1})	3297.0 ± 4	3246.0 ± 0.46	3230.1 ± 0.5	3270.4 ± 1.9	3184.7 ± 1.7	3282.0 ± 1.8
Asymmetry s or $2t/3$ or $2r$	0.0000 (fixed)	0.1245 ± 0.0001	0.1248 ± 0.0002	0.0565 (fixed)	0.098 ± 0.002	0.119 ± 0.003
Parameters of the singlet–triplet electronic transition (Gaussian) $(^3\Pi_{0u} \leftarrow {}^1\Sigma_g^+)$						
Amplitude	237.5 ± 8	382.1 ± 0.8	384.8 ± 0.8	304.0 ± 3.4	323.1 ± 1.6	385.2 ± 6
Centre V_0 (cm^{-1})	23275 ± 50	24602.3 ± 7.5	24612 ± 8	23777 ± 22	24016 ± 14	24413 ± 50
Width β/V'_e (cm^{-1})	1612 ± 68	2608.9 ± 6	2614.7 ± 6	1969 ± 30	2178 ± 16	2375 ± 40
Fit residual						
Chi^2/DoF	6281	12.155	12.152	1239	62	580

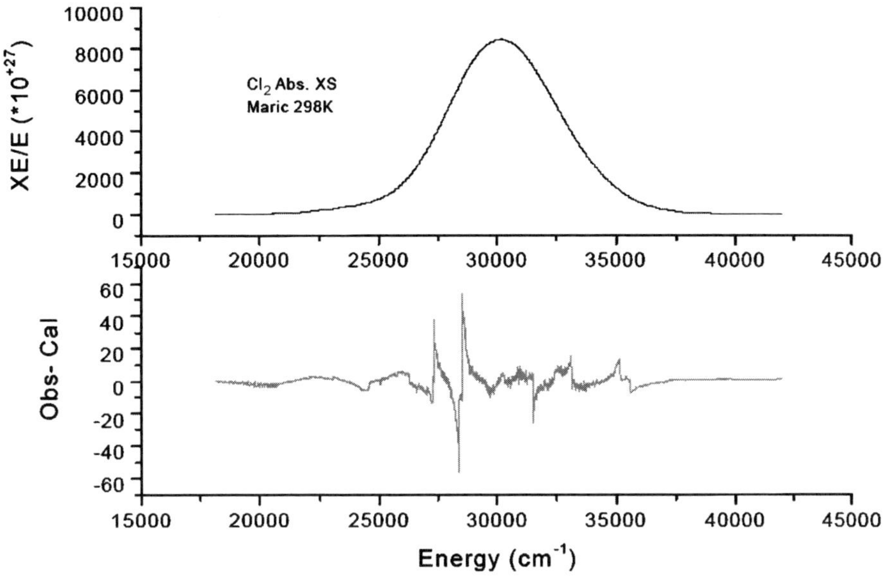

FIGURE 6.2 Upper: plot of the Cl_2 XS/E of Maric et al. The fitted curve almost superimposes with the experimental and cannot be discriminated. The lower part shows the difference between the experimental and fitted curves. Note that the vertical scale is magnified by about a factor 70. The discontinuities in the lower part (present but not visible in upper part!) show that the experimental spectrum results from a concatenation of adjacent parts of the whole XS. The largest step around 28000 cm^{-1} represents about 1% of the maximum amplitude. At least five others discontinuities can be seen. When the experimental XS (or XS/E) is smoothed, these discontinuities can hardly be seen.

mula (29). In association with each of these six fitted functions, the 3 parameters of a Gaussian describing the weak S–T transition located around 24000 cm^{-1} were simultaneously fitted because these two transitions strongly overlap (see Figure 6.1). Globally, each fit involves either 6 (3 + 3) (for fits 1 and 4) or 7 (4 + 3) fitted parameters (for fits 2, 3, 5, 6). A description of six slightly different fits (Fit-1 to Fit-6) is given below and the corresponding parameters of these fits are given in Table 6.2. Figure 6.2 shows the result of Fit-3 which corresponds to Formula (12′) without prefactor or to Formula (29) and which is the best fit. In the upper part of Figure 6.2 the fitted curve overlaps perfectly the experimental $\sigma(E)/E$ and cannot be visually discriminated. The residual is shown on the lower part of Figure 6.2 with a magnification factor of about 100. This residual displays some tiny discontinuities (which cannot be seen on the upper part) and which are probably due to the concatenation procedure, because the spectrometer cannot cover the whole range with the same detector and beam splitter. In any case, the residual is significantly lower than the quoted experimental uncertainty and any numerical improvement may be meaningless.

To ease the comparison between r (when using (12′)), t (when using (27′)), and s (when using (29)), we give, in Table 6.2, either $2r$ (when using (12′)), or $2t/3$ (when using (27′)) or s (when using (29)). Below, the six sets of fitted parameters given in Table 6.2 are described. The first column (Fit-1) of Table 6.2 gives the 6 fitted parameters when two Gaussians are used, describing respectively the S–S and the S–T transitions. For the 5 other fits (Fit-2 to Fit-6) the 3 parameters describing the weak S–T transitions are given in the lower part of Table 6.2. The second column (Fit-2) gives the 4 parameters of Formula (12′) describing the S–S transition. The third column (Fit-3) gives the 4 parameters of Formula (12′) when the prefactor $(1 - rX + r^2X^2)$ is removed. This Fit-3 corresponds also to the use of Formula (29). The comparison between Fit-2, *with* the prefactor of (12′), and Fit-3, *without* this prefactor, shows that its does not play a significant role. Conversely, this means that most of the asymmetry of the Cl_2 cross section ($\sigma(E)/E$) is described by the odd term, $-2rX$ in (12′) or $-(2/3)t$ in (27′), in the argument of the exponential. The fourth column (Fit-4) gives the 3 fitted parameters of Formula (27′) when t is constrained to $t = (1/2)\hbar\omega_e/(V_e'/\beta)$. This Fit-4 has only 6 *free* parameters because the parameter t is proportional to the inverse of (V_e'/β) (the third parameter) and because $\hbar\omega_e$ is known and fixed at 544.6 cm^{-1}. The fifth column (Fit-5) gives the 4 fitted parameters of Formula (27′) when t is free. The sixth (last) column gives the fitted parameters when the prefactor of (27′) is removed. Again, these two fits show that the prefactor does not play a dominant role in describing the asymmetry of the experimental $\sigma(E)/E$ of Cl_2. However the Chi^2/DoF of Fit-5 and Fit-6 are significantly larger than Fit-3, even if the fitted parameters are similar to those obtained with Fit-2 and Fit-3. It should be noted that the only difference between Fit-3 and Fit-6 (both without prefactor) is between the term $\{1 - 2rX + 3r^2X^2\}$ in Fit-3 and the extra term $\{1 - (2/3) \cdot tX + t^2X^2\}$ in Fit-6. If we set $s = 2r = 2t/3$, we get respectively $\{1 - sX + (3/4) \cdot s^2X^2\}$ and $\{1 - sX + (9/4) \cdot s^2X^2\}$. Then, the difference lies only in coefficients 3/4 and 9/4, the first (corresponding to Fit-3) giving a much better fit than the second (corresponding to Fit-6). This observation is consistent with the fact that below we will find a dominant contribution of the curvature of the upper potential (Formulae (12′)) and not of the quantum effect described by (27′). The global comparison of Fit-1 using a Gaussian with the five other fits shows the dramatic effect of the asymmetry parameter which can be $2r$ (using (12′)) or $2t/3$ (using (27′)) or s (using (29)). This is true with or without the prefactor which does not play an important role in (12′) or (27′).

To summarize, the observed asymmetry of the Cl_2 $\sigma(E)/E$ is due to two physical origins: the curvature of the upper potential and/or the "quantum effect" described in Section 3. So, the question is: what are the relative contributions of these two coexisting effects? Fit-4 shows clearly that the "quantum effect" is not able to describe on its own the observed curvature because the value of $2t/3$ derived from $t = \frac{1}{2}\hbar\omega_e/(V_e'/\beta)$ using the fitted value of V_e'/β is only 0.0565, about half the values obtained when the parameter describing the asymmetry is fitted. These fitted values, $2r$ or $2t/3$, given in Table 6.2 (Fit-2, Fit-3, Fit-5 and Fit-6), have a weighted average close to 0.12. This means that, when t is fixed to $= \frac{1}{2}\hbar\omega_e/(V_e'/\beta)$, the model (27′), is *not* able to quantitatively describe the asymmetry of the experimental $\sigma(E)/E$ of Cl_2. Consequently, there is a dominant contribution of the curvature of the upper potential to the asymmetry of the $\sigma(E)/E$ of Cl_2. We can consider that the contribution of the "quantum effect" presented in Section 3 is well characterized by $t = \frac{1}{2}\hbar\omega_e/(V'_e/\beta)$. Since the fitted value of (V'_e/β) is well determined around $3240 \pm 10\ \text{cm}^{-1}$ (see Table 6.2) for Fits 2 and 3, and $\frac{1}{2}\hbar\omega_e$ is well known at $277\ \text{cm}^{-1}$ (see above), we can conclude that $2t/3$ is well known at 0.057 ± 0.001. From this, using $s = 0.12 \pm 0.004$, we derive $r = 0.0515 \pm 0.0025$ when using the tentative Formula (28). The value of r for Cl_2 can be derived from the values of V'_e (parameter C) and V''_e (parameter D) published by Burkholder [4] in his Table 6.3, for his Abs. XS and using $\beta^{-1} = 0.058$ Å. These values give $r = 2D/\beta C = 0.055 \pm 0.005$. The agreement between these two determinations of r may seems fortuitous because Burkholder also takes into account the dependence of the transition dipole moment on $R - R_e$ (which is assumed to be constant in the present model) but the relative variation of this transition dipole moment on the relevant range of $(R - R_e)$, estimated from $b'_u/\beta b_u$, is only 3.5%. Moreover, the uncertainty given by Burkholder on the coefficient b'_u describing the variation of the transition dipole moment is about 70%. The weak dependence of the transition dipole moment and its large uncertainty indicate that the effect of the dependence of transition dipole moment is much smaller than the curvature effect and than the quantum effect described in Section 3. Using $2r = 0.11 \pm 0.01$ and $2t/3 = 0.057 \pm 0.001$ we derive $s = 0.123 \pm 0.01$ from Formula (28), in good agreement with the fitted value of $s = 0.1245$ (Fit-2) or $s = 0.1248$ (Fit-3).

To conclude, most of the asymmetry observed in the singlet–singlet transition in the Abs. XS of Cl_2 can be ascribed to the curvature of the $^1\Pi_{1u}$ upper potential and not to the quantum effect described in Section 3, nor to the dependence of the transition dipole moment. However, this is probably not always the case and the three effects (the curvature effect (Section 2), the quantum effect (Section 3) and the nuclear dependence of the transition dipole moment) may contribute to the asymmetry of any XS is analyzed. It is important to note that the asymmetry due to the "quantum effect" detailed in Section 3 does not require an additional parameter in the fit. Consequently, the first (and easy!) step is to compare two fits of a $\sigma(E)/E$ both using Formula (27′): the first fit with t constrained to $t = \frac{1}{2} \cdot \hbar\omega/(V'_e/\beta)$ (Formula (26)) and a second fit with t as a free parameter. This comparison allows us to estimate if the quantum effect is dominant or not and then, to know whether another contribution is important, e.g., the curvature of the upper potential. In addition, the contribution of the hot bands discussed by Burkholder [4] and Alder-Golden [24] can also contribute to observed asymmetry.

5. A 3D VERSION OF THE MODEL AND ITS APPLICATION TO TRIATOMIC MOLECULES

In this section we apply the model presented in Section 3 to the XS of triatomic molecules such as O_3, SO_2, CO_2. In line with Heller [5], Schinke, on pages 115–116 of his book [6], proposed a 2D version of the simple reflection model but without the curvature and quantum effects taken into account in Sections 2 and 3. For triatomic molecules the quantities corresponding to V'_0/β and V''_e or r (see Formula (12)) defined in Section 2 and corresponding to t (see Formula (27)) defined in Section 3 are difficult to determine. For triatomic molecules, there are 3 components of the gradient to be substituted into V_0', one along each normal coordinate of the ground state, combined with the three corresponding β (defining the three widths of the initial wavefunction), and 6 components of the Hessian corresponding to V_e''. Analogously, the equivalent of t (see Section 3) for triatomic molecules involves an integral overlap between a three dimensional Gaussian and a three dimensional Airy function which appears to be very difficult (may even be impossible) to calculate analytically.

Nevertheless, some simplifications may occur according to the symmetries of the lower and upper PESs of the electronic transition under consideration. For example, when we consider an $A(C_{2v}) \leftarrow X(C_{2v})$ electronic transition of a C_{2v} (ABA) molecule, two of the three normal modes (the symmetric stretch and the bending) belong to a_1 symmetry and the third mode, the antisymmetric stretch, belongs to b_2 symmetry. Then, the 3D calculation of the XS/E splits into a 2D problem linked to the bending and symmetric stretch coordinates and a 1D problem along the antisymmetric stretch for which the equilibrium geometry is the same for both C_{2v} PESs. These symmetry considerations may lead to some tractable simplifications but the corresponding analytic treatment is heavy and out of the scope of this paper. Nevertheless, we mention Formulae (6.18) and (6.19) given by Schinke, page 116 of his book [6], which define the term replacing V_e'/β when two of the three coordinates of a triatomic molecule are involved in the reflection method. These formulae can be used and extended to include the curvatures (the components of the Hessian) of the upper PES.

Alternatively, Formulae (12′), (27′) or (29) can be used *numerically* to fit the $\sigma(E)/E$ of triatomic molecules, even if the interpretation of the fitted parameters is not yet possible. The results presented below show that the numerical improvement obtained by using Formulae (12), (27) or (29) (all have 4 parameters and are able to describe the asymmetry of a $\sigma(E)/E$) is comparable with the improvement observed for Cl_2 (the Chi^2/DoF is reduced by typically up two orders of magnitude; see Section 4). Here, it is essential to note that the reflection models are only able to describe the envelope of the XS, (corresponding to very short time evolution ($t < 10$ fs (femtosecond)) of the wavepacket after the photon absorption) and *not* the vibronic structures which are specific to each molecule and correspond to some vibrational (and or vibronic) oscillations at a time scale of several hundred femtoseconds.

This explains why, below, for each XS, the first step of the analysis is to smooth the XS in order to blur all the vibronic structures. The second step is to divide each smoothed XS by E in order to characterize the asymmetry of the $\sigma(E)/E$.

As mentioned above, the interpretation of the asymmetry of a $\sigma(E)/E$ for a triatomic molecule is much more difficult than for a diatomic molecule. However, the asymmetry is a relevant phenomenological parameter which characterizes the XS in addition to the centre, the width and the amplitude. The isotopic (isotopologue) dependence of these four parameters is of special interest when dealing with the photodissociation processes [9,15]. In fact, this aspect was our initial motivation when starting this study and will be treated in the future. Below, we characterize the XS asymmetry of three important molecules: O_3, SO_2 and CO_2.

5.1 The Abs. Cross Section of the Hartley band of O_3

Several published XSs of the Hartley band were compared with special attention to their 4 fitted parameters obtained when Formula (29) is used. According to our fits, the best XS is the one of Bogumil *et al.* [25]. This XS, more precisely $\sigma(E)/E$ which is the smoothed XS divided by E, is plotted in the upper part of Figure 6.3. The fitted models with four parameters (Formula (29)), overlap the experimental curve and they cannot be discriminated. This is why the fit residual (the difference between the Exp. and the Cal $\sigma(E)/E$) is shown on the lower part of Figure 6.3. The XS of Burrows *et al.* [26] is very similar to the one of Bogumil. In contrast, the XS of Voigt-2001 [27] is significantly different from the two others (and is probably incorrect) above 40000 cm^{-1} as shown on Figure 6.4. We have compared the fit with a Gaussian with the fits using several versions of the 4 parameter models discussed above. These parameters are given in Table 6.3 only when Formula (29) is used. The fit residuals (see the lower part of Figure 6.3) are typically less than 0.3% of the $\sigma(E)/E$ maxima. This is significantly less than the experimental uncertainty and any further improvement may be meaningless. The Chi^2/DoF decreases by about two orders of magnitude (compared with a fitted Gaussian) when an asymmetric parameter (r, t or s) is introduced. This strong improvement demonstrates

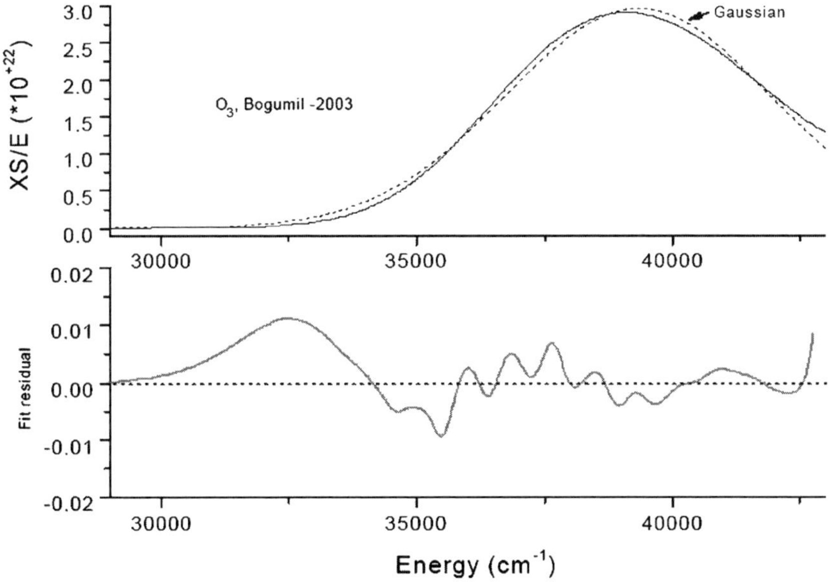

FIGURE 6.3 Fit of the ozone XS ($\sigma(E)/E$) of Bogumil (2003) (Ref. [25]). The upper part shows the smoothed $\sigma(E)/E$ and a fitted Gaussian (dashed). When fitted with Formula (29) the experimental and fitted curves overlap and cannot be discriminated. In the lower part, the difference between the exp. and fitted curves is shown. Note that this difference is, at most, 0.3% of the maximum amplitude while the max difference with the Gaussian (dashed curve of the upper part) reach 7%. The superimposed oscillating structure is due to the vibronic structure of the Exp. Abs. XS which has not been completely smoothed.

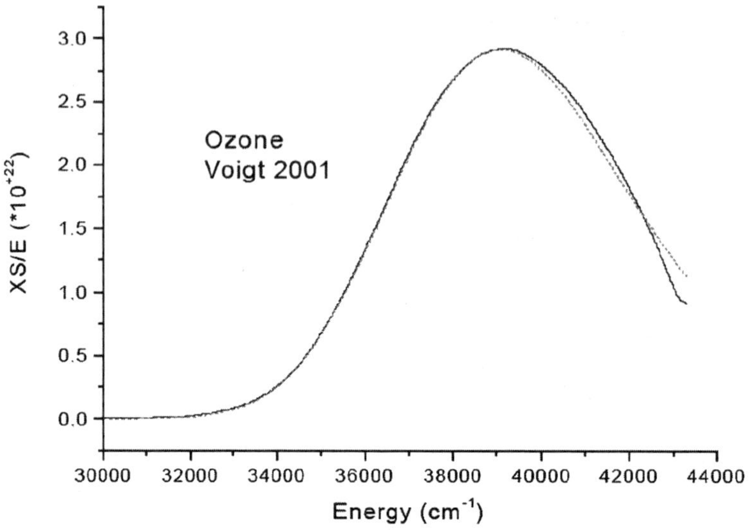

FIGURE 6.4 The smoothed XS/E of Voigt [27] and, (dashed) the XS/E of Bugumil [26]. A visual comparison shows a very good agreement below 39000 cm^{-1} and significant discrepancies above 39000 cm^{-1}.

TABLE 6.3 The fitted parameters[a] of the O_3 Hartley Band. Two fits are compared: in the left column the 3 parameters of a fitted Gaussian are given with the resulting Chi^2/DoF. This fit corresponds to $s = 0$ in Formula (29). In the right column, the four fitted parameters of Formula (29) are given. The Chi^2/DoF decreases by almost two orders of magnitude compared with the Gaussian fit. The uncertainties are 2 S.D (Standard Deviation). Note that *s*, the dimensionless asymmetry parameter, is defined to within 0.3%

Model parameters	XS of Bogumil (2003) [25] fitted range: 28900–42780 cm^{-1}	
	Gaussian model (3 parameters)	Asymmetric model[b] (4 parameters)
Amplitude ($\times 10^{+22}$) ($cm^2/(mol\,cm)$)	2.98 ± 0.01	$2.919\ (\pm 0.001)$
Centre (V_0) (cm^{-1})	39281.0 ± 11	$3\,9071.1 \pm 1.8$
V'_0/β (cm^{-1})	3597.4 ± 16	3768.9 ± 1.6
Asymmetry (s)	0.000 (fixed)	0.2362 ± 0.0013
Chi^2/DoF	3.8×10^{-3}	3×10^{-5}

a These parameters are for the XS/E and not for the XS. For $s = 0$, the HWHM $= 0.83255\ V'_0/\beta$.

b When $s \neq 0$, these parameters do not provide the true amplitude, centre and HWHM of the XS/E but, instead, they characterize the asymmetric Gaussian.

the relevance of the asymmetric Gaussian model which gives a very precise and compact analytic description of the Abs. XS. In contrast with the analysis of the Cl_2 XS, we have not been able to analyze the relative contributions of the curvature ef-

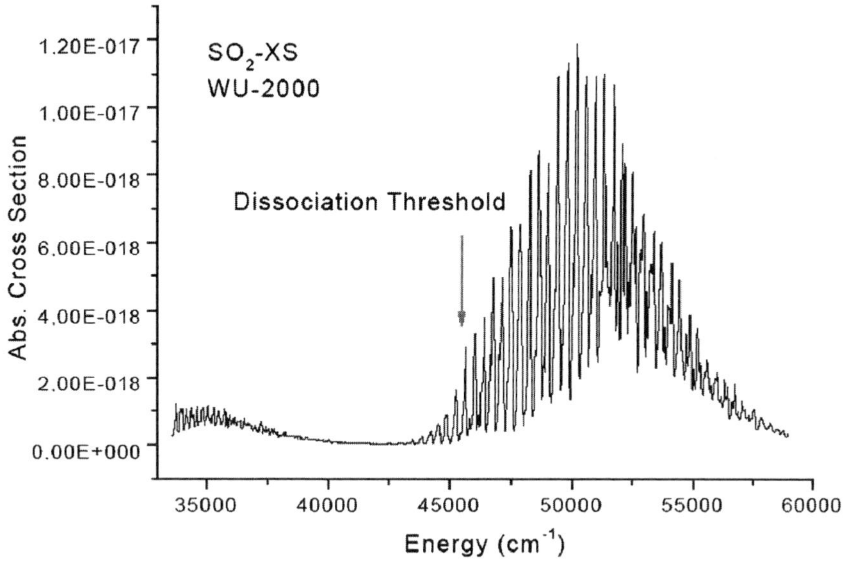

FIGURE 6.5 The Abs. XS of SO_2 from Ref. [28]. This XS is highly vibronically structured. The 42000–59000 cm^{-1} energy range of this XS has been smoothed and divided by E (in cm^{-1}) and is plotted on Figure 6.6.

fect (the Hessian for a 3D potential) and the quantum effect discussed in Sections 2 and 3. Note that the experimental temperature dependence of the parameters can also be analyzed in the future. This may require inclusion of the contribution of the hot bands.

5.2 The SO_2 Abs. Cross Section

We used the XS of Wu [28] which covers a wide energy range and seems to be the best existing SO_2 XS. This XS displays a strong vibronic structure as shown on Figure 6.5. After a smoothing, the XS/E of Figure 6.6 shows contributions from two electronic transitions. The main electronic transition has a centre at 51110 cm^{-1} and has been fitted with Formula (29) while the second transition, which is about 10 times weaker, has a centre at 46408 cm^{-1}. This weak transition was fitted with a Gaussian (the 3 parameters of the left column in Table 6.4) because the asymmetry parameter is not well determined. The 7 fitted parameters are given in Table 6.4. The residual of the fit shown on Figure 6.7 is of the order of 1%, significantly less than the experimental error and any further numerical improvement of the fit may be meaningless.

5.3 The CO_2 Abs. Cross Section

We concatenated the XS, or $\sigma(E)$, of Yoshino [29] between 118 and 163 nm and the XS of Parkinson [30] between 163 and 200 nm. As shown in Figure 6.8, this

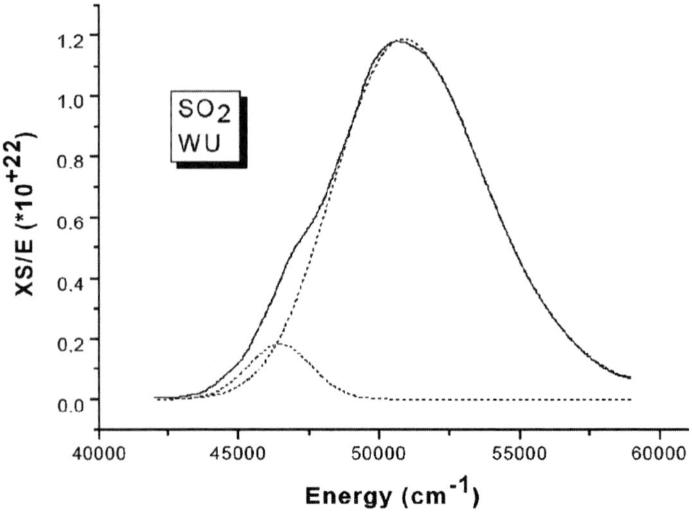

FIGURE 6.6 A plot of the smoothed exp. SO_2 $\sigma(E)/E$ and its decomposition in two electronic transitions (dashed). These two components have been obtained from a fit of the exp. $\sigma(E)/E$ with a sum of two XSs, the strongest being described with a 4 parameter Formula (29) and the weaker one described by a three parameter Gaussian. The corresponding parameters are given in Table 6.4.

TABLE 6.4 The fitted parameters[a] of the smoothed SO_2 Abs. XS/E from a single fit involving two electronic transitions. The lower electronic transition is fitted with a Gaussian (3 parameters, first column) and the second electronic transition is fitted with Formula (29) (4 parameters, second column). The uncertainties are 2 S.D. (Standard Deviation). Note that the uncertainty of the *s* parameter of the upper transition is only 1%. Note that the lower (weak) electronic transition is much narrower than the upper strong electronic transition because the widths are proportional to the slopes of the PES of these two states at the G.S. (Ground State) equilibrium geometry; see Formula 6.18 and 6.19 of Ref. [15]

	Lower electronic transition	Upper electronic transition
Amplitude $(\times 10^{+20})^{b}$ $(\mathrm{cm}^2/(\mathrm{mol\,cm}))$	0.182 ± 0.003	1.188 ± 0.001
Centre[b] (cm^{-1})	46488 ± 12	50922 ± 3
Width V'_e/β^{b} (cm^{-1})	1596 ± 15	3734.5 ± 8
Asymmetry: s	0.000 (fixed)	0.197 ± 0.002

a These parameters are for the XS/E and not for the XS. For $s = 0$, the HWHM $= 0.83255\ V'_0/\beta$.

b When $s \neq 0$, these parameters do not provide the true amplitude, centre and HWHM of the XS/E but, instead, they characterize the asymmetric Gaussian.

concatenation allows for a global description of two overlapping electronic transitions. After smoothing and division by E, the $\sigma(E)/E$, which is the sum of two electronic transitions, has been fitted, each electronic transition being described

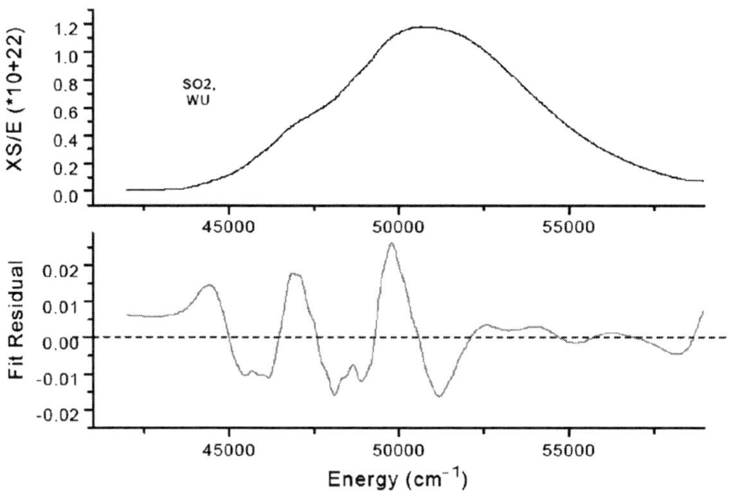

FIGURE 6.7 Comparison between the smoothed exp. and the fitted $\sigma(E)/E$ of SO_2. Upper part: the smoothed exp. $\sigma(E)/E$ of SO_2. On this upper part, the fitted curve is not superimposed on the exp. curve because these two curves are too close. The two components of this XS have been fitted simultaneously: the strong component is described with Formula (29) (4 parameters) and the weak component with a Gaussian (3 parameters). The fitted parameters are given in Table 6.4. Lower part: the fit residual which is typically 1% of the max amplitude. Note that the typical experimental uncertainty is significantly larger than 1%. Then the residual in not necessarily due to the model but may be mostly due to the experimental XS.

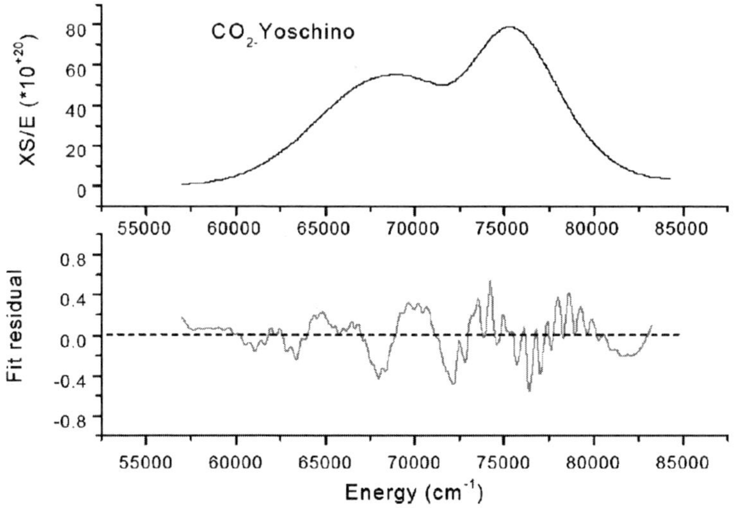

FIGURE 6.8 A comparison between the exp. and the fitted $\sigma(E)/E$ of CO_2, Refs. [29,30]. This XS is made of two components. These two components have been fitted simultaneously, each components being described with the Formula (29). The fit residual is at most about 0.5% of the max amplitude. The superimposed fast oscillations are due to the vibronic structures which have not been completely smoothed.

TABLE 6.5 The fitted parameters[a] of the smoothed CO_2 Abs. XS/E from a single fit involving two electronic transitions. Each electronic transition is fitted with a Formula (29) (4 parameters each). The uncertainties are 2 S.D. (Standard Deviation). Note that the uncertainties in the *s* parameters are only 0.2% and 1.3% for the upper and lower transitions. Note that the lower electronic transition is significantly steeper than the upper strong electronic transition because the widths are proportional to the slopes of the PES of these two states at the G.S. (Ground State) equilibrium geometry; see Formula 6.18 and 6.19 of Ref. [15]

	Lower electronic transition	Upper electronic transition
Amplitude $(\times 10^{+20})^{b}$ $(cm^2/(mol\,cm))$	55.56 ± 0.01	62.09 ± 0.04
Centre^{b} (cm^{-1})	68742.8 ± 2	75779.4 ± 0.8
Width V'_e/β^{b} (cm^{-1})	5954 ± 6	3189.4 ± 2.2
Asymmetry: s	0.078 ± 0.001	0.2455 ± 0.0004

a These parameters are for the XS/E and not for the XS. For $s = 0$, the HWHM $= 0.83255\, V'_0/\beta$.

b When $s \neq 0$, these parameters do not provide the true amplitude, centre and HWHM of the XS/E but, instead, they characterize the asymmetric Gaussian.

with the 4 parameter function defined by Formula (29). The fitted parameters are given in Table 6.5. Again the fit residual is typically less than 0.5% of the maximum amplitude. This is significantly less than the experimental uncertainty and any further improvement of the fit may be meaningless.

6. CONCLUSIONS AND PERSPECTIVES

New models, able to describe precisely the *shape* of the low resolution visible–UV Absorption Cross Section(s) of diatomic and triatomic molecules, have been designed. These models involve mostly three parameters, describing the upper electronic state potential: its vertical energy, slope and curvature. These three parameters, combined with the equilibrium geometry and the force constant of the ground state, allow for a precise and quantitative description of the *shape* of the Abs. Cross Section. However, the amplitude of the Cross Section, which is mostly determined by the transition dipole moment between the ground and the upper electronic state, was only considered as an additional free parameter (independent of the nuclear distance). These models, which enable us to predict the isotopologue (i.e. the mass) dependence of the Absorption Cross Section, can be used to determine isotope ratios after photodissociation [4,9]. An important (and still open) question is whether these isotope ratios are mass independent (MIF type) or not. We guess that the answer to this question is specific to each molecule and each type of photon irradiation. The actinic flux is obviously the most relevant irradiation profile but, in the very important UV range, it depends on the altitude, mostly because of the ozone layer.

From a dynamical (and/or spectroscopic) perspective, we may ask ourselves how to describe and predict the vibronic structures which are superimposed on many low resolution Abs. Cross Sections. These vibronic structures are deeply linked to the time evolution of the wavepacket, after the initial excitation, over typical times of a few hundreds of femtoseconds as discussed by Grebenshchikov *et al.* [31]. In 1D, for a diatomic molecule, the time evolution is rather simple when only one upper electronic state is involved. In contrast, for triatomic molecules the 3D character of the PESs makes the wavepacket dynamics intrinsically complex. So, for most of the polyatomic molecules, the quantitative interpretation of the vibronic structures superimposed to the absorption cross section envelope remains a hard task for two main reasons: first because it requires high accuracy PESs in a wide range of nuclear coordinates and, second, it is not easy to follow the ND ($N = 3$ for triatomic molecules) wavepacket over several hundred femtoseconds, i.e., the timescale of some vibrational periods.

For these reasons, the experimental determination of the Absorption Cross Sections of polyatomic molecules will not be supplanted for a long time by theoretical calculations. This is also true when the Cross Sections of various isotopologues need to be compared but, then, some significant progress can be predicted, mostly because, within the Born-Oppenheimer approximation, the same PESs are involved for the various isotopologues. So, *differential* effects due to various isotope substitutions in triatomic molecules seem easier to characterize than the differences between various molecules.

ACKNOWLEDGEMENTS

This work would not have been possible without the initial stimulations of Mark Thiemens and Greg. Michalski and, more recently, by Joel Savarino and others. Some aspects of this paper would be missing without the suggestions of Reinhard Schinke and Mark Child. I warmly thank Matt Johnson for a careful language and scientific proofreading. This work is supported be the CNRS, PICS contract. R. Jost belongs to ENSPG-INP Grenoble.

APPENDIX A. A POLYNOMIAL VERSION OF FORMULA (12′)

The Formulae (12) or (12′) can be approximated as product of a Gaussian, $\sigma_0(E)$, by a correction function, $\gamma_c(E)$ where the subscript "c" stands for "curvature" because Formula (12′) and then $\gamma_c(E)$ describes the effect of the curvature of the upper PES:

$$\sigma(E)/E = \big(\sigma_g(E)/E\big)\gamma_c(E), \tag{A.1}$$

where

$$\sigma_g(E)/E = A' \exp\big[-X^2\big]. \tag{A.2}$$

(A.2) is the same as (3″), and, $\gamma_c(E)$, limited to second order in r and of fourth order in X, is:

$$\gamma_c(E) = 1 - rX + r^2X^2 + 2rX^3 - 5r^2X^4. \tag{A.3}$$

In these expressions X is defined as in (4): $X = \beta(E - V_0)/V'_e$.

Note that $\sigma_g(E)$ only depends on three independent parameters, A, V_0 and V'_e/β, and that $\gamma_c(E)$ depends on these three parameters, and also on r, which characterizes the asymmetry of $\sigma(E)/E$.

More precisely, the asymmetry of $\sigma(E)/E$ is described by the two odd terms in X, $-rX$ and $+2rX^3$ of $\gamma_c(E)$, this second term being necessary when a large energy range of the XS around V_0 is considered.

Note that the analytic formula of $\gamma(E)$ given by Liang *et al.* [15] is incorrect because they have neglected the contribution of their η parameter in their denominator, $(V'_0 + \eta)$, of the second line of their Formula (7). This denominator gives a linear contribution $-rX$ in Formula (A.3) which is dominant for small X (for E close to V_0). This linear contribution is significant on Figures 6.2, 6.4 and 6.7 of Ref. [15].

APPENDIX B. A POLYNOMIAL VERSION OF FORMULA (27′)

The Formula (27′) can be approximated, a product of a Gaussian, $\sigma_0(E)$, by a correction function, $\gamma_q(E)$ where the subscript "q" stands for "quantum" because Formula (27′) and then $\gamma_q(E)$ describes the quantum effect linked to the Airy function of the final state of the photon absorption:

$$\sigma(E)/E = (\sigma_g(E)/E)\gamma_q(E), \tag{B.1}$$

where

$$\sigma_g(E)/E = A' \exp[-X^2]. \tag{B.2}$$

(B.2) is the same as (3″)), and, $\gamma_q(E)$, limited to the second order in t and of the fourth order in X, is:

$$\gamma_q(E) = 1 - 2tX + 6t^2X^2 + (2/3)tX^3 - (7/3)t^2X^4. \tag{B.3}$$

In (B.2) and (B.3), X is defined as above in (4): $X = (E - V_0)/(V'_e/\beta)$.

Note that $\sigma_g(E)$ depends only on three independent parameters, A, V_0 and V'_e/β, and that $\gamma_q(E)$ depends on these three parameters, and also on t, which characterizes the asymmetry of $\sigma(E)/E$ due to the quantum effect describe in Section 3. More precisely, the asymmetry of $\sigma(E)/E$ is described by the two odd terms in X, $-2tX$ and $+(2/3)tX^3$ of $\gamma_q(E)$, this second term being necessary if a large energy range of the XS around V_0 is considered.

A comparison of (A.3) and (B.3) shows that these two formulae are similar but not equivalent: if we set $r = 2t$ in order to equalize the linear term in X of these formulae, then the terms of higher orders are not proportional. Numerically, the contributions of the two effects are difficult to discriminate. This is why Formula (29) is prefered.

APPENDIX C. THE TEMPERATURE DEPENDENCE OF THE ABSORPTION CROSS SECTION

The treatments presented in Sections 2 and 3 assume the initial state to be a 1D Gaussian for diatomics and 3D for triatomics. This is correct at 0° K but, at a given temperature T, contributions from hot band(s) should be considered. In 1D the contribution of hot bands can be calculated using the wavefunction for $v = 1$ and the harmonic approximation:

$$\Phi_1(x) = \left[(4/\pi)(\mu\omega/\hbar)^3\right]^{1/4} x \exp\left[-(\mu\omega/2\hbar)x^2\right], \tag{C.1}$$

where ω is the vibrational frequency and μ the reduced mass.

Note that, with this notation, the ground state, $v = 0$, given by (2), should be written as:

$$\Phi_0(x) = [\mu\omega/\pi\hbar]^{1/4} \exp\left\{-(\mu\omega/2\hbar)x^2\right\}. \tag{C.2}$$

For most rigid diatomic molecules, ρ, the ratio of population in $v = 1$ to $v = 0$, is quite small, ($\rho \ll 1$) and can be approximated by $\rho \cong \exp[-\hbar\omega/kT]$. Then, the population of $v = 0$ is close to $(1 - \rho)$ and the higher vibrational levels ($v > 1$) can be neglected. As a result, Formula (12′) should be modified as follows:

$$\sigma(E)/E = A\left(1 - \rho + 2\rho X^2\right)\left(V_e'(1 + rX)\right)^{-1} \exp\left[-\left(X^2/(1 + rX)^2\right)\right]. \tag{C.3}$$

The new factor, $1 - \rho + 2\rho X^2$, can be interpreted as follows: the factor $(1 - \rho)$ is the contribution of the population of $v = 0$ and the term $2\rho X^2$ is the contribution from $v = 1$. The term $2X^2$ comes from the square of x in Formula (C.1) because the XS is proportional to the probability distribution along the x axis which is the square of $\Phi_1(x)$. The factor of 2 arises from the square of the ratio of normalization factors in (C.1) and (C.2).

The hot band contributions in Formula (27′) have not been calculated analytically.

For Cl_2, we have $\hbar\omega/hc = 554\ \text{cm}^{-1}$ and $\rho \cong 0.063$ at 300 K ($\cong 200\ \text{cm}^{-1}$). For heavy and or weakly bound molecules the contribution of hot bands can be dominant in the Abs. XS.

For triatomic molecules, the contribution of hot bands cannot be expressed as a function of energy alone (see (5)) and therefore cannot be expressed in a compact analytic formula like Formula (C.3). However, for rigid triatomic molecules like CO_2, NO_2, SO_2, O_3 and N_2O, the contribution of hot bands is weak at room temperature (and below) because $\hbar\omega \ll kT$ for all normal mode frequencies. Note that the width of the contribution to the Abs. XS associated with each excited vibrational level (hot bands) is proportional to the slope of the upper PES along the normal mode of the ground electronic corresponding to each excited (thermally populated) vibrational level. This fact explains why numerical models (e.g. using ground state normal coordinates) are able to calculate the Abs. XS. These calculations are of Frank–Condon type.

REFERENCES

[1] E.U. Condon, *Phys. Rev.* **32** (1928) 858.
[2] E.A. Gislason, *J. Chem. Phys.* **58** (1973) 3702.
[3] R.J. LeRoy, R.G. MacDonald, Burns, *J. Chem. Phys.* **65** (1976) 1485.
[4] J.B. Burkholder, E.J. Bair, *J. Phys. Chem.* **87** (1983) 1859.
[5] E.J. Heller, *J. Chem. Phys.* **68** (1978) 2066.
[6] R. Schinke, *Photodissociation Dynamics*, Cambridge Univ. Press, Cambridge, 1993.
[7] K. Mauersberger, P. Lämmerzahl, D. Krankowsky, *Geophys. Res. Lett.* **28** (16) (2001) 3155; K. Mauersberger, B. Erbacher, D. Krankowsky, J. Günter, R. Nickel, *Science* **283** (1999) 370; K. Mauersberger, D. Krankowsky, C. Janssen, R. Schinke, *Adv. At. Mol. Opt. Phys.* **50** (2005) 1.
[8] G.S. Selwyn, H.S. Johnston, *J. Chem. Phys.* **74** (1981) 3791.
[9] Y.L. Yung, C.E. Miller, *Science* **278** (1997) 1778.
[10] M.S. Johnson, G. Billing, A. Gruodis, M. Janssen, *J. Phys. Chem. A* **105** (2001) 8672.
[11] C. Parisse, J. Brion, J. Malicet, *Chem. Phys. Lett.* **248** (1996) 31.
[12] P. von Hessberg, J. Kaiser, M.B. Enghoff, C.A. McLinden, S.L. Sorensen, T. Röckmann, M.S. Johnson, *Atmos. Chem. Phys.* **4** (2004) 1237.
[13] A.J. Colussi, F.Y. Leung, M.R. Hoffmann, *Environ. Chem.* **44** (2004) 1.
[14] The XSs of $^{14}N^{18}O_2$, $^{15}N^{16}O_2$, $^{15}N^{18}O_2$, $^{18}O^{14}N^{16}O$ have been recorded recently at ULB by Ann-Carine Vandaele and Sophie Fally. A manuscript is under preparation.
[15] M.C. Liang, G.A. Blake, Y.L. Yung, *J. Geophys. Res. D* **109** (2004) 10308; See also G.A. Blake, M.C. Liang, C.G. Morgan, Y.L. Yung, *J. Geophys. Res. Lett.* **30** (2003) 1656; S. Nanbu, M.S. Johnson, *J. Chem. Phys.* **108** (2004) 8905; M.K. Prakash, J.D. Weibel, R. Marcus, *J. Geophys. Res. Atmos. D* **110** (2005) 21315.
[16] R.T. Pack, *J. Chem. Phys.* **65** (1976) 4765.
[17] J.K. Cohen-Tannoudji, B. Diu, F. Laloë, *Mechanique Quantique*, vol. 1, Hermann, Paris, 1977, p. 499.
[18] M.S. Child, *Semiclassical Mechanics with Molecular Applications*, Oxford Science Publications, Clarendon Press, Oxford, 1991, see p. 123.
[19] J. Tellinghuisen, The Franck–Condon principle in bound free transitions, in: K.P. Lawley (Ed.), *Photodissociation and Photoionization*, Willey, New York, 1985.
[20] M. Abramowitz, I.A. Stegun, *Handbook of Mathematical Functions*, Dover, London, 1970.
[21] K.F. Freed, Y.B. Band, in: E.C. Lim (Ed.), *Excited States*, vol. 3, Academic Press, New York, 1977.
[22] D. Maric, J.P. Burrows, R. Meller, G.K. Moortgat, *J. Photochem. Photobiol. A Chem.* **70** (1993) 205.
[23] D. Maric, J.P. Burrows, G.K. Moortgat, A study of the UV–visible absorption spectra of Br_2 and BrCl, *J. Photochem. Photobiol. A Chem.* **83** (1994) 179–192.
[24] S.M. Alder-Golden, *Chem. Phys.* **64** (1982) 421.
[25] K. Bogumil, J. Orphal, T. Homann, S. Voigt, P. Spietz, O.C. Fleischmann, A. Vogel, M. Hartmann, H. Bovensmann, J. Frerick, J.P. Burrows, *J. Photochem. Photobiol. A Chem.* **157** (2003) 167.
[26] J.P. Burrows, A. Richter, A. Dehn, B. Deters, S. Himmelmann, S. Voigt, J. Orphal, *J. Quant. Spectrosc. Radiat. Transfer* **61** (1999) 509.
[27] S. Voigt, J. Orphal, K. Bogumil, J.P. Burrows, *J. Photochem. Photobiol. A Chem.* **143** (2001) 1.
[28] C.Y.R. Wu, B.W. Yang, F.Z. Chen, D.L. Judge, J. Caldwell, L.M. Trafton, *Icarus* **145** (2000) 289.
[29] K. Yoshino, J.R. Esmond, Y. Sun, W.H. Parkinson, K. Ito, T. Matsui, *J. Quant. Spectrosc. Radiat. Transfer* **55** (1996) 53; High-resolution (0.005 nm) data from http://www.atmosphere.mpg.de/Spectra/Quick_Search_5so.html.
[30] W.H. Parkinson, J. Rufus, K. Yoshino, *Chem. Phys.* **290** (2003) 251; High-resolution (0.005 nm) data from http://www.atmosphere.mpg.de/Spectra/Quick_Search_5so.html.
[31] S. Grebenshchikov, *et al.*, *Phys. Chem. Chem. Phys.* **9** (2007) 2044.

CHAPTER 7

Isotope Effects in Photodissociation: Chemical Reaction Dynamics and Implications for Atmospheres

Solvejg Jørgensen*, **Mette M.-L. Grage*,****, **Gunnar Nyman**** and **Matthew S. Johnson**[1*]

Contents

1. Introduction 102
2. Electronic Structure Calculations 103
   2.1 Computing adiabatic potential energy surfaces 104
   2.2 Diabatization 105
   2.3 Electronic transition dipole moment 106
   2.4 Complete basis set extrapolation 106
   2.5 Interpolation/fitting 107
3. Construction of the Time-Independent Hamiltonian Operator 108
4. Time-Independent Methods 109
5. Time-Dependent Methods 111
   5.1 Methodology 111
   5.2 Propagation schemes 113
6. Examples of Photodissociation 115
   6.1 Photolysis of HCl 115
   6.2 Photolysis of N_2O 119
   6.3 Photolysis of OCS 123
   6.4 Photolysis of HCHO 125
7. Perspective 128
Acknowledgements 129
References 129

* Copenhagen Center of Atmospheric Research, Department of Chemistry, University of Copenhagen, Universitetsparken 5, DK-2100 Copenhagen Ø, Denmark
** Department of Chemistry, University of Gothenburg, S-412 96 Göteborg, Sweden
1 Corresponding author. E-mail: msj@kiku.dk

Advances in Quantum Chemistry, Vol. 55
ISSN 0065-3276, DOI: 10.1016/S0065-3276(07)00207-9

Abstract Obtaining the absorption and/or photodissociation cross section is a threefold challenge: computing the electronic potential energy surfaces, interpolating the potentials, and finding the cross section either by time-dependent or time-independent methods. We review electronic structure methods used for computing accurate potential energy surfaces for the electronic ground and accessible excited state as well as coupling between them (electronic transition dipole moments and diabatic coupling). Methods used for interpolation are discussed. The time-independent methods are based on the reflection principle and implicitly involve the short time approximation. In the time-dependent methods the time-dependent Schrödinger equation is solved exactly and the method considers the effect of dynamics away from the Franck–Condon region. We illustrate the presented methods using small molecules (HCl, N_2O, OCS and HCHO) and their isotopic analogues.

1. INTRODUCTION

The general chemistry of the atmosphere depends critically on the concentrations of gases present at trace levels, with mole fractions of parts per thousand to parts per billion. These gases are governed by the mass balance equation, which states that the change in concentration with time equals the difference between the rates of production and loss, $dC/dt = r_P - r_L$. Most atmospheric gases have distributions that are further characterized according to their isotopologues and/or isotopomers. Isotopologues are molecules that differ according to the number of isotopic substitutions, for example CH_4 and CH_3D. Isotopomers differ according to the position of the substitution, for example symmetric and asymmetric ozone, $^{16}O^{18}O^{16}O$ and $^{16}O^{16}O^{18}O$. These molecules are special because of the presence of relatively rare stable isotopes of which ^2H (D), ^{13}C, ^{15}N, ^{17}O, ^{18}O, ^{33}S and ^{34}S are the most important. The additional information obtained from the construction of an isotope budget can be used to investigate atmospheric photochemistry and constrain emission and deposition budgets. In order to do this, both the isotopic enrichment or depletion of the sources (often a combination of natural and anthropogenic) and the isotopic fractionation of the sinks (often chemical reaction or photolysis) must be quantified [1,2]. In this review the emphasis will lie on using quantum calculations to explore the differences caused by isotopic substitution in a given molecule.

Photons are emitted by the sun. When a molecule absorbs a photon it may dissociate breaking one or more chemical bonds. The rate of a reaction in the atmosphere, *e.g.* $ABC + h\nu \rightarrow A + BC$, depends on the absorption cross section of the molecule ABC, the photolysis quantum yield and the actinic solar flux, shown in Figure 7.1, all of which are wavelength dependent.

The probability of photodissociation depends on the spectral and intensity distribution of the light flux, the absorption cross section and the photolysis branching ratio(s). The key molecular property is the transition dipole that governs the intensity or oscillator strength of the transition. Due to violation of space and spin

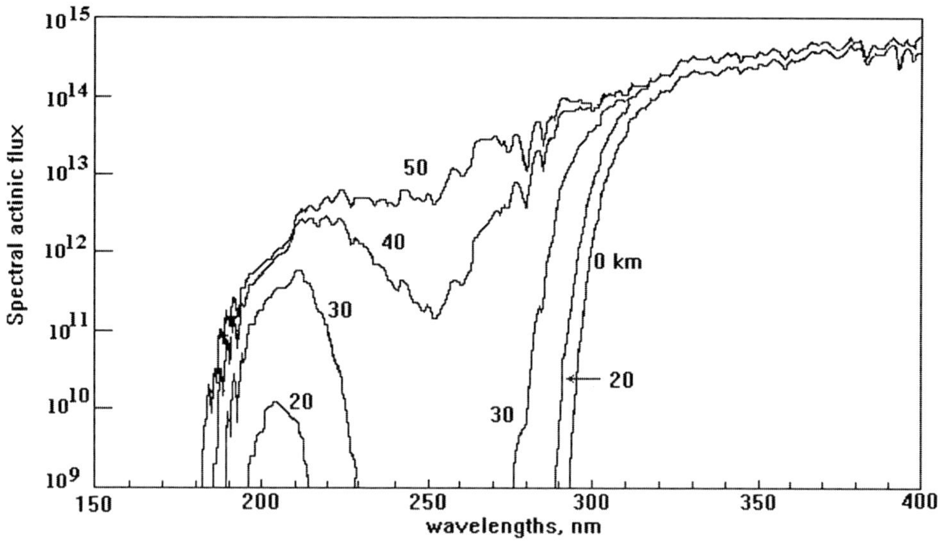

FIGURE 7.1 The solar actinix flux photons $cm^{-2}\,s^{-1}\,nm^{-1}$.

symmetry, certain electronic transitions are forbidden. Upon excitation from the ground state to an excited electronic state the dissociation is governed by the topology of the potential energy surfaces. If the excited state potential is repulsive along the reaction coordinate the molecule will undergo direct dissociation leading to a bell-shaped Gaussian type absorption cross section, whereas if the excited state potential has a barrier it will lead to peaks in the spectrum depending on the lifetime of the resonant states. However, most photodissociation reactions involve avoided crossings between the potential energy surfaces of the electronic excited states. These enable the molecule to be transferred from one excited state potential energy surface to another. The exit channel for the new electronic state may differ from that of the old [3]. In order to gain insight into the dynamics of the photodissociation a microscopic description is needed. This requires computing the potential energy surfaces of the electronic ground and the accessible excited electronic states, and their couplings. The evolution of the dynamics of the photodissociation can be followed once the time-dependent Schrödinger equation (TDSE) has been solved. Often there are several product reaction channels either on the same potential energy surface, or accessible through coupling (vibronic, spin orbit) to other potential energy surfaces, making it interesting to compute state-to-state cross sections and branching ratios.

2. ELECTRONIC STRUCTURE CALCULATIONS

In general, a molecule with N atoms has $3N - 6$ internal degrees of freedom denoted by $R = (r_1, r_2, \ldots, r_{3N-6})$. However, if the molecule is linear it has $3N - 5$ internal degrees of freedom. The potential energy surfaces are functions of the internal degrees of freedom. The key to understanding photodissociation dynamics is obtaining accurate electronic potential energy surfaces as well as the electronic coupling elements between them. Quantitative insight about the photodissociation dynamics can be obtained by considering the topology of the potential energy surface. If the electronic excited state potential is repulsive along the reaction coordinate the molecule will undergo direct dissociation, whereas for a barrier the molecule will dissociate indirectly. Avoided crossings between the potential energy surfaces of the electronic excited states along the reaction coordinate will enable the molecule to transfer from one excited state potential energy surface to another.

2.1 Computing adiabatic potential energy surfaces

Computing accurate potential energy surfaces is a challenging task. In order to describe photolysis the electronic Schrödinger equation must be solved for a large set of geometries for the ground and excited electronic states. The accuracy depends on the selection of an approximate method for solving the electronic Schrödinger equation, and on the choice of basis set. In the following the fact that acquiring the excited state potential surfaces is essential will be reflected in the choice of which methods to present.

In order to properly describe the electronic ground and excited states a highly correlated level of theory and many electronic configurations are typically required. We note that multi-configurational techniques are needed to describe bond breakings. The multi-configurational self consistent field (MCSCF) approach yields a flexible wave function based on a linear combination of configuration state functions (CSFs), which are spin-adapted eigenfunctions [4]. The usual difficulty with MCSCF calculations is selecting the configuration space. A successful approach for this is the complete active space self consistent field (CASSCF) method [5,6]. The orbitals are then classified as inactive, active or virtual. The inactive ones are always doubly occupied and the virtual ones are unoccupied. The remaining ones are the active ones and within these the active electrons are distributed in all possible ways, yielding the complete active space (CAS). Usually the core orbitals are chosen as the inactive orbital space. In many cases it is not possible to include all valence orbitals in the active space. Therefore, and in particular for systems containing more than two atoms, part of the valence space must also be included in the inactive space. CASSCF calculations may be improved by second order Møller–Plesset perturbation theory (CASPT2) [7,8].

The CASSCF wavefunctions are often used to find the reference space for multi-reference configuration interaction (MRCI) [9] calculations. In the MRCI approach a suitable reference space can be obtained from MCSCF calculations by employing state-averaged CASSCF wave functions, which have the advantage of being orthonormal. Usually only single and double excitations from each reference configuration are included. The Davidson correction (MRCI + Q) may be used to approximate the contributions of higher excitations and for size-extensive energies. The choice of method depends on the size of the system. For systems with four atoms or less, highly correlated methods such as MRCI + Q have been used in many cases, for example O_3 [10–12] and ClO [13,14].

If several electronically excited states are relevant for describing the photodissociation then one or more of the Rydberg orbitals of the molecule must be included in the (CAS) [13]. As the number of orbitals and electrons increases in the CAS, the computational time increases dramatically. In order to obtain accurate potential energy surfaces for the excited electronic states, one must include diffuse functions in the basis set [4]. For heavier atoms, a relativistic effective core potential (ECP) can be used to treat the scalar relativistic effects. The ECP basis sets have been developed by several research groups [15,16] and have been implemented in most of the standard electronic structure programs.

2.2 Diabatization

Avoided crossings are often found between adiabatic electronic potential energy surfaces, indicating the breakdown of the Born–Oppenheimer approximation, which assumes that the motion of the electrons is much faster than the nuclear motion and that therefore the nuclear and electronic degrees of freedom are decoupled. On the contrary the two involved potential energy surfaces are coupled around an avoided crossing. The non-adiabatic coupling between two electronic states can be described by including the first and the second derivative couplings that are neglected in the Born–Oppenheimer approximation [17–21],

$$\left\langle \Xi_k^{\mathrm{a}}(\mathbf{r};\mathbf{R}) \left| \frac{\partial}{\partial \mathbf{R}} \right| \Xi_j^{\mathrm{a}}(\mathbf{r};\mathbf{R}) \right\rangle \quad \text{and} \tag{1a}$$

$$\left\langle \Xi_k^{\mathrm{a}}(\mathbf{r};\mathbf{R}) \left| \frac{\partial^2}{\partial \mathbf{R}^2} \right| \Xi_j^{\mathrm{a}}(\mathbf{r};\mathbf{R}) \right\rangle, \tag{1b}$$

where $\Xi^{\mathrm{a}}(\mathbf{r};\mathbf{R})$ are the adiabatic electronic wave functions calculated at fixed coordinate $\mathbf{R}$.

The sudden changes in the adiabatic wavefunctions near avoided crossings make it more convenient to use diabatic potential energy surfaces when simulating photodissociation dynamics. The adiabatic potentials, usually constructed from electronic structure calculation data, should therefore be transformed to diabatic potentials. The adiabatic–diabatic transformation yields diabatic states for which the derivative couplings above approximately vanish. The diabatic potential energy surfaces are obtained from the adiabatic ones by a unitary orthogonal transformation [22,23]

$$\begin{pmatrix} \Xi_k^{\mathrm{d}}(\mathbf{r};\mathbf{R}) \\ \Xi_j^{\mathrm{d}}(\mathbf{r};\mathbf{R}) \end{pmatrix} = \begin{pmatrix} \cos(\theta(\mathbf{R})) & -\sin(\theta(\mathbf{R})) \\ \sin(\theta(\mathbf{R})) & \cos(\theta(\mathbf{R})) \end{pmatrix} \begin{pmatrix} \Xi_k^{\mathrm{a}}(\mathbf{r};\mathbf{R}) \\ \Xi_j^{\mathrm{a}}(\mathbf{r};\mathbf{R}) \end{pmatrix}, \tag{2}$$

where Ξ^{d} are the new diabatic electronic wave functions and $\theta(\mathbf{R})$ is a coordinate-dependent mixing angle. In the diabatic representation the kinetic energy matrix operator is approximately diagonal as opposed to the adiabatic representation. Instead the potential energy matrix operator is non-diagonal in the diabatic representation. The diagonal elements describe the diabatic potential energy surfaces whereas the off-diagonal elements describe the coupling between them.

2.3 Electronic transition dipole moment

Photodissociation involves electronic transitions initiated by the absorption of light. The key molecular property that mediates the interaction with light is the transition dipole moment. The electronic transition dipole moment between the jth and kth electronic state is defined by the integral over the electronic degrees of freedom for the operator where the dipole moment is sandwiched between the two electronic wave functions given by

$$\mu_{jk}(\mathbf{R}) = \int \Xi_k^{a}(\mathbf{r};\mathbf{R})\Big(\sum_i(-e\hat{\mathbf{d}}\cdot\mathbf{r}_i)\Big)\Xi_j^{a}(\mathbf{r};\mathbf{R})\,dr. \tag{3}$$

The electronic transition dipole moment depends on the nuclear coordinates. Certain transitions are forbidden by symmetry, for example $\Delta \leftarrow \Sigma$ in CO_2 and N_2O. Vibrational bending motion, however, breaks the symmetry and the transition becomes allowed [24,25].

2.4 Complete basis set extrapolation

The goal is to obtain the potential energy surfaces, electronic transition moment(s) and intersystem coupling(s) at the complete basis set (CBS) limit for a given level of theory. As the number of basis functions increases the computational time increases tremendously. To overcome this difficulty, methods have been developed to extrapolate the calculated energies to the CBS limit [26–33], based on the correlation consistent basis sets denoted cc-pVnZ by Dunning and co-workers [34–37]. The symbol n is the cardinal number labeling the basis set, *e.g.* $n = 2, 3, 4$ for the aug-cc-pVDZ, aug-cc-pVTZ and aug-cc-pVQZ basis sets respectively. In *ab initio* calculations the correlation energy converges much slower than the Hartree–Fock (HF) energy, and Halkier *et al.* [32] suggest a particularly simple form to extrapolate the dynamical correlation energy

$$E_{\mathrm{CBS}}^{\mathrm{corr}} = E^{\mathrm{corr}}(n) + \frac{A}{n^3}, \tag{4}$$

where $E_{\mathrm{CBS}}^{\mathrm{corr}}$ is the estimated dynamical correlation energy at the CBS limit and A is a linear parameter. By performing calculations for two cardinal numbers, n_1 and n_2, A can be eliminated and $E_{\mathrm{CBS}}^{\mathrm{corr}}$ can then be obtained from (4) as

$$E_{\mathrm{CBS}}^{\mathrm{corr}}(n_1, n_2) = \frac{E^{\mathrm{corr}}(n_1)n_1^3 - E^{\mathrm{corr}}(n_2)n_2^3}{n_1^3 - n_2^3}. \tag{5}$$

In single reference problems, where HF is a good first approximation, only the dynamic correlation energy is retrieved in subsequent correlation treatments. In multi-reference situations the static (or nondynamic) correlation energy, which relates to near-degeneracy, should also be considered. The reference space must then be carefully considered and the orbitals and the configurations should be optimized simultaneously [4], as in, *e.g.*, a CASSCF calculation. If the active space in the CASSCF calculations is adequate, the static correlation energy is included [38]. The dynamical correlation energy can then be considered to be the difference between the CASSCF energy and, *e.g.*, the CASPT2 or MRCI energy.

2.5 Interpolation/fitting

Solving the time-dependent nuclear Schrödinger equation using a wave packet approach implies a grid representation. It is necessary to be able to estimate the potential energies at any given grid point. If *m ab initio* points are needed to obtain a proper description of the potential energy functions for each degree of freedom, we will need $(3N - 6)^m$ *ab initio* points to establish a global fit of the potential energy surfaces. Not only the potential energy surfaces but also the electronic transition dipole moment and the electronic diabatic coupling elements must be fitted/interpolated.

Murrel and co-workers [39] have developed a functional form in which the global analytic potential energy surface is written as a sum of N-body terms

$$V_{\mathrm{AB}\ldots N}(\mathbf{R}) = \sum_{\mathrm{A}} V_{\mathrm{A}}^{(1)} + \sum_{\mathrm{AB}} V_{\mathrm{AB}}^{(2)}(R_{\mathrm{AB}}) + \cdots + V_{\mathrm{AB}\ldots N}^{(N)}(R), \tag{6}$$

where the one-body term $V_{\mathrm{A}}^{(1)}$ describes the energy of atoms A, the two-body term $V_{\mathrm{AB}}^{(2)}(R_{\mathrm{AB}})$ is the potential energy curve of the diatom AB, and so on. Other functional methods such as the London–Eyring–Polanyi–Sato (LEPS) [40–42], diatomic-in-molecule (DIM) [43–45], bond-energy-bond-order (BEBO) [46,47], rotated-Morse-curve-spline (RMCS) [48–50], and large angle generalization of rotating bond order (LAGROBO) [51] have been used for triatomic systems and also for several larger systems. The disadvantage of the functional forms is that they require good insight concerning the nature of the interactions of molecular/atomic fragments in the molecule. Furthermore there are few adjustable parameters to tune the potential energy surface, which may lead to poor accuracy.

Spline functions have been used successfully to fit potential energy surfaces in one and two dimensions [52,53]. However for higher dimensional surfaces unphysical oscillations [53] can easily occur, and densely distributed *ab initio* points may be required [53,54].

The reproducing kernel Hilbert space [55–57], modified Shepard interpolation [58–63], interpolation moving least squares (IMLS) [64–67], distributed approximating functional [68,69], generalized discrete variable representation [70] and neural network algorithms [71,72] are well known interpolation methods. The advantage of these methods is that tedious fitting of the functional form is not needed. IMLS is based on a polynomial expansion. The Shepard interpolation is a zero-degree polynomial IMLS method. In the Shepard interpolation the potential energy and the corresponding gradient and Hessian matrix are needed at the geometry points, whereas in IMLS only the potential energy is needed. Obtaining the gradient and the Hessian requires additional electronic structure calculation but reduces the number of data points required. In the Shepard interpolation method there are different approaches for selecting the geometries for the electronic structure calculations.

Shepard interpolation has been applied to several systems. Wu *et al.* have used it to construct the ground potential energy surface for the reaction $CH_4 + H \rightarrow CH_3 + H_2$ [63]. Neural networks can be described as general, non-linear fitting functions that do not require any assumptions about the functional form of the underlying problem. The evaluation of the potential energy at the grid points is cheap when employing a neural network. Lorenz *et al.* have used a 2×2 potential energy surface generated by a neural network to represent a potassium covered Pd(100) surface [71].

3. CONSTRUCTION OF THE TIME-INDEPENDENT HAMILTONIAN OPERATOR

The Hamiltonian operator consists of the potential energy operator and the kinetic energy operator. These operators have to be expressed in the coordinate system to be used. The potential energy operator is not a problem in single surface calculations. All that is required is to convert between whatever coordinates the potential energy routine was written in and the coordinates to be used in the dynamics, although obtaining an accurate potential energy surface itself may be a major obstacle. The kinetic energy operator on the other hand requires that the Laplace operator be expressed in the relevant coordinate system, which should be done such that the operator is Hermitian. It is also important in numerical calculations to choose proper basis sets so that the Hermitian property is numerically retained [73]. Podolsky has described a well known procedure for deriving the kinetic energy operator [74].

For a diatomic molecule (AB) there is only one degree of freedom namely the bond length between the two atoms denoted r. The Hamiltonian without rotational contributions is given by

$$\hat{H}(r) = -\frac{\hbar^2}{2\mu_{\mathrm{AB}}}\frac{\partial^2}{\partial r^2} + V(r), \tag{7}$$

where μ_{AB} is the reduced mass of the diatomic molecule. Rotation can be included by adding the centrifugal potential.

For a non-linear triatomic molecule (ABC) there are three degrees of freedom. The representation of the three degrees of freedom depends on how many reaction channels are open/relevant. A review of the choice of coordinates and the corresponding Hamiltonian is given in Ref. [75]. It is common to use either the Jacobi or the hyper-spherical coordinate representation.

If only one reaction channel is relevant, *e.g.* $\mathrm{ABC} + h\nu \rightarrow \mathrm{AB} + \mathrm{C}$, Jacobi coordinates are suitable since one bond (AB) dominates and is not broken. The Jacobi coordinates are the set (R, r, θ), where r is the diatomic internuclear distance (A–B) and R is the distance between the atom (C) and the centre of the mass of the diatom (AB). θ is the Jacobi angle between R and r. Following Balint-Kurti *et al.* the non-rotating Hamiltonian in Jacobi coordinates is given by

$$\begin{aligned}\hat{H}(R, r, \theta) = &-\frac{\hbar^2}{2\mu_{\mathrm{AB\text{-}C}}}\frac{\partial^2}{\partial R^2} - \frac{\hbar^2}{2\mu_{\mathrm{AB}}}\frac{\partial^2}{\partial r^2} \\ &+ \left(\frac{1}{2\mu_{\mathrm{AB\text{-}C}}R^2} + \frac{1}{2\mu_{\mathrm{AB}}r^2}\right)\frac{1}{\sin\theta}\frac{\partial}{\partial\theta}\sin\theta\frac{\partial}{\partial\theta} + V(R, r, \theta),\end{aligned} \tag{8}$$

where $\mu_{AB\text{–}C} = \frac{(m_A+m_B)m_C}{m_A+m_B+m_C}$ is the reduced mass of the of the C–AB system and μ_{AB} is the reduced mass of the diatomic fragment AB [76].

The hyperspherical coordinate Hamiltonian is suitable for photodissociation with two or more arrangement channels *e.g.* ABC + $h\nu$ $\rightarrow$ AB + C or A + BC. The three degrees of freedom are represented by (ρ, θ, ϕ). Employing Johnson's modified hyperspherical coordinates the transformation from hyperspherical to Jacobi coordinates is given by [77,78]:

$$R = \frac{1}{\sqrt{2}}\frac{1}{d_1}\rho\sqrt{1 - \sin\theta\cos\phi}, \tag{9}$$

$$r = \frac{1}{\sqrt{2}}d_1\rho\sqrt{1 + \sin\theta\cos\phi} \tag{10}$$

$$\text{and} \quad \cos\theta_{\text{Jacobi}} = \frac{\sin\theta\cos\phi}{\sqrt{1 - \sin^2\theta\cos^2\phi}}, \tag{11}$$

where

$$d_1 = \frac{m_C}{\mu_{ABC}}\left(1 - \frac{1}{m_A + m_B + m_C}\right) \quad \text{and} \quad \mu_{ABC} = \left(\frac{m_A m_B m_C}{m_A + m_B + m_C}\right)^{1/2}.$$

The non-rotating Hamiltonian in these coordinates becomes [78]:

$$\hat{H}(\rho, \theta, \phi) = -\frac{\hbar^2}{2\mu_{ABC}}\left(\frac{1}{\rho^5}\frac{\partial}{\partial\rho}\rho^5\frac{\partial}{\partial\rho} + \frac{4}{\rho^2}\left(\frac{1}{\sin 2\theta}\frac{\partial}{\partial\theta}\sin 2\theta\frac{\partial}{\partial\theta} + \frac{1}{\sin^2\theta}\frac{\partial^2}{\partial\phi^2}\right)\right) + V(\rho, \theta, \phi). \tag{12}$$

Approximations are usually used for systems that comprise four or more atoms. Derivation of the Hamiltonian operator is facilitated by symbolic algebra packages like Mathematica and Maple. Recently Evenhuis *et al.*, employing Mathematica, derived an exact kinetic operator for AB_3 molecules in curvilinear coordinates together with a series of simpler, progressively more approximate kinetic energy operators [79]. The coordinates used were chosen to work well in multi-configuration time-dependent Hartree (MCTDH) calculations [80] and therefore to give as small correlation as possible between the coordinates.

4. TIME-INDEPENDENT METHODS

In this section we review three simple approximate time-independent methods which are specifically designed for calculating absorption spectra. Time-independent quantum scattering calculations, which can give exact results, will not be discussed. Such methods have been reviewed for instance by Nyman and Yu 2000 [81] and Althorpe and Clary 2004 [82]. In Section 6 we instead describe the time-dependent wavepacket approach, which can also give exact results. In the present section we review (i) a zero point energy model, (ii) the simple reflection principle model and (iii) the reflection principle model [3]. The accuracy of these models, which should only be applied to direct or near-direct reactions, will

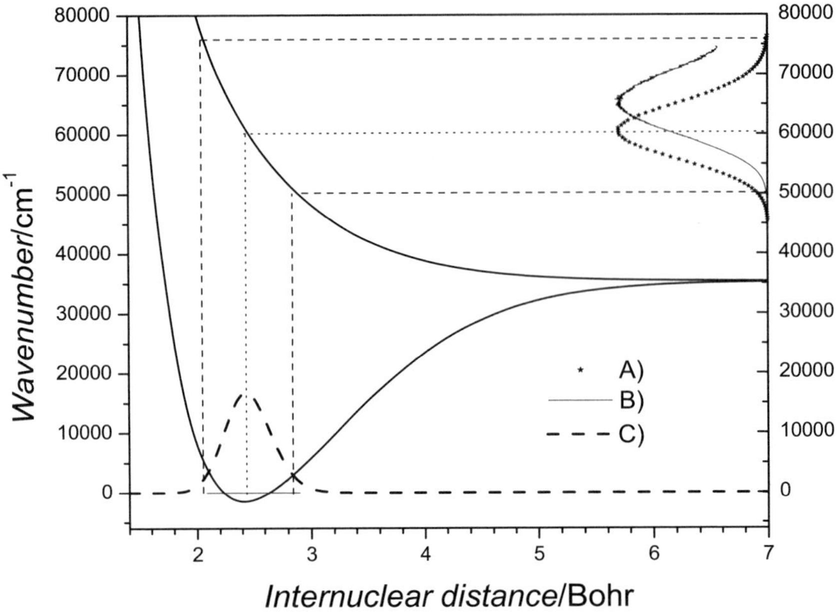

FIGURE 7.2 (A) Prediction of HCl spectrum using the simple reflection principle. (B) Experimental HCl spectrum. (C) Ground state wavepacket.

be illustrated in Section 6.1 by comparison with experimental spectra and exact calculations for the absorption spectra of HCl and DCl.

The zero point energy (ZPE) model assumes that the experimental spectrum of the isotopologue with the largest natural abundance is known. The zero point energies for the isotopologues/isotopomers of interest are then determined. The change in ZPE for a given isotopologue is used to shift the UV absorption spectra by the corresponding amount. Thus the intensities and widths of the spectra of all isotopically substituted species are the same in this simple model, which was introduced by Yung and Miller [83] for nitrous oxide photodissociation in the stratosphere. The model is not reliable in general as exemplified in Section 6.1.

In the simple reflection principle model the vibrational wave function is determined for the lower (ground) electronic state. It is then reflected using the upper electronic potential onto the energy axis as sketched in Figure 7.2. The width of the spectrum is related to the width of the vibrational wave function in the ground state and to the slope of the dissociative potential energy curve. The differences between the model spectrum and the experimental spectrum in Figure 7.2 will be discussed in Section 6.1.

It is usually assumed that the interaction between the electromagnetic field and the absorbing matter only involves two electronic states and that the interaction is weak and of long duration. This simplifies the expressions for the absorption spectrum $\sigma(\omega)$ from an initial state (i, ν) which may be approximated as,

$$\sigma(\omega) = \frac{\pi\omega}{c\hbar\varepsilon_0}\sum_{\nu'}\delta(\omega_{f\nu',i\nu} - \omega)\left|\langle\psi_{\nu'}(\mathbf{R})|\mu_{if}(\mathbf{R})|\psi_v(\mathbf{R})\rangle\right|^2. \tag{13}$$

Here ω is the angular frequency of the incoming light, c is the velocity of light and ε_0 is the vacuum permittivity. f, i, ν' and ν are quantum numbers for the final (f, ν') and initial (i, ν) electronic (f, i) and vibrational (ν', ν) states. The electronic transition dipole moment $\mu_{if}(\mathbf{R})$ is defined in equation (3). $\psi(\mathbf{R})$ is a vibrational wave function. Solving (13) requires knowledge of the vibrational states of the upper electronic state. These can be difficult to obtain, which motivates the reflection principle approximation, which is explained next.

In the reflection principle, the kinetic energy is approximated by $(1/2)E_{\mathrm{ZPE}}$, where E_{ZPE} is the harmonic ZPE of the ground electronic state. The absorption cross section in (13) can then be further simplified,

$$\sigma(\omega) \cong \frac{\pi\omega}{3c\hbar\varepsilon_0}\langle\psi_v(\mathbf{R})|\mu_{if}^{*}\delta\big(E_{iv}/\hbar + \omega - (1/2)E_{\mathrm{ZPE}}/\hbar - V_f(\mathbf{R})/\hbar\big)\mu_{if}|\psi_v(\mathbf{R})\rangle \tag{14a}$$

$$= \frac{\pi\omega}{3c\hbar\varepsilon_0}\int d\mathbf{R}\left|\psi_v(\mathbf{R})\right|^2\left|\mu_{if}(\mathbf{R})\right|^2\delta\big(E_{iv}/\hbar + \omega - \big(V_f(\mathbf{R}) + (1/2)E_{\mathrm{ZPE}}\big)/\hbar\big). \tag{14b}$$

Here E_{iv} is the energy of the initial state and $\mathbf{R}$ is the nuclear geometry. The division by 3 in (14) comes from orientational averaging. In this form, calculation of the absorption cross section requires the initial vibrational wave function, the transition dipole moment surface and the excited state potential. The reflection principle can be employed for direct or near direct photodissociation. It is again an approximation where the ground state wave function is reflected off the upper potential curve or surface. Prakash *et al.* and Blake *et al.* [84–86] have used this theory to calculate isotope effects in N_2O photolysis.

5. TIME-DEPENDENT METHODS

5.1 Methodology

A general description of the time-dependent wave packet approach is presented for N' variables (degrees of freedom) represented by the vector $\mathbf{R} = \{r_1, r_2, \ldots, r_{N'}\}$. The time-dependent wave packet method is based on the solution of the TDSE describing the coupling between M electronic states

$$i\hbar\frac{\partial\boldsymbol{\Psi}(\mathbf{R},t)}{\partial t} = \hat{\mathbf{H}}_{\mathrm{mol}}(\mathbf{R})\boldsymbol{\Psi}(\mathbf{R},t). \tag{15}$$

The TDSE is solved by propagating the wave function forward in time. This can be done by several methods to be discussed later in this section.

The Hamiltonian $\hat{\mathbf{H}}_{\text{mol}}(\mathbf{R})$ is given by

$$\hat{\mathbf{H}}_{\text{mol}}(\mathbf{R}) = \begin{bmatrix} \hat{\mathbf{H}}_X(\mathbf{R}) & \mathbf{V}_{X1}(\mathbf{R}) & \mathbf{V}_{X2}(\mathbf{R}) & \cdots & \mathbf{V}_{XM}(\mathbf{R}) \\ \mathbf{V}^*_{X1}(\mathbf{R}) & \hat{\mathbf{H}}_1(\mathbf{R}) & \mathbf{V}_{12}(\mathbf{R}) & & \\ \mathbf{V}^*_{X2}(\mathbf{R}) & \mathbf{V}^*_{12}(\mathbf{R}) & \hat{\mathbf{H}}_2(\mathbf{R}) & & \\ \vdots & & & \ddots & \\ \mathbf{V}^*_{XM}(\mathbf{R}) & & & & \hat{\mathbf{H}}_M(\mathbf{R}) \end{bmatrix}. \tag{16}$$

The diagonal elements $\hat{\mathbf{H}}_i(\mathbf{R})$ represent the individual Hamiltonian of the ith electronic state. The construction of the Hamiltonian was described in the previous section. The off-diagonal elements $\mathbf{V}_{ij}$ $(\mathbf{R})$ describe the diabatic electronic coupling between the ith and jth electronic states. In cases where the electromagnetic field is explicitly treated couplings due to the field also appear as off-diagonal elements. The subscript X represents the electronic ground state, and subscripts $\{1, 2, \ldots, M\}$ represent the electronic excited states. The number of electronically excited states treated in the description of the photodissociation dynamics depends on the system.

The TDSE requires an initial condition, *i.e.* one must specify the wave packet at $t = 0$. Based on the assumption that the nuclei do not move during an electronic transition but only the electrons, the photo-excitation can be described as a Franck–Condon transition where the wave packet is excited vertically to the excited electronic state. Here $\boldsymbol{\mu}_{Xj}$ is the electronic transition dipole moment of the transition between the ground state X and the jth electronically excited state, whereas $\boldsymbol{\chi}_X$ is a wave function of the electronic ground state, typically the lowest vibrational state. Assuming a vertical electronic transition the initial wave packet on the excited state PES is given by

$$\boldsymbol{\Psi}(\mathbf{R}, t = 0) = \begin{pmatrix} \boldsymbol{\Phi}_X(\mathbf{R}, t = 0) \\ \boldsymbol{\Phi}_1(\mathbf{R}, t = 0) \\ \vdots \\ \boldsymbol{\Phi}_M(\mathbf{R}, t = 0) \end{pmatrix} = \begin{pmatrix} \mathbf{0} \\ \boldsymbol{\mu}_{X1}(\mathbf{R})\boldsymbol{\chi}_X(\mathbf{R}) \\ \vdots \\ \boldsymbol{\mu}_{XM}(\mathbf{R})\boldsymbol{\chi}_X(\mathbf{R}) \end{pmatrix}. \tag{17}$$

The evolution of the wave packet is governed by the TDSE. The cross section for absorption onto the jth electronic surface is given by the Fourier transform of the autocorrelation function:

$$\sigma_j(\nu) = \frac{\pi \nu}{3c\varepsilon_0 \hbar} \int_0^\infty \exp\bigl(i(E_X + h\nu)t/\hbar\bigr)\bigl\langle \boldsymbol{\Phi}_j(\mathbf{R}, t = 0)\big|\boldsymbol{\Phi}_j(\mathbf{R}, t)\bigr\rangle|\, dt. \tag{18}$$

The total cross section becomes

$$\sigma^{\text{tot}}(\nu) = \frac{\pi \nu}{3c\varepsilon_0 \hbar} \int_0^\infty \exp\bigl(i(E_X + h\nu)t/\hbar\bigr)\bigl\langle \boldsymbol{\Psi}(\mathbf{R}, t = 0)\big|\boldsymbol{\Psi}(\mathbf{R}, t)\bigr\rangle|\, dt. \tag{19}$$

It can be decomposed into contributions from the different electronic states. The factor $1/3$ enters in the expression for the cross section due to orientational averaging. The energy of the incident photon is $h\nu$.

In order to evaluate partial photodissociation cross sections (vibrational and rotational product distributions) *i.e.* $ABC + h\nu \rightarrow A + BC(n, K)$ the wave function can be projected onto the different rovibrational eigenstates of the molecular fragment BC at fixed distance R' between the two fragments. The chosen R' should be on the asymptote of the potential energy surface where the two fragments do not interact. Balint-Kurti *et al.* [87] have shown that the partial cross section is given by

$$\sigma_{nK,j}(R_\infty, E) \approx \frac{\nu k_{nj}}{4\pi^2} \left| \int_0^\infty \exp(iEt/\hbar)\langle\psi_{nK}(r)|\mathbf{\Phi}_j(R' = R_\infty, t)\rangle_r dt \right|^2, \tag{20}$$

where

$$k_{nj} = \sqrt{2\mu_{BC}(E - \varepsilon_{nj})/\hbar^2} \tag{21}$$

is the wave number corresponding to the relative kinetic energy E between the atom (A) and the diatom (BC) for the given internal state of the diatomic. The rovibrational eigenfunction of the diatomic BC is denoted $\psi_{nj}(r)$ with the corresponding energy ε_{nj}.

5.2 Propagation schemes

Normally the TDSE cannot be solved analytically and must be obtained numerically. In the numerical approach we need a method to render the wave function. In time-dependent quantum molecular reaction dynamics, the wave function is often represented using a discrete variable representation (DVR) [88–91] or Fourier Grid Hamiltonian (FGH) [92,93] method. A Fast Fourier Transform (FFT) can be used to evaluate the action of the kinetic energy operator on the wave function. Assuming the Hamiltonian is time independent, the solution of the TDSE may be written

$$\mathbf{\Psi}(\mathbf{R}, t + \Delta t) = \exp\left(-i\hat{\mathbf{H}}\Delta t/\hbar\right)\mathbf{\Psi}(\mathbf{R}, t). \tag{22}$$

The time evolution operator $\exp(-i\hat{\mathbf{H}}\Delta t/\hbar)$ acting on $\mathbf{\Psi}(t)$ propagates the wave function forward in time. A number of propagation methods have been developed and we will briefly describe the following: the split operator method [91,94,95], the Lanzcos method [96] and the polynomial methods such as Chebychev [93,97], Newtonian [98], Faber [99] and Hermite [100,101]. A classical comparison between the three first mentioned methods was done by Leforestier *et al.* [102].

The idea behind split operator propagation [91,94,95] is to split the action of the time evolution operator such that the kinetic energy operator $\hat{\mathbf{T}}$ and the potential energy operator $\hat{\mathbf{V}}$ are separated into different exponentials, causing a small error since $\hat{\mathbf{T}}$ and $\hat{\mathbf{V}}$ do not commute. In second order potential referenced split operator propagation the evolution operator can be approximated as

$$\exp\left(-i\hat{\mathbf{H}}\Delta t/\hbar\right) \approx \exp\left(-i\hat{\mathbf{T}}\Delta t/2\hbar\right)\exp\left(-i\hat{\mathbf{V}}\Delta t/\hbar\right)\exp\left(-i\hat{\mathbf{T}}\Delta t/2\hbar\right) + O\left((\Delta t)^3\right). \tag{23}$$

This approximation is numerically stable since it is unitary. The method is suitable for short time dynamics and time-dependent Hamiltonian operators. To reduce the error, higher orders of the split operator propagator have been developed. For very small time steps the split propagator methods become numerically unstable.

For multidimensional systems it is necessary to split the kinetic energy operator into several parts, in order to decouple the degrees of freedom in the kinetic energy operator. The split operator propagation has been implemented in *e.g.* spherical [91,94] and hyper spherical coordinates [95].

The time evolution operator can be expanded in any classical orthogonal polynomial such as Chebychev [93,97], Newtonian [98], Faber [99], Hermite [100,101] and others. In general, the evolution operator can be written in polynomial series

$$\exp(-i\hat{\mathbf{H}}\Delta t/\hbar) \approx \sum_{j=0}^{M} a_j P_j\big(-i\hat{\mathbf{H}}\Delta t/\hbar\big), \tag{24}$$

where a_j are the expansion coefficients of the terms P_j in the polynomial. The polynomial is determined using recursive algorithms in which the elementary mapping of the Hamiltonian on the wave function *e.g.* $\hat{\mathbf{H}}\boldsymbol{\Psi}$ is used.

Each of the polynomial methods has advantages and disadvantages. For Hermitian time-independent Hamiltonians the Chebychev algorithm is stable and reliable. Because the Chebychev polynomials are defined using the unit circle, the convergence error is uniform and vanishes exponentially. The introduction of a boundary condition that absorbs the dissociating wave function makes the Hamiltonian complex and non-Hermitian and can cause trouble for the Chebychev propagation, but usually it goes well if the complex eigenvalues are small. Large complex eigenvalues can cause severe instability. The Newtonian and Faber propagation methods were developed to handle complex eigenvalues, which are especially useful for solving the Liovuille von Neumann equation for dissipative systems. The Chebychev, Newtonian and Faber methods were developed for the time-independent Hamiltonian but they can be used for time-dependent Hamiltonians if the time step is sufficiently small that the time variation of the Hamiltonian between steps is negligible.

The Lanczos propagator is, like the polynomial methods, based on an expansion of the time evolution operator. In this case the expansion makes use of the Lanczos vectors [103], which are generated iteratively from a recursion relation. The vectors are orthogonal in theory, but in numerical implementations orthogonality is lost after a few tens of iterations. The Lanczos vectors can be reorthogonalized, but this is numerically costly. The Lanczos propagator is therefore used as a short time propagator, employing rather few iterations whereby the need to reorthogonalize is avoided [102]. The Chebychev method, on the other hand, is known to be stable for many iterations and is therefore useful for long-time propagations.

The Hermite propagator can be used for any length of time step. As the Hermite propagator separates into a time part and a recursive Hamiltonian the integrals involving the time parameter can be carried out analytically, implying essentially infinite time propagation. This feature makes the method suitable for spectral filter problems [101]. A complex absorbing boundary can be used to prevent transmission and reflection of the wave function at large distances for dissociative degrees of freedom [104,105].

6. EXAMPLES OF PHOTODISSOCIATION

6.1 Photolysis of HCl

Isotope effects provide a very sensitive test of theoretical methods. Here we illustrate the three models described in Section 4 with results for the UV absorption spectra of HCl and DCl [106]. The transition considered is from the electronic ground state X $^1\Sigma^+$ to an electronic excited state A $^1\Pi$. The MRCI calculated potential energy curves are taken from Alexander *et al.* [107].

We first note the poor agreement between the ZPE model and the experimental results shown in Figure 7.3. The ZPE model predicts that DCl should absorb at slightly higher energies (419 cm^{-1}) than HCl because of the lower zero point energy of DCl. In contrast, as can be seen in Figure 7.3, the experimental absorption peak of HCl is about 1500 cm^{-1} higher than that of DCl.

In the simple reflection principle, illustrated for HCl in Figure 7.2, the shape of the predicted spectrum is in reasonable agreement with the experimental spectrum. However the energy of the experimental spectrum is approximately 5000 cm^{-1} higher than predicted by the model. The reasons for this inaccuracy

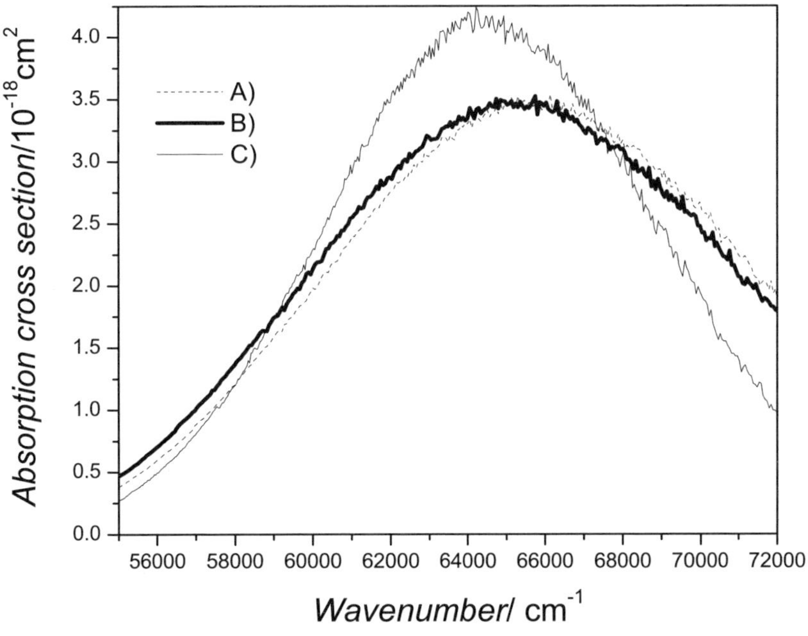

FIGURE 7.3 (A) ZPE model prediction of the DCl spectrum (*i.e.* the experimental HCl spectrum shifted 419 cm^{-1}). (B) Experimental HCl spectrum. (C) Experimental DCl spectrum.

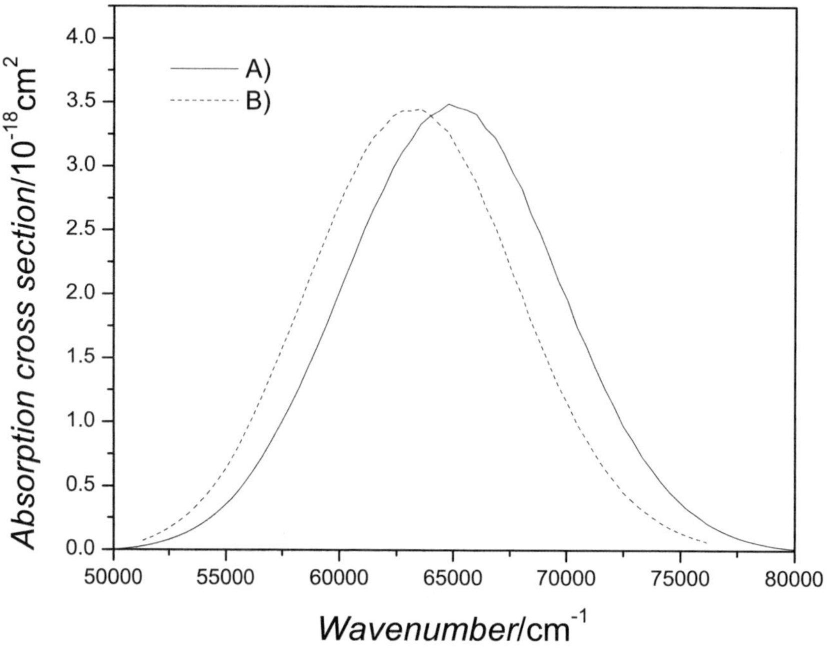

FIGURE 7.4 Wavepacket absorption spectra of HCl, using (A) a transition dipole moment curve (max absorption at approximately 65300 cm^{-1}) and (B) a constant transition dipole moment (max absorption at approximately 63500 cm^{-1}).

are the variation of the transition dipole moment with bond length, and the dynamics on the upper potential, which have been neglected. Setting the transition dipole moment to be independent of the internuclear distance accounts for approximately 1800 cm^{-1} of the error, as illustrated in Figure 7.4.

Figures 7.5 and 7.6 show the experimental absorption spectra of HCl and DCl, respectively, together with the spectra calculated using wavepacket propagation and the reflection principle. It can be seen that the wavepacket method is in good agreement with the experimental results. For both HCl and DCl the wavepacket propagation method yields the correct frequency for the absorption peak. The wavepacket propagation method is exact and the deviation from the experimental spectrum must be attributed to the use of only two electronic states and/or inaccurate transition dipole moments and/or potential energy curves and/or not treating the rotations of the molecule. The experimental results are quite accurate.

The HCl and DCl spectra calculated with the reflection principle model have too low intensities in comparison to the wavepacket results, which are for the same conditions. The differences in the results demonstrate that the dynamics of the system play an important role even for direct photodissociation reactions like HCl and DCl.

The autocorrelation functions of both HCl and DCl are given in Figure 7.7, and it can be seen that the greater reduced mass of DCl results in the disso-

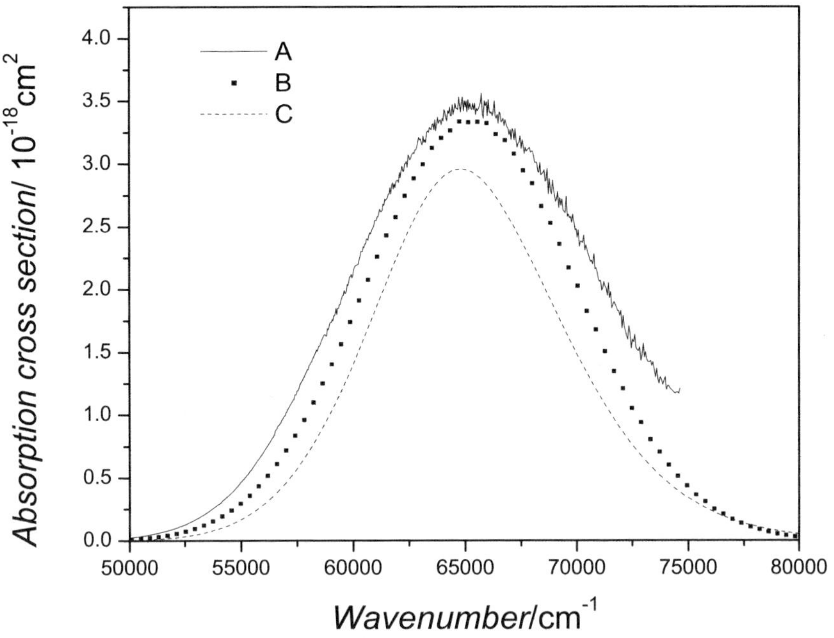

FIGURE 7.5 (A) Experimental, (B) wavepacket and (C) reflection principle absorption spectra of HCl.

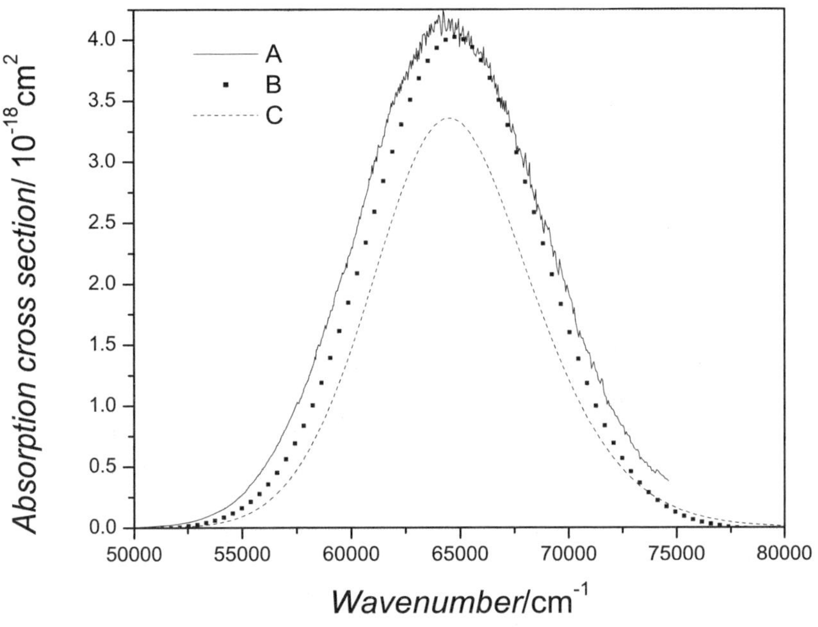

FIGURE 7.6 Experimental, (B) wavepacket and (C) reflection principle absorption spectra of DCl.

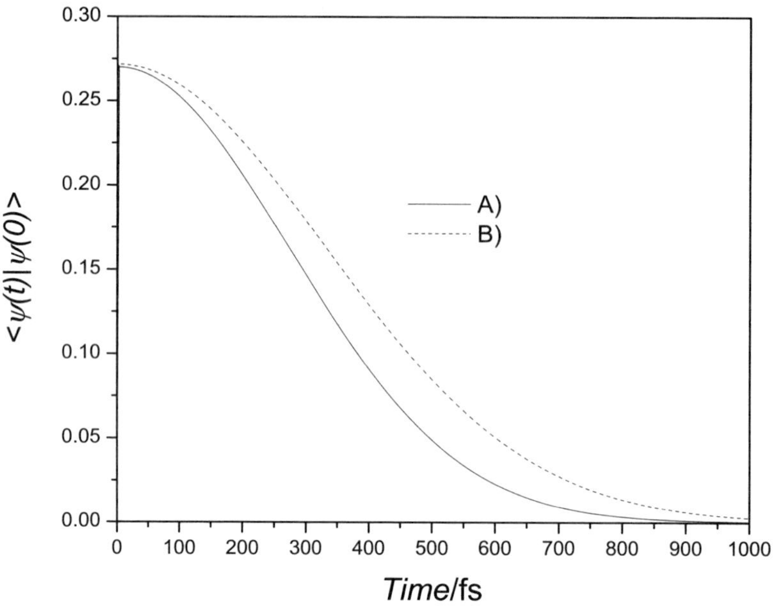

FIGURE 7.7 Autocorrelation functions for (A) HCl and (B) DCl.

ciation process being slower than for HCl. This is reflected in the higher peak intensity and narrower width of the DCl spectrum compared to the HCl spectrum.

For most molecules, electronic absorptions shift to higher energy upon deuteration. This is not the case however for hydrochloric acid. Some authors have explained this in terms of mixing in of a triplet electronic state into the ground state wave function [108]. However, the triplet state is only mixed in to 0.5% and it is therefore unlikely that this transition would have the intensity required to dramatically shift the position of the maximum. An alternative explanation is the change in the transition dipole as a function of bond length, combined with dynamic effects.

We have seen the limitations of the ZPE model and the simple reflection principle. The reflection principle is a reasonable approximation for predicting the absorption spectra of molecules photolysed by direct dissociation. However, because dynamics are excluded it results in significant errors even for one of the simplest direct photodissociation processes, HCl + $h\nu$. For systems that undergo indirect as opposed to direct photodissociation the reflection principle should not be applied, since phenomena such as curve crossing, multiple crossings and semi-bound potentials may give rise to interferences. These cases may be described in the framework of propagating wavepackets and can result in recurring oscillations on the upper potential surface(s) which transform into spikes in the absorption spectrum, as is seen for instance for ClO_2, NO_2 and N_2O [95,109,110].

6.2 Photolysis of N_2O

Apart from molecular nitrogen, nitrous oxide (N_2O) is the most abundant nitrogen-containing species in the atmosphere. It is a greenhouse gas [111] with a warming potential per molecule 200–300 times that of CO_2. It is therefore of great interest to monitor and regulate emissions, which turns out to be difficult. There are uncertainties of about 50% in the source and sink fluxes, the primary reason being very diffuse sources. In cases like this, isotopic analysis makes a valuable contribution by constraining the atmospheric budget. Isotopic distributions [112–115] have been characterized for the main sources, of which about a third stem from human activity, mainly agricultural fertilizer and biomass burning. The most important isotopologues are $^{14}N^{14}N^{16}O$ (abbreviated 446), 456, 546, 556, 447 and 448. In contrast the sink mechanisms are simple, mainly photolysis in the stratospheric UV window at around 205 nm (90%):

$$N_2O(1^1\Sigma) + h\nu \rightarrow N_2(1^1\Sigma_g^+) + O(^1D) \tag{25}$$

and oxidation by $O(^1D)$ (10%) [116,117].

Several approaches have been used to calculate the effect of isotopic substitution on the absorption cross section of N_2O. The zero point energy (ZPE) model [83] as described in Section 4, 2D and 3D wavepacket dynamics [110,118], a semi-empirical model [85,86] and an extended reflection principle model [84] have been used to explain the differences in the absorption that stem from isotopic substitution. In the following the emphasis lies on the enhanced understanding of the isotopic differences in the photodissociation reactions, which arises from employing wavepacket propagation calculations for the various isotopologues.

In the 2D case the relative photoabsorption cross sections of the isotopologues 446, 447, 448, 456, 546 and 556 were calculated using a Hermite time propagator [101] and potential energy surfaces (in Jacobi coordinates) from the literature [119]. The initial ground state wave function was found in a "local mode" as a Morse wave function in the N–O bond multiplied by a harmonic oscillator function in the bending angle, transformed into Jacobi coordinates.

The main conclusions from this first study in which dynamical effects and the transition dipole moment surface were taken into account were: (1) a prediction that the isotopic enrichment would increase with decreasing temperature, (2) the enrichment would be "mass-independent" as defined by the mass spectroscopy community and (3) that the heavier isotopologues would exhibit an overall smaller absorption cross section because of a smaller bending angle. This last point was counteracted by the observation that one would observe larger populations in the higher excited bending states of the heavier isotopologues. Here we attend to the third point first.

We note that N_2O is a linear molecule, where an orbital transition to the upper state of the UV transition is forbidden, whereas it is vibronically allowed for a bent system. Furthermore a bending in the ground state will lead to a more favorable Franck–Condon factor with the upper state. In Figure 7.8 are shown the cross sections $\sigma_{ijk,n}$ of the ground, first and second levels of the ν_2 bending excitation. There is a progression towards a relatively higher probability of excitation to the

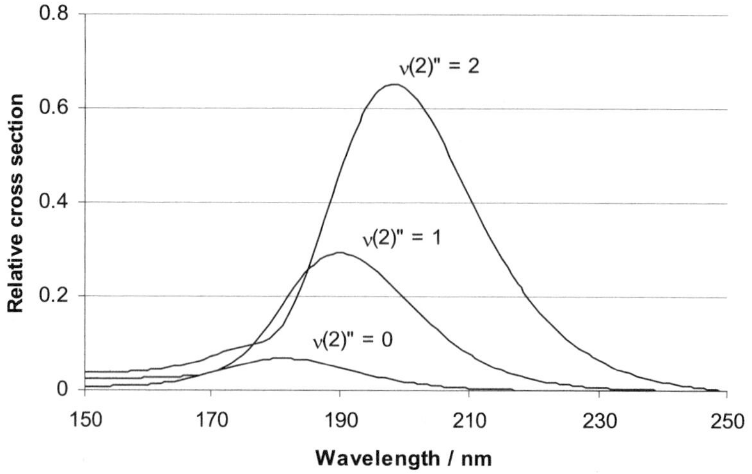

FIGURE 7.8 The relative cross section as a function of the bending vibrational quantum number in the ground state.

TABLE 7.1 Band centers of the fundamental vibrational frequencies of N_2O, together with the zero point energy (ZPE), and the difference of the ZPE from the ZPE of the 446 isotopologue (ΔZPE). The uncertainties are indicated in parentheses. The unit is cm^{-1}

ijk	ν_1	ν_2	ν_3	ZPE	ΔZPE
446	1284.903369(21) [155]	588.76798(1) [156]	2223.756764(16) [157]	2343.10	0.00
447	1264.704263(65) [155]	586.362(1) [158]	2220.073460(38) [157]	2328.75	−14.35
448	1246.884557(35) [155]	584.225(1) [158]	2216.711180(35) [157]	2316.02	−27.08
456	1280.354121(42) [155]	575.4340(3) [156]	2177.656805(25) [157]	2304.44	−38.66
546	1269.891964(70) [155]	585.3121(4) [156]	2201.605289(30) [157]	2321.06	−22.04
556	1265.33383(42) [155]	571.894(1) [158]	2154.72590(20) [157]	2281.92	−61.17

$\nu_2 = 1$ and $\nu_2 = 2$ vibrational states as we move to greater mass j the vibrations are shown in Table 7.1. At the same time the intensity factor $f_{ijk,n}$, which is the square of the product of the vibrational frequency and the transition dipole surface, is decreasing, as shown in Table 7.2. This results in a decrease in the total cross section for heavier isotopologues. The total cross section is governed by

$$\sigma_{ijk}(T) \propto \frac{\sum_n \sigma_{ijk,n} g_n f_n \exp((-n\nu_{ijk}^{\text{bend}})/(kT))}{Q}, \tag{26}$$

where g_n, is the (pseudo-)degeneracy of the angular momentum of the degenerate bending mode. Each contribution is scaled with the Boltzman population factor for the actual bending frequency (listed in Table 7.1), and the sum is divided by the partition function, Q. Figure 7.9 shows the weighted contribution from each vibrational state to the total cross section together with experiment.

TABLE 7.2 The intensity factors $f_{ijk,m}$ of the isotopomers in the ground, first and second bending vibrational states are shown. The factor accounts for the product of the isotopomer dependent vibrational wave function with the transition dipole surface

Isotopomer ijk	$f_{ijk,0}$	$f_{ijk,1}$	$f_{ijk,2}$
446	0.01357392	0.03464763	0.06346950
447	0.01351649	0.03446404	0.06315070
448	0.01346522	0.03430018	0.06286494
456	0.01330919	0.03372219	0.06178399
546	0.01351304	0.03439177	0.06301048
556	0.01325017	0.03347542	0.06133432

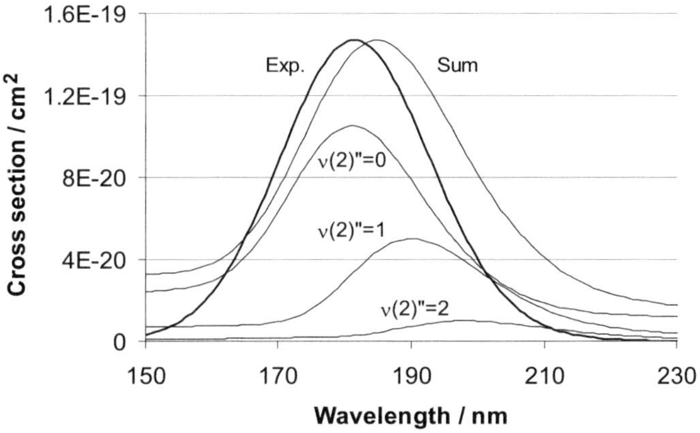

FIGURE 7.9 Contribution from each vibrational state to the total cross section calculated using the degeneracy factor and Boltzmann weight at 298 K and comparison to experiment. The cross sections were scaled by a factor of 1.7×10^{-18} cm^{-2}.

The enrichment factor is defined as $\varepsilon = \alpha - 1$, where α is the ratio of the reaction rate, or, as in this case, the photolysis rate of the isotopologue to the reference, *e.g.* $\varepsilon_{456} = (\sigma_{456}/\sigma_{446}) - 1$. Bearing in mind the analysis of the absorption cross sections it is not surprising that the total enrichment increases as the temperature drops, as can be viewed when comparing Figure 7.10 ($T = 298$ K) with Figure 7.11 ($T = 220$ K). At lower temperatures the population of the higher vibrational states will diminish, and this will influence the absolute cross sections of the heavier isotopologues more. Enrichment factors observed in the earth's stratosphere at 205 nm [120] fit well with the model results at the temperature of 233 K, which is a representative temperature for the stratosphere.

In 2004 Nanbu and Johnson [110] re-examined the isotope effects in NNO photolysis using a three dimensional potential energy surface, the added degree of freedom being NN vibration. This allowed an investigation of the effect of this degree of freedom on the dynamics, and of the $\nu_1 = 1$ vibrational state, both of which impact the absorption cross section. The NN bond can absorb some of the impulse

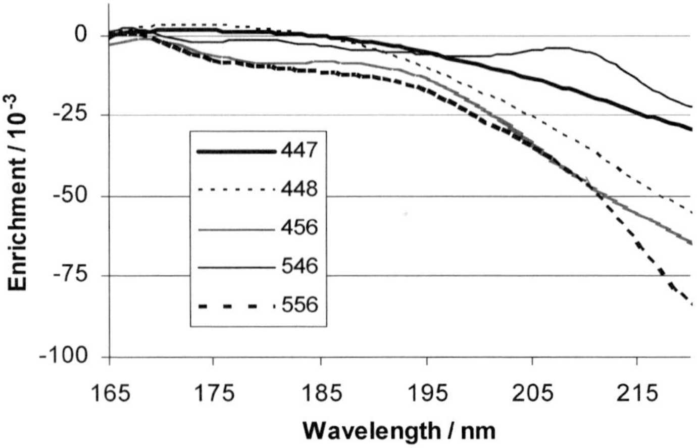

FIGURE 7.10 Wavelength dependent enrichment coefficients for the nitrous oxide isotopologues in the region of stratospheric photolysis. Temperature is set to 298 K.

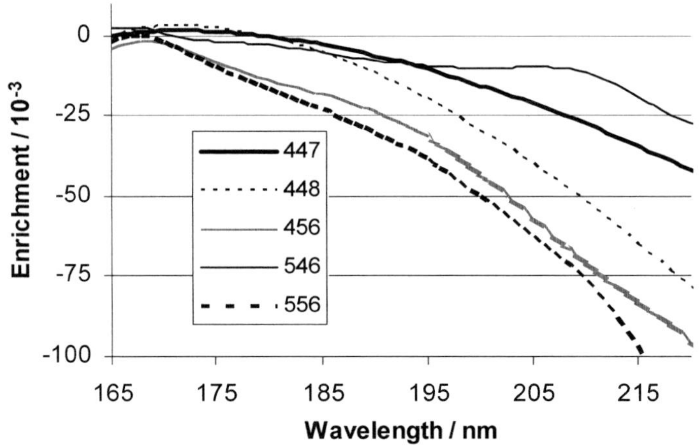

FIGURE 7.11 Wavelength dependent enrichment coefficients for the nitrous oxide isotopologues in the region of stratospheric photolysis. Temperature is set to 220 K.

released by photoexcitation, changing the width of the absorption, as noted by Marcus [84]. *Ab initio* molecular orbital CI calculations were performed using aug-cc-pVTZ (diffusion-function-augmented, correlation consistent, polarized valence, triple-ζ) basis set. The molecular orbitals used in the CI calculations were determined using CASSCF calculations; the natural orbitals found for the excited states were used in MRCI calculations. The transition dipole moments were ascertained from the CI calculations.

The wavepacket dynamics were carried out using the Lanczos method with real wavepackets [121] employing Jacobi coordinates to describe the relative positions of the three nuclei in the body fixed plane. As mentioned the novel results of the 3D calculations included a test of the effect of the NN vibrations. The vibrational quanta of the symmetric vibration (resembling NO stretching), bending and antisymmetric stretching (resembling the NN stretching) are denoted ν_1, ν_2 and ν_3, respectively. The rovibrational quantum number J is denoted $|l|$, as the rotational quantum number can take the values $|l|$, $|l| + 1$ and so forth, where l is the almost good quantum number for the bending and takes on the values $-\nu_2, -\nu_2 + 2, \ldots, \nu_2 - 2, \nu_2$. As noted earlier an orbital transition to the upper state is forbidden, so the initial wave packet was obtained for the $(0, 1^1, 0)$ level where $J = 1$. For the ν_1 and ν_2 vibrations the theoretical and experimental results were in good accordance, whereas the results for ν_3 had a larger quantitative deviance. This was quite to be expected, as the correlation energy in the MRCI was not sufficient to describe the PES well enough in the case of the NN bond. As a matter of fact anti bonding MOs were not included for NN in the CAS space, something that could lead to an underestimation of the ν_3 frequencies. One point that was noted was that the contribution of the $(1, 0^0, 0)$ and the $(0, 2^2, 0)$ levels are predicted to be larger than the contribution of the $(0, 1^1, 0)$ level on the low energy side. Of particular interest is the wavelength region around 205 nm, where field data have been recorded [120].

Enrichment is said to be mass dependent when it can be directly related to the difference in mass, as in the cases of evaporation and diffusion. The isotopic composition of a sample is characterised by the value $\delta = R_i/R_{std} - 1$, where R_i is the isotope ratio e.g. $[^{13}C]/[^{12}C]$ in the sample and R_{std}—a reference standard respectively. In a sample of N_2O the enrichment will be mass dependent if the three-isotope factor $\rho = \delta^{17}/\delta^{18} = 0.51$ (*cf.* Ref. [1]). In atmospheric nitrous oxide there is an excess of ^{17}O relative to the isotope's typical environmental concentration, and the enrichment here is mass independent [122,123]. Whether the photolytic fractionation is mass independent or mass dependent is a matter of debate. Kaiser *et al.* [124,125], Prakash and Marcus [126] and Nanbu and Johnson [110] assert that it is mass dependent whereas Johnson *et al.* [118] and McLinden *et al.* [116] find that the photolytic enrichment probably may be non-mass dependent.

6.3 Photolysis of OCS

It is important to understand the sources and loss mechanisms of stratospheric sulfate aerosols. These aerosols are linked to the decrease in ozone at mid-latitudes because they hydrolyse N_2O_5, reducing the amount of NO_x that would otherwise limit the efficiency of chlorine-catalysed ozone depletion. In addition these aerosols scatter light, cooling the planet [127]. Their concentration increases dramatically following major volcanic eruptions however they are always present at background levels. The source of these background aerosols is a matter of debate. In 1976 Paul Crutzen presented the idea that sulfate aerosols result from the photolysis of carbonyl sulphide [128]:

$$OCS + h\nu\ (\lambda \sim 205\ \text{nm}) \rightarrow CO + S. \tag{27}$$

The remaining source is believed to result from the transport of SO_2 from the troposphere.

Leung, Colussi and Hoffmann have used isotopic analysis in an attempt to constrain the amount of sulfate that could be produced by Crutzen's mechanism [129,130]. The first study retrieved the concentration profiles of $OC^{32}S$ and $OC^{34}S$ from infrared transmission spectra of the atmosphere recorded by the NASA MkIV balloon-borne interferometer. They derived an enrichment factor of 73.8 ± 8.6‰ defined such that photolytically generated sulfur would be enriched in ^{34}S. An isotopic budget based on this result shows that OCS photolysis cannot be a significant source of sulfate aerosol, since the ^{34}S enrichments of OCS, sulfuric acid aerosol and SO_2 are known to be small [131]. A later laboratory study by the same group came to the conclusion that stratospheric photolysis results in an enrichment of 67 ± 7‰.

The OCS photodissociation has a number of interesting features. The stratospheric photolysis window lies to the blue of the absorption maximum, not to the red as in nitrous oxide. The two systems are similar in that the transition is forbidden for the linear molecule and therefore the transition intensity is expected to grow significantly with bending excitation. The absorption cross section of the OCS isotopologues was calculated using a method analogous to that used for N_2O [118]. The 2D potential energy surface of Suzuki, Nanbu and co-workers was used [132,133]. The Jacobi coordinates involved the carbon monoxide fragment and the sulfur atom. A bending force constant was taken from the literature [134]. The Hermite propagator technique was used to calculate the absorption cross section from a series of isotopologues. Representing $^{16}O^{12}C^{32}S$ by 622, UV cross sections of the first three bending states of 622, 623, 624, 632, 722 and 822 were calculated. As an example the relative cross sections of the first three bending states are shown in Figure 7.12. The fractionation factor $\varepsilon = \sigma_{624}/\sigma_{622} - 1$ is shown in Figure 7.13.

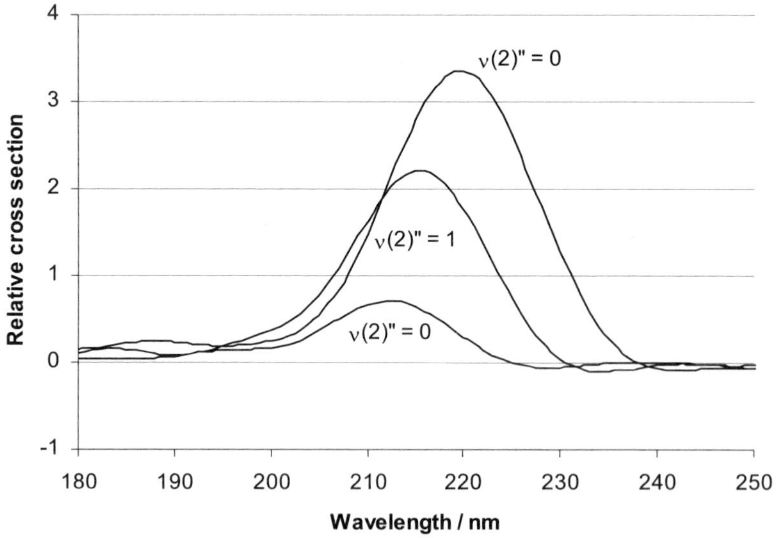

FIGURE 7.12 Relative cross section of the three lowest bending states of the parent isotopologue of carbonyl sulfide.

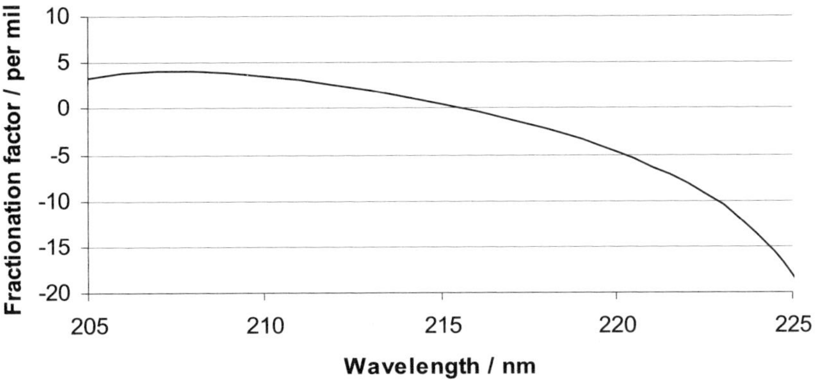

FIGURE 7.13 Fractionation factor for $OC^{34}S$ vs. $OC^{32}S$ photolysis, calculated at 298 K.

The simulation predicted modest fractionations for all of the isotopologues, in contrast to the large values obtained by Leung and co-workers. We recommend that the experimental results of Colussi *et al.* be reinvestigated [129] in another study [129].

The carbonyl sulfide calculations were not as straightforward as the 2D nitrous oxide work. One issue is that the transition involves three transition dipoles to various states, whereas in nitrous oxide the absorption is dominated by a single transition. The calculation for OCS only considered a single transition. In addition the quality of the potential energy surfaces was not as high, and the 2D approximation not as good, for OCS relative to NNO. An illustration of this is that the predicted OCS spectrum has a maximum at 214 nm while the experimental spectrum has a maximum at 223 nm. In comparison the difference in peak location for NNO was only 3 nm.

6.4 Photolysis of HCHO

Formaldehyde is formed in the atmospheric degradation of virtually all hydrocarbons. Its photolysis has an important effect on the atmosphere's oxidation capacity since it is a significant source of HO_X radicals in the middle and upper troposphere, and in polluted regions [135]:

$$\mathrm{HCHO} + h\nu\ (\lambda < 340\ \mathrm{nm}) \rightarrow \mathrm{H} + \mathrm{HCO}. \tag{28}$$

In addition the photolysis of formaldehyde produces half of the molecular hydrogen found in the atmosphere:

$$\mathrm{HCHO} + h\nu\ (\lambda < 360\ \mathrm{nm}) \rightarrow \mathrm{H_2} + \mathrm{CO}. \tag{29}$$

About half of formaldehyde reacts through each of these processes under atmospheric conditions. The absorption spectrum of formaldehyde shows extensive vibrational structure as shown in Figure 7.14, which also shows the photolysis quantum yields into the radical and molecular channels [136].

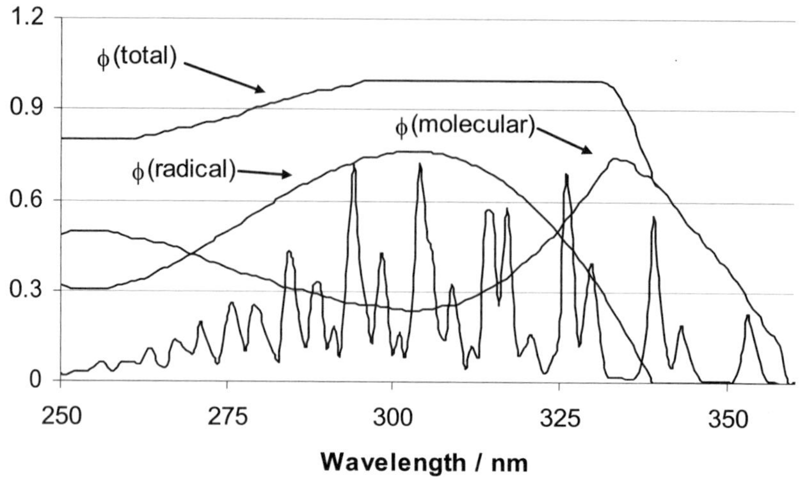

FIGURE 7.14 The cross section and photolysis quantum yields for formaldehyde [136]. The cross sections where divided by 10^{-19} cm^2.

Formaldehyde photolysis thus provides an important link between the atmosphere's carbon and hydrogen cycles. Isotopic analysis has been used to explore this link in a series of articles from several laboratories [137–140]. The main precursors of formaldehyde are methane and isoprene and the amount of deuterium in atmospheric hydrogen will depend on its concentration in the starting material and on the fractionation through the intermediate reaction steps, *e.g.* methane CH_4, methyl CH_3, methyl peroxyl CH_3O_2, methoxyl CH_3O, formaldehyde HCHO and finally hydrogen [138,140–142]. Most of atmospheric formaldehyde is formed from the following reaction

$$CH_3O + O_2 \rightarrow HCHO + HO_2. \tag{30}$$

A recent study has shown that this reaction enriches deuterium in formaldehyde relative to methoxy [142]. Moore and co-workers have reviewed the considerable literature concerning the photodissociation dynamics of formaldehyde [143–145]. In addition the so-called roaming atom pathway has recently been discovered by Bowman and co-workers. This exciting theoretical and experimental effort has demonstrated a new photodissociation path in which one hydrogen atom takes a long trajectory around the HCO radical before abstracting the other H atom [146–151]. Recently Troe has analysed the photodissociation and thermal dissociation yields of formaldehyde [152–154].

The photodissociation of formaldehyde is associated with significant isotope effects [137,138]. These are summarized in Table 7.3.

It is seen that DCDO is photolysed the most slowly in natural sunlight, followed by HCDO, and the ^{13}C and ^{18}O isotopologues. Symmetry and the roaming atom pathway are likely to play a role in these isotope effects. Further evidence for this appears in the channel-specific photolysis rates, which show that HCDO

TABLE 7.3 Isotope effects in the photodissociation of formaldehyde [137,138,159,160][a]

Isotopologue	Photolysis rate relative to HCHO
DCDO	0.34 ± 0.03
HCDO	0.633 ± 0.012
$H^{13}CHO$	0.894 ± 0.006
$HCH^{18}O$	0.911 ± 0.011

a In addition the relative photolysis rates in the molecular and radical channels have been measured for the pair HCHO and HCDO: $j_{HCHO \to H_2+CO}/j_{HCDO \to HD+CO} = 1.82 \pm 0.07$, $j_{HCHO \to H+HCO}/(j_{HCDO \to H+DCO} + j_{HCDO \to D+HCO}) = 1.10 \pm 0.06$ [138].

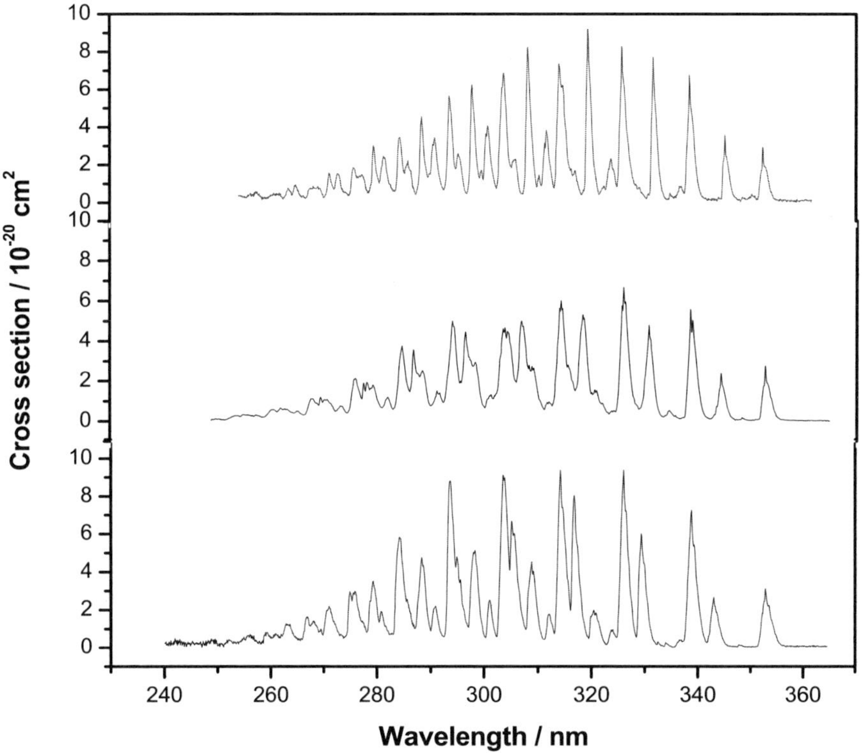

FIGURE 7.15 From top to bottom, the UV absorption cross sections of DCDO, HCDO and HCHO.

is quite bad at producing HD as compared to the production of H_2 from HCHO. This difference is not paralleled in the radical channel [138].

The isotope effects do not arise from differences in the absorption cross sections themselves. In fact the integrated UV absorption cross sections of HCHO, HCDO and DCDO are indistinguishable (Figure 7.15). The key then to their different photolysis behaviour must lie in the photodissociation dynamics. Further study is clearly needed.

7. PERSPECTIVE

In this presentation we have only considered single photon transitions since multi-photon transitions are not relevant to the atmosphere, however the time-dependent framework can easily be extended for this purpose. This requires computing the diabatic transition dipole moment for the multi-photon transitions. Furthermore, the couplings between the time-dependent external electric field and the molecular system would change.

The photodissociation dynamics of di- and triatomic molecules are well established. The great difficulty lies in obtaining highly accurate potential energy surfaces and their diabatic couplings. This can be an especially challenging task if many electronic excited states are available in the Franck–Condon region.

For 4-atom systems such as formaldehyde the photodissociation dynamics have yet to be established. A fully quantal description of the photodissociation is still very expensive computationally. So far 6-dimensional wave packet studies have been applied only to one electronic surface, for example for scattering of a molecule or atoms on surfaces or collisions of molecules. Currently, for molecular systems with more than four atoms, a reduced dimensionality model must be used. This means that certain degrees of freedom are fixed throughout the dynamical simulation.

Light interaction with molecules drives much of the chemistry of our atmosphere. Photodissociation thereby plays an important role. Theoretically this is a most challenging task as by necessity one must include excited potential energy surfaces, the calculation of which is demanding. In addition, the dynamics are prone to nonadiabatic effects and therefore more than one potential energy surface must be included. This means that a complete study of the photodissociation dynamics, even of a diatomic, can be nontrivial. Despite this, substantial progress has been made and we have given some examples here. There are however a vast number of reactions that have received little attention. We conclude this paper by mentioning some that we consider to be of high priority to investigate.

The reflection principle approach (see article by Jost in this issue) produces essentially perfect agreement with experiment but it is not able to provide vibrational structure. Therefore the time-dependent techniques are important for indirect photodissociation, of which there are many examples in atmospheric chemistry, including HCHO, SO_2, NO_2, O_2, CO, HCl, H_2O and O_3. Many studies have shown how isotopic analysis is able to provide valuable insight concerning atmospheric photochemical reactions, and emissions sources and loss mechanisms. One of the largest uncertainties in projections of future climate is the ability to predict greenhouse gas concentrations, which depend on accurate knowledge of their sources, sinks and atmospheric photochemistry.

ACKNOWLEDGEMENTS

Financial support from the Swedish Research Council and funding of CCAR by The Danish Natural Science Research Council, and the Villum Kann Rasmussen fund are gratefully acknowledged. S.J. is supported by the Danish Natural Science Research Council (Project no. 272-05-0230). We gratefully thank Stefan Andersson for helpful discussions.

REFERENCES

[1] M.S. Johnson, *et al.*, Isotopic processes in atmospheric chemistry, *Chem. Soc. Rev.* **31** (6) (2002) 313–323.

[2] C.A.M. Brenninkmeijer, *et al.*, Isotope effects in the chemistry of atmospheric trace compounds, *Chem. Rev.* **103** (12) (2003) 5125–5161.

[3] R. Schinke, *Photodissociation Dynamics: Spectroscopy and Fragmentation of Small Polyatomic Molecules, Cambridge Monographs on Atomic, Molecular and Chemical Physics*, Cambridge University Press, Cambridge, 1993.

[4] T. Helgaker, P. Jørgensen, J. Olsen, *Molecular Electronic-Structure Theory*, Wiley, Chichester, 2000.

[5] H.J. Werner, P.J. Knowles, A 2nd order multiconfiguration SCF procedure with optimum convergence, *J. Chem. Phys.* **82** (11) (1985) 5053–5063.

[6] P.J. Knowles, H.J. Werner, An efficient 2nd-order MC SCF method for long configuration expansions, *Chem. Phys. Lett.* **115** (3) (1985) 259–267.

[7] H.J. Werner, Third-order multireference perturbation theory—The CASPT3 method, *Mol. Phys.* **89** (2) (1996) 645–661.

[8] P. Celani, H.J. Werner, Multireference perturbation theory for large restricted and selected active space reference wave functions, *J. Chem. Phys.* **112** (13) (2000) 5546–5557.

[9] H.J. Werner, P.J. Knowles, An efficient internally contracted multiconfiguration reference configuration-interaction method, *J. Chem. Phys.* **89** (9) (1988) 5803–5814.

[10] Z.W. Qu, *et al.*, The Huggins band of ozone: A theoretical analysis, *J. Chem. Phys.* **121** (23) (2004) 11731–11745.

[11] Z.W. Qu, *et al.*, The photodissociation of ozone in the Hartley band: A theoretical analysis, *J. Chem. Phys.* **123** (7) (2005) 074305.

[12] E. Baloitcha, G.G. Balint-Kurti, Theory of the photodissociation of ozone in the Hartley continuum: Potential energy surfaces, conical intersections, and photodissociation dynamics, *J. Chem. Phys.* **123** (1) (2005), Art. No. 014306.

[13] A. Toniolo, M. Persico, D. Pitea, An ab initio study of spectroscopy and predissociation of ClO, *J. Chem. Phys.* **112** (6) (2000) 2790–2797.

[14] S. Jørgensen, Unpublished.

[15] P.J. Hay, W.R. Wadt, Ab initio effective core potentials for molecular calculations—Potentials for the transition-metal atoms Sc to Hg, *J. Chem. Phys.* **82** (1) (1985) 270–283.

[16] A. Nicklass, *et al.*, Ab-initio energy-adjusted pseudopotentials for the noble-gases Ne through Xe—Calculation of atomic dipole and quadrupole polarizabilities, *J. Chem. Phys.* **102** (22) (1995) 8942–8952.

[17] G.A. Worth, L.S. Cederbaum, Beyond Born–Oppenheimer: Molecular dynamics through a conical intersection, *Annual Rev. Chem. Phys.* **55** (127) (2004) 127–158.

[18] M. Baer, *Beyong Born–Oppenheiner: Electronic Nonadiabatic Coupling Terms and Conical Intersection*, Wiley, New York, 2006.

[19] H. Lischka, *et al.*, Analytic evaluation of nonadiabatic coupling terms at the MR-CI level. I. Formalism, *J. Chem. Phys.* **120** (16) (2004) 7322–7329.

[20] M. Dallos, *et al.*, Analytic evaluation of nonadiabatic coupling terms at the MR-CI level. II. Minima on the crossing seam: Formaldehyde and the photodimerization of ethylene, *J. Chem. Phys.* **120** (16) (2004) 7330–7339.

[21] H.J. Werner, B. Follmeg, M.H. Alexander, Adiabatic and diabatic potential-energy surfaces for collisions of CN ($X^2\Sigma^+ \rightarrow A^2\Pi$) with He, *J. Chem. Phys.* **89** (5) (1988) 3139–3151.
[22] D. Simah, B. Hartke, H.J. Werner, Photodissociation dynamics of H_2S on new coupled ab initio potential energy surfaces, *J. Chem. Phys.* **111** (10) (1999) 4523–4534.
[23] J.A. Klos, *et al.*, Ab initio calculations of adiabatic and diabatic potential energy surfaces of $Cl(^2P)\ldots HCl(^1\Sigma^+)$ van der Waals complex, *J. Chem. Phys.* **115** (7) (2001) 3085–3098.
[24] R. Donovan, *et al.*, General discussion, *Faraday Discuss.* (1997) 427–467.
[25] Tannor, *et al.*, General discussion, *Faraday Discuss.* (1999) 455–491.
[26] D. Feller, Application of systematic sequences of wave-functions to the water dimer, *J. Chem. Phys.* **96** (8) (1992) 6104–6114.
[27] M.P. de Lara-Castells, *et al.*, Complete basis set extrapolation limit for electronic structure calculations: Energetic and nonenergetic properties of HeBr and $HeBr_2$ van der Waals dimers, *J. Chem. Phys.* **115** (22) (2001) 10438–10449.
[28] J.M.L. Martin, Ab initio total atomization energies of small molecules—Towards the basis set limit, *Chem. Phys. Lett.* **259** (5–6) (1996) 669–678.
[29] J.M.L. Martin, P.R. Taylor, Benchmark quality total atomization energies of small polyatomic molecules, *J. Chem. Phys.* **106** (20) (1997) 8620–8623.
[30] D. Feller, J.A. Sordo, A CCSDT study of the effects of higher order correlation on spectroscopic constants. I. First row diatomic hydrides, *J. Chem. Phys.* **112** (13) (2000) 5604–5610.
[31] K.A. Peterson, D.E. Woon, T.H. Dunning, Benchmark calculations with correlated molecular wave-functions. 4. The classical barrier height of the $H + H_2 \rightarrow H_2 + H$ reaction, *J. Chem. Phys.* **100** (10) (1994) 7410–7415.
[32] A. Halkier, *et al.*, Basis-set convergence in correlated calculations on Ne, N_2, and H_2O, *Chem. Phys. Lett.* **286** (3–4) (1998) 243–252.
[33] A. Tajti, *et al.*, HEAT: High accuracy extrapolated ab initio thermochemistry, *J. Chem. Phys.* **121** (23) (2004) 11599–11613.
[34] T.H. Dunning, Gaussian-basis sets for use in correlated molecular calculations. 1. The atoms boron through neon and hydrogen, *J. Chem. Phys.* **90** (2) (1989) 1007–1023.
[35] R.A. Kendall, T.H. Dunning, R.J. Harrison, Electron-affinities of the 1st-row atoms revisited—Systematic basis-sets and wave-functions, *J. Chem. Phys.* **96** (9) (1992) 6796–6806.
[36] D.E. Woon, T.H. Dunning, Gaussian-basis sets for use in correlated molecular calculations. 3. The atoms aluminum through argon, *J. Chem. Phys.* **98** (2) (1993) 1358–1371.
[37] A.K. Wilson, *et al.*, Gaussian basis sets for use in correlated molecular calculations. IX. The atoms gallium through krypton, *J. Chem. Phys.* **110** (16) (1999) 7667–7676.
[38] B.O. Roos, The complete active space self-consistent field method and its applications in electronic structure calculations, *Adv. Chem. Phys.* **69** (1987) 399.
[39] J.T. Murrel, S. Carter, S.C. Farantos, P. Huxley, A.J.C. Varandas, *Molecular Potential Energy Functions*, Wiley, Chichester, 1988.
[40] F. London, *Z. Elecktrochem.* **35** (1929) 552.
[41] H. Eyring, M. Polanyi, Uber einfache Gasreaktionen, *Z. Phys. Chem. Abt. B* (12) (1931), Heft 4.
[42] S. Sato, On a new method of drawing the potential energy surface, *J. Chem. Phys.* **23** (3) (1955) 592–593.
[43] F.O. Ellison, A method of diatomics in molecules. 1. General theory and application to H_2O, *J. Am. Chem. Soc.* **85** (22) (1963) 3540–3544.
[44] P.J. Kuntz, in: R.B. Bernstein (Ed.), *Atom–Molecular Collision Theory*, Plenum, New York, 1979.
[45] J.C. Tully, *Adv. Chem. Phys.* **42** (1980) 63.
[46] H.S. Johnston, C. Parr, Activation energies from bond energies. 1. Hydrogen transfer reactions, *J. Am. Chem. Soc.* **85** (17) (1963) 2544–2551.
[47] E. Garcia, A. Lagana, A new bond-order functional form for triatomic-molecules a fit of the BETH potential-energy, *Mol. Phys.* **56** (3) (1985) 629–639.
[48] F.T. Wall, R.N. Porter, General potential-energy function for exchange reactions, *J. Chem. Phys.* **36** (12) (1962) 3256–3260.
[49] J.M. Bowman, A. Kuppermann, Semi-numerical approach to construction and fitting of triatomic potential-energy surfaces, *Chem. Phys. Lett.* **34** (3) (1975) 523–527.
[50] J.N.L. Connor, W. Jakubetz, J. Manz, Exact quantum transition-probabilities by state path sum method—Collinear $F + H_2$ reaction, *Mol. Phys.* **29** (2) (1975) 347–355.

[51] A. Lagana, G.O. de Aspuru, E. Garcia, The largest angle generalization of the rotating bond order potential: Three different atom reactions, *J. Chem. Phys.* **108** (10) (1998) 3886–3896.
[52] Dr. McLaughl, D.L. Thompson, Ab-initio dynamics: $HeH^+ + H_2 \rightarrow He + H_3^+$ (C_{2v}) classical trajectories using a quantum-mechanical potential-energy surface, *J. Chem. Phys.* **59** (8) (1973) 4393–4405.
[53] N. Sathyamurthy, L.M. Raff, Quasiclassical trajectory studies using 3d spline interpolation of ab initio surfaces, *J. Chem. Phys.* **63** (1) (1975) 464–473.
[54] J.M. Bowman, J.S. Bittman, L.B. Harding, Ab initio calculations of electronic and vibrational energies of HCO and HOC, *J. Chem. Phys.* **85** (2) (1986) 911–921.
[55] T.S. Ho, H. Rabitz, A general method for constructing multidimensional molecular potential energy surfaces from ab initio calculations, *J. Chem. Phys.* **104** (7) (1996) 2584–2597.
[56] T.S. Ho, *et al.*, A global H_2O potential energy surface for the reaction $O(^1D) + H_2 \rightarrow OH + H$, *J. Chem. Phys.* **105** (23) (1996) 10472–10486.
[57] T. Hollebeek, T.S. Ho, H. Rabitz, A fast algorithm for evaluating multidimensional potential energy surfaces, *J. Chem. Phys.* **106** (17) (1997) 7223–7227.
[58] J. Ischtwan, M.A. Collins, Molecular-potential energy surfaces by interpolation, *J. Chem. Phys.* **100** (11) (1994) 8080–8088.
[59] M.J.T. Jordan, K.C. Thompson, M.A. Collins, Convergence of molecular-potential energy surfaces by interpolation: Application to the $OH + H_2 \rightarrow H_2O + H$ reaction, *J. Chem. Phys.* **102** (14) (1995) 5647–5657.
[60] T. Ishida, G.C. Schatz, Automatic potential energy surface generation directly from ab initio calculations using Shepard interpolation: A test calculation for the H_2+H system, *J. Chem. Phys.* **107** (9) (1997) 3558–3568.
[61] K.C. Thompson, M.J.T. Jordan, M.A. Collins, Polyatomic molecular potential energy surfaces by interpolation in local internal coordinates, *J. Chem. Phys.* **108** (20) (1998) 8302–8316.
[62] K. Thompson, T.J. Martinez, Ab initio interpolated quantum dynamics on coupled electronic states with full configuration interaction wave functions, *J. Chem. Phys.* **110** (3) (1999) 1376–1382.
[63] T. Wu, H.J. Werner, U. Manthe, Accurate potential energy surface and quantum reaction rate calculations for the $H + CH_4 \rightarrow H_2 + CH_3$ reaction, *J. Chem. Phys.* **124** (16) (2006) 164307.
[64] T. Ishida, G.C. Schatz, A local interpolation scheme using no derivatives in quantum-chemical calculations, *Chem. Phys. Lett.* **314** (3–4) (1999) 369–375.
[65] G.G. Maisuradze, *et al.*, Interpolating moving least-squares methods for fitting potential energy surfaces: Detailed analysis of one-dimensional applications, *J. Chem. Phys.* **119** (19) (2003) 10002–10014.
[66] G.G. Maisuradze, D.L. Thompson, Interpolating moving least-squares methods for fitting potential energy surfaces: Illustrative approaches and applications, *J. Phys. Chem. A* **107** (37) (2003) 7118–7124.
[67] G.G. Maisuradze, *et al.*, Interpolating moving least-squares methods for fitting potential energy surfaces: Analysis of an application to a six-dimensional system, *J. Chem. Phys.* **121** (21) (2004) 10329–10338.
[68] A.M. Frishman, *et al.*, Distributed approximating functional approach to fitting and predicting potential surfaces. 1. Atom–atom potentials, *Chem. Phys. Lett.* **252** (1–2) (1996) 62–70.
[69] D.K. Hoffman, A. Frishman, D.J. Kouri, Distributed approximating functional approach to fitting multi-dimensional surfaces, *Chem. Phys. Lett.* **262** (3–4) (1996) 393–399.
[70] H.G. Yu, S. Andersson, G. Nyman, A generalized discrete variable representation approach to interpolating or fitting potential energy surfaces, *Chem. Phys. Lett.* **321** (3–4) (2000) 275–280.
[71] S. Lorenz, A. Gross, M. Scheffler, Representing high-dimensional potential-energy surfaces for reactions at surfaces by neural networks, *Chem. Phys. Lett.* **395** (4–6) (2004) 210–215.
[72] J.B. Witkoskie, D.J. Doren, Neural network models of potential energy surfaces: Prototypical examples, *J. Chem. Theory Comput.* **1** (1) (2005) 14–23.
[73] I. Tuvi, Y.B. Band, Hermiticity of the Hamiltonian matrix in a discrete variable representation, *J. Chem. Phys.* **107** (21) (1997) 9079–9084.
[74] B. Podolsky, Quantum mechanically correct form of Hamiltonian function for conservative systems, *Phys. Rev.* **32** (5) (1928) 812–816.
[75] G. Katz, *et al.*, The Fourier method for triatomic systems in the search for the optimal coordinate system, *J. Chem. Phys.* **116** (11) (2002) 4403–4414.

[76] G.G. Balint-Kurti, R.N. Dixon, C.C. Marston, Grid methods for solving the Schrödinger-equation and time-dependent quantum dynamics of molecular photofragmentation and reactive scattering processes, *Int. Rev. Phys. Chem.* **11** (2) (1992) 317–344.
[77] B.R. Johnson, The classical dynamics of 3 particles in hyperspherical coordinates, *J. Chem. Phys.* **79** (4) (1983) 1906–1915.
[78] J.T. Muckerman, R.D. Gilbert, G.D. Billing, A classical path approach to reactive scattering. 1. Use of hyperspherical coordinates, *J. Chem. Phys.* **88** (8) (1988) 4779–4787.
[79] C. Evenhuis, G. Nyman, U. Manthe, *J. Chem. Phys.* **127** (14) (2007) 144302.
[80] M.H. Beck, *et al.*, The multiconfiguration time-dependent Hartree (MCTDH) method: A highly efficient algorithm for propagating wavepackets, *Phys. Rep.-Rev. Sect. Phys. Lett.* **324** (1) (2000) 1–105.
[81] G. Nyman, H.G. Yu, Quantum theory of bimolecular chemical reactions, *Rep. Prog. Phys.* **63** (7) (2000) 1001–1059.
[82] S.C. Althorpe, D.C. Clary, Quantum scattering calculations on chemical reactions, *Annu. Rev. Phys. Chem.* **54** (2003) 493–529.
[83] Y.L. Yung, C.E. Miller, Isotopic fractionation of stratospheric nitrous oxide, *Science* **278** (5344) (1997) 1778–1780.
[84] M.K. Prakash, J.D. Weibel, R.A. Marcus, Isotopomer fractionation in the UV photolysis of N_2O: Comparison of theory and experiment, *J. Geophys. Res.-Atmos.* **110** (D21) (2005) D21315.
[85] G.A. Blake, *et al.*, A Born–Oppenheimer photolysis model of N_2O fractionation, *Geophys. Res. Lett.* **30** (12) (2003), Art. No. 1656.
[86] M.C. Liang, G.A. Blake, Y.L. Yung, A semianalytic model for photo-induced isotopic fractionation in simple molecules, *J. Geophys. Res.-Atmos.* **109** (D10) (2004) D10308.
[87] G.G. Balint-Kurti, R.N. Dixon, C.C. Marston, Time-dependent quantum dynamics of molecular photofragmentation processes, *J. Chem. Soc. Faraday Transact.* **86** (10) (1990) 1741–1749.
[88] J.C. Light, I.P. Hamilton, J.V. Lill, Generalized discrete variable approximation in quantum-mechanics, *J. Chem. Phys.* **82** (3) (1985) 1400–1409.
[89] Z. Basic, J.C. Light, Theoretical methods for rovibrational states of floppy molecules, *Annu. Rev. Chem. Phys.* **40** (1989) 469–498.
[90] S.E. Choi, J.C. Light, Use of the discrete variable representation in the quantum dynamics by a wave packet propagation—Predissociation of $NaI(^1\Sigma_0^+) \to NaI(0^+) \to Na(^2S) + I(^2P)$, *J. Chem. Phys.* **90** (5) (1989) 2593–2604.
[91] M.R. Hermann, J.A. Fleck, Split-operator spectral method for solving the time-dependent Schrödinger-equation in spherical coordinates, *Phys. Rev. A* **38** (12) (1988) 6000–6012.
[92] D. Kosloff, R. Kosloff, A Fourier method solution for the time-dependent Schrödinger-equation as a tool in molecular-dynamics, *J. Comput. Phys.* **52** (1) (1983) 35–53.
[93] R. Kosloff, Time-dependent quantum-mechanical methods for molecular-dynamics, *J. Phys. Chem.* **92** (8) (1988) 2087–2100.
[94] M.D. Feit, J.A. Fleck, Solution of the Schrödinger-equation by a spectral method. 2. Vibrational-energy levels of triatomic-molecules, *J. Chem. Phys.* **78** (1) (1983) 301–308.
[95] G. Barinovs, N. Markovic, G. Nyman, Split operator method in hyperspherical coordinates: Application to CH_2I_2 and OClO, *J. Chem. Phys.* **111** (15) (1999) 6705–6711.
[96] A. Nauts, R.E. Wyatt, New approach to many-state quantum dynamics—The recursive-residue-generation method, *Phys. Rev. Lett.* **51** (25) (1983) 2238–2241.
[97] H. Talezer, R. Kosloff, An accurate and efficient scheme for propagating the time-dependent Schrödinger-equation, *J. Chem. Phys.* **81** (9) (1984) 3967–3971.
[98] G. Ashkenazi, *et al.*, Newtonian propagation methods applied to the photodissociation dynamics of I_3^-, *J. Chem. Phys.* **103** (23) (1995) 10005–10014.
[99] W. Huisinga, *et al.*, Faber and Newton polynomial integrators for open-system density matrix propagation, *J. Chem. Phys.* **110** (12) (1999) 5538–5547.
[100] G.D. Billing, Quantum corrections to the classical path equations: Multitrajectory and Hermite corrections, *J. Chem. Phys.* **107** (11) (1997) 4286–4294.
[101] A. Vijay, R.E. Wyatt, G.D. Billing, Time propagation and spectral filters in quantum dynamics: A Hermite polynomial perspective, *J. Chem. Phys.* **111** (24) (1999) 10794–10805.
[102] C. Leforestier, *et al.*, A comparison of different propagation schemes for the time-dependent Schrödinger-equation, *J. Comput. Phys.* **94** (1) (1991) 59–80.

[103] C. Lanczos, An iteration method for the solution of the eigenvalue problem of linear differential and integral operators, *J. Res. Nat. Bur. Stand.* **45** (4) (1950) 255–282.
[104] A. Vibok, G.G. Balintkurti, Parametrization of complex absorbing potentials for time-dependent quantum dynamics, *J. Phys. Chem.* **96** (22) (1992) 8712–8719.
[105] D.E. Manolopoulos, Derivation and reflection properties of a transmission-free absorbing potential, *J. Chem. Phys.* **117** (21) (2002) 9552–9559.
[106] M.M.L. Grage, G. Nyman, M.S. Johnson, HCl and DCl: A case study of different approaches for determining photo fractionation constants, *Phys. Chem. Chem. Phys.* **8** (41) (2006) 4798–4804.
[107] M.H. Alexander, B. Pouilly, T. Duhoo, Spin–orbit branching in the photofragmentation of HCI, *J. Chem. Phys.* **99** (3) (1993) 1752–1764.
[108] B.M. Cheng, *et al.*, Quantitative spectral analysis of HCl and DCl in 120–220 nm: Effects of singlet-triplet mixing, *J. Chem. Phys.* **117** (9) (2002) 4293–4298.
[109] G. Barinovs, Time dependent wavepackets—Applications to spectroscopy, eigenstates and ultrafast dynamics, *Physical Chemistry*, Department of Chemistry, Göteborg University, Göteborg, 2002.
[110] S. Nanbu, M.S. Johnson, Analysis of the ultraviolet absorption cross sections of six isotopically substituted nitrous oxide species using 3D wave packet propagation, *J. Phys. Chem. A* **108** (41) (2004) 8905–8913.
[111] Y.L. Yung, W.C. Wang, A.A. Lacis, Greenhouse effect due to atmospheric nitrous-oxide, *Geophys. Res. Lett.* **3** (10) (1976) 619–621.
[112] F. Turatti, *et al.*, Positionally dependent ^{15}N fractionation factors in the UV photolysis of N_2O determined by high resolution FTIR spectroscopy, *Geophys. Res. Lett.* **27** (16) (2000) 2489–2492.
[113] N. Yoshida, S. Matsuo, Nitrogen isotope ratio of atmospheric N_2O as a key to the global cycle of N_2O, *Geochem. J.* **17** (5) (1983) 231–239.
[114] T. Rockmann, *et al.*, Isotopic enrichment of nitrous oxide ($(NNO)^{15}N^{14}N$, $(NNO)^{14}N^{15}N$, $(NNO)^{14}N^{14}N^{18}O$) in the stratosphere and in the laboratory, *J. Geophys. Res.-Atmos.* **106** (D10) (2001) 10403–10410.
[115] K.R. Kim, H. Craig, ^{15}N and ^{18}O characteristics of nitrous-oxide—A global perspective, *Science* **262** (5141) (1993) 1855–1857.
[116] C.A. McLinden, M.J. Prather, M.S. Johnson, Global modeling of the isotopic analogues of N_2O: Stratospheric distributions, budgets, and the $^{17}O^{18}O$ mass-independent anomaly, *J. Geophys. Res.-Atmos.* **108** (D8) (2003) 4233.
[117] K. Minschwaner, R.J. Salawitch, M.B. McElroy, Absorption of solar-radiation by O_2—Implications for O_3 and lifetimes of N_2O, $CFCl_3$, and CF_2Cl_2, *J. Geophys. Res.-Atmos.* **98** (D6) (1993) 10543–10561.
[118] M.S. Johnson, *et al.*, Photolysis of nitrous oxide isotopomers studied by time-dependent Hermite propagation, *J. Phys. Chem. A* **105** (38) (2001) 8672–8680.
[119] A. Brown, P. Jimeno, G.G. Balint-Kurti, Photodissociation of N_2O. I. Ab initio potential energy-surfaces for the low-lying electronic states $\widetilde{X}^1A'$, $2^1A'$, and $1^1A''$, *J. Phys. Chem. A* **103** (50) (1999) 11089–11095.
[120] D.W.T. Griffith, *et al.*, Vertical profiles of nitrous oxide isotopomer fractionation measured in the stratosphere, *Geophys. Res. Lett.* **27** (16) (2000) 2485–2488.
[121] S.K. Gray, G.G. Balint-Kurti, Quantum dynamics with real wave packets, including application to three-dimensional ($J = 0$)$D + H_2 \rightarrow HD + H$ reactive scattering, *J. Chem. Phys.* **108** (3) (1998) 950–962.
[122] S.S. Cliff, M.H. Thiemens, The $^{18}O/^{16}O$ and $^{17}O/^{16}O$ ratios in atmospheric nitrous oxide: A mass-independent anomaly, *Science* **278** (5344) (1997) 1774–1776.
[123] S.S. Cliff, C.A.M. Brenninkmeijer, M.H. Thiemens, First measurement of the $^{18}O/^{16}O$ and $^{17}O/^{16}O$ ratios in stratospheric nitrous oxide: A mass-independent anomaly, *J. Geophys. Res.-Atmos.* **104** (D13) (1999) 16171–16175.
[124] J. Kaiser, T. Rockmann, C.A.M. Brenninkmeijer, Contribution of mass-dependent fractionation to the oxygen isotope anomaly of atmospheric nitrous oxide, *J. Geophys. Res.-Atmos.* **109** (D3) (2004), D03305.
[125] T. Rockmann, *et al.*, The origin of the anomalous or "mass-independent" oxygen isotope fractionation in tropospheric N_2O, *Geophys. Res. Lett.* **28** (3) (2001) 503–506.

[126] M.K. Prakash, R.A. Marcus, Three-isotope plot of fractionation in photolysis: A perturbation theoretical expression, *J. Chem. Phys.* **123** (17) (2005), Art. No. 174308.
[127] G. Myhre, *et al.*, The radiative effect of the anthropogenic influence on the stratospheric sulfate aerosol layer, *Tellus Series B Chem. Phys. Meteorology* **56** (3) (2004) 294–299.
[128] P.J. Crutzen, Possible importance of CSO for sulfate layer of stratosphere, *Geophys. Res. Lett.* **3** (2) (1976) 73–76.
[129] A.J. Colussi, F.-Y. Leung, M.R. Hoffmann, Electronic spectra of carbonyl sulfide sulfur isotopologues, *Environ. Chem.* **1** (2004) 44–48.
[130] F.Y.T. Leung, *et al.*, Isotopic fractionation of carbonyl sulfide in the atmosphere: Implications for the source of background stratospheric sulfate aerosol, *Geophys. Res. Lett.* **29** (10) (2002), doi:10.1029/2001GL013955.
[131] S.O. Danielache, *et al.*, Ab-initio study of sulfur isotope fractionation in the reaction of OCS with OH, *Chem. Phys. Lett.* **450** (2008) 214–220.
[132] T. Suzuki, *et al.*, Nonadiabatic bending dissociation in 16 valence electron system OCS, *J. Chem. Phys.* **109** (14) (1998) 5778–5794.
[133] T. Suzuki, S. Nanbu, Non-adiabatic bending dissociation of OCS induced by orbital unlocking, *Low-Lying Potential Energy Surfaces, ACS Symposium Series* **828** (2002) 300–313.
[134] G. Herzberg, *Molecular Spectra and Molecular Structure II. Infrared and Raman Spectra of Polyatomic Molecules*, D. Van Nostrand Company, New York, 1945.
[135] A. Stickler, *et al.*, Influence of summertime deep convection on formaldehyde in the middle and upper troposphere over Europe, *J. Geophys. Res.-Atmos.* **111** (D14) (2006) D14308.
[136] S.P. Sander, *et al.*, Chemical kinetics and photochemical data for use in atmospheric studies: Evaluation number 15, in JPL Publication 06-02, 2006, Jet Propulsion Laboratory, California Institute of Technology: Pasadena, CA.
[137] K.L. Feilberg, *et al.*, Relative tropospheric photolysis rates of HCHO, $(HCHO)^{13}C$, $(HCHO)^{18}O$, and DCDO measured at the European photoreactor facility, *J. Phys. Chem. A* **109** (37) (2005) 8314–8319.
[138] K.L. Feilberg, *et al.*, Relative tropospheric photolysis rates of HCHO and HCDO measured at the European photoreactor facility, *J. Phys. Chem. A* **111** (2007) 9034–9046.
[139] K.L. Feilberg, M.S. Johnson, C.J. Nielsen, Relative reaction rates of HCHO, HCDO, DCDO, $(HCHO)^{13}C$, and $(HCHO)^{18}O$ with OH, Cl, Br, and NO_3 radicals, *J. Phys. Chem. A* **108** (36) (2004) 7393–7398.
[140] T.S. Rhee, C.A.M. Brenninkmeijer, T. Röckmann, Hydrogen isotope fractionation in the photolysis of formaldehyde, *Atmos. Chem. Phys. Discuss.* **7** (2007) 12715–12750.
[141] A. Gratien, *et al.*, UV and IR absorption cross-sections of HCHO, HCDO and DCDO, *J. Phys. Chem. A* **111** (45) (2007) 11506–11513.
[142] E. Nilsson, *et al.*, Atmospheric deuterium fractionation: HCHO and HCDO yields in the $CH_2DO + O_2$ reaction, *Atmos. Chem. Phys.* **7** (2007) 5873–5881.
[143] C.B. Moore, J.C. Weisshaar, Formaldehyde photochemistry, *Annu. Rev. Phys. Chem.* **34** (1983) 525–555.
[144] C.B. Moore, D.J. Bamford, State-selected photodissociation dynamics of formaldehyde, *Laser Chem.* **6** (2) (1986) 93–102.
[145] C.B. Moore, A spectroscopist's view of energy states, energy transfers, and chemical reactions, *Annu. Rev. Phys. Chem.* **58** (2007) 1–33.
[146] X.B. Zhang, *et al.*, A global ab initio potential energy surface for formaldehyde, *J. Phys. Chem. A* **108** (41) (2004) 8980–8986.
[147] X.B. Zhang, J.L. Rheinecker, J.M. Bowman, Quasiclassical trajectory study of formaldehyde unimolecular dissociation: $H_2CO \rightarrow H_2 + CO, H + HCO$, *J. Chem. Phys.* **122** (11) (2005) 114323.
[148] H.M. Yin, *et al.*, Signatures of H_2CO photodissociation from two electronic states, *Science* **311** (5766) (2006) 1443–1446.
[149] D. Townsend, *et al.*, The roaming atom: Straying from the reaction path in formaldehyde decomposition, *Science* **306** (5699) (2004) 1158–1161.
[150] S.A. Lahankar, *et al.*, The roaming atom pathway in formaldehyde decomposition, *J. Chem. Phys.* **125** (4) (2006) 044303.
[151] J.M. Bowman, X.B. Zhang, New insights on reaction dynamics from formaldehyde photodissociation, *Phys. Chem. Chem. Phys.* **8** (3) (2006) 321–332.

[152] J. Troe, Theory of multichannel thermal unimolecular reactions. 2. Application to the thermal dissociation of formaldehyde, *J. Phys. Chem. A* **109** (37) (2005) 8320–8328.
[153] J. Troe, Refined analysis of the thermal dissociation of formaldehyde, *J. Phys. Chem. A* **111** (19) (2007) 3862–3867.
[154] J. Troe, Analysis of quantum yields for the photolysis of formaldehyde at $\lambda > 310$ nm, *J. Phys. Chem. A* **111** (19) (2007) 3868–3874.
[155] R.A. Toth, Frequencies of N_2O in the 1100 cm^{-1} to 1440 cm^{-1} region, *J. Opt. Soc. Am. B Opt. Phys.* **3** (10) (1986) 1263–1281.
[156] K. Jolma, J. Kauppinen, V.M. Horneman, Vibration–rotation spectrum of N_2O in the region of the lowest fundamental ν_2, *J. Mol. Spectrosc.* **101** (2) (1983) 278–284.
[157] R.A. Toth, N_2O vibration–rotation parameters derived from measurements in the 900–1090 cm^{-1} and 1580–2380 cm^{-1} regions, *J. Opt. Soc. Am. B Opt. Phys.* **4** (3) (1987) 357–374.
[158] C. Amiot, Vibration–rotation bands of $^{15}N_2{}^{16}O$, $^{14}N_2{}^{18}O$, *J. Mol. Spectrosc.* **59** (3) (1976) 380–395.
[159] K.L. Feilberg, *et al.*, Relative tropospheric photolysis rates of HCHO, $(HCHO)^{13}C$, $(HCHO)^{18}O$, and DCDO measured at the European photoreactor facility, *J. Phys. Chem. A* **109** (2005) 8314–8319.
[160] E. Nilsson, *et al.*, Relative tropospheric photolysis rates of HCHO and DCDO measured at the European photoreactor facility, *Phys. Chem. Chem. Phys.*, (2008), in press.

CHAPTER 8

Atmospheric Photolysis of Sulfuric Acid

Henrik G. Kjaergaard[*,**], **Joseph R. Lane**[*], **Anna L. Garden**[*], **Daniel P. Schofield**[*], **Timothy W. Robinson**[*] and **Michael J. Mills**[***]

Contents

1. Introduction 138
2. Vibrational Transitions 141
   2.1 Local mode model and computational details 142
   2.2 Vibrational cross sections 145
3. Electronic Transitions 149
   3.1 Computational details 150
   3.2 Electronic cross sections 151
4. Atmospheric Simulations 153
5. Conclusion 155
Acknowledgements 156
References 156

Abstract

We describe theoretical methods for the calculation of vibrational and electronic transitions in sulfuric acid, from which absorption cross sections can be obtained in the infrared through to the vacuum ultraviolet region. In the absence of experimental cross sections these calculations provide invaluable input for the assessment of the atmospheric photolysis of sulfuric acid. The vibrational model is based on a local mode model that includes the OH-stretching and SOH-bending vibrations, while the electronic transitions are calculated with coupled cluster response theory. These approaches are sufficient to describe the dominant vibrational transitions in

[*] Department of Chemistry, University of Otago, P.O. Box 56, Dunedin, New Zealand
[**] Lundbeck Foundation Center for Theoretical Chemistry, Department of Chemistry, Aarhus University, DK-8000 Aarhus, Denmark
[***] LASP/PAOS, University of Colorado, Boulder, Colorado, 80309, USA

Advances in Quantum Chemistry, Vol. 55
ISSN 0065-3276, DOI: 10.1016/S0065-3276(07)00208-0

the near infrared and visible regions, the lowest lying electronic transitions in the ultraviolet region and the higher energy electronic transitions in the region of Lyman-α radiation. We highlight the influence quantum mechanical calculations have had in the recent discussion of the atmospheric photolysis of sulfuric acid, and show that theoretical calculations can provide absorption cross sections of an accuracy that is useful in atmospheric science.

1. INTRODUCTION

In the last 150 years the anthropogenic emission of sulfur has increased dramatically, primarily due to combustion processes [1]. In the 1950s anthropogenic emission surpassed natural emission and the atmospheric sulfur cycle is one of the most perturbed biogeochemical cycles [1,2]. The oceans are the largest natural source of atmospheric sulfur emissions, where sulfur is emitted in a reduced form, predominantly as dimethyl sulfide (DMS) and to a much lesser extent carbonyl sulfide (OCS) and carbon disulfide (CS_2) [3]. Ocean emitted DMS and CS_2 are initially oxidised to OCS, which diffuses through the troposphere into the stratosphere where further oxidation to sulfur dioxide (SO_2), sulfur trioxide (SO_3) and finally sulfuric acid (H_2SO_4) occurs [1–4].

The fate of sulfur in the Earth's middle atmosphere has been the subject of recent scientific investigation due to its roles in global climate change. While most tropospheric sulfur is short-lived, and thereby has its greatest impact on local and regional environments (i.e. acid rain), sulfur in the stratosphere is long-lived and affects both climate and chemistry [2,3]. In the stratosphere and mesosphere, sulfuric acid is the dominant form of sulfur. At altitudes above ~35 km sulfuric acid is present in gaseous form, whereas at lower altitudes it exists in aerosol form as part of the global stratospheric aerosol layer.

This aerosol layer is enhanced many-fold following volcanic eruptions that enter the stratosphere and are high in sulfur content. Such volcanic aerosol has been known to scatter enough solar radiation back into space to cool the Earth's surface for years following certain eruptions. The June 1991 eruption of Mt. Pinatubo offset global warming for several years [5,6], and the 1815 eruption of Tambora in Indonesia has been implicated in the devastating "year without a summer" in 1816 in Europe and North America [7]. The sulfate layer has also been shown to modulate anthropogenic ozone depletion at mid- and high-latitudes via heterogeneous chemistry on the surface of sulfate [8,9].

The fate of sulfuric acid in the upper stratosphere and mesosphere is key to understanding the role of sulfur in the upper atmosphere [10]. In this regard the photolysis of sulfuric acid (H_2SO_4) via the reaction

$$H_2SO_4 + h\nu \rightarrow SO_3 + H_2O, \tag{1}$$

is an important reaction and the focus of this paper. In the stratosphere, H_2SO_4 is photolysed to sulfur trioxide (SO_3), which is subsequently photolysed to sulfur dioxide (SO_2) [11]. Observation of SO_2 concentrations in the upper stratosphere, which increase with altitude [12], could be explained by a photolytic decomposition of stratospheric H_2SO_4. Modeling studies supported this explanation, provided sufficiently large absorption cross sections were assumed [13]. The anomalously large concentration of small condensation nuclei (CN) measured at the top of the aerosol layer in polar spring or in midlatitude air of recent polar origin [14, 15] could also be modeled assuming a photolytic breakdown of sulfuric acid in the stratosphere [13,16]. Thus modeling studies suggested that light initiated H_2SO_4 photolysis must occur to explain observations [13,16,17] and hence knowledge of the absorption spectrum of vapor phase sulfuric acid is crucial to assess its role in the atmosphere.

In early studies, the photolysis of H_2SO_4 was assumed to occur via electronic excitation in the UV region [12,13,16,18], however experimental attempts to measure the electronic spectrum of H_2SO_4 from 330 nm to 140 nm have been unsuccessful [10,19]. Supported by *ab initio* calculations of the lowest lying electronic transitions [19–21], these experimental results led to the suggestion that electronic transitions could not be the primary pathway for photolysis of H_2SO_4 in the atmosphere [22]. In addition, modeling studies using the measured upper limits for the H_2SO_4 UV absorption cross section could not explain the CN layer observations [13,23]. The wavelength of the lowest lying electronic transition in H_2SO_4 was calculated to occur at $\sim$144 nm [21]. This transition led to an estimated cross section in the spectral region relevant to atmospheric photolysis (photons with wavelengths shorter than 179 nm are very limited in the stratosphere) that is even lower than the measured upper limits [21]. Thus UV driven photolysis is likely negligible in the stratosphere and another mechanism had to be proposed.

Vaida *et al.* proposed that excitation of a high OH-stretching vibrational overtone in H_2SO_4 could provide sufficient energy for photolysis [22]. The energy required for H_2SO_4 dissociation via Eq. (1), has been calculated at the MP2 level to be $\sim$40 kcal/mol [24]. Based on the measured fundamental and first two OH-stretching overtone transitions [19], anharmonic local mode calculations were used to predict that sufficient energy for dissociation of H_2SO_4 is available if the third OH-stretching overtone ($v = 0$ to $v = 4$ excitation) or higher is excited [22]. Thus rather than electronic excitations, high vibrational OH-stretching overtone transitions could provide the energy necessary for atmospheric photolysis of H_2SO_4 [22]. The third and fourth OH-stretching overtones are excited by photons with wavelengths in the visible region, photons that are abundant throughout the atmosphere.

Vaida *et al.* calculated the atmospheric *J*-values (the integrated value of the cross section, photon flux and quantum yield) for the vibrational overtone initiated visible photolysis and UV photolysis. The *J*-values were obtained assuming the local mode calculated third and fourth OH-stretching overtone cross sections for the vibrational transitions and the experimental upper limit for the UV cross section across the 179–224 nm range, and a unitary quantum yield for both regions [22]. These *J*-values showed that at altitudes below $\sim$45 km, the OH-stretching overtone initiated photolysis was the predominant process, whereas for higher altitudes the visible and UV photolysis rates were comparable. Atmospheric modeling studies using these *J*-values confirmed that photolysis of sulfuric acid by visible light could be responsible for the formation of the springtime CN layer, and is sufficient to explain the vertical increase of SO_2 mixing ratio in the upper stratosphere [23]. In addition it was shown that enough H_2SO_4 survives this weak photolysis mechanism to produce significant sulfate aerosol area in the mesosphere [25]. Mesospheric sulfate may act as nuclei for polar mesospheric clouds, themselves possible indicators of global climate change [26]. Recently, the calculated cross sections of the OH-stretching overtone transitions have been supported by cavity ring down measurements [27]. Classical trajectory simulations of the photolysis reaction in Eq. (1) have also shown that this reaction is viable and occurs on a picosecond timescale [28], significantly faster than the collision rate in the stratosphere [29].

Theoretical and experimental studies of the hydration reaction of SO_3, the reverse of Eq. (1), have shown that this reaction occurs significantly faster in the presence of an additional H_2O molecule [24,30,31]. The dissociation barrier of the hydrated complex H_2SO_4–H_2O is significantly lower than the barrier of the analogous dissociation of H_2SO_4 following Eq. (1) [24]. Hence the role of the hydrated complex was suggested in the atmospheric photolysis of H_2SO_4 [22,32]. However, recent dynamics simulations have shown that the hydrated complex, H_2SO_4–H_2O, dissociates to H_2SO_4 and H_2O monomers upon excitation of a high OH-stretching overtone [28].

The possibility and importance of OH-stretching overtone induced photolysis mechanisms have been shown in both field and laboratory measurements for atmospheric radical production from, for example, hydrogen peroxide [33], nitric acid [34,35], peroxynitric acid [36,37] and hydroxymethyl hydroperoxide [38] with recent reviews providing a fuller discussion of this including additional references [39,40]. However, there has been no experimental verification of the overtone initiated dissociation reaction of H_2SO_4, and we rely on theoretical calculations. The reaction in Eq. (1) is difficult to observe in the flow cell environment because the products are the same as the reactants from which the H_2SO_4 is generated [10,19]. It has recently been suggested that the photolysis reaction of H_2SO_4 could be validated by proxy if the photolysis of the more stable fluorosulfonic acid (FSO_3H) was instead followed [41]. *Ab initio* calculations showed that FSO_3H dissociation, to form HF and SO_3, is energetically possible via excitation of the fourth OH-stretching overtone [41].

In the present paper, we show that it is possible to calculate both vibrational and electronic transitions of H_2SO_4 with an accuracy that is useful in atmospheric simulations. We calculate the absorption cross sections from the infrared to the vacuum UV region. In Section 2 we describe the vibrational local mode model used to calculate OH-stretching and SOH-bending vibrational transitions as well as their combinations and overtones [42–44]. This model provides frequencies and intensities of the dominant vibrational transitions from the infrared to the visible region. In Section 3 we present vertical excitation energies and oscillator strengths of the electronic transitions calculated with coupled cluster response theory. These coupled cluster calculations provide us with an accurate estimate of the lowest lying electronic transitions from which we can assess the accuracy of the previously given upper limits for the UV cross section. We also calculate higher energy electronic transitions to estimate the cross section in the Lyman-α region around 121.6 nm. The photon flux is very high in the narrow Lyman-α wavelength region and penetrates down into the mesopause. In Section 4 we illustrate what the effect the newly calculated cross sections in the Lyman-α region have, by modeling high altitude atmospheric SO_2 concentrations.

Our theoretical results in Sections 2 and 3 should provide a useful reference for absorption cross sections of H_2SO_4 in the region of interest to atmospheric modeling. The presented results show that it is possible to provide theoretical estimates that are of sufficient accuracy to alleviate the need to employ speculative cross sections in atmospheric simulations [10,12,13,18].

2. VIBRATIONAL TRANSITIONS

The vibrational spectroscopy of vapor phase H_2SO_4 has been studied in the infrared (IR) [45,46], near infrared (NIR) [19,47,48] and very recently visible region [27]. The fundamental vibrations were assigned in the early IR spectra, with additional combination and overtone transitions identified in the more recent IR/NIR spectra, which also provide absolute intensities of the dominant vibrational transitions. Harmonic frequencies and intensities calculated with the B3LYP/6-311++G(2d,2p) method were in reasonable agreement with the observed fundamental frequencies and intensities [19]. The measured relative intensities of the OH-stretching fundamental and first two overtones were reproduced well with a one dimensional (1D) anharmonic oscillator local mode model using experimental local mode parameters and a B3LYP/6-311++G(2d,2p) calculated dipole moment function (DMF) [19]. The results for the 1D local mode calculations were also in reasonable agreement with the more recent measurements of the third and fourth OH-stretching overtones [27]. High resolution (0.05 cm^{-1}) vapor phase spectra in the 1200–10000 cm^{-1} region, and comparison with $HDSO_4$ and D_2SO_4 spectra, facilitated the identification of additional weak combination bands [48]. In addition to the *ab initio* harmonic frequency calculations, a correlation consistent vibrational self-consistent field (VSCF) method was used to calculate anharmonic frequencies of fundamental, combination and first overtone transitions [49].

In the present paper we calculate frequencies and intensities of OH-stretching and SOH-bending vibrational transitions as well as their combinations and overtones—the dominant vibrational transitions from $\sim$1000 cm^{-1} to $\sim$20000 cm^{-1}. The vibrational calculation is based on the harmonically coupled anharmonic oscillator (HCAO) local mode model [42–44] combined with *ab initio* calculated dipole moment functions [50]. This local mode method has been successful in the calculation of OH- and CH-stretching overtone spectra [19,51,52]. The local mode parameters, frequency and anharmonicity, are obtained either from the observed experimental transitions or calculated *ab initio* [53–56].

2.1 Local mode model and computational details

We express absorption intensities in terms of oscillator strength, f, which for a transition from the ground state to a vibrational or electronic excited state is given by [50,57]

$$f_{\mathrm{eg}} = 4.702 \times 10^{-7}[\mathrm{cm\,D^{-2}}]\tilde{\nu}_{\mathrm{eg}}|\vec{\mu}_{\mathrm{eg}}|^2, \tag{2}$$

where $\tilde{\nu}_{\mathrm{eg}}$ is the frequency of the transition in wavenumbers and $\vec{\mu}_{\mathrm{eg}} = \langle e|\vec{\mu}|9\rangle$ is the transition dipole moment matrix element in Debye (D). We calculate the oscillator strength of electronic transitions using the electric dipole and Born-Oppenheimer approximations. Within these approximations, symmetry and spin forbidden transitions have zero intensity.

Our local mode model includes the OH-stretching and SOH-bending modes in the vibrational Hamiltonian [58]. The zeroth order Hamiltonian includes the two vibrational modes as uncoupled Morse oscillators according to

$$\left(H^0 - E^0_{|00\rangle}\right)/hc = v_{\mathrm{s}}\tilde{\omega}_{\mathrm{s}} - \left(v_{\mathrm{s}}^2 + v_{\mathrm{s}}\right)\tilde{\omega}_{\mathrm{s}}x_{\mathrm{s}} + v_{\mathrm{b}}\tilde{\omega}_{\mathrm{b}} - \left(v_{\mathrm{b}}^2 + v_{\mathrm{b}}\right)\tilde{\omega}_{\mathrm{b}}x_{\mathrm{b}}, \tag{3}$$

where $E^0_{|00\rangle}$ is the energy of the vibrational ground state, v_{s} and v_{b} are the vibrational quantum numbers of the OH-stretching and SOH-bending modes, respectively, and $\tilde{\omega}_i$ and $\tilde{\omega}_i x_i$ are the local mode frequencies and anharmonicities of the oscillators. We include a Fermi resonance type (1:2) stretch–bend coupling through the perturbation

$$H^1/hc = fr'\left(a_{\mathrm{s}}^+ a_{\mathrm{b}} a_{\mathrm{b}} + a_{\mathrm{s}} a_{\mathrm{b}}^+ a_{\mathrm{b}}^+\right), \tag{4}$$

where fr' is the effective stretch–bend coupling constant. The harmonic oscillator creation (a^+) and annihilation (a) operators are used as an approximation within the HCAO local mode model [50,59]. The indices on a and a^+ denote operation on either the stretching (s) or bending (b) coordinate. The fr' coupling constant contains both kinetic and potential energy coupling, and is expressed by [58,60,61]

$$fr' = \frac{1}{hc}\left\{\left(\frac{\partial G_{\mathrm{bb}}}{\partial q_{\mathrm{s}}}\right)_{\mathrm{e}}\left(\frac{-q_{\mathrm{cs}}p_{\mathrm{cb}}^2}{4\sqrt{2}}\right) + \left(\frac{\partial G_{\mathrm{sb}}}{\partial q_{\mathrm{b}}}\right)_{\mathrm{e}}\left(\frac{p_{\mathrm{cs}}q_{\mathrm{cb}}p_{\mathrm{cb}}}{2\sqrt{2}}\right) + F_{\mathrm{sbb}}\left(\frac{q_{\mathrm{cs}}q_{\mathrm{cb}}^2}{4\sqrt{2}}\right)\right\}, \tag{5}$$

where

$$q_{ci} = \hbar\sqrt{\frac{G_{ii}}{\tilde{\omega}_i hc}} \tag{6}$$

and

$$p_{ci} = \sqrt{\frac{\tilde{\omega}_i hc}{G_{ii}}} \tag{7}$$

are the classical turning points and corresponding momenta [62]. F_{sbb} is the mixed cubic derivative of the potential. The expressions for the G-matrix elements and their derivatives are given in Table 8.1.

In addition, based on the approximately 1:3 ratio of the OH-stretch to SOH-bend frequency [19] we also add a 1:3 stretch-bend resonance perturbation term to

TABLE 8.1 *G*-matrix elements, their derivatives and potential energy coupling terms for the S–O–H group in H_2SO_4[a]

Term	CCSD(T)
$G_{ss} = \frac{1}{m_H} + \frac{1}{m_O}$	$1.0548\ \text{amu}^{-1}$
$G_{bb} = \frac{1}{r_{SO}^2 m_S} + \frac{1}{r_{OH}^2 m_H} + \frac{1}{m_O}\left(\frac{1}{r_{SO}^2} + \frac{1}{r_{OH}^2} - \frac{2\cos\phi}{r_{SO} r_{OH}}\right)$	$1.1866\ \text{amu}^{-1}\ \text{Å}^{-2}$
$G_{sb} = \frac{-\sin\phi}{m_O r_{SO}}$	$-0.0374\ \text{amu}^{-1}\ \text{Å}^{-1}$
$\left(\frac{\partial G_{bb}}{\partial q_s}\right)_e = \frac{2\cos\phi}{m_O r_{SO} r_{OH}^2} - \frac{2}{m_H r_{OH}^3} - \frac{2}{m_O r_{OH}^3}$	$-2.347\ \text{amu}^{-1}\ \text{Å}^{-3}$
$\left(\frac{\partial G_{sb}}{\partial q_b}\right)_e = \frac{-\cos\phi}{m_O r_{SO}}$	$0.0123\ \text{amu}^{-1}\ \text{Å}^{-1}$
$\left(\frac{\partial G_{sb}}{\partial q_b^2}\right)_e = \frac{\sin\phi}{m_O r_{SO}}$	$0.0374\ \text{amu}^{-1}\ \text{Å}^{-1}$
$\left(\frac{\partial^2 G_{bb}}{\partial q_b \partial q_s}\right)_e = \frac{-2\sin\phi}{m_O r_{OH} r_{SO}^2}$	$-0.0797\ \text{amu}^{-1}\ \text{Å}^{-3}$
F_{sbb}	$-0.0816\ \text{au}\ \text{Å}^{-1}$
F_{sbbb}	$0.190\ \text{au}\ \text{Å}^{-1}$

a Calculated at the CCSD(T)/aug-cc-pV(T+d)Z optimized geometry. The angle dependence in F_{sbb} and F_{sbbb} is in radians.

the Hamiltonian:

$$H^2/hc = tr'\left(a_s^+ a_b a_b a_b + a_s a_b^+ a_b^+ a_b^+\right), \tag{8}$$

where tr' is the effective 1:3 stretch-bend coupling constant that can be expressed as:

$$tr' = \frac{1}{hc}\left\{\left(\frac{\partial^2 G_{bb}}{\partial q_s \partial q_b}\right)_e \left(\frac{-q_{cs} q_{cb} p_{cb}^2}{16}\right) - \left(\frac{\partial^2 G_{sb}}{\partial q_b^2}\right)_e \left(\frac{p_{cs} p_{cb} q_{cb}^2}{8}\right) + F_{sbbb}\left(\frac{q_{cs} q_{cb}^3}{24}\right)\right\}, \tag{9}$$

where F_{sbbb} is the mixed quartic derivative of the potential.

The local mode parameters, $\tilde{\omega}_i$ and $\tilde{\omega}_i x_i$, can be determined from a Birge–Sponer fit of the observed pure local mode transitions to

$$\frac{\tilde{\nu}_{v0}}{v} = (\tilde{\omega}_i - \tilde{\omega}_i x_i) - v\tilde{\omega}_i x_i, \tag{10}$$

where $\tilde{\nu}_{v0}$ is the wavenumber of the $v = 0$ to v transition. Alternatively, $\tilde{\omega}_i$ and $\tilde{\omega}_i x_i$ can be determined directly from *ab initio* calculated potential energy curves along the local mode coordinates. The local mode frequency is given by

$$\tilde{\omega}_i = \frac{\sqrt{G_{ii}F_{ii}}}{2\pi c}, \tag{11}$$

where F_{ii} is the second order derivative of the potential energy (force constant). The anharmonicity has in the past been obtained from [53,63]

$$\tilde{\omega}_i x_i = \frac{hG_{ii}}{72\pi^2 c}\left(\frac{F_{iii}^2}{F_{ii}}\right), \tag{12}$$

where F_{iii} is the third order derivative of the potential energy. The results obtained with use of Eqs. (11) and (12), are reasonable at lower *ab initio* levels, provided some scaling is applied [64]. This approach has been used successfully to predict OH-stretching and HOH-bending overtone vibrations of the water dimer [53,65]. However, an improved expression for the anharmonicity can be derived if the fourth (F_{iv}) order derivative of the potential energy is included [54,66]

$$\tilde{\omega}_i x_i = \frac{hG_{ii}}{64\pi^2 c F_{ii}}\left(\frac{5F_{iii}^2}{3F_{ii}} - F_{iv}\right). \tag{13}$$

We have recently found that use of Eq. (13) with force constants derived from an *ab initio* calculated potential energy curve at the coupled cluster including singles, doubles and perturbative triples (CCSD(T)) level of theory with an aug-cc-pVTZ basis set gives local mode frequencies and anharmonicities that are in good agreement with experimental observations, thereby alleviating the need for scaling [54,56].

Values of the F_{ii}, F_{iii}, and F_{iv} force constants are determined from an *ab initio* potential energy curve calculated as a function of displacements of the local mode coordinate from equilibrium. The potential energy curves extend from −0.2 Å to 0.2 Å in 0.05 Å steps for the stretching coordinate, and from −20° to 20° in 5° steps for the bending coordinate. These nine point grids are fitted with an eighth order polynomial to determine the diagonal force constants. The mixed force constants F_{sbb} and F_{sbbb} are found using standard numerical techniques as derivatives of the potential energy from a two dimensional (2D) 9×9 grid (81 points) calculated with the same displacements as the diagonal grids.

We approximate the dipole moment function by a series expansion in the internal OH-stretching and SOH-bending displacement coordinates about the calculated equilibrium geometry. The dipole moment function is therefore written as

$$\vec{\mu}(q_{\mathrm{s}}, q_{\mathrm{b}}) = \sum_{i,j} \vec{\mu}_{ij} q_{\mathrm{s}}^i q_{\mathrm{b}}^j, \tag{14}$$

where the coefficients $\vec{\mu}_{ij}$ are given by

$$\vec{\mu}_{ij} = \frac{1}{i!j!}\frac{\partial^{i+j}\vec{\mu}}{\partial q_{\mathrm{s}}^i \partial q_{\mathrm{b}}^j}\bigg|_{\mathrm{e}}. \tag{15}$$

The summation in Eq. (14) is limited to sixth-order diagonal terms, and fourth order mixed terms. The diagonal dipole moment derivatives are determined by fitting sixth order polynomials to 9 point grids of the dipole moment, similar to the energy grids. The off-diagonal terms are determined by standard numerical techniques from the 9×9 grid [53].

The equilibrium geometry and all grid points were obtained with the CCSD(T)/aug-cc-pV(T+d)Z method. The aug-cc-pV(T+d)Z basis set includes additional tight *d* functions on the sulfur atom. These additional basis functions have been shown to significantly improve the description of molecules containing second row elements [67–69]. The CCSD(T) geometry optimizations were performed with the default convergence criteria in Molpro 2002.6 [70]. The bond lengths and angles of the CCSD(T)/aug-cc-pV(T+d)Z optimized geometry are: $R_{OH} = 0.969$ Å, $R_{SO} = 1.588$ Å and $\theta_{SOH} = 108.3°$ [41], in close agreement with the experimental microwave values: $R_{OH} = 0.97(1)$ Å, $R_{SO} = 1.574(10)$ Å and $\theta_{SOH} = 108.5(15)°$ [71,72]. The dipole moments were found using a finite field method with a field strength of ± 0.025 au [70].

2.2 Vibrational cross sections

The CCSD(T)/aug-cc-pV(T+d)Z calculated values of the G-matrix and force constant parameters used in Eqs. (5) and (9) are given in Table 8.1. From these parameters the effective stretch-bend coupling constants are calculated to be: $fr' = 31.5$ cm^{-1} and $tr' = 1.0$ cm^{-1}. The coupling between the two OH-stretching modes is very small, as expected [73] and evident from the experimentally observed small splitting ($\sim$1 cm^{-1}) of the symmetric and asymmetric fundamental OH-stretching vibrations [48]. Hence an isolated OH-stretching local mode model is a reasonable approximation. The experimentally derived and CCSD(T) *ab initio* calculated local mode frequencies and anharmonicities of the OH-stretching and SOH-bending modes are compared in Table 8.2. For the bending mode only the fundamental and first overtone transitions have been observed. The two bending local modes are coupled more than the OH-stretching modes, leading to symmetric and asymmetric vibrations, observed at 1141 and 1157 cm^{-1} in the IR spectrum [19,45,50]. Harmonic frequency calculations also suggest that the bending modes couple

TABLE 8.2 Local mode parameters (in cm^{-1}) of sulfuric acid

	Expt.[a]	Calc.[b]
$\tilde{\omega}_s$	3769.3	3778.5
$\tilde{\omega}_s x_s$	79.75	79.71
$\tilde{\omega}_b$	1169.3	1187.6
$\tilde{\omega}_b x_b$	10.1	8.57

a Determined from Birge–Sponer fit to the $\Delta v_{OH} = 1$–3 transitions for the OH-stretching local mode and the $\Delta v_{SOH} = 1$–2 transitions for the SOH-bending mode.

b Calculated with the CCSD(T)/aug-cc-pV(T+d)Z *ab initio* method.

TABLE 8.3 Calculated 2D OH-stretching and SOH-bending vibrational frequencies (in cm^{-1}) and oscillator strengths in sulfuric acid

State	Calc.[a]		Calc.[b]		Expt.[c]	
	$\tilde{\nu}$	f	$\tilde{\nu}$	f	$\tilde{\nu}$	f
$\|0\rangle_s\|1\rangle_b$	1149.1	3.8×10^{-5}	1170.5	3.8×10^{-5}	1140 1157.1	1.8×10^{-5}
$\|0\rangle_s\|2\rangle_b$	2276.5	1.7×10^{-6}	2322.2	1.7×10^{-6}	2278	6.5×10^{-7}
$\|1\rangle_s\|0\rangle_b$	3611.3	3.7×10^{-5}	3620.7	3.7×10^{-5}	3609.6	3.4×10^{-5}
$\|1\rangle_s\|1\rangle_b$	4763.3	9.7×10^{-7}	4794.2	9.5×10^{-7}	4760.8	4.1×10^{-7}
$\|2\rangle_s\|0\rangle_b$	7063.7	1.4×10^{-6}	7079.2	8.0×10^{-7}	7060.6	1.2×10^{-6}
$\|1\rangle_s\|3\rangle_b$	7000.4	8.5×10^{-9}	7086.5	5.5×10^{-7}	–	–
$\|2\rangle_s\|1\rangle_b$	8219.4	1.7×10^{-8}	8260.3	1.7×10^{-8}	8163	3.4×10^{-8}
$\|3\rangle_s\|0\rangle_b$	10356.6	3.9×10^{-8}	10385.2	3.8×10^{-8}	10350.3	4.4×10^{-8}
$\|3\rangle_s\|1\rangle_b$	11514.5	7.1×10^{-10}	11566.5	7.5×10^{-10}	–	–
$\|4\rangle_s\|0\rangle_b$	13491.5	2.1×10^{-9}	13529.8	2.1×10^{-9}	13490	3.3×10^{-9}
$\|4\rangle_s\|1\rangle_b$	14657.5	5.4×10^{-11}	14718.1	5.3×10^{-11}	–	–
$\|5\rangle_s\|0\rangle_b$	16468.5	1.9×10^{-10}	16516.9	1.9×10^{-10}	16494	2.7×10^{-10}
$\|5\rangle_s\|1\rangle_b$	17643.5	5.2×10^{-12}	17714.9	5.0×10^{-12}	–	–
$\|6\rangle_s\|0\rangle_b$	19289.5	2.6×10^{-11}	19348.6	2.5×10^{-11}	–	–

a Calculated with the CCSD(T)/aug-cc-pV(T+d)Z dipole moment function and experimental local mode parameters.

b Calculated with the CCSD(T)/aug-cc-pV(T+d)Z dipole moment function and *ab initio* local mode parameters.

c Experimental values taken from Hintze *et al.* [19,48] and Feierabend *et al.* [27].

significantly to the SO-stretching modes [19]. In the first SOH-bending overtone region only one transition is observed. We use the midpoint of the symmetric and asymmetric fundamental transitions and the first overtone transition in the Birge–Sponer fit of the $\Delta v_{SOH} = 1$–2 transitions to obtain the local mode parameters for the SOH-bending mode. The experimental parameters for the OH-stretching mode are obtained from a Birge–Sponer fit of the $\Delta v_{OH} = 1$–3 transitions. We find that the *ab initio* calculated local mode parameters are in reasonable agreement with the experimentally derived local mode frequency and anharmonicity.

In Table 8.3 we present frequencies and intensities of the dominant OH-stretching and SOH-bending transitions up to 20 000 cm^{-1}. The calculated results are obtained using a Hamiltonian that includes the terms in Eqs. (3), (4) and (8) with either the experimental or *ab initio* calculated local mode parameters and a CCSD(T)/aug-cc-pV(T+d)Z calculated DMF. As expected the frequencies of transitions obtained with the *ab initio* local mode parameters are slightly higher than those obtained with the experimentally fitted parameters. The intensities of the two calculations are in general similar. The exception is near resonances such as between the $|2\rangle_s|0\rangle_b$ and $|1\rangle_s|3\rangle_b$ states. In the calculation with the *ab initio* local mode parameters, the coupling causes significant mixing of the two states and the intensity of the inherently weak $|1\rangle_s|3\rangle_b$ state increases dramatically. However, the

TABLE 8.4 Calculated 1D OH-stretching vibrational frequencies (in cm^{-1}) and oscillator strengths in sulfuric acid

State	Morse[a]		Numeric[b]		Expt.[c]	
	$\tilde{\nu}$	f	$\tilde{\nu}$	f	$\tilde{\nu}$	f
$\|1\rangle_S$	3619	3.74×10^{-5}	3621	3.72×10^{-5}	3609.6	3.4×10^{-5}
$\|2\rangle_S$	7079	1.32×10^{-6}	7086	1.44×10^{-6}	7060.6	1.2×10^{-6}
$\|3\rangle_S$	10379	3.71×10^{-8}	10398	5.19×10^{-8}	10350.3	4.4×10^{-8}
$\|4\rangle_S$	13520	2.03×10^{-9}	13559	3.75×10^{-9}	13490	3.3×10^{-9}
$\|5\rangle_S$	16501	1.87×10^{-10}	16571	4.27×10^{-10}	16494	2.7×10^{-10}
$\|6\rangle_S$	19323	2.59×10^{-11}	19434	5.56×10^{-11}	–	–

a Calculated with the CCSD(T)/aug-cc-pV(T+d)Z dipole moment function and *ab initio* local mode parameters from Table 8.2.
b Calculated with the CCSD(T)/aug-cc-pV(T+d)Z dipole moment function and the numeric potential.
c Experimental values taken from Hintze *et al.* [19,48] and Feierabend *et al.* [27].

combined intensity of the two transitions is very similar in the two calculations and is in good agreement with the experimental intensity of the band observed in this region. The agreement between the calculated and observed transitions is reasonably good, especially considering that no scaling is employed, that the frequencies span 15000 cm^{-1} and that the intensities range about five orders of magnitude. The OH-stretching frequencies calculated with the *ab initio* local mode parameters are within 40 cm^{-1} of the measured frequencies. Not surprisingly, the discrepancy in intensities is somewhat larger as they are more difficult to both measure and calculate. The calculated intensities of the lower overtone OH-stretching transitions are within 20% of the observed values, and for the $\Delta v = 4$ and 5 transitions the calculated intensities are within a factor of 2 of the experimental values. The increased discrepancy with increasing v is not surprising and suggests the need for a more suitable potential than the Morse potential for higher energy transitions [41,74]. For the bending states and the stretch-bend combination states the accuracy is less good, with calculated frequencies within 100 cm^{-1} and calculated intensities within about a factor of 2 of the experimental values.

In Table 8.4 we compare OH-stretching frequencies and intensities calculated using a 1D Morse oscillator model (with *ab initio* local mode parameters) with those obtained using wavefunctions and energies from a numeric solution of the 1D Schrödinger equation with an *ab initio* calculated potential [41,74,75]. For the numeric solution the CCSD(T) potential energy is calculated with the OH bond displaced from equilibrium in steps of 0.025 Å from −0.3 Å to +0.6 Å for a total of 37 points. For the fundamental and first overtone transitions the Morse, numeric and experimental frequencies and intensities are all in good agreement. For $\Delta v = 3$–5, the Morse and numeric intensities differ significantly, with the numeric higher and the Morse lower than the experimental intensities, however both sets of calculations are within a factor of 2 of the experimental intensities. We find that the oscillator strengths of the $\Delta v = 4$ and 5 OH-stretching overtone transitions calculated with a numerical potential are approximately twice those calculated with a Morse potential, in agreement with similar investigations on other molecular systems [41,74]. Comparison of the 1D results in Table 8.4 with the 2D results in Table 8.3, shows that the 1:2 and 1:3 coupling between the OH-stretching and HOH-bending mode has only a small effect on the dominant OH-stretching transitions. The coupling reduces the OH-stretching frequency by up to 30 cm^{-1} and has almost no effect on the intensities. The exception is the aforementioned resonance between the $|2\rangle_s|0\rangle_b$ and $|1\rangle_s|3\rangle_b$ states.

It is possible that some of the discrepancy between the calculated and experimental results presented is due to coupling to other modes [19]. Miller *et al.* have previously used the correlation consistent VSCF method to calculate vibrational frequencies including all vibrational modes and coupling between these [49]. Their approach provides good agreement for most fundamental vibrations [49]. However, for the OH-stretching vibration the local mode approach seems more appropriate. The correlation consistent VSCF calculation predicts the asymmetric and symmetric OH-stretching vibration of H_2SO_4 at 3500 and 3590 cm^{-1} with a splitting much larger than the observed splitting of less than 1 cm^{-1} [49]. A small splitting is expected for two OH-stretching bonds not sharing a common atom [73]. For the stretch–bend combination transitions the agreement with experimental results is similar between the local mode and correlation consistent VSCF methods.

In Figure 8.1 we present the CCSD(T) calculated and experimental cross section in the IR to visible region. Each calculated and observed vibrational transition has been convoluted with a Gaussian function with a half width at half maxi-

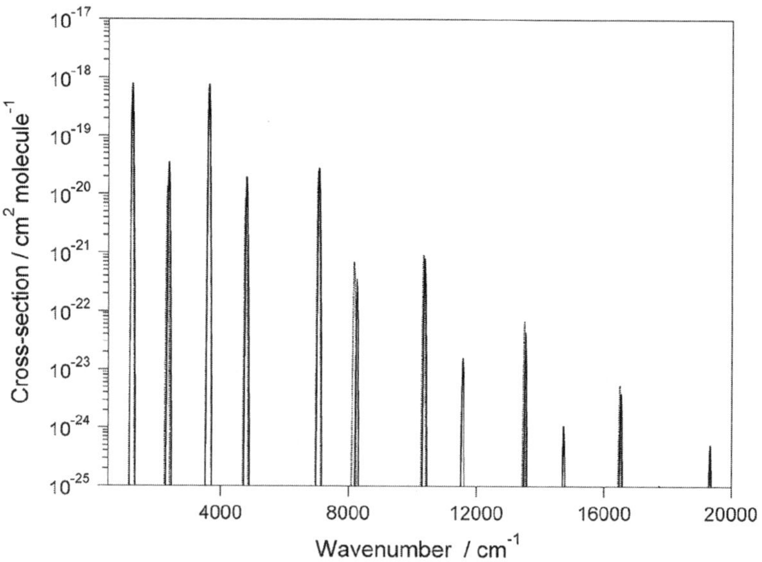

FIGURE 8.1 Absorption cross sections simulated from calculated (solid) and experimental (dotted) vibrational transitions from Table 8.3. Each vibrational transition has been convoluted with a Gaussian HWHM width of 20 cm^{-1}. The calculated transitions are obtained purely *ab initio*.

mum (HWHM) of 20 cm^{-1}. We have used the calculated results obtained with the *ab initio* local mode parameters to illustrate the agreement with experimental results that is possible from a purely calculative approach without any empirical scaling. Overall the agreement is reasonable with the largest uncertainty in frequency being 100 cm^{-1}. Such a difference is significant in overtone spectroscopy where accuracy better than 1 cm^{-1} is routinely obtained. However, in atmospheric modeling, the spectroscopic region is often separated into bins that typically have widths of ~150 cm^{-1} [23,25]. In addition the solar flux in the visible region is near constant. Thus an uncertainty in the frequency of the transitions of ~100 cm^{-1} will have little impact on the atmospheric modeling of OH-stretching photolysis.

3. ELECTRONIC TRANSITIONS

Electronic absorption transitions in H_2SO_4 have so far not been observed [10,19]. However, experimental efforts have led to the following upper limits on the absorption cross section up to 140 nm (8.6 eV): 10^{-21} cm^2 $molecule^{-1}$ in the region 330–195 nm; 10^{-19} cm^2 $molecule^{-1}$ in the region 195–160 nm and 10^{-18} cm^2 $molecule^{-1}$ in the region 160–140 nm. These upper limits are shown schematically as dotted lines in Figure 8.2. Previously, the lowest lying electronic transitions in H_2SO_4 have been calculated with the configuration interaction including singles (CIS) *ab initio* method and the lowest energy transition was found to occur

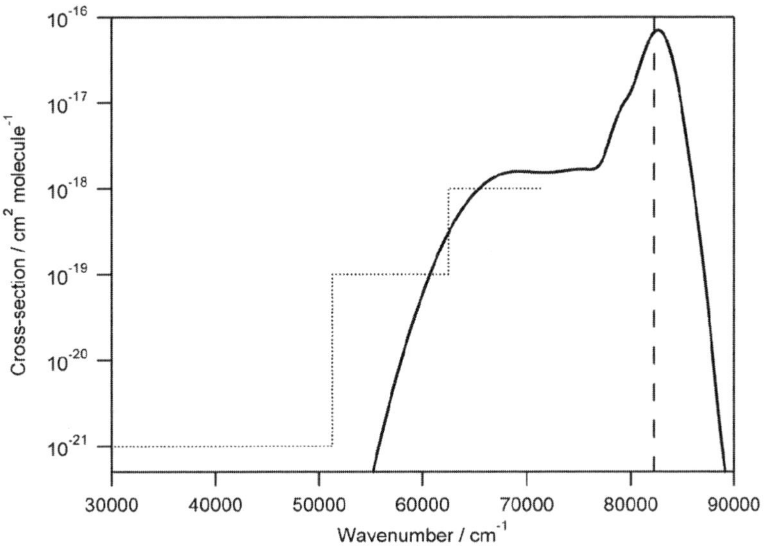

FIGURE 8.2 Cross sections simulated from electronic transitions. Each CCSD calculated electronic transition has been convoluted with a Gaussian with a HWHM of 3800 cm^{-1} for the transitions below 9.5 eV (76600 cm^{-1}) and 1200 cm^{-1} for the higher energy transitions. Experimental upper limits have been indicated with connected dotted lines. The Lyman-α energy is indicated by a vertical dashed line.

at 111 nm (11.2 eV) with the CIS/6-31G(d) method and at 122 nm (10.2 eV) with the CIS/6-311++G(d,p) method [19,20]. The CIS method is known to give electronic transition energies that are too high. The TDDFT method with the B3LYP functional and aug-cc-pVTZ basis set gives the lowest energy transition at 168 nm (7.4 eV). [21] More accurate, multi reference configuration interaction (MRCI) calculations with the aug-cc-pVTZ basis set found that the lowest energy transition occurs at 144 nm with an oscillator strength of about 0.03, in support of the experimental upper limits [21].

For reactions at high altitude the cross section in the wavelength region around Lyman-α radiation (121.6 nm, $\sim$10.2 eV) is important [23]. To our knowledge, no cross section measurement has been attempted for H_2SO_4 in the vacuum UV region around 121.6 nm. To determine the cross sections in this wavelength region it is necessary to calculate transitions to higher lying electronic states. Previously, we extended the MRCI calculations of Robinson *et al.* [21] to include additional transitions at higher energy and used these to determine a cross section in the Lyman-α region that could be used in atmospheric simulations [23,25]. This extension used the previous complete active space (CAS) which was formed from valence molecular orbitals and was not well suited to describe the higher energy electronic transitions in H_2SO_4. The electronic transitions in H_2SO_4 have been shown to have significant Rydberg character, that is, the electronic excited states are dominated by excitations to principally atomic orbitals [76].

Given that the electronic structure of H_2SO_4 is well described by a single Slater determinant and does not have significant multi reference character, an alternative approach is to use coupled cluster (CC) response functions to calculate the electronic transitions [76]. We will show in this section that CC calculations can provide calculated electronic transitions in H_2SO_4 and thus a cross section up to and including the Lyman-α region.

3.1 Computational details

We have calculated vertical excitation energies and oscillator strengths of H_2SO_4 using CC response functions. The vertical excitation energies (E) are given in eV (1 eV $=$ 8065.544 cm^{-1}). All electronic transitions were calculated at the experimentally determined geometry of H_2SO_4 [71].

We calculate electronic transitions with the coupled cluster singles (CCS), second order approximate coupled cluster singles and doubles (CC2), the coupled cluster singles and doubles (CCSD) and the third order approximate coupled cluster singles, doubles and triples (CC3) response functions [77,78], using the aug-cc-pV(D+d)Z and aug-cc-pV(T+d)Z basis sets. To ensure saturation with diffuse basis functions for some of the highly delocalized Rydberg excited states we have also constructed a series of molecule-centered primitive basis functions, originating from the center of mass. These basis functions were generated according to the procedure by Kaufmann *et al.* [79]. We have chosen a set of 3s3p3d functions with the "semi-quantum numbers" 2.0, 2.5 and 3.0. The aug-cc-pV(D+d)Z and aug-cc-pV(T+d)Z basis sets augmented with this 3s3p3d set are denoted aug-cc-pV(D+d)Z+3 and aug-cc-pV(T+d)Z+3, respectively. Basis sets of this type have

TABLE 8.5 CCS, CC2, CCSD and CC3 calculated vertical excitation energies (in eV) and oscillator strengths[a]

State	CCS		CC2		CCSD		CC3	
	E	*f*	*E*	*f*	*E*	*f*	*E*	*f*
2A	10.51	0.0051	8.18	0.0012	8.66	0.0018	8.49	0.0010
3A	10.92	0.0119	8.72	4×10^{-5}	9.21	0.0095	9.09	0.0081
1B	10.11	0.0159	7.74	0.0079	8.24	0.0094	8.08	0.0081
2B	10.22	0.0034	7.90	0.0018	8.35	0.0013	8.20	0.0016

a Calculated with the aug-cc-pV(D+d)Z basis set at the experimental geometry.

been previously shown to be well converged for the $n = 3$, 4 and 5 Rydberg states of s, p, d type in furan and pyrrole [80,81]. The twin hierarchy of coupled cluster response functions in combination with correlation consistent basis sets provides results that converge and hence provides an estimate of the accuracy, which is essential in the absence of experimental spectra. In all coupled cluster calculations the oxygen 1s, sulfur 1s, 2s, and 2p orbitals were assumed to be core. All coupled cluster response calculations were performed using a local version of Dalton 2.0 [82].

3.2 Electronic cross sections

In Table 8.5, we compare results for the 4 lowest energy excited states, calculated with the hierarchy of coupled cluster methods (CCS, CC2, CCSD, and CC3) and the aug-cc-pV(D+d)Z basis set. We find that vertical excitation energies calculated with the CCS method (identical to CIS energies) are overestimated by 1–2 eV and as such can be considered useful for qualitative interpretation only. For the four lowest energy excited states we find that the energy differences from CC2 to CCSD are about 0.5 eV and from CCSD to CC3 about 0.15 eV. The CC3 method has been benchmarked for smaller molecules and with a suitably large basis set it is able to predict vertical excitation energies to ~0.05 eV accuracy [78,83]. The difference in oscillator strengths between CC2, CCSD, and CC3 is less than a factor of two for each transition, except for the 3A state calculated with CC2.

In Table 8.6 we present the results obtained with the CCSD method and the aug-cc-pV(T+d)Z and aug-cc-pV(T+d)Z+3 basis sets for the first 12 excited states (the CC3 method is prohibitively expensive for this many states with this basis set). The aug-cc-pV(T+d)Z+3 basis set contains additional very diffuse basis functions necessary for the adequate description of Rydberg states. Comparison of results obtained with the aug-cc-pV(D+d)Z+3 basis set and those obtained with the much larger aug-cc-pV(D+d)+7 basis set showed that H_2SO_4 is saturated with diffuse basis functions with the aug-cc-pV(D+d)Z+3 basis set [76]. For the 4 lowest energy excited states shown in Table 8.6, we see that the difference in energy between the aug-cc-pV(T+d)Z and aug-cc-pV(T+d)Z+3 basis sets with the CCSD

TABLE 8.6 CCSD calculated vertical excitation energies (in eV) and oscillator strengths[a]

State	aug-cc-pV(T+d)Z		aug-cc-pV(T+d)Z+3	
	E	*f*	*E*	*f*
2A	8.83	0.0016	8.83	0.0017
3A	9.41	0.0074	9.40	0.0061
4A	9.46	0.0069	9.44	0.0077
5A	10.16	0.0744	10.12	0.0445
6A	10.31	0.0195	10.21	0.0476
7A	10.57	0.0051	10.38	0.0003
1B	8.43	0.0103	8.42	0.0104
2B	8.57	0.0019	8.56	0.0019
3B	9.87	0.0230	9.86	0.0214
4B	10.24	0.0648	10.10	0.0048
5B	10.35	0.0712	10.27	0.1241
6B	10.73	0.0042	10.50	0.0039

a Calculated at the experimental geometry.

method is small. However for some of the higher energy states, e.g. 4B and 6B, the effect of the extra diffuse basis function is significant. Investigations of the second moments of electronic charge distribution also show that these two states are significantly more diffuse than, for example, the 5B state [76]. A recent comparison of CCSD vertical excitation energies for H_2SO_4 obtained with the aug-cc-pV(X+d)Z basis sets found that energies of all the excited states increased about 0.2 eV from X = D to X = T and about 0.05 eV from X = T to X = Q [76]. This basis set dependence is indicative of significant Rydberg character in all electronic transitions for H_2SO_4. We estimate that the CCSD/aug-cc-pV(T+d)Z+3 calculated transition energies are converged to within ~0.15 eV (~2 nm) and intensities to within a factor of 2. The lowest energy CCSD/aug-cc-pV(T+d)Z+3 calculated transition occurs at 8.42 eV (147 nm) with an oscillator strength of 0.01. This result is similar to the previous MRCI/aug-cc-pVTZ result of 8.61 eV (144 nm) with an oscillator strength of 0.03 [21].

We estimate the cross section of H_2SO_4 by convoluting each of the CCSD/aug-cc-pV(T+d)+3 calculated transitions from Table 8.6 with a Gaussian band shape and illustrate this graphically in Figure 8.2. We use a HWHM of 3800 cm^{-1} for the lowest energy transitions below 9.5 eV (2A–4A and 1B–2B) and 1200 cm^{-1} for the higher energy transitions (5A–7A and 3B–6B), respectively. The 3800 cm^{-1} HWHM width is similar to the width observed in the SO_3 electronic spectrum around 165 nm [11,19] and the 1200 cm^{-1} HWHM is approximately the observed width of electronic transitions in the spectrum of SO_2 around 120 nm [84].

As seen in Figure 8.2, our simulated cross section supports the upper limits established by previous experimental investigations up to 140 nm [10,19]. It is worth noting that the cross section in the region up to 55900 cm^{-1} (179 nm) is likely significantly less than the 10^{-21} cm^2 $molecule^{-1}$ upper limit of the cross section previously used (226–179 nm) to estimate the *J*-value of UV driven photolysis [22] and in the atmospheric simulations [23,25].

We simulate a relatively large cross section of 6.3×10^{-17} cm^2 $molecule^{-1}$ around 83000 cm^{-1}, which coincides with the window of Lyman-α radiation. The dominant transitions in the Lyman-α region are 5A/6A and 4B/5B. The variation in vertical transition energy is small, of the order of 0.1 eV, between the aug-cc-pV(T+d)Z and aug-cc-pV(T+d)Z+3 calculations (Table 8.6), however the difference in intensities is more substantial, likely due to changes in state mixing. Despite this, the total intensity envelope in this region does not change substantially between the aug-cc-pV(T+d)Z and aug-cc-pV(T+d)Z+3 basis sets. If we simulate the cross section at Lyman-α we find very similar results using either the aug-cc-pV(T+d)Z and aug-cc-pV(T+d)Z+3 results.

We have estimated the sensitivity of the Lyman-α cross section to the band width used, by evaluating the cross section using half and double the assigned HWHM of the higher energy transitions. The low energy transitions have little effect on the cross section in the Lyman-α region. If we use a HWHM of 600 cm^{-1} we estimate the Lyman-α cross section to be 8.6×10^{-17} cm^2 $molecule^{-1}$ whereas a 2400 cm^{-1} HWHM leads to a 3.9×10^{-17} cm^2 $molecule^{-1}$ cross section. Irrespective of which width is used, it is clear that the cross section in the Lyman-α region is significantly larger than the original speculative estimate based on the HCl absorption spectrum [10,12,13]. We have also estimated the sensitivity to the uncertainty of our calculated vertical excitation energies and oscillator strengths of the electronic transitions. A shift of ± 0.1 eV in vertical excitation energy of the electronic transitions in the Lyman-α region lowers the cross section at Lyman-α by up to a third. A doubling or halving of the calculated oscillator strengths produce a cross section at Lyman-α in the range from $\sim 2 \times 10^{-16}$ to $\sim 2 \times 10^{-17}$ cm^2 $molecule^{-1}$, clearly the most sensitive parameter.

4. ATMOSPHERIC SIMULATIONS

If we compare the simulated cross sections in Figures 8.1 and 8.2, we find that at energies higher than required for photolysis via Eq. (1) ($\sim$14000 cm^{-1}) the cross section from vibrational transitions is very small. Similarly, in the region where there are photons in the stratosphere, up to $\sim$55900 cm^{-1}, the cross section from electronic transitions is also very low. Both these regions are likely to contribute to the atmospheric photolysis of H_2SO_4, the region that dominates will depend on the actual cross sections, the altitude (photon flux) and the dynamics of the dissociation reactions.

We have investigated the atmospheric implications of our newly calculated absorption cross sections with the Garcia–Solomon 2D dynamical/chemical model [85,86], to which we have added sulfur chemistry and aerosol microphysics [87]. The model spans 56 pressure levels from $\sim$2 to $\sim$112 km above sea level, and 36 latitudes from 89.5°S to 89.5°N. Further details of the sulfur chemistry and

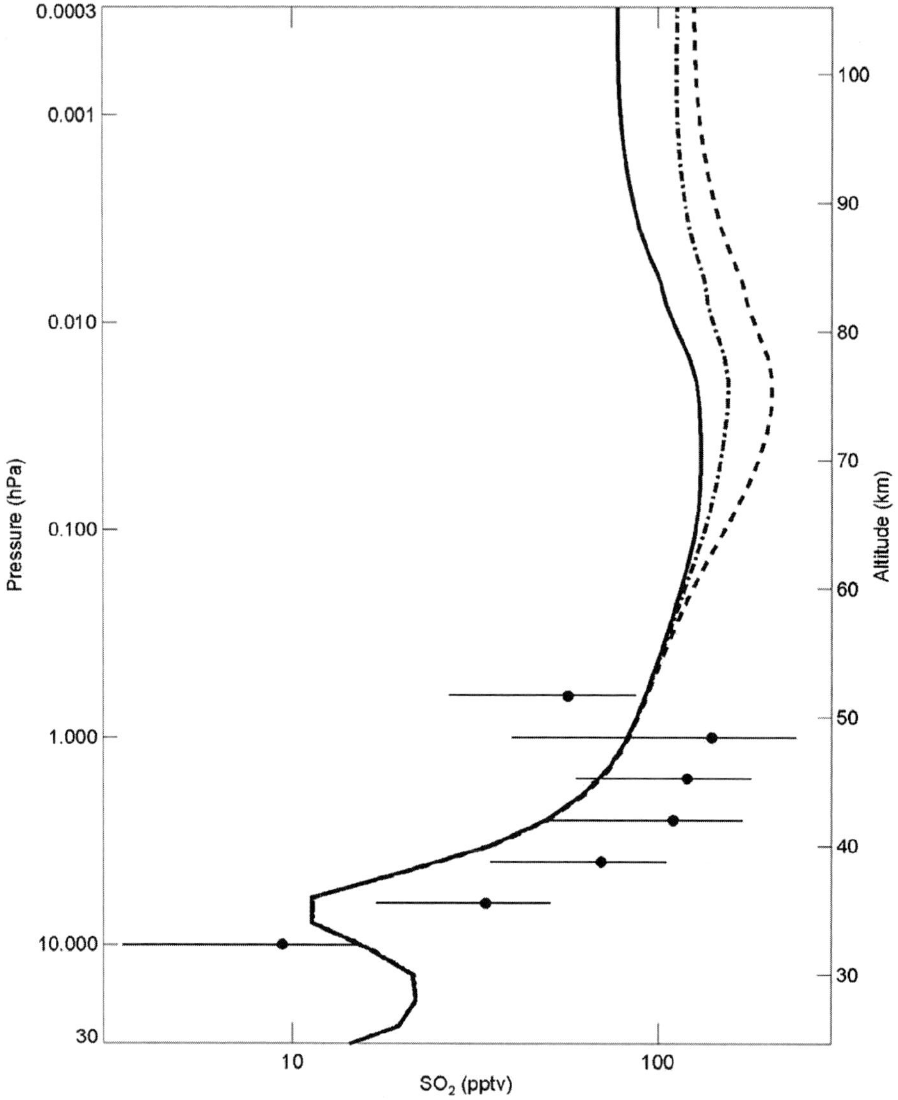

FIGURE 8.3 Calculated and observed SO_2 mixing ratios for the April–May period between 26°N to 32°N latitude. 2D model calculations incorporate our calculated vibrational transitions in the visible region and three values for the Lyman-α cross section: no Lyman-α cross section (solid), the Lyman-α cross section of 6.4×10^{-18} cm^2 molecule^{-1} (dot dash) and Lyman-α cross section of 6.3×10^{-17} cm^2 molecule^{-1} (dashed). Observations from Spacelab 3 (circles with horizontal lines) from 1985 [12].

stratospheric aerosol microphysics employed can be found in the previous paper: [23] here we simulate the atmospheric SO_2 mixing ratios.

The previously calculated vibrational transitions for H_2SO_4 are similar to the ones we calculate here and as expected there is very little change in the simulated

SO_2 mixing ratios due to this slight change in calculated vibrational cross sections in the visible region [23]. The change in the calculated electronic transition around the Lyman-α region is more significant and we discuss its effect on the simulated SO_2 mixing ratios. Lyman-α radiation is effective at dissociating H_2O, CO_2, CH_4, and HCl, among other gases, in the upper part of the middle atmosphere due to the high intensity of the solar Lyman-α line, its position in an atmospheric window, and the high energy of the photons. Strong atmospheric absorption attenuates this high-energy radiation at lower altitudes. The Lyman-α region is very narrow ($\sim$10.2 $\pm$ 0.02 eV) and the cross section is dependent on the calculated transition energy and oscillator strength of the electronic transitions as well as the width and shape of the associated bands.

Due to the lack of measured and calculated H_2SO_4 cross sections, the early atmospheric modeling of H_2SO_4 photodissociation used a cross section taken from the HCl spectrum, with a cross section of 2.0×10^{-18} cm^2 molecule^{-1} in the Lyman-α region [10,12,13]. More recently a cross section of 6.4×10^{-18} cm^2 molecule^{-1}, based on MRCI transitions, was used in the Lyman-α region [23,25]. Here we use the CCSD/aug-cc-pV(T+d)+3 calculated transitions convoluted with Gaussian band shapes as described in the previous section and obtain a significantly larger cross section of 6.3×10^{-17} cm^2 molecule^{-1} in the Lyman-α region. Based on our uncertainty estimates in the previous section this value of the cross section seems reasonable.

In Figure 8.3 we show the average calculated SO_2 mixing ratios for the April–May period in the region from 26°N to 32°N latitude. Our new calculations for Lyman-α produce a significant increase in calculated SO_2 concentration above 60 km. The greatest increase in SO_2 concentration is calculated near 75 km, where SO_2 mixing ratios for our CCSD Lyman-α cross section are more than 30% greater than those obtained with a cross section of 6.4×10^{-18} cm^2 molecule^{-1}, used in the recent simulation [23]. Compared to calculations neglecting Lyman-α absorption, the SO_2 concentration increases by about 60% at altitudes above 75 km. If we increase or decrease the cross section at Lyman-α by 30%, the SO_2 concentration changes by less than 7%. As seen previously our calculations are in good agreement with the observations in the upper stratosphere from the same period in 1985 and the same latitude region [12].

5. CONCLUSION

We have calculated the absorption cross sections of sulfuric acid from the infrared to vacuum ultraviolet region ($\sim$1000 to 85000 cm^{-1}) including vibrational and electronic transitions. We have calculated vibrational transitions with a local mode model that includes OH-stretching and SOH-bending vibrations, described by Morse oscillators. We find that calculated transitions obtained using a high level CCSD(T)/aug-cc-pV(T+d)Z potential energy surface and dipole moment function are in good agreement with experiment. The calculated intensity of the fundamental and lower overtone OH-stretching transitions are in good agreement with experiment. For the higher OH-stretching overtones, the use of a simple Morse potential becomes less appropriate, and we find that the intensity of the $\Delta v = 4$ and 5 OH-stretching overtone transitions calculated with a numerical potential are approximately twice that of those calculated with a Morse potential.

We have calculated electronic transitions with a range of coupled cluster response functions and correlation consistent basis sets. This twin hierarchy provides results that converge and hence provide a good estimate of the accuracy of the calculations, which is essential in the absence of experimental spectra. We estimate that our vertical excitation energies have an accuracy of 0.15 eV and our oscillator strengths are reliable within a factor of two. With the CCSD/aug-cc-pV(T+d)Z+3 method we calculate the lowest energy transition to occur at 8.42 eV with an oscillator strength of 0.01. This corroborates the earlier proposal that photolysis of sulfuric acid in the stratosphere is dominated by the vibrational overtone initiated photolysis in the visible region. Our calculations also included electronic excited states of sufficiently high energy to describe absorption in the region of Lyman-α radiation. The CCSD/aug-cc-pV(T+d)Z+3 method gives an estimate of the cross section at the Lyman-α region of $\sim 6 \times 10^{-17}$ cm^2 molecule^{-1}. This cross section is a factor 30 larger than the speculative cross section used in the early atmospheric sulfur simulations. Not surprisingly the effect on the calculated SO_2 mixing ratio is significant at high altitudes where the Lyman-α radiation is appreciable.

We have shown that modern theoretical methods are capable of calculating cross sections for sulfuric acid purely *ab initio*. This includes vibrational transitions, low lying electronic transitions and transitions to high energy Rydberg states. The results obtained are of an accuracy that is suitable as input parameters for atmospheric modeling of the photolysis of sulfuric acid.

ACKNOWLEDGEMENTS

We would like to thank Veronica Vaida, R. Benny Gerber, Yifat Miller, Daryl L. Howard and Poul Jørgensen for helpful discussions. We thank Jeppe Olsen for the use of his onedim program and Rolando Garcia and Susan Solomon for the use of the Garcia–Solomon model. T.W.R, D.P.S and J.R.L are grateful to the Foundation for Research, Science and Technology for Bright Futures scholarships. The Marsden Fund administrated by the Royal Society of New Zealand, the Lundbeck Foundation, and the Research Foundation at Aarhus University have provided funding for this research.

REFERENCES

[1] H. Berresheim, P.H. Wine, D.D. Davis, *et al.*, in: H.B. Singh (Ed.), *Composition, Chemistry, and Climate of the Atmosphere*, Van Nostrand Reinhold, New York, 1995, p. 251.
[2] R.J. Charlson, J.E. Lovelock, M.O. Andreae, S.G. Warren, *Nature* **326** (1987) 655.
[3] P. Warneck, *Chemistry of the Natural Atmosphere*, 2nd ed., Academic Press, San Diego, 2000.
[4] B.J. Finlayson-Pitts, J.N. Pitts Jr., *Chemistry of the Upper and Lower Atmosphere*, Academic Press, San Diego, 2000.

[5] M.P. McCormick, L.W. Thomason, C.R. Trepte, *Nature* **373** (1995) 399.
[6] P. Minnis, E.F. Harrison, L.L. Stowe, G.G. Gibson, F.M. Denn, D.R. Doelling, W.L. Smith Jr., *Science* **259** (1993) 1411.
[7] H. Stommel, E. Stommel, *Volcano Weather: The Story of 1816, the Year Without a Summer*, Seven Seas Press, Newport, 1983.
[8] S. Solomon, R.W. Portmann, R.R. Garcia, L.W. Thomason, L.R. Poole, M.P. McCormick, *J. Geophys. Res.* **101** (1996) 6713.
[9] R.W. Portmann, S. Solomon, R.R. Garcia, L.W. Thomason, L.R. Poole, M.P. McCormick, *J. Geophys. Res.* **101** (1996) 22.
[10] J.B. Burkholder, M. Mills, S. McKeen, *Geophys. Res. Lett.* **27** (2000) 2493.
[11] J.B. Burkholder, S. McKeen, *Geophys. Res. Lett.* **24** (1997) 3201.
[12] C.P. Rinsland, M.R. Gunson, M.K.W. Ko, D.W. Weisensten, R. Zander, M.C. Abrams, A. Goldman, N.D. Sze, G.K. Yue, *Geophys. Res. Lett.* **22** (1995) 1109.
[13] M.J. Mills, O.B. Toon, S. Solomon, *Geophys. Res. Lett.* **26** (1999) 1133.
[14] J.M. Rosen, D.J. Hofmann, *J. Geophys. Res.* **88** (1983) 3725.
[15] D.J. Hofmann, J.M. Rosen, *Geophys. Res. Lett.* **12** (1985) 13.
[16] J.X. Zhao, O.B. Toon, R.P. Turco, *J. Geophys. Res.* **100** (1995) 5215.
[17] S. Bekki, J.A. Pyle, *J. Geophys. Res.* **99** (1994) 18861.
[18] R.P. Turco, P. Hamill, O.B. Toon, R.C. Whitten, C.S. Kiang, *J. Atmos. Sci.* **36** (1979) 699.
[19] P.E. Hintze, H.G. Kjaergaard, V. Vaida, J.B. Burkholder, *J. Phys. Chem. A* **107** (2003) 1112.
[20] S.J. Wrenn, L.J. Butler, G.A. Rowland, C.J.H. Knox, L.F. Phillips, *J. Photochem. Photobiol. A* **129** (1999) 101.
[21] T.W. Robinson, D.P. Schofield, H.G. Kjaergaard, *J. Chem. Phys.* **118** (2003) 7226.
[22] V. Vaida, H.G. Kjaergaard, P.E. Hintze, D.J. Donaldson, *Science* **299** (2003) 1566.
[23] M.J. Mills, O.B. Toon, V. Vaida, P.E. Hintze, H.G. Kjaergaard, D.P. Schofield, T.W. Robinson, *J. Geophys. Res.* **110** (2005) D08201.
[24] K. Morokuma, C. Muguruma, *J. Am. Chem. Soc.* **116** (1994) 10316.
[25] M.J. Mills, O.B. Toon, G.E. Thomas, *J. Geophys. Res.* **110** (2005) D24208.
[26] G.E. Thomas, J. Olivero, *Adv. Space Res.* **28** (2001) 937.
[27] K.J. Feierabend, D.K. Havey, S.S. Brown, V. Vaida, *Chem. Phys. Lett.* **420** (2006) 438.
[28] Y. Miller, R.B. Gerber, *J. Am. Chem. Soc.* **128** (2006) 9594.
[29] Y. Miller, R.B. Gerber, V. Vaida, *Geophys. Res. Lett.* **34** (2007) L16820.
[30] C.E. Kolb, J.T. Jayne, D.R. Worsnop, M.J. Molina, R.F. Meads, A.A. Viggiano, *J. Am. Chem. Soc.* **116** (1994) 10314.
[31] E.R. Lovejoy, D.R. Hanson, L.G. Huey, *J. Phys. Chem.* **100** (1996) 19911.
[32] V. Vaida, H.G. Kjaergaard, K.J. Feierabend, *Int. Rev. Phys. Chem.* **22** (2003) 203.
[33] T.R. Rizzo, C.C. Hayden, F.F. Crim, *Faraday Discuss. Chem. Soc.* **75** (1983) 223.
[34] A. Sinha, R.L. Vander Wal, F.F. Crim, *J. Chem. Phys.* **92** (1990) 401.
[35] D.J. Donaldson, G.J. Frost, K.H. Rosenlof, A.F. Tuck, V. Vaida, *Geophys. Res. Lett.* **24** (1997) 2651.
[36] C.M. Roehl, S.A. Nizkorodov, H. Zhang, G.A. Blake, P.O. Wennberg, *J. Phys. Chem. A* **106** (2002) 3766.
[37] P.O. Wennberg, R.J. Salawitch, D.J. Donaldson, T.F. Hanisco, E.J. Lanzendorf, K.K. Perkins, S.A. Lloyd, V. Vaida, R.S. Gao, E.J. Hintsa, R.C. Cohen, W.H. Swartz, T.L. Kusterer, D.E. Anderson, *Geophys. Res. Lett.* **26** (1999) 1373.
[38] J.L. Fry, J. Matthews, J.R. Lane, C.M. Roehl, A. Sinha, H.G. Kjaergaard, P.O. Wennberg, *J. Phys. Chem. A* **110** (2006) 7072.
[39] D.J. Donaldson, A.F. Tuck, V. Vaida, *Chem. Rev.* **103** (2003) 4717.
[40] V. Vaida, *Int. J. Photoenergy* **7** (2005) 61.
[41] J.R. Lane, H.G. Kjaergaard, *J. Phys. Chem. A* **111** (2007) 9707.
[42] B.R. Henry, *Acc. Chem. Res.* **20** (1987) 429.
[43] L. Halonen, *Adv. Chem. Phys.* **104** (1998) 41.
[44] B.R. Henry, H.G. Kjaergaard, *Can. J. Chem.* **80** (2002) 1635.
[45] K. Stopperka, F. Kilz, *Z. Anorg. Allg. Chem.* **370** (1969) 49.
[46] S.M. Chackalackal, F.E. Stafford, *J. Am. Chem. Soc.* **88** (1966) 723.
[47] D.K. Havey, K.J. Feierabend, V. Vaida, *J. Mol. Struct. (THEOCHEM)* **680** (2004) 243.

[48] P.E. Hintze, K.J. Feierabend, D.K. Harvey, V. Vaida, *Spectrochim. Acta A* **61** (2005) 559.
[49] Y. Miller, G.M. Chaban, R.B. Gerber, *J. Phys. Chem. A* **109** (2005) 6565.
[50] H.G. Kjaergaard, H. Yu, B.J. Schattka, B.R. Henry, A.W. Tarr, *J. Chem. Phys.* **93** (1990) 6239.
[51] H.G. Kjaergaard, B.R. Henry, *J. Chem. Phys.* **96** (1992) 4841.
[52] H.G. Kjaergaard, *J. Phys. Chem. A* **106** (2002) 2979.
[53] G.R. Low, H.G. Kjaergaard, *J. Chem. Phys.* **110** (1999) 9104.
[54] D.L. Howard, P. Jørgensen, H.G. Kjaergaard, *J. Am. Chem. Soc.* **127** (2005) 17096.
[55] T. Helgaker, T.A. Ruden, P. Jørgensen, J. Olsen, W. Klopper, *J. Phys. Org. Chem.* **17** (2004) 913.
[56] D.P. Schofield, Ph.D. thesis, University of Otago, 2005.
[57] P.W. Atkins, R.S. Friedman, *Molecular Quantum Mechanics*, 3rd ed., Oxford University Press, Oxford, 1997.
[58] D.P. Schofield, H.G. Kjaergaard, J. Matthews, A. Sinha, *J. Chem. Phys.* **123** (2005) 134318.
[59] O.S. Mortensen, B.R. Henry, M.A. Mohammadi, *J. Chem. Phys.* **75** (1981) 4800.
[60] L. Halonen, T. Carrington Jr., *J. Chem. Phys.* **88** (1998) 4171.
[61] H.G. Kjaergaard, B.R. Henry, H. Wei, S. Lefebvre, T. Carrington Jr., O.S. Mortensen, M.L. Sage, *J. Chem. Phys.* **100** (1994) 6228.
[62] A. Messiah, *Quantum Mechanics*, Wiley, New York, 1961.
[63] M.G. Sowa, B.R. Henry, Y. Mizugai, *J. Phys. Chem.* **95** (1991) 7659.
[64] G.R. Low, Ph.D. thesis, University of Otago, 2002.
[65] D.P. Schofield, H.G. Kjaergaard, *Phys. Chem. Chem. Phys.* **5** (2003) 3100.
[66] G. Herzberg, *Molecular Spectra and Molecular Structure I. Spectra of Diatomic Molecules*, D. Van Nostrand Company, Inc., Princeton, NJ, 1950.
[67] T.H. Dunning Jr., K.A. Peterson, A.K. Wilson, *J. Chem. Phys.* **114** (2001) 9244.
[68] A.K. Wilson, T.H. Dunning Jr., *J. Chem. Phys.* **119** (2003) 11712.
[69] R.D. Bell, A.K. Wilson, *Chem. Phys. Lett.* **394** (2004) 105.
[70] R.D. Amos, A. Bernhardsson, A. Berning, P. Celani, D.L. Cooper, M.J.O. Deegan, A.J. Dobbyn, F. Eckert, C. Hampel, G. Hetzer, P.J. Knowles, T. Korona, R. Lindh, A.W. Lloyd, S.J. McNicholas, F.R. Manby, W. Meyer, M.E. Mura, A. Nicklass, P. Palmieri, R. Pitzer, G. Rauhut, M. Schütz, U. Schumann, H. Stoll, A.J. Stone, R. Tarroni, T. Thorsteinsson, H.-J. Werner, MOLPRO, A package of *ab initio* programs designed by H.-J. Werner and P.J. Knowles, 2002.6 ed. (2002).
[71] R.L. Kuczkowski, R.D. Suenram, F.J. Lovas, *J. Am. Chem. Soc.* **103** (1981) 2561.
[72] J. Demaison, M. Herman, J. Liévin, H.D. Rudolph, *J. Phys. Chem. A* **111** (2007) 2602.
[73] H.G. Kjaergaard, J.D. Goddard, B.R. Henry, *J. Chem. Phys.* **95** (1991) 5556.
[74] D.P. Schofield, J.R. Lane, H.G. Kjaergaard, *J. Phys. Chem. A* **111** (2007) 567.
[75] J. Olsen, ONEDIM program, private communication.
[76] J.R. Lane, H.G. Kjaergaard, in preparation.
[77] O. Christiansen, A. Halkier, H. Koch, P. Jørgensen, *J. Chem. Phys.* **108** (1998) 2801.
[78] O. Christiansen, H. Koch, P. Jørgensen, J. Olsen, *Chem. Phys. Lett.* **256** (1996) 185.
[79] K. Kaufmann, W. Baumeister, M. Jungen, *J. Phys. B At. Mol. Opt. Phys.* **22** (1989) 2223.
[80] O. Christiansen, P. Jørgensen, *J. Am. Chem. Soc.* **120** (1998) 3423.
[81] O. Christiansen, J. Gauss, J.F. Stanton, P. Jørgensen, *J. Chem. Phys.* **111** (1999) 525.
[82] C. Angeli, K.L. Bak, V. Bakken, O. Christiansen, R. Cimiraglia, S. Coriani, P. Dahle, E.K. Dalskov, T. Enevoldsen, B. Fernandez, C. Hättig, K. Hald, A. Halkier, H. Heiberg, T. Helgaker, H. Hettema, H.J.A. Jensen, D. Jonsson, P. Jørgensen, S. Kirpekar, W. Klopper, R. Kobayashi, H. Koch, A. Ligabne, O.B. Lutnæs, K.V. Mikkelsen, P. Norman, J. Olsen, M.J. Packer, T.B. Pedersen, Z. Rinkevicius, E. Rudberg, T.A. Ruden, K. Ruud, P. Salek, A. Sanchez de Meras, T. Saue, S.P.A. Sauer, B. Schimmelpfennig, K.O. Sylvester-Hvid, P.R. Taylor, O. Vahtras, D.J. Wilson, H. Ågren, *DALTON, A Molecular Electronic Structure Program*, Release 2.0 (2005), see http://www.kjemi.uio.no/software/dalton/dalton.html, 2005.
[83] H. Larsen, K. Hald, J. Olsen, P. Jørgensen, *J. Chem. Phys.* **115** (2001) 3015.
[84] R. Feng, Y. Sakai, Y. Zheng, G. Cooper, C.E. Brion, *Chem. Phys.* **260** (2000) 29.
[85] R.R. Garcia, S. Solomon, *J. Geophys. Res.* **99** (1994) 12937.
[86] R.R. Garcia, F. Stordal, S. Solomon, J.T. Kiehl, *J. Geophys. Res.* **97** (1992) 12967.
[87] M.J. Mills, Ph.D. thesis, University of Colorado, 1996.

CHAPTER 9

Computational Studies of the Thermochemistry of the Atmospheric Iodine Reservoirs HOI and $IONO_2$

Paul Marshall*

Contents

1. Introduction 160
2. Methodology and Results 161
3. Discussion 165
   3.1 Possible uncertainties 165
   3.2 Thermochemistry of HOI 167
   3.3 Thermochemistry of $IONO_2$ 168
   3.4 Comparison with halogen analogs 170
   3.5 $IONO_2$ kinetics 171
4. Conclusions 173
Acknowledgements 173
References 174

Abstract

CCSD(T) theory with aug-cc-pVTZ-PP and aug-cc-pVQZ-PP basis sets has been applied to hypoiodous acid and iodine nitrate, two potential reservoirs for atmospheric iodine. The results are employed in bond-conserving reactions and extrapolated to the complete basis set limit, to yield $\Delta_f H_{298}(\mathrm{HOI}) = -59.2 \pm 3.3$ kJ mol^{-1} and $\Delta_f H_{298}(\mathrm{IONO_2}) = 37.4 \pm 3.9$ kJ mol^{-1}. For iodine nitrate the bond dissociation enthalpies $DH_0(\mathrm{IO{-}NO_2}) = 113.6 \pm 3.1$ kJ mol^{-1} and $DH_0(\mathrm{I{-}NO_3}) = 141.6 \pm 3.9$ kJ mol^{-1} are derived. $DH_0(\mathrm{IO{-}NO_2})$ is used in Troe's unimolecular formalism to yield the 298 K low-pressure limiting rate constant for $IO + NO_2$ addition as $(5.3\text{–}13.3) \times 10^{-31}$ cm^6 molecule^{-1} s^{-1} for N_2 bath gas, depending on the approach taken to define the rotational term F_{rot}. This

* Center for Advanced Scientific Computing and Modeling, Department of Chemistry, University of North Texas, P.O. Box 305070, Denton, Texas 76203-5070, USA
E-mail: marshall@unt.edu

Advances in Quantum Chemistry, Vol. 55
ISSN 0065-3276, DOI: 10.1016/S0065-3276(07)00209-2

range is in good accord with measured values. At 1 atm N_2 and 298 K, the lifetime for $IONO_2$ with respect to thermal dissociation is predicted to be of the order of 6 h, with an uncertainty of a factor of 3.5.

1. INTRODUCTION

Small quantities of iodine compounds, typically iodoalkanes and/or I_2 [1,2], are emitted from biomass combustion over land and by marine plankton and algae. Photolysis of these iodine compounds rapidly yields atomic iodine, which reacts with ozone:

$$I + O_3 \rightarrow IO + O_2. \tag{1}$$

Because a cycle can be closed by fast processes such as

$$Cl + O_3 \rightarrow ClO + O_2, \tag{2}$$

$$IO + ClO \rightarrow I + Cl + O_2, \tag{3}$$

even traces of iodine have been proposed to have a significant influence on stratospheric ozone concentrations [3,4]. The reaction

$$IO + HO_2 \rightarrow HOI + O_2 \tag{4}$$

followed by photolysis of HOI back to I + OH has been suggested, together with (1), to form a significant ozone depletion cycle in the troposphere [5]. The efficiency of such cycles depends in part on how much active iodine (I and IO) is temporarily in the form of less reactive reservoirs such as HOI and iodine nitrate, formed via

$$IO + NO_2 + M \rightarrow IONO_2 + M. \tag{5}$$

The reverse of reaction (5) may be a nighttime source of IO. Iodine nitrate may be formed by the reaction

$$NO_3 + I_2 \rightarrow IONO_2 + I, \tag{6}$$

which, followed by reaction (1), may also act as an atmospheric source of IO at night [2]. Oxidation of IO may yield OIO, which has been detected in the marine boundary layer [6] and is implicated in the formation of aerosol particles by polymerization of iodine oxides in the marine boundary layer [7,8], which can act as cloud condensation nuclei. The broader significance of these studies concerns the influence of ozone and cloud cover on climate.

The aim of the present work is to derive the enthalpies of formation of HOI and $IONO_2$ via *ab initio* methods. The latter is used to investigate the kinetics of reaction (5). The NASA-JPL recommendation [9] for $\Delta_f H$(HOI) is based on calculations by Hassanzadeh and Irikura [10] and previous work from our laboratory on OH + CF_3I kinetics [11]. In the intervening decade there have been revisions to most of the ancillary thermochemical and kinetic data used in these calculations, and advances in computational technology and methodology make it appropriate to revisit these molecules. In particular, the systematic extrapolation of coupled-cluster theory to the infinite basis set limit has proven to be of generally high accuracy [12], and the development of new basis sets by Peterson and coworkers has permitted sophisticated analysis of iodine-containing compounds [13,14]. The case of $IO + NO_2$ is one of several examples where Golden has noted difficulties in rationalizing the kinetics of halogen/nitrogen oxide systems with their thermochemistry [15] and found that an $IO–NO_2$ bond strength higher than values from most previous studies was needed to match RRKM theory with observation for reaction (5) [16]. A focus of the present study is to derive this bond strength independently and to explore the association and dissociation kinetics involving $IONO_2$.

2. METHODOLOGY AND RESULTS

We employed the aug-cc-pVTZ-PP and aug-cc-pVQZ-PP basis sets of Peterson *et al.* [14], which incorporate a relativistic pseudopotential (effective core potential) that largely accounts for scalar relativistic effects in iodine. For O and Cl we employed the aug-cc-pVTZ and aug-cc-pVQZ basis sets [17]. First, the optimized geometry, energy and harmonic frequencies for each molecule of interest were obtained with the triple zeta (TZ) basis sets and spin-unrestricted coupled-cluster theory with single and double excitations and quasiperturbative triples, CCSD(T), based on spin-restricted wavefunctions. Then the geometries were refined with the quadruple zeta (QZ) basis sets to obtain the QZ energy. All calculations were carried out with the Molpro series of programs [18]. The results are shown in Figure 9.1 and Table 9.1.

The thermochemistry of HOI was initially derived via the isodesmic process

$$IO^- + HOCl \rightarrow ClO^- + HOI, \tag{7}$$

which conserves the number and type of bonds. Experimental 0 K enthalpies of formation of the halogen monoxide anions were obtained from the enthalpies of the neutral oxides plus the energies of electron attachment (that is, minus the electron affinities). At 0 K the ion and electron conventions are equivalent. The closed-shell anions were chosen because first-order spin–orbit coupling (a vector relativistic effect) is eliminated, unlike for open-shell radicals with degenerate ground states, and second-order spin–orbit effects are negligible [19]. Bedjanian *et al.* [20] determined $\Delta_f H_{298}(IO) = 115.6 \pm 5.0\,\text{kJ mol}^{-1}$ from the ratio of forward and back rate constants for $ClO + IO = I + OClO$, which equals the equilibrium constant K_{eq} and therefore yields ΔG_{298}, which in combination with a separately estimated ΔS_{298} leads to ΔH_{298}. As noted by Peterson *et al.* [14], a subsequent revision of $\Delta_f H_{298}(OClO)$ to $98.4 \pm 1.0\,\text{kJ mol}^{-1}$ by Davis and Lee [21] implies $\Delta_f H_{298}(IO) = 120.5 \pm 2.4\,\text{kJ mol}^{-1}$. This value is employed here. It is consistent with the estimate of 119.2 kJ mol^{-1} by Kim *et al.* [22] who combined their new BrO thermochemistry with the $Br + IO = BrO + I$ equilibrium data of Bedjanian *et al.* [23,24]. For comparison, Hassanzadeh and Irikura computed

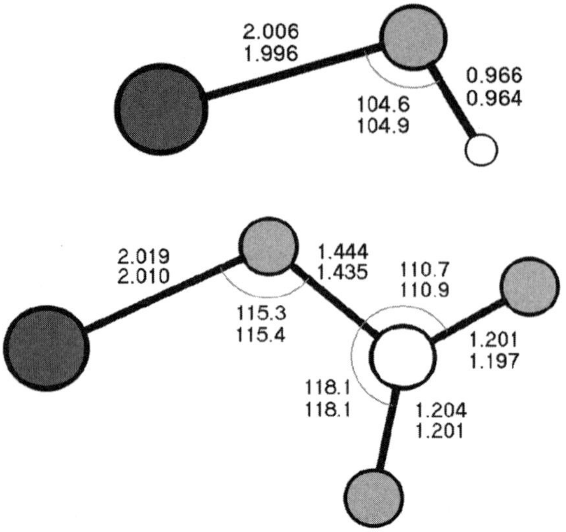

FIGURE 9.1 Structures of planar molecules computed with CCSD(T)/aug-cc-pVTZ-PP (upper numbers) and CCSD(T)/aug-cc-pVQZ-PP (lower numbers) theory. Distances are in 10^{-10} m and angles are in degrees. From top to bottom, HOI $^1A'$ and $IONO_2$ $^1A'$.

118.8 ± 7.6 kJ mol^{-1}, and Peterson *et al.* computed 125.1 ± 2.5 kJ mol^{-1} [10,14]. $\Delta_f H_{298}(\text{IO}) = 120.5 \pm 2.4$ kJ mol^{-1} in turn leads to $\Delta_f H_0(\text{IO}) = 122.6 \pm 2.4$ kJ mol^{-1} [14]. Again for comparison, we note that Lee used similar arguments to reach $\Delta_f H_0(\text{IO}) = 120.4 \pm 2.6$ kJ mol^{-1} [25], and that our chosen thermochemistry for IO lies midway between the recent calculations of Peterson and the re-evaluation of Lee. Addition of the electron affinity of 2.378 ± 0.006 eV (1 eV = 96.485 kJ mol^{-1}) measured by Gilles *et al.* [26] gives $\Delta_f H_0(\text{IO}^-) = -106.8 \pm 2.6$ kJ mol^{-1}. Coxon and Ramsay [27] measured the 0 K bond dissociation enthalpy of ClO to be 265.37 ± 0.03 kJ mol^{-1}. With enthalpy data for the atoms and elements, this yields $\Delta_f H_0(\text{ClO}) = 101.05 \pm 0.10$ kJ mol^{-1} and $\Delta_f H_{298}(\text{ClO}) = 100.97 \pm 0.10$ kJ mol^{-1}. Using the electron affinity of 2.2775 ± 0.0013 eV measured by Distelrath and Boesl [28], we arrive at $\Delta_f H_0(\text{ClO}^-) = -118.69 \pm 0.16$ kJ mol^{-1}. Joens has reassessed $\Delta_f H_0(\text{HOCl})$ as −73.99 ± 0.12 kJ mol^{-1}, a value which takes account of recent revision in the thermochemistry of OH [29].

The CCSD(T) energy changes ΔE for reaction (7) are shown in Table 9.2. The energy changes were extrapolated to the complete basis limit using the relation of Halkier *et al.* [30]:

$$\Delta E_{\text{CBS}} = \frac{4^3 \Delta E_{\text{QZ}} - 3^3 \Delta E_{\text{TZ}}}{4^3 - 3^3}.$$

Further correction for changes in vibrational zero-point energy (ZPE) yields the reaction enthalpy at 0 K. Because of anharmonicity, the observed fundamental frequency for the transition $v = 0$ to $v = 1$, ν_0, is less than ν_e, the harmonic frequency at the bottom of the well. The ZPE is between $\frac{1}{2}h\nu_0$ and $\frac{1}{2}h\nu_e$ for each mode, and

TABLE 9.1 Vibrational frequencies and *ab initio* energies

Species	TZ energy (au)[a]	QZ energy (au)[b]	Measured ν_0 (cm^{-1})					$\frac{1}{2}\Sigma h\nu_0$ (kJ mol^{-1})	Calculated TZ ν_e (cm^{-1})					$\frac{1}{2}\Sigma h\nu_e$ (kJ mol^{-1})
HOI	−370.529772	−370.567556	576	1070	3626			31.53	579	1095	3793			32.69
ClO^-	−534.830449	−534.872357	658					3.93	647					3.87
HOCl	−535.409072	−535.450074	724	1239	3609			33.32	726	1263	3779			34.49
IO^-	−369.953785	−369.992599	572					3.42	581					3.47
OH^-	−75.709425	−75.730336	3556					21.26	3723					22.26
H_2O	−76.342326	−76.363588	1595	3657	3756			53.87	1646	3810	3919			56.06
H	−0.499821	−0.499948	0					0.00	0					0.00
O	−74.978823	−74.995132	0					0.00	0					0.00
I	−294.799661	−294.815813	0					0.00	0					0.00
$IONO_2$	−574.734656	−574.824148	(97)[c]	(185)[c]	(356)[c]	580	(711)[c]	–	101	192	370	604	739	16.46
			(718)[c]	815	1276	1673			746	822	1296	1706		
$ClONO_2$	−739.612951	−739.705569	124	273	434	563	711	17.25	122	260	443	579	724	17.47
			780	809	1293	1737			794	814	1312	1767		

a CCSD(T)/aug-cc-pVTZ-PP.
b CCSD(T)/aug-cc-pVQZ-PP.
c Estimated from harmonic data (see text).

TABLE 9.2 Reaction energetics

Reaction	ΔE_{TZ} ($kJ\,mol^{-1}$)[a]	ΔE_{QZ} ($kJ\,mol^{-1}$)[b]	ΔE_{CBS} ($kJ\,mol^{-1}$)[c]	ΔE_{ZPE} ($kJ\,mol^{-1}$)[d]	ΔH_0 ($kJ\,mol^{-1}$)
$IO^- + HOCl \rightarrow ClO^- + HOI$ (7)	6.9	7.2	7.5	−1.3	6.1
$IO^- + H_2O \rightarrow OH^- + HOI$ (8)	149.4	153.1	155.7	−4.5	151.1
$HOI \rightarrow H + O + I$ (9)	660.2	673.9	683.8	−32.1	620.5[e]
$IO^- + ClONO_2 \rightarrow ClO^- + IONO_2$ (10)	4.3	4.4	4.4	−1.4	3.0

a Energy change computed at CCSD(T)/aug-cc-pVTZ-PP level.
b Energy change computed at CCSD(T)/aug-cc-pVQZ-PP level.
c Energy change computed with molecular CCSD(T) energies extrapolated to the complete basis set limit.
d Zero-point vibrational energy change (see text).
e Includes atomic spin–orbit corrections.

we employ the mean of these two quantities, which should be a good estimate [31] (as discussed below, our results are rather insensitive to the extent of anharmonicity). We find that this estimate of ZPE is equal to 0.981 $\frac{1}{2}\Sigma h\nu_e$ for the seven known molecules in Table 9.1. This relation is used to derive the ZPE for the molecule $IONO_2$, for which a complete set of ν_0 has not been measured. The ν_0 data shown in Table 9.2 were taken from the NIST compilation [32].

The computed enthalpy change for reaction (7) is $\Delta_r H_0 = \Sigma\,\Delta_f H_0(\text{products}) - \Sigma\,\Delta_f H_0(\text{reactants})$, and with all $\Delta_f H_0$ except for HOI established above, $\Delta_f H_0(\text{HOI})$ is obtained. From Table 9.2, $\Delta_r H_0$ is 6.1 $kJ\,mol^{-1}$ and thus $\Delta_f H_0(\text{HOI}) = -53.8\,kJ\,mol^{-1}$. The enthalpy correction $H_{298} - H_0$ was calculated via standard methods, with the JANAF [33] convention for polyatomic molecules of employing ν_0 for the frequencies. In the case of $IONO_2$, where a complete set of observed ν_0 is unavailable, we use the relation observed $\nu_0 = (0.962 \times \text{calculated } \nu_e)$, derived from our TZ results in Table 9.1, to fill in the missing data. Together with $H_{298} - H_0$ for the elements in their reference state [33], the enthalpy of formation at 298 K is obtained as $\Delta_f H_{298} = \Delta_f H_0 + H_{298} - H_0(\text{HOI}) - \Sigma H_{298} - H_0(\text{elements})$.

As a check, the alternative isodesmic reaction

$$IO^- + H_2O \rightarrow OH^- + HOI \quad (8)$$

was employed, with $\Delta_f H_0(OH^-) = -139.2 \pm 0.1\,kJ\,mol^{-1}$ derived from $\Delta_f H_0(OH) = 37.1 \pm 0.1\,kJ\,mol^{-1}$ from Joens [29] and the electron affinity of 176.3 $kJ\,mol^{-1}$ from Smith *et al.* [34]. Reaction (8) yields $\Delta_f H_0(\text{HOI}) = -55.2\,kJ\,mol^{-1}$. In principle, bond-conserving reactions of neutral IO, such as reaction (4), could also be analyzed. However there is some ambiguity in the spin–orbit splitting in IO, necessary for computation of reaction enthalpies, where the value of $25.0 \pm 0.5\,kJ\,mol^{-1}$ measured by Gilles *et al.* [26] differs somewhat from the 21.2 $kJ\,mol^{-1}$ recently computed by Peterson *et al.* [14]. The atomization reaction

$$HOI \rightarrow H + O + I \quad (9)$$

is also used for comparison, although the large changes in the electronic structure of the atoms eliminate the favorable error cancellation. There is a potential advantage that the enthalpies of formation of the atoms are very well defined. The energies of the O and I atoms were corrected downwards by 0.9 and 30.3 kJ mol^{-1}, respectively, based on the measured spin–orbit splitting in these atoms. This atomization approach yields $\Delta_f H_0(\text{HOI}) = -50.5$ kJ mol^{-1}.

Iodine nitrate was analyzed via the congeneric exchange reaction

$$IO^- + ClONO_2 \rightarrow ClO^- + IONO_2 \tag{10}$$

in a similar manner. The NASA-JPL recommendation [9] for $\Delta_f H_{298}(\text{ClONO}_2) = 22.9 \pm 2.0$ kJ mol^{-1} is based on the work of Anderson and Fahey, who analyzed forward and reverse kinetics for $ClO + NO_2 = ClONO_2$ [35]. This is inconsistent with a prior determination of 26.4 ± 0.8 kJ mol^{-1} by Alqasmi *et al.* [36]. It is noted that the reaction enthalpy from Anderson and Fahey, combined with modern values for $\Delta_f H_{298}$ of ClO and NO_2 [9], yields $\Delta_f H_{298}(\text{ClONO}_2) = 24.3 \pm 2.0$ kJ mol^{-1} which appears to eliminate the disagreement. This value leads to $\Delta_f H_0(\text{ClONO}_2) = 30.9 \pm 2.0$ kJ mol^{-1}, which is employed here. The computed $\Delta_r H_0$ of 3.0 kJ mol^{-1} for reaction (10) (see Table 9.2) corresponds to $\Delta_f H_0(\text{IONO}_2) = 45.8$ kJ mol^{-1}. The enthalpy change $H_{298} - H_0$ is calculated to be 15.6 kJ mol^{-1} and, with similar data for the elements in their standard states [33], leads to $\Delta_f H_{298}(\text{IONO}_2) = 37.4$ kJ mol^{-1}. The uncertainty in this result is discussed in Section 3.

The present values of $\Delta_f H_0(\text{IO})$ and $\Delta_f H_0(\text{IONO}_2)$, together with $\Delta_f H_0(\text{NO}_2) = 36.8 \pm 0.1$ kJ mol^{-1} from the Active Thermochemical Tables as quoted by Burcat and Ruscic [37], yield the bond dissociation enthalpy $DH_0(\text{IO–NO}_2) = 113.6 \pm 3.1$ kJ mol^{-1}. It should be noted that because $\Delta_f H_0(\text{IO})$ is also used as the basis for the anion and thence $\Delta_f H_0(\text{IONO}_2)$, the bond dissociation enthalpy derived here is independent of the value chosen for $\Delta_f H_0(\text{IO})$, and there is no contribution from its uncertainty. NO_3 has a complex vibronic structure, that arises in part from Jahn–Teller distortion and coupling between the $X\,^2A_2'$ and $A\,^2E''$ states [38]. $\Delta_f H_{298}(\text{NO}_3) = 74.6 \pm 0.7$ kJ mol^{-1} from the Active Thermochemical Tables as quoted by Burcat and Ruscic [37] plus data for atomic I [33] leads to $DH_{298}(\text{I–NO}_3) = 144.0 \pm 3.9$ kJ mol^{-1}. The vibronic levels tabulated by Stanton [38] yield $H_{298} - H_0(\text{NO}_3) = 11.8$ kJ mol^{-1}, and thus $DH_0(\text{I–NO}_3) = 141.6 \pm 3.9$ kJ mol^{-1}.

3. DISCUSSION

3.1 Possible uncertainties

Neither the geometry of HOI nor that of $IONO_2$ has been determined experimentally. For comparison, the geometry of neutral OIO is also computed. At the TZ and QZ levels of theory r_{IO} and a_{OIO} equal 1.826×10^{-10} m and 111.0°, and 1.815×10^{-10} m and 110.5°, respectively. These parameters may be compared to the estimated equilibrium parameters of 1.800×10^{-10} m and 109.8°, from microwave spectroscopy by Miller and Cohen [39]. For neutral IO, r_{IO} is computed as 1.900×10^{-10} and 1.890×10^{-10} m at the TZ and QZ levels of theory, respectively, which may be compared to the state-resolved equilibrium values of

1.868×10^{-10} m ($^2\Pi_{3/2}$) and 1.885×10^{-10} m ($^2\Pi_{1/2}$) measured by Miller and Cohen [40]. This level of accord suggests the QZ geometries in Figure 9.1 may be reliable to around $\sim 0.02 \times 10^{-10}$ m and $\sim 1°$. We note that for I–O stretching, such bond length errors correspond to energy errors of about 0.4 kJ mol^{-1} per bond.

The computed harmonic frequencies shown in Table 9.1 are, as expected, slightly higher than the observed fundamental frequencies. The scale factor of 0.962 given above corresponds to a root mean square deviation from the observed ν_0 values of 22 cm^{-1}. The observed fundamental frequency of ClO^- is atypically *above* the computed harmonic one, however, there is agreement when the ± 25 cm^{-1} experimental uncertainty [26] is taken into account. The scale factor of 0.981 proposed for ZPE leads to a small root-mean-square deviation in ZPE for the data set in Table 9.1, of 0.06 kJ mol^{-1}.

The important consideration for the present analysis is uncertainty in the overall energy changes for the bond conserving reactions. The uncertainty from geometry errors is expected to cancel because of similar bonding in reactants and products, with residual errors perhaps as large as 0.5 kJ mol^{-1}. The greatest contribution of ZPE to the reactions used, (7), (8) and (10), is less than 5 kJ mol^{-1}. Thus the difference between the computed harmonic and observed fundamental contributions of $\Sigma \frac{1}{2}h\nu$ to $\Delta_r H_0$ is less than 0.2 kJ mol^{-1}. Therefore the impact of any uncertainty in the scaling of ZPE has a negligible impact.

Next the effect of basis set size is considered, i.e., potential uncertainty in the CBS extrapolation. For reaction (7) there is little difference (0.3 kJ mol^{-1}) between TZ and QZ reaction energies, while this difference is 3.6 kJ mol^{-1} for reaction (8). Because ΔE is already close to the convergence limit and is extrapolated by only 0.3 and 2.6 kJ mol^{-1}, respectively, the CBS extrapolation introduces little uncertainty, less than 0.2 kJ mol^{-1}. For reaction (10) the ΔE results are already converged before extrapolation (see Table 9.2).

The level of electron correlation employed influences the computed reaction energies. For reaction (7), at the HF/aug-cc-pVQZ-PP level of theory, i.e., with complete neglect of correlation, $\Delta E = -1.7$ kJ mol^{-1}, indicating a fair degree of independence of ΔE from the level of correlation treatment. In other words, $\Delta E_{QZ}(\mathrm{HF}) - \Delta E_{QZ}(\mathrm{CCSD(T)}) = -8.9$ kJ mol^{-1}. Comparison with CCSD results yields $\Delta E_{QZ}(\mathrm{CCSD}) - \Delta E_{QZ}(\mathrm{CCSD(T)}) = -1.8$ kJ mol^{-1}, indicating that even before inclusion of the perturbative triples (T) contribution, ΔE is almost converged with respect to the level of correlation treatment. For reaction (8), the difference $\Delta E_{QZ}(\mathrm{HF}) - \Delta E_{QZ}(\mathrm{CCSD(T)})$ is 17.1 kJ mol^{-1}. The result from reaction (8), by the arguments of Hassanzadeh and Irikura [10], is therefore less reliable. CCSD data, i.e., with neglect of the triple excitation contribution, yield ΔE only 1.1 kJ mol^{-1} higher, so the result is insensitive to the level of correlation. For reaction (10) the impact is larger, where neglect of the (T) term reduces ΔE by 6.6 kJ mol^{-1}. The effect of extrapolation of electron correlation to the full configuration interaction limit was explored by Peterson *et al.*, who found for ClO^- and IO^- that the dissociation energies were increased by 1.5 and 1.8 kJ mol^{-1}, respectively [14]. Because these terms appear on opposite sides of the bond conserving reactions they largely cancel here. An estimated uncertainty of about 1 kJ mol^{-1} is allowed for here.

For the atomization reaction (9) the potential errors are much larger. The difference between ΔE at the TZ and QZ levels of theory, the size of the energy extrapolation, and the sensitivity to the correlation treatment as defined as the difference between CCSD(T) and HF results, are all much larger than for the other reactions at 13.7, 9.9 and -319.8 kJ mol^{-1}, respectively, and there is no ZPE or geometry error cancellation. The thermochemistry obtained via reaction (9) is therefore the least reliable contribution towards $\Delta_f H_0$(HOI). The best estimate of $\Delta_f H_0$(HOI) is -54.5 kJ mol^{-1}, which comes from averaging the results obtained via reactions (7) and (8) that differ by only 1.4 kJ mol^{-1}. The corresponding $\Delta_f H_{298}$(HOI) is -59.2 kJ mol^{-1}. The propagated uncertainty of the ancillary thermochemistry, combined in quadrature, is 2.6 kJ mol^{-1}. An approximate estimate of the total uncertainty arising from the CBS extrapolation, ZPE difference, geometry errors and incomplete electron correlation is $\sim$2 kJ mol^{-1}. Combination in quadrature with the uncertainty of the ancillary thermochemistry leads to an overall uncertainty of $\sim$3.3 kJ mol^{-1}. Similar computational uncertainties are expected for the result obtained via reaction (10), where the propagated error limits in the auxiliary experimental data are 3.3 kJ mol^{-1}, with final error limits of around 3.9 kJ mol^{-1}.

3.2 Thermochemistry of HOI

The present result for $\Delta_f H$(HOI) is shown in Table 9.3, along with prior values of the thermochemistry. Early values based on extrapolated empirical trends in the series HOCl–HOBr–HOI [41,42] are seen to be too positive by 10–20 kJ mol^{-1}.

TABLE 9.3 Thermochemistry of HOI

$\Delta_f H_0$ (kJ mol^{-1})	$\Delta_f H_{298}$ (kJ mol^{-1})	Method
-33.5 to -37.7		Empirical trend[a]
-42.7		Empirical trend[b]
-44.7	-48.9	*Ab initio*[c]
-55.2 ± 6.9	-59.9 ± 6.9	*Ab initio*[d]
-64.9 ± 5.4	-69.6 ± 5.4	E_a difference[e]
-60 ± 12		*Ab initio*[e]
-68.4	-73.1	DFT[f]
-54.5 ± 3.3	-59.2 ± 3.3	This work

a Ref. [42].
b Ref. [41].
c Ref. [43].
d Ref. [10].
e Ref. [11].
f Ref. [44].

The result of Glukhovtsev *et al.* [43], who applied G2 theory to reaction (9), is also somewhat high, but there is particularly close accord with the results of Hassanzadeh and Irikura [10]. Begović *et al.* [44] used density functional theory (DFT) to obtain values that are too negative by 14 $kJ\,mol^{-1}$. The value of Berry *et al.* [11], which is significantly below the present value, was derived from the difference between the measured forward activation energy E_a for

$$OH + CF_3I \rightarrow CF_3 + HOI \quad (11)$$

of 11.3 $kJ\,mol^{-1}$ and a predicted $E_a = 0 \pm 5\ kJ\,mol^{-1}$ for the reverse process. This prediction was based on the absence of a barrier to the reverse process in a scan along the reaction coordinate at the G2(MP2) level of theory. E_a in the forward direction has since been redetermined as $16.6 \pm 1.2\ kJ\,mol^{-1}$ [45] which, with the new thermochemistry plus data on CF_3, CF_3I [46] and OH [29] that yield $\Delta_r H_{298} = 23.9 \pm 4.4\ kJ\,mol^{-1}$, implies $E_a = -7.3 \pm 4.6\ kJ\,mol^{-1}$ for the back reaction. This is consistent with the earlier *ab initio* analysis of reaction (11), and is similar to the activation energy measured, for example, for the $(CH_3)_3C$ + HI reaction where $E_a = -6.3 \pm 0.8\ kJ\,mol^{-1}$ [47].

3.3 Thermochemistry of $IONO_2$

Table 9.4 summarizes prior data for $IONO_2$. Some of these have been reviewed recently by Golden [16], and are briefly discussed here. Complicating factors have included uncertainty in the heats of formation of IO and NO_3, where errors in these quantities will impact any $IONO_2$ thermochemistry derived via bond strengths. For this reason the 0 K bond strengths DH_0(IO–NO_2) and DH_0(I–NO_3), where

TABLE 9.4 Thermochemistry of $IONO_2$

$\Delta_f H_0$ ($kJ\,mol^{-1}$)	$\Delta_f H_{298}$ ($kJ\,mol^{-1}$)	DH_0(IO–NO_2) ($kJ\,mol^{-1}$)	DH_0(I–NO_3) ($kJ\,mol^{-1}$)	Method
		134 ± 13		Semi-empirical calculation[a]
70 ± 16		94.9	102.8	DFT[b]
		105		RRKM fit[b]
29.5 or 36.1 or 42.4		146.0 or 131.4	157.3	*Ab initio*[c]
		96	116	DFT[d]
		150		RRKM fit[e]
45.8 ± 3.9	37.4 ± 3.9	113.6 ± 3.1	141.6 ± 3.9	This work

a Ref. [48].
b Ref. [49].
c Ref. [53].
d Ref. [54].
e Ref. [16].

derived previously, are also tabulated. These are fundamentally related to the kinetics of dissociation and recombination as well.

As seen in Table 9.4, the 1993 semiempirical (PM3) calculation of $DH_0(IO–NO_2)$ [48], which did not benefit from isodesmic reactions to minimize errors, is in the middle of the range of bond strengths which have subsequently appeared in the literature. Allan and Plane [49] applied DFT (B3LYP/6-311 + G(2d,p)) to the two bond strengths in $IONO_2$, which both led to the same heat of formation at 0 K, of 70 kJ mol^{-1}. This convergence is fortuitous because they used 0 K enthalpies of formation of IO and NO_3 of 132 and 68.8 kJ mol^{-1}, respectively, whereas the recommendations discussed above correspond to 122.6 and 80.2 kJ mol^{-1}, respectively. Allan and Plane carried out an RRKM fit to their measured rate constants k_5 for reaction (5) with $DH_0(IO–NO_2)$ as one of six adjustable parameters [49]. This is discussed in more detail in the next section. An important claim is that $IONO_2$ dissociation can be observed at 473 K. The ratio of their forward and reverse rate constants yields $K_{eq,5} = k_5/k_{-5} = 1.8 \times 10^{-16}$ cm^3 molecule^{-1}, so that $\Delta G_{473} = -31.2$ kJ mol^{-1}. With entropies at that temperature of 274.1 [10], 258.4 and 358.0 J K^{-1} mol^{-1} for IO, NO_2 and $IONO_2$, respectively, one derives $DH_{473}(IO–NO_2) = 114.2$ kJ mol^{-1} and hence $DH_0(IO–NO_2) = 107.9$ kJ mol^{-1}. This approach has the advantage of independence from any details of an RRKM analysis, but in this case the k_{-5} information is highly uncertain. It corresponds to a small deviation from the observed decay of IO (see Figure 7 of Ref. [49]) with k_{-5} set to zero, comparable to the scatter in the IO fluorescence signal. If k_{-5} was smaller by, say, a factor of two, there would still be a match with the IO decay shown, and the corresponding $DH_0(IO–NO_2)$ would increase by 2.7 kJ mol^{-1}, to within the error limits of the present value. Similarly, Dillon *et al.* [50] set a lower limit $DH_0(IO–NO_2) > 107$ kJ mol^{-1} which is consistent with the above analysis and the present *ab initio* result. The present $DH_{298}(I–NO_3) = 144.0 \pm 3.9$ kJ mol^{-1} result given above, combined with $DH_{298}(I–I) = 151.1$ kJ mol^{-1} [33], implies that reaction (6) is endothermic by 7.1 ± 3.9 kJ mol^{-1} at room temperature. The lower end of this range is more consistent with the arguments of Chambers *et al.* [51], who observed a zero activation energy for this process and thereby deduced that reaction (6) is not endothermic. However, as emphasized recently, this is not strictly true [52]. In other words, the reverse of reaction (6) is predicted to show a negative activation energy. Such a measurement would be a good test of the thermochemistry proposed here.

Papayannis and Kosmas made single-point CCSD(T) calculations at B3LYP geometries, using a modest-sized LANL2DZ basis set augmented with two *d* and one *f* polarization function [53]. Some complications with their results have been discussed previously [16]: uncertainty in the spin–orbit splitting in IO gives alternative thermochemistry, and the bond strengths are not entirely consistent with each other when correct enthalpies for the fragments are employed. Their thermochemistry based on $DH_0(I–NO_3)$ appears to give the closest accord with the new values, but this was based on $\Delta_f H_0(NO_3) = 62.4$ kJ mol^{-1}. The current Active Thermochemical Tables value of 80.2 kJ mol^{-1} would increase their $\Delta_f H_0(IONO_2)$ to 60.2 kJ mol^{-1}.

Plane *et al.* made further DFT calculations on the OIO + NO system [54]. These included bond-additivity corrections for I_2 and IO, absorbed into "spin–orbit" corrections for I and IO. The bond strengths shown in Table 9.4 are derived from the differences between the computed reaction enthalpies reported by Plane *et al.* [54], as noted by Golden [55], and are reasonably close to the present values, both falling about 8 kJ mol^{-1} smaller.

3.4 Comparison with halogen analogs

The thermochemistry of HOCl and $ClONO_2$ has been discussed above, and data for HOBr were reanalyzed by Joens [29]. Orlando and Tyndall studied the decomposition of $BrONO_2$ to BrO + NO_2, and from the equilibrium constant deduced the BrO–NO_2 bond strength. This is combined here with a new determination of $\Delta_f H_{298}$(BrO) by Kim *et al.* [22] to obtain $\Delta_f H_{298}(BrONO_2) = 39.4 \pm 6.3$ kJ mol^{-1}. Alternatively, Hanson *et al.* studied the equilibrium $BrONO_2 + H_2O = HOBr + HNO_3$ and deduced $\Delta_r H_{298} = 5.4 \pm 2.9$ kJ mol^{-1} [56], which together with information about HOBr, H_2O and HNO_3 [29,37] implies $\Delta_f H_{298}(BrONO_2) = 44.1 \pm 3.4$ kJ mol^{-1}. These two estimates are in accord, and are provisionally summarized here as 42 ± 5 kJ mol^{-1}. These results are shown in Table 9.5. It may be seen that bond dissociation enthalpies are generally similar across the series, given the uncertainties. An exception is halogen–oxygen bonding, which is distinctly stronger for chlorine than bromine or iodine, in both the hypohalous acids and the halogen nitrates. The bonding to bromine and iodine is of similar strength in both systems. Differences between the behavior of halogen species depends significantly on differences in their reaction kinetics and also photochemistry: the typical photolysis of iodine compounds in the visible spectrum means that much of atmospheric iodine is in chemically active radical forms during daylight. An example where nighttime chemistry is impacted by the thermochemistry of iodine nitrate is discussed below.

TABLE 9.5 Comparison of HOX and $XONO_2$ thermochemistry (kJ mol^{-1})

Halogen X	$\Delta_f H_{298}$ (HOX)	DH_0 (H–OX)	D_0 (HO–X)	$\Delta_f H_{298}$ ($XONO_2$)	DH_0 (XO–NO_2)	DH_0 (X–ONO_2)
Cl	-76.9 ± 0.1^{a}	391.1 ± 0.2^{a}	230.75 ± 0.02^{a}	$24.3 \pm 2.0^{b,c}$	$168.8 \pm 2.1^{b,c}$	$107.0 \pm 2.0^{b,c}$
Br	-58.2 ± 1.8^{a}	$394.7 \pm 1.8^{a,b}$	202.7 ± 1.8^{a}	$42 \pm 5^{b,d,e}$	$142 \pm 5^{b,d,e}$	$111 \pm 5^{b,d,e}$
I	-59.2 ± 3.3^{f}	393.1 ± 2.3^{f}	198.8 ± 3.5^{f}	37.4 ± 3.9^{f}	141.6 ± 3.9^{f}	113.6 ± 3.1^{f}

a Ref. [29].
b See text.
c Ref. [35].
d Ref. [62].
e Ref. [56].
f This work.

3.5 $IONO_2$ kinetics

A particular application of $DH_0(IO–NO_2)$ is interpretation of the $IO + NO_2$ recombination reaction (5) at the low-pressure limit. Golden [16] has noted that the collisional energy transfer parameters employed previously [49] are unusually large. This will tend to increase the low-pressure limiting rate constant k_0, while with smaller collisional parameters Golden needed $DH_0(IO–NO_2)$ to be as large as 150 kJ mol^{-1}. Golden also reported that Plane was subsequently able to fit the k_5 data with $DH_0(IO–NO_2) = 128$ kJ mol^{-1} and smaller collisional parameters than previously used [16]. Thus a lower bond strength offsets a higher collisional stabilization efficiency. Here, the new $DH_0(IO–NO_2)$ is kept fixed in RRKM calculations. Troe's unimolecular formalism [57] is employed, which for the addition reaction (5) at the low pressure limit may be cast in the form [58]:

$$k_0 = \beta_c Z_{LJ} \frac{\rho(E_0)RT}{Q_{vib}(IONO_2)} F_E F_{anh} F_{rot} \frac{Q(IONO_2)}{Q(IO) \cdot Q(NO_2)}, \tag{12}$$

where $\rho(E_0)$ is the vibrational density of states of $IONO_2$ at the threshold energy E_0 for dissociation to $IO + NO_2$, i.e., $E_0 = DH_0(IO–NO_2)$. The Qs are the partition functions, derived via the rigid rotor-harmonic oscillator approximation. F_E, F_{anh} and F_{rot} are factors to account for the energy dependence of $\rho(E_0)$, for the effects of vibrational anharmonicity, and for centrifugal barriers, respectively. $\beta_c Z_{LJ}$ is the weak-collision stabilization rate constant. The Lennard-Jones parameters chosen by Golden imply $Z_{LJ} = 7.3 \times 10^{-10}$ cm^3 molecule^{-1} s^{-1} [16]. Here it is assumed that the efficiency of energy transfer between excited $IONO_2$ and N_2 bath gas is similar to that for excited INO_2, for which $\langle \Delta E \rangle = -4.1$ kJ mol^{-1} [57]. The corresponding β_c is 0.46, and so $\beta_c Z_{LJ} = 3.4 \times 10^{-10}$ cm^3 molecule^{-1} s^{-1}. A difference from the various RRKM calculations reported to date (except the initial work of Rayez and Destriau [48] who also applied the Troe formalism) is the explicit inclusion of an anharmonicity correction via $F_{anh} = 1.38$. A crucial term is F_{rot}, which in essence is determined by the ratios of the moments of inertia of the structure at the dissociation transition state and of equilibrium $IONO_2$, I^+/I. For a loose transition state, as assumed here, with no potential energy barrier to association and located at the centrifugal maximum, F_{rot} depends sensitively upon the (unknown) long-range potential between IO and NO_2. Two different approaches are examined here. First, Troe's simplest treatment, based on a quasi-diatomic van der Waals potential [57], is employed here as Model A. This yields the inertia ratio via

$$\left(\frac{I^+}{I}\right)^2 = \frac{ABC}{\pi(RT)^3}\left(\frac{RT}{B_{eff}}\right)\left(\frac{24E_0}{B_{eff}}\right), \tag{13}$$

where A, B and C are the moments of inertia of $IONO_2$, expressed in J mol^{-1}, and B_{eff} is the mean of the two smallest values, B and C. Table 9.6 lists the input parameters for the Troe formalism, and these yield $I^+/I = 58.2$. There is an alternative choice [57] for the C_6 parameter that defines the $-C_6/r_6$ van der Waals potential, which replaces the 24 in Eq. (13) by 12. If this alternative is applied, then $I^+/I = 41.2$. This uncertainty in C_6 has little im-

TABLE 9.6 Parameters for Troe unimolecular analysis at 298 K[a]

$IONO_2$ frequencies 97, 185, 356, 580, 711, 718, 815, 1276, 1673 cm^{-1}; $E_{ZN} = 38.3$ kJ mol^{-1}
IO and NO_2 frequencies 673 and 750, 1318, 1618 cm^{-1}; $E_{ZP} = 26.1$ kJ mol^{-1}
NO_2 product of moments of inertia 15.36×10^{-138} $kg^3\,m^6$
IO moment of inertia 8.230×10^{-46} kg m^2
$IONO_2$ rotational constants $A, B, C = 12.43, 1.346, 1.214$ GHz; $B_{eff} = (B + C)/2 = 0.511$ J mol^{-1}
Critical energy for dissociation $E_0 = 113.6$ kJ mol^{-1}; $\Delta E_Z = 8.7$ kJ mol^{-1}
Reaction coordinate stretch 580 cm^{-1}; Morse $\beta = 4.06 \times 10^{10}$ m^{-1} and $D = 125.8$ kJ mol^{-1}
Looseness parameter $\alpha = 1 \times 10^{10}$ m^{-1}; $a_1 = 0.119$; $a_2 = 0.00666$
Centrifugal barriers $E_0(J) - E_0(J = 0) = 0.243$ J mol^{-1} $[J(J + 1)]^{1.009}$
$F_E = 1.149$; $F_{anh} = 1.381$; $I^+/I = 8.16$; $F_{rot\,max} = 20.2$; $\rho(E_0) = 387$ $(J\,mol^{-1})^{-1}$; $Q_{vib}(IONO_2) = 6.40$; $Q_{rot}(IONO_2) = 1.92 \times 10^5$
$Q(IONO_2) = 3.10 \times 10^{39}$ m^{-3}; $Q(NO_2) = 1.37 \times 10^{36}$ m^{-3}; $Q(IO) = 2.09 \times 10^{36}$ m^{-3}

a See Refs. [57,59].

pact on F_{rot}, which would drop from 15.2 to 13.8. Model B is a more sophisticated treatment of F_{rot}, derived by Troe via his statistical adiabatic channel model [59]. Centrifugal maxima are located in channel potentials of the form

$$V(z) = D\left[1 - \exp(-z)\right]^2 + \Delta E_Z \exp\left(-\frac{\alpha}{\beta}z\right) + E_{ZP} + J(J+1)\frac{B_{eff}}{1 + a_1 z + a_2 z^2}, \tag{14}$$

with $z = \beta(q - q_e)$, where q is the IO–NO_2 separation, and D and β are Morse parameters (see Table 9.6). The second term describes the decay of zero-point energy in terms of a looseness parameter α, with a standard value of $\alpha = 1 \times 10^{-10}$ m assumed here. The final term describes the rotational energy, with a_1 and a_2 derived in the standard manner [59]. The rotational barriers E_0 as a function of J are fitted to the form $E_0(J) - E_0(J = 0) = C_\nu J(J + 1)^\nu$, from which I^+/I may be deduced [57]. This second procedure leads to $F_{rot} = 6.02$, which is about a factor of 2.5 smaller than that derived via Model A. This difference emphasizes the inherent uncertainty in rotational effects.

Combining these data through the unimolecular formulation (12) yields $k_0 = 13.3 \times 10^{-31}$ (Model A) or 5.3×10^{-31} (Model B) cm^6 $molecule^{-1}$ s^{-1} for N_2 at 298 K.

These estimates bracket the NASA-JPL and IUPAC recommendations of 6.5×10^{-31} and 7.7×10^{-31} cm^6 $molecule^{-1}$ s^{-1} [9,60]. It is therefore possible to reconcile the thermochemistry proposed here with the observed IO + NO_2 recombination kinetics while employing reasonable input parameters for the unimolecular model. Nevertheless it must be stressed, as emphasized earlier [16], that there is considerable uncertainty in some of the input parameters to an RRKM analysis, especially the F_{rot} term. It is of interest to compare the present kinetic calculations with the Multiwell [61] Master Equation calculations on this system by Golden [16]. He used a Morse potential to locate the centrifugal maximum, and from the bond extension $F_{rot} \sim 2.1$ is derived, about 1/7 of that used here. On the other hand, the higher E_0 value yields a density of states larger by a factor of ~6, and these two factors largely cancel.

The present computed equilibrium constant for reaction (5) at 298 K is 8.9×10^{-8} cm^3 $molecule^{-1}$. For 1 atm of N_2, the effective second-order $k_5 \sim 4 \times 10^{-12}$ cm^3 $molecule^{-1}$ s^{-1} [49]. Thus the corresponding first-order dissociation rate constant is $k_{-5} \sim 4.5 \times 10^{-5}\,s^{-1}$ at these conditions, i.e., the lifetime is ~6 h. This estimate of k_{-5} is about two orders of magnitude smaller than that of Allan and Plane [49], but still high enough for appreciable $IONO_2$ dissociation to take place overnight. The dissociation rate is very sensitive to DH_0(IO–NO_2): the present error limits correspond to an uncertainty of a factor of 3.5, and so a reliable measurement of k_{-5} would allow reliable experimental determination of the thermochemistry.

4. CONCLUSIONS

Hypoiodous acid and iodine nitrate have been investigated with coupled cluster theory and the results extrapolated to the complete basis set limit. Together with revised thermochemistry for several ancillary molecules, the enthalpy changes of working reactions yields new thermochemistry for HOI and $IONO_2$. The latter data, employed in unimolecular rate theory, appear to be consistent with kinetic measurements on the IO + NO_2 reaction to within the uncertainties of the kinetic analysis.

ACKNOWLEDGEMENTS

Drs. D.M. Golden, K.A. Peterson and N.J. de Yonker are thanked for valuable discussions and advice. This work was supported by the Robert A. Welch Foundation (Grant B-1174) and the UNT Faculty Research Fund, and computational facilities were provided at the National Center for Supercomputing Applications (Grant CHE000015N), at the Research Cluster operated by UNT Academic Computing Services, and purchased with funding from the National Science Foundation (Grant CHE-0342824).

REFERENCES

[1] L.J. Carpenter, W.T. Sturges, S.A. Penkett, P.S. Liss, B. Alicke, K. Hebestreit, U. Platt, *J. Geophys. Res. Atmos.* **104** (1999) 1679.
[2] A. Saiz-Lopez, J.M.C. Plane, *Geophys. Res. Lett.* **31** (2004) L04112.
[3] S. Solomon, J.B. Burkholder, A.R. Ravishankara, R.R. Garcia, *J. Geophys. Res.* **99** (1994) 20929.
[4] S. Solomon, R.R. Garcia, A.R. Ravishankara, *J. Geophys. Res. D* **99** (1994) 20491.
[5] W.L. Chameides, D.D. Davis, *J. Geophys. Res.* **85** (1980) 7383.
[6] B.J. Allan, J.M.C. Plane, G. McFiggans, *Geophys. Res. Lett.* **28** (2001) 1945.
[7] J.B. Burkholder, J. Curtius, A.R. Ravishankara, E.R. Lovejoy, *Atmos. Chem. Phys.* **4** (2004) 19.
[8] R.W. Saunders, J.M.C. Plane, *Environ. Chem.* **2** (2005) 299.
[9] S.P. Sander, R.R. Friedl, D.M. Golden, M.J. Kurylo, G.K. Moortgat, H. Keller-Rudek, P.H. Wine, A.R. Ravishankara, C.E. Kolb, M.J. Molina, B.J. Finlayson-Pitts, R.E. Huie, V.L. Orkin, Chemical Kinetics and Photochemical data for Use in Stratospheric Modeling. Evaluation Number 15. JPL Publication 06-2 (http://jpldataeval.jpl.nasa.gov), JPL, Pasadena, 2006.
[10] P. Hassanzadeh, K.K. Irikura, *J. Phys. Chem. A* **101** (1997) 1580.
[11] R.J. Berry, J. Yuan, A. Misra, P. Marshall, *J. Phys. Chem. A* **102** (1998) 5182.
[12] W. Klopper, J. Noga, *ChemPhysChem* **4** (2003) 32.
[13] B.C. Shepler, N.B. Balabanov, K.A. Peterson, *J. Phys. Chem. A* **109** (2005) 10363.
[14] K.A. Peterson, B.C. Shepler, D. Figgen, H. Stoll, *J. Phys. Chem. A* **110** (2006) 13877.
[15] D.M. Golden, *J. Phys. Chem. A* **111** (2007) 6772.
[16] D.M. Golden, *J. Phys. Chem. A* **110** (2006) 2940.
[17] T.H. Dunning Jr., *J. Chem. Phys.* **90** (1989) 1007.
[18] H.-J. Werner, P.J. Knowles, R. Lindh, M. Schütz, P. Celani, T. Korona, F.R. Manby, G. Rauhut, R.D. Amos, A. Bernhardsson, A. Berning, D.L. Cooper, M.J.O. Deegan, A.J. Dobbyn, F. Eckert, C. Hampel, G. Hetzer, A.W. Lloyd, S.J. McNicholas, W. Meyer, M.E. Mura, A. Nicklaß, P. Palmieri, R. Pitzer, U. Schumann, H. Stoll, A.J. Stone, R. Tarroni, T. Thorsteinsson, *Molpro Quantum Chemistry Package*, Birmingham, UK, 2006.
[19] A. Tajti, P.G. Szalay, A.G. Császár, M. Kállay, J. Gauss, E.F. Valeev, B.A. Flowers, J. Vázquez, J.F. Stanton, *J. Chem. Phys.* **121** (2004) 11599.
[20] Y. Bedjanian, G. Le Bras, G. Poulet, *J. Phys. Chem. A* **101** (1997) 4088.
[21] H.F. Davis, Y.T. Lee, *J. Chem. Phys.* **105** (1996) 8142.
[22] H. Kim, K.S. Dooley, E.R. Johnson, S.W. North, *J. Chem. Phys.* **124** (2006) 134304.
[23] Y. Bedjanian, G.L. Bras, G. Poulet, *Chem. Phys. Lett.* **266** (1997) 233.
[24] Y. Bedjanian, G.L. Bras, G. Poulet, *J. Phys. Chem. A* **102** (1998) 10501.
[25] S.Y. Lee, *J. Phys. Chem. A* **108** (2004) 10754.
[26] M.K. Gilles, M.L. Polak, W.C. Lineberger, *J. Chem. Phys.* **96** (1992) 8012.
[27] J.A. Coxon, D.A. Ramsay, *Can. J. Phys.* **54** (1976) 1034.
[28] V. Distelrath, U. Boesl, *Faraday Discuss.* **115** (2000) 161.
[29] J.A. Joens, *J. Phys. Chem. A* **105** (2001) 11041.
[30] A. Halkier, T. Helgaker, P. Jorgensen, W. Klopper, H. Koch, J. Olsen, A.K. Wilson, *Chem. Phys. Lett.* **286** (1998) 243.
[31] R.S. Grev, C.L. Janssen, H.F. Schaefer III, *J. Chem. Phys.* **95** (1991) 5128.
[32] P.J. Linstrom, W.G. Mallard, NIST Chemistry WebBook, NIST Standard Reference Database Number 69 (http://webbook.nist.gov), National Institute of Standards and Technology, Gaithersburg MD, 2005.
[33] M.W. Chase Jr. (Ed.), *NIST-JANAF Thermochemical Tables*, 4th edition, American Chemical Society and the American Institute of Physics, Woodbury, NY, 1998.
[34] J.R. Smith, J.B. Kim, W.C. Lineberger, *Phys. Rev. A* **55** (1997) 2036.
[35] L.C. Anderson, D.W. Fahey, *J. Phys. Chem.* **94** (1990) 644.
[36] R. Alqasmi, H.D. Krauth, D. Rohlack, *Ber. Bunsen-Ges.* **82** (1978) 217.
[37] A. Burcat, B. Ruscic, Ideal Gas Thermochemical Database with updates from Active Thermochemical Tables, http://garfield.chem.elte.hu/Burcat/burcat.html, July 5, 2007.
[38] J.F. Stanton, *J. Chem. Phys.* **126** (2007) 134309.
[39] C.E. Miller, E.A. Cohen, *J. Chem. Phys.* **118** (2003) 1.

[40] C.E. Miller, E.A. Cohen, *J. Chem. Phys.* **115** (2001) 6459.
[41] Z. Zhang, P.S. Monks, L.J. Stief, J.F. Liebman, R.E. Huie, S.-C. Kuo, R.B. Klemm, *J. Phys. Chem.* **100** (1996) 63.
[42] B. Ruscic, J. Berkowitz, *J. Chem. Phys.* **101** (1994) 7795.
[43] M.N. Glukhovtsev, A. Pross, L. Radom, *J. Phys. Chem.* **100** (1996) 3498.
[44] N. Begović, Z. Marković, S. Anić, L. Kolar-Anić, *J. Phys. Chem. A* **108** (2004) 651.
[45] M.K. Gilles, R.K. Talukdar, A.R. Ravishankara, *J. Phys. Chem. A* **104** (2000) 8945.
[46] B. Ruscic, J.V. Michael, P.C. Redfern, L.A. Curtiss, K. Raghavachari, *J. Phys. Chem. A* **102** (1998) 10889.
[47] J.A. Seetula, J.J. Russell, D. Gutman, *J. Am. Chem. Soc.* **112** (1990) 1347.
[48] M.T. Rayez, M. Destriau, *Chem. Phys. Lett.* **206** (1993) 278.
[49] B.J. Allan, J.M.C. Plane, *J. Phys. Chem. A* **106** (2002) 8634.
[50] T.J. Dillon, M.A. Blitz, D.E. Heard, *J. Phys. Chem. A* **110** (2006) 6995.
[51] R.M. Chambers, A.C. Heard, R.P. Wayne, *J. Phys. Chem.* **96** (1992) 3321.
[52] Y. Gao, I.M. Alecu, P.-C. Hsieh, B.P. Morgan, P. Marshall, L.N. Krasnoperov, *J. Phys. Chem. A* **110** (2006) 6844.
[53] D.K. Papayannis, A.M. Kosmas, *Chem. Phys. Lett.* **398** (2004) 75.
[54] J.M.C. Plane, D.M. Joseph, B.J. Allan, S.H. Ashworth, J.S. Francisco, *J. Phys. Chem. A* **110** (2006) 93.
[55] D.M. Golden, 2007, personal communication.
[56] D.R. Hanson, A.R. Ravishankara, E.R. Lovejoy, *J. Geophys. Res. [Atmos.]* **101** (1996) 9063.
[57] J. Troe, *J. Phys. Chem.* **83** (1979) 114.
[58] Y. Shi, P. Marshall, *J. Phys. Chem.* **95** (1991) 1654.
[59] J. Troe, *J. Chem. Phys.* **75** (1981) 226.
[60] R. Atkinson, R.A. Cox, J.N. Crowley, J.R.F. Hampson, R.G. Hynes, M.E. Jenkin, J.A. Kerr, M.J. Rossi, J. Troe, Summary of Evaluated Kinetic and Photochemical Data for Atmospheric Chemistry, Section IV - FO_X, BrO_X and IO_X Reactions, http://www.iupac-kinetic.ch.cam.ac.uk, 2006.
[61] J.R. Barker, *Int. J. Chem. Kinet.* **33** (2001) 232.
[62] J.J. Orlando, G.S. Tyndall, *J. Phys. Chem.* **100** (1996) 19398.

CHAPTER **10**

Theoretical Investigation of Atmospheric Oxidation of Biogenic Hydrocarbons: A Critical Review

Jun Zhao* and **Renyi Zhang[1]***

Contents

1. Introduction 177
2. Theoretical Approaches in Atmospheric Hydrocarbon Oxidation Research 178
3. Theoretical Investigation of Biogenic Hydrocarbon Oxidation 183
   - 3.1 Isoprene 183
   - 3.2 Pinenes 199
   - 3.3 Other monoterpenes and sesquiterpenes 206
4. Conclusions and Future Research 207

Acknowledgements 209

References 209

1. INTRODUCTION

Non-methane hydrocarbons (NMHCs) are emitted into the atmosphere from natural and anthropogenic sources and represent an important fraction of volatile organic compounds (VOCs) in the atmosphere [1–4]. Globally, the annual emission of VOCs from biogenic sources (primarily from vegetation) is estimated to be about 1150 Tg C, more than half of which (~55%) is isoprene and monoterpenes [1]. Biogenic VOCs dominate over the anthropogenic counterparts by an order magnitude, which only account for ~10% of the total VOC budget, although the anthropogenic fraction is higher in urban atmosphere (~20–30%) [5,6].

Once emitted into the atmosphere, the VOCs involve in several chemical and physical processes, leading to transformation and removal from the atmosphere [6]. Chemical transformation of VOCs in the atmosphere occurs in sev-

* Department of Atmospheric Sciences and Department of Chemistry, Texas A&M University, College Station, TX 77843, USA

1 Corresponding author. Fax: +1 979 862 4466. E-mail: zhang@ariel.met.tamu.edu

Advances in Quantum Chemistry, Vol. 55
ISSN 0065-3276, DOI: 10.1016/S0065-3276(07)00210-9

eral ways including photolysis, thermal decomposition, and most importantly the reactions with atmospheric oxidants, which include hydroxyl radical OH, nitrate radical NO_3, ozone O_3, and halogen atoms (e.g., Cl, Br or I) [5,6]. Physical removal of VOCs includes dry deposition and wet deposition, eventually depositing them to the Earth's surface. The lifetimes of VOCs in the troposphere depend on their reactivity and the abundance of oxidants, ranging from seconds to years [4–6]. In particular, photochemical oxidation of hydrocarbons in the atmosphere leads to ozone, toxic compound, and secondary organic aerosol (SOA) formation, with major implications for air quality, human health, and climate change [7–10].

Oxidation of hydrocarbons has long been considered as a fundamental problem to atmospheric chemists, both from experimental and theoretical points of view, because of the inherent complexity. The reaction kinetics and mechanism of atmospheric hydrocarbons have been the focuses of numerous researches in both experimental and theoretical aspects. Although advances have been made in elucidation of the VOC oxidation mechanisms, large uncertainty and tremendous numbers of unexplored reactions still remain. Several review articles on the atmospheric degeneration of VOCs have been published [4,11–14]. In this review, recent advances in the application of theoretical methods to the atmospheric oxidation of biogenic hydrocarbons are discussed. We will introduce the backgrounds on the quantum chemical calculations and kinetic rate theories, recent progress on theoretical studies of isoprene and α-, β-pinenes, and studies on other monoterpenes and sesquiterpenes.

2. THEORETICAL APPROACHES IN ATMOSPHERIC HYDROCARBON OXIDATION RESEARCH

The chemistry of atmospheric hydrocarbon oxidation is highly complex. Despite nearly decades of research conducted in this field, there remains an inadequate understanding of this class of reaction mechanisms at the fundamental molecular level. Oxidation of hydrocarbons is initiated by various radical species and proceeds through multiple reaction pathways and steps. At each stage of the chain reactions intermediate organic radicals are produced to propagate or terminate the oxidation process, and the organic radicals play a key role in determining the final product distribution. Currently, the detailed kinetics and mechanism of hydrocarbon oxidation remain highly uncertain, hindering numerical simulations of the VOC oxidation in atmospheric chemical transport models [15]. Most of organic intermediate radicals arising from oxidation of atmospherically important hydrocarbons have not been detected directly in the gas phase. While the enormous chemical complexity in the hydrocarbon oxidation poses an insurmountable obstacle for current analytical techniques in resolving the intermediate steps and isomeric branching, theoretical calculations promise great advantages in providing insights into the intermediate processes.

With rapid development of computer techniques and algorithms to improve the computational capacities and efficiencies over the last few decades, quantum chemical calculations have been employed as a useful and accessible tool in the hydrocarbon oxidation research and demonstrated for the important roles in elucidating the oxidation mechanisms of atmospheric hydrocarbons. Quantum chemical calculations utilize a variety of computational methods, ranging from semi-empirical, density functional methods to *ab initio* calculations [16]. The latter approaches are extensively employed in the field of chemistry of hydrocarbon oxidation, which involves primarily gas-phase reactions. Density functional methods have been widely adopted in many fields of chemistry by merit of their relatively moderate computational cost and high accuracy [17,18]. Geometries obtained from density functional methods, especially those employing hybrid functionals (e.g. B3LYP, B3PW91) are often as reliable as those from Moller–Plesset perturbation (MP2) theory. However, density functional methods in some cases fail to provide accurate energetic information. For example, it has been shown density functional methods fail to predict the potential energy surface in some reaction systems (e.g., in the case of a van der Waals complex and loose transition state) [19–21]. Hence, high levels of *ab initio* calculations are required to obtain more accurate reaction energies and activation energies and other thermochemical parameters. Methods incorporating high level of electron correlation (e.g., MP2 method, CCSD(T) and QCISD(T)) have been widely used in the hydrocarbon oxidation community [22–27]. In recent years, composite methods (e.g., Gaussian series of theory (G1, G2, G3 and G4) and the complete basis set (CBS)) have been applied to the hydrocarbon oxidation and have provided intriguing and highly accurate thermochemical data (e.g., bonding energy, heat of formation and proton affinity) [28,29]. Those methods are based on a well-defined single-configuration wave function. It should be pointed out that one can obtain the multi-reference character of a molecular system by analyzing the single-configuration wave function using the T1 diagnostic method. For example, the single reference method (e.g., CCSD) is not reliable if the T1 diagnostic value of the wave function is larger than 0.044 [30,31]. Many reaction systems are essential multi-reference configurations so that multi-configuration interaction theories are required to accurately describe the reaction systems. The most efficient approach is the complete active space self-consistent field (CASSCF), which has been applied to investigate the complex oxidation mechanisms of the atmospheric hydrocarbons [32–34]. The choice of level of theory and basis set in quantum chemical calculations requires consideration of the best compromise between the accuracy and computational cost. Since the hydrocarbon oxidation systems contain both open and close shell species, caution needs to be excised regarding basis set related error and spin contamination, which can be critical factors in many calculations [35]. The above-mentioned methodologies are discussed extensively in the literature [16–18,22–33,36–40].

Chemical kinetic rate methods including conventional transition state theory (TST), canonical variational transition state theory (CVTST) and Rice–Ramsperger–Kassel–Marcus in conjunction with master equation (RRKM/ME) and separate statistical ensemble (SSE) have been successfully applied to the hydrocarbon oxidation. Transition state theory has been developed and employed in many disciplines of chemistry [41–44]. In the atmospheric chemistry field, conventional transition state theory is employed to calculate the high-pressure-limit unimolecular or bimolecular rate constants if a well-defined transition state (i.e., a tight transition state) exists in the potential energy surface along the reaction coordinate; that is, there exists a surface dividing the reactants from the products and passing through the saddle point of the potential energy surface. The rate constants is expressed by [44]

$$k_{\mathrm{TST}} = \kappa \frac{k_{\mathrm{b}}T}{h} \frac{Q^{\neq}}{Q_{\mathrm{A}}} \exp\left(-\frac{E_{\mathrm{a}}}{RT}\right) \quad \text{(for unimolecular reaction)}, \tag{1}$$

$$k_{\mathrm{TST}} = \kappa \frac{k_{\mathrm{b}}T}{h} \frac{Q^{\neq}}{Q_{\mathrm{A}}Q_{\mathrm{B}}} \exp\left(-\frac{E_{\mathrm{a}}}{RT}\right) \quad \text{(for bimolecular reaction)}, \tag{2}$$

where $Q^{\neq}$ is the partition function of the transition state with the vibrational frequency corresponding to the reaction coordinate removed, Q_{A} and Q_{B} are the partition functions of the reactants. k_{b} and h are Boltzmann constant and the Planck constant, respectively. T is the temperature and κ is the tunneling effect correction factor. The partition functions are calculated from the rotational constants and frequency information by performing the frequency calculations. The activation energy E_{a} is determined from the energy difference between the transition state and the reactants with zero-point energy corrected. If the dividing surface does not pass through the saddle point, a well-defined transition state does not exist (i.e., a loose transition state). The transition state can only be located by varying the dividing surface to minimize the one-way flux coefficient. In this case, the high-pressure limit rate constants are evaluated using CVTST [45,46]. For a bimolecular reaction: A + B → AB, the unimolecular rate is given by

$$k_{\mathrm{uni}} = \frac{kT}{h} \frac{Q_{AB}^{\neq}}{Q_{AB}} \exp\left(-\frac{\Delta E}{kT}\right), \tag{3}$$

where ΔE is the zero-point corrected transition state energy relative to the separated reactants. The unimolecular rates (k_{uni}) are converted to the bimolecular rates (k_{rec}) via the equilibrium constant,

$$\frac{k_{\mathrm{rec}}}{k_{\mathrm{uni}}} = K_{\mathrm{eq}} = \frac{Q_{\mathrm{AB}}}{Q_{\mathrm{A}}Q_{\mathrm{B}}} \exp\left(\frac{\Delta E'}{kT}\right), \tag{4}$$

where $\Delta E'$ is the zero-point corrected reaction energy. The partition functions in Eqs. (3) and (4) are calculated by treating the rotational and translational motion classically and treating vibrational modes quantum mechanically. Unscaled vibrational frequencies and moments of inertia are taken from the frequency calculations. The conserved modes of the transition state are assumed to assemble the product modes. The dependence of the transitional mode frequencies on reaction coordinate is modeled using [47]

$$\upsilon(r) = \upsilon_0 \exp\left[-a(r - r_{\mathrm{e}})\right] + b, \tag{5}$$

where υ_0 is the vibrational frequency in the reactant molecule, r_{e} is the equilibrium bond distance in the reaction coordinate, b is the sum of the rotational constants of the individual reactant molecule. The coefficient a is a constant which can be determined by performing constrained optimization along the reaction coordinate.

Moments of inertia at fixed geometries are calculated by changing only the bond distance in the reaction coordinate. The potential energy surface along the reaction coordinate is modeled by a Morse function including the centrifugal barrier [48]

$$V(r) = \mathrm{De}\left[1 - \exp(-\beta(r - r_e))\right]^2 + B_{ext}(r)J(J+1), \tag{6}$$

where D_e is the bond dissociation energy, $B_{ext}(r)$ is the external rotational constant determined by assuming that the molecular is a symmetric top, and J is assumed to be the average rotational quantum number of a Boltzmann distribution calculated using the external constant of the molecule at the equilibrium configuration. The β parameter for the Morse function is expressed as $\beta = (2\pi^2\mu/D_e)^{1/2}\nu$, where μ is the reduced mass of the bond atoms, D_e is the bond dissociation energy, ν is the vibrational frequency of the reaction coordinate in parent molecule.

Many hydrocarbon oxidation reactions are strongly exothermic and the excessive internal energy is retained and partitioned over the products, leading to formation of chemically excited intermediated species. The statistical-dynamical master equation in combination with standard RRKM or variational RRKM (vRRKM) theory has been employed to assess the fate of the vibrationally excited species governed by competition between collisional stabilization and consecutive chemical reactions [48–50]. The excited species are stabilized by transferring their energy to the surrounding bath gas. An exponential model is employed to model the collision energy transfer with a specified average energy in the RRKM calculations. An example to illustrate the application of kinetic methods to elucidate the oxidation mechanism of atmospheric hydrocarbons is presented in Figure 10.1, showing the reaction diagram for the OH–isoprene alkoxy radicals formed in the reaction of peroxy radical with NO. The reaction of peroxy radical (RO_2) with NO leads to the formation of a vibrationally excited peroxynitrite (ROONO*) with an internal energy inherited from the thermal reactants [51,52]. The nascent molecular population of the ROONO is described by a shifted thermal distribution [53],

$$f^{\mathrm{ROONO}}(E) = \frac{W_{\mathrm{ROONO}}(E - E_{(-x)})\exp[-(E - E_{(-x)})/k_bT]}{\int_0^\infty W_{\mathrm{ROONO}}(\varepsilon)\exp(\varepsilon/k_bT)\,d\varepsilon} \tag{7}$$

for $E \geqslant E_{(-x)}$ with the energy being counted from the ground state of the ROONO. W_{ROONO} is the sum of the states of the ROONO, $E_{(-x)}$ is the barrier height of the reverse reaction of x. Since the entrance channel is barrierless, a Morse potential including the centrifugal barrier (as mentioned above) can be used to variationally locate the transition states for NO addition to the hydroxyperoxy radical as a function of energy. The collisional energy loss of the ROONO* prior to its dissociation into RO + NO_2 modifies its internal energy distribution to $f^{\mathrm{RO}}(E)$ at dissociation by assuming only the energy in excess of the activation barrier partitioning statistically between the dissociating fragments [53,54],

$$f^{\mathrm{RO}}(E) = \frac{\rho_{\mathrm{RO}}(E)W_{\mathrm{RO}}(E_{tot} - E)}{\int_0^{E_{tot}} \rho_{\mathrm{RO}}(\varepsilon)W_{\mathrm{RO}}(E_{tot} - \varepsilon)\,d\varepsilon}, \tag{8}$$

where E_{tot} is the overall disposable energy and ρ_{RO} is the density of states of the RO, that is the number of electron states per unit energy interval. The dissociation of the ROONO* and partitioning of the disposable excess energy over two fragments (RO + NO_2) and over the degrees of freedom of their relative motion can be calculated according to the separate statistical ensemble (SSE) theory [55,56]. The probability of formation of RO radical with in a given vibrational (rotational) level i is given by [55,56]

$$P_{RO_i}^{E_{tot}}(E_{RO}) = \left(N_{RO_i}(E_{RO}) \int_0^{E_{tot}-E_{RO}} \left[N_{NO_2}(E_{NO_2}) N_{rel\,mot}(E_{tot} - E_{RO} - E_{NO_2}) \right] dE_{NO_2} \right)$$
$$\times \left(\int_0^{E_{tot}} \left\{ \left(\sum_i N_{RO_i}(E_{RO}) \right) \int_0^{E_{tot}-E_{RO}} \left[N_{NO_2}(E_{NO_2}) N_{rel\,mot} \right.\right.\right.$$
$$\left.\left.\left. \times (E_{tot} - E_{RO} - E_{NO_2}) \right] dE_{NO_2} \right\} dE_{RO} \right)^{-1}, \quad (9)$$

where $N_{RO_i}(E_{RO})$, $N_{NO_2}(E_{NO_2})$, and $N_{rel\,mot}(E_{tot} - E_{RO} - E_{NO_2})$ are the density of state of the ith energy level for RO, the density of state of NO_2, and the density of state of relative motion, respectively. For the relative motion, $N_{rel\,mot}(E)$ varies with E^2 since the relative motion is treated as unhindered motion which $N(E)$ varies with its degrees freedom y as $E^{(y/2)-1}$.

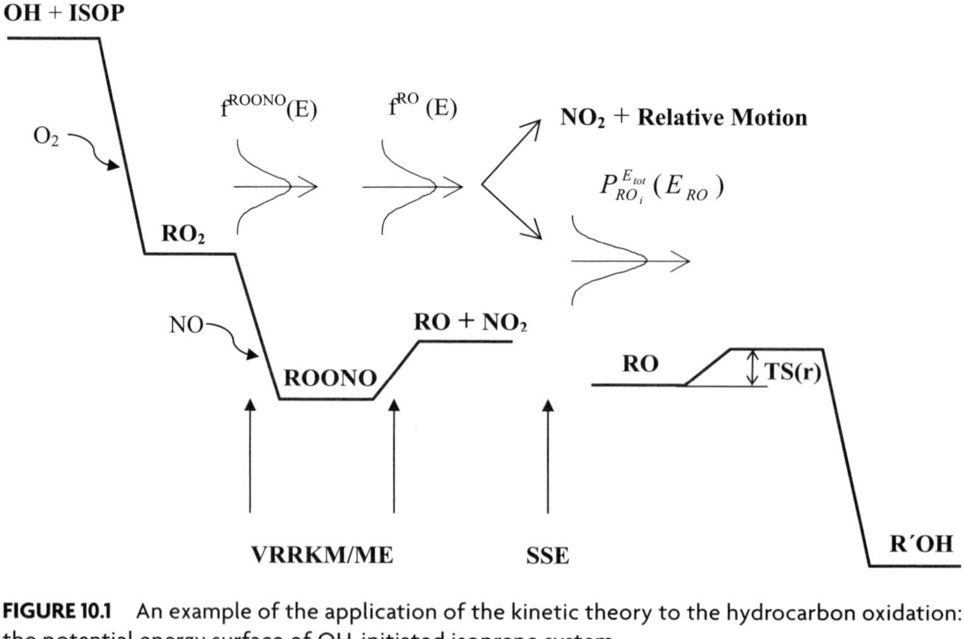

FIGURE 10.1 An example of the application of the kinetic theory to the hydrocarbon oxidation: the potential energy surface of OH-initiated isoprene system.

The microcanonical rate constant at energy E can be expressed using the RRKM theory [48],

$$k_r(E) = \frac{W_r(E - E_{TS(r)})}{h\rho(E)}, \quad (10)$$

where W_r is the sum of states of the transition state for reaction r. The concentration of a relevant species is related to a balance over all gain and loss processes for a given energy level i according to statistical dynamic master equation [48],

$$\frac{dn_i}{dt} = Rf_i - \omega n_i + \sum_i P_{ij}n_j - \sum_l k_{li}n_i, \quad (11)$$

where n_i is the concentration of the species having internal energy E_i, R is the overall formation rate, f_i is the normalized energy distribution, P_{ij} is the energy transfer probability from j to i, k_{li} is the reaction rate constant for pathway l, and ω is the collision frequency. Assuming the steady-state condition, Eq. (11) becomes

$$RF = \left[\omega(I - P) + \sum K_r\right]N^s \equiv JN^s, \quad (12)$$

where the vector/matrix symbols correspond to those in Eq. (11) and I denotes the unit matrix. The steady-state population N^s follows from $N^s = RJ^{-1}F$, and the rate of reaction r is $D_r = \sum(K_rN^s)_i$. The relative yields are then expressed by $Dr/R = \sum(K_rJ^{-1}F)_i$, and the stabilization fraction follows from $S/R = 1 - \sum D_r/R$. It should be pointed out that while most of the reactions can be solved using traditional RRKM approach, experimental and theoretical studies have shown that non-RRKM dynamics is important for moderate to large-sized molecules with various barriers for unimolecular dissociation [57,58]. In these cases, non-RRKM behavior needs to be taken into account and direct chemical dynamic simulation is suggested to serve this purpose [58].

3. THEORETICAL INVESTIGATION OF BIOGENIC HYDROCARBON OXIDATION

3.1 Isoprene

Isoprene (2-methyl-1,3-butadiene, $CH_2{=}C(CH_3){-}CH{=}CH_2$) is one of the most abundant hydrocarbons emitted by the terrestrial biosphere with a global average production rate of 450 Tg yr^{-1} and is sufficiently reactive to influence oxidation levels over large portions of the continental troposphere [59,60]. An accurate and complete knowledge of the atmospheric chemistry of isoprene is critical to elucidate chemical mechanisms of atmospheric hydrocarbons in urban and regional environments [15]. Atmospheric oxidation reactions of isoprene are initiated by attacks from the atmospheric oxidants, including OH, O_3, NO_3, and halogen atoms.

3.1.1 OH-initiated oxidation

Since isoprene is emitted from vegetation only during daylight hours, the reaction with OH is expected to be the dominant tropospheric removal pathway. The reaction between isoprene and OH occurs almost entirely by OH addition to the >C=C< bonds, yielding four possible hydroxyalkyl radicals. Under atmospheric conditions, the hydroxyalkyl radicals react primarily with oxygen molecules to form hydroxyalkyl peroxy radicals (RO_2). In the presence of nitric oxide NO, the subsequent reactions of the hydroxyalkyl peroxy radicals lead to the formation of hydroxyalkoxy radicals (RO). Alternatively, a small fraction of peroxy radicals will react with NO to produce hydroxylalkyl nitrates ($RONO_2$). Under low NO_x conditions, the peroxy radicals will undergo permutation reactions (e.g., self reactions and cross reactions) or reaction with HO_2 to produce a variety of products (e.g., hydoxyalkoxy radicals, hydroperoxide, carbonyls, alcohol and carboxylic acids) [61–64]. The dominant tropospheric reaction of β-hydroxyalkoxy radicals is believed to be decomposition, leading to the formation of various oxygenated (e.g., methyl vinyl ketone, methacrolein, 3-methyl furan) and nitrated organic compounds while for δ-hydroxyalkoxy radicals, the isomerization pathway has been shown to be dominant over the decomposition, leading to the formation of hydroxycarbonyls, which have been detected and quantified experimentally [65–68]. Figure 10.2 depicts the mechanistic diagram of the OH-initiated oxidation of isoprene [69].

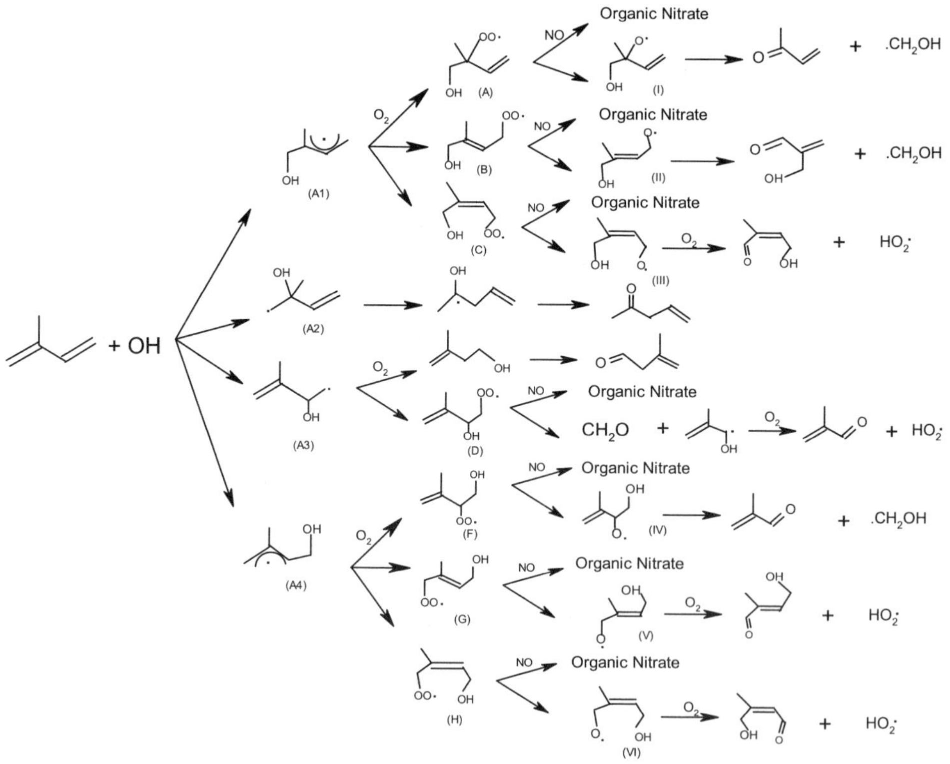

FIGURE 10.2 Mechanistic diagram of OH-initiated isoprene oxidation: detailed reaction pathways and major products.

Systematic theoretical studies of OH–isoprene oxidation have been conducted in recent years. Although the initial step of this reaction is relatively well understood, some discrepancies exist for the potential energy surface of the initial step. In the previous work by Zhang, North, and co-workers (e.g., [51,69–72]), a series of theoretical calculations have been performed to elucidate the mechanisms of the OH–isoprene reactions. Isomeric branching ratios of the initial step have been determined using density functional theory (DFT) and *ab initio* methods coupled to kinetic theories, e.g., CVTST and RRKM/ME. Density functional and *ab initio* methods obtained geometry, frequency and energetic information of the OH–isoprene adducts, which were then incorporated into CVTST to calculate the branching ratios. Four adduct isomers have been globally located along the potential energy surface by gradient corrected density functional theory (DFT) in conjunction with a split valence polarized basis set (B3LYP/6-31G(d,p)). This level of theory has been evaluated for many hydrocarbon oxidation systems and has been shown to be quite robust in geometry optimization and frequency calculations. High levels of *ab initio* methods (e.g., CCSD(T) and MP2 with various basis sets) were then employed to perform single energy calculations to reevaluate the relative energies and activation energies for the relevant reaction systems. A correction factor (CF) term was introduced and defined as the energy difference between the two MP2 levels: MP2/6-311++G(d,p) and MP2/6-31G(d). Hence, the highest level of theory corresponded to CCSD(T)/6-31G(d) + CF for most of *ab initio* calculations. A branching ratio of 0.56:0.02:0.05:0.37 for the isomers A1–A4 is obtained (labeling refers to Figure 10.2) and a high-pressure limit reaction rate of 1.0×10^{-10} cm^{-3} $molecule^{-1}$ s^{-1} is calculated according to the CVTST calculations [70,71]. In addition, the results also show that the inter-conversions (e.g., A1 $\leftrightarrow$ A2 and A3 $\leftrightarrow$ A4) between OH–isoprene adducts are hindered due to very high activation energies (Figure 10.3) [70]. The calculated rate constant is in agreement with the experimental values [69,73–77], confirming the accuracy of the selected levels of theory. Kinetic calculations employing RRKM/ME method show a consistence of fall-off behavior for OH–isoprene reaction with the recent experimental measurements at low pressure [70]. The branching ratio at the low-pressure limit was predicted by Stevens [75], with values of 0.72 and 0.28 for isomer A1 and A4, respectively, and less than 1% for both isomer A2 and A3. The β-hydroxyl alkyl radical from internal addition of OH has been suggested to undergo cyclization to produce α-hydroxy radical [78] and result in the formation of C5 carbonyls, whose yields have been estimated recently [68]. While Lei *et al.* [70,72] found that addition of OH to isoprene proceeds without an activation barrier, Francisoco-Marquez *et al.* [79] using both density functional theory and Moller–Plesset perturbation theory to the second-order (MP2) showed the formation of pre-reactive complexes and large activation energies for the internal carbon addition. They concluded that the internal addition pathways are unlikely responsible for the formation of 3-methyl furan; the likely pathways for the formation 3-methyl furan are via external addition according to their recent theoretical calculations [80]. Francisoco-Marquez *et al.* [80] also predicted negative activation energies for addition at the terminal carbon atoms for the OH–isoprene reaction. In general, activation energies obtained using DFT are rather unreliable.

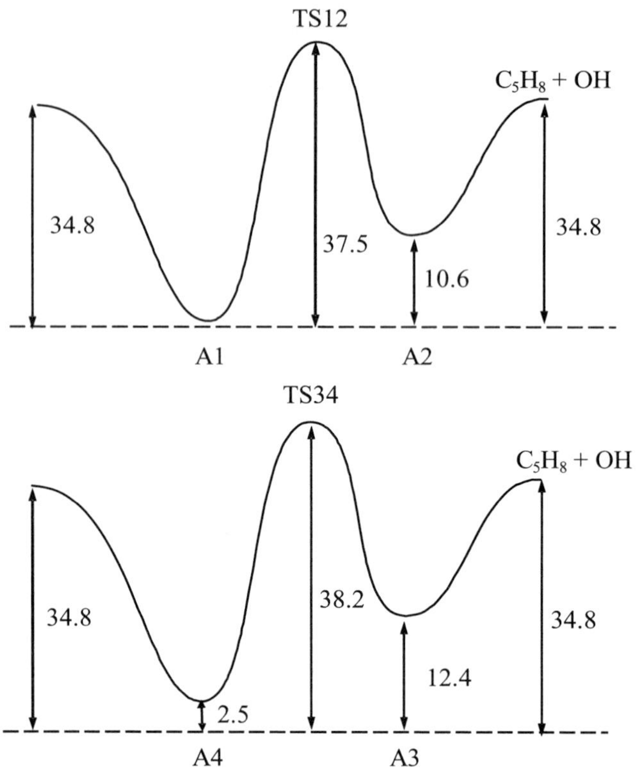

FIGURE 10.3 Schematic potential energy surfaces for the inter-conversions of OH–isoprene adduct reactions calculated at the CCSD(T)/6-311G(d,p)//B3LYP/6-31G(d,p) level of theory.

Peroxy radicals involve in many reactions as key intermediates in propagating catalytic cycles, which lead to ozone formation in the troposphere. The reactivity of the peroxy radicals depends strongly on the structure of the radicals. The geometries of six isoprene peroxy radical isomers (without Z-configuration isomers) are obtained and the peroxy radicals are more stable than the separated reactants (isoprene, OH and O_2) by 47–53 kcal mol^{-1} at CCSD(T)/6-31G(d) level of theory [81]. The CVTST calculations predict branching ratios of 0.60:0.40 for isomer A and B and 0.78:0.22 for isomer F and G [81]. The strong propensity for isomers A, B and F is predicted in good agreement with the model of Jenkin and Hayman [82], but qualitatively different from the model of Paulson and Seinfeld [83]. Jenkin and Hayman suggested the branching ratios based on the assumption that the peroxy radical center would form at the more substituted site. The predicted high-pressure limit rate constants are significantly faster than the corresponding rates associated with O_2 addition to aromatic–OH adducts, but slower than the reaction of O_2 with alkyl and hydroxyalkyl radicals. It is speculated that the calculated branching ratios and rate constants of the peroxy radicals are mostly dependent on their binding energies and the nature of the transition states, not on their relative stability. Peroxy radicals have been detected and a rate constant of 7×10^{-13} cm^3 $molecule^{-1}$ s^{-1} for the reaction of OH–isoprene adducts with O_2 has been determined from the decay of the peroxy radicals detected by the chemical ionization mass spectrometry (CIMS) [84], while Park *et al.* [85] reported a rate constant of 2×10^{-12} cm^3 $molecule^{-1}$ s^{-1} for the same reactions using laser photolysis/laser-induced fluorescence technique.

The reaction of hydroxy peroxy radicals (RO_2) with NO represents one of the most crucial tropospheric processes, leading to terrestrial ozone formation or NO_x removal and chain termination. The C–C bond fission of the resultant OH–isoprene alkoxy radicals was investigated by Dibble [65]. It is found that the low barrier heights (0.7–2.1 kcal mol^{-1} at B3LYP/6-311G(2df, 2p) level) for the decomposition of four β-hydroxyalkoxy radicals allow this channel to be dominant, while in contrast decomposition of δ-hydroxyalkoxy radicals possesses high barrier heights and endothermic characteristics (16–20 kcal mol^{-1}), rendering this pathway unimportant in the atmosphere. The results from theoretical calculations provide useful information to assess the potential detection of these alkoxy radicals in the laboratory based on the stabilities and it has been pointed out that the β-hydroxyalkoxy radical is difficult to be probed due to their short lifetime (picoseconds). Lei and Zhang [66] carried out higher levels of *ab initio* calculations to investigate the geometries and energetics of the alkoxy radicals and their decomposition pathways. They found that activation energies for the preferable decomposition pathway fall in the range of 6–7 kcal mol^{-1} for the β-hydroxyalkoxy isomers and 18–21 kcal mol^{-1} for δ-hydroxyalkoxy isomers, providing more accurate and reliable activation energies. The activation barriers for decomposition of alkoxy radicals calculated by Dibble are lower than those calculated by Lei *et al.*, and the difference between the two studies is explained since the B3LYP level of theory is notorious for under-prediction of barrier height. Thermal decomposition of hydroxylalkoxy I and IV is mainly responsible for the formation of the major isoprene-OH products: methyl vinyl ketone (MVK) and methacrolein (MACR) (Figure 10.4) [66]. Ramirez-Ramirez *et al.* [86] investigated the formation and interconversion of cis- and trans-isomer of the MVK from the alkoxy radical I at MP2/6-31G(d)//B3LYP/6-31G(d,p) level and found that the barrier heights are comparable to the ones reported by Lei and Zhang [66] at the similar level. Park *et al.* performed kinetic calculations using RRKM/ME method to investigate the temperature and pressure dependent rate constants between 220 and 310 K and found similar features in the fall-off region for the β-hydroxyalkoxy radicals [87]. The high-pressure limit rate constants for the β-hydroxyalkoxy radicals obtained from RRKM/ME calculations are close to the values from TST calculations [66]. The rapid decomposition of the energized β-hydroxyalkoxy radicals may play an important role in the overall decomposition pathways since significant fraction of them (except for isomer I) was found to undergo prompt reactions. However, a rate constant of 3×10^4 s^{-1} was reported using laser photolysis/laser-induced fluorescence technique [88], which is significant lower than those values from Lei and Zhang [66] and Dibble [65], in the range of 10^8–10^{10} s^{-1}. A rate constant of 9×10^{-12} cm^3 $molecule^{-1}$ s^{-1} for reactions of the peroxy radicals with NO was reported from the decay of the peroxy radicals using the CIMS detection [84], which is consistent with the predicted rate constant using the CVTST [66]. A higher

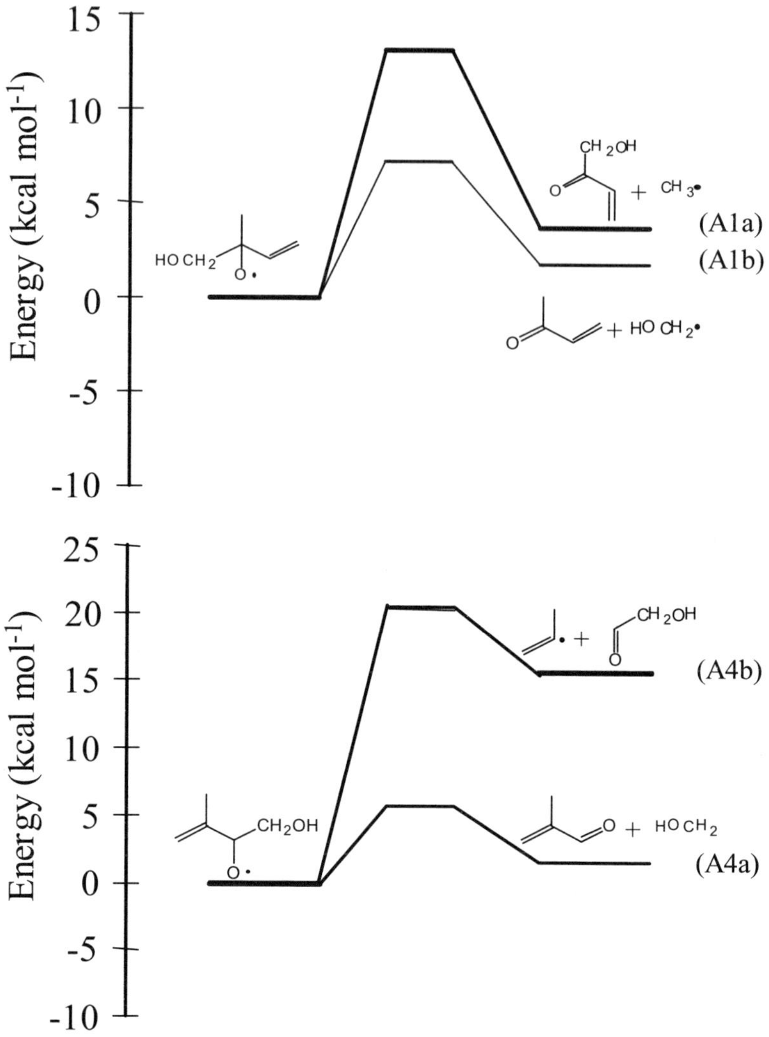

FIGURE 10.4 Potential energy surfaces for the C–C bond fission of radicals I and IV obtained at the CCSD(T)/6-31G(d) + CF//B3LYP/6-31G(d,f) level of theory.

value of 2.5×10^{-11} cm^3 $molecule^{-1}$ s^{-1} was obtained using laser photolysis/laser-induced fluorescence technique [88]. For the β-hydroxy-peroxy radicals with an unsaturated C=C bond, addition of one more oxygen molecule leads to unimolecular ring-closure (Figure 10.5) (forming a bridge between the oxygen of radical center and the unsaturated terminal carbon), forming a 6-membered ring structure radicals (cycloperoxide-peroxy radicals) with high oxygenated content, which potentially alter the final product distribution of the OH–isoprene system and have important implication for secondary organic aerosol (SOA) formation. For example, Vereecken and Peeters presented a theoretical study of such reactions and

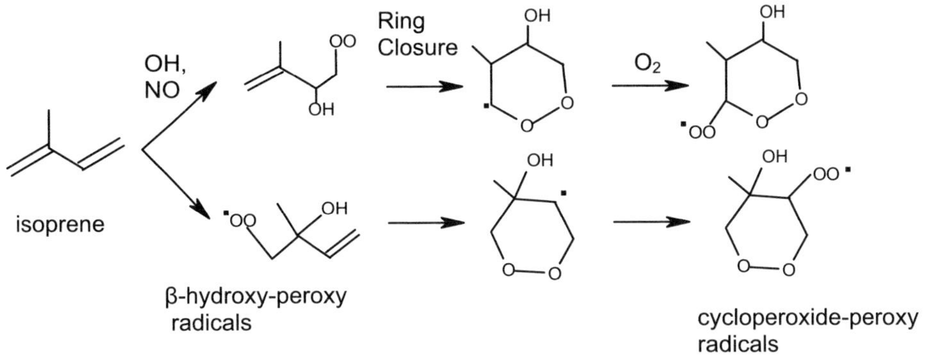

FIGURE 10.5 Mechanistic representation of the ring-closure reactions for the β-hydroxy-peroxy radicals arising from the OH–isoprene reactions.

found that in low NO_x or even moderately polluted atmosphere (e.g., 1 ppbv NO) the rates of ring closure dominate or are comparable to the reaction with NO [89].

The reactions of hydroxy peroxy radicals (RO_2) with NO lead to major formation of alkoxy radicals and minor production of organic nitrates. It has been proposed that these two channels are bridged via a common hydroxy peroxy nitrite intermediate. Similar to the scheme in Figure 10.1, the entrance channel of the RO_2–NO reaction is exoergic by about 20 kcal mol^{-1}, leading to a vibrationally excited hydroxyperoxy nitrite (ROONO*), which subsequently reacts via unimolecular reactions or collisional stabilization [51]. The excited or thermalized nitrite undergoes two possible prompt unimolecular reactions, isomerization to the isoprene nitrate ($RONO_2$) or decomposition to a hydroxyalkoxy radical (RO) and NO_2. The high-pressure limit rate constants for the formation of the ROONO isomers at 300 K were calculated using CVTST method, with values of $3\text{–}10 \times 10^{-12}$ cm^3 molecule^{-1} s^{-1} [51], consistent with the two experimental results mentioned above (9×10^{-12} and 2.5×10^{-11} cm^3 molecule^{-1} s^{-1}) [84,88]. The fate of the excited hydroxyperoxy nitrites was assessed using the steady-state master equation (ME) formalism in conjunction with the variational-RRKM (vRRKM/ME) method [51]. The entrance channel and the exit channel (dissociation of the nitrite to RO and NO_2) were found to be barrierless according to the quantum chemical calculations because the calculated energy relative to the equilibrium peroxynitrite increased monotonically when the O–O bond length was successively increased at the B3LYP/6-31G(d,p) level. The results indicated that at all energies above the entrance channel (20.2 kcal mol^{-1}) the dissociation rates are larger than the collision rate and the formation of thermalized nitrite is insignificant under ambient conditions. On average, loss of the nitrite internal energy due to collisions is insignificant (1–4 kcal mol^{-1}).

Rearrangement of the peroxynitrite to nitrate has been a long outstanding issue. A previous theoretical study suggested that this rearrangement proceeds via a three-centered transition state, which for HOONO corresponds to an activation energy of about 60 kcal mol^{-1} [89–92]. However, the three-centered transition state is likely hindered by the substitute of a carbon chain to the hydrogen atom, and

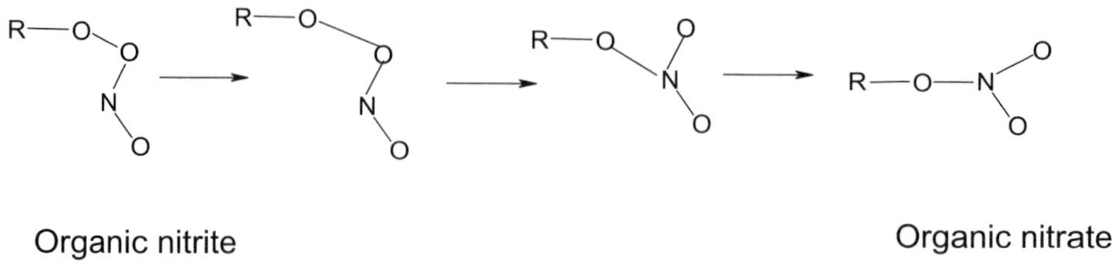

FIGURE 10.6 Schematic representation of isomerization of organic nitrite to nitrate.

the rearrangement of ROONO to $RONO_2$ may go through a partial O–O bond fission followed by RO and NO_2 recombination at elongated O–O bond lengths (Figure 10.6). Several experimental studies have reported the nitrate yield from the OH-initiated oxidation of isoprene, ranging from 4 to 12% [93]. Using the vRRKM/ME analysis and the experimentally determined nitrate yield, Zhang *et al.* [51] predicted that the activation barrier of ROONO isomerization to $RONO_2$ is between 0.4 and 1.1 kcal mol^{-1} below and 3.3–3.7 kcal mol^{-1} above the dissociation energy of ROONO, significantly lower than the barrier height for the three-center transition state in HOONO [89–92].

For both thermalized and activated β-hydroxyalkoxy radicals, the dominant fate is decomposition, leading to the formation of various oxygenated organic compounds. For the δ-hydroxyalkoxy radicals, however, the decomposition channel is minor due to the high barrier height, and other channels (e.g., H-abstraction and H-migration) are more important. Zhao *et al.* [94] investigated the competing pathways of H-abstraction by oxygen molecules and 1,5 H-shift of the δ-alkoxy using multiple level calculations (B3LYP, CCSD(T), and MPW1K with various basis sets) to obtain the geometries and energetic information of these two reactions. The B3LYP significantly underestimates the activation barriers, while the CCSD(T) and MPW1K give comparable results in predicting the activation barriers. Kinetic calculations employing vRRKM/ME formalism and separate statistical ensemble (SSE) theory showed that a significant fraction of the chemically excited alkoxy radicals undergo prompt 1,5 H-shift. The results revealed that 1,5 H-shift of thermalized δ-alkoxy radicals dominates over H-abstraction by O_2 (see Figure 10.7). The consecutive reactions of the dihydroxy radical intermediates with O_2 form hydroxycarbonyls which have been qualitatively observed [95,96] and quantified [68], with a yield of about 19%. Ramirez-Ramirez *et al.* [97] performed *ab initio* calculations of subsequent reactions of the dihydroxyl radicals and found that these reactions occur quickly in atmospheric conditions and proceed without an energy barrier, providing an explanation for the observed formation of hydroxylcarbonyls. Alternatively, the resonant allylic dihydroxyl radicals can form dihydroxyperoxy radicals by O_2 addition and subsequently react with NO to produce the corresponding alkoxy radicals [98]. These second-generation of peroxy and alkoxy radicals exhibit two intramolecular hydrogen bonds and the simultaneous transfer of two H-atoms across the hydrogen bonds with a barrier of only about 5 kcal mol^{-1} in the alkoxy radicals, but about 20 kcal mol^{-1} in the peroxy radicals. These double H transfer reactions in the alkoxy radicals are exother-

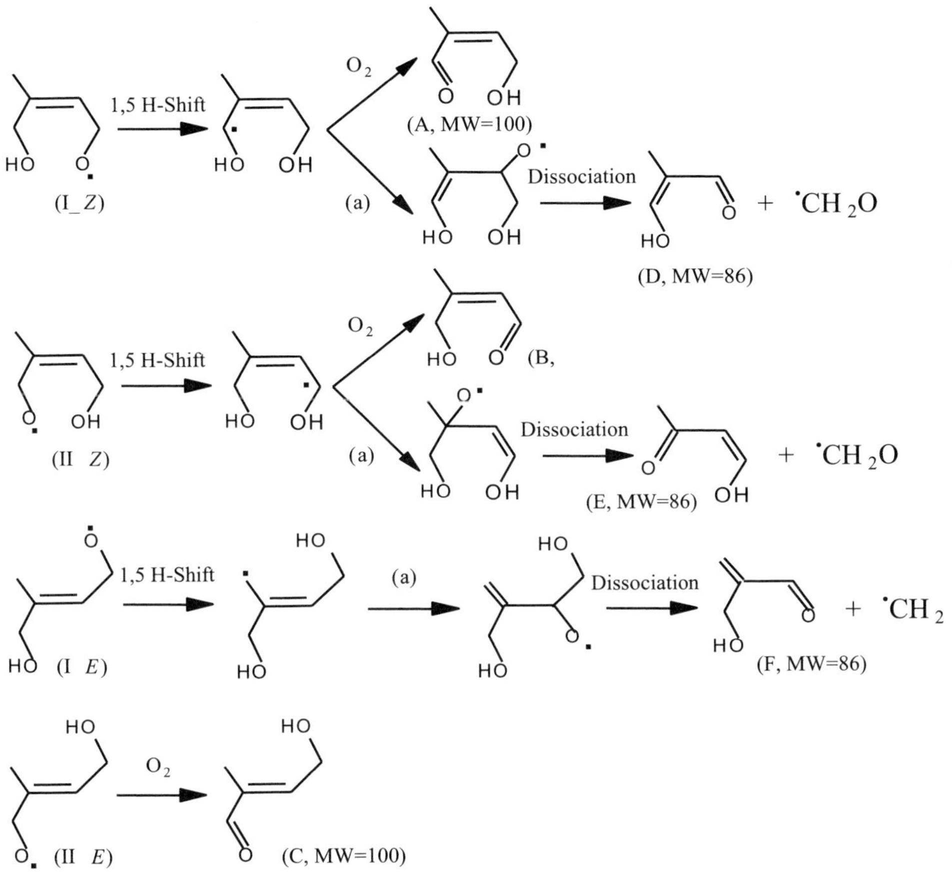

FIGURE 10.7 Reaction mechanisms of the δ-alkoxy radicals arising from the OH–isoprene reactions.

mic and their fates are governed by competition between collisional stabilization and prompt decomposition. The thermalized alkoxy radicals are further decomposed to form C4-hydroxycarbonyls, which has been detected and quantified recently [68].

3.1.2 O_3-initiated oxidation

While reaction with OH is the dominant tropospheric removal pathway for isoprene, a substantial amount of isoprene is also degraded by reaction with ozone in the troposphere. There exists increasing evidence that the OH radicals are formed during the course of ozonolysis of isoprene, providing an important source of nighttime OH radicals on the regional scale [99]. Ozonolysis of isoprene may also play an important role in SOA formation since low volatile organic compounds generated from this process are potential precursors of SOA [100,101]. The initial ozonolysis reaction occurs primarily by cycloaddition of O_3 to either one of the >C=C< double bonds of isoprene, resulting in the formation of primary ozonide

(POZ). The reaction enthalpy is retained as the internal energy of the products, resulting in formation of the vibrationally excited ozonide, which subsequently undergoes unimolecular decomposition to yield a chemically activated biradical, known as the carbonyl oxide or Criegee intermediate (CI), and an aldehyde (e.g., MVK, MACR or formaldehyde). A total of nine carbonyl oxides (methyl vinyl carbonyl oxide, derived from 1,2-ozonide; isopropyl carbonyloxide, derived from 3,4-ozonide and formaldehyde oxide, H_2COO from both ozonide) are formed and a large fraction of these carbonyl oxides with ample internal energy facilitates prompt unimolecular reactions or stabilization. Two primary reaction pathways exist for the vibrationally excited carbonyl oxide, ring closure to form dioxirane or H-migration to form a hydroperoxide intermediate. There are several pathways leading to the formation of OH radicals: (1) the resulting hydroperoxide decomposes to form OH and RCO radicals; (2) decomposition of dioxirane may lead to formation of vibrationally excited organic acids which may also dissociate to form OH; and (3) the cleavage of the O–O bond in the primary ozonides results in formation of OH radicals via an unstable hydroperoxide intermediate. Figure 10.8 [69] summarizes the mechanistic diagram of the ozonolysis of isoprene, which only consider the major reaction pathways. The OH formation yield has been measured experimentally by several methods (e.g., indirect scavenger method and direct laser-induced fluorescence detection) with values in a wide range of 0.19–0.53 [102–106]. Theoretical investigation of isoprene ozonolysis fo-

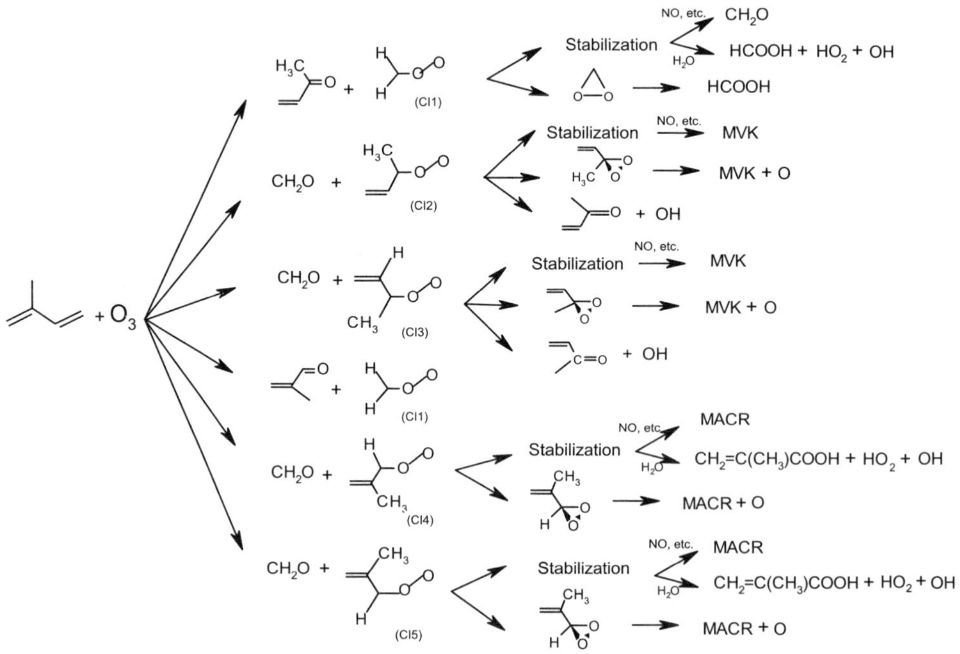

FIGURE 10.8 Mechanistic diagram of ozonolysis of isoprene.

cuses on the formation of OH radicals and the subsequent reactions of the carbonyl oxides.

The reaction of ozone with isoprene involves a series of reactions and intermediates, with excessive reaction energy propagating among the intermediates and products. The reaction mechanism of ozone with isoprene is not as well understood as that of OH with isoprene in both experimental and theoretical aspects. In addition, since the lifetimes of most of the intermediates are very short, experimental detection of these intermediate species is extremely difficult. Hence, theoretical methods provide an alternative tool to investigate the formation of the intermediates and their fates in the ozone–isoprene system. Gutbrod *et al.* [107] performed both experimental and theoretical investigation of the formation of various syn and anti stereoisomers of the carbonyl oxides from ozonolysis of isoprene and reported their structures and energies at the B3LYP/6-31G(d,p) level of theory. An OH formation yield of 19% was obtained and the OH radicals were primarily from the decomposition of carbonyl oxides depending on a syn-positioned methyl (alkyl) group and the interaction with the terminal O atom of a carbonyl oxide. Quantum chemical calculations showed that the formation of hydroperoxide is more favorable than isomerization to dioxirane for carbonyl oxides, with a barrier of 15.5 verse 23 kcal mol^{-1} for the two channels at level of CCSD(T)/6-31G(d,p)//B3LYP/6-31G(d,p), respectively. Zhang and Zhang [108] performed density functional theory and *ab initio* calculations to investigate the formation and unimolecular reactions of primary ozonides and carbonyl oxides from isoprene ozonolysis and found no significantly preferential branching for the initial step of O_3 cycloaddition to the two double bonds of isoprene. Cleavage of thermalized primary ozonides to form carbonyl oxides occurs with barriers of 11–16 kcal mol^{-1} above the ground state of the primary ozonides. However, the collisional stabilizations of the primary ozonides are negligible under atmospheric conditions. Prompt dissociation of the excited primary ozonides leading to formation of syn and anti conformations of carbonyl oxides should be considered due to the large exothermicity associated with the formation of the primary ozonides. The OH formation is shown to occur primarily via decomposition of the carbonyl oxides with the syn-positioned methyl (alkyl) group, which is more favorable than isomerization to form dioxirane (by 1.1–3.3 kcal mol^{-1}). An OH formation yield of 0.25 from prompt and thermal decomposition of the carbonyl oxides was estimated using the transition state theory and master equation formalism. A calculated rate constant of 1.58×10^{-17} cm^3 $molecule^{-1}$ s^{-1} for the O_3-isoprene is in agreement with the experimental value (1.4×10^{-17} cm^3 $molecule^{-1}$ s^{-1}). In another study, Zhang *et al.* [109] evaluated the fate of primary ozonides and carbonyl oxides using a statistical–dynamical master equation and transition state theory and the results showed that excited carbonyl oxides promptly dissociate to produce OH (11%) or isomerize to form dioxirane (32%), while the remaining (57%) are collisionally stabilized. An OH formation yield of 0.25 obtained from previous study [108] was distributed to prompt (0.11) and thermal (0.14) decomposition of the carbonyl oxides. The results reveal slow thermal decomposition of the carbonyl oxide to form OH due to high activation barriers. Except for OH radicals, the yields of the stable products formaldehyde, MACR, and MVK are estimated to be 0.67, 0.21, and 0.12, respectively. A very recent study by Kuwata *et al.* [110] provided a new insight into the isoprene ozonolysis using CBS-QB3 calculations and RRKM/master equation simulations. Kuwata *et al.* investigate the formation of the methyl vinyl carbonyl oxides and the interconversion among four possible conformers, and their isomerization. They predicted an OH formation yield of 19%, quantitatively consistent with the experiment value assuming that all thermalized syn-methyl carbonyl oxides form OH. A predicted MVK formation yield of 0.17 is in agreement with the current recommended experimental MVK yield of 0.16 [110–112]. Natural bond order (NBO) analysis reveals that the effect of the vinyl group has profound impact on the chemistry of the methyl vinyl carbonyl oxide by lowering the barrier to rotation about the C=O bond and allowing the carbonyl oxide to cyclize to form a dioxole [110], which was not considered previously. Dioxole formation with an activation barrier of 13.9 kcal mol^{-1} at CBS-QB3 level is predicted to be favored over other channels (e.g., vinyl hydroperoxide formation, dioxirane formation, and collisional stabilization). Kuwata *et al.* [110] also concluded that two dioxole derivatives, 1,2-epoxy-3-butanone and 3-oxobutanal, are major products of isoprene ozonolysis, which need to be investigated experimentally in the future study. Hence, there still exists discrepancy among the above studies: it needs to be verified that the complete basis set (CBS) employed in Kuwata *et al.* provide more accurate prediction of the branching ratios and hence the OH yield than those in the two previous studies.

An important fraction of carbonyl oxides are collisionally stabilized and further react with other atmospheric species (e.g., water, sulfuric dioxide and sulfuric acid [113]). The reaction with water vapor represents one of the most important processes for the atmospheric degradation of the stabilized carbonyl oxides and is known to lead to the formation of α-hydroxy hydroperoxide compounds, organic acids, aldehydes, and H_2O_2 [114–118]. Figure 10.9 shows the schematic mechanisms for the reactions between carbonyl oxides and water. Two reaction pathways have been proposed: (1) An α-hydroxy hydroperoxide is formed by addition of the water molecule to the carbonyl oxide; (2) Water-assisted hydrogen migration occurs to the terminal oxygen of the COO group and subsequent OO bond cleavage lead to the formation OH radicals. Aplincourt and Anglada [119] presented a

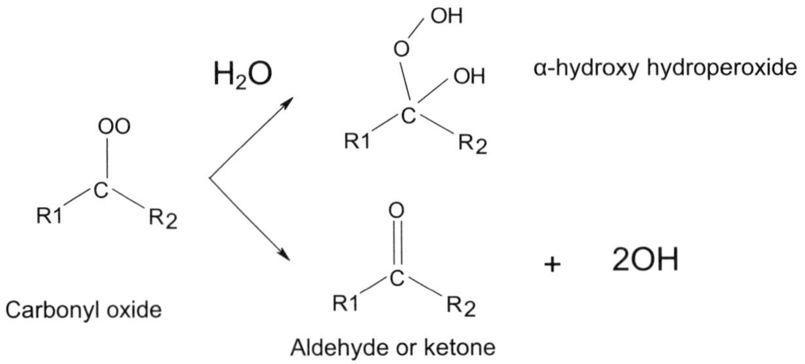

FIGURE 10.9 Schematic representation of reactions of carbonyl oxides with water.

theoretical investigation of the stabilized isoprene carbonyl oxides with water and found formation of a stable hydrogen-bond complex (about 6 kcal mol^{-1} more stable than the reactants). The water addition involve in a five-membered ring transition state and is the most favorable reaction channel with the activation enthalpies of about 14 kcal mol^{-1} and 9 kcal mol^{-1} at G2M-RCC5//B3LYP/6-311 + G(2d,2p) level (a modified Gaussian 2 theory) of theory for carbonyl oxides derived from 1,2- and 3,4-ozonide, respectively. The water-assisted hydrogen migration path corresponds to the water catalyzed reaction of the unimolecular decomposition pathway involving in a seven-membered ring transition state and is a possible source of atmospheric OH radicals. Furthermore, the water-catalyzed effect reduces the activation enthalpy by ~7 kcal mol^{-1} (at CCSD(T)/6-31G(d) + CF level) compared to the decomposition pathways reported by Zhang and Zhang [108]. However, the resultant activation barriers are still higher than those of the water addition reaction, rendering this pathway unimportant compared to the water addition reaction. Kinetic calculations employing transition state theory showed that only two water-catalyzed reaction channels involving the hydrogen transfer from the β-hydrogen (with respect to the COO group) in syn position to the terminal oxygen of the COO group are accessible with a branching ratio of 13–20% at temperature 273–298 K. Another theoretical study by Aplincourt and Anglada [120] showed that the unimolecular decomposition of both α-hydroxy hydroperoxides requires very high activation enthalpies (~44 kcal mol^{-1}), rendering this pathway implausible in the atmosphere. On the other hand, lower activation enthalpies have been found for the water-assisted decomposition of these α-hydroxy hydroperoxides. The reaction between α-hydroxy hydroperoxides derived from 1,2-ozonide and H_2O leads to the formation of MVK and H_2O_2 exclusively. The reaction between α-hydroxy hydroperoxides derived from 3,4-ozonide and H_2O has two possible reaction channels, MACR+H_2O_2 and methacrylic acid (MTA)+H_2O, respectively. However, the activation enthalpies and the computed unimolecular reaction constants reveal that the reaction is almost exclusively MACR + H_2O_2. Ryzhkov and Parisa [121,122] presented theoretical studies of the reactions of carbonyl oxides with water and its clusters. They found that the most kinetically favorable pathway is the reaction with water dimer, which involves a seven-member ring transition state. The authors also found that the relative humidity may influence the rates and reaction mechanisms. For example, the rate constants of the carbonyl oxides and water clusters are increasing with relative humidity. The results indicate that the reaction mechanisms may be different under different humidities.

3.1.3 NO_3-initiated oxidation

The nitrate radical, formed by the reaction of NO_2 and ozone, is a major tropospheric nighttime oxidant and its mixing ratio can reach up to several hundred ppt at night [123]. The nighttime reaction between isoprene and nitrate radical contributes significantly to the degradation of isoprene [124,125]. The reaction between isoprene and NO_3 occurs by NO_3 addition to the $>C=C<$ bonds, forming thermodynamically favored nitrooxyalkyl radicals. Under atmospheric conditions, the nitrooxyalkyl radicals react primarily with oxygen molecules to form

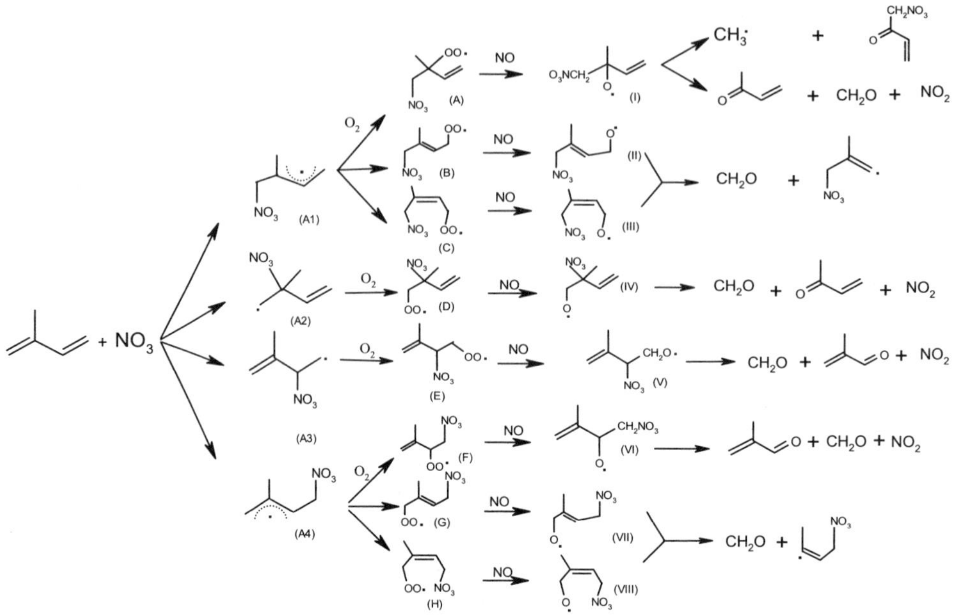

FIGURE 10.10 Mechanistic diagram of NO_3-initiated oxidation of isoprene.

nitrooxyalkyl peroxy radicals. Similarly to the OH–isoprene reaction, addition of O_2 occurs only at the carbons β to the NO_3 position for the NO_3–isoprene adducts of internal NO_3 (C2- and C3-position) addition, but takes place at two centers (β or δ to the NO_3 position) for terminal NO_3 (C1- and C4-position) addition, leading to the formation of β- and δ-nitrooxyalkyl peroxy radicals. The δ-nitrooxyperoxy isomers include both *E* and *Z* configurations. The nitrooxyalkyl peroxy radicals can further react with NO or engage in a self-reaction or cross-reaction with other peroxy radicals to form nitrooxyalkoxy radicals, or react with HO_2 to form nitrooxyperoxide. The nitrooxyalkoxy radicals can undergo isomerization or decomposition, or abstraction by O_2, leading to various oxygenated or nitrated organic species. Figure 10.10 represents the mechanistic diagram of the NO_3-initiated oxidation of isoprene. Current, only a few theoretical studies have been reported in the literature on the NO_3 initiated isoprene oxidation. Suh *et al.* [126] reported the structures and energies of the NO_3–isoprene adduct isomers employing density functional theory and *ab initio* calculations and the rate constants of the formation of the NO_3–isoprene adducts and their isomeric branching have been calculated using the canonical variational transition state theory. The calculated isomeric branching suggested preferential addition of NO_3 to C1 position (a ratio of 0.84), consistent with previous product studies of the NO_3–isoprene reaction system [127,128]. Zhang and Zhang [129] investigated unimolecular decomposition of nitrooxyalkyl radicals from NO_3–isoprene reaction using quantum chemical calculations, CVTST, and master equation analysis. The results indicated that the channel leading to the formation of the two oxiranes proposed by Berndt and Boge [128] is negligible. Instead, the chemically activated nitrooxyalkyl radicals become stabilized by collision and the thermal nitrooxyalkyl radicals mainly react with O_2 to form the nitrooxyalkyl peroxy radicals, which contribute to the formation of MACR and MVK. Zhao and Zhang [130] presented the structures and energetics of the eight peroxy radicals arising from NO_3–isoprene reactions using density functional theory and *ab initio* methods and the rate constants for addition of O_2 to the NO_3–isoprene adducts using CVTST. The results provided the isomeric branching ratios between the NO_3–O_2–isoprene peroxy radicals and predicted a ratio of 7.4 of 1,2-addition to 4,3-addition, in agreement with the experimental value reported by Berndt and Boge [128], almost two fold higher than the experimental values by Skov *et al.* [127]. No explanation was given for the discrepancy between the two experimental results. Zhao and Zhang [131] also investigated the structures and energetics of the alkoxy radicals and their decomposition pathways. For both thermalized and activated β-alkoxy radicals the dominant fate is decomposition, but for δ-hydroxy alkoxy radicals the decomposition channel is minor due to a large barrier and the H-abstraction and H-migration channels are more important, similar to the alkoxy radicals of OH–isoprene systems.

3.1.4 Cl-initiated oxidation

The reaction between isoprene and chlorine atoms may play an important role both in the continental troposphere and in the marine boundary layer [132–135]. The initial reaction between isoprene and Cl is believed to proceed mainly by Cl addition to the C=C bond, forming a Cl–isoprene adduct radical. Alternatively, Cl can abstract one hydrogen atom from methyl group of isoprene to form HCl and an allylic radical. Under the tropospheric conditions, the adduct reacts primarily with oxygen molecules to form the β-chloroalkenylperoxy radicals. In the presence of nitric oxide NO, initial Cl radical addition at the C1-, C2-, C3-, or C4-positions with subsequent addition of O_2 at the C2-, C1-, C4-, and C3-positions, respectively, leads to the formation of β-chloroalkenylalkoxy radicals. The dominant tropospheric reaction of β-chloroalkenylalkoxy radicals is unimolecular decomposition or bimolecular reaction with oxygen molecule, leading to the formation of various oxygenated and nitrated organic compounds (Figure 10.11). A rate constant of $(3.4–4.6) \times 10^{-10}$ cm^3 molecule^{-1} s^{-1} has been determined experimentally for the reaction between Cl and isoprene [136–140] and the branching ratio for the initial addition and abstraction reaction has been inferred from the HCl formation yield, with values between 13% and 17% [136,137]. Various final products have been identified including CO, CO_2, formyl chloride, formic acid, methylglyoxal, hydrogen chloride, and 1-chloro-3-methyl-3-buten-2-one (CMBO, a tracer for chlorine radicals) [136,137]. However, direct experimental data concerning intermediate processes of the oxidation reactions of isoprene with Cl are very limited, due to the lack of sensitive detection schemes for these species. Lei and Zhang [141] reported the first quantum chemical calculations on the Cl–isoprene oxidation and found the isomers with Cl addition to the terminal C1-position are the most energetically favorable. The activation energies for interconversion between isomers are low ($\sim$2.8–4.7 kcal mol^{-1}), indicating that thermal equilibrium between the isomers is easily established. The formation rates and isomeric branching ratios were

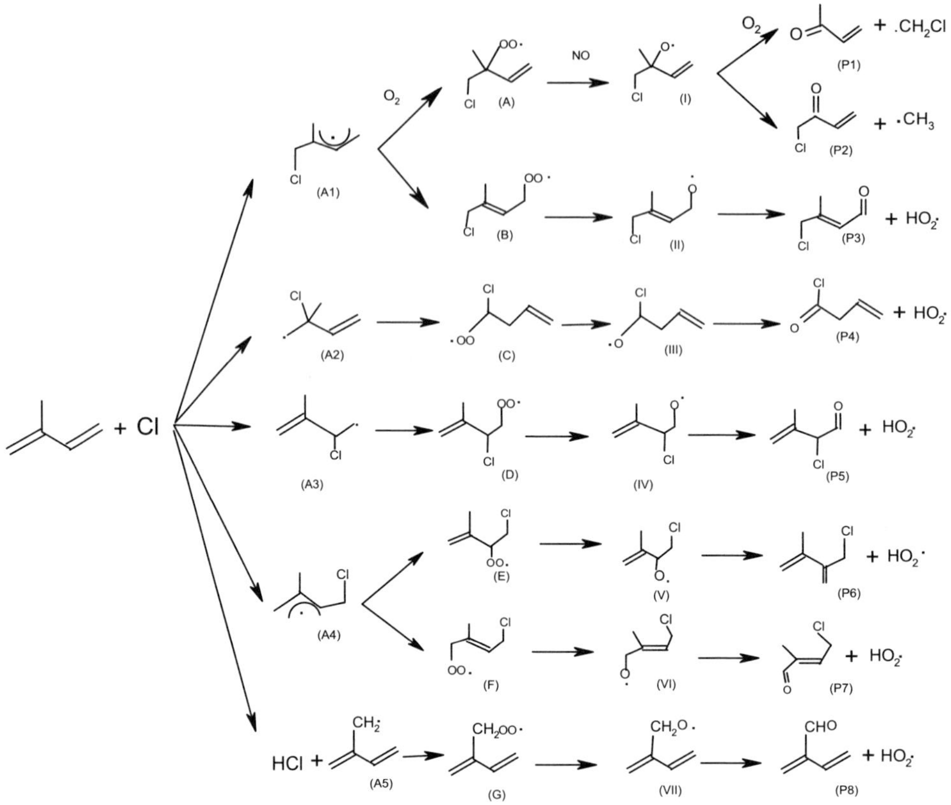

FIGURE 10.11 Mechanistic diagram of Cl-initiated oxidation of isoprene.

calculated using CVTST [142]: a rate constant of 4.2×10^{-10} cm^3 molecule^{-1} s^{-1} is determined and a strong preference is found for Cl addition to the terminal carbon, with an amount of ~90% for the terminal addition. RRKM/ME calculations show that almost all of the adducts from the terminal addition are collision stabilized, while adducts from internal addition undergo exclusively prompt isomerization and the collisional stabilization can be negligible under atmospheric conditions. No activation barrier is found for Cl addition to isoprene and the Morse potential well represents the energetics along the reaction coordinate. Brana and Sordo [143] investigated the abstraction channel and found that the reaction proceeds through an association-elimination mechanism in which a weakly bound intermediate is formed via a six-membered ring transition state, followed by HCl elimination leading to the final the allylic radical formation. DFT and *ab initio* molecular orbital calculations have been performed to investigate the structures and energetics of the Cl–O_2–isoprene peroxy radicals: the peroxy radicals are ~39 to 43 kcal mol^{-1} more stable than the separated reactants at the CCSD(T)/6-31G(d) level of theory [144]. The O_2 addition is found to be barrierless and the rate constants are calculated using CVTST based on the Morse potential to describe the reaction coordinate, in agreement with the experimental values [145]. The results indicate that the two β-chloroalkenylperoxy radicals with initial Cl addition at C1 and C4 positions and subsequent O_2 addition at C2 and C3 positions, respectively, play an important role in determining the reaction pathways and final product distributions of the Cl–isoprene reaction system. The predicted branching ratio facilitates the application of CMBO as a tracer to deduce chlorine radical concentrations in the urban and regional atmosphere. Density-functional theory and *ab initio* calculations have been employed to explore the structures and energetics of the chloroalkenyl alkoxy radicals and their decomposition pathways [146]. The activation and reaction energies of C–C bond scission of the alkoxy radicals are in the ranges of 12–25 and 23–22 kcal mol^{-1}, respectively. The results indicate that C–C bond decomposition of the chloroalkenyl alkoxy radicals is rather slow and likely plays a minor role in the Cl–isoprene reactions, consistent with laboratory studies which show small MVK and MACR yields but significant CMBO formation. A yield of 30% is predicted for CMBO formation from Cl-initiated oxidation of isoprene. Considering the potential importance of the Cl–isoprene reactions, more studies of Cl-initiated isoprene oxidation are needed to assess the product distribution of the Cl–isoprene reactions in the future.

Recently, the explicit oxidation mechanism of isoprene initiated by OH, O_3, NO_3, and Cl, incorporating the most recent laboratory and theoretical studies, has been evaluated using a box model [69]. The updated mechanism provides explicit reaction steps and detailed intermediates for isoprene oxidation and facilitates more accurate modeling of isoprene photochemistry in the atmosphere.

3.2 Pinenes

Monoterpenes account for ~10% of the total biogenic hydrocarbons globally, with α- and β-pinene being the most abundant monoterpenes emitted in North America [1]. The pinenes (α- and β-pinene) are removed from the atmosphere primarily by reactions with OH radicals, NO_3 radicals and O_3 [12]. Due to their high reactivity toward oxidants, atmospheric oxidation of the pinenes contributes significantly to ozone and SOA formation in the troposphere. The OH reactions dominate during daytime while reactions with NO_3 and O_3 are significant at night. The OH reactions proceed mainly by initial addition to the endocyclic (for α-pinene) or exocyclic (for β-pinene) C=C double bond, resulting in the formation of β-hydroxylalkyl radicals. Subsequent reaction of β-hydroxylalkyl radicals with other atmospheric constituents (e.g., O_2 and NO) or isomerization leads to multifunctional products (e.g., aldehydes, ketones and carboxylic acids) in both gas phase and particulate phase. Alternatively, H-abstraction can account for ~10–20% of the total reaction rate, forming allyl resonance stabilized α-pinenyl radicals. Figure 10.12 shows an overview of the OH-initiated oxidation of α-pinene.

The initial step of the OH–pinenes reaction has been investigated using density functional theory and *ab initio* methods by Fan *et al.* [147] (for α- and β-pinenes) and Ramirez-Ramirez *et al.* [148] (for β-pinene). Both studies considered site-specific and stereo-specific (anti and syn position relative to the $-C(CH_3)_2-$ bridge) addition of OH to the pinene C=C bond, forming four adducts for each pinene.

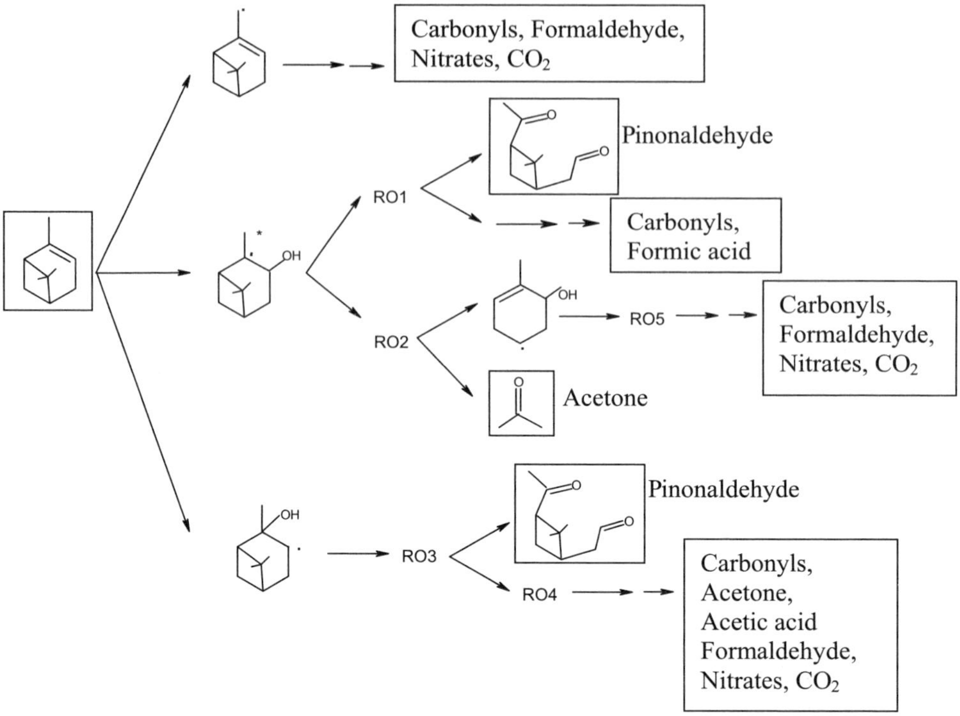

FIGURE 10.12 Mechanistic diagram of the reactions between OH and α-pinene (RO1–RO5 represent the various alkoxy radicals formed in the reactions).

Fan *et al.* [147] found that the OH addition to both pinenes proceeds without an activation barrier, in contrast to a well-defined transition state (for β-pinene only) with level-of-theory dependent barriers reported by Ramirez-Ramirez *et al.* [148]. Fan *et al.* [147] pointed out that the bond lengths (2.7–2.8 Å) of the reaction coordinate in the loose transition states determined by the CVTST is much longer than those (2.1–2.2 Å) in the tight transition states reported by Ramirez-Ramirez *et al.* for β-pinene [148]. The transition states reported by Ramirez-Ramirez *et al.* suffer from high spin contamination so that the geometries of the transition states and the activation energies of the reactions are less reliable. Fan *et al.* reported a rate constant of 5.0×10^{-11}, 6.1×10^{-11} cm^3 $molecule^{-1}$ s^{-1} at B3LYP/6-31G(d,p) level for α- and β-pinene respectively, consistent with the recommended experimental rate constants [149–152]. Peeters *et al.* [153] presented a detailed mechanism for the OH-initiated oxidation of α-pinene in the presence of NO_x using various theoretical methods, e.g., structure–activity relationship (SAR) (It is wide used to model the correlation between the chemical activity and the structure of a family compounds. Such models can be used to predict the activity of an unknown compound based on its structure), quantum chemical methods, transition state theory (TST), and RRKM/master equation analysis. OH addition to the C=C bond accounts for 88% of the reaction, leading to two chemically activated β-hydroxylalkyl radicals with nearly equal branching ratios. H-abstraction channel is found to be a minor route (~12%), but contributes significantly to the overall yield of formaldehyde. The overall product distribution under the simulation conditions are predicted on a molar basis: 35.7% pinonaldehyde, 18.8% formaldehyde, 19% organic nitrates, 17.9% acetone, 25.9% other (hydroxy)carbonyls, 17.2% C1 and C2 carboxylic acids, and 30.7% CO_2, in general agreement with experimentally measured yields. However, under low NO_x conditions the theoretically predicted yields differ significantly, with 59.5% pinonaldehyde, 12.6% formaldehyde, 13.1% organic nitrates, 11.9% acetone, 16.4% other (hydroxy) carbonyls, 8.7% CO_2, and a negligible fraction of C1 and C2 carboxylic acids, possible due to the dominance self- and cross-reactions of the peroxy radicals under low NO_x conditions, which current mechanisms are not able to accurately represent. This theoretical study outlined a comprehensive mechanism and provided a detailed product distribution for the OH-initiated oxidation of α-pinene.

The gas-phase products and mechanisms of OH-initiated oxidation of the pinenes have been explored by numerous experimental studies. Pinonaldehyde from α-pinene and nopinone from β-pinene are observed and quantified as the most abundant products in the gas phase, with a yield of 6% to 87% for pinonaldehyde [154–162] and 17% to 25% for nopinone [160,161,163], respectively. Significant discrepancy exists regarding the reported pinonaldehyde yield from α-pinene, with different analytic techniques and photochemical conditions. The mechanisms leading to the formation of pinonaldehyde and nopinone have been proposed. The resultant β-hydroxylalkyl radicals from addition of OH radicals to the C=C double bond of α-pinene results react rapidly with O_2 and subsequent reactions with NO form β-hydroxylaloxy radicals. The C–C bond fragmentation and subsequent abstraction by O_2 produce pinonaldehyde. For nopinone from β-pinene, a primary and a tertiary β-hydroxylaloxy radical are formed in a similar way as α-pinene, followed by C–C bond scission and subsequent abstraction by O_2, leading to the formation of nopinone, with formaldehyde as co-product. Several studies have reported theoretically predicted yields of pinonaldehyde, with a value of 67% by Fan *et al.* [147], 60% by Peeters *et al.* [153]. A nopinone yield of 23% is predicted by Fan et al [147]. The subsequent oxidations of pinonaldehyde and nopinone by OH radicals have been investigated theoretically. Fantechi *et al.* [163] developed a detailed mechanism of product formation in atmospheric conditions for six H-abstraction sites of pinonaldehyde–OH reaction using SARs, quantum chemistry methods, and TST or RRKM/ME analysis. They found that the overall predicted product yields under high-NO conditions are 22.9% (4-hydroxynorpinonaldehyde), 9.9% (acetone), 12.9% (formaldehyde), 30.3% (organic nitrates), 73.8% (CO_2), 11.4% (HC(O)OH), 16.6% (norpinonaldehyde and CO), and 16.6% (other (hydroxy)(poly)carbonyls). Under lower NO conditions, the theoretically predicted yields differ only regarding 4-hydroxynorpinonaldehyde (38.2%), HC(O)OH (0.3%), and nitrates (26.2%). Lewis *et al.* [164] presented a theoretical study of OH–nopinone reaction and obtained branching ratios of H-abstraction from seven different positions. They found that the most preferable abstraction is at the bridgehead position, with a branching ratio of 23%, inconsistent with the one derived from SARs, which suggests much less oxidation in this position. However, the results are in agreement with available experimental evidence [165], suggesting formation of significant amounts of products such as 1-hydroxynopinone during terpene oxidation. Calculated rate coefficients using density functional theory with the KMLYP functional are surprisingly in good agreement with the experimental results, possibly because the KMLYP method is a hybrid DFT method with a combination of exchange and correlation functional to minimize the errors for the hydrogen atoms, while preserving the correct asymptotic values.

Another important product from both pinenes oxidation is acetone, an important source of HO_x in the upper troposphere [166] and low stratosphere and also an important precursor for peroxyacetyl-nitrate (PAN) in the atmosphere [167]. Acetone formation from both pinenes has been experimentally determined, with a yield of 5–18% from α-pinene [157–159,168–170] and 2–13% from β-pinene [159, 160,168–170]. Several mechanisms have been proposed for the formation of acetone from OH-initiated oxidation of α- and β-pinene. A theoretical study by Vereecken and Peeters [171] showed that the initial OH-addition reaction leads to the formation of chemically activated tertiary β-hydroxylalkyl radicals with an initial internal energy of 32 kcal mol^{-1}. Subsequent prompt isomerization of the activated radicals to 6-hydroxymenthen-8-yl is comparable to the thermal stabilization according to the kinetic calculations using RRKM/ME. The resultant radical (6-hydroxymenthen-8-oxy) reacts with O_2 and NO and subsequent decomposition of the β C=C bond provide one possible route for formation of acetone. Figure 10.13 shows the major formation mechanism of acetone from OH–α-pinene reactions. The authors reported an acetone yield of 8.5%, in agreement with available experimental data [157–159,168–170]. This channel is also confirmed by Dibble [172] in a theoretical study: the activation energy of ~8-12 kcal mol^{-1} is obtained for the rearrangement of the tertiary β-hydroxylalkyl radical, which is crucial for subsequent formation of acetone. An alternative pathway for acetone formation has been proposed, involving C–C (either one of C–C bonds connected to the radical center) scission of the tertiary β-hydroxylalkoxy radical, followed by alkoxy radical formation via reaction with O_2 and NO and subsequent decomposition, leading to the formation of acetone. A less possible pathway for the formation of acetone is suggested via OH abstraction, which has been demonstrated to be implausible in a theoretical study by Vereecken and Peeters [171]

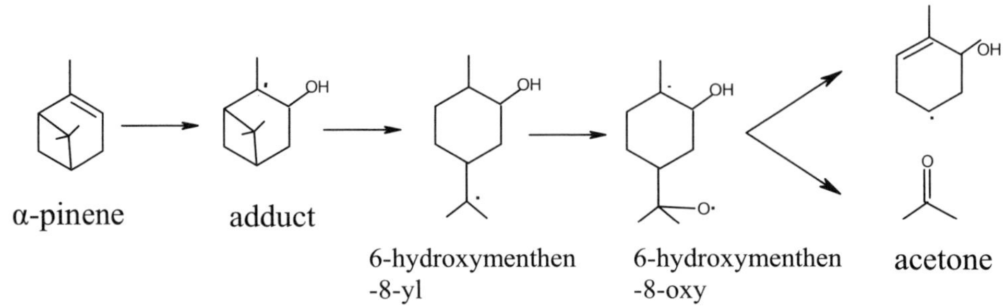

FIGURE 10.13 Major channel of acetone formation from OH–α-pinene reactions.

due to the constraints of the bicyclo skeleton. Fan *et al.* [147] predicted a yield of 9.5% for acetone from OH–α-pinene reaction. Similar mechanisms for the formation of acetone from β-pinene have been proposed via OH-addition channel (tertiary OH–β-pinene adduct). Several other gas-phase products (e.g., formaldehyde, nitrates, hydroxycarbonyls) are also observed and quantified with various yields. For example, formaldehyde is quantified with a molar yield of 19% and 23% [158,170] for α-pinene and 45% and 54% for β-pinene [155,170]. Organic nitrates have been quantified only in two experiments, with the yield of 1% [154] and 18% [157], showing significant disagreement. The mechanisms leading to these products have been proposed, although remain to be scrutinized. A detailed gas-phase mechanism [173] for the degradation of α-pinene by OH has been proposed based on box model simulations of laboratory measurements performed in the presence as well as in the absence of NO by Noziere *et al.* [158]. The results showed that the levels of OH, NO, NO_2 are well reproduced in the model and the model succeeds in reproducing the average apparent yields of pinonaldehyde, acetone, total nitrates and total PANs in the experiments performed in the presence of NO. In the absence of NO, pinonaldehyde is fairly well reproduced, but acetone is largely underestimated. The results also showed that the main oxidation channels differ largely according to photochemical conditions. The pinonaldehyde yield is estimated to be about 10% in the remote atmosphere and up to 60% in very polluted areas. This finding explained the large existing discrepancy for the measured pinonaldehyde yield reported previously [154–161].

As previously mentioned in the OH–isoprene reaction, the ring-closure reaction in both unsaturated peroxy- and oxy radicals are expected to be important pathways in the atmospheric chemistry of the OH–pinenes system. Vereecken and Peeters [89] predicted a peroxy ring-closure rate constant of 2.6 s^{-1} at 298 K, well above the rate of the NO reaction even at 1 ppbv NO and those with HO_2/RO_2 at the atmospheric concentrations. The formation of ring closure peroxide not only changes the product distribution, e.g., reducing acetone formation, but also these peroxides are highly oxygenated and hence low volatile, contributing to SOA formation. For the unsaturated structures discussed above, the resulting oxy radical RO likely undergoes ring closure, forming cyclic ether structures. For example, the oxy radicals form from α-pinene after opening of the 4-membered ring and subsequent reactions with O_2 and NO. The oxy radical site can attack the double bond, forming a cyclic ether radical which can in turn react quickly with O_2 to form a cycloetherperoxy radical. However, this new channel is not currently well-understood and more systematic investigation is needed in order to better understand this unique channel.

Ozonolysis of pinenes contributes to OH radical and SOA formation. The latter has been speculated to be responsible for the formation of "blue haze". Hence, it is of vital importance to quantitatively understand the roles of ozonolysis of α- and β-pinenes on both OH and SOA formation in the troposphere. The O_3 reaction with both pinenes follows the Criegee mechanism, similar to that of isoprene. The initial step proceeds through cycloaddition of O_3 to the C=C double bond of each pinene, forming a primary ozonide (POZ). The available reaction energy is retained as the internal energy of the product, resulting in formation

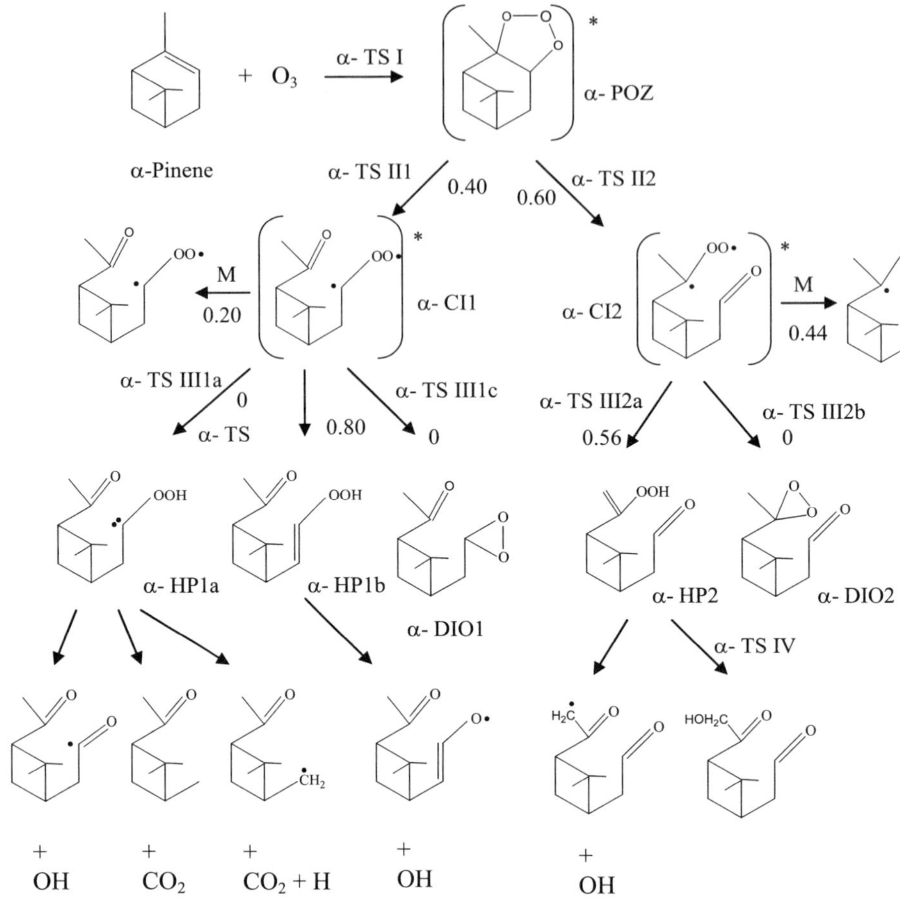

FIGURE 10.14 Mechanistic diagram of ozonolysis of α-pinene.

of the vibrationally excited primary ozonide. The excited primary ozonide subsequently undergoes unimolecular isomerization (for α-pinene) or decomposition (for β-pinene) to yield chemically activated carbonyl oxides and/or aldehydes (for β-pinene). A large fraction of the carbonyl oxides have ample internal energy and are subjected to prompt unimolecular reactions or collisional stabilization. Two reaction pathways were proposed for the carbonyl oxide, ring closure to form dioxirane or H-migration to produce a hydroperoxide intermediate. The hydroperoxide subsequently undergoes isomerization or decomposition, leading to formation of OH, carbonyls, CO_2, and a variety of other products, some of which are potential SOA precursors. Thermally stabilized CIs may react with other atmospheric compounds (e.g., H_2O, SO_2, HCHO, H_2SO_4). Numerous experiments have been carried out to investigate the products and SOA formation from ozonolysis of pinenes [174–179]. However, only one theoretical study has been reported for ozonolysis of pinenes. Zhang and Zhang [180] presented a comprehensive theoretical study of the reactions of ozone with α- and β-pinenes using combined quantum-chemical

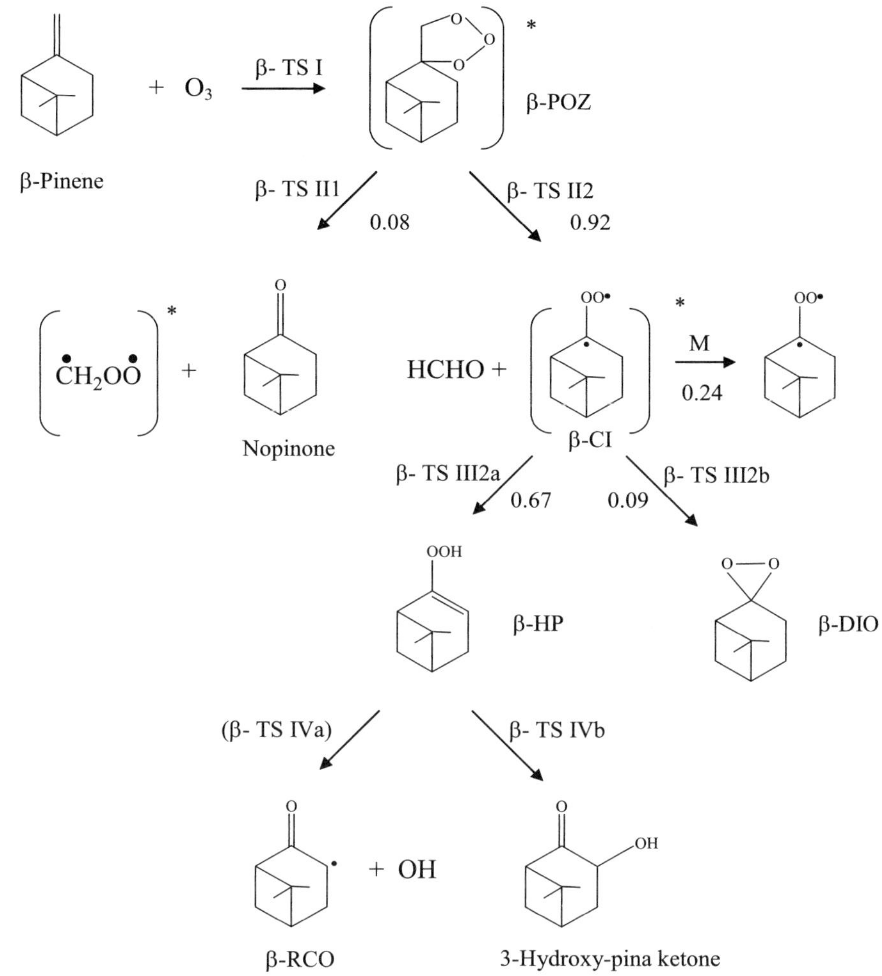

FIGURE 10.15 Mechanistic diagram of ozonolysis of β-pinene.

methods (e.g., DFT and CCSD(T)) and RRKM/ME to determine the structures and energetics of relevant species in the reaction systems and to predict the kinetics and mechanisms of pinenes ozonolysis. Figures 10.14 and 10.15 represent the mechanistic diagrams of the ozonolysis of α-, β-pinene, respectively. The formation of POZs occurs with the activation barriers of 0.3 kcal mol^{-1} for α-POZ and 2.0 kcal mol^{-1} for β-POZ. The addition reactions are highly exothermic, with the reaction energies of 55.1 and 51.1 kcal mol^{-1} for α- and β-pinenes, respectively. The results indicate a negligible collision stabilization of the POZs. Cleavage of both POZs forms two carbonyl oxides along with aldehyde or ketone, respectively. For both activated and stabilized CIs from α- and β-pinene ozonolysis, H-migration to hydroperoxides dominates over ring closure into dioxirane. Significant fractions of CIs are stabilized with 0.34 for α-CIs and 0.22 for β-CI, consistent with the experimental observations. The theoretical calculations suggest that stabilized CIs contribute significantly to previously measured high yield of OH for α-pinene, but stabilized CIs contribute negligibly to the experimental OH yield for β-pinene because of the reaction with formaldehyde.

Reaction with NO_3 radicals represents another important gas phase loss pathway for α-pinene during the night time. Under elevated NO_x conditions, reaction with NO_3 can also be significant during the daytime and contribute substantially to the monoterpene decay. However, only a few studies have been performed on NO_3-initiated oxidation of pinenes [181–184]. To our knowledge, no theoretical studies have been performed to investigate the mechanistic behaviors of these reactions.

3.3 Other monoterpenes and sesquiterpenes

The α- and β-pinenes with other monoterpenes ($C_{10}H_{16}$) naturally-emitted into the troposphere from vegetation is comparable to or can even exceed the emissions of nonmethane organic compounds from anthropogenic sources on a regional and global scale. There are evidences that show significant amounts of OH radicals formed from reaction of monoterpenes with ozone. The monoterpene is very reactive toward OH radical with a typical lifetime of about several hours in the atmosphere. The intermediates and products produced from monoterpene oxidation also contribute to the ozone and secondary aerosol formation in the troposphere. Theoretical studies on other monoterpenes than pinenes are sparse in the literature and only a few are discussed in this review. Carrasco *et al.* [185] performed both experimental and theoretical study to investigate gas phase OH-initiated sabinene reaction. Three primary carbonyl products (acetone, sabinaketone and formaldehyde) have been observed and quantified in the absence and presence of NO_x. Quantum chemical DFT-B3LYP calculations reveal that both abstraction and addition channel need to be taken into account for acetone production and sabinaketone and formaldehyde are the mainly products of the addition channels. The results also indicate that DFT is an appropriate level of theory with reasonable reliability and computational cost for such a large reaction system. Their previous studies showed that barrier heights calculated at the DFT-B3LYP are good agreement with experimental results and the high level *ab initio* calculations. Semi-empirical parameters obtained from correlation between Cl and simple monoalkene and dienes were used to predict the rate constants of Cl atom and a series of monoterpenes including α-pinene, 2-carene, 3-carene, myrcene, and γ-terpinene, based on the framework of perturbation molecular orbital (PFMO) theory [186]. The HOMO energies were calculated using *ab initio* and DFT quantum chemical methods. The calculated rate constants were found to be generally in agreement with available literature values. Ramirez-Ramirez and Nebot-Gil [187] carried out *ab initio* calculations to investigate the initial step of the gas-phase OH–*d*-limonene reaction by considering eight different possibilities for OH addition to the endo and exo C=C double bonds with syn or anti configuration (Figure 10.16). Activation energies calculated at the QCISD(T)/6-31G(d)//UMP2/6-31G(d) level show that OH is preferably added to endocyclic C=C bond under atmospheric

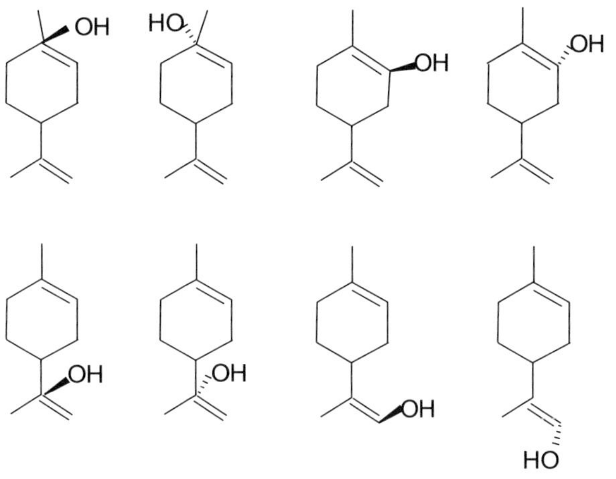

FIGURE 10.16 Schematic diagram of the eight OH–*d*-limonene adduct isomers.

conditions. The gas-phase basicity (GB) and proton affinity (PA) of limonene were calculated based on the density functional methods with best predicted values of 869.6 and 873.9 kJ mol^{-1} for PA at B3PW91/6-31G* and BLYP/6-31G* respectively, in agreement with the measure value (875.5 kJ mol^{-1}) [188], providing valuable information for the measurement of this compound in the atmosphere.

In recent years, increasing interests have arisen for sesquiterpenes ($C_{15}H_{24}$) due to the fact that oxidation of these terpenes has potentially significant contribution to biogenic SOA formation. However, theoretical studies of sesquiterpenes are very limited and only one study concerning the molecular conformations of β-caryophyllene has been reported [189]. Four conformations of β-caryophyllene ($\alpha\alpha$, $\alpha\beta$, $\beta\alpha$, and $\beta\beta$) were investigated by means of *ab initio* calculations at several levels of theory (e.g., HF/6-31G*, MP2/6-31G*, and B3LYP/6-31G*) to investigate their relative thermodynamic stabilities. The results predicted that the $\alpha\alpha$ conformer corresponds to the most stable geometry, in agreement with low-temperature NMR measurements.

4. CONCLUSIONS AND FUTURE RESEARCH

During the past two decades there has been significant progress in the application of quantum chemical calculations to the field of atmospheric chemistry. In this article, we have presented an overview of the application of quantum chemical calculations and kinetic methods to elucidate the mechanism of atmospheric biogenic hydrocarbon oxidations. Recent advances in isoprene and α-, β-pinenes are discussed in this review. The mechanism of isoprene oxidation is better understood than those of other biogenic hydrocarbons. The initial isoprene-oxidant adducts, the peroxy radicals, the alkoxy radicals and the first generation products are investigated by means of quantum chemical calculations. The roles of the intermediate species are assessed using kinetic methods and the overall picture of the isoprene oxidation mechanism is obtained. For pinenes, monoterpenes, and sesquiterpenes, there have been limited theoretical studies on the kinetics and mechanism of these systems initiated by oxidants such as OH and O_3. Consequently, large uncertainty exists on the atmospheric oxidation of these biogenic VOCs.

Typically, several aspects need to be considered for application of quantum chemical calculations and kinetic methods to the atmospheric biogenic hydrocarbons:

(1) The accuracy of the geometries of the species. A reliable potential energy surface depends on reliable geometries of reactants, transition states, intermediates, and products. For biogenic hydrocarbons (e.g., isoprene, α-, β-pinenes), it is difficult or formidable to determine their geometries experimentally. In most of the hydrocarbon oxidation systems, density functional methods have been proven to be an efficient approach, which accurately predicts structural parameters (e.g., bond length, bond angle, and dihedral angle) considering the relatively low computational cost of the methods and have been validated in many applications. Higher level theories have been employed to obtain more accurate structural requirement by balancing between accuracy and computational efficiency. The MP2 and G2 levels are also frequently employed in geometry optimization of species from atmospheric hydrocarbons oxidation.
(2) The accuracy of the reaction energy and activation energy. Quantum chemical calculations provide a simple way to compute the reaction energy/enthalpy and activation energy for a reaction in the biogenic hydrocarbon oxidation system. The theoretical results are needed to be compared with available experimental values. In most cases, higher levels of calculations than DFT calculations (e.g., Gaussian series of theory) are required to obtain accurate energy profiles, especially for the activation energy, which is needed in predicting the reaction rate constant. Once the reaction energy and the activation energy are accurately defined, a reliable potential energy surface is established. For reactions proceeding without barriers, accurate reaction energy, geometry, vibrational frequencies are required to model the potential energy surface using variational transition state theory.
(3) Other thermochemical data (e.g., heat of formation and proton affinity). These thermochemical data are very helpful to atmospheric modeling to understand the complex chemical reactions. Thermochemical data can be obtained by high levels of calculations, e.g., CCSD(T), Gn series, CBS, QCI, etc.
(4) Kinetic theories. The high-pressure limit rate constants can be determined using classic kinetic theory, e.g., TST for tight transition states and CVTST for loose transition states. Kinetic methods such as RRKM or vRRKM can be employed to investigate the temperature and pressure dependent behaviors in fall-off region. RRKM (or vRRKM) in conjunction with master equation is used to evaluate the competition between prompt chemical reactions and collisional stabilization of the excited species. Separate statistical ensemble (SSE) approach can be used to determine the energy partitioning from the parent species into its fragments. The accuracies of the rate constants determined from various kinetic rate methods strongly depend on the accuracies of the

predicted barrier height and vibration frequencies of the transition states and the reactants.

(5) The fates of chemically activated intermediate species. For many oxidation reactions encountered in biogenic hydrocarbon oxidation, especially ozonolysis of isoprene and monoterpene, large excessive energy is released, resulting in the formation of highly excited intermediate species. The fates of these species can be evaluated by RRKM or vRRKM/ME formulism and the energy can be further partitioned into the decomposition products according to the SSE theory.

As the late Nobel laureate John A. Pople [190] pointed out almost ten years ago that "the reaction mechanisms with pollutants in the atmosphere could be studied using computational chemistry". Today, more studies have been undertaken by exploiting quantum chemical calculations and chemical kinetic methods to investigate the oxidation mechanisms of volatile organic compounds in the atmosphere. Quantum chemical calculations and theoretical kinetic models serve as the guidance and support for experimental research. The theoretical framework has been developed over many decades and the maturation of concepts, methods, and algorithms have been validated in the field of atmospheric chemistry. High-level calculations for large organic molecular systems are still computationally demanding and extremely challenging for atmospheric theoretical investigators. However, with the development of high-performance computational facility and better mathematical algorithms, such difficulties and challenges will be overcome in the near future.

ACKNOWLEDGEMENTS

This work was supported by the Robert A. Welch Foundation (Grant A-1417). Additional support was provided by the Texas A&M University Supercomputing Facilities. The authors also acknowledge the use of the Laboratory for Molecular Simulations at Texas A&M University.

REFERENCES

[1] A. Guenther, C.N. Hewitt, D. Erickson, R. Fall, C. Geron, T. Graedel, P. Harley, L. Klinger, M. Lerdau, W.A. McKay, T. Pierce, B. Scholes, R. Steinbrecher, R. Tallamraju, J. Taylor, P.J. Zimmermann, *J. Geophys. Res.* **100** (1995) 8873.

[2] A. Guenther, C. Geron, T. Pierce, B. Lamb, P. Harley, R. Fall, *Atmos. Environ.* **34** (2000) 2205.

[3] R. Fall, in: C.N. Hewitt (Ed.), *Reactive Hydrocarbons in the Atmosphere*, Academic Press, San Diego, 1999, pp. 41–96.

[4] J.D. Fuentes, M. Lerdau, R. Atkinson, D. Baldocchi, J.W. Bottenheim, P. Ciccioli, B. Lamb, C. Geron, L. Gu, A. Guenther, T.D. Sharkey, W. Stockwell, *Bull. Am. Meteorol. Soc.* **81** (2000) 1537.

[5] J. Finlayson-Pitts, J.N. Pitts Jr., *Chemistry of the Upper and Lower Atmosphere: Theory, Experiments, and Applications*, Academic Press, San Diego, CA, 2000.

[6] J.H. Seinfeld, S.N. Pandis, *Atmospheric Chemistry and Physics: From Air Pollution to Climate Change*, Wiley, New York, 1997.

[7] R. Zhang, W. Lei, X. Tie, P. Hess, *Proc. Natl. Acad. Sci. USA* **101** (2004) 6346.

[8] R. Zhang, I. Suh, J. Zhao, D. Zhang, E.C. Fortner, X. Tie, L.T. Molina, M.J. Molina, *Science* **304** (2004) 1487.

[9] W. Lei, R. Zhang, X. Tie, P. Hess, *J. Geophys. Res.* **109** (2004) D12301, doi:10.1029/2003JD004219.

[10] J. Fan, R. Zhang, G. Li, J. Nielsen-Gammon, Z. Li, *J. Geophys. Res.* **110** (2005) D16203, doi: 10.1029/2005JD005805.

[11] R. Atkinson, J. Arey, *Atmos. Environ.* **37** (2003) S197.

[12] R. Atkinson, *Atmos. Environ.* **24** (1990) 1.

[13] R. Atkinson, *J. Phys. Chem. Ref. Data* **26** (1997) 215.

[14] R. Atkinson, D.L. Baulch, R.A. Cox, J.N. Crowley, R.F. Hampson, R.G. Hynes, M.E. Jenkin, M.J. Rossi, J. Troe, *Atmos. Chem. Phys.* **6** (2006) 3625.

[15] G. Li, R. Zhang, J. Fan, X. Tie, *J. Geophys. Res.* **112** (2007) D10309, doi:10.1029/2006JD007924.

[16] F. Jensen, *Introduction to Computational Chemistry*, 2nd ed., Wiley, Hoboken, NJ, 2007.

[17] W. Koch, M.C. Holthausen, *A Chemist's Guide to Density Functional Theory*, Willey–VCH, New York, 2000.

[18] M.A.L. Marques, E.K.U. Gross, *Annu. Rev. Phys. Chem.* **55** (2004) 427.

[19] R.O. Jones, O. Gunnarsson, *Rev. Mod. Phys.* **61** (1989) 681.

[20] S. Kristyan, P. Pulay, *Chem. Phys. Lett.* **229** (1994) 175.

[21] Y. Anderson, D.C. Langreth, C.A. Gonzales, P.M.W. Gill, J.A. Pople, *Chem. Phys. Lett.* **221** (1994) 100.

[22] J. Cizek, *J. Chem. Phys.* **45** (1966) 4256.

[23] R. Krishnan, J.A. Pople, *Int. J. Quantum Chem.* **14** (1978) 91.

[24] R. Krishnan, M.J. Frisch, J.A. Pople, *J. Chem. Phys.* **72** (1980) 4244.

[25] J.A. Pople, M. Headgordon, K. Raghavachari, *J. Chem. Phys.* **87** (1987) 5968.

[26] K. Raghavachari, J.A. Pople, E.S. Replogle, M. Headgordon, *J. Phys. Chem.* **94** (1990) 5579.

[27] J.R. Bartlett, Couple Cluster theory: An overview of recent developments, in: D.R. Yarkony (Ed.), *Modern Electronic Structure Theory*, Part II, World Scientific, Singapore, 1995.

[28] J.W. Ochterski, G.A. Petersson, J.A. Montgomery, *J. Chem. Phys.* **104** (1996) 2598, and references therein.

[29] L.A. Curtiss, P.C. Redfern, K. Raghavachari, *J. Chem. Phys.* **126** (2007), Art. No. 084108, and references therein.

[30] T. P Lee, P.R. Taylor, *Int. J. Quantum Chem. Suppl.* **23** (1989) 199.

[31] J.C. Rienstra-Kiracofe, W.D. Allen, H.F. Schaefer III, *J. Phys. Chem. A* **104** (2000) 9823.

[32] M. Hotokka, B. Roos, P. Siegbahn, *J. Am. Chem. Soc.* **105** (1983) 5263.

[33] O. Roos, The complete active space self-consistent field method and its applications in electronic structure calculations, in: K.P. Lawley (Ed.), *Ab Initio Methods in Quantum Chemistry*, Part II, Wiley, New York, 1987, pp. 399–445.

[34] R. Shapard, The mutliconfiguration self-consistent field method, in: K.P. Lawley (Ed.), *Ab Initio Methods in Quantum Chemistry*, Part II, Wiley, New York, 1987, pp. 63–200.

[35] Y.Y. Chuang, E.L. Coitino, D.G. Truhlar, *J. Phys. Chem. A* **104** (2000) 446.

[36] R.G. Parr, W. Yang, *Density Functional Theory of Atoms and Molecules*, Oxford University Press, New York, 1989.

[37] H. Nakano, K. Hir, *Bull. Korean Chem. Soc.* **24** (2003) 812.

[38] M. Alcami, O. Mo, M. Yanez, *Mass Spectrom. Rev.* **20** (2001) 195.

[39] O. Christiansen, *Theor. Chem. Acc.* **116** (2006) 106.

[40] M. Garavelli, *Theor. Chem. Acc.* **116** (2006) 87.

[41] K.J. Laidler, M.C. King, *J. Phys. Chem.* **87** (1983) 2657.

[42] G. Truhlar, W.L. Hase, J.T. Hynes, *J. Phys. Chem.* **87** (1983) 2664.

[43] G. Truhlar, B.C. Garrett, S.J. Klippenstein, *J. Phys. Chem.* **100** (1996) 12771.

[44] J.I. Steinfeld, J.S. Francisco, W.L. Hase, *Chemical Kinetics and Dynamics*, Prentice Hall, Upper Saddle River, NJ, 1999.

[45] D.G. Truhlar, B.C. Garrett, *Acc. Chem. Res.* **13** (1980) 440.

[46] D.G. Truhlar, B.C. Garrett, *Annu. Rev. Phys. Chem.* **35** (1984) 159.

[47] W.L. Hase, *Chem. Phys. Lett.* **139** (1987) 389.

[48] K. Holbrook, M. Pilling, S. Robertson, *Unimolecular Reactions*, Wiley, Chichester, UK, 1996.

[49] R.G. Gilbert, S.C. Smith, *Theory of Unimolecular and Recombination Reactions*, Blackwell Scientific, Oxford, UK, 1990.

[50] T. Baer, W.L. Hase, *Unimolecular Reaction Dynamics: Theory and Experiments*, Oxford University Press, New York, 1996.
[51] D. Zhang, R. Zhang, J. Park, S.W. North, *J. Am. Chem. Soc.* **124** (2002) 9600.
[52] J.Y. Zhang, T. Dransfield, N.M. Donahue, *J. Phys. Chem. A* **108** (2004) 9082.
[53] M. Olzmann, E. Kraka, D. Cremer, R. Gutbrod, S. Andersson, *J. Phys. Chem. A.* **101** (1997) 9421.
[54] J.H. Kroll, S.R. Sahay, J.G. Anderson, K.L. Demerjian, N.M. Donahue, *J. Phys. Chem. A* **105** (2001) 4446.
[55] C. Wittig, I. Nadler, H. Reisler, M. Nobel, J. Catanzarite, G. Radhakrishnan, *J. Chem. Phys.* **83** (1985) 5581.
[56] J.J. Orlando, G.S. Tyndall, L. Vereecken, J. Peeters, *J. Phys. Chem. A* **104** (2000) 11578.
[57] L. Bunker, W.L. Hase, *J. Chem. Phys.* **59** (1973) 4621.
[58] G. Vayner, S.V. Addepalli, K. Song, W.L. Hase, *J. Chem. Phys.* **125** (2006) 014317.
[59] M. Trainer, E.J. Williams, D.D. Parrish, M.P. Buhr, E.J. Allwine, H.H. Westberg, F.C. Fehsenfeld, S.C. Liu, *Nature* **329** (1987) 705.
[60] R.A. Rasmussen, M.A. Khalil, *J. Geophys. Res.* **93** (1988) 1417.
[61] S. Madronich, J.G. Calvert, *J. Geophys. Res.* **95** (1990) 5697.
[62] R.D. Lightfoot, R.A. Cox, J.N. Crowley, M. Destriau, G.D. Hayman, M.E. Jenkin, G.K. Moortgat, F. Zabel, *Atmos. Environ.* **26** (1992) 1805.
[63] M.E. Jenkin, S.M. Saunders, M.J. Pilling, *Atmos. Environ.* **31** (1997) 81.
[64] M.E. Jenkin, A.A. Boyd, R. Lesclaux, *J. Atmos. Chem.* **29** (1998) 267.
[65] T.S. Dibble, *J. Phys. Chem. A* **103** (1999) 8559.
[66] W.F. Lei, R.Y. Zhang, *J. Phys. Chem. A* **105** (2000) 3808.
[67] T.S. Dibble, *J. Phys. Chem. A* **108** (2004) 2208.
[68] J. Zhao, R.Y. Zhang, E.C. Fortner, S.W. North, *J. Am. Chem. Soc.* **126** (2004) 2686.
[69] J. Fan, R. Zhang, *Environ. Chem.* **1** (2004) 140.
[70] S. McGivern, I. Suh, A.D. Clinkenbeard, R. Zhang, S.W. North, *J. Phys. Chem. A* **104** (2000) 6609.
[71] W. Lei, R. Zhang, W.S. McGivern, A. Derecskei-Kovacs, S.W. North, *Chem. Phys. Lett.* **326** (2000) 109.
[72] W. Lei, A. Derecskei-Kovacs, R. Zhang, *J. Chem. Phys.* **113** (2000) 5354.
[73] R. Zhang, W. Lei, *J. Chem. Phys.* **113** (2000) 8574.
[74] P. Campuzano-Jost, M.B. Williams, L. D'Ottone, A.J. Hynes, *Geophys. Res. Lett.* **27** (2000) 693.
[75] P.S. Stevens, E. Seymour, Z.J. Li, *J. Phys. Chem. A* **104** (2000) 5989.
[76] R. Zhang, I. Suh, W. Lei, A.D. Clinkenbeard, S.W. North, *J. Geophys. Res.* **105** (2000) 24627.
[77] B. Chuong, P.S. Stevens, *J. Phys. Chem. A* **104** (2000) 5230.
[78] J. Park, C.G. Jongsma, R. Zhang, S.W. North, *Phys. Chem. Chem. Phys.* **5** (2003) 3638.
[79] M. Francisco-Marquez, J.R. Alvarez-Idaboy, A. Galano, A. Vivier-Bunge, *Phys. Chem. Chem. Phys.* **5** (2003) 1392.
[80] M. Francisco-Marquez, J.R. Alvarez-Idaboy, A. Galano, A. Vivier-Bunge, *Environ. Sci. Technol.* **39** (2005) 8797.
[81] W. Lei, R. Zhang, W.S. McGivern, A. Derecskei-Kovacs, S.W. North, *J. Phys. Chem. A* **105** (2001) 471.
[82] M.E. Jenkin, G.D. Hayman, *J. Chem. Soc. Faraday Trans.* **91** (1995) 1911.
[83] S.E. Paulson, J.H. Seinfeld, *J. Geophys. Res.* **97** (1992) 20703.
[84] D. Zhang, R. Zhang, C. Church, S.W. North, *Chem. Phys. Lett.* **343** (2001) 49.
[85] J. Park, C.G. Jongsma, R. Zhang, S.W. North, *J. Phys. Chem. A* **108** (2004) 10688.
[86] V.M. Ramirez-Ramirez, I. Nebot-Gil, *Int. J. Quantum Chem.* **105** (2005) 518.
[87] J. Park, J.C. Stephens, R. Zhang, S.W. North, *J. Phys. Chem.* **107** (2003) 6408.
[88] J.E. Reitz, W.S. McGivern, M.C. Church, M.D. Wilson, S.W. North, *Int. J. Chem. Kinet.* **34** (2002) 255.
[89] L. Vereecken, J. Peeters, *J. Phys. Chem. A* **108** (2004) 5197.
[90] K.N. Houk, K.R. Condroski, W.A. Pryor, *J. Am. Chem. Soc.* **118** (1996) 13002.
[91] R. Cameron, A.M.P. Borrejo, B.M. Bennett, G.R.J. Thatcher, *Can. J. Chem.* **73** (1995) 1627.
[92] B.S. Jursic, L. Klasine, S. Pecur, W.A. Pryor, *Nitric Oxide* **1** (1997) 494.
[93] J.M. O'Brien, E. Czuba, D.R. Hastie, J.S. Francisco, P.B. Shepson, *J. Phys. Chem. A* **102** (1998) 8903.
[94] J. Zhao, R. Zhang, S.W. North, *Chem. Phys. Lett.* **369** (2003) 204.

[95] J. Yu, H.E. Jeffries, R.M. Le Lacheur, *Environ. Sci. Technol.* **29** (1995) 1923.
[96] S. Kwok, R. Atkinson, J. Arey, *Environ. Sci. Technol.* **29** (1995) 2467.
[97] V.M. Ramirez-Ramirez, I. Nebot-Gil, *Chem. Phys. Lett.* **406** (2005) 404.
[98] T.S. Dibble, *J. Phys. Chem. A* **108** (2004) 2199.
[99] T.A. Biesenthal, J.W. Bottenheim, P.B. Shepson, P.C. Brickell, *J. Geophys. Res.* **103** (1998) 25487.
[100] J.H. Kroll, N.L. Ng, S.M. Murphy, R.C. Flagan, J.H. Seinfeld, *Geophys. Res. Lett.* **32** (2005), Art. No. L18808.
[101] T.E. Kleindienst, M. Lewandowski, J.H. Offenberg, M. Jaoui, E.O. Edney, *Geophys. Res. Lett.* **34** (2007), Art. No. L01805.
[102] R. Atkinson, S.M. Aschmann, J. Arey, B. Shorees, *J. Geophys. Res.* **97** (1992) 6065.
[103] S.M. Aschmann, J. Arey, R. Atkinson, *Atmos. Environ.* **30** (1996) 2939.
[104] R. Gutbrod, S. Meyer, M.M. Rahman, R.N. Schindler, *Int. J. Chem. Kinet.* **29** (1997) 717.
[105] S.E. Paulson, M. Chung, A.D. Sen, G. Orzechowska, *J. Geophys. Res.* **103** (1998) 25333.
[106] P. Neeb, G.K. Moortgat, *J. Phys. Chem. A* **103** (1999) 9003.
[107] R. Gutbrod, E. Kraka, R.N. Schindler, D. Cremer, *J. Am. Chem. Soc.* **119** (1997) 7330.
[108] D. Zhang, R.Y. Zhang, *J. Am. Chem. Soc.* **124** (2002) 2692.
[109] D. Zhang, W.F. Lei, R.Y. Zhang, *Chem. Phys. Lett.* **358** (2002) 171.
[110] K.T. Kuwata, L.C. Valin, A.D. Converse, *J. Phys. Chem. A* **109** (2005) 10710.
[111] S.M. Aschmann, R. Atkinson, *Environ. Sci. Technol.* **28** (1994) 1539.
[112] D. Grosjean, E.L. Williams II, E. Grosjean, *Environ. Sci. Technol.* **27** (1993) 830.
[113] T. Kurten, B. Bonn, H. Vehkamaki, M. Kulmala, *J. Phys. Chem. A* **111** (2007) 3394.
[114] K.H. Becker, K.J. Brockmann, J. Bechara, *Nature* **346** (1990) 256.
[115] K.H. Becker, J. Bechara, K. Brockmann, *J. Atmos. Environ.* **27A** (1993) 57.
[116] R. Simonaitis, K.J. Olsyna, J.F. Meagher, *Geophys. Res. Lett.* **18** (1991) 9.
[117] P. Neeb, F. Sauer, O. Horie, G.K. Moortgat, *Atmos. Environ.* **31** (1997) 417.
[118] F. Sauer, C. Schafer, P. Neeb, O. Horie, G.K. Moortgat, *Atmos. Environ.* **33** (1999) 229.
[119] P. Aplincourt, J.M. Anglada, *J. Phys. Chem. A* **107** (2003) 5798.
[120] P. Aplincourt, J.M. Anglada, *J. Phys. Chem. A* **107** (2003) 5812.
[121] A.B. Ryzhkov, P.A. Ariya, *Phys. Chem. Chem. Phys.* **6** (2004) 5042.
[122] A.B. Ryzhkov, P.A. Ariya, *Chem. Phys. Lett.* **419** (2006) 479.
[123] A. Geyer, B. Alicke, S. Konrad, T. Schmitz, J. Stutz, U. Platt, *J. Geophys. Res.* **106** (2001) 8013.
[124] D.D. Riemer, P.J. Milne, C.T. Farmer, R.G. Zika, *Chemosphere* **28** (1994) 837.
[125] T.K. Starn, P.B. Shepson, S.B. Bertman, D.D. Riemer, R.G. Zika, K. Olszyna, *Geophys. Res. Lett.* **103** (1998) 22437.
[126] I. Suh, W. Lei, R. Zhang, *J. Phys. Chem. A* **105** (2001) 6471.
[127] H. Skov, J. Hjorth, C. Lohse, N.R. Jenen, G. Restelli, *Atmos. Environ.* **26A** (1992) 2771.
[128] T. Berndt, O. Boge, *Int. J. Chem. Kinet.* **29** (1997) 755.
[129] D. Zhang, R. Zhang, *J. Chem. Phys.* **116** (2002) 9721.
[130] J. Zhao, R.Y. Zhang, *Atmos. Environ.*, in press.
[131] J. Zhao, R.Y. Zhang, *J. Phys. Chem. A*, in preparation for publication.
[132] B.T. Jobson, H. Niki, Y. Yokouchi, J. Bottenheim, F. Hopper, R. Leaitch, *J. Geophys. Res.* **99** (1994) 25355.
[133] S. Solberg, N. Schmidbauer, A. Semb, F. Stordal, O. Hov, *J. Atmos. Chem.* **23** (1996) 301.
[134] W. Wingenter, M.K. Kubo, N.J. Blake, T.W. Smith, F.S. Rowland, *J. Geophys. Res.* **101** (1996) 4331.
[135] A.P. Pszeeny, W.C. Keene, D.J. Jacob, S. Fan, J.R. Maben, M.P. Zetwo, M. Springer-Young, J.N. Galloway, *Geophys. Res. Lett.* **20** (1993) 699.
[136] L. Ragains, B.J. Finlayson-Pitts, *J. Phys. Chem.* **101** (1997) 1509.
[137] G. Fantechi, N.R. Jensen, O. Saastad, J. Hjorth, J. Peeters, *J. Atmos. Chem.* **31** (1998) 247.
[138] E. Canosa-Mas, H.R. Hutton-Squire, M.D. King, D.J. Stewart, K.C. Thompson, R.P. Wayne, *J. Atmos. Chem.* **34** (1999) 163.
[139] A. Notario, G. Le Bras, A. Mellouki, *Chem. Phys. Lett.* **281** (1997) 421.
[140] Y. Bedjanian, G. Laverdet, G. Le Bras, *J. Phys. Chem.* **102** (1998) 953.
[141] W. Lei, R. Zhang, *J. Chem. Phys.* **113** (2000) 153.
[142] W. Lei, D. Zhang, R. Zhang, L.T. Molina, M.J. Molina, *Chem. Phys. Lett.* **357** (2002) 45.
[143] P. Brana, J.A. Sordo, *J. Am. Chem. Soc.* **123** (2001) 10348.

[144] W. Lei, R. Zhang, L.T. Molina, M.J. Molina, *J. Phys. Chem.* **106** (2002) 6415.
[145] I. Suh, R. Zhang, *J. Phys. Chem.* **104** (2000) 6590.
[146] D. Zhang, R. Zhang, D.T. Allen, *J. Chem. Phys.* **118** (2003) 1794.
[147] J. Fan, J. Zhao, R.Y. Zhang, *Chem. Phys. Lett.* **411** (2005) 1.
[148] V.M. Ramirez-Ramirez, J. Peiro-Garcıa, I. Nebot-Gil, *Chem. Phys. Lett.* **391** (2004) 152.
[149] A.M. Winer, A.C. Lloyd, K.R. Darnall, J.N. Pitts, *J. Phys. Chem.* **80** (1976) 1635.
[150] R. Atkinson, S.M. Aschmann, J.N. Pitts, *Int. J. Chem. Kinet.* **18** (1986) 287.
[151] T.E. Kleindienst, G.W. Harris, J.N. Pitts, *Environ. Sci. Technol.* **16** (1982) 844.
[152] J. Gill, R.A. Hites, *J. Phys. Chem. A* **106** (2002) 2538.
[153] J. Peeters, L. Vereecken, G. Fantechi, *Phys. Chem. Chem. Phys.* **3** (2001) 5489.
[154] J. Arey, R. Atkinson, S.M. Aschmann, *J. Geophys. Res.* **95** (1990) 18539.
[155] S. Hatakeyama, K. Izumi, T. Fukuyama, H. Akimoto, N. Washida, *J. Geophys. Res.* **96** (1991) 947.
[156] H. Hakola, J. Arey, S. Aschmann, R. Atkinson, *J. Atmos. Chem.* **18** (1994) 75.
[157] C. Vinckier, F. Compernolle, A. Saleh, M.N. Van Hoof, I. Van Hees, *Fresenius Environ. Bull.* **7** (1998) 361.
[158] B. Nozière, I. Barnes, K.-H. Becker, *J. Geophys. Res.* **104** (1999) 23645.
[159] A. Wisthaler, N.R. Jensen, R. Winterhalter, W. Lindinger, J. Hjorth, *Atmos. Environ.* **35** (2001) 6181.
[160] R. Larsen, D. di Bella, M. Glasius, R. Winterhalter, N.R. Jensen, J. Hjorth, *J. Atmos. Chem.* **38** (2001) 231.
[161] S.M. Aschmann, R. Atkinson, J. Arey, *J. Geophys. Res.* **107** (2002), doi:10.1029/2001JD001098.
[162] V. Librando, G. Tringali, *J. Environ. Manage.* **75** (2005) 275.
[163] G. Fantechi, L. Vereecken, J. Peeters, *Phys. Chem. Chem. Phys.* **4** (2002) 5795.
[164] P.J. Lewis, K.A. Bennett, J.N. Harvey, *Phys. Chem. Chem. Phys.* **7** (2005) 1643.
[165] R. Atkinson, S.M. Aschmann, *J. Atmos. Chem.* **16** (1993) 337.
[166] S.A. McKeen, T. Gierczak, J.B. Burkholder, P.O. Wennberg, T.F. Hanisco, E.R. Keim, R.-S. Gao, S.C. Liu, A.R. Ravishankara, D.W. Fahey, *Geophys. Res. Lett.* **24** (1997) 3177.
[167] K.-H. Wohlfrom, T. Hauler, F. Arnold, H. Singh, *Geophys. Res. Lett.* **26** (1999) 2849.
[168] S.M. Aschmann, A. Reissell, R. Atkinson, J. Arey, *J. Geophys. Res.* **103** (1998) 25553.
[169] A. Reissell, C. Harry, S.M. Aschmann, R. Atkinson, J. Arey, *J. Geophys. Res.* **104** (1999) 13869.
[170] J.J. Orlando, B. Nozière, G.S. Tyndall, G.E. Orzechowska, S.E. Paulson, Y. Rudich, *J. Geophys. Res.* **105** (2000) 11561.
[171] L. Vereecken, J. Peeters, *J. Phys. Chem. A* **104** (2000) 11140.
[172] T.S. Dibble, *J. Am. Chem. Soc.* **123** (2001) 4228.
[173] M. Capouet, J. Peeters, B. Noziere, J.-F. Muller, *Atmos. Chem. Phys.* **4** (2004) 2285.
[174] S. Hatakeyama, K. Izumi, T. Fukuyama, H. Akimoto, *J. Geophys. Res.* **94** (1989) 13013.
[175] D. Grosjean, E.L. Williams, E. Grosjean, J.M. Andino, J.H. Seinfeld, *Environ. Sci. Technol.* **27** (1993) 2754.
[176] T. Hoffmann, R. Bandur, U. Marggraf, M. Linscheid, *J. Geophys. Res.* **103** (1998) 25569.
[177] R. Winterhalter, P. Neeb, D. Grossmann, A. Kolloff, O. Horie, G. Moortgat, *J. Atmos. Chem.* **35** (2000) 165.
[178] W. Hoppel, J. Fitzgerald, G. Frick, P. Caffrey, L. Pasternack, D. Hegg, S. Gao, R. Leaitch, N. Shantz, C. Cantrell, T. Albrechcinski, J. Ambrusko, W. Sullivan, *J. Geophys. Res.* **106** (2001) 27603.
[179] Y. Iinuma, O. Boge, T. Gnauk, H. Herrmann, *Atmos. Environ.* **38** (2004) 761.
[180] D. Zhang, R.Y. Zhang, *J. Chem. Phys.* **122** (2005) Art. No. 114308.
[181] T. Berndt, O. Boge, *J. Chem. Soc. Faraday Trans.* **93** (1997) 3021.
[182] M. Hallquist, I. Wangberg, E. Ljungstrom, I. Barnes, K.H. Becker, *Environ. Sci. Technol.* **33** (1999) 553.
[183] B. Bonn, G.K. Moortgat, *Amos. Chem. Phys.* **2** (2002) 183.
[184] M. Spittlera, I. Barnes, I. Bejan, K.J. Brockmann, Th. Benter, K. Wirtz, *Atmos. Environ.* **40** (2006) S116.
[185] N. Carrasco, M.T. Rayez, J.C. Rayez, J.F. Doussin, *Phys. Chem. Chem. Phys.* **8** (2006) 3211.
[186] Q.K. Timerghazin, P.A. Ariya, *Phys. Chem. Chem. Phys.* **3** (2001) 3981.
[187] V.M. Ramirez-Ramirez, I. Nebot-Gil, *Chem. Phys. Lett.* **409** (2005) 23.
[188] M. Fernandez, C. Williams, R.S. Mason, B.J.C. Cabral, *J. Chem. Soc. Faraday Trans.* **94** (1998) 1427.
[189] M. Clericuzio, G. Alagona, C. Ghio, L. Toma, *J. Org. Chem.* **65** (2000) 6910.
[190] J.A. Pople, *R&D Mag.* **41** (1999) 44.

CHAPTER 11

Computational Study of the Reaction of *n*-Bromopropane with OH Radicals and Cl Atoms

Claudette M. Rosado-Reyes*, **Mónica Martínez-Avilés***, and **Joseph S. Francisco**[1*]

Contents

1. Introduction 216
2. Computational Methods 219
3. Results and Discussion 220
   3.1 Mechanistic analysis 221
   3.2 Energetics of the reaction of *n*-bromopropane with OH radical and Cl atom 237
4. Atmospheric Implications 241
5. Conclusion 242
Acknowledgements 243
References 243

Abstract

Ab initio molecular orbital theory is utilized to study the hydrogen abstraction reaction of *n*-bromopropane with hydroxyl radical and chlorine atom. The stability of the *trans* and *gauche* isomers of *n*-bromopropane is explored. The potential energy surface of both reactions is characterized by pre- and post-reactive complexes, as well as transition state structures in both *trans* and *gauche* isomeric forms. The importance of these two reactions relies on the ultimate product distribution from both reactions. Differences in the reactivity of 1-bromopropane toward OH and Cl are observed. The reaction of *n*-bromopropane with OH radical favors the abstraction of β hydrogen atoms while the reaction with Cl atoms favors the abstraction of hydrogen atoms at the α and β carbon sites.

* Department of Chemistry and Department of Earth and Atmospheric Sciences, Purdue University, West Lafayette, IN 47907, USA
1 Corresponding author.

Advances in Quantum Chemistry, Vol. 55
ISSN 0065-3276, DOI: 10.1016/S0065-3276(07)00211-0

1. INTRODUCTION

Under the agreements of the Montreal Protocol on Substances that Deplete the Ozone Layer, the production of chlorofluorocarbons (CFCs), halons and several other halocarbons has been prohibited [1,2]. Consequently, there is an interest in replacing these compounds [3]. As part of the development of such replacing compounds, it is necessary to consider and evaluate the potential environmental effects of their use, especially on stratospheric ozone [2].

Although the emphasis has been on chlorine-containing compounds, other halogenated compounds, such as bromine-containing compounds, play a role in the interaction between tropospheric and stratospheric chemistry [4,5]. Both natural and anthropogenic processes contribute to the brominated organics emitted into the troposphere [5]. Through sea-salt aerosols produced by bursting of small air bubbles near the sea surface, the ocean contributes a great part of the halogenated substances that are present in the atmosphere [6]. Marine biological activity is believed to be the major natural source of brominated organics producing methyl bromide, the major bromine-containing organic in the remote troposphere [7,8]. Anthropogenic sources include soil fumigants (CH_3Br), gasoline additives (C_2H_4Br), and flame retardants (e.g. $CBrF_3$) [5,9].

Gas phase reactions involving bromine are related to ozone decrease at the ground level [10,11]. Measurements in the Arctic have shown the destruction of boundary-layer ozone from 30–40 ppb to undetectable levels in less than a day [12–16]. This destruction of ozone is strongly linked to high levels of filterable bromine occurring since the beginning of polar sunrise until April [12,17]. Mechanisms, involving aerosol particles or reactions on the snow pack, have been proposed for the regeneration of photochemically active bromine [12,18–20]. McConnell and coworkers suggested in 1992 that the concentration of sea-salt Br^- in the snow pack is high during the long polar night, which is released into the atmosphere during polar sunrise as Br_2 [12]. Numerical simulations were performed proposing that Br_2 was released to the atmosphere by heterogeneous reactions among scavengers (i.e. ambient aerosols and ice crystals) and bromine containing compounds (i.e. HBr and brominated organic compounds (BOCs)). The authors claim that this bromine cycling between the aerosol and the gas phase, maintains the high levels of Br atoms and BrO radicals necessary to destroy ozone, as observed previously in Arctic measurements [12–16]. Fan and Jacob [19] proposed a mechanism for such aerosol-phase production of active bromine species, since McConnell did not specified one. In 1996, Sander and Crutzen developed a model describing the chemistry of the marine boundary layer at midlatitudes, where autocatalytic cycles involving sea salt particles generate photochemically active gases such as Br_2, BrCl and Cl_2 [21]. These types of reactions in the condensed phase lead to the formation of photochemically active halogens in the gas phase.

Bromine compounds, when diffused to and photolyzed in the stratosphere, can produce Br atoms that participate in catalytic ozone destruction that can reach efficiencies that are 40–50 times more effectively than Cl atoms [9,22–24]. The stratospheric chemistry of brominated organic compounds is similar to that of chlorine-containing organics with a few exceptions: (1) the abstraction of a hydrogen atom from H_2 and CH_4 is too slow to be of interest in the stratosphere, as it is a highly endothermic reaction; (2) radical–radical processes play a more important role in Br_x chemistry than in Cl_x chemistry; and (3) the photolysis rate of BrO is $\sim 10^3$ faster than for ClO [25].

An important aspect of stratospheric bromine chemistry is the possibility of synergistic interactions between bromine and chlorine cycles via the following reaction,

$$\text{BrO} + \text{ClO} \rightarrow \text{Cl} + \text{Br} + \text{O}_2.$$

It has been suggested that this ClO_x–BrO_x interaction could lead to enhanced ozone destruction in the lower stratosphere [2,25].

Long lived compounds in the troposphere are transported into the stratosphere, where they decompose providing a source of inorganic stratospheric bromine [25]. Even short-lived halocarbons such as bromopropane can have significant ozone depletion potentials (ODP) due to rapid transport to the stratosphere by tropical convection [26]. Bromopropane molecules release bromine atoms 2–3 times more effectively than some CFCs would release chlorine atoms in the lower stratosphere [27].

Reactions with hydroxyl radicals are considered one of the most efficient ways used by the atmosphere to remove natural and anthropogenic trace gases in the atmosphere [28,29]. Previous kinetics [27,30–32] and modeling studies [2, 33–35] have shown that the primary atmospheric sink of bromopropane is the reaction with OH and that it has an atmospheric lifetime of 10–16 days. Further studies have determined that the lifetime of short-lived species, such as 1-bromopropane, is dependent on the geographic distribution of surface emissions; therefore it depends mostly on local concentrations of the scavenger rather than its tropospheric average [27,31,34,35]. Based on this assertion, Wuebbles *et al.* [35] divided the modeling study in three regions and found the atmospheric lifetime of 1-bromopropane (nPB) to be 9 days and an ODP range of 0.087 to 0.105 for the emissions in the tropical Southeast Asia, 19 days and 0.021 to 0.028 for the emissions of the industrialized regions of the Northern Hemisphere, and 19 days and an ODP range of 0.033 to 0.040 for the global emissions over landmasses, respectively [35]. On the contrary, Bridgeman *et al.* [34] determined the value to be 19–33 days and an ODP range of 0.0109–0.0033. The calculated ODP for 1-bromopropane (1-BP) is 0.027 using a fixed mixing ratio boundary condition [27,33]. Previously Wuebbles and coworkers [2], as well as Teton *et al.* [32], determined that photolysis has an insignificant effect on the residence time of 1-bromopropane in the atmosphere. Also, Wuebbles *et al.* [2] arrived at the same conclusion for an estimated ocean sink.

Kinetic studies of the reaction of 1-bromopropane with OH radicals have been previously performed [27,30,32,36,37]. Donaghy *et al.* [30], using the relative rate technique, found a rate constant of $(11.8 \pm 3.0) \times 10^{-13}$ cm^3 molecule^{-1} s^{-1} while using c-C_6H_{12} as the reference compound. The obtained atmospheric lifetime was 11–16 days. Teton *et al.* [32] via pulsed laser photolysis followed by laser induced fluorescence obtained a temperature-dependent rate constant of $5.28 \times 10^{-12} e^{-0.91/RT}$ cm^3 molecule^{-1} s^{-1} at a 100 torr pressure and 233–372 K temperature range. The authors calculated an atmospheric lifetime of 10 days. Using the same technique and total pressure, Herndon *et al.* [36] found the temperature-dependent rate constant for the OH reaction to be $9.1 \times 10^{-14} T^{0.5} e^{-157/T}$ cm^3 molecule^{-1} s^{-1} for a temperature range of 230–298 K. Kozlov *et al.* [31] used flash photolysis followed by resonance fluorescence detection in order to report a temperature-dependent rate constant measurement of $3.03 \times 10^{-12} e^{-0.66/RT}$ cm^3 molecule^{-1} s^{-1} for a temperature range of 210–480 K. The results led to a 14 days estimation of the atmospheric lifetime of 1-bromopropane. Nelson *et al.* [27] determined a temperature-dependent rate constant of $5.75 \times 10^{-12} e^{-1.00/RT}$ cm^3 molecule^{-1} s^{-1}, while performing the studies in a range of 1.1 to 2.3 torr of pressure and 271 to 263 K with a discharge flow followed by laser induced fluorescence detection. They estimated a tropospheric lifetime of 15 days for 1-bromopropane, and with that a global warming potential of 1.0, 0.3 and 0.1 for integration time horizons of 20, 100, and 500 years, respectively [27].

Gilles and coworkers [37], via pulsed laser photolysis followed by laser induced fluorescence, found the temperature-dependent rate constant for the OH reaction at 50 torr pressure to be $6.6 \times 10^{-18} T^2 e^{-154/T}$ cm^3 molecule^{-1} s^{-1} for a temperature range of 230–360 K. The group developed a fit for all the published data previously mentioned [27,30–32,36,37], resulting in the expression $k(T) = (7.40 \pm 1.07) \times 10^{-18} T^2 \exp[-(-140 \pm 45)/T]$. They recommend this expression for atmospheric modeling. In their paper, the rate coefficient and product branching ratios for the reaction were reported [37,38]. The β abstraction was determined the most favorable reaction pathway, with a 56% yield at room temperature. The α abstraction from the substituted carbon had a 32% yield, while the γ abstraction had a 12% yield [37,38]. The experiments showed bromoacetone and propionaldehyde as stable degradation products of both the reaction in discussion and in the reaction with Cl atom. For the reaction with Cl atom, no rate constant was reported in the studies of Gilles *et al.* [37]. The only experimental study is that of Donaghy *et al.* [30], where a rate constant of $(61 \pm 18) \times 10^{-12}$ molecule^{-1} s^{-1} relative to C_2H_6 was determined for the reaction of 1-bromopropane with Cl atoms at room temperature. An atmospheric lifetime of 95 days was estimated. It was observed that the rate constant value is lowered by the presence of a halogen substituent. The reaction of Cl atoms with non-methane hydrocarbons (NMHC) is a significant atmospheric process. Despite the low abundance of Cl atoms in the atmosphere ($\sim 10^3$ molecules cm^{-3}), Singh and Kasting [39] showed that 20 to 40% of the NMHC oxidation in the troposphere (0–10 km) and 40 to 90% of NMHC oxidation in the lower stratosphere (10–20 km) is caused by Cl atoms. The photochemical activity of NMHC not only depends on their reaction with OH radicals but also on Cl atom reactions.

Ab initio calculations were performed to obtain the bond strength and the relative rate coefficients, using the Gaussian-3 methodology previously extended to brominated compounds by Curtiss *et al.* [40,41]. The results of the calculations showed that, by adding Br to the α site, the C–H bond energies decreased at the other sites. As well, the calculated energies of the transition states decreased as suggested by the experimental observations ($\gamma > \alpha > \beta$).

Hydrogen abstraction reactions in the atmosphere, initiated by OH radicals and Cl atoms, are some of the mechanism to trigger the production of free photochemically active bromine from bromine-containing compounds. To gain a better understanding of the fate of n-bromopropane as an atmospheric species [10], and to establish global warming and climate change potentials, this paper approaches the atmospheric impact of n-bromopropane which is proposed as a replacement for CFCs used as the active component of industrial cleaning solvents [2,27,31,37], more specifically HCFC-141b [36]. The following work intends to relate the previously discussed experimental results [27,30–32,36,37] with theory, by performing *ab initio* calculations of the 1-bromopropane reaction with OH radicals and with Cl atoms to determine energetics of both reactions.

2. COMPUTATIONAL METHODS

Gaussian 03 suites of programs [42] were utilized to perform *ab initio* molecular orbital calculations on n-bromopropane, radicals, transition states, and complexes structures. Geometry optimizations and frequency calculations were performed using the levels of theory of Hartree–Fock (HF) and second-order Møller–Plesset perturbation (MP2) [43], with 6-31G(d) basis set to find equilibrium geometries, vibrational frequencies and zero point energies. All optimizations were carried out to better than 0.001 Å for bond lengths and 0.1° for angles, with a self-consistent field convergence criteria of at least 10^{-7} on the density matrix and a residual rms force of less than 10^{-4} atomic units. Harmonic vibrational frequencies, obtained with analytical second derivatives for the optimized minimum energy geometries, confirmed true minima and first-order saddle points. Global minima had all positive frequencies. One imaginary frequency was found for each transition state, confirming their location as maxima in one reaction coordinate. The proper connectivity between reactants, pre- and post-reactive complexes, transition states, and products were verified for all pathways by intrinsic reaction coordinate calculations (IRC) [44]. To refine energy values of equilibrium geometries, two single-point calculations of high level electron correlation of coupled cluster including singles and doubles with approximate triples were performed using CCSD(T) with 6-311G(2d,2p) and 6-311++G(2df,2p) basis sets. The degree of spin contamination was monitored because of the possible production of inaccurate total energies. For open shell systems the $\langle S^2 \rangle$ value did not have major deviations from 0.75.

Gaussian-2 [45–47] and Gaussian-3 [40,41] calculations were also performed for calibration purposes. G2 and G3 theories are composite techniques in which a series of *ab initio* molecular orbital calculations, optimization/frequency and single-points, is performed to arrive at a total energy of a given molecular species. Since third-row heavy atoms have not been defined in the G3 theory basis sets in the Gaussian 03 set of programs, basis sets defining bromine atom were incorporated in the *ab initio* sequence of calculations. The G3 energy is defined as,

$$\begin{aligned} E_0(\text{G3}) = {} & \text{MP4/6-31G(d)} + [\text{QCISD(T)/6-31G(d)} - \text{MP4/6-31G(d)}] \\ & + [\text{MP4/6-31+G(d)} - \text{MP4/6-31G(d)}] \\ & + [\text{MP4/6-31G(2df,p)} - \text{MP4/6-31G(d)}] \\ & + [\text{MP2(FU)/G3Large} - \text{MP2/6-31G(2df,p)} \\ & - \text{MP2/6-31+G(d)} + \text{MP2/6-31G(d)}] \\ & + \text{E(SO)} + \text{E(HLC)} + \text{E(ZPE)}. \end{aligned}$$

The zero-point energy correction (ZPE) [48] is determined at the HF/6-31G(d) level of theory with a harmonic frequency scaling factor of 0.8929. In this study we have modified the procedure by calculating the ZPE at the MP2(FU)/6-31G(d). The harmonic frequencies were scaled by 0.9427 to account for any deficiencies at this level [49].

Enthalpies of reaction ($\Delta_{\text{rxn}}H_{298}$) and energy barriers ($E_a$), corrected with zero-point energies (ZPE), were calculated to characterized the energetics of primary, secondary, and tertiary hydrogen abstractions during the reaction of $BrCH_2CH_2CH_3$ with OH radical and Cl atoms. The $BrCH_2CH_2CH_3 + OH$ reaction serves as a calibration method for the characterization reaction of *n*-bromopropane with Cl, since little is known about the latter.

3. RESULTS AND DISCUSSION

The potential energy surface of hydrogen abstractions during the reaction of *n*-bromopropane with OH radicals and Cl atoms is explored in detail. 1-Bromopropane contains three different reactive sites from where hydrogen atoms can be abstracted: α, β, and γ carbon atoms. The α carbon (primary) atom is defined as the carbon to which the substituent, in this case the bromine atom, is attached. The adjacent carbon atom is called the β-carbon. The three resulting reaction channels proceed as follows, where X = OH/Cl,

$$\begin{aligned} BrCH_2CH_2CH_3 + X & \xrightarrow{\alpha} BrCHCH_2CH_3 + HX \\ & \xrightarrow{\beta} BrCH_2CHCH_3 + HX \\ & \xrightarrow{\gamma} BrCH_2CH_2CH_2 + HX. \end{aligned}$$

The results will be discussed and analyzed in terms of these three reaction branches.

Table 11.1 shows the enthalpies of reaction and activation barriers for the formation of primary radicals and transition state structures. Binding energies for the formation of pre- and post-reactive complexes are listed in Table 11.2. The minimum energy structures (closed- and open-shell systems) are illustrated in

TABLE 11.1 Enthalpies of isomerization (gauche → trans) and formation of the 1-bromopropane—*trans* isomer

Level of theory	$\Delta H_{0\ \mathrm{T\text{-}G}}$ [a]	$\Delta H_{0,f\ n\mathrm{BP}\text{-}trans}$ [a,b]
HF/6-31G(d)	−0.0067	−20.2
MP2(full)/6-31G(d)	0.47	−22.4
CCSD(T)/6-311G(2d,2p)	−0.12	−21.5
CCSD(T)/6-311++G(2df,2p)	−0.065	−21.4
G2	−0.088	−21.5
Experimental value	–	−20.1 ± 0.1[59]

a In kcal mol^{-1}.

b Based on isodesmic reaction: $BrCH_2CH_2CH_3 + CH_4 \rightarrow CH_3CH_2CH_3 + CH_3Br$.

Figure 11.1 (n-bromopropane and primary radicals), Figure 11.2 (nBP + OH complexes), Figure 11.3 (nBP + OH transition states), Figure 11.4 (nBP + Cl complexes), and Figure 11.5 (nBP + Cl transition states). The energetics of the potential energy surface are depicted in Figures 11.6A, 11.6B, and 11.6C for the reaction of bromopropane + OH radical and in Figures 11.7A, 11.7B, and 11.7C for the reaction of bromopropane with Cl atom.

3.1 Mechanistic analysis

3.1.1 Stability of 1-bromopropane and bromopropyl alkyl radicals

In this study, we explored the stability of the n-bromopropane *trans* and *gauche* isomers by calculating the isomerization enthalpy at the levels of theory of interest. The results are summarized in Table 11.1. The values vary from −0.12 to 0.47 kcal mol^{-1}. Only the MP2/6-31G(d) calculation predict that the *gauche* isomer is the most stable. The MP2 determination (0.47 kcal mol^{-1}) seems overestimated. Only high-level quantum calculations made predictions in accordance with experimental findings.

The heat of formation of n-bromopropane is determined in this study via the isodesmic reaction of $BrCH_2CH_2CH_3 + CH_4 \rightarrow CH_3CH_2CH_3 + CH_3Br$. Isodesmic reactions are those in which the number of bonds of equal type is conserved [50]. The energies of isodesmic reactions are usually well reproduced, which allow for high-accuracy determination of reaction enthalpies. The heat of formation of n-bromopropane is determined from literature values of experimental heat of formations ($\Delta H_{0,f}$) for the remaining species in the reaction, and these are: −16.0 ± 0.1 kcal mol^{-1} (CH_4) [51], −8.96 ± 0.36 kcal mol^{-1} (CH_3Br) [52], −25.0 ± 0.15 kcal mol^{-1} ($CH_3CH_2CH_3$) [53]. n-Bromopropane was found to have a heat of formation at 0 K of −21.4 kcal mol^{-1} at the CCSD(T)/6-311++G(2df,2p) level of theory. The literature value of −20.2 ± 0.1 kcal mol^{-1} deviates from ours by 6%.

TABLE 11.2 Enthalpies of reaction ($\Delta H_{r,0}$) and energy barriers (EB) for the reaction of bromopropane with OH radicals and Cl atoms (kcal mol^{-1})†

Theory	$BrCH_2CH_2CH_3 + OH \rightarrow$										
	$BrCHCH_2CH_3 + H_2O$			$BrCH_2CHCH_3 + H_2O$				$BrCH_2CH_2CH_2 + H_2O$			
	$\Delta H_{r,0}$	EB^{1a}	EB^{1b}	$\Delta H_{r,0}$	EB^{2a}	EB^{2b}	EB^{2c}	$\Delta H_{r,0}$	EB^{3a}	EB^{3b}	EB^{3c}
HF/6-31G(d)	−0.02	26.2	26.4	−3.6	24.6	24.8	24.7	0.7	27.5	27.1	26.7
MP2/6-31G(d)	−13.2	6.9	6.9	−16.5	6.2	5.0	5.6	−11.6	9.6	8.8	7.8
CCSD(T)/6-311G(2d,2p)	−16.3	1.8	2.4	−18.8	1.4	1.6	1.4	−13.8	5.2	4.2	3.8
CCSD(T)/6-311++G(2df,2p)	−18.9	0.9	1.2	−21.6	0.2	0.4	0.3	−16.5	3.2	3.1	2.7
G2	−19.3	0.8	1.2	−22.0	0.3	0.4	0.2	−16.8	3.2	3.0	2.9
G3	−15.7	0.4	0.5	−18.3	−0.2	−0.6	−0.3	−13.0	3.0	2.7	2.0
G3 (Gilles *et al.*)	–	–	1.0	–	–	–	0.0	–	–	2.9	2.4

Theory	$BrCH_2CH_2CH_3 + Cl \rightarrow$										
	$BrCHCH_2CH_3 + HCl$			$BrCH_2CHCH_3 + HCl$				$BrCH_2CH_2CH_2 + HCl$			
	$\Delta H_{r,0}$	EB^{1a}	EB^{1b}	$\Delta H_{r,0}$	EB^{2a}	EB^{2b}	EB^{2c}	$\Delta H_{r,0}$	EB^{3a}	EB^{3b}	EB^{3c}
HF/6-31G(d)	6.2	19.6	19.4	2.6	17.6	17.6	–	6.9	18.6	18.3	20.5
MP2/6-31G(d)	5.6	7.1	6.7	2.3	6.7	6.2	–	7.2	9.8	8.8	10.4
CCSD(T)/6-311G(2d,2p)	−2.7	−1.2	−1.0	−5.2	−1.0	−0.8	–	−0.2	1.5	1.4	3.0
CCSD(T)/6-311++G(2df,2p)	−3.3	−2.2	−1.9	−6.0	−2.0	−1.9	–	−0.9	0.7	0.5	2.0
G2	−3.8	−2.8	−2.6	−6.6	−2.4	−2.2	–	−1.3	0.4	0.3	1.7
G3	−5.7	−4.8	−4.6	−8.3	−4.4	−4.4	–	−3.0	−1.3	−1.7	−0.5

† Include MP2(full)/6-31G(d) zero point energy correction.

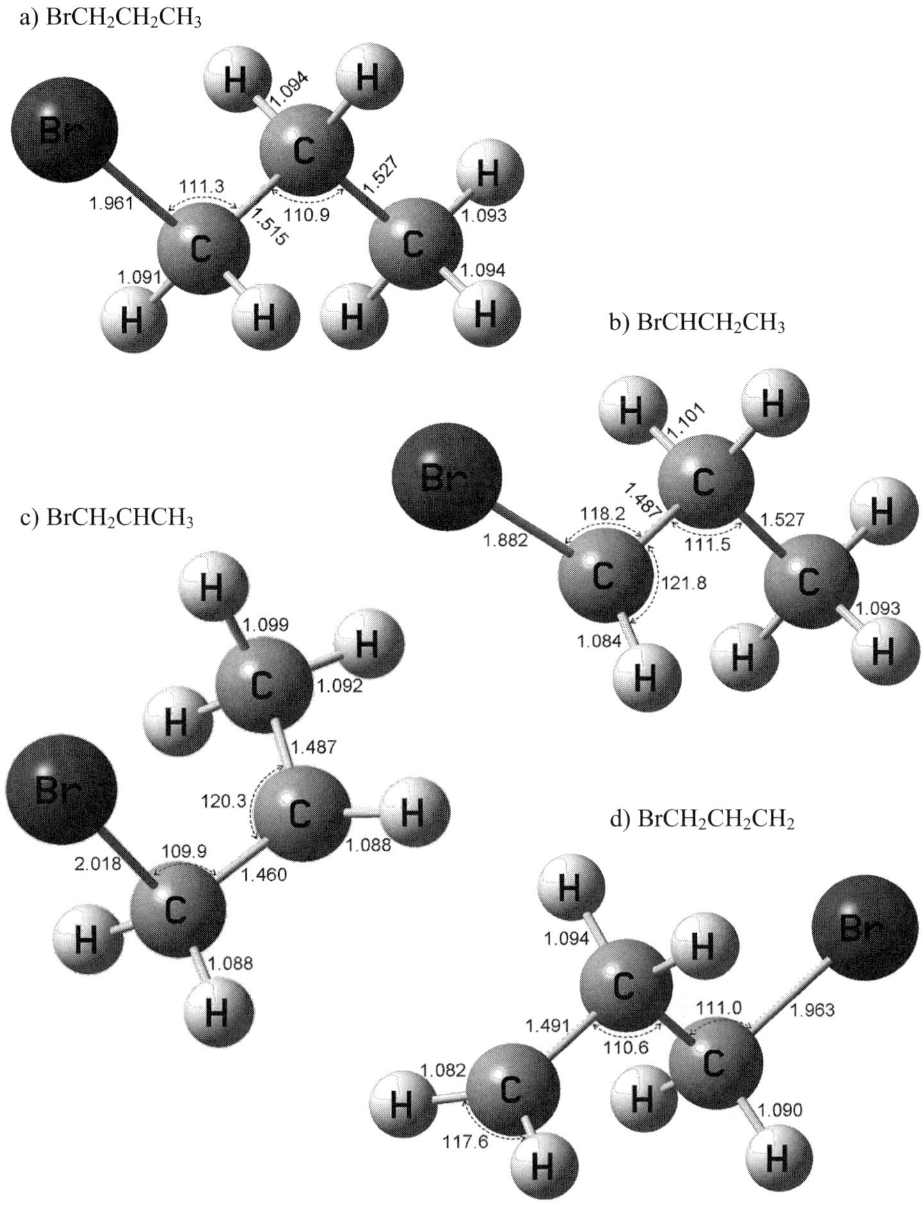

FIGURE 11.1 Geometries of n-bromopropane, primary, secondary, and tertiary alkyl radicals. Parameters are calculated at the MP2(full)/6-31G(d) level of theory.

The geometry of n-bromopropane in the *trans* conformation is illustrated in Figures 11.1a. The resulting alkyl radicals from the hydrogen abstraction reactions are shown in Figures 11.1b, 11.1c, and 11.1d. The α radical ($BrCHCH_2CH_3$, Figure 11.1b), keeps the same C-atom backbone and H-atom arrangement. At the radical site (αC), the remaining H atom forms an angle of 121.8° with the adjacent C-atom, being stretched by ~10°, and the Br atom is positioned closer with respect to the αC at 1.882 Å, after the transition of *n*-bromopropane geometry from a close to an open-shell structure. Figure 11.1c illustrates the β-alkyl radical. With the removal of a secondary hydrogen atom the C-atom backbone forms an angle of 120.3° and the C–C bonds shorten by roughly 0.05 Å. Also the Br atom bends outside the C–C–C plane while the C–Br bond elongates by ~0.06 Å. Figure 11.1d illustrates the γ *n*-bromopropyl alkyl radical ($BrCH_2CH_2CH_2$). The remaining two terminal H atoms accommodate 90.0° out of plane with respect to C–C–C backbone, forming an angle of 117.6° with the γC.

3.1.2 *n*-Bromopropane with OH radical reaction

The most recent experimental study of the OH + *n*-bromopropane reaction, supported with *ab initio* calculations, was performed by Gilles *et al.* [37]. The determination of the activation barriers at the G3 level was performed with respect to the *n*-bromopropane—*trans* isomer. We have explored the stability of both *trans* and *gauche* conformers of *n*-bromopropane, and only high-level *ab initio* calculations predicted that the *gauche* form was the most stable by 0.15 kcal mol^{-1}. For comparison and calibration purposes with the Gilles *et al.* [37] study, the energetics in this study were determined with respect to the –*trans* form. On the other hand, since all experimental studies to date have agreed in observing both isomeric forms in liquid and gas phases, we anticipate that alkyl radicals, complexes, and transition state structures in the *trans* and *gauche* forms are participating simultaneously in the dynamics of the reactions under study. In the following sections, we show how the *trans* and *gauche* forms of the different species contribute to the overall reaction in very good agreement with experimental determinations.

Hydrogen abstraction reactions by OH radicals are known to be characterized by strongly bonded complexes where the hydrogenated compound approached by OH radical is stabilized by hydrogen bonds. This trend is not an exception for *n*-bromopropane. Figure 11.2 shows the pre- and post-reactive complexes (preceding and succeeding the formation of the transition state, respectively) throughout the reaction of *n*-bromopropane with OH radicals. The respective transition structures are shown in Figure 11.3. In total, eight transition states were found and fully optimized. Gilles *et al.* [37] previously found and reported four of these eight transition states. The removal of the α hydrogen is described by two different transition state structures, TS1a (Figure 11.3a) and TS1b (Figure 11.3b). The main difference between these two is the orientation of the Br atom and the leaving H atoms with respect to the C-atom backbone. TS1a resembles more closely the *trans* conformation of *n*-bromopropane on its global minimum structure, where the Br atom is coplanar to the carbon atoms. On the other hand, the C–Br coordinate has rotated 120° away from the carbon atoms plane in the TS1b structure and the leaving H atom is now oriented coplanar to the C-atom chain. In both structures, TS1a and TS1b, the OH group is oriented perfectly coplanar with the C–Br bond. The interaction of Br atoms with the H atom on OH radicals seems to stabilize the transition states. The complex structure preceding TS1a (pre-CX1a) is shown in Figure 11.2a.

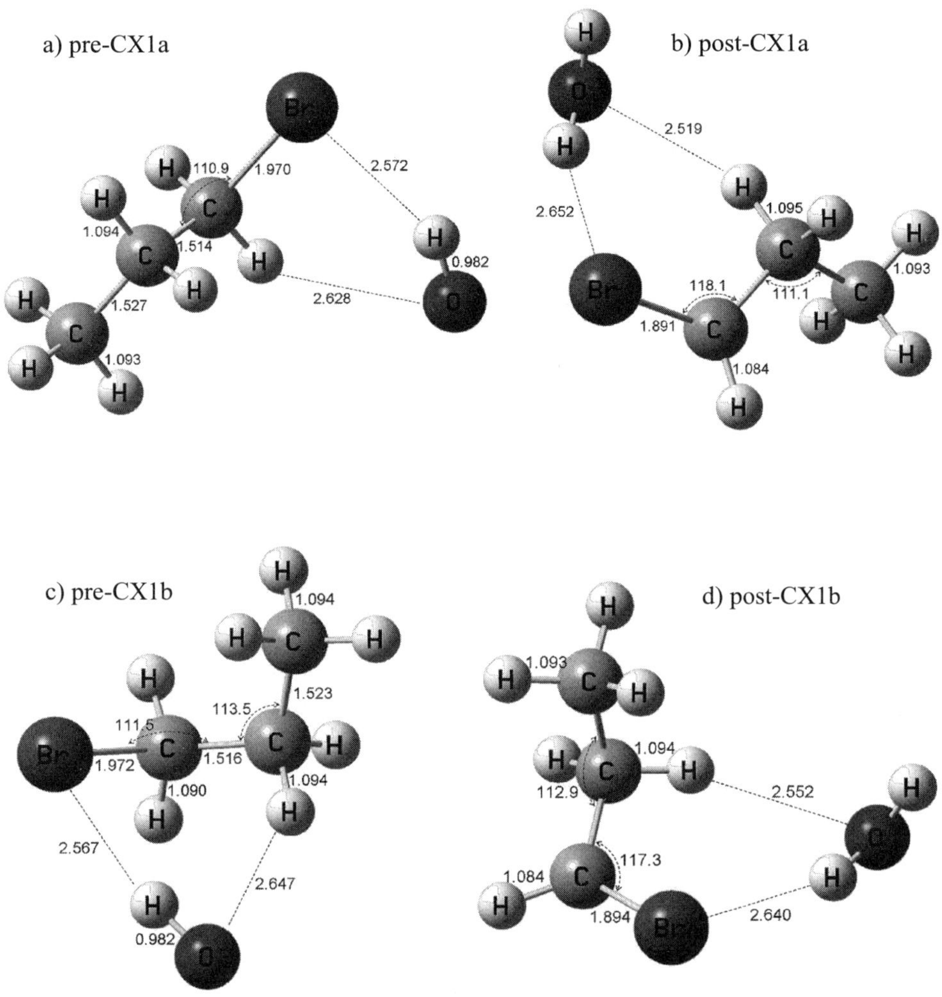

FIGURE 11.2 Geometries of pre- and post-reactive complexes involved in the $BrCH_2CH_2CH_3 + OH$ reaction. Parameters are calculated at the MP2(full)/6-31G(d) level of theory.

The hydrogen bonding interaction is taking place between OH radical, Br atom, and one of the α hydrogen atoms. In contrast, pre-CX1b, complex structure preceding TS1b, shows that OH radical prefers stabilization with the secondary β hydrogen atoms (Figure 11.2c). After the removal of the α H atom through any of the two different conformations, water molecule stabilizes the formed primary radical, by forming hydrogen bonds of approximately 2.6 and 2.5 Å with bromine and secondary hydrogen atoms, respectively.

The abstraction of secondary β hydrogen atoms proceeds via three different transition states, TS2a, TS2b, and TS2c. The transition state TS2c was originally proposed by Gilles *et al.* [37] and it is illustrated in Figure 11.3e. It involves the

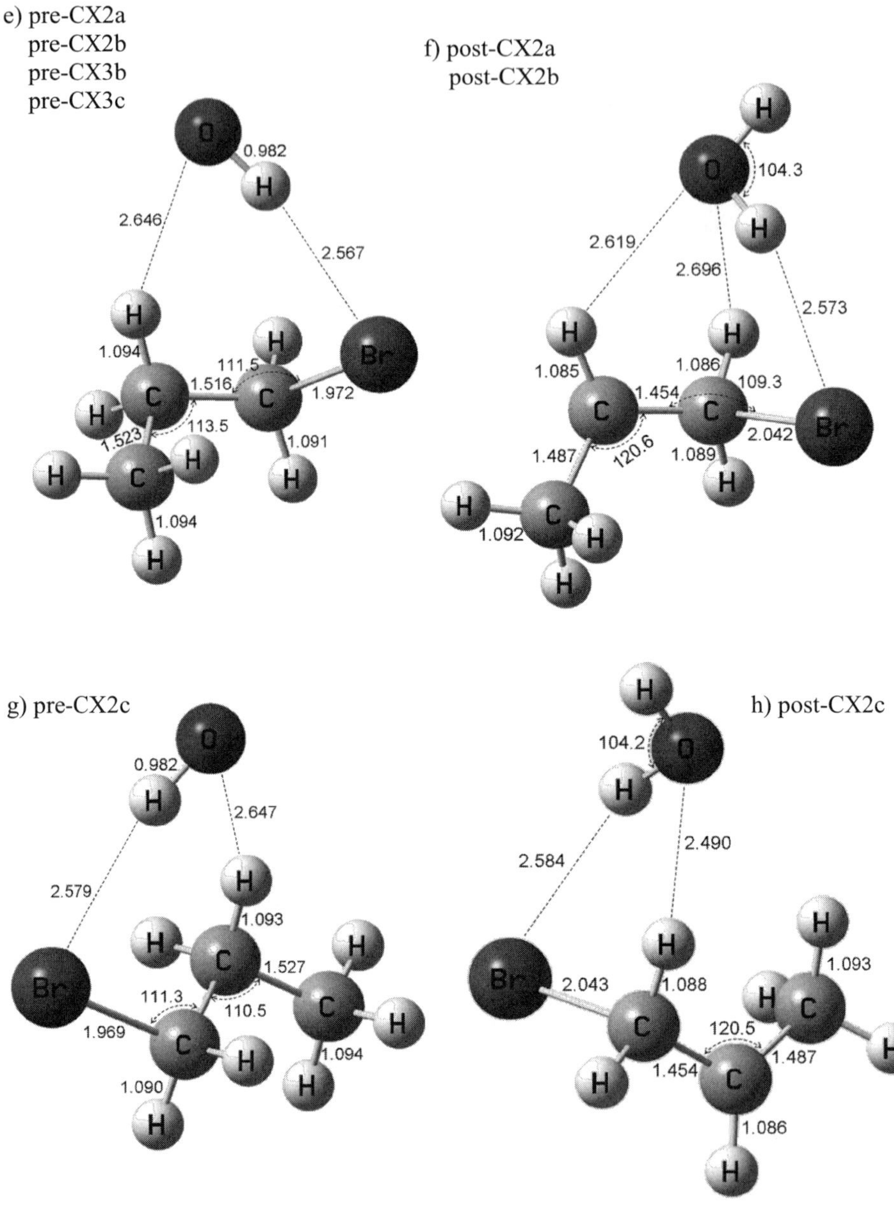

FIGURE 11.2 *(Continued.)*

abstraction of the β hydrogen from bromopropane at its minimum energy configuration. The OH radical is oriented toward the Br atom. This orientation is preserved from the configuration of the pre-reactive complex (Figure 11.2g) where also the OH radical is stabilizing the precursor via the interaction with the H atom subjected to be abstracted. The transition state structures found in this study

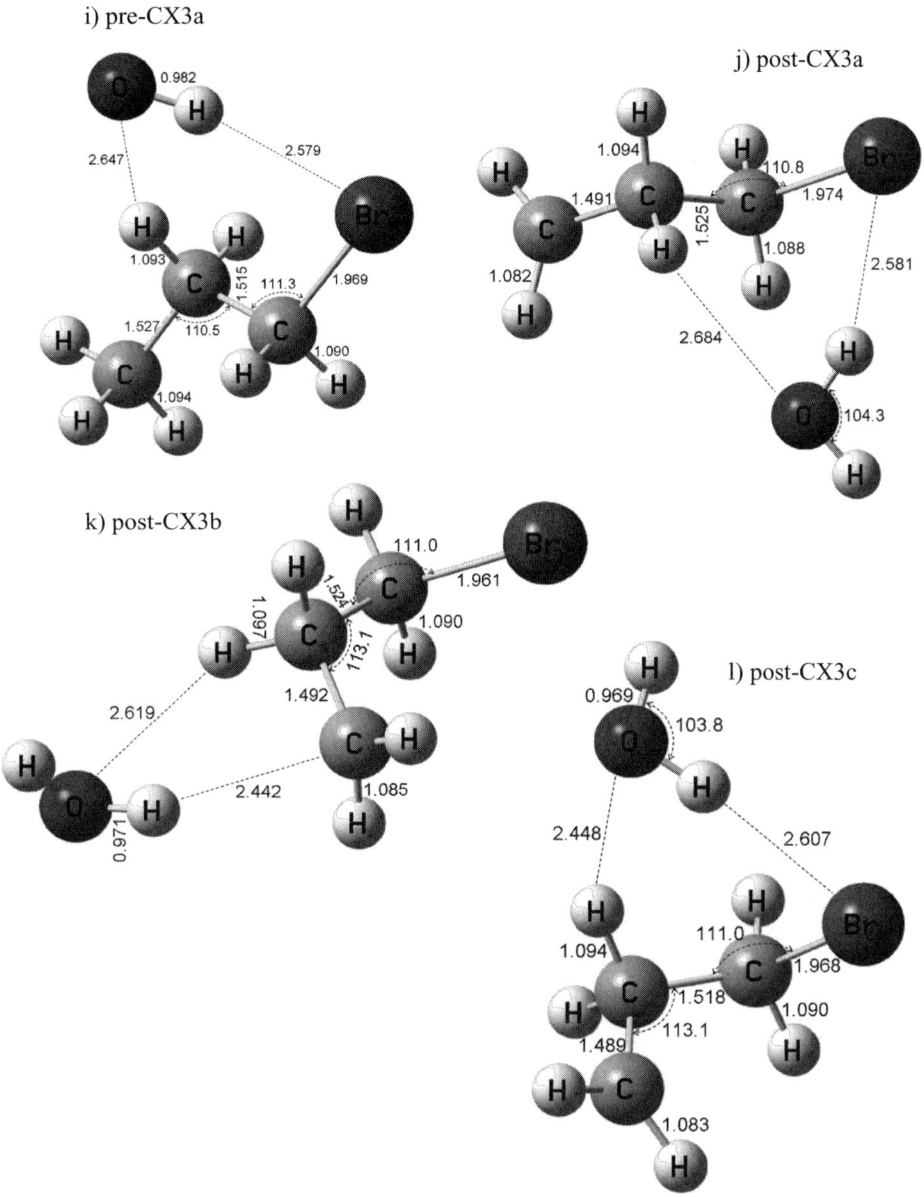

FIGURE 11.2 *(Continued.)*

for the β abstraction are illustrated in Figures 11.3c (TS2a) and 11.3d (TS2b). In both configurations the bromine atom is oriented in a dihedral angle of ~120° with respect to the C-atom chain. OH radicals are removing the secondary H-atom oriented in opposite direction to Br in TS2a, while removing the H-atom oriented toward the Br in TS2b. The post-reactive complexes for the β abstrac-

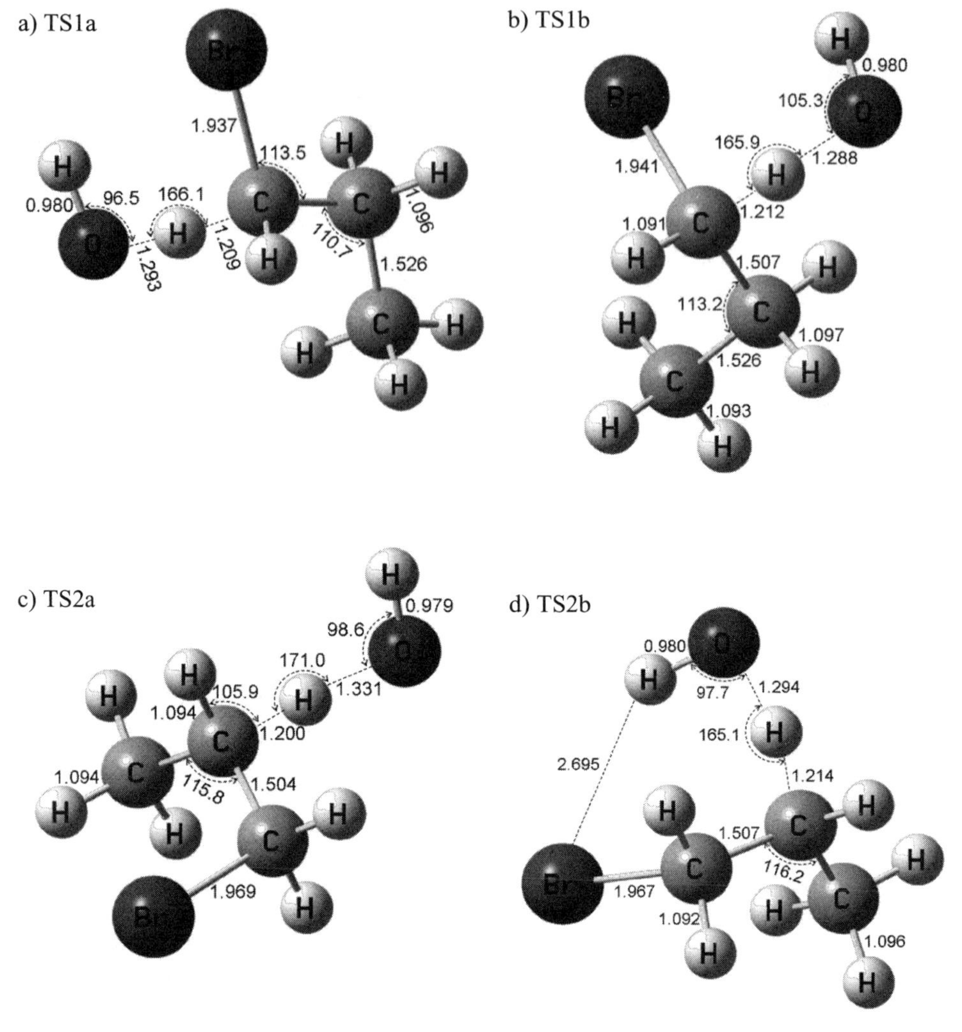

FIGURE 11.3 Transition state structures involved in the $BrCH_2CH_2CH_3 + OH$ reaction. Parameters are determined at the MP2(full)/6-31G(d) level of theory.

tions resemble similar structures, as shown in Figures 11.2f (post-CX2a and 2b) and 11.2h (post-CX2c), where water complexes with bromine, αH and βH in post-CX2a and 2b while the H-bonding interaction takes place via Br atom and αH in post-CX2c.

Three different transition states describe the abstraction of γ hydrogen atoms, by TS3a (Figure 11.3f) proposed in this study, and TS3b & 3c (Figures 11.3g and 11.3h) previously proposed by Gilles *et al.* [37]. TS3a involves the removal of the hydrogen atom in the *trans* orientation with respect to the C-chain and *cis* with the bromine atom. TS3b and 3c involves the abstraction of the *trans* and *cis* (with respect to Br) H atoms from bromopropane on its 120° Br–C–C–C

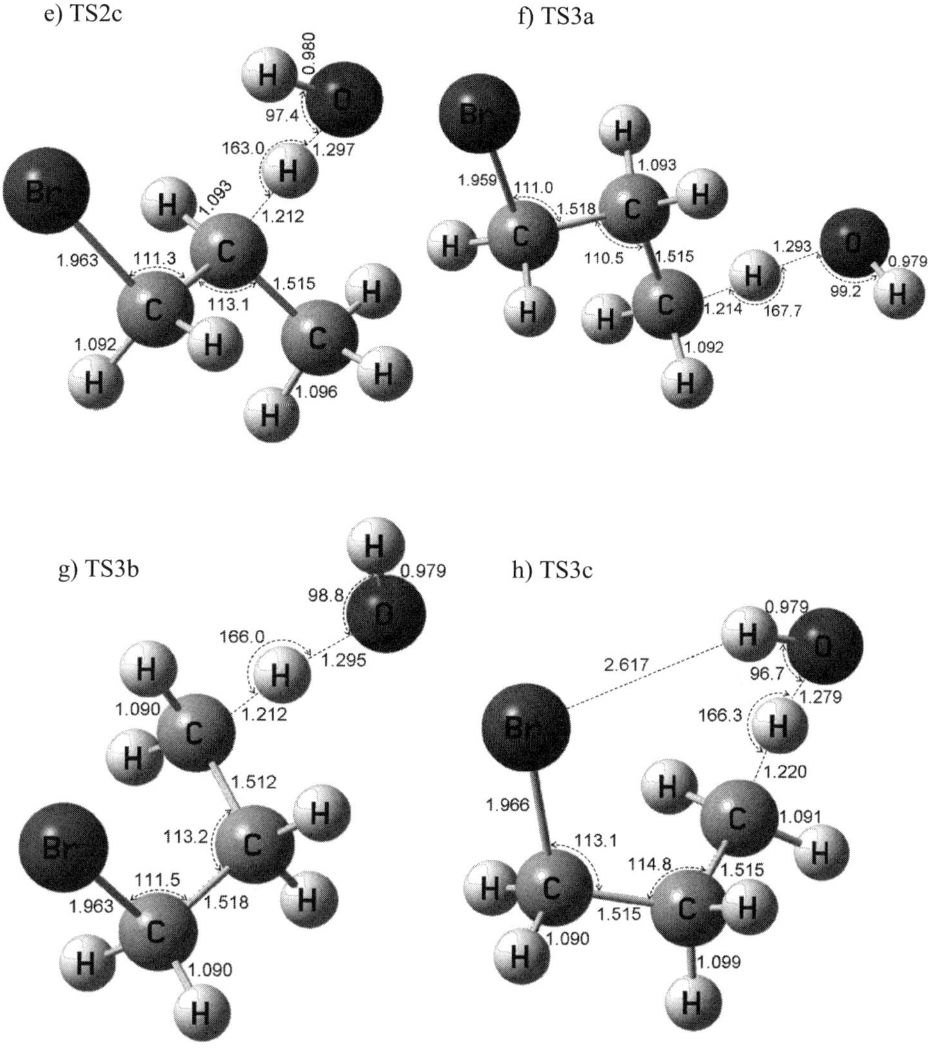

FIGURE 11.3 (*Continued.*)

dihedral angle configuration, respectively. TS3c is stabilized by an intramolecular hydrogen bond 2.617 Å in length between bromine and the H atom on OH radical. The complex structure preceding TS3a is shown in Figure 11.2i, where OH radical is stabilizing the precursor through interaction via bromine atom and a β hydrogen atom. Figure 11.2e also resembles the pre-reactive complex preceding TS3b and TS3c. Figures 11.2j and 11.2l shows complex structures following the γ hydrogen abstraction reaction where H_2O stabilizes the resulting radical via bromine and β hydrogen atoms on bromopropane two different conformations, while it is also stabilized via γC and a βH as illustrated in Figure 11.2k.

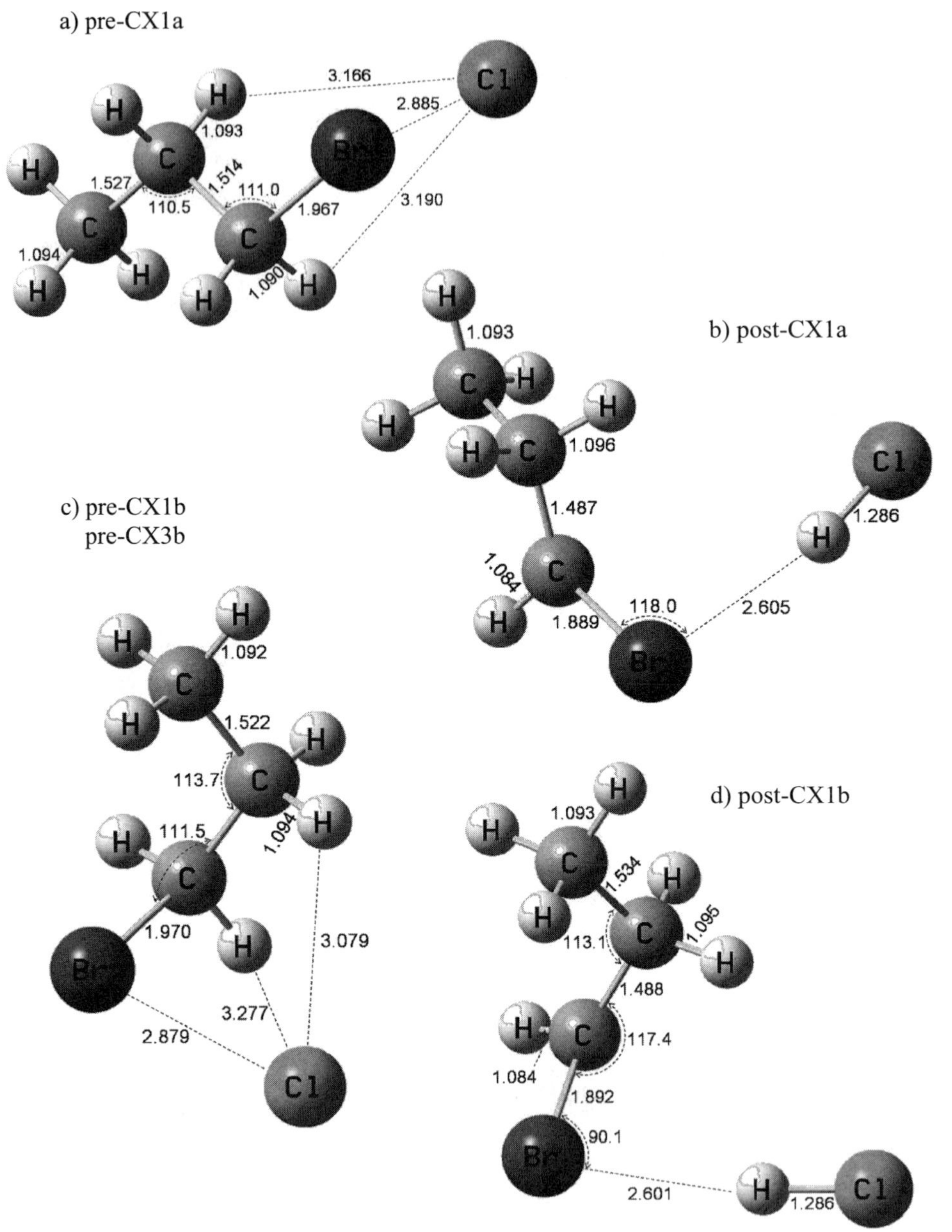

FIGURE 11.4 Geometries of pre- and post-reactive complexes involved in the $BrCH_2CH_2CH_3 + Cl$ reaction. Parameters are calculated at the MP2(full)/6-31G(d) level of theory.

3.1.3 *n*-Bromopropane with Cl radical reaction

Figure 11.4 shows the pre- and post-reactive complexes (preceding and succeeding the formation of the transition state, respectively) throughout the reaction of

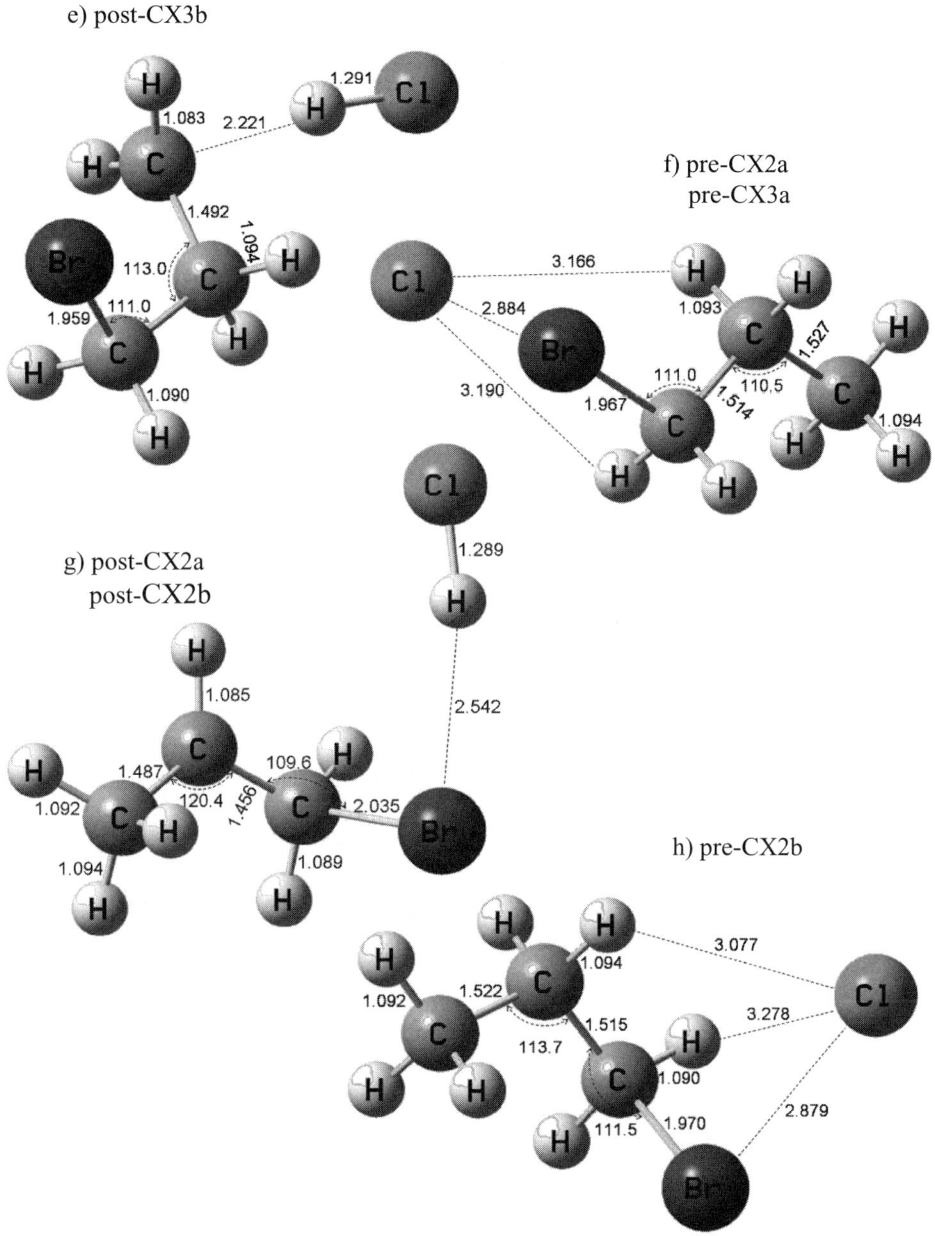

FIGURE 11.4 (*Continued.*)

n-bromopropane with Cl atoms. The respective transition structures are shown in Figure 11.5. In total, seven transition states were found and fully optimized, inspired by those obtained from the reaction with OH radical. The removal of the α hydrogen is described by two different transition state structures, $TS1a_{Cl}$

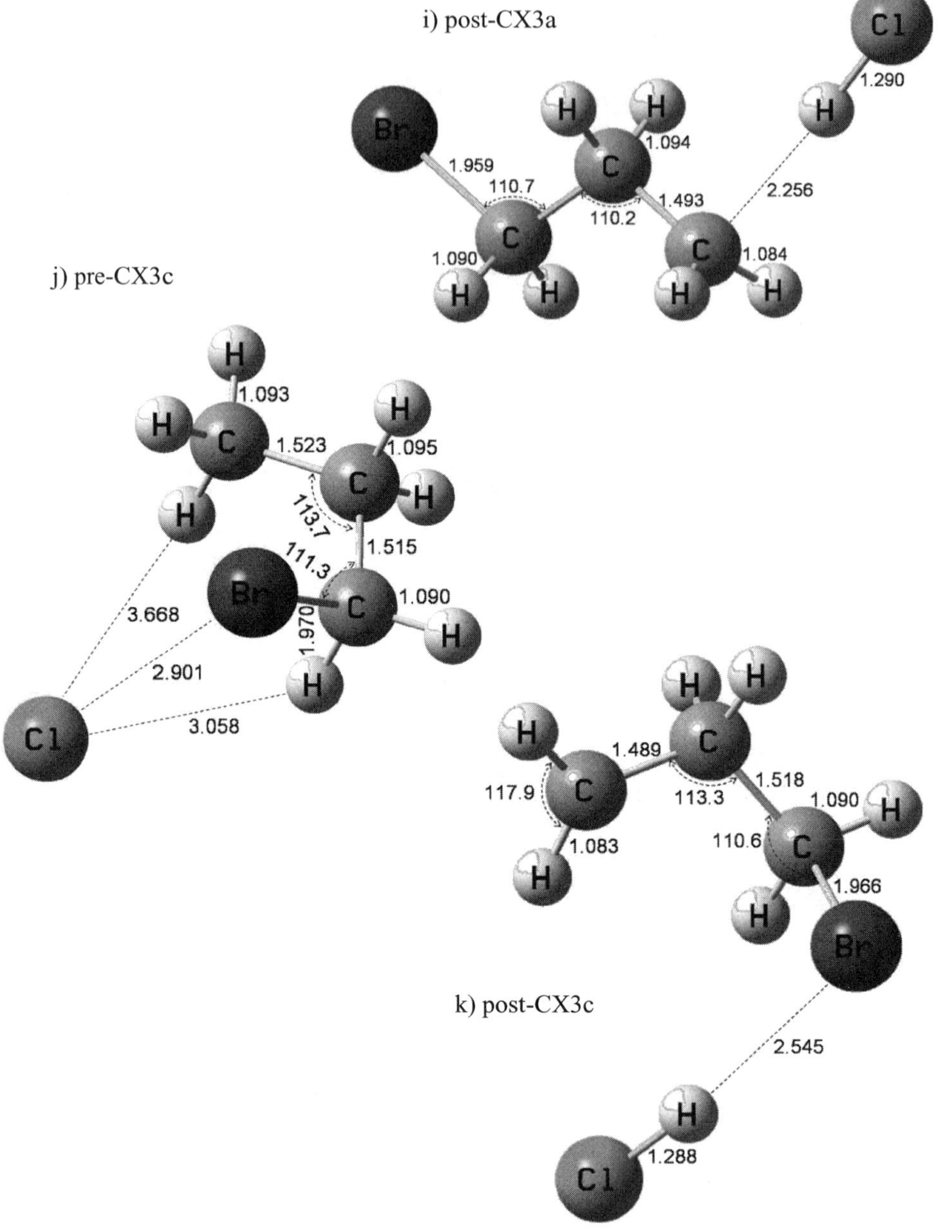

FIGURE 11.4 (*Continued.*)

(Figure 11.5a) and TS1b$_{Cl}$ (Figure 11.5b). In any case, the abstraction of the hydrogen atom takes place in a C–H–Cl angle of approximately 180°. The pre- and post-reactive complexes for TS1a are shown in Figures 11.4a and 11.4b. Cl atom complexes around Br atom, αH, and βH before it abstract the αH. The produced

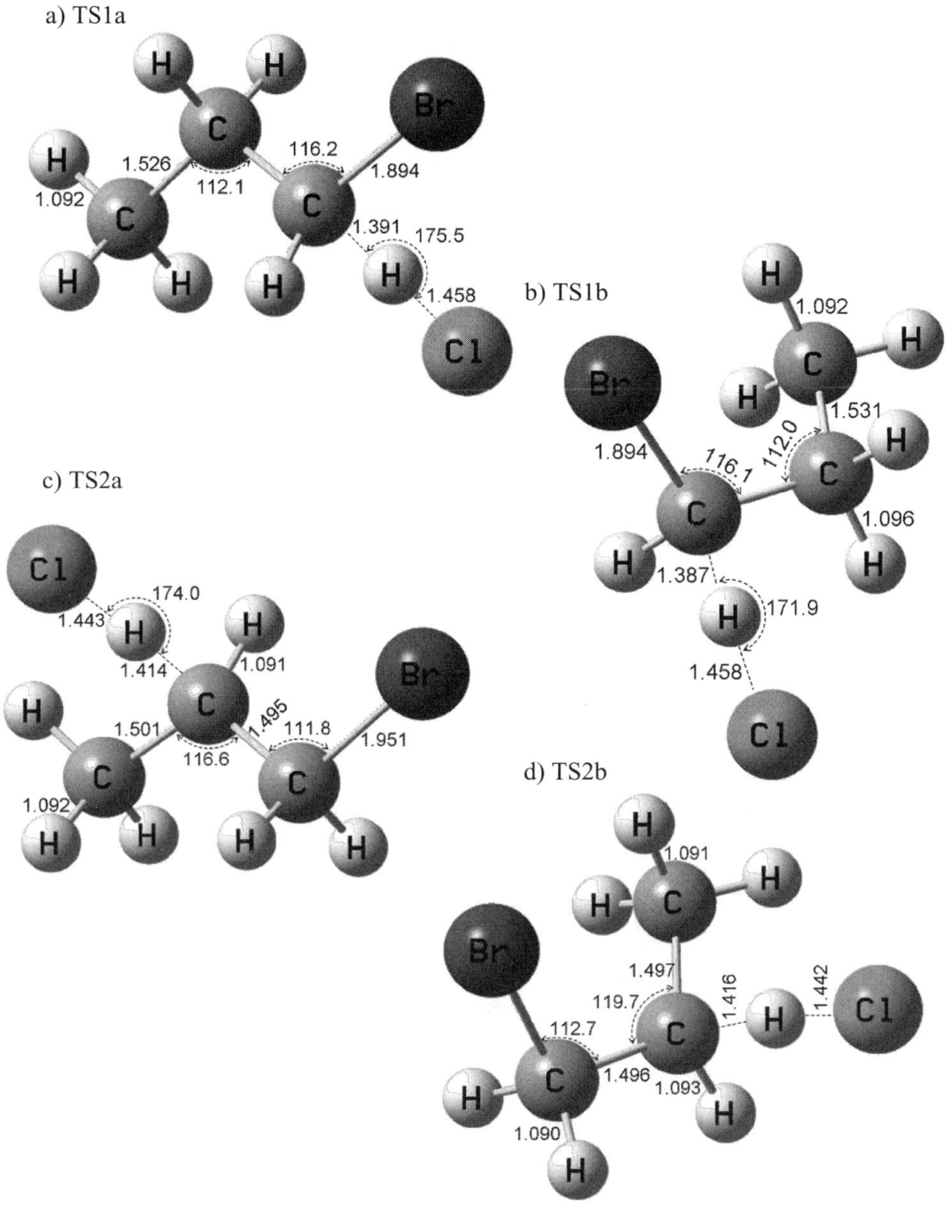

FIGURE 11.5 Transition state structures involved in the $BrCH_2CH_2CH_3$ + Cl reaction. Parameters are determined at the MP2(full)/6-31G(d) level of theory.

HCl forms a hydrogen bond with Br atom of 2.605 Å. Figures 11.6c and 11.6d show the pre- and post-reactive complexes predicted by the IRC calculation of TS1b. Both complexes describe the reaction being bromopropane in the gauche conformation, complexing Cl at the Br-αH-βH side of bromopropane in the pre-complex

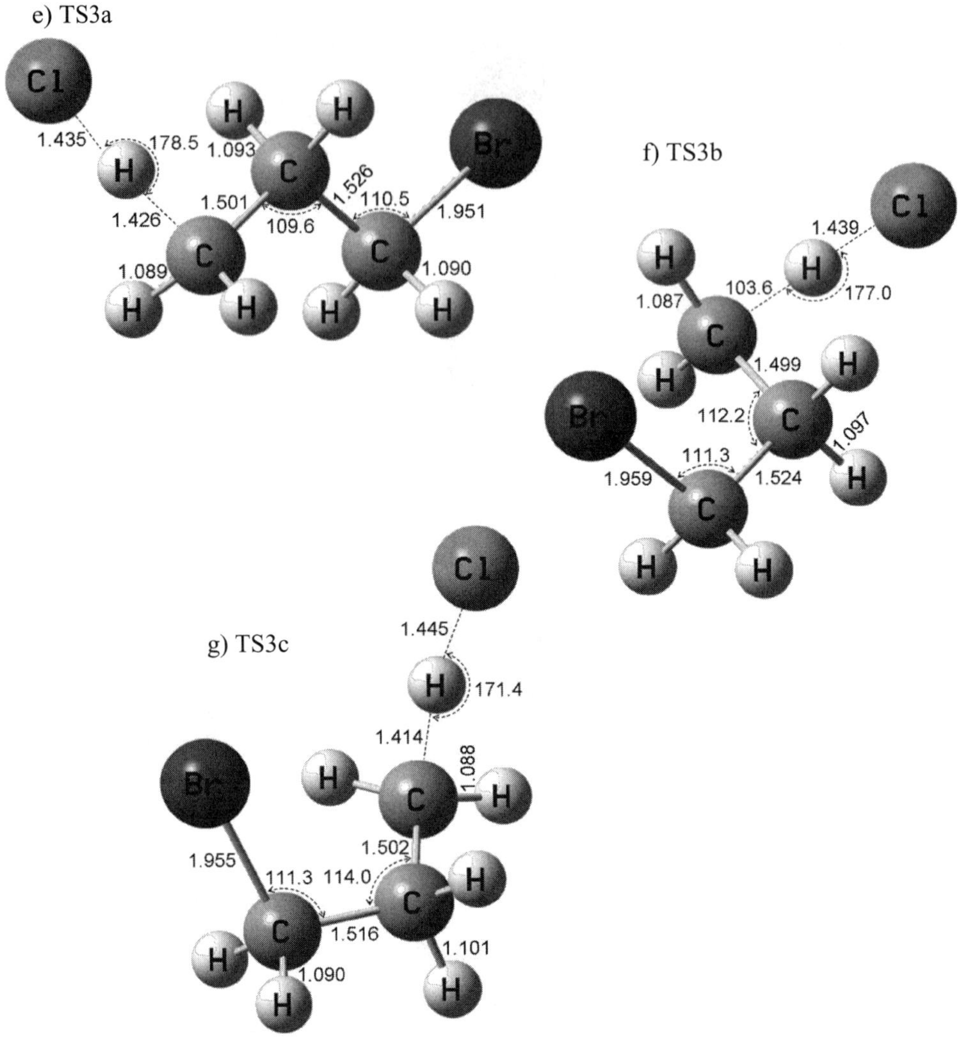

FIGURE 11.5 *(Continued.)*

and HCl toward Br atom by hydrogen bonding in the post-reactive complexes. These stabilizing interactions are very similar in nature. Interestingly, their distances are kept within the same range throughout the reaction.

The abstraction of β hydrogen atoms is represented by TS2a and TS2b in Figures 11.5c and 11.5d. In the *gauche* conformation (TS2b) the C–C–C angle in the *gauche* conformation is found to be larger than that in the *trans* conformation by approximately three degrees due to the steric interaction between Br atom and the methyl group in the *gauche* rearrangement. Figures 11.4f and 11.4h show the complex structures preceding transition states TS2a and TS2b. The stabilization interactions are basically the same, Cl atom complexing with Br atom, αH, and

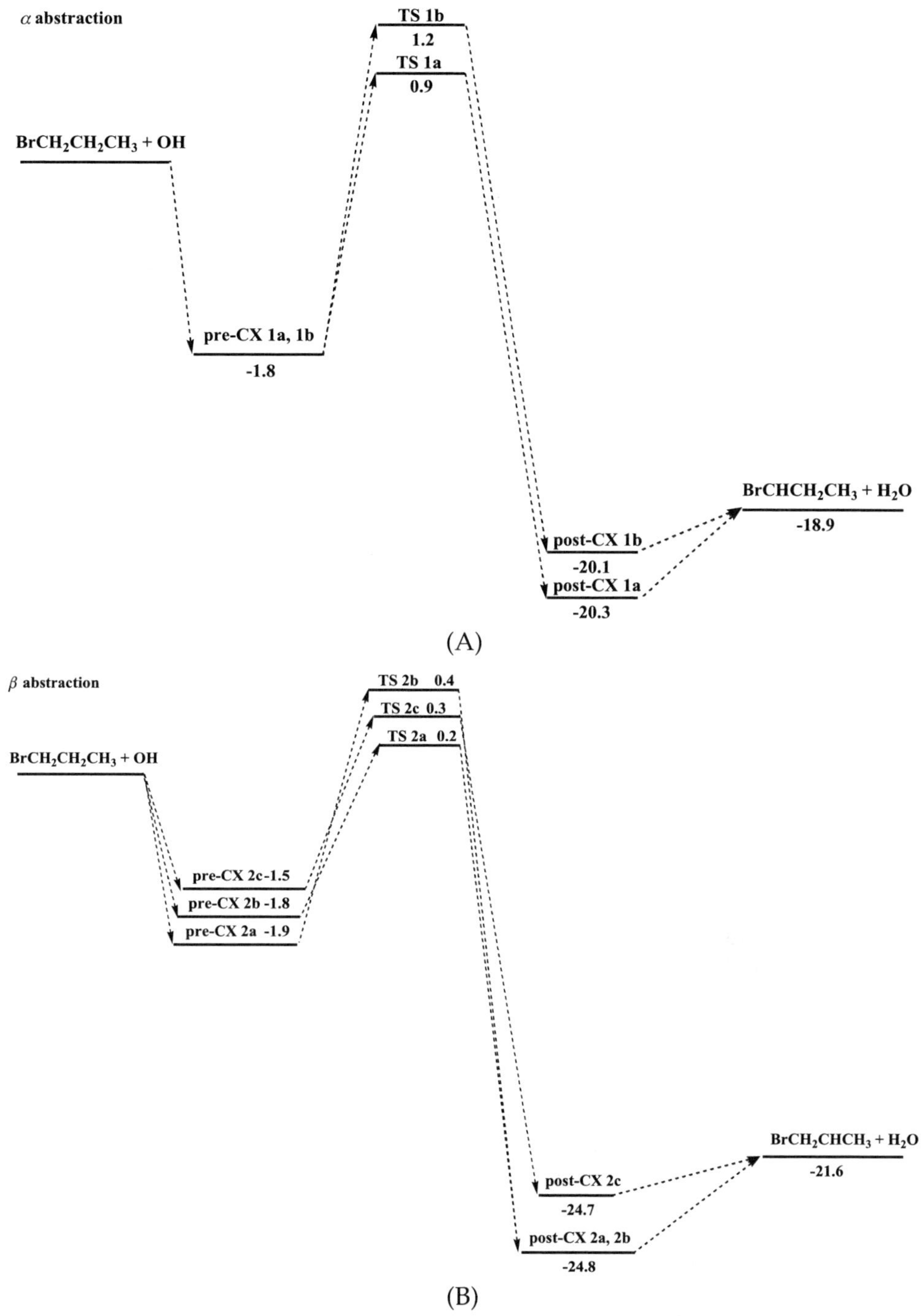

FIGURE 11.6 Potential energy surface of the (A) α, (B) β, and (C) γ hydrogen abstraction during the reaction of 1-bromopropane with OH radicals (kcal mol^{-1}).

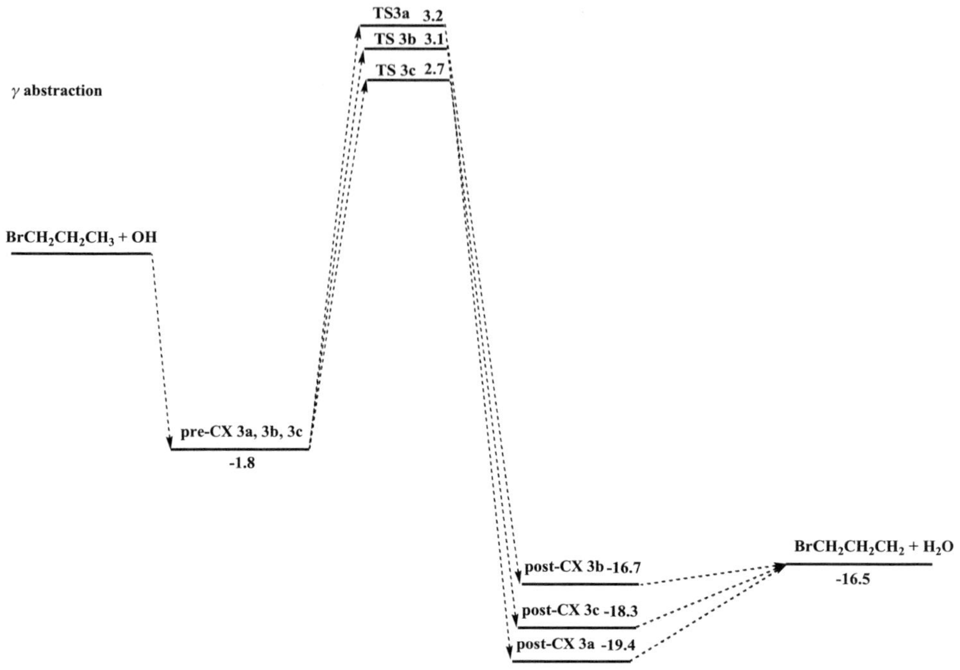

(C)

FIGURE 11.6 *(Continued.)*

βH in the two different *trans* and *gauche n*-bromopropane conformations, with the closest distance headed for the Br atom. Both transition states share a common post-reactive complex since the β radical will be produced in its most stable configuration, the *gauche* (Figure 11.4g). Again, HCl will complex toward bromine atom.

The γ hydrogen abstraction is described by TS3a, 3b, and 3c, shown in Figures 11.5e, 11.5f, and 11.5g, respectively. Structural differences between the *trans* (TS3a) and the *gauche* (TS3b and TS3c) conformations are observed, mainly based on steric hindrance. The C–Br bond distances of the *gauche* conformations are longer than that of the *trans* conformation. This is because the steric effect makes the C–Br bond less stable. For the similar reason as in α and β abstractions, the C–C–C angle in *gauche* is larger by 3 to 5 degrees than in the *trans* conformation. The complex prior to TS3a (Figure 11.4f) resembles the structure of all pre-reactive complexes where the transition state under discussion is on the *trans* configuration. Regarding the pre-reactive complexes in the *gauche* conformation (Figures 11.4c and 11.4j), Cl atom lies on top and bottom of the Br–C–C plane on CX3b and CX3c, respectively. The post-reactive complexes CX3a and CX3b show that HCl stabilizes the radical in the direction of the γC atom with an interaction distance of 2.256 Å and 2.221 Å, respectively. On the other hand, post-reactive complex CX3c shows that the stabilization interaction is taking place between HCl toward bromine atom at a distance of 2.545 Å.

The complexation commonly occurs on the bromine atom. This type of adduct formation in halogenated systems has also been observed during the reaction of Cl with iodoethane, in which the $Cl{\cdot}ICH_2CH_3$ adduct structure where Cl atom complexes ICH_2CH_3 toward iodide, has been identified at the B3LYP/ECP level of theory [54]. It should be noted that similar Cl–I distances of ca. 2.9 Å are involved in the Cl + ICH_2CH_3 reaction as to the Cl–Br distance for the Cl + $BrCH_2CH_3$ reaction. Because iodine is larger than bromine, this implies a weaker Cl interaction with Br. Consistent with this is the calculated Cl–I bond strength [54] of ca. 14 kcal mol^{-1}, which is twice that found for Cl–Br in this work.

3.2 Energetics of the reaction of *n*-bromopropane with OH radical and Cl atom

3.2.1 $BrCH_2CH_2CH_3$ + OH

Figures 11.6A, 11.6B and 11.6C summarize the energetics of the potential energy surface for the $BrCH_2CH_2CH_3$ + OH reaction determine at the CCSD(T)/6-311++G(2df,2p)//MP2(full)/6-31G(d). Table 11.2 shows the calculated enthalpies of reaction and activation barriers, corrected for zero-point energies, for the various reaction channels for the abstraction of a hydrogen atom from bromopropane by OH radicals. The binding energies of pre- and post-reactive complexes are listed in Table 11.3. To recall, the hydrogen abstraction from bromopropane can take place at the α (primary), β (secondary), γ (tertiary) carbon atoms. A description of the potential energy surface of the abstraction of a α hydrogen from n-bromopropane by OH radicals is shown in Figure 11.6A. The CCSD(T)/6-311++G(2df,2p) calculations predict that the α hydrogen abstraction proceeds via the formation two different pre-reactive complexes, CX1a and CX1b, both with binding energies of −1.8 kcal mol^{-1}, respectively, as shown in Table 11.3. The reactants can rearrange into two different transition structures, $TS1a_{OH}$ and TS1b, with energy barriers of 0.8 and 1.2 kcal mol^{-1}. Both transition states are first order saddle-points with imaginary frequencies at 2242i cm^{-1} for TS1a and 2289i cm^{-1} for TS1b. After the α hydrogen is transfer from n-bromopropane to OH, the products stabilize each other in a complex arrangement with binding energies (with respect to the reactants) of −20.3 (post-CX1a) and −20.1 (post-CX1b) kcal mol^{-1}. The hydrogen abstraction at the α carbon is exothermic by −18.9 kcal mol^{-1}. The G3 energy barriers determined at 0 K by Gilles *et al.* [37] are included in Table 11.2. These values represent approximately an average between our CCSD(T)/6-311++G(2df,2p) and G3 values, and are within 0.2–0.3 kcal mol^{-1} from the G3 energy barriers determined in this study.

The β hydrogen abstraction pathway (Figure 11.6B) is exothermic by −21.6 kcal mol^{-1} at the CCSD(T)/6-311++G(2df,2p)//MP2(full)/6-31G(d) level of theory. Three transition state structures were located for this secondary hydrogen abstraction reaction. TS2a, 2b, and 2c, are positioned at 0.2, 0.4, and 0.3 kcal mol^{-1} above the reactants level, respectively. One imaginary frequency was found for each transition state at 1858i cm^{-1} (TS2a), 2120i cm^{-1} (TS2b), and 2101i cm^{-1} (TS2c), frequencies that characterized the C–H stretching motion against OH radical. Pre- and post-reactive complexes were located from each transition state by

TABLE 11.3 Binding energies (BE) in kcal mol^{-1} of complexes

Complex	X = OH, Cl	
	BE_{nBP+OH} [a]	BE_{nBP+Cl} [a]
$BrCH_2CH_2CH_3 + X \rightarrow BrCHCH_2CH_3 + HX$		
pre-CX1a	−1.8	−5.7
post-CX1a	−20.2	−5.3
pre-CX1b	−1.8	−5.7
post-CX1b	−20.0	−5.2
$BrCH_2CH_2CH_3 + X \rightarrow BrCH_2CHCH_3 + XH$		
pre-CX2a	−1.9	−5.7
post-CX2a	−24.8	−9.3
pre-CX2b	−1.8	−5.7
post-CX2b	−24.8	−8.6
pre-CX2c	−1.5	–
post-CX2c	−24.7	–
$BrCH_2CH_2CH_3 + X \rightarrow BrCH_2CH_2CH_2 + XH$		
pre-CX3a	−1.8	−5.7
post-CX3a	−19.4	−2.4
pre-CX3b	−1.8	−5.7
post-CX3b	−16.7	−2.0
pre-CX3c	−1.8	−5.3
post-CX3c	−18.3	−3.5

a Determined at CCSD(T)/6-311++G(2df,2p)//MP2(full)/6-31G(d).

IRC calculations. For this reaction, the *gauche*-like structures, CX2a and CX2b, have the largest binding energies. Preceding the transition states, these binding energies were determined to be −1.8 and −1.9 kcal mol^{-1}; and −24.8 kcal mol^{-1} after the transition barrier. The *trans* structure, CX2c, have smaller binding energies, −1.5 kcal mol^{-1} and −24.7 kcal mol^{-1}, before and after passing through the transition barrier. The abstraction of β hydrogen atoms results in the most kinetically and thermodynamically favorable pathway, specifically through TS2a. Gilles *et al.* [37] predicted an activation barrier of 0.0 kcal mol^{-1} for TS2c at the G3 level of theory, compared to −0.3 kcal mol^{-1} predicted by this study at the G3 level with MP2 scaled zero-point energy correction.

The potential energy surface of the abstraction of the terminal methyl (γ) hydrogen is depicted in Figure 11.6C. The enthalpy of reaction is predicted to be −16.50 kcal mol^{-1} at the CCSD(T)/6-311++G(2df,2p)//MP2(full)/6-31G(d). The reaction proceeds via three possible transition states. Vibrational frequency analysis shows that TS3a is a first-order saddle point characterized by one negative frequency of 2129i cm^{-1}. The formation of a pre-reactive complex, pre-CX3a, is observed with binding energy of −1.8 kcal mol^{-1}. A product-like complex, post-

CX3a, is formed involving hydrogen bond interaction between water and bromine atom and a γ-hydrogen atom with binding energy of -19.4 kcal mol^{-1}. TS3b and TS3c were originally predicted by Gilles *et al.* [37], whose activation barriers were predicted to be 2.9 and 2.4 kcal mol^{-1} at the G3 level. The frequency analysis in this study attributed one imaginary frequency to TS3b and TS3c, of values of 2093i and 2254i cm^{-1}, respectively, indicating that they are first-order saddle points. Given the magnitude of the imaginary frequencies for TS3b and TS3c, for example, suggest that these barriers could be quite narrow, and hence tunneling could be important factors in these reactions. Pre- and post-reactive complexes CX3b resemble the gauche-like conformation of TS3b, with binding energies of -1.8 and -16.7 kcal mol^{-1}, respectively. TS3c pathway has the smallest barrier height of 2.7 kcal mol^{-1}. The formation of the respective reactant-like and product-like complexes is also observed with binding energies of -1.8 and -18.3 kcal mol^{-1}. Among the three possible γ branches, the abstraction of terminal hydrogen will proceed via TS3c. Its lowest barrier height is attributed to the fact that the transition state structure is stabilized by an intramolecular hydrogen bond between OH radical and Br atom with a value of 2.617 Å. The γ hydrogen abstraction reaction involves the higher energy barriers and smaller reaction enthalpy, making this channel the least kinetically and thermodynamically favorable.

3.2.2 $BrCH_2CH_2CH_3 + Cl$

The potential energy surface of the $BrCH_2CH_2CH_3$ + Cl is depicted in Figures 11.7A, 11.7B, and 11.7C. The α hydrogen abstraction is characterized by two different branches as shown in Figure 11.7A. The enthalpy of reaction is predicted to be -3.3 kcal mol^{-1} at the CCSD(T)/6-311++G(2df,2p)//MP2(full)/6-31G(d) level of theory. The pre-reactive complexes that were found, *trans* and *gauche*, have

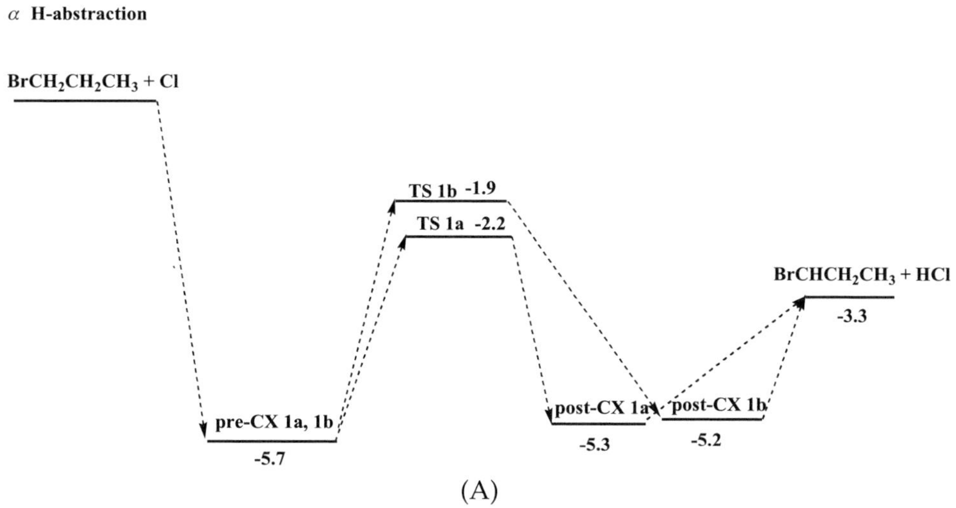

FIGURE 11.7 Potential energy surface of the (A) α, (B) β, and (C) γ hydrogen abstraction during the reaction of 1-bromopropane with Cl atoms (kcal mol^{-1}).

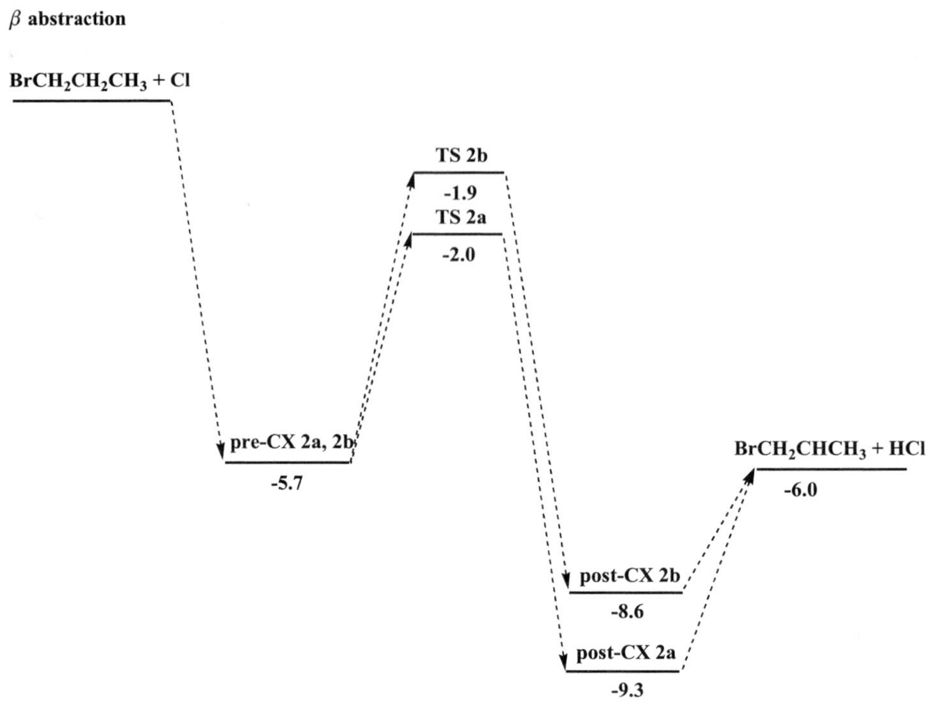

(B)

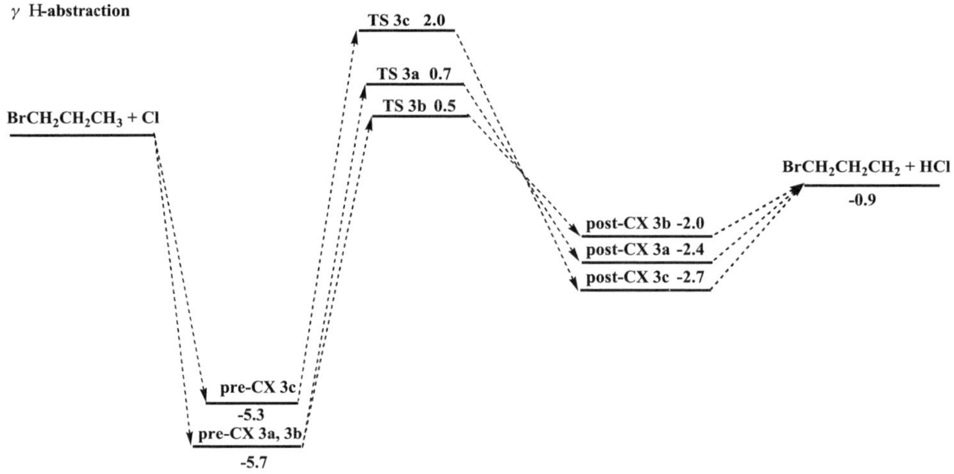

(C)

FIGURE 11.7 *(Continued.)*

similar relative energetics. Despite the fact that there are conformational differences, both have a binding energy of -5.7 kcal mol^{-1}. On the other hand, such differences lead to two differences transition states, TS1a (*trans*) and TS1b (*gauche*), with activation barriers of −1.9 and −2.2 kcal mol^{-1}, respectively. The respective post-reactive complexes keep the same structural features along the chemical reaction. Post-CX1a and 1b differ in their binding energies by 0.07 kcal mol^{-1}, being post-CX1a with the largest stabilization.

Figure 11.7B describes the energetics of the abstraction of β hydrogen atoms from *n*-bromopropane by chlorine atoms. The βH as well as the αH abstraction reaction share the same reactant-like complexes, with binding energy of 5.7 kcal mol^{-1}. Two transition state structures characterized the energetics of the β hydrogen abstraction. TS2a and TS2b lie at −2.0 kcal mol^{-1} and −1.9 kcal mol^{-1} below the reactant level, respectively. TS2a structure resembles a *trans* configuration while TS2b involves the transference of the hydrogen atom in a *gauche* conformation. The post-reactive binding energies are in accordance with those observed at the reaction barriers, the complex proceeding TS2a in the *trans* conformation is more stable than the gauche complex, which proceeds TS2b, by 0.7 kcal mol^{-1}. The overall reaction is slightly exothermic by −6.0 kcal mol^{-1}.

Figure 11.7C shows the energy profile along the reaction path of the abstraction of γ hydrogen from *n*-bromopropane by chlorine atoms. Three different branches were explored. To recall, the pre-reactive complexes CX3a and CX3b involve stabilization within Br and β hydrogen atoms, in *trans* and *gauche* conformations, respectively. These features result in the same binding energy relative to the free reactants, −5.7 kcal mol^{-1}. On the other hand, pre-reactive complex CX3c involves the stabilization of bromine and one γ hydrogen atom in a gauche configuration with Cl atom. Pre-CX3c is less stable than pre-CX3a and pre-CX3b by 0.4 kcal mol^{-1}. True first-order saddle point transition state structures were located, TS3a, TS3b, and TS3c, with imaginary frequencies of 1168i cm^{-1}, 1076i cm^{-1}, and 1291i cm^{-1}, and activation barriers of 0.7, 0.5, and 2.0 kcal mol^{-1}, respectively. TS3b represents the most kinetically favorable structure for the abstraction of γ hydrogen atoms, and to recall, it involves the arrangement of Br and Cl atoms opposite to each other. TS3b results in the formation of the least stable product-like complex, with binding energy of −2.0 kcal mol^{-1}. The remaining post-complexes are 0.4 kcal mol^{-1} and 0.7 kcal mol^{-1} below post-CX3b. It is not surprising that post-CX3c has the largest stabilization energy since HCl interacts with the alkyl radical in the direction of the bromine atom, in contrast to the other two complexes where the stabilization between HCl and the gamma alkyl radical occurs toward the radical γ carbon atom. The γ hydrogen abstraction reaction is slightly exothermic by only −0.9 kcal mol^{-1}.

4. ATMOSPHERIC IMPLICATIONS

Bromine-containing compounds have the potential to released Br atoms upon degradation in the atmosphere. Once in the stratosphere, Br atoms are about 45 times more effective than chlorine in destroying stratospheric ozone [22]. An important example of Br-containing compounds is 1-bromopropane, which is currently utilized as a industrial solvent and have been proposed as a replacement for CFCs, controlled under the agreements of the Montreal Protocol.

The stability of *n*-bromopropane has also been explored by *ab initio* methods. The majority of the ab initio methods predicted that, between the *trans* and *gauche* bromopropane isomers, the *trans* isomer is the most stable. Only high-level calculations predict that the gauche form is the most stable, in accordance with experimental observation where an abundance of approximately 64% (*gauche*) and 36% (*trans*) have been determined of the two isomers.

Two major bromopropane atmospheric degradation routes have been explored: hydrogen abstraction reactions initiated by OH radical and Cl atoms. The energetics of the potential energy surface of both reactions are described by pre- and post-reactive complexes and transition states. In total, eight and seven transition state structures supported the characterization of the *n*BP + OH and *n*BP + Cl reactions, respectively. The transition state structures resemble both the *trans* and *gauche* forms. The *trans* isomeric forms of the transition states contributed the most to the α and β hydrogen abstraction in the reactions of *n*-bromopropane with OH radical and Cl atom, while the *gauche* form contributed significantly to the abstraction of γ-hydrogen atoms in both reactions, with an internal stabilizing Br–OH(Cl)–H interaction. From the perspective of the energy barriers, the reaction of *n*-bromopropane with OH radicals will proceed via the abstraction of β hydrogen atoms, while the reaction with Cl atoms should proceed via the abstraction of α and β hydrogen atoms. A difference of at most 3 kcal mol^{-1} between the energy barriers in the reaction of *n*BP + OH versus those in the *n*BP + Cl reaction was observed. This energy barrier difference suggests that the *n*BP + Cl should be faster than the *n*BP + OH reaction. Although it is expected for the reaction with Cl atoms to be faster, the atmospheric concentrations of Cl atoms (10^4 molecules cm^{-3}) are approximately two orders of magnitude smaller than those of OH (10^6 molecules cm^{-3}). These data suggest that the reaction of *n*-bromopropane with Cl atoms is just as important as the reaction with OH radicals. Even though the nature of the reaction is the same, the *n*-bromopropane oxidation chemistry, whether initiated by OH radicals or Cl atoms, will lead to different by-products. According to previous modeling studies by Wuebbles *et al.* [35] the removal of the α hydrogen will produce $BrCH_2CH_2C(O)H$. Removal of the β hydrogen yields $CH_3C(O)CH_2Br$, $CH_3CH{=}CH_2$ and Br are expected as products. $CH_3CH_2C(O)H$ and Br are proposed to be produced from the abstraction of γ hydrogen atoms from *n*-bromopropane.

5. CONCLUSION

Results from this study confirms the experimental findings that the reaction of *n*-bromopropane with OH radicals should be slower than the reaction with Cl atoms. The present results show that pre- and post-reaction complexes are important in the hydrogen abstraction reactions. A detailed study of these rate constants that incorporates the contribution of pre-reactive complexes, the multi-channel nature of these reactions, and temperature dependence is necessary. The results also find that there are subtle reaction preferences for the abstraction of site specific hydrogen on *n*-bromopropane. While knowledge of the dominant products of the

OH reaction is experimentally known, the information for the Cl reaction is new. For the OH reaction, the β hydrogen atom is preferred while the reaction with Cl atoms both α and β hydrogen are favored. This suggests that the oxidation by-products of n-bromopropane is going to be different for the OH versus Cl initiation chemistry. The results presented in this computational study should provide useful insight for modeling n-bromopropane global cycle in atmospheric reactivity models.

ACKNOWLEDGEMENTS

The authors greatly thank Dr. Paul C. Redfern, Chemistry Division from Argonne National Laboratory, Argonne, IL 60439 for providing the Br atom basis sets and for assistance in making the G3 calculations possible.

REFERENCES

[1] X. Yu, G. Ichihara, J. Kitoh, Z. Xie, E. Shibata, M. Kamijima, N. Asaeda, Y. Takeuchi, *J. Occup. Health* **40** (1998) 234.
[2] D.J. Wuebbles, A.K. Jain, K.O. Patten, P.S. Connell, *Atmos. Environ.* **32** (1997) 107.
[3] M.-C. Su, S.S. Kumaran, K.P. Lim, J.V. Michael, A.F. Wagner, D.A. Dixon, J.H. Keifer, J. DiFelice, *J. Phys. Chem.* **100** (1996) 15827.
[4] R.J. Cicerone, *Rev. Geophys. Space Phys.* **19** (1981) 123.
[5] B.J. Finlayson-Pitts, J.N. Pitts, *Atmospheric Chemistry: Fundamentals and Experimental Techniques*, Wiley, New York, 1986.
[6] B.J. Finlayson-Pitts, J.N. Pitts, *Chemistry of the Upper and Lower Atmosphere: Theory, Experiments and Applications*, Academic Press, New York, 2000.
[7] H.B. Singh, L. Salas, H. Shigeishi, A. Crawford, *Atmos. Environ.* **11** (1997) 819.
[8] J.E. Lovelock, *Nature* **256** (1975) 193.
[9] D.A. Dixon, W.A.D. Jong, K.A. Peterson, J.S. Francisco, *J. Phys. Chem. A* **106** (2002) 4725.
[10] S.J. Oltmans, W.D. Kohyr, *J. Geophys. Res.* **91** (1986) 5229.
[11] J.W. Bottenheim, A.G. Gallant, K.A. Brice, *Geophys. Res. Lett.* **13** (1986) 113.
[12] J.C. McConnell, G.S. Henderson, L. Barrie, J. Bottenheim, H. Niki, C.H. Langford, E.M.J. Templeton, *Nature* **355** (1992) 150.
[13] S.J. Oltmans, R.C. Schnell, P.J. Sheridan, R.E. Peterson, S.-M. Li, J.W. Winchester, P.P. Tans, W.T. Sturges, J.D. Kahl, L.A. Barrie, *Atmos. Environ.* **23** (1989) 2431.
[14] R.E. Mickel, J.W. Bottenheim, W.R. Leaitch, W. Evans, *Atmos. Environ.* **23** (1989) 2443.
[15] L.A. Barrie, J.W. Bottenheim, R.C. Schnell, P.J. Crutzen, R.A. Rasmussen, *Nature* **334** (1988) 138.
[16] L.A. Barrie, G. den Hartog, J.W. Bottenheim, S. Landsberger, *J. Atmos. Chem.* **9** (1989) 101.
[17] B.J. Finlayson-Pitts, F.E. Livingston, H.N. Berko, *Nature* **343** (1990) 622.
[18] T. Tang, J.C. McConnell, *Geophys. Res. Lett.* **23** (1996) 2633.
[19] S.-M. Fan, D.J. Jacob, *Nature* **359** (1992) 522.
[20] A. Aranda, G. Le Bras, G. La Verdet, G. Poulet, *Geophys. Res. Lett.* **24** (1997) 2745.
[21] R. Sander, P.J. Crutzen, *J. Geophys. Res.* **101** (1996) 9121.
[22] J.S. Daniel, S. Solomon, R.W. Portmann, R.R. Garcia, *J. Geophys. Res.* **104** (D19) (1999) 23871.
[23] M.A. Kamboures, J.C. Hansen, J.S. Francisco, *Chem. Phys. Lett.* **353** (2002) 335.
[24] J.G. Anderson, W.H. Brune, S.A. Lloyd, D.W. Toohey, S.P. Sander, W.L. Starr, M. Loewenstein, J.R. Podolske, *J. Geophys. Res.* **94** (1989) 11480.
[25] Y.L. Yung, J.P. Pinto, R.T. Watson, S.P. Sander, *J. Atmos. Sci.* **37** (1980) 339.
[26] D.J. Wuebbles, A. Jain, J. Edmonds, D. Harvey, K. Hayhoe, *Environ. Pollut.* **100** (1999) 57.
[27] D.D. Nelson Jr., J.C. Wormhoudt, M.S. Zahniser, C.E. Kolb, M.K.W. Ko, D.K. Weisenstein, *J. Phys. Chem. A* **101** (1997) 4987.

[28] H. Levy II, *Planet. Space Sci.* **20** (1972) 919.

[29] J. Matthews, A. Sinha, J.S. Francisco, *Proc. Nat. Acad. Sci.* **102** (2005) 7449.

[30] T. Donaghy, I. Shanahan, M. Hande, S. Fitzpatrick, *Int. J. Chem. Kinet.* **25** (1993) 273.

[31] S.N. Kozlov, V.L. Orkin, R.E. Huie, M.J. Kurylo, *J. Phys. Chem. A* **107** (2003) 1333.

[32] S. Teton, A.E. Boudali, A. Mellouki, *J. Chim. Phys.* **93** (1996) 274.

[33] D.J. Wuebbles, R. Kotamarthi, K.O. Patten, *Atmos. Environ.* **33** (1999) 1641.

[34] C.H. Bridgeman, J.A. Pyle, D.E. Shallcross, *J. Geophys. Res.* **105** (2000) 26493.

[35] D.J. Wuebbles, K.O. Patten, M.T. Johnson, *J. Geophys. Res.* **106** (2001) 14551.

[36] S.C. Herndon, T. Gierczak, R.K. Talukdar, A.R. Ravishankara, *Phys. Chem. Chem. Phys.* **3** (2001) 4529.

[37] M.K. Gilles, J.B. Burkholder, T. Gierczak, P. Marshall, A.R. Ravishankara, *J. Phys. Chem. A* **106** (2002) 5358.

[38] J.B. Burkholder, M.K. Gilles, T. Gierczak, A.R. Ravishankara, *Geophys. Res. Lett.* **29** (2002) 1822.

[39] H.B. Singh, J.F. Kasting, *J. Atmos. Chem.* **7** (1988) 261.

[40] L.A. Curtiss, K. Raghavachari, P.C. Redfern, V. Rassolov, J.A. Pople, *J. Chem. Phys.* **109** (1998) 7764.

[41] L.A. Curtiss, P.C. Redfern, V. Rassolov, G. Kedziora, J.A. Pople, *J. Chem. Phys.* **114** (2001) 9287.

[42] M.J. Frisch, G.W. Trucks, H.B. Schlegel, G.E. Scuseria, M.A. Robb, J.R. Cheeseman, J.A. Montgomery Jr., T. Vreven, K.N. Kudin, J.C. Burant, J.M. Millam, S.S. Iyengar, J. Tomasi, V. Barone, B. Mennucci, M. Cossi, G. Scalmani, N. Rega, G.A. Petersson, H. Nakatsuji, M. Hada, M. Ehara, K. Toyota, R. Fukuda, J. Hasegawa, M. Ishida, T. Nakajima, Y. Honda, O. Kitao, H. Nakai, M. Klene, X. Li, J.E. Knox, H.P. Hratchian, J.B. Cross, C. Adamo, J. Jaramillo, R. Gomperts, R.E. Stratmann, O. Yazyev, A.J. Austin, R. Cammi, C. Pomelli, J.W. Ochterski, P.Y. Ayala, K. Morokuma, G.A. Voth, P. Salvador, J.J. Dannenberg, V.G. Zakrzewski, S. Dapprich, A.D. Daniels, M.C. Strain, O. Farkas, D.K. Malick, A.D. Rabuck, K. Raghavachari, J.B. Foresman, J.V. Ortiz, Q. Cui, A.G. Baboul, S. Clifford, J. Cioslowski, B.B. Stefanov, G. Liu, A. Liashenko, P. Piskorz, I. Komaromi, R.L. Martin, D.J. Fox, T. Keith, M.A. Al-Laham, C.Y. Peng, A. Nanayakkara, M. Challacombe, P.M.W. Gill, B. Johnson, W. Chen, M.W. Wong, C. Gonzalez, J.A. Pople, *Gaussian 03*, Revision C. 02, Gaussian, Inc., Wallingford CT, 2004.

[43] A.L. Leach, *Molecular Modeling: Principles and Applications*, second ed., Pearson Education, Dehli, 2001.

[44] J.B. Foresman, A. Frisch, *Exploring Chemistry with Electronic Structure Methods*, second ed., Gaussian, Inc., Pitsburgh, PA, 1996.

[45] J.A. Pople, M. Head-Gordon, D.J. Fox, K. Raghavachari, L.A. Curtiss, *J. Chem. Phys.* **90** (1989) 5622.

[46] L.A. Curtiss, K. Raghavachari, G.W. Trucks, J.A. Pople, *J. Chem. Phys.* **94** (1991) 7221.

[47] L.A. Curtiss, K. Raghavachari, J.A. Pople, *J. Chem. Phys.* **98** (1993) 1293.

[48] J.I. Steinfeld, J.S. Francisco, W.L. Hase, *Chemical Kinetics and Dynamics*, second ed., Prentice Hall, Upper Saddle River, NJ, 1999.

[49] A.P. Scott, L. Radom, *J. Phys Chem.* **100** (1996) 16502.

[50] *Encyclopedia of Computational Chemistry*, vol. 2, Wiley, New York, 1998.

[51] L.A. Curtiss, K. Raghavachari, P.C. Redfern, J.A. Pople, *J. Chem. Phys.* **106** (1997) 2498.

[52] P. Fowell, J.R. Lancher, J.D. Park, *Trans. Faraday Soc.* **61** (1965) 1324.

[53] S.G. Lias, J.E. Bartmess, J.F. Liebman, J.L. Holmes, R.D. Levin, W.G. Mallard, *J. Phys. Chem. Ref. Data* **17** (1988) Suppl. 1.

[54] J.J. Orlando, C.A. Piety, J.M. Nicovich, M.L. McKee, P.H. Wine, *J. Phys. Chem. A* **109** (2005) 6659.

CHAPTER 12

Atmospheric Reactions of Oxygenated Volatile Organic Compounds + OH Radicals: Role of Hydrogen-Bonded Intermediates and Transition States

Annia Galano[*,**] and **J. Raúl Alvarez-Idaboy**[***]

Contents

1. Introduction 246
2. Kinetics 247
   - 2.1 Conventional Transition State Theory (CTST) 249
   - 2.2 Canonical Variational Theory (CVT) 249
3. Energies 251
   - 3.1 The B//A approach 251
   - 3.2 On the Basis Set Superposition Error (BSSE) dilemma 251
4. Aliphatic Alcohols 252
   - 4.1 Structure–activity relationship 254
5. Aldehydes 256
6. Ketones 258
7. Carboxylic Acids 264
8. Multifunctional Oxygenated Volatile Organic Compounds 266
9. Concluding Remarks 268

References 270

[*] Instituto Mexicano del Petróleo, Eje Central Lázaro Cárdenas 152, 07730, México D.F., México
E-mail: agalano@prodigy.net.mx

[**] Departamento de Química, Universidad Autónoma Metropolitana-Iztapalapa, San Rafael Atlixco 186, Col. Vicentina, Iztapalapa, C.P. 09340, México D.F., México

[***] Facultad de Química, Universidad Nacional Autónoma de México, Ciudad Universitaria, 04510, México D.F., México
E-mail: jidaboy@servidor.unam.mx

Advances in Quantum Chemistry, Vol. 55
ISSN 0065-3276, DOI: 10.1016/S0065-3276(07)00212-2

Abstract In the last few decades there has been an increasing concern for the environment, worldwide. The scientific community has actively contributed to the efforts to understand the chemical processes that impact our surroundings, as the vast amount of publications in this area shows. There is considerable current interest in the high levels of oxygenated compounds which are both emitted and formed in the troposphere. Their main sink involves in most cases their reactions with OH radicals, at least at daytimes. Since the OH radical has a rather large electric dipole moment it is clearly capable of forming strong hydrogen bonds. In this work, we examine the relevance of such interactions in the reactions of this radical with different oxygenated volatile organic compounds (VOCs). According to the available data, at the present time, it seems well established that these reactions take place through a complex mechanism, with a first step involving the formation of reactant complexes, caused by hydrogen bond like interactions. These interactions can also be present in the transition states and play a relevant role in the branching rations of the studied reactions. The reliability of Quantum Chemistry to predict and reproduce the kinetics of oxygenated VOCs reactions, as well as its capability to explain unusual experimental findings, is discussed.

1. INTRODUCTION

Nowadays *ab initio* quantum chemical calculations can provide results approaching benchmark accuracy for small molecules in the gas phase [1], and they have proven to be very useful to complement experimental studies. Small molecules in the gas phase are typically addressed by high-level methods such as CCSD(T), QCISD(T) and MRCI, which in many cases are equally accurate than experiments [2]. A wide variety of properties such as: structures [3]; thermochemistry [4]; spectroscopic quantities [5,6], and kinetics [7] can be effectively computed.

At the same time, in the last few decades there has been an increased interest on different environmental issues. One of them is the degradation of volatile organic compounds (VOCs) in the troposphere leading to the production of a wide range of secondary pollutants such as ozone, peroxyacyl nitrates, and secondary organic aerosols VOCs are directly emitted into the atmosphere from anthropogenic and biogenic sources [8–12]. In urban air oxygenated compounds represent about 20% of the emitted VOCs [13,14]. They are also formed in situ in the atmosphere from the oxidation of other VOCs [15].

There is a variety of processes that act to remove a trace gas from either the troposphere or the stratosphere. For oxygenated VOCs the main tropospheric loss occurs via gas phase oxidation reactions involving OH, O_3, NO_x and Cl radicals, and photolysis. However, the hydroxyl radical is the most important oxidizing species in the global troposphere [16–19]. As a result of its role in initiating the majority of oxidation reaction chains, the OH radical is the primary cleansing agent for the lower atmosphere and has been called the "tropospheric vacuum cleaner" [20]. The dominant production cycle for tropospheric OH involves the reaction of $O(^1D)$, produced from the photolysis of O_3, with H_2O:

$$O_3 + h\nu \rightarrow O(^1D) + O_2, \tag{R1}$$

$$O(^1D) + H_2O \rightarrow 2OH. \tag{R2}$$

Its subsequent reactions with O_3, CO, and other VOCs, in the presence of nitrogen oxides (NO_x) lead to the formation of HO_2, hydrogen peroxide, alkyl peroxy radicals, and ultimately to the production of ozone via the conversion of NO_2 to NO [8,21].

The reactions of oxygenated VOCs with the OH radical proceed mainly by H-atom abstraction from C–H and, to a much lesser extent, from O–H bonds, or by addition to the carbon atoms of any C=C bonds [19,22]. Many of these reactions exhibit apparent negative activation energies, especially at temperatures below room temperature, i.e. the rate constants are observed to decrease with increasing temperature

The behavior of such reactions has been successfully described by Mozurkewich and Benson [23,24] for systems at low pressures, and by Singleton and Cvetanovic [25] for systems at high pressures. In this case, several explanations have been proposed, which are summarized in Ref. [26]. Three of them support the idea of an elementary reaction but suggest a modification of the pre-exponential factor in the Arrhenius equation including a $T^{-1.5}$ term. On the other hand Singleton and Cvetanovic [25] propose a complex mechanism and explain the occurrence of such negative activation energies by the reversible formation of a loosely bound reactant complex (RC) which is formed without activation energy, followed by a second reaction, which is irreversible, and whose transition state energy is lower than the energy of the separated reactants.

Reactant complexes seem to be common in radical–molecule reactions, and they are mainly formed through long-range Coulombic interactions between the reactant molecules. Recent theoretical studies [27–32] have proposed that oxygenated and unsaturated compounds react with OH radicals through a complex mechanism, involving a reactant complex. The role of these intermediates in bimolecular reactions has been recently reviewed [33,34]. Reactant complexes involving OH radicals have also been studied experimentally [35,36]. Moreover, it has been established that the presence of an attractive well at the entrance channel of a potential energy surface can influence the dynamics, and hence the course of the reaction [33]. If the reaction occurs at high enough pressures for these complexes to be collisionally stabilized, and if the overall energy barriers are small, they are likely to play an important role. In the present work the role of such complexes as well as the influence of hydrogen bond like interactions in transition states are reviewed for tropospheric VOCs + OH radicals reactions. The reactions reviewed in this work include the following VOCs: aliphatic alcohols, aldehydes, ketones, carboxylic acids and multifunctional VOCs.

2. KINETICS

Since the presence of reactant complexes (RC) and stabilizing interactions in transition states can directly affect the kinetic parameters of the reactions, we present here a brief overview on the methodology used to compute the rate constants in the examples that are going to be discussed later. For a chemical reaction that occurs through a stepwise mechanism involving the formation of a reactant complex in the entrance channel at least two steps must be considered in the kinetics, namely: (1) the formation or the reactant complex from the isolated reactants, and (2) the formation of the products from the reactant complex. For VOCs + OH reactions that mechanism can be written as:

$$\mathrm{VOC} + \mathrm{OH}^{\bullet} \underset{(-1)}{\overset{(1)}{\leftrightarrows}} [\mathrm{VOC}\cdots\mathrm{OH}]^{\bullet} \overset{(2)}{\rightarrow} \mathrm{VOC}^{\bullet}_{-\mathrm{H}} + \mathrm{H_2O}.$$

Provided that $k_{-1} + k_2 > k_1$, i.e., the complex rapidly disappears, a steady-state analysis leads to a rate coefficient for each overall reaction channel which can be written as:

$$k = \frac{k_1 k_2}{k_{-1} + k_2}. \tag{1}$$

For the studied reactions in general the energy barrier for k_{-1} is about the same size as that for k_2, in terms of enthalpy. However, the entropy change is much larger in the reverse reaction than in the formation of the products. The activation entropy ΔS_2 is small and negative because the transition state structure is tighter than the reactant complex, while ΔS_{-1} is large and positive because six vibrational degrees of freedom are converted into three translational plus three rotational degrees of freedom. This leads to k_{-1} values that are much larger than k_2. Based on this assumption, first considered by Singleton and Cvetanovic [25], the overall rate coefficient (k) can be rewritten as:

$$k = \frac{k_1 k_2}{k_{-1}} = K_{\mathrm{eq}} \cdot k_2, \tag{2}$$

where k_2 is the rate constant corresponding to the second step of the mechanism, i.e., transformation of the reactant complex into products; and K_{eq} is the equilibrium constant between the isolated reactants and the reactant complex. Applying basic statistical thermodynamic principles the equilibrium constant (k_1/k_{-1}) of the fast pre-equilibrium between the reactants and the reactant complex may be obtained as:

$$K_{\mathrm{eq}} = \frac{Q_{\mathrm{RC}}}{Q_{\mathrm{R}}} \exp\left[(E_{\mathrm{R}} - E_{\mathrm{RC}})/RT\right], \tag{3}$$

where Q_{RC} and Q_{R} represent the partition functions corresponding to the reactant complex and the isolated reactants, respectively.

Let us emphasize that when it is claimed that the reactant complexes are lower in energy than the isolated reactant, this statement is made on terms of enthalpy or ZPE corrected electronic energies. On the other hand, the Gibbs free energy associated with step 1, i.e. the formation of the reactant complex ($\Delta G_1 = G_{\mathrm{RC}} - G_{\mathrm{R}}$), for the formation of radical–molecule complexes is always positive in a wide range of temperatures, around 300 K. The expression for the equilibrium constant can also be written as:

$$K_{eq} = \frac{k_1}{k_{-1}} = \exp(-\Delta G_1/RT). \tag{4}$$

From this expression, it is evident that $K_{eq} < 1$ for endergonic processes ($\Delta G_1 > 0$), i.e. $k_{-1} > k_1$, which validates the steady state hypothesis.

On the other hand k_2 can be calculated within the frame of the Transition State Theory.

2.1 Conventional Transition State Theory (CTST)

From a phenomenological point of view, numerous experiments have shown that the variation of the rate constant with temperature can be described by the Arrhenius equation [37]:

$$k = A\exp(-E_a/RT), \tag{5}$$

where E_a is the activation energy and A is the pre-exponential or frequency factor, which may have a weak dependence on temperature. If a reaction obeys the Arrhenius equation, then the Arrhenius plot ($\ln k$ versus $1/T$) should be a straight line with the slope and the intercept being $-E_a/R$ and A, respectively.

In a unimolecular process, under high-pressure conditions, an equilibrium distribution of reactants is established and the Transition State Theory formula can be applied [38] to calculate k_2:

$$k_2 = \sigma\kappa_2 \frac{k_B T}{h} \frac{Q_{TS}}{Q_{RC}} \exp\big[(E_{RC} - E_{TS})/RT\big], \tag{6}$$

where σ is the symmetry factor, which accounts for the number of equivalent reaction paths, κ_2 is the tunneling factor, k_B and h are the Boltzmann and Planck constants, respectively, Q_{TS} is the transition state partition function, and the energy difference includes the ZPE corrections. In this approach the reactant complex is assumed to be in its vibrational ground state.

In a classical treatment the influence of the complex exactly cancels in Eq. (2) and the overall rate coefficient depends only on the properties of OH, VOCs, and the transition states. However, when there is a possibility of quantum mechanical tunneling, the existence of the complex means that there are extra energy levels from where tunneling may occur so that the tunneling factor, κ, increases. Since it has been assumed that a thermal equilibrium distribution of energy levels is maintained, the energy levels from the bottom of the well of the complex up to the barrier might contribute to tunneling.

2.2 Canonical Variational Theory (CVT)

The Canonical Variational Theory [39] is an extension of the Transition State Theory (TST) [40,41]. This theory minimizes the errors due to recrossing trajectories [42–44] by moving the dividing surface along the minimum energy path (MEP) so as to minimize the rate. The reaction coordinate (s) is defined as the distance along the MEP, with the origin located at the saddle point and is negative on the reactants side and positive on the products side of the MEP. For a canonical ensemble at a given temperature T, the canonical variational theory (CVT) thermal rate constant is given by:

$$k^{\mathrm{CVT}}(T,s) = \min\{k^{\mathrm{GT}}(T,s)\}, \tag{7}$$

where s is the reaction coordinate, $k^{\mathrm{GT}}(T,s)$ is the rate constant for the passage through the generalized transition state (GTS) [39,45–48] that intersects the MEP at s:

$$k^{\mathrm{GT}}(T,s) = \sigma(s)\frac{k_{\mathrm{B}}T}{h}\frac{Q_{\mathrm{GT}}(T,s)}{Q_{\mathrm{R}}(T)}\exp\left[-\frac{V_{\mathrm{MEP}}(s)}{k_{\mathrm{B}}T}\right]. \tag{8}$$

In this expression, Q_{GT} and Q_{R} are the partition functions of the generalized transition state and of the reactants, and $V_{\mathrm{MEP}}(s)$ is the potential energy of the MEP at s.

The vibrational (q_{v}) and rotational (q_{r}) partition functions may be calculated within the harmonic and rigid rotor approximations, respectively. In addition, in this work, the q_{v} values were corrected by replacing some of the large amplitude vibrations by the corresponding hindered internal rotations, when necessary.

The quantum mechanical effect on the motion along the reaction coordinate is included in the kinetics calculations by multiplying the CVT rate constant by a temperature-dependent transmission coefficient $\kappa(T)$ which accounts for tunneling and non-classical reflexion. Therefore, the final expression for the rate constant is given by:

$$k(T) = \kappa(T)k^{\mathrm{CVT}}(T,s), \tag{9}$$

where $k(T)$ may be computed using the Small-Curvature Tunneling (SCT) method [49,50], which constitutes a generalization of the Marcus–Coltrin method [51]. In SCT it is assumed that the tunneling path is displaced from the MEP to a concave-side vibrational turning point in the direction of the internal centrifugal force. In this method, the probability that a system be transmitted through the ground-state level of the transition state is approximated by the Centrifugal-Dominant Small-Curvature Semiclassical Adiabatic Ground-state method (CD-SCSAG) [49,50]. The SCT transmission coefficient includes the effect of the reaction path curvature on the ground-state transmission probability at total energy E, $P(E)$, which is calculated as:

$$P(E) = 1/\{1+\exp[2\theta(E)]\}, \tag{10}$$

where $\theta(E)$ is the imaginary action integral evaluated along the tunneling path:

$$\theta(E) = \frac{2\pi}{h}\int_{S1}^{S2}\sqrt{2\mu_{\mathrm{eff}}(s)|E - V_{\mathrm{a}}^{\mathrm{G}}(s)|}\,ds. \tag{11}$$

The integration limits S_1 and S_2 are the reaction coordinate classical turning points; μ_{eff} is the reduced mass, which introduces the reaction path curvature; and $V_{\mathrm{a}}^{\mathrm{G}}(s)$ is the adiabatic ground-state potential.

3. ENERGIES

3.1 The B//A approach

To the MEP of chemical reactions there is a procedure for that has become common in the study of polyatomic systems because it is relatively inexpensive from a computational point of view and it usually reproduces correctly the main features of the reaction path. It is known as B//A approach, and it consists of geometry optimizations at a given level (A) followed by single point calculations, without optimization, at a higher level (B). Espinosa-Garcia and Corchado [52] argue that, when the MEP is constructed using the B//A approach, the energy maximum is artificially located away from the saddle point corresponding to the level of optimization (A). This shift, that is simply a numerical effect, could be mistaken with a variational effect and mislead the kinetic calculations. Consequently, in our calculations we have used the modification proposed by Espinosa-Garcia and Corchado [52], which consists of simply moving the maximum of the single-point calculation curve, B//A, to its original position ($s = 0$) at the A//A level. It should be noticed that according to this procedure the frequencies are not shifted, i.e., to each geometry at the A//A level there corresponds a set of original frequencies (calculated at the A//A level) and a shifted energy (calculated at the B//A level). Based on our previous experience, the use of B//A approach at BHandHLYP//CCSD(T) level of theory properly describes the energetic and kinetics features of VOCs+OH hydrogen abstraction reactions. In addition, for this kind of reactions it has been proved that the differences in geometries between several DFT methods compared to CCSD and QCISD are minimal for BHandHLYP [53]. However in some of our earlier works other methods were used.

3.2 On the Basis Set Superposition Error (BSSE) dilemma

One issue of concern when modeling weak bonded systems is the BSSE, which is caused by the truncation of the basis set. From a theoretical point of view its existence has been well established [54–56]. It is especially important in weakly bond complexes such as van der Walls and hydrogen bonded complexes and transition states because for such systems the BSSE and the binding energies are of the same order. The most widely used and simplest way to correct BSSE is the counterpoise procedure (CP) [57,58]. In CP, for a dimer system formed by two interacting monomers: CP^2, the BSSE is corrected by calculating each monomer with the basis functions of the other (but without their nuclei or electrons), using the so-called "ghost orbitals". However, the use of this method is polemic since several authors have proposed that it overestimates the BSSE [59–65].

It has been pointed out by Dunning [66] that "It is quite possible and even probably that the binding energies computed without the counterpoise correction are closer to the complete basis set limit than the uncorrected values. This situation is due to the fact that BSSE and basis set convergence error are often of opposite sign". Since it cannot be established *a priori* if that is the case, we have modeled the closest system to our reactant complexes with known experimental binding energy: the water dimer. The experimental value for the electronic dissociation energy (DE) of water dimer is 5.4 kcal/mol. We have modeled the water dimer at CCSD(T)//BHandHLYP/6-311++G(d,p) and DE values of 5.2 and 3.9 kcal/mol are obtained without and with CP^2 corrections, respectively. The CP uncorrected value is closer to the experimental one, suggesting that for the water dimer, at this level of calculation, the BSSE and Basis set truncation errors cancel each other. Since the reactant complexes between oxygenated VOCs and OH radicals are formed in an equivalent chemical way, i.e they are formed trough the interaction of the H atom in the OH radical and an oxygen atom in the VOCs, and as it will be discussed later they also have electronic binding energies that are very similar in magnitude to that of the water dimer, it is reasonable to assume that the same cancellation of errors occurs in our systems.

4. ALIPHATIC ALCOHOLS

There are over 70 alcohols in the atmosphere as a result of biogenic and anthropogenic emissions [67]. For example methanol and ethanol [68–70] have been used as fuels additives to reduce automobile emissions of carbon monoxide and hydrocarbons [71], in particular ethanol has been used in Brazil as a fuel for over 20 years [72]. 1-Propanol is widely used as a solvent in the manufacturing of different electronic components. The high volatility of these compounds causes their relative abundance in the troposphere and makes it relevant to determine their degradation pathways. During daytime the major loss process for alcohols is their reaction with OH radicals [68]. Accordingly, several experimental [69,70,73–84] and theoretical [85–88] kinetic studies of alcohols + OH reactions have been performed.

Two of the theoretical works [85,88], dealing with alcohols + OH reactions have not included weakly bonded complexes in the modeling. In the work on methanol + OH reaction, by Jodkowski *et al.* [86], it has been proposed the existence of a complex in the exit channel but none in the entrance channel. On the other hand reactant complexes have been modeled by Galano *et al.* [87]. The relevance of such inclusion on the tunneling corrections can be shown by comparing the corresponding results from both approaches [85,87], for 2-propanol. When the hydrogen abstractions take place from beta sites the tunneling corrections are about 3 to 6 [85] and 30 [87], and for abstractions from the OH group in the alcohol the values of κ_{298} are about 3 [85] and 96 [87], respectively. These values illustrate the relevance of reactant complexes in H abstraction reactions.

The ZPE corrected energies of the complexes, relative to isolated reactants, reported for different aliphatic alcohols [87] (R–OH) are equal to −5.73, −6.55, −6.29, −6.50, and −6.00 kcal/mol for R = CH_3, C_2H_5, n-C_3H_7, i-C_3H_7 and n-C_4H_9, respectively. These energies were obtained within the B//A approach, at CCSD(T)//BHandHLYP/6-311G(d,p) level of theory. Their large magnitude supports the formation of stable reactant complexes (RC) and unambiguously shows that the mechanism is complex with a first step leading to the RC formation, previous to the yielding of the corresponding radical and water. The structures of the

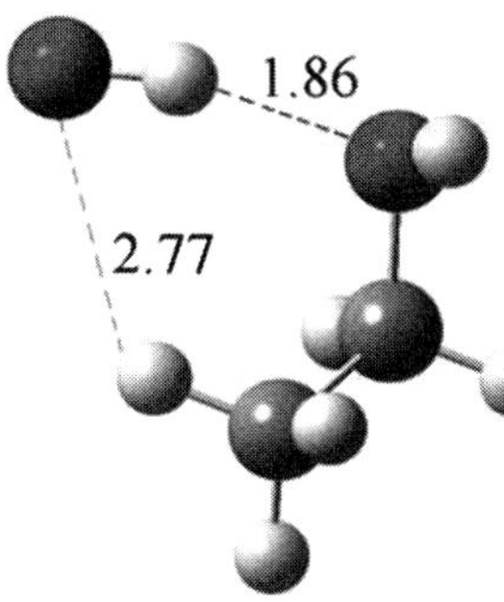

FIGURE 12.1 Structure of ethanol···OH complex.

complexes [87] show that for all the studied reactions the main stabilizing interaction involves the H in the OH radical and the O atom in the alcohols, with O···H distances of 1.84, 1.86, 1.85, 1.86 and 1.85 Å for R = CH_3, C_2H_5, n-C_3H_7, i-C_3H_7 and n-C_4H_9, respectively. A less important interaction was also found between the O atom in the OH radical and one of the hydrogens in the side chain of the alcohols with interaction distances from 2.6 to 2.9 Å. To illustrate this kind of complex, the structure corresponding to ethanol···OH complex is shown in Figure 12.1.

Intramolecular interactions in the transitions states (TS) are also relevant to properly predict or reproduce experimental rate constants, since they directly affect the TSs energies and small variations in reaction barriers have relative large impact on k since they enter exponentially in the rate constant equation. A detailed discussion on such interactions, in the TSs of different H abstraction paths, for 2-propanol + OH reaction has been provided by Luo *et al.* [85]. These authors have also discussed the influence of the interactions on the reaction barriers and rate coefficients.

In the case of 1-butanol + OH reactions the situation is even more drastic: the intramolecular hydrogen bond like interactions in the transition states structures are responsible for the branching ratios. It was found that while for methanol, ethanol, 1-propanol and 2-propanol the main reaction channels is that involving alpha abstractions, for 1-butanol the rate coefficient corresponding to the gamma channel (k_γ) is larger than those of the other competing channels over a temperature range from 290 to 350 K, while the alpha channel was found predominant over the range 350–500 K [87]. The alpha (TS-1B_α), beta (TS-1B_β) and gamma (TS-1B_γ) transition states showed hydrogen bond like stabilizations. However, the shortest distance in TS-1B_γ (Figure 12.2) causes a larger stabilization, i.e.: it shows a lower energy barrier and a larger rate coefficient.

The gamma predominance at room temperatures, for 1-butanol + OH reaction, also suggests an additional explanation to the findings of Oh and Andino [77], that the presence of ammonium sulfate aerosols promote the reactions of OH radicals only with aliphatic alcohols containing fewer than four carbon atoms. The fact that the preferred site for the hydrogen abstraction, at this temperature, is the gamma one, implies that the interaction of the aerosol with the functional group of the alcohols is apart enough to provoke any appreciable effect on the rate coefficient.

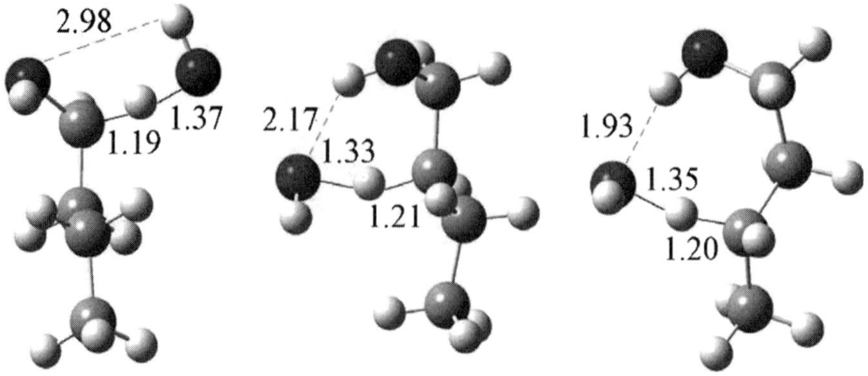

FIGURE 12.2 Intramolecular hydrogen bond like interactions in the transition state structures of channels α, β and γ of the 1-butanol + OH reaction.

TABLE 12.1 Calculated rate coefficients (cm^3/(molecule s)) for alcohols + OH radical gas phase reactions at 298 K, compared to the recommended ones

	Methanol	Ethanol	1-Propanol	2-Propanol	1-Butanol
Ref. [85]	–	–	–	8.2×10^{-12}	–
Ref. [86]	6.5×10^{-16}	–	–	–	–
Ref. [87]	6.82×10^{-13}	2.40×10^{-12}	4.04×10^{-12}	4.19×10^{-12}	6.88×10^{-12}
Ref. [88]	1.98×10^{-14}	3.08×10^{-13}	5.50×10^{-12}	–	8.56×10^{-12}
Recomm. [89]	9.30×10^{-13}	3.20×10^{-12}	5.50×10^{-12}	5.10×10^{-12}	8.10×10^{-12}
Δk [89]	±25%	±25%	±30%	±22%	±30%

A comparison between the overall rate coefficients, calculated in different ways, and those recommended by Atkinson *et al.* [89] is provided in Table 12.1 and Figure 12.3. All the theoretical calculations were performed within the CTST approach. As these results show, taking into account the reactant complexes and the intramolecular interactions in the transition structures in the modeling of alcohols + OH radical reactions in gas phase, improve the agreement of the theoretical calculations with the experimental data.

4.1 Structure–activity relationship

The higher reactivity of 1-butanol gamma site [87] is in line with the results of Wallington and Kurylo [90] who have proven that in ketones there is a significant enhancement of the group reactivity toward OH for both CH_3 and CH_2 when moved from the alpha to beta positions. This has been taking into account in the Structure-Activity Relationship (SAR) method [91,92] by factors of $F[\text{–}CH_2C(O)] = F[>CHC(O)] = F[\gg CC(O)] = 3.9$ [22]. In this improved version of

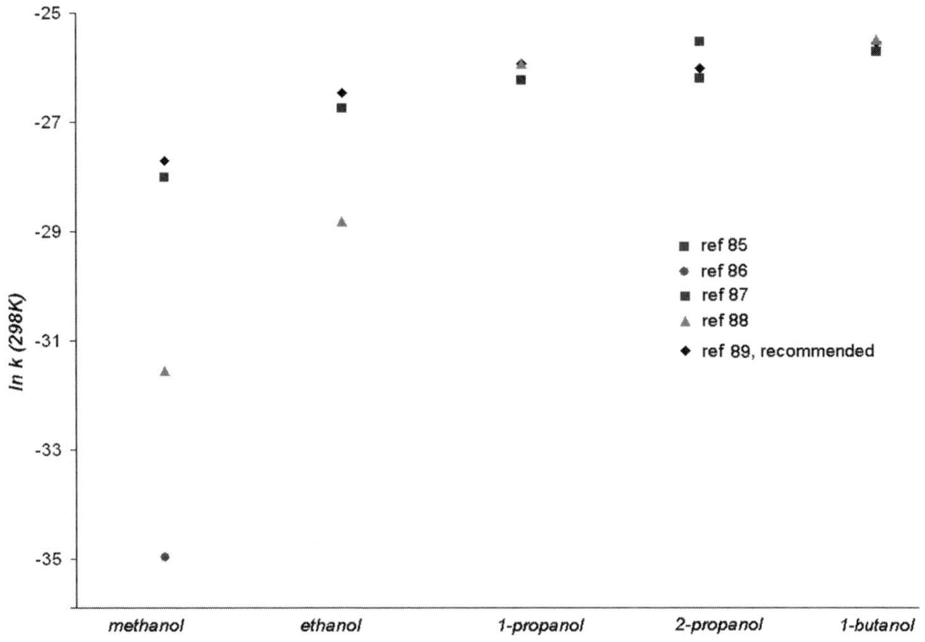

FIGURE 12.3 Comparison between experimental and theoretical rate coefficients.

SAR [22] also a new factor was included for abstractions from beta sites in alcohols: $F[-CH_2OH] = 1.6$. Beta carbons in ketones and gamma carbons in alcohols are both two carbon atoms apart from the oxygen atom, in fact the distance $H_\beta \cdots O$ in ketones and $H_\gamma \cdots O$ in alcohols were reported to be equal to 2.69 and 2.63 Å [87], respectively, in BHandHLYP/6-311G(d,p) fully optimized geometries. Based on this similarity, Galano *et al.* [87] proposed the following new corrections to SAR for alcohols. The proposed values are $F[-CH_2CH_2OH] = F[>CHCH_2OH] = F[\gg CCH_2OH] = 3.9$, which are equivalent factors to those for beta sites in ketones. Another set of SAR substitution factors for hydroxyl compounds have also been proposed by Bethel *et al.* [93], assuming that beta sites are more reactive, towards OH radicals, than gamma sites. These authors considered the effects of the OH substituents on H atom abstraction at the α and β positions, with new factors of $F[-OH] = 2.9$ and $F[-CH_2OH] = F[>CHOH] = F[\gg COH] = 2.6$.

Very recently Mellouki *et al.* [83] have determined the overall rate coefficient for the reaction of 3-methyl-1-butanol (3M1B), and 3-methyl-2-butanol (3M2B). Taking into account this new data an improved corrections for gamma sites in alcohols is proposed in this work: $F[-OH] = 3.3$, $F[-CH_2OH] = F[>CHOH] = F[\gg COH] = 1.5$ and $F[-CH_2CH_2OH] = F[>CHCH_2OH] = F[\gg CCH_2OH] = 3.7$. The rate constants estimates by different SAR approaches are compared to the most recent experimental values in Table 12.2. As these estimated values show, the best agreement with the experimental data is obtained when the sets of factors proposed in this work are used. The relevant sets of factors used in the different approaches are reported in Table 12.3. The main structural difference between

TABLE 12.2 Comparison between rate constants estimated using SAR method with experimental values

	1B	3M1B	3M2B
Ref. [92]	7.21×10^{-12}	8.68×10^{-12}	1.17×10^{-11}
Ref. [22]	6.92×10^{-12}	8.33×10^{-12}	1.14×10^{-11}
Ref. [93]	7.81×10^{-12}	9.21×10^{-12}	1.28×10^{-11}
Ref. [87]	8.73×10^{-12}	1.28×10^{-11}	1.07×10^{-11}
This work	9.30×10^{-12}	1.32×10^{-11}	1.22×10^{-11}
Exp[a]	9.31×10^{-12}	1.45×10^{-11}	1.23×10^{-11}

a Average from the available experimental data.

TABLE 12.3 Sets of SAR substitution factors for OH abstractions from hydroxyl compounds

Factors	Ref. [92]	Ref. [22]	Ref. [93]	Ref. [87]	This work
$-CH_3$	1.00	1.00	1.00	1.00	1.00
$-CH_2-$	1.29	1.23	1.23	1.23	1.23
>CH–	1.29	1.23	1.23	1.23	1.23
>C<	1.29	1.23	1.23	1.23	1.23
–OH	3.40	3.50	2.90	2.90	3.30
$-CH_2-OH$	1.29	1.23	2.60	1.23	1.50
>CH–OH	1.29	1.23	2.60	1.23	1.50
⋟C–OH	1.29	1.23	2.60	1.23	1.50
$-CH_2-CH_2-OH$	1.29	1.23	1.23	3.90	3.70
$>CH-CH_2-OH$	1.29	1.23	1.23	3.90	3.70
⋟$C-CH_2-OH$	1.29	1.23	1.23	3.90	3.70

these compounds is that in 1B the gamma and beta carbons are secondary, while the tertiary carbon is gamma in 3M1B and beta in 3M2B. Accordingly, the fact that *k*(3M1B) [83] is actually almost twice *k*(1B) [89] supports the higher reactivity of gamma sites in alcohols.

Please notice that the set of SAR substitution factors proposed in Ref. [87], and those from the present work, are meant for OH abstractions from alcohols with central chain formed by four or more carbon atoms, since in these cases gamma carbons are secondary o tertiary and their contributions to the overall rate coefficient become significant.

5. ALDEHYDES

Aldehydes are of crucial importance in atmospheric chemistry since they are primary biogenic and anthropogenic pollutants and also secondary pollutants due to partial oxidation of VOCs [26]. Moreover the reaction of aldehydes with OH radical are among the fastest initial oxidation of VOCs [94–101], leading to a fast oxidation and consequently to very short lifetimes.

A few years ago there were several open questions about aldehydes + OH reactions:

- There was not a clear explanation to the close to zero and negative Arrhenius activation energies, experimentally observed for formaldehyde and acetaldehyde, respectively [95–97].
- It was not clear if they occur by hydrogen abstraction, by addition or if both channels significantly contribute to the overall reaction.
- It was unknown if the mechanism was elementary or involves a reactant complex formation.
- The structure of this hypothetical complex was unknown.

Quantum Chemical calculations played an important role in the elucidations of these questions.

Pioneering theoretical work on the formaldehyde + OH reaction has been reported by Dupuis and Lester [102] using multiconfiguration self-consistent-field (MCSCF) and configuration interaction (CI) wave functions. These authors [102] predicted a large positive activation barrier of 5.5 kcal/mol. Using the *ab initio* Møller–Plesset method up to fourth order (MP4), Francisco [103] determine the barriers and energetic of this reaction. He obtained a small but positive barrier of 1.2 kcal/mol and a rate constant in very good agreement with experiment. The formation of a reactant complex was not considered in that work. The weakly bound complexes of the hydroxyl radical with formaldehyde and acetaldehyde were calculated later [29] using a density functional approach, but the mechanistic implications were not discussed.

Taylor *et al.* [104] investigated the reaction of hydroxyl radicals with acetaldehyde in a wide temperature range using a quantum RRK model to describe the competition between addition and abstraction. They conclude that different reaction mechanisms occur, depending on the temperature, and that OH addition followed by CH_3 elimination is the dominant reaction pathway between 295 and 600 K. Moreover, they claimed that the H-atom elimination pathway is largely insignificant, except possibly at the lowest temperatures.

A complex mechanism, involving the formation of a reactant complex, was postulated for addition reactions to C=C double bonds several years ago to explain negative activation energies [25]. The possible role of such reactant complexes, for the reaction of OH radical with aldehydes, was studied for the first time at CCSD(T)/6-311++G(d,p)//MP2/6-311++G(d,p) level of theory [27]. The proposed structure of the reactant complexes for formaldehyde and acetaldehyde are shown in Figure 12.4. They are weakly bonded complexes formed through hydrogen bond like interactions between the hydrogen atom in the OH radical and the oxygen in the carbonyl group of the aldehyde. In the same paper it was concluded that the magnitude of the stabilization energy is large enough to assure that the reaction is complex at 298.15 K. The apparent barrier of reaction, calculated as the difference in energy between the transition state and isolated reactants ($E_{TS} - E_R$),

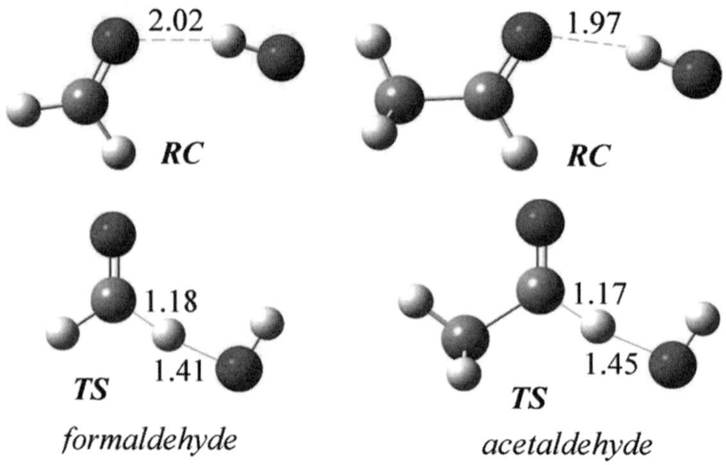

FIGURE 12.4 Optimized geometries of reactant complexes and transition states of the reaction of formaldehyde and acetaldehyde with OH radical.

was found to be close to zero for formaldehyde: 0.03 kcal/mol, and slightly negative: −1.71 kcal/mol for acetaldehyde.

The calculated rate coefficients at 298 K are 1.10 and 1.45 $\times$ 10^{-11} L/(cm^3 s) for formaldehyde and acetaldehyde, respectively [27]. These values are in excellent agreement with the experimental data. The structure of the transition states are shown in Figure 12.4. Both are early transition states consistent with reactions with very low barriers. The transition state of OH hydrogen abstraction from acetaldehyde is earlier than the one of formaldehyde. Additionally, the possibility of the addition channel was excluded due to the large barrier associated to this process. The temperature dependence of the rate coefficient was not studied. In a more recent work [105] the level of the calculations was increased, the temperature dependence of the rate coefficient and the role of direct abstraction were studied. All the main conclusions from these articles are now accepted in recent works [106–108], and it is well known that the hydrogen abstraction is the main reaction channel if not unique [108,109].

6. KETONES

The experimental evidence [22,89,90,110–115] suggests that ketones react with OH radicals via a hydrogen abstraction mechanism, leading to a water molecule and a new radical. Nevertheless, there is a peculiarity in the ketones + OH reactions: hydrogen atoms attached to carbon atoms in beta positions to the carbonyl group are the most likely to be abstracted [110–113,115]. However, if the beta carbon is a primary carbon, its contribution to the total reaction is much less important. The contribution is about 66% [116] to 67% [117] for secondary beta carbons, while it is only about 11% [116] to 17% [117] for primary ones. To explain the large contribution of the beta abstractions, Wallington and Kurylo [90] have proposed a complex mechanism that involves the formation of a short-lived six-member ring complex. Alternatively, a seven-member ring has been proposed by Klamt [117].

Another important experimental finding for the propanone+OH reaction [118] is that the temperature dependence is not described by a simple Arrhenius expression but by $k_{202\text{–}395\,\mathrm{K}} = 8.8 \times 10^{-12} \exp(-1320/T) + 1.7 \times 10^{-14} \exp(423/T)\ \mathrm{cm^3/s}$, indicating that a simple H atom abstraction may not be the only reaction path. The first term is assumed to correspond to a direct abstraction and the second one, with small negative activation energy, to a complex mechanism. According to the results from that work, the complex mechanism should proceed through an association complex prior to final reaction which is absolutely correct as will be shown latter. However, it was also assumed that this complex evolves to addition products (to the carbon atom in the carbonyl group) in the same way that OH radical reacts with C=C double bond. The C=O double bond is not as electron rich as C=C due to the large electronegativity of oxygen atom. For this reason ketones react with nucleophiles instead of electrophiles in most of their reaction in solution. When they react with electrophiles, as in acid catalyzed reactions, the target is the negative charged O atom, creating a high energy intermediate that later or concertedly stabilizes due to a nucleophilic attack to the carbon atom.

Quantum mechanical calculations show that the barrier of the OH addition to C=O double bond is positive and larger than that corresponding to H abstraction [27,119] in several kcal/mol, which is in contradiction with the small negative activation energy proposed for the complex reaction [118]. This indicates that assuming that the reaction is complex because the addition path also contributes to the overall reaction is not a valid hypothesis. Since it is a fact that the mechanism is complex, there must be another path, in addition to direct H abstraction, contributing to the overall reaction. The obvious candidate for this mechanism is the complex hydrogen abstraction previously proposed by Wellington and Kurylo [90].

This alternative has been addressed theoretically in three kinetic studies. Two of them deals only with propanone [120,121], and were performed using the state of the art for quantum chemical calculations and variational transition state theory. In the third one a series of ketones was studied using high level methods (CCSD(T)/6-311g(d,p)//BHandHLYP/6-311g(d,p)) and conventional transition state theory [122]. Masgrau *et al.* [121] have used the low pressure limit case to calculate the rate constant for propanone. Despite of the high level of theory used, they did not reproduce the temperature dependence, especially at low temperatures. On the other hand, Yamada *et al.* [120,123] properly reproduced the experimental temperature dependence of the rate coefficient by using the high pressure approach, CVT, and almost the same level of calculation that was used by Masgrau *et al.* [121]. Figure 12.5 shows Arrhenius plots from the experimental data [118], compared to different theoretical approaches, assuming low and high pressure limits. There is also new data from our group, in Figure 12.5, that was not previously published. The main difference between high and low pressure approaches is that in the first one tunneling is calculated from the vibrational ground state of the reactant complex whereas in the low pressure approach it is calculated from the isolated reactants.

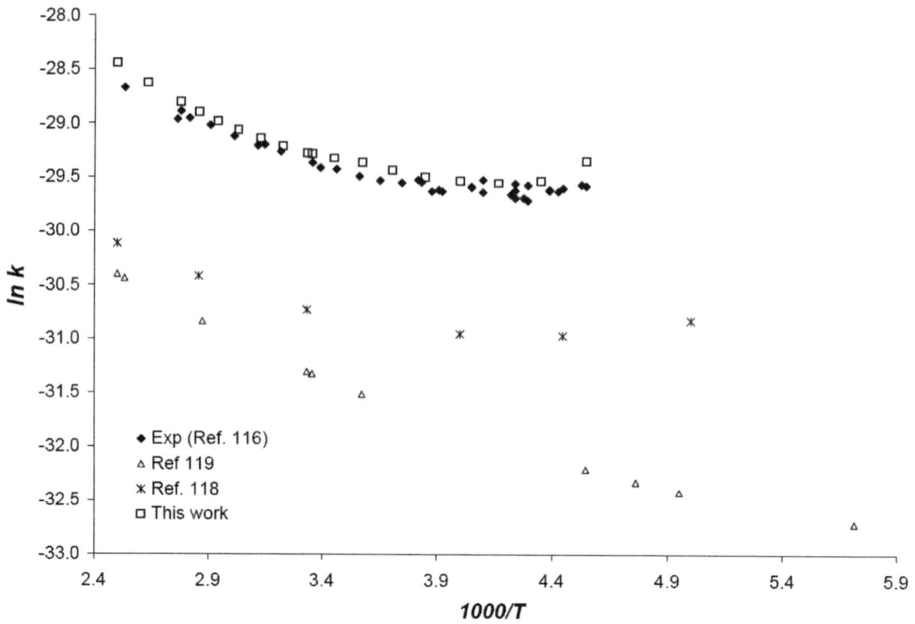

FIGURE 12.5 Comparison between experimental and theoretical Arrhenius plots for propanone + OH reaction.

It is clear from Figure 12.5 that both "high pressure" approaches (reference 120 and this work) adequately reproduce the temperature dependence of *k*. This strongly suggests that the reactant complex is in its vibrational ground state, even if it is not collisionally stabilized, which was recently suggested for this kind of complex reactions [124]. It seems surprising that the theoretical results that best agree with the experimental data are those obtained from conventional TST and not those from variational transition state theory and higher level of electronic calculations. There are several reasons for that outcome. The highly sophisticated method used in reference 120 for electronic calculations (quadratic complete basis set, CBS-QB3 [125,126]) was developed to predict excellent thermodynamic heats of reactions, not kinetic parameters. The one we used was selected among a large number of trials for frequency/geometry/energy combinations aiming the best reproduction of hydrogen abstraction reactions by OH radical. Concerning rate constants calculations, there are two major disadvantages related to CTST/Eckart scheme. The first one is the recrossing effect which leads to overestimations of *k*, mainly at high temperatures. The second one is related to tunneling corrections, which are larger than those from CVT/SCT approach, and this is especially important at low temperatures. However, around 298 K none of these drawbacks are critical. Figure 12.5 shows that our values actually lay above the experimental ones. It is significant that the Arrenius plots from reference 120 and from this work are parallel, i.e. the same activation energies are predicted by both approaches. Accordingly their differences are related to the pre-exponential factors. Since in the variational TST the dividing surface is moved to the maximum in Gibbs free en-

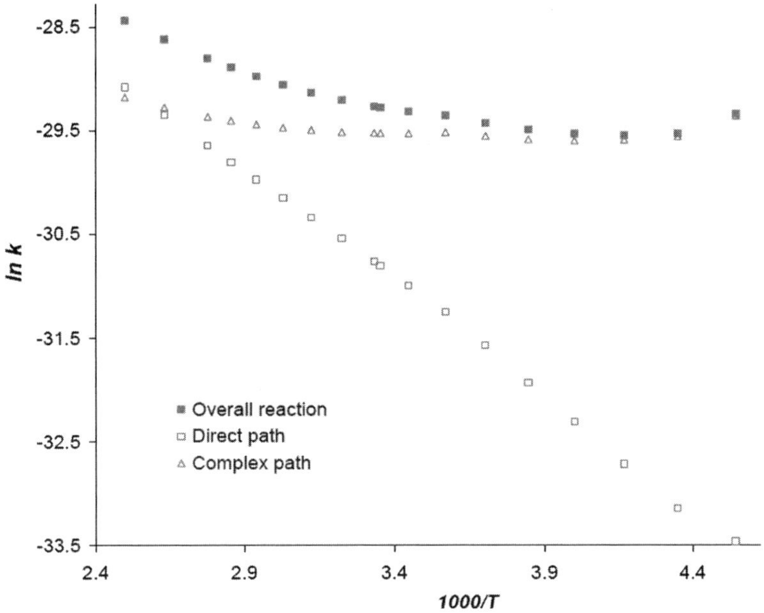

FIGURE 12.6 Arrhenius plots showing the contribution of direct and complex abstraction channels to the overall propanone + OH reaction.

ergy, the entropy of activation increases and the pre-exponential factor decreases. Evidently, in the conventional approach there is a fortunate cancellation of errors that leads to k values closer to the experimental ones. It should be taken into account that thermodynamic corrections were calculated at similar levels of theory [120] and in this work.

Figure 12.6 shows the Arrhenius plots for the different reaction channels contributing to propanone + OH overall reaction. None of the rate constants of the individual paths matches the experimental ones in the studied temperature range. Only at the highest temperature (440 K) the contribution of the direct abstraction is larger than that of the complex abstraction. At the lowest temperatures the direct abstraction is negligible; while at intermediate temperatures both are important. This is in excellent agreement with the proposal of two channels: one complex and one direct H abstraction.

Let us insist on the nature of the path used to calculate the previous rate coefficients. Direct abstractions refer to elementary reactions with the reactants are converted into abstraction products without any intermediate step. On the other hand, in the complex path a weakly bonded complex is formed in the entrance channel. For H abstraction from alpha sites in ketones it is important to distinguish between eclipsed and alternated hydrogens [122]. The transition state of the abstractions from eclipsed alpha sites are the only one directly connected to the reactant complex.

2-Pentanone and methyl butanone were also studied, in addition to propanone [122]. All the reactant complexes show ring-like structures, and they are

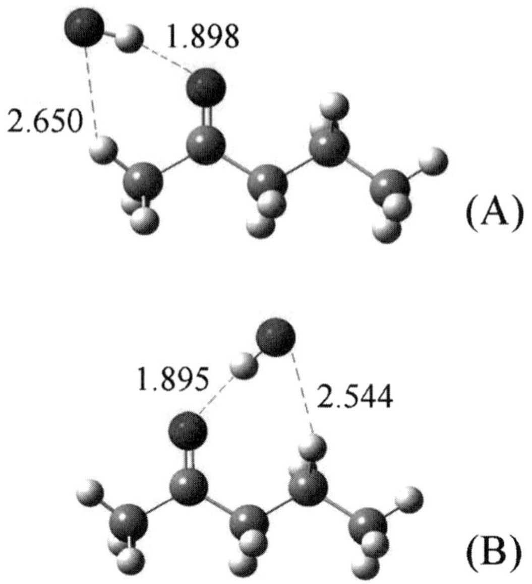

FIGURE 12.7 Fully optimized geometry of 2-pentanone· · ·OH reactant complexes, involved in primary alpha (A) and secondary beta (B) abstractions.

caused by two attractive interactions. The strongest one occurs between the H atom in the OH radical and the O atom in the carbonyl group (Figure 12.7). The weakest one is formed between the O in OH and one of the hydrogens in the ketone. For non-symmetric ketones two reactant complexes were found: one for eclipsed alpha abstraction and another involving beta hydrogens (A and B in Figure 12.7).

The eclipsed transition states which correspond to the complex mechanism, present an intramolecular interaction between the H atom in the OH radical and the O atom in the carbonyl group (Figure 12.8A). This interaction stabilizes the TS that present such interactions and lowers the reaction barrier. Additionally since for this channel the tunneling correction is calculated from the reactant complex the tunneling corrections are larger than for direct abstractions. Both effects are especially important at low (tropospheric) temperatures. However, these transition states are tighter than those corresponding to direct hydrogen abstractions (Figure 12.8B), leading to a larger entropy loss. At high temperatures the entropy loss becomes more important and the direct abstraction channel turns out to be the dominant path.

The strongest intramolecular interaction was found for transition states corresponding to beta sites [122]. This feature was found responsible for the "anomalous" higher reactivity of beta sites. The ratios between the complex/direct normalized rate coefficients were found to be 7 and 20 for alpha and beta abstractions, respectively, at room temperature (Table 12.4). As a rough approximation, this means that tunneling is responsible for increasing k in a factor of 7 and the additional stabilization of the TS increases the rate coefficient in a factor of 3.

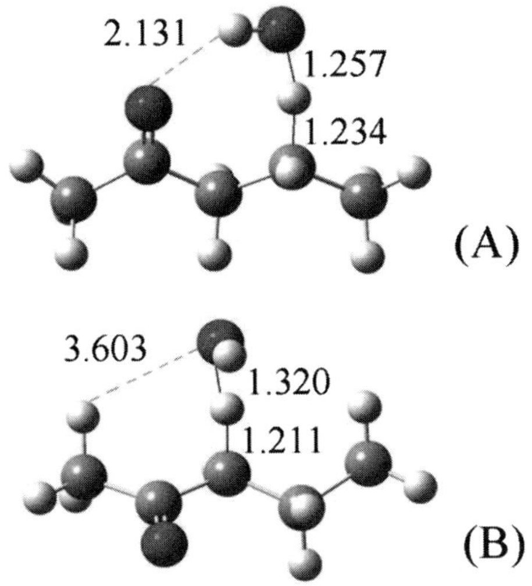

FIGURE 12.8 Fully optimized geometry of selected transition states for 2-pentanone + OH reaction, corresponding to secondary beta (A) and secondary alpha (B) abstractions.

TABLE 12.4 Comparison between eclipsed and alternated calculated rate coefficients at 298 K, corresponding to mixed abstractions, normalized to one H atom

Site	Ketone	Rate coefficients (cm^3/(molecule s))	
		Eclipsed	Alternated
Primary alpha	Propanone	7.51×10^{-14}	1.05×10^{-14}
	2-Pentanone	9.06×10^{-14}	1.26×10^{-14}
	Methyl-butanone	3.09×10^{-13}	4.32×10^{-14}
Primary beta	Methyl-butanone	4.02×10^{-13}	1.97×10^{-14}

Finally, although the reactant complexes formation enhances the tunneling factor and increases the rate coefficients, diminishing the expected deactivating effect of the carbonyl group, this effect was found in all the studied paths and cannot be accounted for the higher reactivity of beta sites. The experimentally observed enhanced reactivity of beta sites was found to be caused by the hydrogen bond interactions in the transition states corresponding to beta abstractions.

A good agreement was obtained between experimental rate coefficients and those calculated assuming a complex/direct mixed mechanism, including TS intramolecular interactions (Table 12.5). The largest discrepancies were of 15%, 52% and 24% for propanone, 2-pentanone and methyl butanone, respectively. This agreement supports the used theoretical approach and validates the relevance of reactant complexes and hydrogen-bond like interactions on ketones + OH reactions.

TABLE 12.5 Experimental and calculated rate coefficients ($cm^3/(molecule\ s)$) of ketones + OH radical gas phase reactions, at 298 K

	Propanone	Ref.	2-Pentanone	Ref.	Methyl butanone	Ref.
k_{exp}	1.90×10^{-13}	[89]				
	2.21×10^{-13}	[114]	4.00×10^{-12}	[69]		
					3.02×10^{-12}	[115]
	1.90×10^{-13}	[115]	4.98×10^{-12}	[112]		
	2.27×10^{-13}	[69]				
k_{calc}	1.92×10^{-13}	[122]	2.38×10^{-12}	[122]	2.30×10^{-12}	[122]

7. CARBOXYLIC ACIDS

It has been reported that formic acid is the most abundant carboxylic acid in the troposphere [127]. In wet deposition, along with acetic acid, it accounts for up to 18% of the total acidity in rain in some areas [128]. The main chemical sink for atmospheric carboxylic acids is their reaction with the hydroxyl radical.

Formic acid + OH reaction has been extensively studied from an experimental point of view [129–134]. It was determined from some of these studies [129–131] that, at room temperature, the reaction pathway involving the abstraction of the acidic hydrogen is dominant over the abstraction of the formyl hydrogen despite of the fact that the C–H bond strength is weaker than the O–H bond strength by about 14 kcal/mol. Accordingly, the need of quantum mechanical modeling of the mechanism was pointed out [130,131], since such studies could help explaining the reason of such apparent contradiction. About the mechanism, Wine *et al.* [129] and Jolly *et al.* [130] proposed a mechanism in which OH forms a hydrogen-bonded complex with formic acid, followed by transfer of the hydroxylic hydrogen within the adduct.

There are also several experimental studies dedicated to the acetic acid + OH reaction [133–138]. According to the primary kinetic isotope effect (KIE) carried out by Singleton *et al.* [136], the preferential pathway in this case is also the H-atom abstraction from the carboxyl group. From the theoretical modeling on this reaction [137,138], it seems that the acidic H-abstraction, is greatly enhanced and largely controlled by the formation of very stable H-bonded reactant complexes.

Two different theoretical approaches have been published to explain the dominance of formyl abstraction in formic acid. Anglada [139] explained it by a mechanism involving a proton coupled electron-transfer process, which can compete with the conventional hydrogen abstraction. On the other hand, Galano *et al.* [140] proposed that large tunneling effects for the acidic path that appears as consequence of a high and narrow barrier, are responsible for the larger rate coefficient of this channel. Stabilizing intramolecular interactions in the transition states were

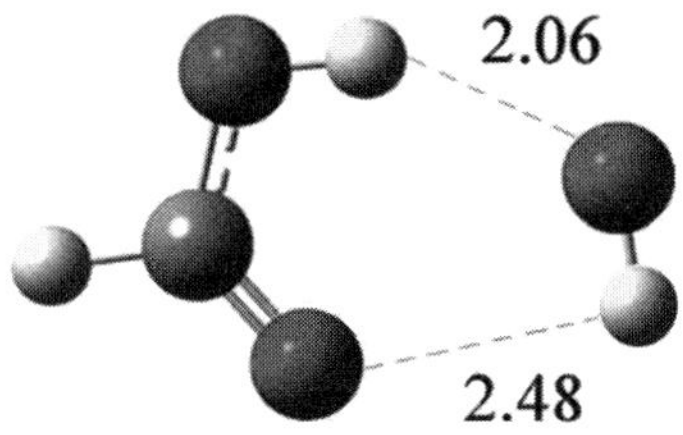

FIGURE 12.9 Structure of the HCOOH···OH complex with the lowest energy. The geometrical parameters are those from Ref. [140].

TABLE 12.6 Overall rate coefficients ($cm^3/(molecule\ s)$) and formyl branching ratio (Γ_{CH}) at 298.15 K, for HCOOH + OH reaction

	$k_{overall}$	Γ_{CH}
Theoretical		
Ref. [139]	6.24×10^{-13}	0.07
Ref. [140]	3.55×10^{-13}	0.08
Experimental		
Ref. [129]	$(4.62 \pm 0.78) \times 10^{-13}$	
Ref. [130]	$(4.93 \pm 0.28) \times 10^{-13}$	
Ref. [131]	$(4.47 \pm 0.28) \times 10^{-13}$	0.08–0.15
Ref. [132]	$(3.70 \pm 0.40) \times 10^{-13}$	
Ref. [142]	$(4.5 \pm 1.8) \times 10^{-13}$	

found in both approaches. Both works also invoke a complex mechanism involving reactant complexes. An exhaustive computational study on the different rearrangements of the possible complexes between formic acid and OH has been provided [141]. The most stable complex from this work shows a six-member ring like structure, where both moieties (HCOOH and OH) act as donor and acceptor simultaneously. This structure actually corresponds to the acidic abstraction in Ref. [140], and explains the high barrier found for the second step of this channel (Figure 12.9).

In summary, both theoretical approaches have shown the importance of including the formation of the reactant complex, as well as the intramolecular interactions in the transition state, to properly describe the experimental behavior of the studied reactions. The experimental and theoretical overall rate constants and formyl branching ratios for HCOOH + OH reaction is shown in Table 12.6. This channel is predicted to occur in significant less extent than that involving the abstraction of the acidic hydrogen. The calculated values not only reproduce this experimental fact but also quantitatively agree with the experimental data.

8. MULTIFUNCTIONAL OXYGENATED VOLATILE ORGANIC COMPOUNDS

It has been established that simple α-dicarbonyl compounds are key intermediate products in the atmospheric oxidation of several hydrocarbons, e.g., OH radical initiated reactions of aromatic compounds and dienes [143–151]. However, there are very few studies concerning their subsequent fate, other than photodissociation. The tropospheric oxidation of dienes leads to the production of multifunctional oxygenated species, which can be highly reactive themselves. Thus, the elucidation of the oxidation rates and mechanisms for these oxygenated species is crucial to the assessment of the overall environmental impact of dienes emissions. Some of these species are glycolaldehyde ($HOCH_2CHO$), methylglyoxal ($CH_3C(O)CHO$), and hydroxyacetone ($HOCH_2C(O)CH_3$) [152,153], which arise from the isoprene oxidation. Unfortunately there is rather scarce data available in the literature dealing with the tropospheric fate of some of these compounds.

From an experimental point of view there is only one work dealing with glyoxal + OH reaction [154] and three with methylglyoxal [154–156], while glycolaldehyde [157–162] and hydroxyacetone [163–167] have attracted more attention. The negative temperature dependence as well as the products yield have been interpreted in terms of formation and stabilization of reactant complexes [161,166–168] and tunneling effects [161]. In addition Kwok and Atkinson [22] have listed glycolaldehyde as one of the molecules for which the measured rate coefficient differs, by more than a factor of two, from the one estimated using structure-activity relationships.

In recent works our group has investigated the reactions of glycolaldehyde [169], glyoxal and methylglyoxal [170] and hydroxyacetone [171] with hydroxyl radicals. Ochando-Pardo *et al.* [172] have also study the mechanism and kinetics of glycolaldehyde + OH reactions using a computational approach. Reactant complexes formed through hydrogen bond interactions have been characterized in all these studies. They show ZPE corrected energies that lies significantly below the energies of the corresponding isolated reactants. The structures of the complexes between these multifunctional oxygenated compounds (MO-VOC) and OH show two different kind of rearrangements. In the first one the complexes are formed through the interaction between the H atom in the OH radical and one of the oxygen atoms in the MO-VOC [169,171,172]. The second rearrangement corresponds to ring-link structures formed by the same interaction just described and another one between the O atom in the OH radical and one of hydrogen atoms in the MO-VOC [169–171]. One example of each type, with MO-VOC = hydroxyacetone, is shown in Figure 12.10.

At his point, it seems worthwhile to revisit the peculiar intramolecular bond that appears in the most abundant isomer of glycolaldehyde, which shows an OO-*s-cis* configuration, since it is relevant to SAR predictions. The glycolaldehyde + OH rate constant that predicted by SAR is 2.3×10^{-11} cm^3/(molecule s), while the average obtained from the available experimental data is 9.25×10^{-12} cm^3/(molecule s). The reaction profiles of the acetaldehyde, ethanol and glycolaldehyde + OH reactions have been modeled at the same level of theory,

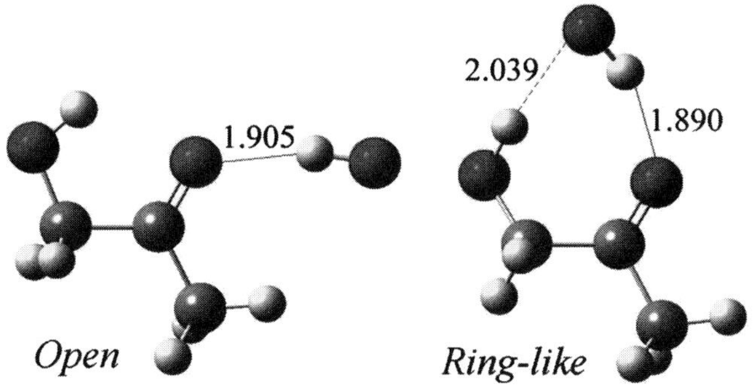

FIGURE 12.10 Two typical rearrangements reported for hydroxyacetona + OH reactant complexes.

TABLE 12.7 Overall rate coefficients (cm^3/(molecule s)) at room temperature for different MO-VOC + OH reactions

	Theoretical	Ref.	Exp[a]
Glyoxal	5.35×10^{-12}	[170]	1.10×10^{-11}
Methylglyoxal	1.35×10^{-11}	[170]	1.25×10^{-11}
Glycolaldehyde	7.29×10^{-12}	[169]	9.25×10^{-12}
	3.83×10^{-11}	[172]	
Hydroxyacetone	3.15×10^{-12}	[171]	3.69×10^{-12}

a Average value, from all available experimental data.

CCSD(T)/6-311++G(d,p)//BHandHLYP/6-311++G(d,p) [169]. The authors found that he ZPE corrected barriers for H abstractions from the alpha position in ethanol and from the aldehydic site in acetaldehyde are 0.2 kcal/mol lower than those corresponding to the equivalent abstractions in glycolaldehyde. This difference is responsible for the decrease in k. In SAR, the deactivation of the $-CH_2$ site by the >C=O substituent is taken into account by introducing a factor $F(>C=O) = 0.75$. However, no deactivation is considered for the aldehydic abstraction. This deactivation occurs because of the intramolecular hydrogen bond in OO-*s-cis* glycolaldehyde, which is a very specific characteristic of this molecule.

The features of the intramolecular hydrogen bonds interactions in the transition states, corresponding to the OH hydrogen abstraction reactions from these compounds, have been discussed in details [169–172]. The associated reaction barriers are quite low and it becomes necessary to use Variational Transition State Theory to avoid significant recrossing effects. The only exception is hydroxyacetone, in this case the barrier of the second step of reaction, that starting at the reactant complex, is high enough to safely use conventional TST. Using the proper approach the theoretical predictions of the rate coefficients of MO-VOCs + OH reaction are in very good agreement with the experimental available data (Table 12.7). By proper approach we mean not only the correct choice between TST and CVT, but also the modeling of the complex mechanism involving reactant complexes in the entrance channel of the MEP. As discussed in the above mentioned references the tunneling effect increases in a significant extent when the RC are taken into account. The rate coefficients calculated for direct reactions are in larger disagreement with the experimental values.

9. CONCLUDING REMARKS

As discussed above, the reactions of OH with oxygenated compounds seem to proceed via formation of complexes that involve one or two hydrogen bonds. Several theoretical studies on these complexes show binding energies large enough to overcome any inaccuracy of the calculation method (Table 12.8).

TABLE 12.8 Binding energies (kcal/mol) of oxygenated VOCs + OH radical complexes, obtained from electronic energies + ZPE corrections

VOC	Binding energy[a]	Ref.
Methanol	−5.73	[87]
Ethanol	−6.55	[87]
1-Propanol	−6.29	[87]
2-Propanol	−6.50	[87]
1-Butanol	−6.00	[87]
Formaldehyde	−3.03	[27]
	−3.0	[29]
Acetaldehyde	−4.18	[27]
	−4.0	[29]
Propanone	−6.06	[122]
	−4.6	[29]
2-Pentanone	−6.08	[122]
Methyl butanone	−6.00	[122]
Formic acid	−4.83	[140]
	−6.12	[139]
	−6.63	[141]
Acetic acid	−7.3	[137]
	−7.3	[138]
Glyoxal	−4.42	[170]
Methylglyoxal	−5.16	[170]
Glycolaldehyde	−4.6	[169]
	−3.89	[172]
Hydroxyacetone	−4.77	[171]

a Corresponding to the highest level of theory, and the most stable adduct, if more than one value is reported.

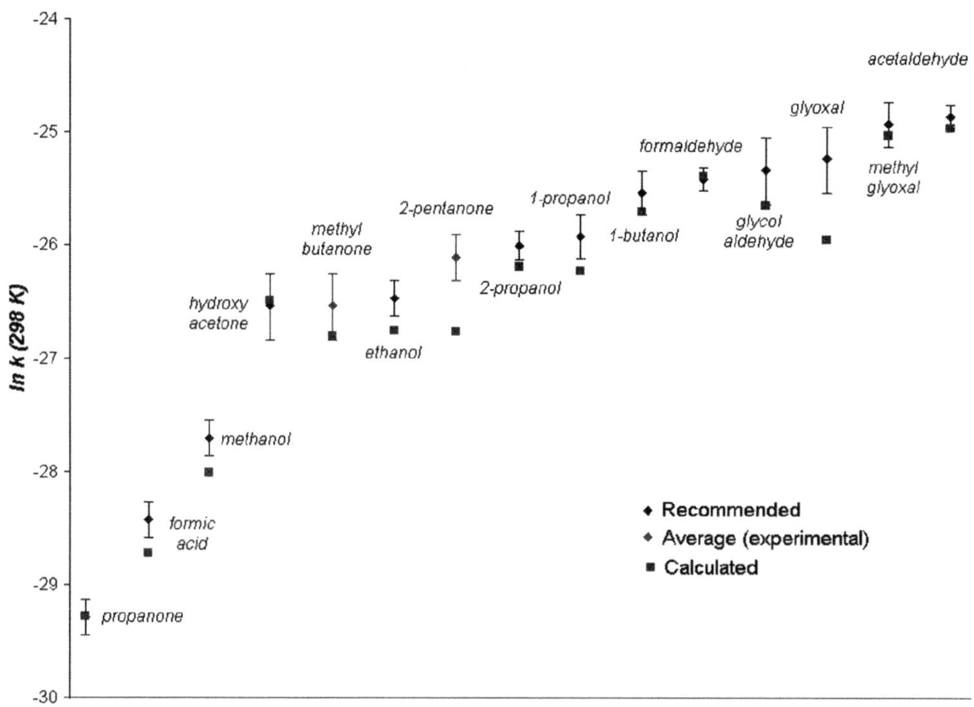

FIGURE 12.11 Comparison between recommended, experimental, and calculated rate coefficients, at 298 K.

The relevance of including such complex in the modeling of oxygenated VOCs + OH reactions has been discussed in detail in this work. Taking them into account, together with any possible intramolecular interaction in the transition states structures, leads to very reliable theoretical values of the rate coefficients. To prove this statement, Figure 12.11 shows a comparison between calculated rate coefficients, at 298 K [27,87,122,140,169–171] for several MO-VOCs + OH reactions and recommended values [89]. For 2-pentanone and methylbutanone there are no recommendations, therefore the average from the experimental values was used instead for 2-pentanone, and the only experimental value available for methylbutanone. As this figure shows, the calculated rate coefficients are in excellent agreement with the available experimental data. The calculated values are within the error range from experimental determinations in most of the studied cases. The largest discrepancies were found for glyoxal and 2-pentanone, which differ from the recommended values in 51 and 48%, respectively. Even for them the agreement is very good, since an error of 1 kcal/mol in energies represent about one order of error in the rate constant, and in all the cases the discrepancies are smaller than that. Accordingly, the calculated results in Figure 12.11 can be considered within the error inherent to the most accurate quantum mechanical calculations, or even smaller.

In summary, the utility of quantum chemical calculations to elucidate the detailed mechanisms of OH radical reactions with oxygenated VOCs has been proven. The importance of including reactant complexes in such modeling, to obtain accurate values of the rate coefficients, has also been shown. The best results are those obtained when it is assumed that such complexes are in their vibrational ground state. The relative site reactivity of the studied compounds towards OH radicals has been shown to be strongly influenced by intramolecular hydrogen-bond-like interactions that arise in the transition states.

Since the main tropospheric sink of oxygenated VOCs seems to be their reactions with OH radicals, at least at daytimes, the mechanistic and kinetic information discussed in this work is relevant to fully understand the tropospheric chemistry of such compounds, as well as their subsequent fate. Hopefully, the large amount of experimental and theoretical work that has been revisited here, which has been devoted to chemical reactions of environmental significance, could contribute in some extent to act in the right directions and prevent more damage to our atmosphere.

REFERENCES

[1] J.M.L. Martin, G. de Oliveira, *J. Chem. Phys.* **111** (1999) 1843.

[2] R.A. Friesner, *Proc. Natl. Acad. Sci.* **102** (2005) 6648.

[3] J.R. Thomas, B.J. Deeleeuw, G. Vacek, T.D. Crawford, Y. Yamaguchi, H.F. Schaefer III, *J. Chem. Phys.* **99** (1993) 403.

[4] V. Guner, K.S. Khuong, A.G. Leach, P.S. Lee, M.D. Bartberger, K.N. Houk, *J. Phys. Chem. A* **107** (2003) 11445.

[5] J.F. Stanton, R.J. Bartlett, *J. Chem. Phys.* **98** (1993) 7029.

[6] J. Gauss, J.F. Stanton, *J. Chem. Phys.* **104** (1996) 2574.

[7] A. Fernandez-Ramos, J.A. Miller, S.J. Klippenstein, D.G. Truhlar, *Chem. Rev.* **106** (2006) 4518, and references therein.

[8] *Scientific Assessment of Ozone Depletion: 2002*, Report No. 47, Global Ozone Research and Monitoring Project, World Meteorological Organization, Geneva, 2003.

[9] A. Guenther, C.N. Hewitt, D. Erickson, R. Fall, C. Geron, T. Graedel, P. Harley, L. Klinger, M. Lerdau, W.A. McKay, T. Pierce, B. Scholes, R. Steinbrecher, R. Tallamraju, J. Taylor, P. Zimmermann, *J. Geophys. Res.* **100** (1995) 8873.

[10] A. Guenther, C. Geron, T. Pierce, B. Lamb, P. Harley, R. Fall, *Atmos. Environ.* **34** (2000) 2205.

[11] R.F. Sawyer, R.A. Harley, S.H. Cadle, J.M. Norbeck, R. Slott, H.A. Bravo, *Atmos. Environ.* **34** (2000) 2161.

[12] M. Placet, C.O. Mann, R.O. Gilbert, M.J. Niefer, *Atmos. Environ.* **34** (2000) 2183.

[13] F.W. Lurmann, H.H. Main, *Analysis of the Ambient VOC Data Collected in the Southern California Air Quality Study*, Final Report to California Air Resources Contract A832-130, Sacramento, CA, 1992.

[14] J.G. Calvert, R. Atkinson, K.H. Becker, R.M. Kamens, J.H. Seinfeld, T.J. Wallington, G. Yarwood, *The Mechanisms of Atmospheric Oxidation of Aromatic Hydrocarbons*, Oxford Univ. Press, New York, 2002.

[15] A. Mellouki, G. Le Bras, H. Sidebottom, *Chem. Rev.* **103** (2003) 5077, and references therein.

[16] H. Levy, *Science* **173** (1971) 500.

[17] D.H. Ehalt, H.P. Dorn, D. Poppe, *Proc. R. Soc. Edinburgh* **97B** (1991) 17.

[18] A.M. Thompson, *Science* **256** (1992) 1157.

[19] R. Atkinson, *Atmos. Environ.* **34** (2000) 2063.

[20] T.E. Graedel, *Chemical Compounds in the Atmosphere*, Academic Press, New York, 1978.

[21] N. Donahue, R. Prinn, *J. Geophys. Res.* **95** (1990) 18387.
[22] E.S.C. Kwok, R. Atkinson, *Atmos. Environ.* **29** (1995) 1685.
[23] M. Mozurkewich, S.W. Benson, *J. Phys. Chem.* **88** (1984) 6249.
[24] M. Mozurkewich, J.J. Lamb, S.W. Benson, *J. Phys. Chem.* **88** (1984) 6435.
[25] D.L. Singleton, R.J. Cvetanovic, *J. Am. Chem. Soc.* **98** (1976) 6812.
[26] B.J. Finlayson-Pitts, N. Pitts, *Atmospheric Chemistry: Fundamentals and Experimental Techniques*, Wiley–Interscience, New York, 1986.
[27] J.R. Alvarez-Idaboy, N. Mora-Diez, R.J. Boyd, A. Vivier-Bunge, *J. Am. Chem. Soc.* **123** (2001) 2018.
[28] J.R. Alvarez-Idaboy, N. Mora-Diez, A. Vivier-Bunge, *J. Am. Chem. Soc.* **122** (2000) 3715.
[29] S. Aloisio, J.S. Francisco, *J. Phys. Chem. A* **104** (2000) 3211.
[30] V. Vasvári, I. Szilágyi, A. Bencsura, S. Dóbe, T. Berces, E. Henon, S. Canneaux, F. Bohr, *Phys. Chem. Chem. Phys.* **3** (2001) 551.
[31] E. Henon, S. Canneaux, F. Bohra, S. Dóbe, *Phys. Chem. Chem. Phys.* **5** (2003) 333.
[32] T. Yamada, P.H. Taylor, A. Goumri, P. Marshall, *J. Chem. Phys.* **119** (2003) 10600.
[33] I.W.M. Smith, A.R. Ravishankara, *J. Phys. Chem. A* **106** (2002) 4798.
[34] J.C. Hansen, J.S. Francisco, *Chem. Phys. Chem.* **3** (2002) 833.
[35] R.A. Loomis, M.I. Lester, *J. Chem. Phys.* **103** (1995) 4371.
[36] M.I. Lester, B.V. Pond, D.T. Anderson, L.B. Harding, A.F. Wagner, *J. Chem. Phys.* **113** (2000) 9889.
[37] S. Arrhenius, *Z. Phys. Chem.* **4** (1889) 226.
[38] M.J. Pilling, P.W. Seakins, *Reaction Kinetics*, Oxford Univ. Press, New York, 1996.
[39] G.K. Schenter, B.C. Garrett, D.G. Truhlar, *J. Chem. Phys.* **119** (2003) 5828.
[40] H. Eyring, *J. Chem. Phys.* **3** (1935) 107.
[41] D.G. Truhlar, W.L. Hase, J.T. Hynes, *J. Phys. Chem.* **87** (1983) 2664.
[42] D.G. Truhlar, B.C. Garrett, *Acc. Chem. Res.* **13** (1980) 440.
[43] D.G. Truhlar, B.C. Garrett, *Annu. Re. Phys. Chem.* **35** (1984) 159.
[44] D.G. Truhlar, A.D. Isaacson, B.C. Garrett, *Generalized Transition State Theory*, vol. 4, CRC Press, Boca Raton, FL, 1985.
[45] B.C. Garrett, D.G. Truhlar, *J. Phys. Chem.* **83** (1979) 1052.
[46] B.C. Garrett, D.G. Truhlar, *J. Phys. Chem.* **83** (1979) 1079.
[47] B.C. Garrett, D.G. Truhlar, *J. Am. Chem. Soc.* **101** (1979) 4534.
[48] B.C. Garrett, D.G. Truhlar, *J. Phys. Chem.* **83** (1979) 3058.
[49] D.-H. Lu, T.N. Truong, V.S. Melissas, G.C. Lynch, R.-P. Liu, B.C. Garrett, R. Steckler, A.D. Isaacson, S.N. Rai, G.C. Hancock, J.G. Lauderdale, T. Joseph, D.G. Truhlar, *Comput. Phys. Commun.* **71** (1992) 235.
[50] Y.-P. Liu, G.C. Lynch, T.N. Truong, D.-H. Lu, D.G. Truhlar, B.C. Garrett, *J. Am. Chem. Soc.* **115** (1993) 2408.
[51] R.A. Marcus, M.E. Coltrin, *J. Chem. Phys.* **67** (1977) 2609.
[52] J. Espinosa-Garcia, J.C. Corchado, *J. Phys. Chem.* **99** (1995) 8613.
[53] M. Szori, C. Fittschen, I.G. Csizmadia, B. Viskolcz, *J. Chem. Theory Comput.* **2** (2006) 1575.
[54] I. Mayer, *Int. J. Quantum Chem.* **23** (1983) 341.
[55] R. Vargas, J. Garza, D.A. Dixon, B.P. Hay, *J. Am. Chem. Soc.* **122** (2000) 4750.
[56] I. Mayer, *J. Phys. Chem.* **100** (1996) 6332.
[57] H.B. Jansen, P. Ros, *Chem. Phys. Lett.* **3** (1969) 140.
[58] S.F. Boys, F. Bernardi, *Mol. Phys.* **19** (1970) 553.
[59] M.J. Frisch, J.E. Del Bene, J.S. Binkley, H.F. Schaefer III, *J. Chem. Phys.* **84** (1986) 2279.
[60] D.W. Schwenke, D.G. Truhlar, *J. Chem. Phys.* **82** (1985) 2418.
[61] K. Morokuma, K. Kitaura, in: P. Politzer (Ed.), *Chemical Application of Atomic and Molecular Electronic Potentials*, Plenum, New York, 1981.
[62] J.C. López, J.L. Alonso, F.J. Lorenzo, V.M. Rayon, J.A. Sordo, *J. Chem. Phys.* **111** (1999) 6363.
[63] S.W. Hunt, K.R. Leopold, *J. Phys. Chem. A* **105** (2001) 5498.
[64] H. Valdés, J.A. Sordo, *J. Comput. Chem.* **23** (2002) 444.
[65] H. Valdés, J.A. Sordo, *J. Phys. Chem. A* **106** (2002) 3690.
[66] T.H. Dunning Jr., *J. Phys. Chem. A* **104** (2000) 9062.
[67] T.E. Graedel, D.T. Hawkings, L.D. Claxton, *Atmospheric Chemical Compounds*, Academic Press, Orlando, 1986.

[68] J.H. Seinfield, J.M. Andino, F.M. Bowman, H.J. Frostner, S. Pandis, *Adv. Chem. Eng.* **19** (1994) 325.
[69] T.J. Wallington, M.J. Kurylo, *Int. J. Chem. Kinet.* **19** (1987) 1015.
[70] B. Picquet, S. Heroux, A. Chebbi, J. Doussin, R. Durand-Jolibois, A. Monod, H. Loirat, P. Carlier, *Int. J. Chem. Kinet.* **30** (1998) 839.
[71] *Guidance on Estimating Motor Vehicle Emission Reduction from the Use of Alternative Fuels and Fuel Blends*, U.S.E.P.A., Report No. 1 APA-AA-TSS-PA-87-4, Ann Arbor, Michigan, 1998.
[72] E. Grosjean, D. Grosjean, R. Gunawardena, R.A. Rasmussen, *Environ. Sci. Technol.* **32** (1998) 736.
[73] R. Overend, G. Paraskevopoulos, *J. Phys. Chem.* **82** (1978) 1329.
[74] A.R. Ravishankara, D.D. Davis, *J. Phys. Chem.* **82** (1978) 2852.
[75] W.P. Hess, F.P. Tully, *J. Phys. Chem.* **93** (1989) 1944.
[76] M. Yujing, A. Mellouki, *Chem. Phys. Lett.* **333** (2001) 63.
[77] S. Oh, J.M. Andino, *Int. J. Chem. Kinet.* **33** (2001) 422.
[78] M. Sorensen, M.D. Hurley, T.J. Wallington, T.S. Dibble, O.J. Nielsen, *Atmos. Environ.* **36** (2002) 5947.
[79] S.A. Chemma, K.A. Holbrook, G.A. Oldershaw, R.W. Walker, *Int. J. Chem. Kinet.* **34** (2002) 110.
[80] F. Cavalli, H. Geiger, I. Barnes, K.H. Becker, *Environ. Sci. Technol.* **36** (2002) 1263.
[81] E. Jimenez, M.K. Gilles, A.R. Ravishankara, *J. Photochem. Photobiol. A Chem.* **157** (2003) 237.
[82] H. Wu, Y. Mu, X. Zhang, G. Jiang, *Int. J. Chem. Kinet.* **35** (2003) 81.
[83] A. Mellouki, F. Oussar, X. Lun, A. Chakir, *Phys. Chem. Chem. Phys.* **6** (2004) 2951.
[84] G. Kovacs, T. Szasz-Vadasz, V.C. Papadimitriou, S. Dobe, T. Berces, F. Marta, *React. Kinet. Catal. Lett.* **87** (2005) 129.
[85] N. Luo, D.C. Kombo, R. Osman, *J. Phys. Chem. A* **101** (1997) 926.
[86] J.T. Jodkowski, M.-T. Rayez, J.-C. Rayez, *J. Phys. Chem. A* **103** (1999) 3750.
[87] A. Galano, J.R. Alvarez-Idaboy, G. Bravo-Perez, M.E. Ruiz-Santoyo, *Phys. Chem. Chem. Phys.* **4** (2002) 4648.
[88] A. Hatipoglu, Z. Cinar, *J. Mol. Struct. THEOCHEM* **631** (2003) 189.
[89] R. Atkinson, D.L. Baulch, R.A. Cox, R.F. Hampson Jr., J.A. Kerr, M.J. Rossi, J. Troe, *J. Phys. Chem. Ref. Data* **28** (1999) 191.
[90] T.J. Wallington, M.J. Kurylo, *J. Phys. Chem.* **91** (1987) 5050.
[91] R. Atkinson, *Chem. Rev.* **86** (1986) 69.
[92] R. Atkinson, *Int. J. Chem. Kinet.* **19** (1987) 799.
[93] H.L. Bethel, R. Atkinson, J. Arey, *Int. J. Chem. Kinet.* **33** (2001) 310.
[94] D.H. Semmes, A.R. Ravishankara, C.A. Gump-Perkins, P.H. Wine, *Int. J. Chem. Kinet.* **17** (1985) 303.
[95] R. Atkinson, D.L. Baulch, R.A. Cox, R.F. Hampson Jr., J.A. Kerr, M.J. Rossi, J. Troe, *J. Phys. Chem. Ref. Data* **26** (1997) 521.
[96] N.I. Butkovskaya, D.W. Setser, *J. Phys. Chem. A* **102** (1998) 9715.
[97] N.I. Butkovskaya, D.W. Setser, *J. Phys. Chem. A* **104** (2000) 9428.
[98] B. D'Anna, O. Andresen, Z. Gefen, C.J. Nielsen, *Phys. Chem. Chem. Phys.* **3** (2001) 3057.
[99] V. Sivakumaran, D. Holscher, T.J. Dillon, J.N. Crowley, *Phys. Chem. Chem. Phys.* **5** (2003) 4821.
[100] V. Sivakumaran, J.N. Crowley, *Phys. Chem. Chem. Phys.* **5** (2003) 106.
[101] K.L. Feilberg, M.S. Johnson, C.J. Nielsen, *J. Phys. Chem. A* **108** (2004) 7393.
[102] M. Dupuis, W.A. Lester Jr., *J. Chem. Phys.* **81** (1984) 847.
[103] J.S. Francisco, *J. Chem. Phys.* **96** (1992) 7597.
[104] P.H. Taylor, M.S. Rahman, M. Arif, B. Dellinger, P. Marshall, *Symp. Int. Comb. Proc.* **26** (1996) 497.
[105] B. D'Anna, V. Bakken, J.A. Beukes, C.J. Nielsen, K. Brudnik, J.T. Jodkowski, *Phys. Chem. Chem. Phys.* **5** (2003) 1790.
[106] J.J. Wang, H.B. Chen, G.P. Glass, R.F. Curl, *J. Phys. Chem. A* **107** (2003) 10834.
[107] E. Vöhringer-Martinez, B. Hansmann, H. Hernandez, J.S. Francisco, J. Troe, B. Abel, *Science* **315** (2007) 497.
[108] G.S. Tyndall, J.J. Orlando, T.J. Wallington, M.D. Hurley, M. Goto, M. Kawasaki, *Phys. Chem. Chem. Phys.* **11** (2002) 2189.
[109] N.I. Butkovskaya, A. Kukui, G. Le Bras, *J. Phys. Chem. A* **108** (2004) 1160.
[110] R. Atkinson, S.M. Aschmann, W.P.L. Carter, J.N. Pitts Jr., *Int. J. Chem. Kinet.* **14** (1982) 839.
[111] P. Dagaut, T.J. Wallington, R. Liu, M.J. Kurylo, *J. Phys. Chem.* **92** (1988) 4375.

[112] R. Atkinson, S.M. Aschmann, *J. Phys. Chem.* **92** (1988) 4008.
[113] R. Atkinson, S.M. Aschmann, *Int. J. Chem. Kinet.* **27** (1995) 261.
[114] W.B. DeMore, S.P. Sander, D.M. Golden, R.F. Hampson, M.J. Kurylo, C.J. Howard, A.R. Ravishankara, C.E. Kolb, M.J. Molina, *Chemical kinetics and photochemical data for use in stratospheric modeling*, Evaluation no. 12, JPL Publication 97-4, 1997.
[115] S. Le Calvé, D. Hitier, G. Le Bras, A. Mellouki, *J. Phys. Chem. A* **102** (1998) 4579.
[116] R. Atkinson, *Int. J. Chem. Kinet.* **19** (1987) 799.
[117] A. Klamt, *Chemosphere* **32** (1996) 717.
[118] M. Wollenhaupt, S.A. Carl, A. Horowitz, J.N. Crowley, *J. Phys. Chem. A* **104** (2000) 2695.
[119] E. Henon, S. Canneaux, F. Bohr, S. Dóbé, *Phys. Chem. Chem. Phys.* **5** (2003) 333.
[120] T. Yamada, P.H. Taylor, A. Goumri, P. Marshall, *J. Chem. Phys.* **119** (2003) 10600.
[121] L. Masgrau, A. González-Lafont, J.M. Lluch, *J. Phys. Chem. A* **106** (2002) 11760.
[122] J.R. Alvarez-Idaboy, A. Cruz-Torres, A. Galano, M.E. Ruiz-Santoyo, *J. Phys. Chem. A* **108** (2004) 2740.
[123] The numerical data from reference 120, which is not in the original article, was kindly provided by Professor Paul Marshall.
[124] Y. Georgievskii, S.J. Klippenstein, *J. Phys. Chem. A* **111** (2007) 3802.
[125] J.A. Montgomery Jr., M.J. Frisch, J.W. Ochterski, A.G. Petersson, *J. Chem. Phys.* **110** (1999) 2822.
[126] J.A. Montgomery Jr., M.J. Frisch, J.W. Ochterski, G.A. Petersson, *J. Chem. Phys.* **112** (2000) 6532.
[127] M. Legrand, M. de Angelis, *J. Geophys. Res.* **100** (1995) 1445.
[128] K. Granby, A.H. Egeløv, T. Nielsen, C. Lohse, *J. Atm. Chem.* **28** (1997) 195.
[129] P.H. Wine, R.J. Astalos, R.L. Mauldin III, *J. Phys. Chem.* **89** (1985) 2620.
[130] G.S. Jolly, D.J. McKenney, D.L. Singleton, G. Paraskevopoulos, A.R. Bossard, *J. Phys. Chem.* **90** (1986) 6557.
[131] D.L. Singleton, G. Paraskevopoulos, R.S. Irwin, G.S. Jolly, D.J. McKenney, *J. Am. Chem. Soc.* **110** (1988) 7786.
[132] P. Dagaut, *Int. J. Chem. Kinet.* **20** (1988) 331.
[133] C. Zetzsch, F. Stuhl, Physico-chemical behavior of atmospheric pollutants, in: B. Versino, H. Ott (Eds.), *Proceedings of the Second European Symposium*, Reidel, Dordrecht, 1982, p. 129.
[134] N.I. Butkovskaya, A. Kukui, N. Pouvesle, G. Le Bras, *J. Phys. Chem. A* **108** (2004) 7021.
[135] Ph. Dagaut, T.J. Wallington, R. Liu, M.J. Kurylo, *Int. J. Chem. Kinet.* **20** (1988) 331.
[136] D.L. Singleton, G. Paraskevopoulos, R.S. Irwin, *J. Am. Chem. Soc.* **111** (1989) 5248.
[137] F. De Smedt, X.V. Bui, T.L. Nguyen, J. Peeters, L. Vereecken, *J. Phys. Chem. A* **109** (2005) 2401.
[138] D. Vimal, P.S. Stevens, *J. Phys. Chem. A* **110** (2006) 11509.
[139] J.M. Anglada, *J. Am. Chem. Soc.* **126** (2004) 9809.
[140] A. Galano, J.R. Alvarez-Idaboy, M.E. Ruiz-Santoyo, A. Vivier-Bunge, *J. Phys. Chem. A* **106** (2002) 9520.
[141] M. Torrent-Sucarrat, J.M. Anglada, *ChemPhysChem* **5** (2004) 183.
[142] W.B. DeMore, S.P. Sander, D.M. Golden, R.F. Hampson, M.J. Kurylo, C.J. Howard, A.R. Ravishankara, C.E. Kolb, M.J. Molina, *JPL Publication* **97** (1997) 4.
[143] B.J. Finlayson-Pitts, J.N. Pitts Jr., *Chemistry of the Upper and Lower Atmosphere. Theory, Experiments, and Applications*, Academic Press, San Diego, 2000.
[144] B.E. Dumdei, D.V. Kenny, P.B. Shepson, T.E. Kleindienst, C.M. Nero, L.T. Culpitt, L.D. Claxton, *Environ. Sci. Technol.* **22** (1988) 1493.
[145] J.M. Andino, J.N. Smith, R.C. Flagan, W.A. Goddard III, J.H. Seinfeld, *J. Phys. Chem.* **100** (1996) 10967.
[146] J. Yu, H.E. Jefries, *Atmos. Environ.* **31** (1997) 2281.
[147] R. Volkamer, U. Platt, K. Wirtz, *J. Phys. Chem. A* **105** (2001) 7865.
[148] I. Magneron, R. Thevenet, A. Mellouki, G. Le Bras, G.K. Moortgat, K. Wirtz, *J. Phys. Chem. A* **106** (2002) 2526.
[149] D.F. Smith, T.E. Kleindienst, C.D. McIver, *J. Atmos. Chem.* **34** (1999) 339.
[150] A. Bierbach, I. Barnes, K.H. Becker, E. Wiesen, *Environ. Sci. Technol.* **28** (1994) 715.
[151] E.C. Tuazon, H. MacLeod, R. Atkinson, W.P.L. Carter, *Environ. Sci. Technol.* **20** (1986) 383.
[152] E.C. Tuazon, R. Atkinson, *Int. J. Chem. Kinet.* **21** (1989) 1141.
[153] E.C. Tuazon, R. Atkinson, *Int. J. Chem. Kinet.* **22** (1990) 591.

[154] C.N. Plum, E. Sanhueza, R. Atkinson, W.P.L. Carter, J.N. Pitts Jr., *Environ. Sci. Technol.* **17** (1983) 479.
[155] T.E. Kleindienst, G.W. Harris, J.N. Pitts Jr., *Environ. Sci. Technol.* **16** (1982) 844.
[156] G.S. Tyndall, T.A. Staffelbach, J.J. Orlando, J.G. Calvert, *Int. J. Chem. Kinet.* **27** (1995) 1009.
[157] H. Niki, P.D. Maker, C.M. Savage, M.D. Hurley, *J. Phys. Chem.* **91** (1987) 2174.
[158] C. Bacher, G.S. Tyndall, J.J. Orlando, *J. Atm. Chem.* **39** (2001) 171.
[159] J. Baker, J. Arey, R. Atkinson, *J. Phys. Chem. A* **108** (2004) 7032.
[160] I. Magneron, A. Mellouki, G. Le Bras, G.K. Moortgat, A. Horowitz, K. Wirtz, *J. Phys. Chem. A* **109** (2005) 4552.
[161] N.I. Butkovskaya, N. Pouvesle, A. Kukui, G. Le Bras, *Phys. Chem. A* **110** (2006) 13492.
[162] R. Karunanandan, D. Holscher, T.J. Dillon, A. Horowitz, J.N. Crowley, L. Vereecken, J. Peeters, *J. Phys. Chem. A* **111** (2007) 897.
[163] P. Dagaut, R. Liu, T.J. Wallington, M.J. Kurylo, *J. Phys. Chem.* **93** (1989) 7838.
[164] J.J. Orlando, G.S. Tyndall, J.M. Fracheboud, E.G. Estupiñan, S. Haberkorn, A. Zimmer, *Atm. Environ.* **33** (1999) 1621.
[165] P.K. Chowdhury, H.P. Upadhyaya, P.D. Naik, J.P. Mittal, *Chem. Phys. Lett.* **351** (2002) 201.
[166] T.J. Dillon, A. Horowitz, D. Hölscher, J.N. Crowley, L. Vereecken, J. Peetersb, *Phys. Chem. Chem. Phys.* **8** (2006) 236.
[167] N.I. Butkovskaya, N. Pouvesle, A. Kukui, Y. Mu, G. Le Bras, *Phys. Chem. A* **110** (2006) 6833.
[168] D.H. Semmes, A.R. Ravishankara, C.A. Gump-Perkins, P.H. Wine, *Int. J. Chem. Kinet.* **17** (1985) 303.
[169] A. Galano, J.R. Alvarez-Idaboy, M.E. Ruiz-Santoyo, A. Vivier-Bunge, *J. Phys. Chem. A* **109** (2005) 169.
[170] A. Galano, J.R. Alvarez-Idaboy, M.E. Ruiz-Santoyo, A. Vivier-Bunge, *ChemPhysChem* **5** (2004) 1379.
[171] A. Galano, *J. Phys. Chem. A* **110** (2006) 9153.
[172] M. Ochando-Pardo, I. Nebot-Gil, A. Gonzalez-Lafont, J.M. Lluch, *J. Phys. Chem. A* **108** (2004) 5117.

CHAPTER 13

Theoretical and Experimental Studies of the Gas-Phase Cl-Atom Initiated Reactions of Benzene and Toluene

A. Ryzhkov[*], P.A. Ariya[*], F. Raofie[*], H. Niki[1]** and **G.W. Harris[**]**

Contents

1. Introduction 276
2. Methods 277
   2.1 Theoretical calculation details 277
   2.2 Experimental methods 277
3. Results and Discussions 279
   3.1 Theoretical calculations 279
   3.2 Experimental section 285
4. Conclusions 292

References 294

Abstract

The reactions of benzene (Bz) and toluene (PhMe) with chlorine atoms in the gas phase have been studied using both theoretical and experimental techniques. Energy and geometry of reaction complexes and transition states were calculated in the Cl-atom initiated reaction of benzene and toluene using modern hybrid functional PBE0 method with the aug-pc1 basis set with an additional CCSD(T)/aug-CC-pVDZ energy single point calculation. Three stationary structures have been found for the Bz···Cl complex: hexahapto-complex, π-complex and σ-complex. The first one is a transition state between two opposite π-complexes. PhMe···Cl has additional structures due to *ipso*-, *ortho*-, *meta*- and *para*-isomerization. The

[*] Department of Chemistry and Department Atmospheric and Oceanic Sciences, McGill University, 801 Sherbrooke St. W., Montreal, Quebec, Canada, H3A 2K6

[**] Centre for Atmospheric Chemistry, York University, 4700 Keele Street, Toronto, Ontario, Canada, M3J 1P3

[1] Corresponding author.

Advances in Quantum Chemistry, Vol. 55
ISSN 0065-3276, DOI: 10.1016/S0065-3276(07)00213-4

stability of all calculated complexes was determined and compared. Two reaction pathways for benzene and toluene with a Cl atom were evaluated: (a) the hydrogen abstraction of benzene and toluene by Cl atom, which is seemingly barrierless and endothermic, and (b) the hydrogen substitution reaction that in contrast has a relatively high energy of activation. Rate coefficients for these same reactions were measured using ethane, n-butane, and chloro-, dichloro- and trichloromethane, as reference compounds, with gas chromatography equipped with mass detection spectrometry and flame ionization detection (GC-MSD and GC-FID). The reaction rates were estimated as $(5.57 \pm 0.15) \times 10^{-11}$ and $\leqslant 1 \times 10^{-15}$ $cm^3/(molecule\ s)$ for benzene and toluene, respectively. Chlorinated products of the reactions were analyzed by GS-MS. Chlorobenzene was the only identified product between a reaction of benzene and the Cl atom. The major products of the PhMe + Cl reaction were chloromethylbenzene with *ortho*- and *para*-chlorotoluenes.

1. INTRODUCTION

Aromatic compounds such as benzene, toluene, and xylene include a significant portion of the volatile organic compounds (VOC) in urban areas [1,2]. The major emission sources of these compounds into the atmosphere are assumed to be motor vehicles [1,2]. Solvents in paints and in polymer manufacturing and biomass burning are amongst other contributors to aromatic emission [3]. Benzene [4] and toluene [5,6] are also shown to be human health hazards. Upon atmospheric oxidation, the reactions of aromatic compounds lead to the formation of secondary organic aerosols [7–11] that can directly and indirectly affect climate by interactions with radiation, and in turn they can affect the contribution of pollutants such as ozone. Due to such environmental impacts atmospheric reactions of aromatic compounds has received much attention during the last few decades. The main oxidation and degradation way of the aromatic compounds is assumed to be the reaction with OH radicals [1], but reactions with ozone, NO_3 and halogens also have significant contribution. Cl-atom high reactivity renders Cl-reaction as an efficient loss pathway for aromatic compounds particularly in the marine boundary layer, polar and coastal areas where relatively high concentrations of reactive halogens are observed [12,13]. Under a high chlorine regime, the degradation rate of VOC by Cl can dominate OH-initiated reactions [14] due to up to two orders of magnitude faster reactions of Cl-atoms with organic compounds in comparison to OH radicals [15].

There are several kinetic and product studies on Cl atom gas phase reactions of hydrocarbons [16–42]. Though the various kinetic studies show more consistency on Cl+toluene reaction rates [41], the reactions of Cl+benzene has been shown to be quite contradictory. Indeed, six existing published studies on the Cl + benzene reaction rates vary by more than 5 orders of magnitudes, ranging from 1.3×10^{-16} to 1.5×10^{-11} $cm^3/(molecule\ s)$ [16,17,19,24,27,42].

The aim of the present work was to evaluate the rate constants of the reactions of Cl initiated oxidation of benzene and toluene, and to provide further mechanistic insights of these two reaction systems using a combined laboratory and theoretical approach. Experiments were performed in N_2 or in dry air at $T = 296 \pm 2$ K, and analyzed using gas chromatograph with the flame ionization detection and with the mass selective detection (GC-FID and GC-MSD, respectively). In addition, we have conducted computational studies of these reactions, and evaluated the geometry and energy of the reactants, intermediates and products. We will herein briefly discuss the importance of these results in understanding of Cl chemistry in troposphere.

2. METHODS

2.1 Theoretical calculation details

The geometries of all molecules, complexes and transition states have been fully optimized using the non-empirical hybrid density functional PBE0 [43,44] with the aug-pc1 polarization consistent basis set [45,46], which was specially developed by Jensen to be used with density functional theory methods. Harmonic vibrational frequencies were used to verify the nature of the stationary points and to estimate zero-point vibration energy at the same level of theory. Additionally, single point CCSD(T) [47–49] energies implementing aug-CC-pV*D*Z correlation consistent basis set [50] were calculated for obtained geometries. All calculations were carried out with the Gaussian 98 [51] and MOLPRO [52] *ab initio* program packages.

2.2 Experimental methods

Experiments have been carried out at York and McGill Universities. In the relative rate technique of bimolecular reactions, we follow two or more reactions, in which two or more molecules will form products with the same reactant, i.e.:

$$\mathrm{Cl} + \mathrm{benzene} \rightarrow \mathrm{products}, \tag{1}$$

$$\mathrm{Cl} + \mathrm{reference} \rightarrow \mathrm{products}. \tag{2}$$

Assuming that such bimolecular reactions initiate solely from the reaction of Cl-atoms with reactants and there is no reformation of the reactant, then we can apply the integrated expression of the second order rate law:

$$\ln\left(\frac{[\mathrm{benzene}]_0}{[\mathrm{benzene}]_t}\right) = \frac{k_1}{k_2}\ln\left(\frac{[\mathrm{reference}]_0}{[\mathrm{reference}]_t}\right), \tag{3}$$

where [benzene] and [reference] are the concentrations of the benzene and reference substance, respectively, and k_1 and k_2 are the rate constants for reactions (1) and (2). However, if there are significant secondary reactions, the plots of $\ln([\mathrm{benzene}]_0/[\mathrm{benzene}]_t)$ vs. $\ln([\mathrm{reference}]_0/[\mathrm{reference}]_t)$ should not yield an straight line.

Cl-atom kinetic reactions of toluene and benzene were carried out using GC-FID and GC-MS. Product studies presented herein were performed using GC-MS.

All experiments were carried out at near atmospheric pressure (about 735 ± 5 Torr) and $T = 296 \pm 2$ K. Cl atoms were generated by UV photolysis of Cl_2 molecules. 2–3 L Pyrex bulbs and 10 L all-Teflon bags were used as reaction chambers. To assure homogeneity, hydrocarbon/Cl_2/N_2 or air mixtures were continuously mixed using a Teflon magnetic rod. 150–200 μl samples were transferred into the GC columns using a gas-tight syringe. Two different columns were used in this study. The experiments involving acetylene and toluene were carried out using a 30 m Agilent PLOT column. Since chlorinated methanes were also used as reference molecules in most benzene-targeted experiments, a DB-624 Agilent column, particularly targeted for separation of chlorinated compounds was employed. The PLOT column was operated at a column head pressure of 30 psi. The pressure was increased at 2.2 psi/min to 50 psi, and maintained for 18 min. The column temperature was increased at 15 °C/min from 30 to 160 °C and maintained for 18 min. The chromatographic retention time ranged from 2.0 min for ethane to 27.0 min for toluene. The DB-624 column head pressure was held constant at 9 psi, while its temperature was increased at 20 °C/min from 30 to 80 °C, and maintained for 10 min. The chromatographic retention time varied from 4.5 min for chloromethane to 10.8 min for benzene. All the reactants were fully resolved under these operating conditions. To avoid any potential chromatographic conflicts between reactants and products, experiments were carried out in which $Cl_2 + N_2$/air mixtures of each hydrocarbon were studied separately, and no interference was observed under these operating conditions.

Mixtures of hydrocarbons in the absence of Cl_2 were photolyzed for c.a. 10 min and no hydrocarbon photolysis was observed. Hydrocarbon/Cl_2 mixtures were aged in dark and within our experimental uncertainties no hydrocarbon decay was observed.

During the measured kinetic reactions, samples were aged to verify the stability of the initial conditions. No such decay was observed in our experiments. The concentration of hydrocarbon reactants and Cl_2 molecules ranged from 10–20 and 20–200 mTorr, respectively. The irradiation time varied from 10 s to 40 min. In GC-MS studies, concentrations of benzene and toluene were monitored by MS detection (quadrupole MSD HP 5973) after separation on a gas chromatograph (HP 6890) equipped with a 0.2 mm i.d. 30 m cross-linked phenyl-methyl-siloxane column (Hp 5-MS). The column was operated at a constant flow (1.5 mL/min) of helium and was kept isothermal at 40 °C for 1 min. During the chromatographic runs, the oven temperature was increased by 25 °C/min from 40 to 150 °C. We performed a full scan of the mass spectrum at the beginning of each kinetic run. Products were observed using the single ion monitoring (SIM) mode. The detailed laboratory kinetics using GC-MS are provided elsewhere [53–56]. In addition to a linear and non-linear least-squares analysis, we used a cumulative error analysis including uncertainties due to sampling, transfer, purification, as well as human and instrument error. A conservative overall accumulative analysis for GC-FID and GC-MS kinetic studies is evaluated to be ±20%.

2.2.1 Long path FTIR studies

The details of experimental conditions and analysis are explained elsewhere [57–59]. Briefly, experiments were performed in a Pyrex photochemical reactor/IR absorption cell with KBr windows coupled to a Mattson FTIR spectrometer (Galaxy 4326C) with a KBr beam splitter and a liquid N_2-cooled Hg/Cd/Te detector at York University. A multipass cell (2 m long with a total 92 m pathlength) was used as the reaction vessel. This cell is 140 L in volume surrounded by 26 Fluorescent lamps (GTEF40/CW for $\lambda \geqslant 400$ nm or GEF 40T12/BLB for $\lambda \geqslant 300$ nm). The IR spectra were collected over the frequency range 500–4000 cm^{-1} at 1/8 cm^{-1} resolution by co-adding 32 scans. Reactants and products were quantified by fitting IR reference spectra of the pure compounds to the observed product spectra using an integrated absorption. Reference spectra for most of the compounds encountered in this study were available from the extensive library of spectra compiled in our laboratories. In all cases, we estimate the potential systematic uncertainties associated with the calibrations of the reference spectra as well as subtractions to be about 5–10% each. This leads to an overall systematic uncertainty of 11% on average, calculated though a propagation of error [60].

2.2.2 Materials and supplies

All reactants, unless otherwise stated, were purchased from Sigma Aldrich (97+%). Trap-by-trap distillation were used for further purifications of organic compounds [53,58,59,61]. Ultra Pure air (containing <1 ppm hydrocarbons) was obtained from Matheson Inc. Carrier gases for GC (hydrogen, helium and air) were also supplied by Matheson. Dimethyldichlorosilane (5% DMDCS in toluene) was obtained from Supelco.

Further purification of benzene The degree of purification of benzene is critically important in the determination of rate constant as well as product study of the reaction of benzene with chlorine. A purification procedure developed by Sokolov *et al.* [27] was used. Cl_2 was dissolved in the deoxygenated liquid benzene (Sigma Aldrich, 99+%) by bubbling gaseous Cl_2 through the sample. Chlorine atoms react first with certain impurities, then with benzene. Thus we convert reactive impurities and a small fraction of the benzene into chlorinate compounds that are eliminated by trap-to-trap distillation.

3. RESULTS AND DISCUSSIONS

3.1 Theoretical calculations

3.1.1 Cl + benzene reaction

According to Tsao *et al.* [62], three possible structures of benzene–chlorine complex were obtained (Figure 13.1). The first one is high symmetric (C_{6v}) hexahapto-complex with chlorine atom over the centre of the benzene ring (Figure 13.1a), which was firstly proposed by Russel [63]. The second and third structures had C_s symmetry; the chlorine atom was located over the carbon–hydrogen bond

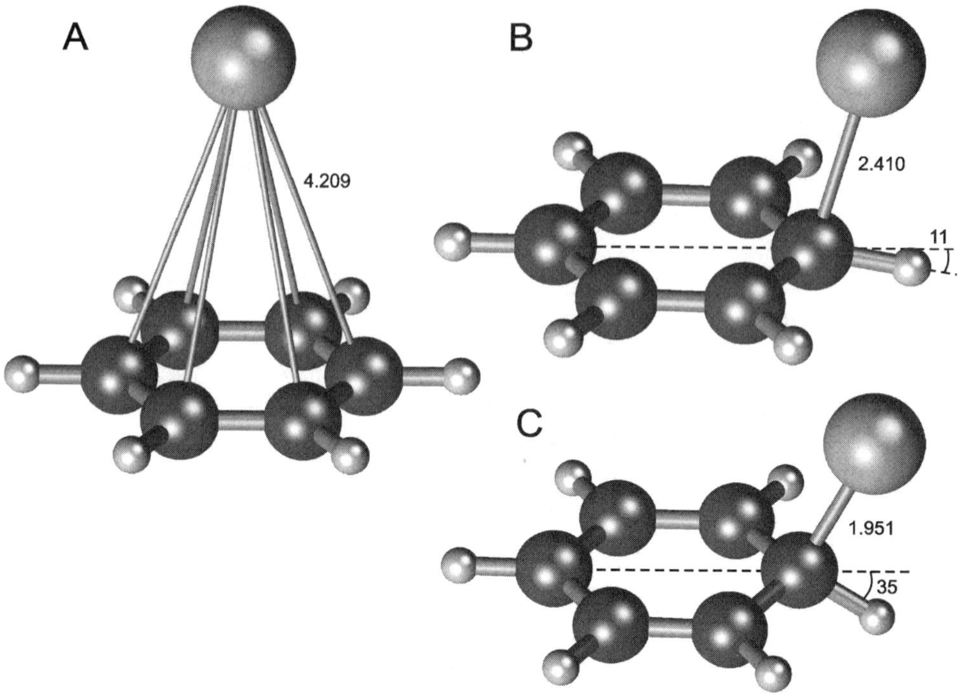

FIGURE 13.1 Structures of benzene–chlorine complexes, (A) C_{6v} symmetric complex, (B) *p*-complex, (C) *s*-complex.

(Figures 13.1b and 13.1c). Similar to Tsao *et al.* [62], all our attempts on finding stable structure for hexahapto-complex were not successful. Any calculated structure with a minimum energy was found to have an unstable wave function. After the re-optimization procedure, a stable wave function was obtained for the structure with minimum energy level. Only one stationary point was obtained, which has two imaginary frequencies, and it is thus a transition state of second order. Scans of potential energy surface at BPE0/6-31+G* (Figure 13.2a) and MP2/6-31+G* (Figure 13.2b) levels of theory in plane perpendicular to the benzene ring, which passes through opposite carbons atoms, indicate that there is no stable structures with C_{6v} symmetry. The transition state connected two minima lying over the carbon–hydrogen bonds (Figure 13.2). These minima for structures with C_s symmetries correspond to so-called π-complexes. They are characterized by relatively long distances of chlorine–carbon and chlorine–hydrogen bonds, 2.410 and 2.567 Å, respectively, and a slightly perturbed geometry of benzene (the hydrogen atom was deflected out from the plane of the benzene ring by 11°, (Figure 13.1b)). The stability of this complex was determined to be 5.0 kcal/mol (Table 13.1). Another possible structure of benzene–chlorine complex is presented in Figure 13.1c with shorter chlorine–carbon and chlorine–hydrogen bonds (1.951 and 2.359 Å, respectively). The hydrogen atom has a larger angle (35°) with respect to the plane of benzene ring. This σ-complex at the chosen level of theory was

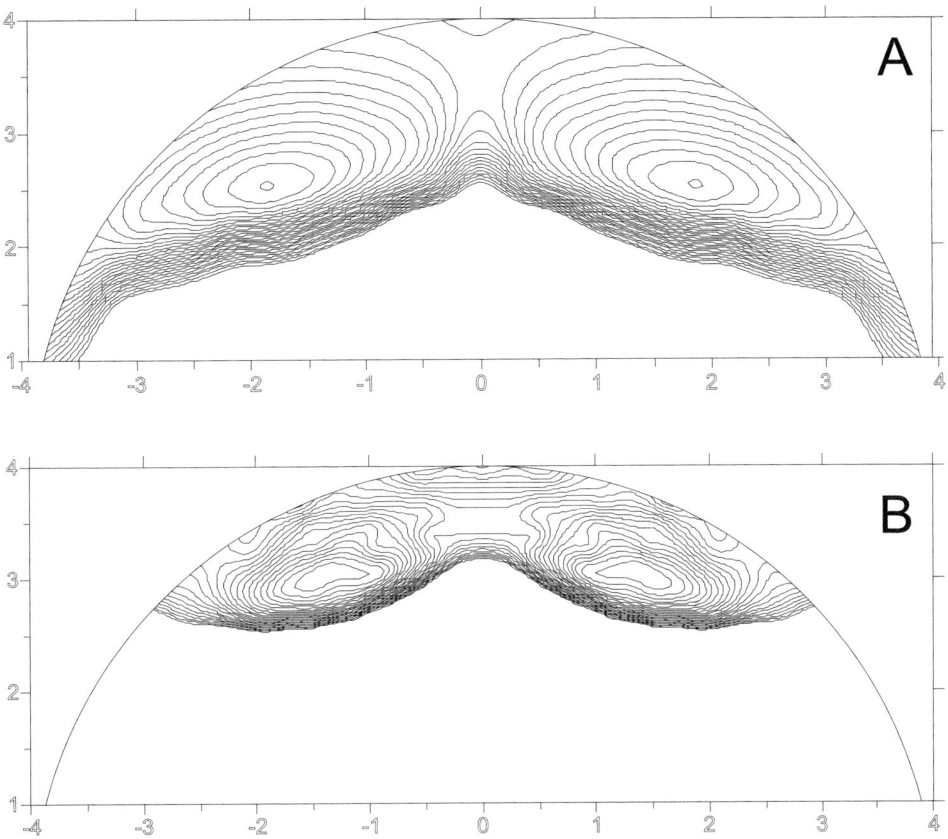

FIGURE 13.2 Potential energy surface scan at (A) BPE0/6-31+G* and (B) MP2/6-31+G* levels of theory of benzene–chlorine system in plane perpendicular to the benzene ring passing through opposite carbons atoms of benzene.

TABLE 13.1 Relative energy, enthalpy and free energy (kcal/mol), and equilibrium constants (cm^3/molecule) of benzene–chlorine complexes, transition state and products of benzene–chlorine reaction

	PBE0/aug-pc1				CCSD(T)/aug-CC-pVDZ			
	ΔE	ΔH (298)	ΔG (298)	K_c (298), 10^{-16}	ΔE	ΔH (298)	ΔG (298)	K_c (298), 10^{-20}
π-C_6H_6–Cl	−10.0	−10.1	−4.7	1.12	−5.0	−5.1	0.3	2.35
σ-C_6H_6–Cl	−10.2	−10.4	−4.3	0.62	−5.4	−5.7	0.4	2.04
C_6H_5 + HCl	2.6	2.8	−0.2		11.5	12.3	8.7	
TS1	32.3	32.1	32.9		32.2	31.9	32.7	
C_6H_5Cl + H	14.1	14.9	14.7		17.9	18.7	18.5	

slightly more stable than π-complexes with stabilisation energies of 5.4 kcal/mol (Table 13.1). Estimated equilibrium constants of these complexes were 2.35×10^{-20} and 2.04×10^{-20} cm^3/molecule (Table 13.1), which are lower than experimentally found upper limit of the constant [27] (1–2) $\times 10^{-18}$ cm^3/molecule. Estimation of the equilibrium constants at the PBE0/aug-pc-1 level yielded to four orders of magnitude increase, 1.12×10^{-16} and 6.18×10^{-17} cm^3/molecule (Table 13.1), respectively.

The chlorine atom reacts with benzene by abstracting a hydrogen atom to form a phenyl radical and HCl:

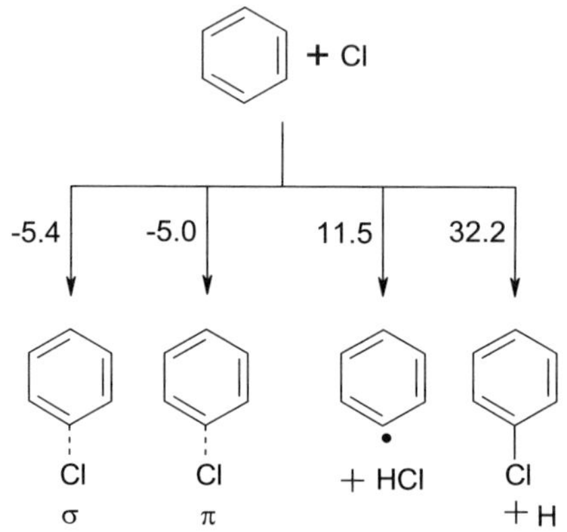

According to our calculations there is no maximum or minimum of energy along the reaction path connecting reagents and products for this reaction. The chlorine atom approaches the hydrogen atom opposite the benzene ring side and forms an H–Cl bond. At the same moment the C–H bond is breaking. Then products,

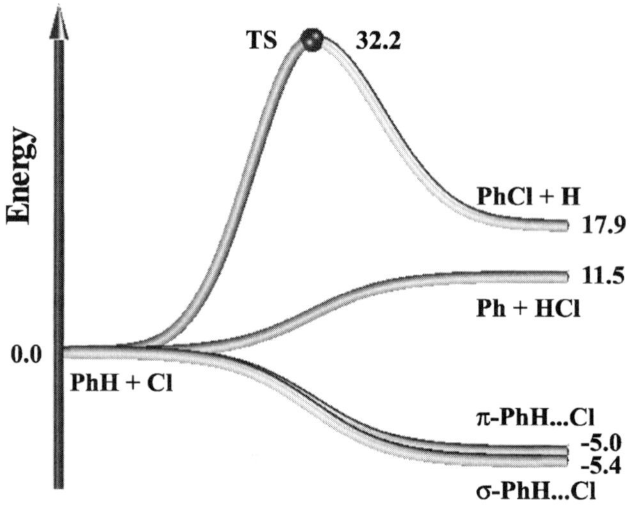

FIGURE 13.3 Energy diagram of chlorine–benzene reaction.

phenyl radical and HCl molecule disperse. Therefore, we suppose that this reaction is seemingly barrierless, where the energy of the products is higher by 11.5 kcal/mol than reactants. Calculated enthalpy of reaction, 12.3 kcal/mol, was higher than the experimental value of 7.9 kcal/mol found by Sokolov *et al.* [27]. Formation of Cl–benzene complexes, which are a more energetically favourable process, can influence the overall reaction. Generally, it leads to lower experimentally observed effective rate constants than for elementary reaction. In another possible reaction pathway, the substitution of a hydrogen atom by chlorine was found to have a transition state, **TS1**. However, the energy barrier for this reaction is very high, 32.2 kcal/mol, hence this channel is expected have no significance (Figure 13.3).

3.1.2 Cl + toluene reaction

There are two possible channels for the Cl atom reaction:

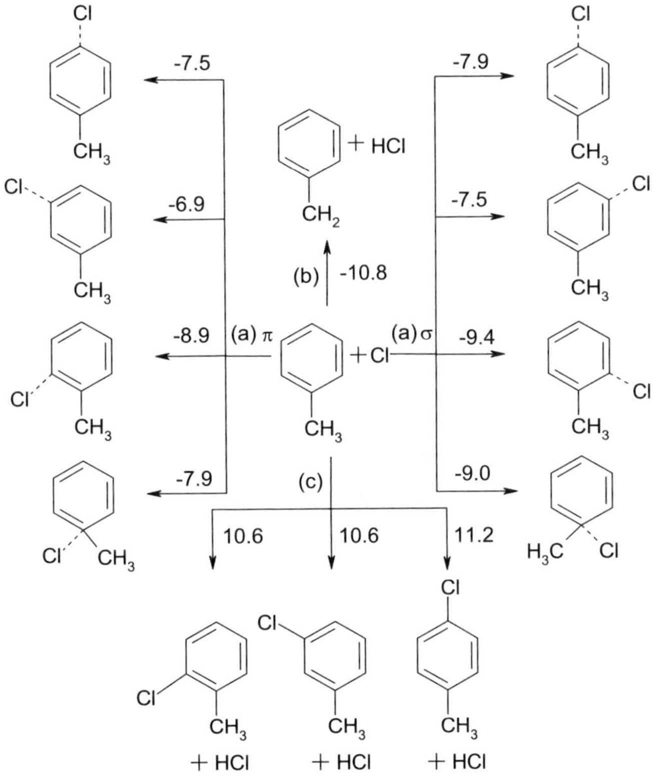

The Cl-atom can add to the benzene ring forming an energized adduct that can (a) be collisionally stabilized, (b) abstract a hydrogen atom from the methyl group, or (c) abstract a hydrogen from the benzene ring itself. In the first scenario, toluene can form two kinds of complexes with the chlorine atom [64], four π-complexes and four σ-complexes, similar to the chlorine–benzene system case. Geometric parameters of the complexes are presented in Figure 13.4. The stabilities of these

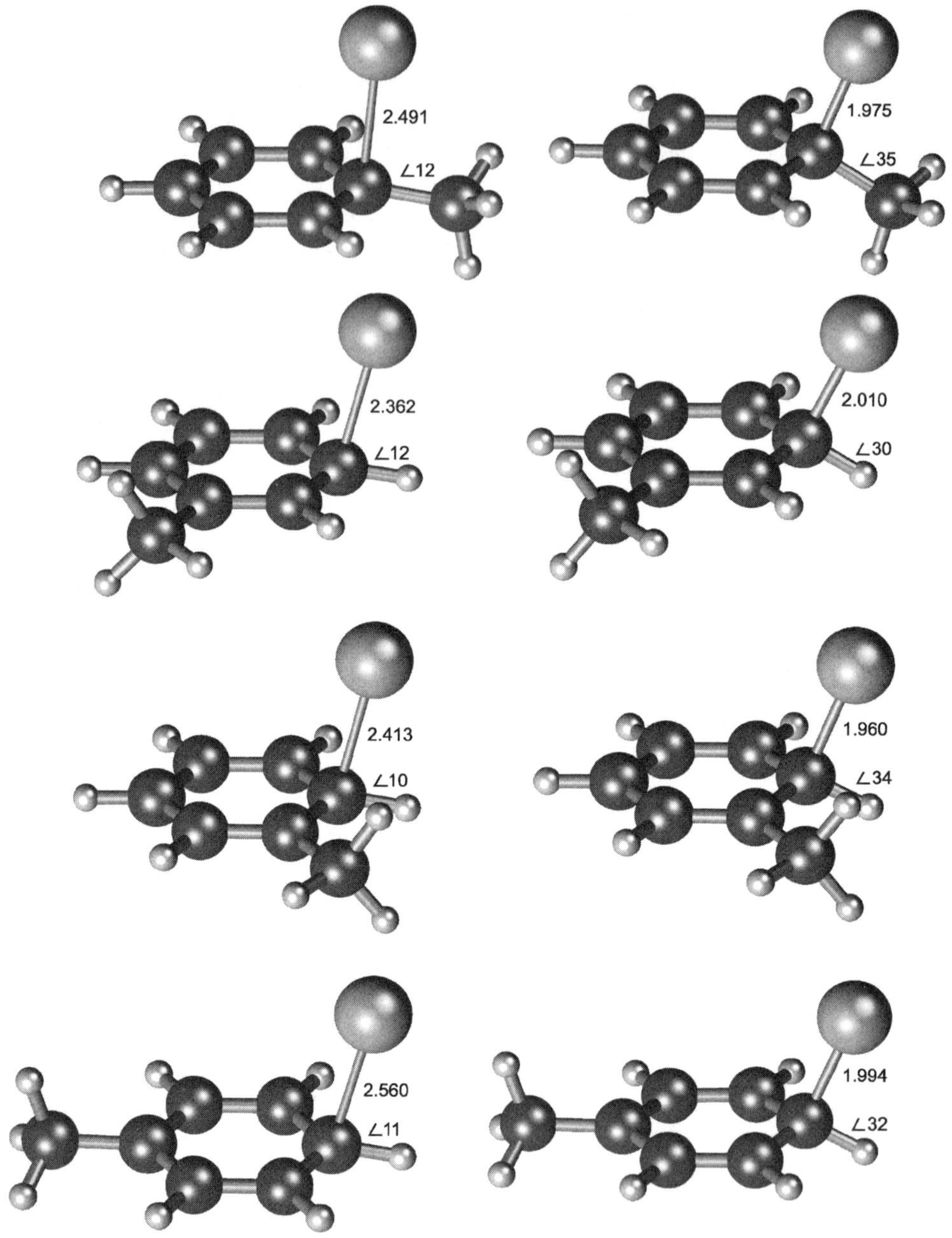

FIGURE 13.4 Structures of toluene–chlorine complexes.

complexes fall in quite narrow ranges of 6.9–8.9 and 7.5–9.4 kcal/mol (Table 13.2) for π- and σ-complexes, respectively. The *ortho* position gives the most stable structures for both kinds of complexes with *ipso* isomers following in second. It is consistent with results of Uc *et al.* [64], who calculated adducts of toluene with

Cl atoms and found the same stabilization energy dependence on structure of the complexes at the QCISD(T)//MP2/6-31G** level of theory.

Another pathway for chlorine atom reactions with toluene, in contrast to the benzene case, is reaction with the methyl group. Abstraction of the hydrogen atom from the methyl group is much more energetically favourable than hydrogen abstraction from benzene ring, −10.8 vs. 11.5 kcal/mol (Table 13.2), and does not have any energy barrier. This leads to an exothermic reaction. Moreover, due to the lower energy of reaction products in comparison to complexes of toluene and chlorine, the complex formation has a little impact on the overall reaction (Figure 13.5).

3.2 Experimental section

3.2.1 Cl + benzene reaction

Compared to toluene, benzene is much less reactive towards Cl-atoms. There are only six studies of this reaction, and these are inconsistent with each other [16,17, 19,24,27,42]. In all studies except one, a relative rate technique was used to determine reaction rate constants at room temperature. Atkinson and Aschmann [16] irradiated Cl_2–benzene–*n*-butane–air mixtures and from the relative loss rates of benzene and *n*-butane obtained the value of $(1.5 \pm 0.9) \times 10^{-11}$ $cm^3/(molecule\ s)$ for the Cl-atom reaction rate of benzene. Wallington *et al.* [17] and Noziere *et al.* [19] estimated an upper limit of 4×10^{-12} cm^3 and 5.1×10^{-16} $cm^3/(molecule\ s)$ correspondingly for this reaction using Cl_2–benzene–methane–N_2 mixtures. Shi *et al.* [24] have measured the rate constant to be $(1.3 \pm 0.3) \times 10^{-15}$ $cm^3/(molecule\ s)$ monitoring the relative loss rates of benzene and CF_2ClH in the air atmosphere. Sokolov *et al.* [27] estimated $(1.3 \pm 1.0) \times 10^{-16}$ $cm^3/(molecule\ s)$ value from the relative rate experiments of benzene–CF_2ClH–Cl_2–N_2 mixtures. A recent study by Alecu *et al.* [42] used a laser flash photolysis resonance fluorescence technique to monitor the Cl concentration. They determined the rate constant to be $6.4 \times 10^{-12} \exp(-18.1\ kJ/mol/RT)$ $cm^3/(molecule\ s)$ over 578–922 K temperature range.

We performed extensive studies to determine the Cl-atom rate constant of benzene. *n*-Butane and ethane were among first molecules that were used as references to measure the relative rate constant of Cl-atom reaction of benzene. By the end of these experiments, the concentration of reference molecules reduced close to the detection limits whereas no decay of benzene outside the uncertainty limits was observed. Although the relative rate technique is particularly powerful in the measurement of reactions with comparable rate constants, for reactions with very different rate constants, this method is less accurate. Since it has become clear that the Cl-atom rate constant of benzene is slow, chloromethane, dichloromethane, and trichloromethane (chloroform) were used as references. These sets of experiments consisted of 20 individual experiments in which the concentration of hydrocarbons ranged from 10 to 15 mTorr and the concentration of Cl_2 varied from 10 to 100 mTorr. The relative rate constants for chloromethane and dichloromethane have been previously measured by Niki *et al.* [65]. Specifically, they were combined as a check on the experimental procedures. No difference in the values of the rate

TABLE 13.2 Relative energy, enthalpy and free energy (kcal/mol) of toluene–chlorine complexes and products of toluene–chlorine reaction

	PBE0/aug-pc1				CCSD(T)/aug-CC-pV*DZ*			
	ΔE	ΔH (298)	ΔG (298)	K_c (298) 10^{-16}	ΔE	ΔH (298)	ΔG (298)	K_c (298) 10^{-19}
Ipso-π-$C_6H_6CH_3$–Cl	−11.6	−11.9	−3.9	0.30	−7.9	−8.2	−0.2	0.61
Ortho-π-$C_6H_6CH_3$–Cl	−13.4	−13.7	−5.7	6.79	−8.9	−9.1	−1.2	3.21
Meta-π-$C_6H_6CH_3$–Cl	−11.7	−11.9	−4.5	0.88	−6.9	−7.1	0.3	0.25
Para-π-$C_6H_6CH_3$–Cl	−12.7	−12.9	−5.6	5.71	−7.5	−7.7	−0.5	0.90
Ipso-σ-$C_6H_6CH_3$–Cl	−11.3	−11.8	−2.7	0.04	−9.0	−9.6	−0.5	0.95
Ortho-σ-$C_6H_6CH_3$–Cl	−13.5	−13.9	−5.6	5.01	−9.4	−9.8	−1.5	5.12
Meta-σ-$C_6H_6CH_3$–Cl	−11.8	−12.2	−3.9	0.31	−7.5	−7.9	0.4	0.20
Para-σ-$C_6H_6CH_3$–Cl	−12.7	−12.9	−5.3	3.22	−7.9	−8.2	−0.6	1.06
$C_6H_5CH_2 + HCl$	−14.3	−14.0	−14.7		−10.8	−10.5	−11.2	
Ortho-$C_6H_4CH_3 + HCl$	7.9	8.5	6.3		10.6	11.2	9.0	
Meta-$C_6H_4CH_3 + HCl$	8.0	8.6	13.1		10.6	11.2	15.7	
Para-$C_6H_4CH_3 + HCl$	8.7	9.2	14.2		11.2	11.7	16.7	
$C_6H_5CH_2Cl + H$	16.0	16.4	17.9		18.4	18.8	20.3	

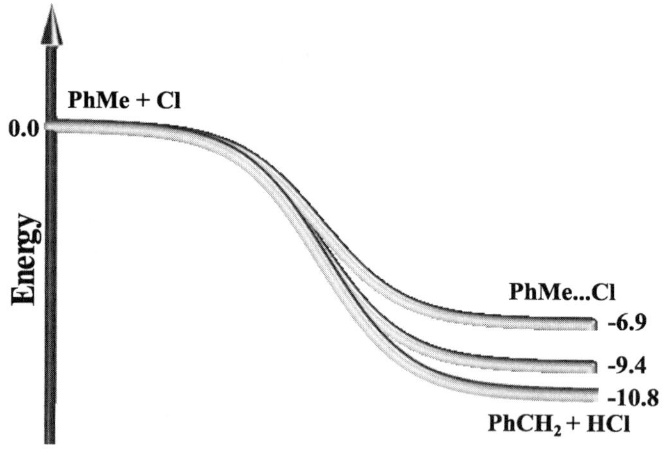

FIGURE 13.5 Energy diagram of chlorine–toluene.

constant ratios for chlorinated methanes was observed, but curvature in the benzene plot was always present. Also, the samples were aged and no dark reaction with Cl_2 molecules resulting in the further decay or reformation of benzene was observed.

Shown in Table 13.3 are the results of these experiments for chlorinated methanes. The ratio of the $k_{\text{Cl-chloromethane}}/k_{\text{Cl-dichloromethane}}$ measured in this study is in excellent agreement with the ratio of 1.31 ± 0.14 obtained by Niki *et al.* [65]

TABLE 13.3 Relative rate constants for the Cl-atom reactions of chlorinated methanes at $T = 297 \pm 2$ K

$k_i/k_{CH_2Cl_2}$	This work	Literature
$i = CH_3Cl$	1.43 ± 0.06	1.31 ± 0.14 [65]
$i = CHCl_3$	0.317 ± 0.016	

using an FTIR technique. These measured ratios can be placed on the absolute basis using the value of $5.15(\pm 0.49) \times 10^{-13}$ $cm^3/(molecule\ s)$ for the Cl-atom reaction coefficient of dichloromethane [66]. Thus the values of $7.36(\pm 0.31) \times 10^{-13}$ and $1.63(\pm 0.08) \times 10^{-13}$ $cm^3/(molecule\ s)$ for Cl-atom reaction rate constant of chloromethane and trichloromethane were obtained. However, using the value of 2.83×10^{-13} $cm^3/(molecule\ s)$ for $Cl + CH_2Cl_2$ derived by Knox [67] leads to the values of $4.05(\pm 0.17) \times 10^{-13}$ and $8.97(\pm 0.05) \times 10^{-14}$ for the Cl-atom reaction coefficients of chloromethane and trichloromethane, respectively.

In absence of secondary reactions, plotting $\ln([HC]_0/[HC]_t)$ against ln([dichloromethane]$_0$/[dichloromethane]$_t$) according to (Eq. (3)) should yield straight lines and zero intercepts for all molecules studied (where HC = chloromethane or trichloromethane). The $\ln([HC]_0/[HC]_t)$ plots are linear with the r^2 of 0.99 and 0.93, respectively. The linearity of these plots suggests that secondary reactions are negligible. In contrast to chlorinated methanes, the benzene plot curves distinctly indicating there are other reactions involved. In the experiments where only chloromethane and benzene were present, the curvature in the benzene plot was always observed. We thus performed a long-path FTIR study using methane as the reference molecule as shown in Figure 13.6.

Experiments were carried out in ca. 700 Torr air and N_2. The experiments were carried out to a maximum conversion of 85% for methane and 24% for benzene. The analysis of IR spectra was performed by analyzing the absorption features of methane in the wave length region of 2844–3202 cm^{-1} and 1199–1429 cm^{-1}. For benzene the wave length regions 3012–3140, 634–714, 1453–1512 and 1005–1068 cm^{-1} were used in order to minimize any potential complications due to the impurities in the benzene. Figure 13.6 shows plots of ln([benzene]$_0$/[benzene]$_t$) vs ln([methane]$_0$/[methane]$_t$) in N_2 and air systems. In N_2 system, only negligible degradation of benzene (<5%) was observed while 85% of methane was reacted. In the air system, a clear degradation of benzene following a formation of a plateau at higher conversion was observed. This behaviour is ascribed to the generation of HO radicals in the system yielding to secondary reactions. The potential formation of HO in the air system, in presence of molecules such as methane has been shown to be feasible due to further reactions of molecules such as CH_3OOH:

$$CH_4 + Cl \rightarrow CH_3 + HCl,$$

$$CH_3 + O_2 \rightarrow CH_3O_2,$$

$$CH_3O_2 + HO_2 \rightarrow CH_3OOH + O_2.$$

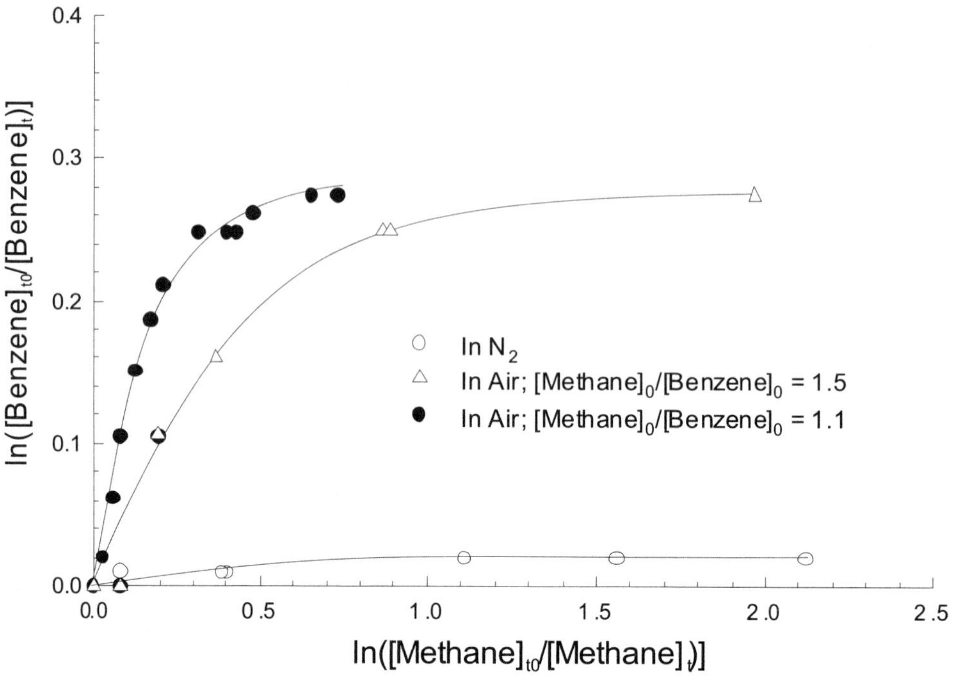

FIGURE 13.6 Plot of observed benzene loss vs. methane loss in air and N_2.

The formation of CH_3OOH in significant amounts has been reported in the Cl-atom initiated oxidation of methane [68]. In the presence of Cl-atoms in the system, there are two possible hydrogen atom abstraction channels from CH_3OOH. Cl-atoms can abstract hydrogen either from OOH or methyl groups. There is evidence in the literature for the importance of both channels. If the abstraction from CH_3 group takes place, CH_2OOH radicals will be formed in the system:

$$CH_3OOH + Cl \rightarrow CH_2OOH + HCl.$$

The fate of this radical is not well established experimentally. If it is unstable, it dissociates to form HCHO and HO [68,69]:

$$CH_2OOH \rightarrow CH_2O + HO.$$

Therefore, in analogy with methane system, in the course of experiments as Cl-atom reacts with chloro or dichloro-methane, there is a potential for formation of HO-radicals. Hence in the initial stages of the reaction at lower rates of methane conversion we expect that HO will primarily attack benzene. As the reaction proceeds, oxidation products of CH_4 and benzene increase in concentration and compete for the HO radicals. Thus, we expect to observe a curvature in relative rate plot of benzene at longer irradiation times, as shown in Figure 13.6. In FTIR studies, for the case of experiments in nitrogen diluent, curvature was observed but substantially lower concentrations of benzene were consumed. It is noteworthy that the nitrogen carrier gas was not free of oxygen and no further purification of benzene besides trap-by-trap distillation was performed. Thus, we suspect that small but observable decrease in benzene concentration is indeed due to the formation of HO radicals formed in the system. Clearly, the extent of the HO formation in nitrogen system is significantly smaller than in air, as depicted in Figure 13.6. We performed a modeling studies using ACCUCHEM [70] kinetic model, and observed that the observed benzene behaviour depicted in Figure 13.6 can likely be due to HO radicals [57].

We performed relative rate experiments using only trichloromethane (chloroform) and benzene in presence of Cl-radicals in N_2 and in air. In nitrogen experiments, the observed benzene decay was insignificant under our experimental uncertainties. In contrast to the air system, where the benzene reactant was not purified using the technique of Sokolov *et al.* [18], benzene decay was observed. However, upon further purification according to Sokolov *et al.*, no significant change in rate was observed. This indicates the low significance of secondary reactions in such relatively slow reactions. Based on our observations, a conservative upper limit estimate of 1×10^{-15} $cm^3/(molecule\,s)$ is estimated. This value is in agreement with Wallington *et al.* [17] and Noziere *et al.* [19] and Sokolov *et al.* [27]. Illustrated in Figure 13.7 is an example of product studies of Cl+benzene reactions using GC-MS analysis. The only identified products in the course of the reaction of benzene with chlorine atoms are chlorobenzene and HCl, in agreement with previous literature data [27] and our theoretical results.

3.2.2 Cl + toluene reaction

We used two reference molecules, ethane and *n*-butane. Equation (3) can be used to determine the relative rate of Cl + toluene with respect to the reference molecules ethane and *n*-butane. In this case, "HC" is toluene and RH is either ethane or *n*-butane. Figures 13.8 and 13.9 are plots of these experiments. All these plots are linear with $r^2 \geqslant 0.99$ and with zero intercepts within 2σ. Depicted in Table 13.4 are the results of these experiments compared with the literature values. This work is in excellent agreement with that of Wallington *et al.* [17] Since our results in air and N_2 are identical within experimental uncertainties, these values are combined together. The ratio of (0.977 ± 0.026) and (0.267 ± 0.01) are obtained using ethane and *n*-butane, respectively. Using the absolute values of 5.7×10^{-11} and 1.94×10^{-10} $cm^3/(molecule\,s)$, we obtain the values of $(5.57 \pm 0.15) \times 10^{-11}$ and $(5.18 \pm 0.19) \times 10^{-10}$ $cm^3/(molecule\,s)$ for the Cl-atom reactions of toluene using ethane and *n*-butane, respectively. It is to be noted that

TABLE 13.4 Relative rate constants for the Cl-atom reactions of toluene at $T = 296 \pm 2$ K[a]

$k_{toluene}/k_j$	In N_2	In air	Literature
j = ethane	0.987 ± 0.040	1.06 ± 0.03	0.981 ± 0.049 [17]
j = *n*-butane	0.259 ± 0.050	0.250 ± 0.006	0.299 ± 0.018 [16]

a Quoted experimental uncertainties are 2σ.

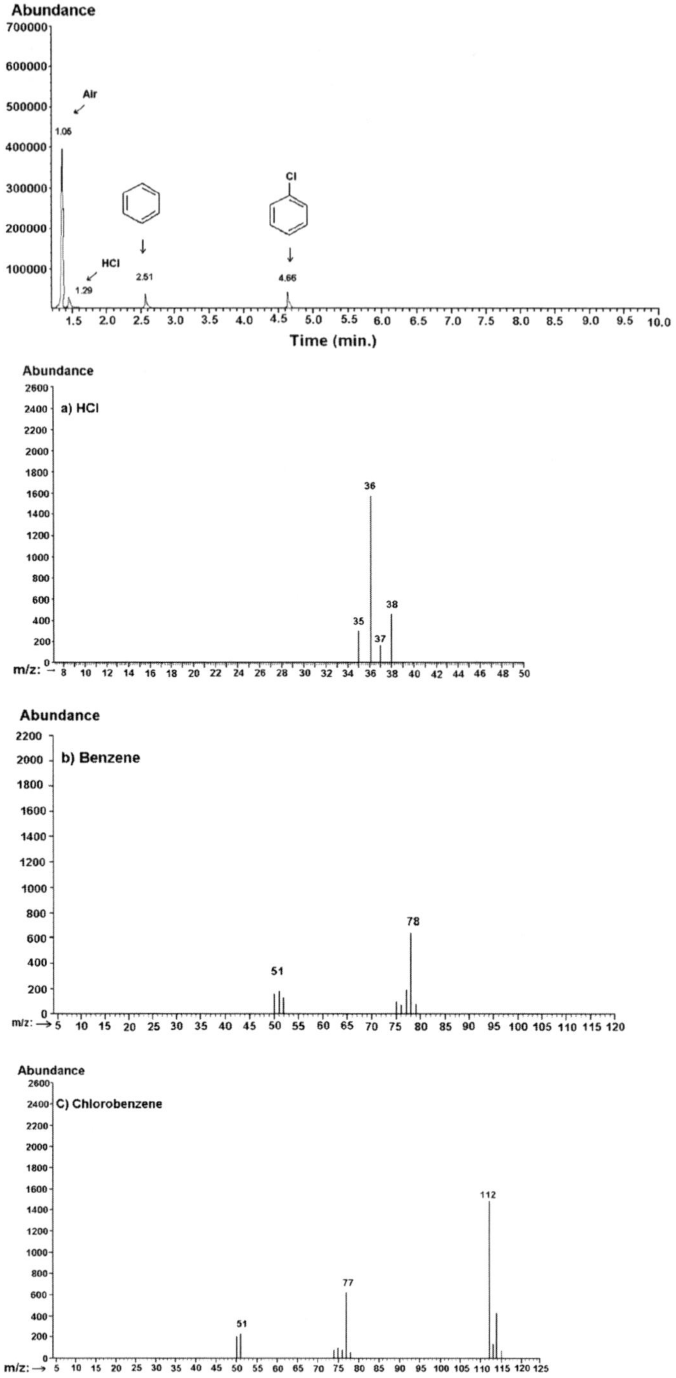

FIGURE 13.7 Gas chromatogram and mass spectra of the product formed in the reaction of benzene with chlorine radicals in the air, at 298 K.

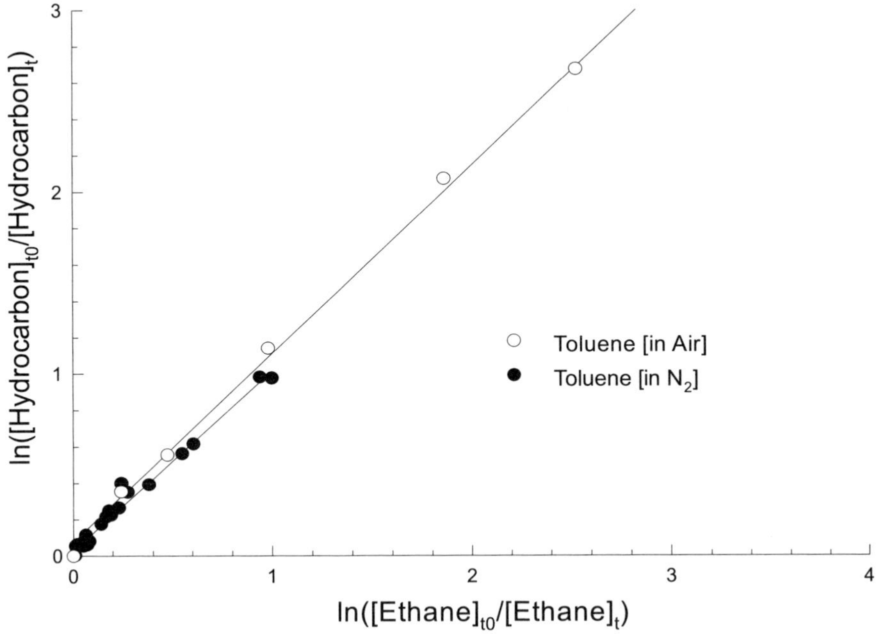

FIGURE 13.8 Plots of observed loss of toluene vs. ethane loss in air and N_2.

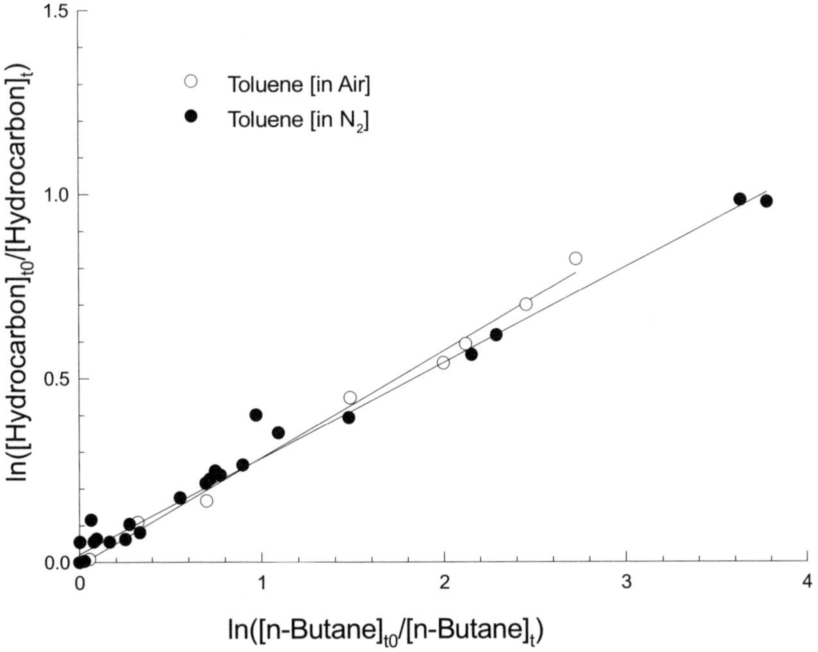

FIGURE 13.9 Plots of observed loss of toluene vs. *n*-butane loss in air and N_2.

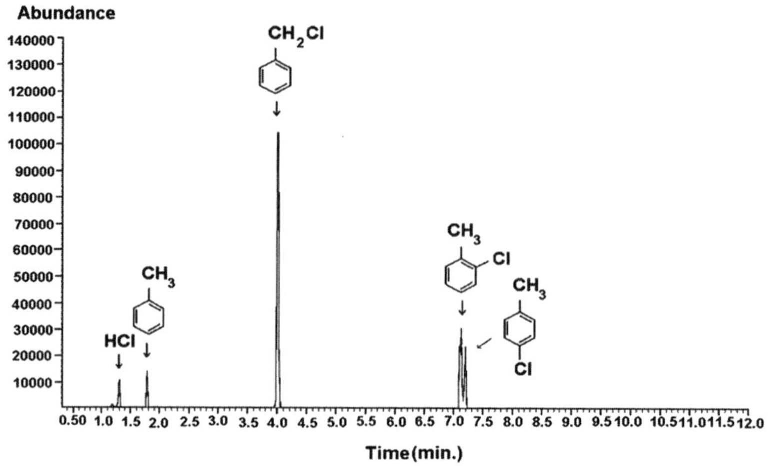

FIGURE 13.10 Gas chromatogram and mass spectra of the product formed in the reaction of toluene with chlorine radicals in the air, at 298 K.

the absolute rate constant of Cl + ethane is well established. However, to our knowledge there is only one published absolute measurement for Cl + *n*-butane. We obtain a value of 1.94 (±11%) × 10^{-10} $cm^3/(molecule\,s)$ from linear regression of measured alkane relative ratios. The Cl-atom rate constants of ethane and toluene are closer together than *n*-butane; hence the ratios measured relative to ethane are more precise. Thus, it is more logical to report the calculated value of $(5.57 \pm 0.15) \times 10^{-11}$ $cm^3/(molecule\,s)$ for the Cl + toluene reaction obtained using the ethane absolute rate constant. We found that chloromethylbenzene is the major product formed in the reaction, with traces of chlorotoluenes (Figure 13.10) from a GC-MS analysis. This is in complete agreement with our aforementioned theoretical calculations. Benzaldehyde and benzyl alcohol, the latter of which has been reported by other authors [19,20,25,41], were not identified in our product study.

4. CONCLUSIONS

We observed consistent theoretical and experimental results on Cl-atom initiated oxidation of benzene and toluene. Based on CCSD(T)/aug-CC-pVDZ//PBE0/aug-pc1 theoretical calculations, three stationary structures have been found for the benzene–chlorine atom complex: hexahapto-complex, π-complex and σ-complex. The first one contains a first order saddle point and is not stable. A potential energy surface scan proves this point and shows that it is a transition state between two opposite π-complexes. Toluene forms eight complexes with the chlorine atom: *ipso-*, *ortho-*, *meta-* and *para*-isomers for both π- and σ-complexes. Benzene and toluene can react with the Cl atom in two ways: a hydrogen abstraction of benzene and toluene by the Cl atom, found to be endothermic and barrierless, or the hydrogen substitution reaction, which by contrast has relatively high energies

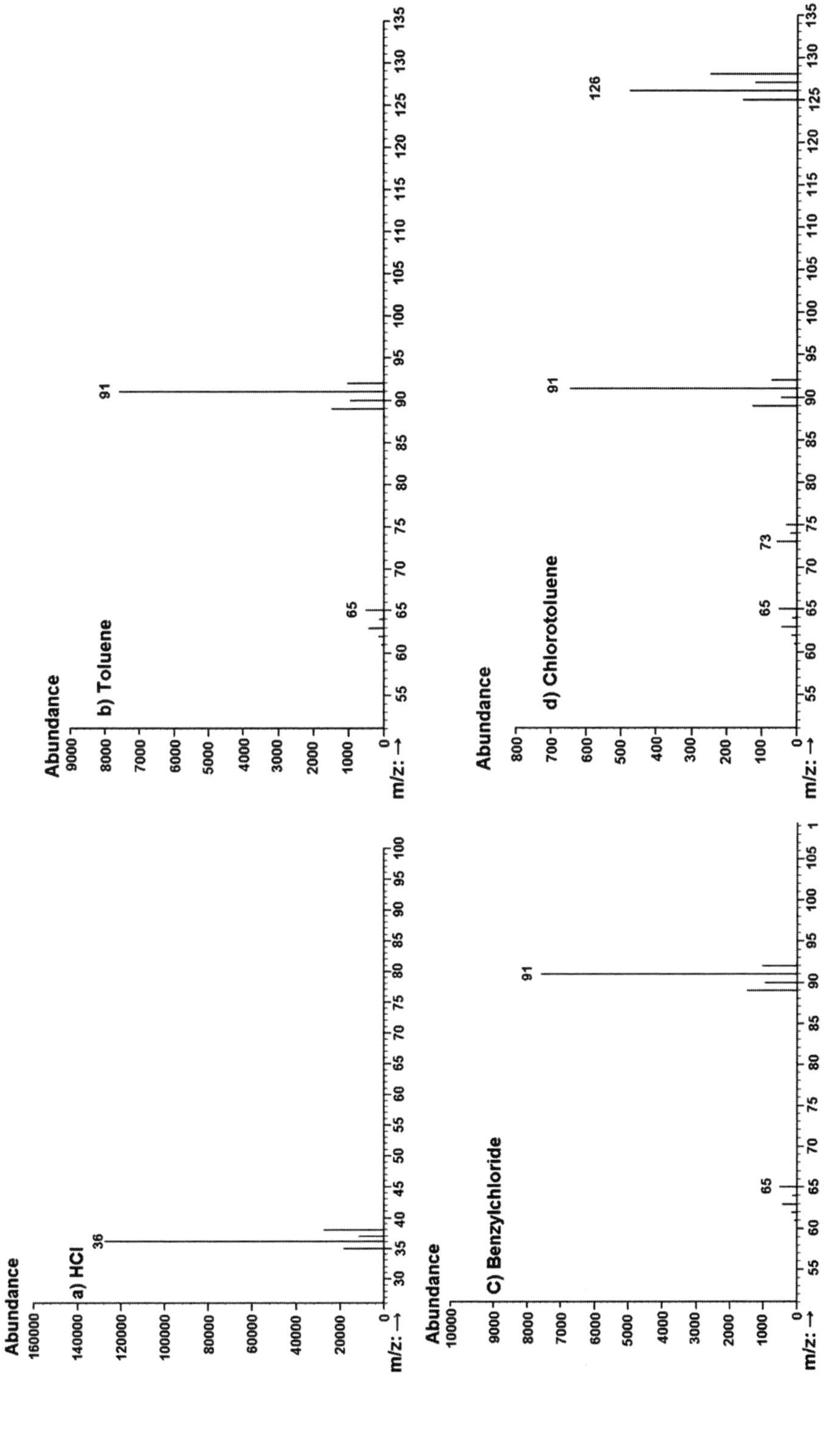

FIGURE 13.10 *(Continued.)*

of activation. Therefore, for benzene and toluene, only the H abstraction channel is significant; our GC-MS product study confirms this conclusion. In benzene, the reaction rate is much slower as Cl may only abstract hydrogen atoms on the benzene ring itself, in contrast to toluene where H-abstraction favors the methyl group. Our experimental results complement our theoretical work in predicting reactions of Cl + toluene are indeed much faster than chlorine initiated reactions of benzene, and agree well with other experimental rate data [16–19,24,25,38]. The established difference between the reactivity of Cl + benzene and Cl + toluene can be employed to distinguish between HO and Cl reactions under particular tropospheric conditions such as Arctic polar sunrise.

REFERENCES

[1] J.G. Calvert, R. Atkinson, K.H. Becker, R.M. Kamens, J.H. Seinfeld, T.H. Wallington, G. Yarwood, *The Mechanisms of Atmospheric Oxidation of the Aromatic Hydrocarbons*, Oxford University Press, New York, 2002.
[2] R. Atkinson, J. Arey, *Chem. Rev. (Washington, DC)* **103** (2003) 4605.
[3] T.M. Sack, D.H. Steele, K. Hammerstrom, J. Remmers, *Atmos. Environ. Part A: Gen. Top.* **26A** (1992) 1063.
[4] T. Eikmann, S. Eikmann, T. Goen, *Umweltmedizin in Forschung und Praxis* **5** (2000) 309.
[5] N.R. Reed, W.A. Reed, K. Weir, I. Encomienda, L.M. Beltran, Health risk assessment and applied action level of toluene, Dep. Environ. Toxicol., Univ. California, Davis, CA, USA, FIELD URL (1989).
[6] V.A. Benignus, *Neurotoxicology* **2** (1981) 567.
[7] D. Johnson, M.E. Jenkin, K. Wirtz, M. Martin-Reviejo, *Environ. Chem.* **2** (2005) 35.
[8] W. Dechapanya, A. Eusebi, Y. Kimura, D.T. Allen, *Environ. Sci. Technol.* **37** (2003) 3671.
[9] W. Dechapanya, A. Eusebi, Y. Kimura, D.T. Allen, *Environ. Sci. Technol.* **37** (2003) 3662.
[10] H. Takegawa, *Toyota Chuo Kenkyusho R&D Rebyu* **38** (2003) 57.
[11] H.J.L. Forstner, R.C. Flagan, J.H. Seinfeld, *Environ. Sci. Technol.* **31** (1997) 1345.
[12] W.C. Keene, *Environ. Chem. (Dordrecht)* **1** (1995) 363.
[13] K.W. Oum, M.J. Lakin, D.O. DeHaan, T. Brauers, B.J. Finlayson-Pitts, *Science (Washington, DC)* **279** (1998) 74.
[14] P.A. Ariya, H. Niki, G.W. Harris, K.G. Anlauf, D.E. Worthy, *J. Atmos. Environ.* **33** (1999) 931.
[15] B.J. Finlayson-Pitts, J.N. Pitts, *Chemistry of the Upper and Lower Atmosphere: Theory, Experiments, and Applications*, Academic Press, San Diego, CA, 1999.
[16] R. Atkinson, S.M. Aschmann, *Int. J. Chem. Kinet.* **17** (1985) 33.
[17] T.J. Wallington, L.M. Skewes, W.O. Siegl, *J. Photochem. Photobiol. A: Chem.* **45** (1988) 167.
[18] F. Markert, P. Pagsberg, *Chem. Phys. Lett.* **209** (1993) 445.
[19] B. Noziere, R. Lesclaux, M.D. Hurley, M.A. Dearth, T.J. Wallington, *J. Phys. Chem.* **98** (1994) 2864.
[20] R. Seuwen, P. Warneck, [Report] EUR European Commission (1994) 137.
[21] S.M. Aschmann, R. Atkinson, *Int. J. Chem. Kinet.* **27** (1995) 613.
[22] T. Nordmeyer, W. Wang, M.L. Ragains, B.J. Finlayson-Pitts, C.W. Spicer, R.A. Plastridge, *Geophys. Res. Lett.* **24** (1997) 1615.
[23] M.L. Ragains, B.J. Finlayson-Pitts, *J. Phys. Chem. A* **101** (1997) 1509.
[24] J. Shi, M.J. Bernhard, *Int. J. Chem. Kinet.* **29** (1997) 349.
[25] G. Fantechi, N.R. Jensen, O. Saastad, J. Hjorth, J. Peeters, *J. Atmos. Chem.* **31** (1998) 247.
[26] A. Mellouki, *J. Chim. Phys. Phys.-Chim. Biol.* **95** (1998) 513.
[27] O. Sokolov, M.D. Hurley, T.J. Wallington, E.W. Kaiser, J. Platz, O.J. Nielsen, F. Berho, M.T. Rayez, R. Lesclaux, *J. Phys. Chem. A* **102** (1998) 10671.
[28] J. Stutz, M.J. Ezell, A.A. Ezell, B.J. Finlayson-Pitts, *J. Phys. Chem. A* **102** (1998) 8510.
[29] B.J. Finlayson-Pitts, C.J. Keoshian, B. Buehler, A.A. Ezell, *Int. J. Chem. Kinet.* **31** (1999) 491.

[30] C.A. Taatjes, *Int. Rev. Phys. Chem.* **18** (1999) 419.
[31] S. Coquet, P.A. Ariya, *Int. J. Chem. Kinet.* **32** (2000) 478.
[32] S. Hewitt, C. Quant, B. Nguyen, C. Frez, M. Luu, D. Robichaud, *Prepr. Ext. Abstr. ACS Nat. Meet., Am. Chem. Soc., Div. Environ. Chem.* **41** (2001) 1185.
[33] Q.K. Timerghazin, P.A. Ariya, *Phys. Chem. Chem. Phys.* **3** (2001) 3981.
[34] M.G. Bryukov, I.R. Slagle, V.D. Knyazev, *J. Phys. Chem. A* **106** (2002) 10532.
[35] M.J. Ezell, W. Wang, A.A. Ezell, G. Soskin, B.J. Finlayson-Pitts, *Phys. Chem. Chem. Phys.* **4** (2002) 5813.
[36] H.-B. Qian, D. Turton, P.W. Seakins, M.J. Pilling, *Int. J. Chem. Kinet.* **34** (2002) 86.
[37] D. Sarzynski, B. Sztuba, *Int. J. Chem. Kinet.* **34** (2002) 651.
[38] J.D. Smith, J.D. DeSain, C.A. Taatjes, *Chem. Phys. Lett.* **366** (2002) 417.
[39] W. Wang, M.J. Ezell, A.A. Ezell, G. Soskin, B.J. Finlayson-Pitts, *Phys. Chem. Chem. Phys.* **4** (2002) 1824.
[40] W. Wang, M.J. Ezell, A.A. Ezell, G. Soskin, B.J. Finlayson-Pitts, in: AMS Annual Meeting, 83rd, Long Beach, CA, United States, Feb. 9–13, 2003, *Combined Preprints CD-ROM* (2003), p. 370.
[41] L. Wang, J. Arey, R. Atkinson, *Environ. Sci. Technol.* **39** (2005) 5302.
[42] I.M. Alecu, Y. Gao, P.C. Hsieh, J.P. Sand, A. Ors, A. McLeod, P. Marshall, *J. Phys. Chem. A* **111** (2007) 3970.
[43] J.P. Perdew, K. Burke, M. Ernzerhof, *Phys. Rev. Lett.* **77** (1996) 3865.
[44] M. Ernzerhof, G.E. Scuseria, *J. Chem. Phys.* **110** (1999) 5029.
[45] F. Jensen, *J. Chem. Phys.* **115** (2001) 9113.
[46] F. Jensen, T. Helgaker, *J. Chem. Phys.* **121** (2004) 3463.
[47] J. Cizek, *Adv. Chem. Phys.* **14** (1969) 35.
[48] G.D. Purvis, R.J. Bartlett, *J. Chem. Phys.* **76** (1982) 1910.
[49] J.A. Pople, M. Headgordon, K. Raghavachari, *J. Chem. Phys.* **87** (1987) 5968.
[50] T.H. Dunning, *J. Chem. Phys.* **90** (1989) 1007.
[51] M.J. Frisch, *et al.*, *GAUSSIAN 98*, Revision A. 7; Gaussian Inc., Pittsburgh, PA, 1998.
[52] H.-J. Werner, P.J. Knowles, R. Lindh, M. Schutz, *et al.*, *MOLPRO*, version 2002.6, a package of *ab initio* programs, 2003.
[53] P.A. Hooshiyar, H. Niki, *Int. J. Chem. Kinet.* **27** (1995) 1197.
[54] P.A. Ariya, A. Khalizov, A. Gidas, *J. Phys. Chem. A* **106** (2002) 7310.
[55] F. Raofie, P.A. Ariya, *Environ. Sci. Technol.* **38** (2004) 4319.
[56] B. Pal, P.A. Ariya, *Phys. Chem. Chem. Phys.* **6** (2004) 572.
[57] P.A. Ariya, Studies of tropospheric halogen chemistry: Laboratory and field measurements, PhD thesis, York University, 1996.
[58] P.A. Ariya, V. Catoire, R. Sander, H. Niki, G.W. Harris, *Tellus B* **49** (1997) 583.
[59] V. Catoire, P.A. Ariya, H. Niki, G.W. Harris, *Int. J. Chem. Kinet.* **29** (1997) 695.
[60] R.J. Cvetanovic, D.L. Singleton, G. Paraskevopoulos, *J. Phys. Chem.* **83** (1979) 50.
[61] J. Chen, V. Young, P.A. Hooshiyar, H. Niki, M.D. Hurley, *J. Phys. Chem.* **99** (1995) 4071.
[62] M.-L. Tsao, C.M. Hadad, M.S. Platz, *J. Am. Chem. Soc.* **125** (2003) 8390.
[63] G.A. Russell, H.C. Brown, *J. Am. Chem. Soc.* **77** (1955) 4031.
[64] V.H. Uc, A. Hernandez-Laguna, A. Grand, A. Vivier-Bunge, *Phys. Chem. Chem. Phys.* **4** (2002) 5730.
[65] H. Niki, P.D. Maker, C.M. Savage, L.P. Breitenbach, *Int. J. Chem. Kinet.* **12** (1980) 1001.
[66] D.D. Davis, W. Braun, A.M. Bass, *Int. J. Chem. Kinet.* **2** (1970) 101.
[67] J.H. Knox, *Transact. Faraday Soc.* **58** (1962) 275.
[68] T.J. Wallington, J.M. Andino, A.R. Potts, P.H. Wine, *Chem. Phys. Lett.* **176** (1991) 103.
[69] D. Shen, A. Moise, H.O. Pritchard, *J. Chem. Soc., Faraday Trans.* **91** (1995) 1425.
[70] W. Braun, J.T. Herron, D.K. Kahaner, *Int. J. Chem. Kinet.* **20** (1988) 51.

CHAPTER 14

Tropospheric Chemistry of Aromatic Compounds Emitted from Anthropogenic Sources

Jean M. Andino[1*] and **Annik Vivier-Bunge**[**]

Contents

1. Introduction	298
2. Reactions	298
2.1 Kinetic rate constants	298
2.2 Mechanisms	299
3. Summary: Areas for Future Work	309
References	309

Abstract

The kinetics and mechanisms associated with the atmospheric photooxidation of aromatic compounds emitted from anthropogenic sources are of seminal importance in the chemistry of the urban and regional atmosphere. Aromatic compounds readily react with hydroxyl radicals to lead to ozone and aerosol formation. However, over the years, difficulties have existed in unambiguously identifying the stable species formed. Thus, only 60–70% of the reacted carbon has been fully accounted for. This article summarizes the major advances that have been made towards elucidating the atmospheric chemistry of anthropogenic aromatic hydrocarbons using computational chemistry. In addition, the computational data are compared to experimental data, and areas for future advances in the community's understanding of aromatic reactions through the use of computational chemistry calculations are discussed.

[*] Arizona State University, Department of Civil & Environmental Engineering and Department of Chemical Engineering, PO Box 875306, Tempe, AZ 85287-5306, USA
E-mail: jean.andino@asu.edu

[**] Área de Química Cuántica, Departamento de Química, Universidad Autónoma Metropolitana, Ixtapalapa, C.P. 09340, Mexico D.F., Mexico
E-mail: annik@xanum.uam.mx

[1] Corresponding author.

Advances in Quantum Chemistry, Vol. 55
ISSN 0065-3276, DOI: 10.1016/S0065-3276(07)00214-6

1. INTRODUCTION

Aromatic compounds are of great abundance in the troposphere, comprising approximately 20% of the composition of non-methane volatile organic carbon [1]. It is well known that these compounds are abundant constituents of motor vehicle emissions [2], and several air sampling studies have been conducted that positively identify aromatic compounds in ambient air [1, and references therein]. Based on the data obtained, it is clear that aromatic compounds comprise a significant portion of the total non-methane organic carbon (NMOC) in many parts of the world. The percentages of aromatic hydrocarbons varies depending on the urban, suburban, or rural nature of the area.

While much research has been conducted in recent years on the reactions of biogenic ring structured hydrocarbons (i.e. α-pinene, and β-pinene, among others) [3–10], in comparison, much less work has been conducted on the reactions of aromatic hydrocarbons (i.e. toluene, *m*-xylene, *p*-xylene, etc.) [1, and references therein]. Recent modeling efforts have underscored the importance of aromatic compounds in relation to the chemistry of the free troposphere [11]. Also, recent chemical mechanism modeling work on aromatic compounds has suggested the need for additional studies to more accurately assess the reactions of aromatic compounds, since existing mechanisms cannot adequately predict the yields of products observed in experimental systems [12–15]. While laboratory data have provided some data to elucidate the mechanisms of these aromatic compounds, computational chemistry efforts have helped to further propel the understanding of these reactions. This article provides a review of the computational work that has been published regarding aromatic compounds, while simultaneously relating this information to existing experimental data to identify data gaps and areas of future need.

2. REACTIONS

2.1 Kinetic rate constants

The primary tropospheric oxidants are $^{\bullet}OH$, O_3, and NO_3, with $^{\bullet}OH$ and O_3 reactions with hydrocarbons dominating primarily during daytime hours, and NO_3 reactions dominating at night. Rate constants for the reactions of many different aromatic compounds with each of the aforementioned oxidants have been determined through laboratory experiments [16]. The rate constant data as well as atmospheric lifetimes for the reactions of toluene, *m*-xylene, *p*-xylene, *m*-ethyltoluene, and 1,2,4-trimethylbenzene appear in Table 14.1. Only these particular aromatic compounds will be discussed in this review paper, since much of the computational chemistry efforts have focused on these compounds. When considering typical atmospheric concentrations of the major atmospheric oxidants, OH, O_3, and NO_3 of 1.5×10^6, 7×10^{11}, and 4.8×10^8, molecules cm^{-3}, respectively [17], combined with the rate constants, it is clear that the major atmospheric loss process for these selected aromatic compounds is reaction with the hydroxyl

TABLE 14.1 Overall rate constants and calculated lifetimes (τ) for the OH, O_3, and NO_3 reactions of selected aromatic compounds. Rate constant data are from Ref. [16]. Lifetimes are calculated assuming atmospheric concentrations for OH, O_3, and NO_3 of 1.5×10^6, 7×10^{11}, and 4.8×10^8 molecules cm^{-3}, respectively [17]

Compound	OH reactions		O_3 reactions		NO_3 reactions	
	k at 298 K (cm^3 $molecule^{-1}$ s^{-1})	τ (days)	k at 298 K (cm^3 $molecule^{-1}$ s^{-1})	τ (days)	k at 298 K (cm^3 $molecule^{-1}$ s^{-1})	τ (days)
Toluene	5.63×10^{-12}	1.37	$<1 \times 10^{-20}$	>1653	7×10^{-17}	344
m-Xylene	23.1×10^{-12}	0.33			2.6×10^{-16}	93
p-Xylene	14.4×10^{-12}	0.54			5.0×10^{-16}	48
m-Ethyltoluene	18.6×10^{-12}	0.41			Not studied	
1,2,4-Trimethyl-benzene	32.5×10^{-12}	0.24			1.8×10^{-15}	13

radical. This same conclusion can be drawn for other aromatic compounds as well as hydrocarbons in general. Thus, much of the past work that has been published on the reactions of aromatic hydrocarbons, either from an experimental or a computational perspective, focuses on their reactions with the OH radical.

2.2 Mechanisms

Despite the prevalence of aromatic compounds in the atmosphere, much work is still required to fully elucidate the atmospheric reaction mechanisms of these compounds. Approximately 60–70% of the reacted carbon can be fully accounted for as products (either in the gas or aerosol phases) in aromatic/OH systems [1,16].

Aromatics react with OH radicals either through an abstraction mechanism to form water and eventually an aromatic aldehyde, or through addition to form an aromatic–OH adduct. The aromatic–OH adduct subsequently reacts with O_2 or NO_2 to form a series of stable ring-retained products or aromatic–OH peroxy radicals. The aromatic–OH peroxy radicals subsequently react to form intermediate radicals that eventually undergo ring fragmentation to form smaller oxygenated organic compounds. Products identified in laboratory studies of the toluene–OH, *m*-xylene–OH, *p*-xylene–OH, and 1,2,4-trimethylbenzene–OH systems appear in Table 14.2.

Several experiments have been conducted to characterize the nature, mechanism, and extent of aerosol formation from gas-to-particle conversion [18–25]. Despite all of these studies, much information has yet to be elucidated. Because of the inability to fully account for all of the reacted carbon through experimental studies, computational studies have been performed to more fully understand the possible reaction mechanisms, thereby providing guidance to experimentalists on the nature of compounds expected to form.

2.2.1 Initial OH attack on the Aromatic ring

The aromatic–OH reaction proceeds initially either by OH addition to or abstraction from the aromatic ring to form a radical (Reaction 1a) or an aromatic–OH adduct (Reaction 1b). Note that the "R" in all of the reactions presented subsequently refers either to a hydrogen atom or an alkyl group. The depiction is meant

TABLE 14.2 Products detected in laboratory studies of the OH-initiated reactions of selected aromatic compounds [1]

Parent aromatic compound	Ring retained products	Ring fragmented products
Toluene	*o*-Cresol	Glyoxal
	m-Cresol	Methyl glyoxal
	p-Cresol	Methylbutenedial
	m-Nitrotoluene	Hydroxymethylbutenedial
	o-Nitrotoluene	Oxoheptadienal
	p-Nitrotoluene	Methylhydroperoxide
	Benzaldehyde	Formaldehyde
	2-Methyl-*p*-benzoquinone	Hexadienyl
	Benzyl alcohol	Hydroxyoxoheptadienyl
	Benzyl nitrate	Maleic anhydride
m-Xylene	2,4-Dimethylphenol	Glyoxal
	2,6-Dimethylphenol	Methylglyoxal
	4-Nitro-*m*-xylene	Formaldehyde
	5-Nitro-*m*-xylene	3-Methyl-5-furanone
	m-Tolualdehyde	4-Oxo-2-pentenal
	3-Methylbenzyl nitrate	
p-Xylene	2,5-Dimethylphenol	Glyoxal
	2-Nitro-*p*-xylene	Methylglyoxal
	p-Tolualdehyde	3-Hexene-2,5-dione
	4-Methylbenzyl nitrate	Formaldehyde
	2,5-Dimethyl *p*-benzoquinone	2-Methylbutenedial
	2,5-Dimethylfuran	
1,2,4-Trimethylbenzene	2,4-Dimethylbenzaldehyde	3-Hexene-2,5-dione
	2,5-Dimethylbenzaldehyde	3-Methyl-3-hexene-2,5-dione
	3,4-Dimethylbenzaldehyde	2-Methylbutenedial
	2,4,5-Trimethylphenol	Biacetyl
	2,3,5-Trimethylphenol	Methylglyoxal
	2,3,6-Trimethylphenol	Glyoxal

to allow for discussion of a generalized mechanism:

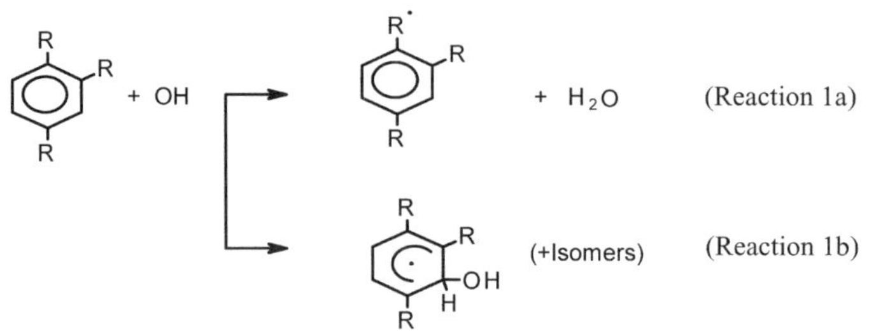

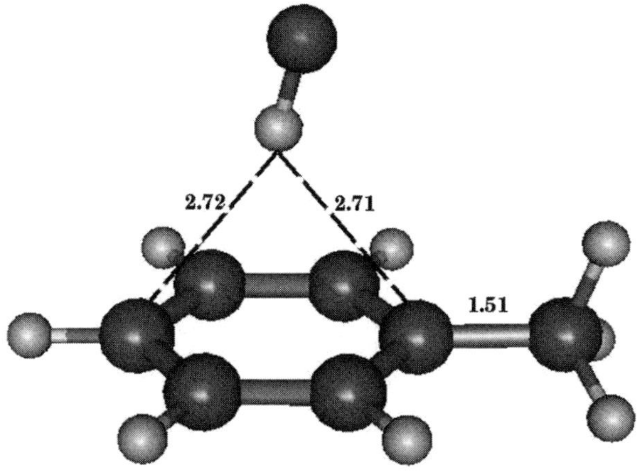

FIGURE 14.1 MP2/6-311G** optimized OH–toluene pre-reactive complex. The indicated interatomic distances are in units of Angstroms.

The OH interaction (either via an addition or abstraction route) with toluene has been studied using computational chemistry [26–29]. Uc *et al.* [28] specifically showed that the mechanism for OH interaction with toluene involves the formation of a stable pre-reactive complex (depicted in Figure 14.1) when the radical approaches the aromatic ring at a van der Waals distance.

Many radical-molecule reaction mechanisms have been shown to be complex and to involve a fast equilibrium between the reactants and the pre-reactive complex, followed by the irreversible formation of products. An example of this mechanism for toluene is shown in Reactions 1bi and 1bii, respectively,

$$C_6H_5CH_3 + OH \underset{k_{-1}}{\overset{k_1}{\rightleftharpoons}} [C_6H_5CH_3\cdots OH], \qquad \text{(Reaction 1bi)}$$

$$[C_6H_5CH_3\cdots OH] \rightarrow C_6H_5(OH)CH_3. \qquad \text{(Reaction 1bii)}$$

The formation of these complexes has often been discussed [30], especially in radical–molecule reactions where their existence is needed to explain observed negative activation energies. It has been shown [31] that pre-reactive complexes are fundamental in that they guide the reaction from the beginning, and also have a strong effect on the reaction barrier height. Pre-reactive complexes may be expected to be responsible for the selectivity of the aromatic–OH adducts formed in (Reaction 1bii). Indeed, considerable spin density is transferred from the radical to the complex [32]. Smith and Ravishankara [33] have described these complexes, which are generally too short-lived to be detected, and whose existence can only be inferred by the way they affect the overall behavior of the reaction. However, in recent work by Davey *et al.* [34], a π-type hydrogen-bonded complex between the hydroxyl radical and acetylene leading to the addition reaction has been stabilized in the reactant channel well. This complex has been characterized by infrared action spectroscopy in the OH overtone region spectrum.

A recent theoretical study by Uc *et al.* [29] on the abstraction pathway for toluene (Reaction 1a), utilized DFT-BHandHLYP calculations with a 6-311++G(d,p) basis set to determine the energetics followed by rate constant determinations using transition state theory. These results indicated a theoretical rate constant for abstraction of an H atom from toluene of 2.4×10^{-13} molecules $cm^3\,s^{-1}$ at 298.15 K. As expected, the results indicated that abstraction of the hydrogen atom preferentially occurred at the substituent methyl group, rather than on the aromatic ring. This calculated theoretical value for abstraction of a hydrogen atom from toluene is consistent with the determination (from experimental studies) that abstraction is a relatively minor pathway for the OH/toluene system, occurring less than approximately 10% of the time for methyl benzenes [1,16,17]. High temperature rate constants for the OH reactions with benzene and toluene have also been obtained theoretically using transition state theory and an energy profile calculated using the G3(MP2)//B3LYP and CBS-QB3 methods [35].

Several computational studies have been conducted on (Reaction 1b) to examine the relative importance of the initial OH addition to the *ipso, ortho, meta,* and *para* sites on the aromatic ring. A summary of the computational approaches used by different authors and the reaction energies calculated for the OH-adduct formation for the case of toluene appear in Table 14.3. The data indicate that (a) there are several possible likely sites of attack, (b) energy differences between the adduct isomers are very small (i.e. in general less than 2 kcal mol^{-1}), (c) the level of theory used to perform the geometry optimizations and energy calculations may influence results considerably, and (d) the inclusion of zero point energy corrections may change the predicted dominance of a particular OH attack site.

Addition at the ipso site had been assumed to be unlikely because of steric hindrance arguments. Therefore, past studies by Bartolotti and Edney [26] and Andino *et al.* [27] had not focused on this pathway. For comparison, *ortho* and *ipso* toluene–OH adducts are represented in Figure 14.2 as calculated at the MP2/6-311++G** level. Several recent calculations have emphasized the possible role of ipso adducts in toluene [29,36,37] in xylenes [38,39], and in another aromatic (phenol [40]). Rather than a steric hindrance, a stabilization due to interactions between the lone pair on the oxygen atom and two methyl hydrogen atoms is observed in some of the recent studies.

To place the results of Table 14.3 into context, it is important to consider experimental data. Based on temperature dependent rate constants, an experimental value for the reaction energy associated with OH radical addition to the toluene ring is -16.5 ± 5 kcal mol^{-1} [41]. With the exception of the data of Bartolotti and Edney [26], all of the computational results fall within the range of values determined in a laboratory setting. It is well known that there is experimental evidence for the formation of *o*-, *m*-, and *p*-cresol from the toluene–OH adduct, with *o*-cresol dominating formation. Given that the formation pathway for the cresols proceeds through the formation of the toluene–OH adduct in (Reaction 1b), as discussed subsequently, the dominance of *o*-cresol suggests the dominance of attack at the *or-*

TABLE 14.3 Reaction energies (kcal mol^{-1}) for the formation of the *ipso*-, *ortho*-, *meta*-, and *para*-toluene/OH adducts

Method basis set	Reaction energies (kcal mol^{-1})				
	Ipso	*Ortho*	*Meta*	*Para*	References
DFT (DMOL) (*Modified basis set*)	Not calculated	43.480*	−41.361*	−41.999*	Bartolotti and Edney [26]
UHF-PM3//B3LYP-DFT *6-31G(d,p)*	Not calculated	−18.4*	−16.8*	−17.2*	Andino *et al.* [27]
B3LYP *6-31G(d,p)*	−19.47*	−20.91*	−19.21*	−19.83*	Uc *et al.* [29]
	−16.33	−17.56	−16.04	−16.71	Uc *et al.* [36], Suh *et al.* [37]
PMP2 *6-31G(d,p)*	−19.65*	−19.03*	−17.13*	−17.37*	Uc *et al.* [29]
CCSD(T)/*6-31G(d)*// B3LYP/*6-31G(d,p)*	−15.54	−14.09	−13.02	−13.33	Suh *et al.* [37]
BHandHLYP *6-311G(d,p)*	−13.41	−13.74	−12.32	−12.57	Uc *et al.* [32]
CCSD/T)/*6-311G(d,p)*// MP2/*6-311G(d,p)*	−14.71	−13.50	−11.95	−12.08	Uc *et al.* [32]

* Numbers marked with an asterisk do not include zero-point corrections.

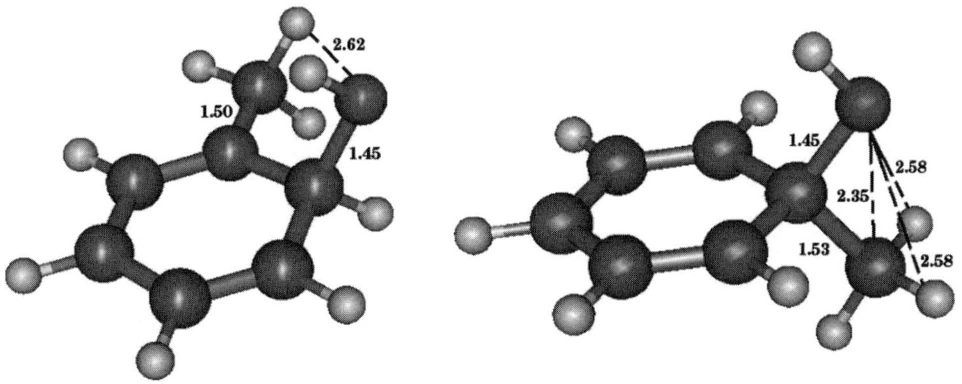

FIGURE 14.2 MP2/6-311++G** optimized OH–toluene *ortho* and *ipso* adducts. Indicated interatomic distances are in Angstroms.

tho site. It is important to note that attack at the ipso site cannot lead to the prompt formation of a cresol. Therefore, simply looking at cresol yields is insufficient to examine the relative importance of addition at each of the sites on the aromatic

TABLE 14.4 Summary of published work on aromatic systems other than toluene studied using computational approaches

System	Predicted initial OH attack site (percentages, if stated)	Method	References
o-Xylene + OH	*Ipso* (0.99)	MP2	Uc *et al.* [38]
	Ipso (0.78), *ortho* (0.07), *para* (0.13)	B3LYP	Uc *et al.* [38]
m-Xylene + OH	*Ortho*	B3LYP	Andino *et al.* [27]
	Ortho (0.61), *ipso* (0.30)	B3LYP and MP2	Uc *et al.* [38]
p-Xylene + OH	*Ortho*	B3LYP	Andino *et al.* [27]
	Ortho (0.23), *ipso* (0.77)	MP2	Uc *et al.* [38]
	Ortho (0.81), *ipso* (0.19)	B3LYP	Uc *et al.* [38]
	Ortho (0.8), *ipso* (0.2)	B3LYP	Fan and Zhang [39]
1,2,4-Trimethyl-benzene + OH	*Ortho*	B3LYP	Andino *et al.* [27]
	Ipso	MP2 and BH and HLYP	Uc [32]
m-Ethyltoluene	*Ortho*	B3LYP	Andino *et al.* [27]

ring. However, it is important to also emphasize that a 100% carbon balance on the reacted aromatic and products has not yet been achieved in laboratory studies for any of the aromatic/OH systems.

New reaction pathways for the *ipso* toluene–OH adduct have been proposed [28]. Elimination of the methyl group on toluene is not an expected route, and, in order to stabilize, the *ipso* adduct probably transforms. Hydroxyl radical addition at the *ipso* site of toluene has been postulated to lead to formation of *o*-cresol through an OH shift from the *ipso* to the *ortho* position on the ring [28]. This formation might be used to explain the relatively large amount of observed products that originate from the *ortho* toluene–OH adduct. Calculations were performed using the MP2/6-311G** method to examine the direct migration of OH from the *ipso* to the *ortho* position on the ring [32]. A large barrier, of 22 kcal mol^{-1} was obtained, suggesting that this is not a likely pathway and that migration, if it occurs, must be catalyzed, perhaps, by another molecule (such as oxygen). Tiecco [42] has suggested that in addition to a shift of the OH from the *ipso* to the *ortho* position, radical *ipso* intermediates could evolve in several ways: *ipso* substitution, return to the starting products, coupling with other radicals, and the rearrangement and fragmentation of a group remote from the *ipso* position. Additional computational work is needed to further examine these theories.

Formation of *o*-cresol is consistent with the primary products that have been identified in toluene–OH chamber studies. It is important to note, however, that cresols quickly react with NO_3, thus leading to the prompt formation of additional ring-retained components. Thus, additional computational studies that specifically examine the formation of products from the subsequent reactions of the primary products with OH and NO_3 are needed.

Computational studies of the OH interactions with other methyl substituted aromatic compounds have been performed by only a few research groups [27,32,38,39]. Table 14.4 details the systems studied and the major results published in the literature.

Based on the data in Table 14.4, OH radical attack appears to occur at either the *ortho* or the *ipso* positions on the aromatic rings. The predicted position of attack appears to vary based on the method used to calculate the energies (i.e. MP2 versus B3LYP), as is clearly evident in the *o*-xylene/OH and *p*-xylene/OH systems. In all cases, the MP2 method tends to favor the *ipso* adduct much more than DFT.

2.2.2 Aromatic–OH adduct reaction with O_2

Under atmospheric conditions, oxygen is expected to rapidly react with the aromatic–OH adduct, forming either an alcohol (Reaction 2a) or a peroxy radical (Reaction 2b):

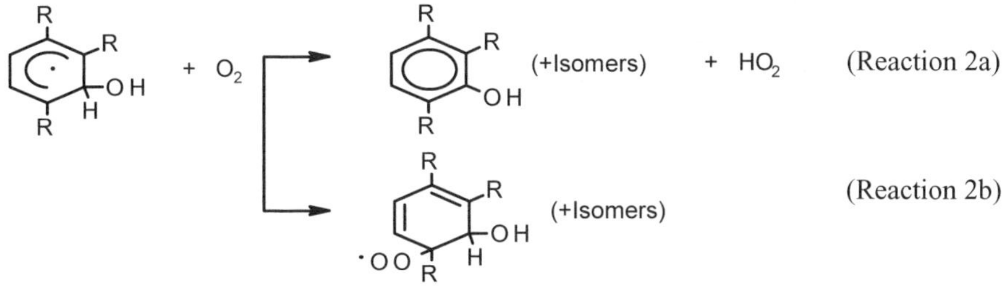

Several computational chemistry studies have focused on the reactions of O_2 with aromatic–OH adducts [27,43–45]. These studies are in good agreement, indicating that the aromatic–OH adduct will add O_2 to form the corresponding peroxy radical, as seen in (Reaction 2b).

2.2.3 Reactions of the OH–aromatic adduct with NO and NO_2

The aromatic OH adduct formed in (Reaction 1b) can react with NO_2 according to a series of different pathways (Reactions 3a–3e). Experimental data have indicated the formation of stable nitroaromatic products (stable product formed in

(Reaction 3a)) from chamber studies:

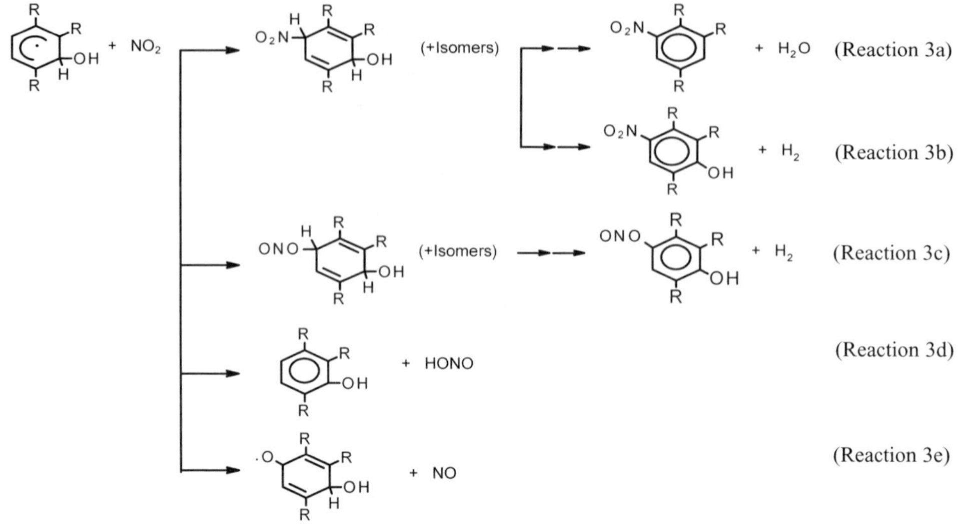

While several studies have also pointed out that nitroaromatic products are more characteristic of laboratory studies where high NO_x mixing ratios are employed rather than the ambient environments where lower NO_x mixing ratios exist, understanding the formation mechanisms of these nitroaromatic compounds is critical to properly defining the reaction mechanisms for aromatic compounds. The only computational study to date that probed the NO_2 reaction with the aromatic–OH adducts is that of Andino *et al.* [27], and their work considered only the reactions of the most likely aromatic–OH adduct, as predicted from the computational results. However, theoretical work involving NO_x tends to be complicated by symmetry problems, so that reaction energy profiles are difficult to obtain.

2.2.4 Reactions of the peroxy radical

The peroxy radical formed in (Reaction 2b) is expected to cyclicize to form a corresponding bicyclic radical (Reaction 4) or react with nitric oxide via an abstraction mechanism to form the corresponding aromatic oxy radicals (Reaction 5a):

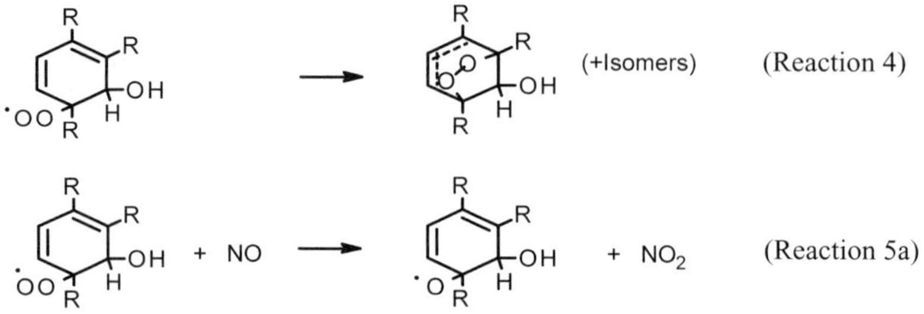

The only studies to date that examined the competing pathways for the peroxy radical are those of Andino *et al.* [27] (focused on the peroxy radicals from toluene, *m*-xylene, *p*-xylene, 1,2,4-trimethylbenzene, and *m*-ethyltoluene), García

Cruz *et al.* [43] and Suh *et al.* [46] (focused on toluene), and Fan and Zhang [39] focused on *p*-xylene). Computational chemistry studies predict the preferential cyclization of the peroxy radical over reaction with NO to form the aromatic oxy radical. Cyclization to specifically form a 5-member ring structure is highly preferred, as might be expected by considering the potential ring strain that would form with a 4-member ring, or the relatively large ring that would form with a 6-member structure. It is important to point out that past studies of the reactions of the aromatic peroxy radicals considered addition of NO to the peroxy radical to form an aromatic nitrate species. For completeness, computational chemistry studies to examine the potential formation of aromatic nitrate species through addition of NO to the peroxy radical according to (Reaction 5b) would be important:

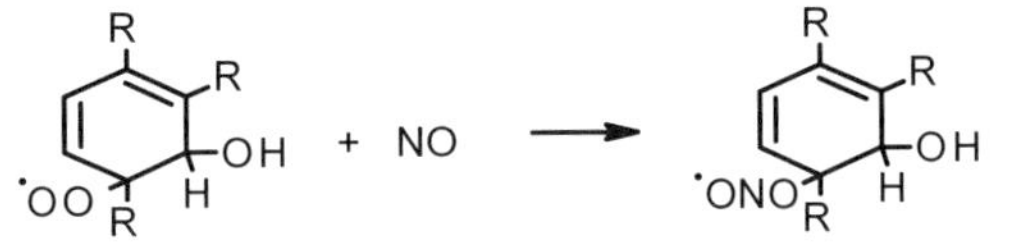

(Reaction 5b)

2.2.5 Reactions of the bicyclic radical

The bicyclic radical is expected to quickly add O_2 (Reaction 6) to form a bicyclic peroxy radical which subsequently reacts with NO to form a bicyclic oxy radical (Reaction 7). It is the subsequent reaction of the bicyclic oxy radical that leads to the formation of ring fragmentation products (Reaction 8):

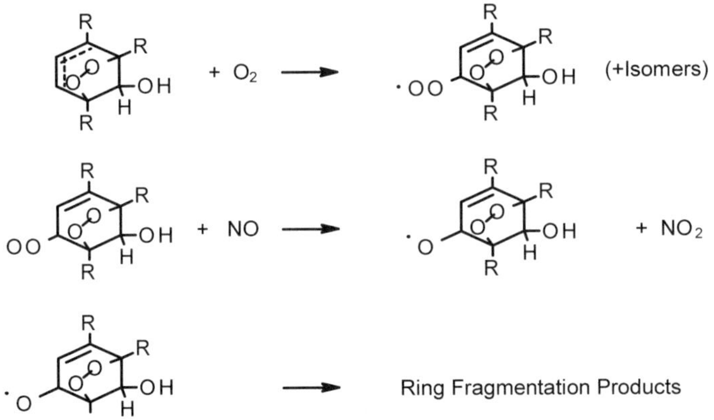

(Reaction 6)

(Reaction 7)

(Reaction 8)

The only fundamental computational chemistry studies of the subsequent reactions of the bicyclic radicals that have been performed are for the toluene–OH and *p*-xylene–OH systems [46,47]. These results examined not only the potential for Reactions 6 and 7 to occur with the toluene/OH and *p*-xylene/OH systems, but the potential for isomerization of the bicyclic radicals to form stable epoxide radicals. In the toluene/OH system, the formation of the epoxide radicals was predicted to be unlikely given that the barriers to formation were relatively high. Computational studies of the *p*-xylene/OH system predicted the likely formation of the epoxide radicals. Clearly, additional aromatic systems must be studied to verify other potential pathways for reaction. This is probably one of the single most needed areas of research required to further elucidate the reactions of aromatic–OH systems, especially considering that recent chemical mechanism modeling work by Bloss *et al.* [12,13] shows poor correlation between predicted results and experimental data for the ring fragmented unsaturated dicarbonyl products.

2.2.6 Formation of secondary organic aerosol

During the gas-phase reactions of aromatic species, numerous products are formed that have relatively low vapor pressures so as to lead to the formation of secondary organic aerosols (SOA) either through nucleation or condensation processes. As indicated previously, laboratory chamber studies have been conducted in the past to identify the nature of the aerosols formed from the gas-phase photooxidation of aromatic species, either in the presence or absence of seed aerosols. Studies have, in general, found that a variety of factors, including the concentration of NO_x, influence the generation of aerosols. However, the chemical mechanisms of SOA formation are not well understood, in part because of the highly reactive nature of the first generation products.

It is expected that SOA formation will originate from a variety of pathways, including cases where:

(I) A gas-phase product of the OH–aromatic reaction simply forms a liquid particle because of its vapor pressure, in the absence of any "seed" aerosol.
(II) A "seed" aerosol exists and the gas is taken up onto, but not into the aerosol.
(III) The same as condition (ii), but now the gas is taken up into the aerosol, thus leaving more room for uptake on the surface.
(IV) Multiple gas-phase products from the OH–aromatic reaction interact to form a critical cluster, and an aerosol is formed.

In general, one might expect the larger ring-retained products that are formed from the reactions of aromatics to have relatively low vapor pressures, and therefore form SOA through mechanism I or IV, whereas the presence of smaller, non-aromatic ring compounds in the aerosol are likely due to mechanisms II or III. Experimental studies exist that detect both larger ring-retained compounds (e.g. benzaldehyde and 2-methyl-4-nitrophenol) and smaller, more volatile non-ring retained compounds (e.g. 3-methyl, 2,5-furanedione, dihydro 2,5-furanedione, and 2,5-furanedione) in the SOA formed from the photooxidation of toluene [18,21]. However, it is important to note that the smaller non-ring retained compounds were not detected in studies where the overall aerosol mass was lower [24], thus providing support for mechanism II (or III). In addition, others have postulated the direct formation of aerosols from aromatic photooxidation that consist of oligomeric forms of polyketone ring-fragmented products [21,24]. These oligomers are believed to decompose to their monomeric forms during the course of typical laboratory analyses.

3. SUMMARY: AREAS FOR FUTURE WORK

Although many advances have been made in understanding the tropospheric reactions of anthropogenic aromatic compounds, additional work is clearly needed. Specific areas of foci for future closely coordinated computational and laboratory-based studies are in the areas of:

- the relative importance of aromatic–OH-addition pathways,
- the gas-phase reactions of ring-retained products,
- secondary organic particle formation mechanisms.

As indicated previously, computational studies have indicated the importance of OH addition to multiple positions on a methylbenzene ring. To further characterize the importance of OH addition to each position, computational chemistry studies that further probe the potential products of addition to all positions (including the ipso position) are needed. Specifically, potential pathways that lead to the generation of ring-retained products that may preferentially form low volatility gas-phase products that partition to form aerosols should be probed.

Of critical need are computational chemistry studies that help to explain the formation of secondary organic aerosols in aromatic–OH systems. Two types of computational studies are needed: ones that focus on the generation of ring-retained semi-volatile products, and those that focus on examining the energies of attraction of a gas-phase molecule to a particle surface. Since computational chemistry studies are now being conducted to examine the properties of nanomaterials [48], it seems plausible that these same techniques could be applied to study the potential for secondary organic aerosol formation and atmospheric transformation, considering the interactions of a gas-phase specie with a spherical particle comprised of elemental carbon, organic carbon (such as ones that might be formed according to Cases I or IV), or inorganic composition (such as ammonium sulfate, sodium chloride, or mineral dust i.e. typical aerosols found in the ambient). In each case, computational chemistry calculations could assist in determining the resulting components on the surface, extent of interaction of additional material with the surface, or the extent of surface coverage of gases onto seed aerosols.

REFERENCES

[1] J.G. Calvert, R. Atkinson, K.H. Becker, R.M. Kamens, J.H. Seinfeld, T. Wallington, G. Yarwood, *The Mechanisms of Atmospheric Oxidation of Aromatic Hydrocarbons*, Oxford University Press, New York, 2002.

[2] J.J. Schauer, M.J. Kleeman, G.R. Cass, B.R. Simoneit, *Environ. Sci. Technol.* **36** (2002) 1169.

[3] S.M. Aschmann, R. Atkinson, J. Arey, *J. Geophys. Res.* **107** (D14) (2002), doi:10.1029/2001JD001098 107.

[4] R.J. Griffin, D.R. Cocker III, R.C. Flagan, J.H. Seinfeld, *J. Geophys. Res.* **104** (1999) 3555.

[5] B. Noziere, I. Barnes, K.H. Becker, *J. Geophys. Res.* **104** (1999) 23645.

[6] A. Reisell, C. Harry, S.M. Aschmann, R. Atkinson, J. Arey, *J. Geophys. Res.* **104** (1999) 13869.

[7] J.J. Orlando, B. Noziere, G.S. Tyndall, G. Orzechowska, S.E. Paulson, Y. Rudich, *J. Geophys. Res.* **105** (2000) 11561.

[8] A. Alvarado, E.C. Tuazon, S.M. Aschmann, R. Atkinson, J. Arey, *J. Geophys. Res.* **103** (1998) 25541.

[9] J. Baker, S.M. Aschmann, J. Arey, R. Atkinson, *Int. J. Chem. Kinet.* **34** (2002) 73.
[10] S.M. Aschmann, J. Arey, R. Atkinson, *Atmos. Environ.* **36** (2002) 4347.
[11] C.L. Heald, D.J. Jacob, R.J. Park, L.M. Russell, B.J. Huebert, J.H. Seinfeld, H. Liao, R.J. Weber, *Geophys. Res. Lett.* **32** (2005) L18809, doi:10.1029/2005GL023831.
[12] V. Bloss, C. Wagner, A. Bonzanini, M.E. Jenkin, K. Wirtz, M. Martin-Reviejo, M.J. Pilling, *Atmos. Chem. Phys.* **5** (2005) 623.
[13] C. Bloss, V. Wagner, M.E. Jenkin, W.J. Bloss, J.D. Lee, D.E. Heard, K. Wirtz, M. Martin-Reviejo, M.J. Pilling, *Atmos. Chem. Phys.* **5** (2005) 641.
[14] V. Wagner, M.E. Jenkin, S.M. Saunders, J. Stanton, K. Wirtz, M.J. Pilling, *Atmos. Chem. Phys.* **3** (2003) 89.
[15] J.F. Hamilton, A.C. Lewis, C. Bloss, V. Wagner, A.P. Henderson, B.T. Golding, K. Wirtz, M. Martin-Reviejo, M.J. Pilling, *Atmos. Chem. Phys.* **3** (2003) 1999–2014.
[16] R. Atkinson, J. Arey, *Chem. Rev.* **103** (2003) 4605.
[17] J.H. Seinfeld, S.N. Pandis, *Atmospheric Chemistry and Physics: From Air Pollution to Climate Change*, Wiley, New York, 2006.
[18] H.J. Forstner, R.C. Flagan, J.H. Seinfeld, *Environ. Sci. Technol.* **31** (1997) 1345.
[19] J.R. Odum, T.P.W. Jungkamp, R.J. Griffin, R.C. Flagan, J.H. Seinfeld, *Science* **276** (1997) 96.
[20] K. Izumi, T. Fukuyama, *Atmos. Environ. A* **24** (1990) 1433.
[21] M.S. Jang, R.M. Kamens, *Environ. Sci. Technol.* **35** (2001) 3626.
[22] H. Takekawa, H. Minoura, S. Yamazaki, *Atmos. Environ.* **37** (2003) 3413.
[23] D.R. Cocker III, B.T. Mader, M. Kalberer, R.C. Flagan, J.H. Seinfeld, *Atmos. Environ.* **37** (2003) 3413.
[24] T.E. Kleindienst, T.S. Conover, C.D. McIver, E.O. Edney, *J. Atmos. Chem.* **47** (2004) 79.
[25] C. Song, K. Na, D.R. Cocker III, *Environ. Sci. Technol.* **39** (2005) 3143.
[26] L.J. Bartolotti, E.O. Edney, *Chem. Phys. Lett.* **245** (1995) 119.
[27] J.M. Andino, J.N. Smith, R.C. Flagan, W.A. Goddard III, J.H. Seinfeld, *J. Phys. Chem.* **100** (1996) 10967.
[28] V.H. Uc, I. Garcia-Cruz, A. Hernandez-Laguna, A. Vivier-Bunge, *J. Phys. Chem. A* **104** (2000) 7847.
[29] V.H. Uc, J.R. Alvarez-Idaboy, A. Galano, I. Garcia-Cruz, A. Vivier-Bunge, *J. Phys. Chem. A* **110** (2006) 10155.
[30] J.R. Alvarez-Idaboy, N. Mora-Diez, R. Boyd, A. Vivier-Bunge, *J. Am. Chem. Soc.* **123** (2001) 2018.
[31] S. Sekušak, K.R. Liedl, A.J. Sabljić, *J. Phys. Chem. A* **102** (1998) 1583.
[32] V.H. Uc, Ph.D. Thesis, Universidad Autónoma Metropolitana, Iztapalapa, México, D.F., México, 2003, also, paper in preparation by V. Uc, J.R. Alvarez-Idaboy, A. Vivier-Bunge.
[33] I.W.M. Smith, A.R. Ravishankara, *J. Phys. Chem. A* **106** (2002) 4798.
[34] J.B. Davey, M.E. Greenslade, M.D. Marshall, M.I. Lester, M.D. Wheeler, *J. Chem. Phys.* **121** (2004) 3009.
[35] T. Seta, M. Nakajima, A. Miyoshi, *J. Phys. Chem. A* **110** (2006) 5081.
[36] V.H. Uc, I. García-Cruz, A. Grand, A. Vivier-Bunge, *J. Phys. Chem.* **105** (2001) 6226.
[37] I. Suh, D. Zhang, R. Zhang, L.T. Molina, M.J. Molina, *Chem. Phys. Lett.* **364** (2002) 454.
[38] V.H. Uc, I. García-Cruz, A. Vivier-Bunge, in: A. Hernández-Laguna, *et al.* (Eds.), *Advanced Problems and Complex Systems*, in: *Quantum Systems in Chemistry and Physics*, vol. II, Kluwer Academic Publishers, Dordrecht, 2000, pp. 241–259.
[39] J.W. Fan, R.Y. Zhang, *J. Phys. Chem. A* **110** (2006) 7728.
[40] M.J. Lundqvist, L.A. Eriksson, *J. Phys. Chem. B* **104** (2000) 848–855.
[41] R.A. Perry, R. Atkinson, J.N. Pitts Jr., *J. Phys. Chem.* **81** (1977) 296.
[42] M. Tiecco, *Acc. Chem. Res.* **13** (1980) 51.
[43] I. García-Cruz, M. Castro, A. Vivier-Bunge, *J. Comput. Chem.* **21** (2000) 1.
[44] A.M. Mebel, M.C. Lin, *J. Am. Chem. Soc.* **116** (1994) 9577.
[45] T.H. Lay, J.H. Bozzelli, J.W. Seinfeld, *J. Phys. Chem.* **100** (1996) 6543.
[46] I. Suh, R. Zhang, L. Molina, M. Molina, *J. Am. Chem. Soc.* **125** (2003) 12655.
[47] I. Suh, J. Zhao, R. Zhang, *Chem. Phys. Lett.* **432** (2006) 313.
[48] A. Galano, *Chem. Phys.* **327** (2006) 159.

CHAPTER 15

Elementary Processes in Atmospheric Chemistry: Quantum Studies of Intermolecular Dimer Formation and Intramolecular Dynamics

Glauciete S. Maciel[*], **David Cappelletti**[**], **Gaia Grossi**[*], **Fernando Pirani**[*] and **Vincenzo Aquilanti**[1*]

Contents

1. Introduction 312
2. Major Atmospheric Components and Their Dimers 313
   2.1 The intermolecular interactions: experimental and theoretical information 314
   2.2 Potential energy surfaces: phenomenological characterization 315
   2.3 Quantum dynamics of bound states and spectral features of clusters 317
3. Aspects of Interactions Involving Minor Atmospheric Components: H_2O and H_2S 319
   3.1 Water–rare gas systems 319
   3.2 The H_2S–rare gas systems 322
4. Peroxides and Persulfides 324
   4.1 Hydrogen peroxide 325
   4.2 Other systems 326
5. Concluding Remarks and Perspectives 328
Acknowledgements 328
References 328

Abstract

The present article provides an account of recent progress in the use of quantum mechanical tools for understanding structure and processes for

[*] Dipartimento di Chimica, Università di Perugia, 06123 Perugia, Italy
[**] Dipartimento di Ingegneria Civile ed Ambientale, Università di Perugia, 06100 Perugia, Italy
[1] Corresponding author. E-mail: aquila@dyn.unipg.it

Advances in Quantum Chemistry, Vol. 55
ISSN 0065-3276, DOI: 10.1016/S0065-3276(07)00215-8

systems of relevance in atmospheric chemistry. The focus is on problems triggered by experimental activity in this laboratory on investigations of intermolecular interactions by molecular beam scattering. Regarding the major components of the atmosphere, results are summarized on dimers (N_2–N_2, O_2–O_2, N_2–O_2) where experimental and phenomenogically derived potential energy surfaces have been used to compute quantum mechanically the intermolecular clusters dynamics. Rovibrational levels and wave functions are obtained, for perspective use in atmospheric modelling, specifically of radiative absorption of weakly bound complexes. Further work has involved interactions of paramount importance, those of water, for which state-of-the-art quantum chemical calculations for its complexes with rare gases yield complementary information on the interaction (specifically the anisotropies) with respect to molecular beam scattering experiments that measure essentially the isotropic forces. Similar approaches and results have been pursued and obtained for H_2S. Stimulated in part by the interesting problem of large amplitude vibrations, such as the chirality change transitions associated with the torsional motions around O–O and S–S bonds, a systematic series of quantum chemical studies has been undertaken on systems that play also roles in the photochemistry of the minor components of the atmosphere. They are H_2O_2, H_2S_2 and several molecules obtained by substitutions of the hydrogens by alkyl groups or halogens. Quantum chemistry is shown to have reached the stage of resolving many previously controversial features regarding these series of molecules (dipole moment, equilibrium geometries, heights of barriers for torsion), which are crucial for intramolecular dynamics. Quantum dynamics calculations are also performed to compute torsional levels and the temperature dependence of their distributions.

1. INTRODUCTION

In atmospheric chemistry, recent attention has been addressed to intermolecular phenomena responsible for the formation of dimers or even of large molecular clusters, leading to measurable effects in the absorption of radiation. A book collects modern contributions relevant to these topics [1]; as amply demonstrated there, it also appears that, besides the major components of the atmosphere (O_2, N_2), also other molecules (e.g. water) need attention from this view point. The focus of our specific investigations on these issues was triggered by combining molecular beam scattering experiments on the determination of intermolecular interactions, and the quantum mechanical treatment of the cluster dynamics for the characterization of the expected spectral features: it has been in part reviewed a few years ago [2,3]. This paper is an account of the role of theory in molecular science as applied to the elucidation of atmospheric phenomena and Section 2 is dedicated to an outline of previous work and its updating. Paradigmatic examples will be the case of molecular oxygen, nitrogen and their dimers, while in Section 3 we will treat from a quantum chemical view point the subject of intermolecular forces, again in connection with recent experiments, regarding the important species H_2O, and also H_2S. Section 4 reports on studies on species like

H_2O_2 and H_2S_2 and their derivatives. For these systems, quantum chemistry has reached recently the status of permitting accurate estimates of geometries, particularly regarding features associated with torsion around the O–O and S–S bonds, and of tackling an issue of interest for chirality changing collisions. In atmospheric chemistry, interest arises in studying effects of substituents of hydrogen atoms and hydrogen peroxide and persulfide, and that of alkyl groups and halogens are being investigated by quantum chemistry. This topic, of great relevance not only for the (photo)dynamics of the atmosphere but also in combustion processes and in biochemistry, will be discussed with reference to mostly unpublished work. The paper ends with conclusion in Section 5.

2. MAJOR ATMOSPHERIC COMPONENTS AND THEIR DIMERS

Weakly interacting molecules leading to collisional complexes, and to either stable or metastable dimers, potentially play an important role in molecular surface physics, in astrophysics, in atmospheric photochemistry and physics, and climate change. High resolution spectra for van der Waals clusters can give insight into the nature of intermolecular forces and of internal dynamics, but results for the case of nonpolar molecules are limited. An account on recent progress of our understanding of the dimers of the major components of the atmosphere is presented in this section. On the basis of potential energy surfaces (PES) which have been characterized by scattering experiments for dimer formation, such as O_2–O_2 [4,5] and N_2–N_2 [6], quantum mechanical predictions of spectral features have been reported [7] by calculations of the dimer bound states. Two further papers have been published on the PES of the O_2–O_2 system [8] in the ground electronic singlet state and more subsequently also on N_2–O_2 [9]. Another paper deals with additional excited PES for these systems [10]. An interesting structural feature is that configurations corresponding to minima in the PES differ in the different cases, while the exact quantum dynamics reveals the role of zero point energy and the floppiness of the dimers.

For such dimers, the interactions involve dispersion forces (induced multipole-induce multipole), which usually are weaker than those arising from electrostatic effects (permanent multipole–permanent multipole), and their complexes rarely have transition moments active in the medium infra-red or microwave ranges. As a consequence, experimental information on the equilibrium geometry and on the bond energy even for diatomic–diatomic complexes is scarce and a full characterization of the interaction is often lacking. Therefore, only recent experiments on molecular beam scattering turned out to be crucial and a simultaneous analysis of different experimental data is recommended. In our laboratory, interference effects in scattering cross sections are measured, obtaining information on interaction potentials, which in the case of anisotropic intermolecular potentials require advanced molecular beam techniques, and possibly a control of mutual orientation. Our current knowledge on the structure of the dimers is summarized in the following, the main focus of this presentation being to emphasize the role of the dynamics and to illustrate how exact quantum mechanics provides insight in cluster formation, specifically for the diatom–diatom case.

Implications for the interpretation of recent laboratory [11,12] and atmospheric [13] spectroscopic observations, as well as current measurements of high pressure behavior of oxygen [14], are amenable to be discussed in this framework. This work provides also the ground for the interpretation of complicated band features in rotational spectra, as exemplified by the case of oxygen. Still appears to be valid the statement [15,16] that "spectral analysis alone is not sufficient to extract information on structure and bonding, and the combined use of scattering and gaseous properties information is therefore confirmed to be crucial".

2.1 The intermolecular interactions: experimental and theoretical information

We give some background experimental information available on the interactions and on theoretical studies devoted to their characterization. Since recent progress has been accounted for in [2,3], we limit ourselves here to updating the list of experimental and theoretical papers [17–37], which have dealt essentially with interactions of O_2 with itself, also in connection with modeling of liquids, solids and nanoagreggates. Some of these papers are also concerned with N_2 aggregates, which are receiving less interest. See however [38–42]. From spectroscopic studies available for N_2–N_2, the infrared spectrum [43] suggests that the equilibrium conformation is T-shaped with the center-of-mass of the monomers approximately separated by 3.7–4.2 Å. The N_2–N_2 dimer has been studied by *ab initio* methods which have provided several alternatives for the interaction potentials [44–52], yielding different results for the equilibrium geometry and the bond energy of the ground state. Previous *ab initio* methods yielded as most stable geometries a T-shaped and a Z one with bond energy D_0 in the range 70–80 cm^{-1} [53], a frequency of the stretching van der Waals mode of 22 cm^{-1} and internal rotation barriers with a maximum at 30 cm^{-1}. These examples show that, for these interactions, quantum chemistry typically gives only fair agreement with experimental results (in this case the venerable ones by Long et al [43]). Spectroscopic and theoretical information on the N_2–O_2 systems is very scarce, see e.g. [53].

Another experimental source of information is from gaseous properties, such as the second virial coefficient. Low temperatures data depend on the full anisotropic potential energy surface, while at high temperature only the spherical part of the interaction is of relevance. Data are available for O_2–O_2 in a wide temperature range [54] and accurate results have been recently derived for the N_2–N_2 systems from acoustic measurements [55]. For the N_2–O_2 systems, for which information is limited, the main features can be indirectly derived as average of those for O_2–O_2 and N_2–N_2, being negligible the correction due to the excess of the second virial coefficients for mixture, and thus exploiting the similarity of the nature of the interaction in N_2–N_2 and in O_2–O_2 [54,56].

The measurement of quantum mechanical interference effects in the velocity dependence of scattering cross-sections provides data which, together with accurate second virial coefficients, yield information on the intermolecular potential and on the dimer structure. In this laboratory the use of this technique was demonstrated for the O_2–O_2 case [4,5], for which not only the singlet, but also the triplet and quintet potential energy surfaces have been characterized. In the velocity dependence of the integral cross-section, the "glory" oscillations, overimposed to a smooth average component, are a probe of the depth of the potential well and of its position, while the absolute value of the cross-section depends on the long range attraction. In our apparatus, experiments can be performed by using "hot" molecules ($T_{rot} = 10^3$ K) as the projectile, and rotationally "cold" molecules ($T_{rot} = 10^2$ K), as the target. At low velocity the measured cross-sections mainly probe the spherical component of the interaction, since the collisional time is larger than the period of a molecular rotation—the critical time needed to induce an angular averaging of the interaction. As the collisional velocity increases, the anisotropy enters into play, since the collisions tend to assume a sudden character. The novel technique developed in our group in 1994 [57], for cooling oxygen to the lowest rovibrational state and for aligning the rotational angular momentum, allows the control of the relative orientation of the colliding molecules and provides more direct information on the interaction anisotropy from the measurement of changes in the smooth component, in the amplitude and in the extreme position of the "glory" pattern, as a function of the alignment degree of projectile molecules.

Along these lines, a simultaneous analysis of our experimental cross-sections and of available second virial coefficients has been carried out to obtain a reliable interaction for O_2–O_2 [5]. The same approach had been applied to the N_2–N_2 and O_2–N_2 systems, although in considerable less detail. Previous results on the N_2–N_2 system [44] have been reanalyzed, and a characterization of the N_2–O_2 complex has been presented a subsequent paper [10]. These results indicate that most of the bonding in the dimers comes from van der Waals (repulsion + dispersion) and electrostatic (permanent quadrupole–permanent quadrupole) forces. On the other hand, chemical (spin–spin) contributions are not negligible for O_2–O_2, which is an open-shell–open-shell system [4,5]. Therefore the geometrical properties of the three dimers have been found to show interesting differences, to be seen in the next paragraph.

2.2 Potential energy surfaces: phenomenological characterization

It has been found useful to represent the interaction potential for a dimer of homonuclear diatomic molecules [4,5,46,58] as a spherical harmonic expansion, separating radial and angular dependencies. The radial coefficients include different types of contributions to the interaction potential (electrostatic, dispersion, repulsion due to overlap, induction, spin–spin coupling). For the three dimers of atmospheric relevance, we provided compact expansions, where the angular dependence is represented by spherical harmonics and truncating the series to a small number of physically motivated terms. The number of terms in the series are six for the N_2–O_2 systems, corresponding to the number of configurations of the dimer (for N_2–N_2 and O_2–O_2 this number of terms is reduced to five and four, respectively).

The radial coefficients were derived from the analysis of experimental data assisted by empirical correlation formulas [59] The leading spherical term has the meaning of a typical isotropic van der Waals interaction, arising from a short-range repulsion, associated to the spherical "size" of both partners, and a long-range dispersion attraction, depending on the spherical polarizability of each of two monomers. Other terms are associated to the orientational anisotropy of each diatom when the other one is considered as a spherical partner: this is the anisotropy expected in the diatomic—"pseudoatom" limit. For N_2–N_2 and O_2–O_2 these terms are undistinguishable, but not in the case of N_2–O_2 where two different limits are possible. To be specific, in the analysis of the scattering of "hot" O_2 (10^3 K) by "cold" N_2 (10^2 K), O_2 acts as a "pseudoatom", and therefore cross sections at high velocity provide information on the potential in this limiting case. Other terms introduce additional corrections to the repulsion arising from the mutual orientation of both molecules, the electrostatic component exclusively depending on the quadrupole–quadrupole interaction, which can be accurately estimated (this term is small in the O_2–O_2 system). Induction contributions due to permanent quadrupole-induced multipole interactions are also included, although they play only a minor role especially in the intermediate range of intermolecular distances, which are those of interest here.

The analysis of the total cross-sections measured as a function of the collisional velocity provides direct information on the isotropic interaction in the low velocity range, and yields information on the strength of the leading anisotropic terms from the "glory" quenching observed at high velocity, which is an attenuation of the amplitude by an interference phenomenon which shows up in molecular beam scattering. The second virial coefficient is affected by all the radial terms in the low temperature range while in the high temperature limit it mainly depends on the isotropic (spherically averaged), components.

From a structural viewpoint it is very interestingly to remark that the three dimers differ both for their geometry and for the nature of the bond. For N_2–N_2 (no electronic spin) the basic feature which determines the equilibrium geometry is the quadrupole–quadrupole interaction, which favors the perpendicular configuration. In the case of oxygen the equilibrium geometry obtained for the ground singlet state is the parallel one. A crucial role is played by the spin–spin interaction, in spite that its contribution to the bond is small (approximately 15% at the equilibrium configuration [4,5]). This leads to the geometry corresponding to the most stable configuration, as the one where the two O_2 molecules are parallel. Indeed, the binding forces are slightly stronger than in the nitrogen case because of this contribution from the spin–spin interaction. Finally a crossed configuration is found to be the most stable for N_2–O_2: here no role is played either by the spin interaction or by the quadrupole–quadrupole interaction, which is not strong enough to stabilize the two possible perpendicular configurations. These potential energy surfaces are considered to be such that their handling and the physical interpretation of their terms make them realistic in the sense that they reproduce micro and macroscopic information. On the other hand, existing spectroscopic information has not been fully analyzed (see e.g. [11,12]), and more spectra may be taken in the future, whose analysis will require theoretical guidance, which will also be needed to study the associated quantum dynamics.

2.3 Quantum dynamics of bound states and spectral features of clusters

A particular effort has been addressed to the study of the dynamics within the dimer and to the characterization of the low lying rovibrational states in view of potential interest for the analysis of spectral features in atmospheric research. Calculations of the bound rovibrational states of the dimers have been performed for rotational states having total angular momentum $J \leqslant 6$ by solving the secular problem over the exact Hamiltonian. We have calculated the rovibrational levels for the potential energy surfaces described above of the dimers N_2–N_2, N_2–O_2 and for all three surfaces (singlet, triplet and quintet) [4,5] of O_2–O_2. A summary of results and their discussion follows. Full account of all available data has been given in [5,6,8–10].

The phenomenological potentials derived as in the previous paragraph enable us to compute rovibrational energy levels of the dimers treating each monomer O_2, N_2 as rigid rotor, decoupling intermonomer (van der Waals) and intramonomer vibrations (which energetically are almost two orders of magnitude larger).

The theory to solve the Schrödinger equation for this Hamiltonian is reported in Ref. [58]. The full quantum mechanical calculations of bound states are carried out using the program BOUND [60], where the intermolecular distance is treated as a scattering coordinate, and a basis set expansion is employed for the remaining degrees of freedom. The coupled equations are then solved using the standard techniques of scattering theory, but with bound state boundary conditions. This method is found to be particularly appropriate for van der Waals complexes, where there is a wide amplitude vibrational motion along the intermolecular coordinate [61]. Previously, the N_2–N_2 and O_2–O_2 dimers had been studied in the body—fixed formulation [45,62] which has some advantages when one neglects Coriolis coupling (an approximation also known as helicity conserving) that we do not make but assess in our study.

It has been found that some of the levels which are degenerate without Coriolis coupling are actually split. Because of Coriolis mixing, there are energy levels (with $J > 0$ but different K) which perturb each other leading to a stabilization of the order of 1 cm^{-1} in some cases. The splitting is of the order of 0.03 cm^{-1} for the $J = 1$ case in some levels, but for higher J can be more than 20 times larger. Computing time for the exact close coupling calculations is only twice than for the helicity decoupling (i.e. neglecting Coriolis) as J increases. But it increases dramatically for higher J. For example, for $J = 6$ it is 40 times larger.

We calculated also other spectral aspects, vibrational frequencies, rotational constants, etc. The results permit comparison with previous works [45,62], where the potential surfaces used are very different: we found that in general although some features may appear similar, zero point and dissociation energies differ notably (compared with corresponding ones in Refs. [45,62]).

Most interesting is the comparison among the three systems. We have seen that the various potential energy surfaces look different topographically: however, similar results are obtained for some spectroscopic observables, in spite of the fact that geometry and values of well depths and positions differ significantly, but what varies is their ability to reproduce other experimental data (integral cross-sections and second virial coefficient). These quantities are therefore most sensitive probes of dimer structure and dynamics.

We also calculated harmonic frequencies through the second derivative of the potential around the equilibrium geometry of the dimers. From these values we obtained the energies of the vibrational levels in the harmonic approximations: the values so calculated are also very different from the exact (close coupling) results, showing that although the interaction forces are very weak, the potential anharmonicity and anisotropy are important.

We can conclude that the range and strength in the bonds of the three dimers N_2–N_2, O_2–O_2, and N_2–O_2 present characteristics more of clusters than of weakly bound molecules and that the interaction between the monomers is very anharmonic.

In the work described here, we have dedicated special emphasis not only to the characterization of the low lying energetic levels, but also to the associated wavefunctions exploiting the availability of realistic potential energy surfaces. This study completes the faithful picture of the internal dynamics of the system involving levels pertaining to the lowest energy states for the dimers.

To obtain such a full characterization of the vibrational states and of related motions within the dimer, a simultaneous analysis of the radial and angular behavior of the total wavefunction is required. However, the program BOUND provides the rovibrational energy levels but not the associated wavefunctions. In order to obtain the latter, we provided the implementation of a method developed by Thornley and Hutson [63]. Specifically, starting from a known eigenvalue of the wavefunction at a particular value of the intermolecular distance we can recover the behavior of the involved wavefunction in a wide range making use of the equations for the propagation of the logarithmic derivative. This mechanism allows us to examine the total wavefunction, facilitating the assignment of the fundamental vibrations within the dimers. Distributions of such probabilities for the ground and second excited state, represented in terms of the coordinates have been given in [3] and references therein.

For O_2–O_2, in contrast with the results that suggest parallel shape conformation for the ground state complex, in the second excited state, the wavefunction presents a maximum probability also for the parallel conformation. Nodes in the angular part are absent and therefore this level exhibits features close to the ground state. However a node in the radial part can be recognized and assigned to an excitation in the stretching intermolecular mode, characterized by a marked anharmonic behavior. Therefore the dimer dynamics is effectively depicted by bound-state calculations of wavefunctions, indicating a poor symmetric distribution of probability as a function of intermolecular distance, most of the corresponding distribution probability being confined at long range. Details on the results outlined here are accessible in previous papers [2–10].

3. ASPECTS OF INTERACTIONS INVOLVING MINOR ATMOSPHERIC COMPONENTS: H_2O AND H_2S

In the context of atmospheric photochemistry and physics, and of climate change, probably the most important interactions of a minor component is that of H_2O and not only with N_2, O_2 and itself, but also with others species, giving rise to a variety of processes that have received little attention and are not well understood. For recent quantum mechanical studies of water interactions with itself and with fluoromethanes, and of water dimers with CO, see papers in [64], which contain references to a large body of literature.

A variety of processes such as homogeneous nucleation of water, water cluster mediated photochemical reactions, and the radiative properties of water vapor and water clusters in the visible/near IR and mid-IR spectral ranges, has attracted attention: as documented in [1], there has been a long-standing controversy in atmospheric physics as to the potential role for dimers especially in explaining the continuum absorption. Another recent challenge concerns our quantitative understanding of the atmospheric absorption in clear skies, in which water clusters may play an important role [65]: following recent reports of the first spectroscopic detection of water dimers in the Earth's atmospheric processes, quantum chemistry and dynamics may contribute to conduct in-depth studies of the physics of weakly interacting molecules and their impact for atmospheric physics and climate.

The following will give a flavour of the accuracy of our experimental technique in the case of water molecules, and a related one, H_2S. More specifically, from the accompanying quantum mechanical studies, we have reported in [66] the test of a recent *ab initio* potential for the H_2O–He dimer [67] on the measured integral cross section data. Since in our experimental conditions water molecules in the beam are kept at 500 K, the molecules rotate so fast to fully average the interaction during the collision and we can therefore calculate the cross sections with the interaction potential properly averaged over all orientations in space. Further work on interactions of water with the rare gases [68] has been reported and is discussed next. Experiments regarding molecules relevant for atmospheric studies (especially H_2, N_2, O_2, CH_4) are currently being planned. This section also reports on the H_2S–rare gas systems.

3.1 Water–rare gas systems

The role of the water molecule in nature is characterized by a peculiar interaction feature, the so-called hydrogen bond, which turns out to be one of the most challenging phenomena to be tackled both experimentally and theoretically [69–71]. A further motivation, of interest in the present context, comes from the evidence that gas-phase complexes of water can play relevant roles in atmospheric chemistry and physics [65,72,73]. Of specific relevance is the need of assessing the relative importance of the various components of the overall interaction, typically electrostatic, induction, dispersion, which operate at long range, and overlap (size repulsion and charge transfer) effects, which act at short range. Their modeling in terms of monomer properties requires appropriate combinations and extensions of experimental and theoretical information on prototypical aggregates.

The simplest water aggregates, amenable to this kind of testing and modeling, are those with the rare gases, in which the contributions to the interaction from electrostatic components are absent. Recently, we demonstrated [68] the incipient hydrogen bond formation for the water complexes with heavier Ar, Kr, and Xe rare gases. In this account, we first focus on the lightest and weakly bound H_2O–He pair, the simplest of the all the possible gas-phase aggregates of water and a benchmark system in the field. In this case, due to the very weak intermolecular electric field, the intermolecular interaction is expected to exhibit a nearly pure van der Waals character.

From the experimental viewpoint, molecular beam scattering measuring differential cross section had been reported by three different laboratories [74–76].

On the theoretical side the H_2O–He systems has a sufficiently small number of electrons to be tackled by the most sophisticated quantum-chemical techniques, and in the last two decades several calculations by various methods of electronic structure theory have been attempted [77–80]. More recently, new sophisticated calculations appeared in the literature: they exploited combined symmetry – adapted perturbation theory SAPT and CCSD(T), purely *ab initio* SAPT [81,82], and valence bond methods [83]. A thorough comparison of the topology, the properties of the stationary points, and the anisotropy of potential energy surfaces obtained with coupled cluster, Möller–Plesset, and valence bond methods has been recently presented [83].

From our laboratory [66], total integral cross section measurements have been recently reported for the scattering of water molecules by helium atoms. The experimental data together with available differential cross section results have allowed a very accurate determination of the isotropic component of the interaction potential for this prototypical system. The spherical average of the most recent and accurate *ab initio* potential energy surfaces has been tested against the new results. The comparison indicates that in the intermediate distance range of 0.33–0.46 nm, the calculations are still not convergent by roughly 10% which is of the order of 0.03 kJ mol^{-1}. Actually, this is quite an acceptable result for state-of-the-art *ab initio* calculations. Better performances are shown at shorter and larger distances. The nature of the interaction in the water–helium weakly bound complex is demonstrated to be essentially of the van der Waals type.

As far as the interactions with the heavier rare gases are concerned, an integrated experimental–theoretical investigation has been also carried out [68]. Illuminating is the direct comparison with results from analogous experiments on O_2–rare gas systems: in such well-established series of partners (prototypical van der Waals interactions [84]), both the long- and medium-range behaviors can be compared with those of the water systems in view of the similarity of oxygen and water in terms of molecular polarizability—the basic property for scaling van der Waals forces in the full distance range [59,85,86].

We have thus provided an accurate determination of the spherically averaged interaction, the leading term in the multipolar expansion. For the water–He and water–Ar systems there is full agreement with previous differential scattering [76] and spectroscopic determination [87], respectively. In detail: (i) For the lighter gases, He and Ne, the coincidence of cross sections for the water and the oxygen systems suggests that a van der Waals interaction is also operative in the water case. Indeed, the involved components (repulsion, dispersion, induction) can be modeled on the basis of the polarizabilities of the partners [59,85,86]. (ii) As we move to the heavier members of the series, the average components of the water cross sections still coincide with those of the oxygen systems and continue to be in accord with polarizability-based models for van der Waals long range attraction [85,86], but this is not so for the "glory" features, which are very sensitive to the depth and position of the potential wells. The observed progressive glory shift towards higher velocities on going towards Xe is thus a precise measure of an increasing binding strength, larger than expected for pure van der Waals forces.

Phenomenological evaluation [85,86] using atomic and molecular polarizabilities and the permanent dipole moment of water suggests that under the condition of the experiment, which involve water at high rotational temperature, the long range attraction is accounted for more than 90% by dispersion forces, while induction contributions play a minor role.

We believe that such an increase in well depths is the manifestation of a short-range reinforcing attractive component to the bond and its role amplifies as distance decreases (as typical of a covalent or charge-transfer contribution): moving towards Xe along the series of rare gases, ionization potentials decrease, proton affinities increase, and there is circumstantial evidence for sharing of external electrons between the partners in the complexes. We looked for confirmation of such a picture and for further insight by extensive state-of-the-art quantum chemical CCSD(T) and MRCI calculations, carried out to determine both energy and structure of the water–rare gas systems (from He to Kr) in the neighborhood of the absolute minimum in the intermolecular interaction potential energy surface. The *ab initio* CCSD(T) method was used with several basis sets: cc-pVDZ, cc-pVTZ, and their augmented versions aug-cc-pVDZ. All geometries were optimized by using analytical gradient procedures. All structures reported in [68] correspond to fully converged geometries with gradients and displacements below the standard thresholds implemented in Gaussian [88]. The multireference method MRCI was used with the aug-cc-pVTZ basis set. An ample set about 600 single potential-energy points on each surface was analyzed. Partial charges and orbital occupations, based on the natural bond orbital (NBO) scheme and partly on Mulliken analysis, have been taken into account to cast light on the nature of these interactions.

It can be stated that experiments and calculations are in a sense complementary here, because quantum chemistry encounters well-known difficulties with absolute estimations of the bond energies of intermolecular forces, but its use is of value in searching for the geometries of most stable structure. In contrast, our experiments probe absolute strengths of interactions, but only orientationally averaged. Specifically, quantum chemical probing was made as the noble gas moves in the plane of the water molecules, and the progressive alignment along the O–H bond is found when heavier rare gases were considered. Such an alignment is known as one of the signatures of the hydrogen bond [89], while van der Waals considerations predict a sideways approach, that is, along the direction of minimum electron density in water. An insurgence of covalence is observed in the density maps of hybrid orbitals in the cases of Ar and Kr.

The relevance of the effect has been discussed quantitatively. In order to summarize the relevant findings, let us take Kr as the case study for comparison of theory and experiments. In [68] we find that the binding energy is about 1 kJ mol^{-1} higher than 1.65 kJ mol^{-1}: this value can be compared with the expected value of 1.37 kJ mol^{-1} for pure van der Waals interactions. The magnitude of the binding energy for Kr and Xe is therefore in the range of the so-called weak hydrogen bonds [69,71] (strong hydrogen bonds can be more than an order of magnitude stronger), and it is important to stress that only a fraction (ca. 50% for Kr) is accounted for by van der Waals forces.

This conclusion establishes how components of different nature contribute to a noncovalent interaction, specifically for a simple systematic study. This type of information is of interest for modeling these interactions to understand the structure and assist in the formulation of the molecular dynamics for the enormous variety of phenomena in which the hydrogen bond plays a major role. Extension of these measurements to systems consisting of water and simple molecules are in progress.

The two systems water–helium and water–argon can be considered as best characterized, also with regard to potential anisotropies. This information opens perspectives for producing and analyzing collisional alignment phenomena in supersonic seeded molecular beams of water, specifically using helium as carrier gas and scattering by argon as probe. This should be in line with our previous work [57,90–93] with diatomic molecules benzene and C_2 hydrocarbons, and would provide the ground for extending this type of measurements to also probe anisotropic effects in the interactions involving water.

3.2 The H_2S–rare gas systems

In conjunction with the experimental characterization of the intermolecular interaction potentials, which are operative in the weakly bound H_2S–Ne, H_2S–Ar, and H_2S–Kr aggregates, we have also carried out *ab initio* calculations, which—as in the previously discussed case of H_2O—provide useful complementary information [94]. The molecular beam apparatus, operating under high resolution conditions in energy and angle, which has been employed, is the same in previous section, and provides integral cross sections, which have been measured in a wide range of collision velocities. The scattering involved velocity selected rotationally "hot" H_2S projectile molecules, emerging from a heated molecular beam source, and rare gases atom targets (Ne, Ar, and Kr) confined into a liquid air cooled chamber. The conditions of the experiments were those appropriate to observe quantum interference effects in the molecular beam scattering, such as the glory pattern in the velocity dependence of cross sections. In fact, in such conditions the projectile molecules rotate sufficiently fast to minimize any quenching and shifting in the quantum interference effects arising from anisotropy components in the intermolecular interaction potential. Therefore, the analysis of the measured cross sections provided well depth, location of the well and long range attraction coefficient, which are representative parameters of an effective interaction supposedly close to the spherical average of the potential energy surface. These results are important to experimentally establish, for the first time, the absolute scale of the binding energy, of the equilibrium distance, and of the long range attraction in the H_2S–rare gases aggregates. For the H_2S–Ar system the determined value for the distance of the minimum (0.405 nm) is in a very good agreement with the value of 0.398 nm, determined by spectroscopy [95].

Theoretical calculations carried out at various levels and in parallel with experiments, provided a set of complementary information especially regarding geometries.

The calculations were performed using the Gaussian 03 program [88]. The choice of the basis sets is the first issue, documented from calculations for the separated monomers (H_2S and the four rare gases) and from preliminary optimization calculations on the equilibrium distance and binding energy of H_2S–rare gas complexes. A total of 13 different basis sets were used at second order MP2 level. Extensive calculations were also carried out at the CCSD(T) level for the same geometries. Interaction energies were only slightly smaller than for MP2 and therefore most of the calculations were done at the less time consuming MP2 level. The results using 6-311++G (3d2f, 3p2d) basis set are of the correct magnitude and exhibit trends for the studied systems, in agreement with the experimental data. Although this basis set has limited diffuse function that are likely important for weak complexes, it has been employed for the subsequent extensive calculations. The interaction energies are determined using the supermolecular approach, defined as the difference energy among that of the H_2S–rare gas complexes and that of the two monomers (H_2S and rare gas). In order to eliminate the basis set superposition error, the full counterpoise Boys and Bernardi method was used.

Specifically, the extensive calculations have demonstrated that in the most stable geometries of all investigated systems the rare gas atom lies in the plane of the H_2S molecule. Several relative minima in the intermolecular potential energy have been localized and barriers between them characterized. All these features relate to in-plane and out-of-plane motions of the rare gas atom, keeping fixed the intermolecular distance. Such calculations also demonstrated that the barrier heights are in general small and of the same order of magnitude or smaller than the binding energy in the most stable geometries. Globally, the potential energy surfaces look quite isotropic and this probably depends on the near symmetric electronic charge distribution around the H_2S molecule. Further calculations showed that pronounced modifications of the small barriers are observable when the atom-molecule distance varies. However, they confirmed, in agreement with previous studies on H_2S–Ar, that nearly isoenergetic and isotropic pathways describe the orbiting of rare gases around H_2S. For these reasons it has not been possible a precise assignment of the most stable configuration for these complexes.

Combining information coming from present experiments and from the reported theoretical calculations, it can be inferred that the binding energy increases by about a factor six, in going from the H_2S–He to the H_2S–Kr system. This work also shows that previous theoretical studies had overestimated the binding energy of about 13% for H_2S–Ar [96] and of about a factor two for H_2S–Ne [97].

A further important effort has been focused on the understanding of the nature of the involved intermolecular potential, i.e., on the relative role of the basic components of the interactions, mainly affecting the intermolecular noncovalent bond in H_2S–rare gas aggregates. Some predictions, concerning the basic intermolecular potential features (well depths, location of the minimum, strength of long range forces) have been obtained considering the combined effect of van der Waals and induction components. The van der Waals features have been anticipated by using correlation formulas, given in terms of polarizability values of the involved partners [84–86]. Corrections due to the role of the induction have been also included [86]. Note that the induction, arising from the permanent dipole-induced dipole interaction and here depending on the permanent dipole of H_2S (about one half of that of water) and on the polarizability of rare gases, provides an increase of about 2.5% of the attraction with respect to the pure dispersion. Anyway its effects has been included in the evaluation. Predicted potential parameters when compared with those obtained from the analysis of measured cross section, show good agreement suggesting that in the present systems the van der Waals component basically controls the intermolecular interaction potential energy. Other components, relevant for the formation of a hydrogen bond, are absent or not measurable. This is to be contrasted with the above results obtained for water. The satisfactory comparison with the parameters for the investigated systems allows us to safely extend the phenomenological evalutation of the same quantities for the H_2S–He and H_2S–Xe cases, for which detail experiments are in progress. Predicted well depth, location of the minimum and long range attraction constants parameters are found to be 0.247 kJ mol^{-1}, 0.388 nm and 11.7 kJ mol^{-1} for He and 2.21 kJ mol^{-1}, 0.428 nm and 191 kJ mol^{-1} nm^{-6} for Xe.

As a conclusion for these systems, we can say that the combination of high resolution experiments with *ab initio* calculations and predictions of empirical correlation formulas sheds light on binding energy, energy barriers, anisotropy of the potential energy surfaces, and on the relative role of the leading interaction components operative in the H_2S–rare gases aggregates. This information is crucial to amplify the phenomenology of noncovalent intermolecular interactions [84], especially to establish when additional components relevant for the formation of the hydrogen bond [68], enter into play.

4. PEROXIDES AND PERSULFIDES

In this section we present a systematic study by quantum mechanical methods of structural and energetic properties of hydrogen peroxide and hydrogen persulfide, and also for a series of substitutions of the hydrogens by alkyl groups and halogens atoms. The emphasis is on the torsion around the peroxidic bond which leads to the chirality changing stereomutation. The dihedral angle dependence of the geometrical features and of the dipole moment is discussed with reference to previous experimental and theoretical information. This is of interest for chiral separation experiments as well as in view of a possible dynamical mechanism for chirality exchange by molecular collisions.

The understanding of the dynamics associated to the relatively weak peroxidic and persulfidic bonds is important, because they play a crucial role in molecules and intermediates of relevance not only in atmospheric science [98–100], but also in other ample fields of chemistry, such as combustion and biochemical processes. These bonds are generally longer and weaker than the organic molecular environment where they occur and this leads them to often exhibit peculiar dynamical features, which manifest as intramolecular and intermolecular phenomena (we leave outside of the scope of this report the otherwise basic issue of the strengths of the bonds and their breakings: see [101,102] and references therein). The *intramolecular* phenomena are kinetically recognized as isomerizations, while spectroscopically are referred to as wide amplitude anharmonic modes, offering the perhaps simplest and neatest paradigm of chirality changing processes, which continue to be an issue of prime interest [103], beyond that of their manifestation through the optical rotatory power [104]. The *intermolecular* aspects have been so far amply examined with respect to hydrogen bonds in adducts such as those with water, and in dimers and larger clusters [105–107]. Interactions with atoms, ions and simple molecules were also investigated [108–110]. In our work, intermolecular interactions turn out to be of specific importance, with reference to the collisional mechanism of chiral effects, such as in scattering from surfaces and gaseous streams and vortices [111].

Regarding this last topic, extensive experimental work in this laboratory has been devoted to the study of intermolecular interactions from scattering measurements, in order to provide ingredients for modelling forces acting in systems involving the components of atmospheres [3]. As noted in previous sections of this article, interesting alignment and orientation effects in the gas phase were experimentally demonstrated, simply due to molecular collisions, and this occurring typically when in a gaseous mixture a "heavier" molecular component is seeded in a lighter one. The detection of aligned oxygen in gaseous streams and further evidence on simple molecules has been extended to benzene and various hydrocarbons [91–93]. Other investigations suggest that chiral effects can be seen in the differential scattering of oriented molecules, in particular from surfaces: this is reviewed in a previous paper [111], where it is further pointed out that it may be of pre-biotical interest a possible mechanism of chiral bio-stereochemistry of oriented reactants, acting through selective collisions whose dynamical treatment requires information on the nature and mechanical properties of the peroxidic bond, and the associated torsional motion.

4.1 Hydrogen peroxide

In a recent paper [112], taking hydrogen peroxide and its isotopomers as the natural starting example, we examined the tools of this investigation, where we employed quantum chemistry to describe the potential energy surface of a molecule with a floppy bond. For this prototypical simple system, we have exploited orthogonal local coordinate systems demonstrating their advantages over the usual representations of intramolecular dynamics. In fact, the modes of a molecule can be seen as stretches, bending and torsions according either to the normal mode pictures or to the bonds and angles of the valence (or local mode) type of description. For H_2O_2, a natural choice for orthogonal vectors turns out to be those joining the two OH radicals plus the Jacobi vectors connecting their centers of mass. Since a further salient feature of the orthogonal local vector representation is the rigorous elimination of intermode coupling in both the classical and the quantum kinetic energy, a model has been recommended for intra and intermolecular treatments of the dynamics of the either thermal or radiation and collision induced mechanisms for chirality interconversion.

In [112] the insight and accuracy of the approach has been demonstrated: it has been extended afterwards to a series of molecules in view of perspective applications to molecular dynamics modeling of collision induced chirality changing processes.

Regarding the methodology of the computational aspects, the issues are the choice of basis sets and theory levels which reproduce known information on features like equilibrium geometry and *cis* and *trans* torsional barriers around the O–O bond [112].

An extensive systematic study has been performed on the choice of method and basis set for H_2O_2. It was found that the results obtained by the B3LYP density functional method using 6-311++G (3df,3pd) basis set and those by the CCSD(T)/aug-cc-pVTZ method gave both consistent geometric parameters and concordance with experimental data. Also the important available experimental features for stereomutation, such as the height of the *trans* barrier, were accurately reproduced, justifying the use of this method and basis set (B3LYP/6-31G++3df, 3pd) for the systematic investigation of substituents, such as alkyl groups and halogens, reported below in this paper. All calculations have been executed by the Gaussian package [88].

The results on torsional levels and their distribution [113] for H_2O_2 have served for the calculation of the dipole moment. For such a calculation, the wavefunctions of the torsional levels are also needed, but they are automatically obtained by our calculation together with their energies. The integrated dipole moments, in Debye, are 1.536 for the two lowest degenerate levels and so this is the value that would apply for the zero absolute temperature limit. At room temperature excited levels contribute, and dipole moments are higher: for the first excited degenerate doublet they are 1.646 D. Our value for the room temperature dipole moments, accounting for the relative population of the torsional levels, is 1.58 D, in close agreement with the accepted experimental value 1.57 ± 0.02 D [114,115].

4.2 Other systems

Among the features which emerge on considering the properties of the alkyl peroxides with respect to H_2O_2, a very significant one is the dihedral angle of the equilibrium configurations. So from 112.5 degrees for H_2O_2, one goes to 115.6 and 114.5 degrees for methyl and ethyl substitution—all other structural properties vary very little from the *cis* to the *trans* barriers, in agreement with much of the established information available, as supplemented also by the recent one [116–118]. Double substitution leads to the disappearance of the *trans* barrier (for the long and controversional history of this effect, see [119]), so that the achiral *trans* structure becomes that corresponding to the equilibrium. This effect is a further manifestation of the floppiness of the O–O bond, where the physical dimension of the substituent groups appears to play a crucial role. The lengths of the O–O bonds (see Ref. [112] for the *cis*, *trans* and equilibrium configurations) shrink in the neighborhood of a right angle torsional configuration for the latter, as showed for hydrogen peroxide, methyl hydroperoxide and dimethyl peroxide, while the H_2O_2 profile had been found in agreement with Ref. [120].

The study reported in [113] had been concerned with the extension of that study presented in the previous paper [112] on quantum chemical exploration of the potential energy surface of hydrogen peroxide—the simplest example of a chiral molecule. Torsional barriers around the O–O bond have been shown to vary enormously by substitution of alkyl groups: a paper in preparation reports a study of the effects by substitution with halogen atoms. Features of interest for the internal and intramolecular dynamics description in terms of stretching, bending and torsional modes are compared with limited experimental and computational information, with special attention to the energy and dipole moment profile upon variation of the dihedral angle.

In these investigations, the steric hindrance of the substituent group has been seen to be a determining fact for the energetic and geometric effects. In the case of halogens as substituent, attention will be needed to more specific features such as electronegativity differences between atoms or groups involved in the bonds. Further work is also being reported on the torsional modes of the S–S bond by substituent groups of the hydrogens in H_2S_2, also of relevance for stereomutation issues [121].

A final remark concerns the descriptions in terms of modes based on the vectors either of the usual valence picture of the molecule and of the orthogonal local representation: in the previous paper [112] we have demonstrated that the orthogonal local representation for H_2O_2 is to be privileged for molecular dynamics simulations since the torsion around the vector joining the center of mass of the two OH radicals mimics very accurately the adiabatic reaction path for chirality changing isomerization, namely the torsional potential energy profile from the equilibrium through the barriers for the *trans* and *cis* configurations. The study of the case that alkyl groups [113], or halogen atoms are substituents, requires further work because the choice of the appropriate local orthogonal systems is not obvious for such systems.

Regarding intermolecular interactions, of specific importance for collisional chirality exchange, a study has been performed on the H_2O_2–rare gas systems [122], for which information should also come from molecular beam experiments in our laboratory (see Sections 2 and 3). This will extend to these systems the joint experimental and theoretical approach already tackled for interactions of H_2O [66, 68] and H_2S [94] with the rare gases.

5. CONCLUDING REMARKS AND PERSPECTIVES

This article has considered quantum mechanical studies of systems of relevance for atmospherical chemistry, illustrating both intramolecular and intermolecular features and pointing out the role of different theoretical approaches.

Regarding the major components of the atmosphere, we have been dealing with the dimers involving N_2 and O_2. Intermolecular potentials needed for describing the dimer structure and dynamics have been obtained from molecular beam scattering experiments and phenomenological analysis. The quantum mechanical tool here has been the solution of the exact diatom–diatom Schrödinger equation generating rovibrational levels and wavefunctions of interest for assisting the spectroscopic probing of atmospheric energy balance.

Regarding minor components, either natural, or pollutants or produced photochemically or during various process of importance in combustion, we have utilized quantum chemical calculations of intermolecular potentials to describe interactions of H_2O and H_2S—in both cases in conjunction with experimental information from molecular beams scattering and other sources. In view of difficulties of quantum chemistry in dealing accurately with intermolecular forces, such synergy with phenomenological approaches is essential. For two series of molecules, H_2O_2 and its alkyl [113] and halogen substituted derivatives [123], and H_2S_2 and the corresponding derivatives, state-of-the art quantum chemistry has been shown to provide accurate description of the structure and the dynamics of torsional modes.

ACKNOWLEDGEMENTS

Collaborations on these researches with Drs M. Bartolomei, A.C.P. Bitencourt, E. Carmona-Novillo and M. Ragni are gratefully acknowledged. This research is supported by FIRB and PRIN grants from the Italian Ministry for University and Scientific and Technological Research, and by ASI.

REFERENCES

[1] C. Camy-Peret, A. Vigasin (Eds.), *Weakly Interacting Molecular Pairs: Unconventional Absorbers of Radiation in the Atmosphere*, Kluwer, Dordrecht, 2003.

[2] V. Aquilanti, M. Bartolomei, D. Cappelletti, E. Carmona-Novillo, E. Cornicchi, M. Moix-Teixidor, M. Sabidò, F. Pirani, in: C. Camy-Peret, A. Vigasin (Eds.), *Weakly Interacting Molecular Pairs: Unconventional Absorbers of Radiation in the Atmosphere*, Kluwer, Dordrecht, 2003, pp. 169–182.

[3] V. Aquilanti, M. Bartolomei, D. Cappelletti, E. Carmona-Novillo, F. Pirani, *Int. J. Photoenergy* **6** (2004) 53.

[4] V. Aquilanti, D. Ascenzi, M. Bartolomei, D. Cappelletti, S. Cavalli, M. De Castro Vitores, F. Pirani, *Phys. Rev. Lett.* **82** (1999) 69.

[5] V. Aquilanti, D. Ascenzi, M. Bartolomei, D. Cappelletti, S. Cavalli, M. De Castro Vitores, F. Pirani, *J. Am. Chem. Soc.* **121** (1999) 10794.

[6] V. Aquilanti, M. Bartolomei, D. Cappelletti, E. Carmona-Novillo, F. Pirani, *J. Chem. Phys.* **117** (2002) 615.

[7] V. Aquilanti, M. Bartolomei, D. Cappelletti, E. Carmona-Novillo, F. Pirani, *Phys. Chem. Chem. Phys.* **3** (2001) 3891.
[8] V. Aquilanti, E. Carmona-Novillo, F. Pirani, *Phys. Chem. Chem. Phys.* **4** (2002) 4970.
[9] V. Aquilanti, M. Bartolomei, E. Carmona-Novillo, F. Pirani, *J. Chem. Phys.* **118** (2003) 22141.
[10] E. Carmona-Novillo, F. Pirani, V. Aquilanti, *Int. J. Quantum Chem.* **99** (2004) 616.
[11] A. Campargue, L. Biennier, A. Kachanov, R. Jost, B. Bussery-Honvault, V. Veyret, S. Churrassy, R. Bacis, *Chem. Phys. Lett.* **288** (1998) 734.
[12] L. Biennier, D. Romanini, A. Kachanov, A. Campargue, B. Bussery-Honvault, R. Bacis, *J. Chem. Phys.* **112** (2000) 6309.
[13] K. Pfeilsticker, H. Bösch, C. Camy-Peyret, R. Fitzenberger, H. Harder, H. Osterkamp, *Geophys. Res. Lett.* **28** (2001) 4595.
[14] F.A. Gorelli, L. Ulivi, M. Santoro, R. Bini, *Phys. Rev. Lett.* **83** (1999) 4093.
[15] C.A. Long, G.E. Ewing, *Chem. Phys. Lett.* **9** (1971) 225.
[16] C.A. Long, G.E. Ewing, *Acc. Chem. Res.* **8** (1975) 185.
[17] F. Cacace, G. de Petris, A. Troiani, *Angew. Chem. Inter. Ed.* **40** (2001) 4062.
[18] C. Coletti, G.D. Billing, *Chem. Phys. Lett.* **356** (2002) 14.
[19] E.N. Timokhina, K.V. Bozhenko, A.E. Gekhman, N.I. Moiseeva, I.I. Moiseev, *Dokl. Phys. Chem.* **384** (2002) 134.
[20] F. Cacace, *Chem.-Eur. J.* **8** (2002) 3839.
[21] J.C.S. Chagas, D.A. Newnham, K.M. Smith, K.P. Shine, *Quartely J. Royal Meteorol. Soc.* **128** (2002) 2377.
[22] R. Hernandez-Lamoneda, M.I. Hernandez, J. Campos-Martinez, *Chem. Phys. Lett.* **368** (2003) 709.
[23] A.C. Pavao, J.C.F. de Paula, R. Custodio, C.A. Taft, *Chem. Phys. Lett.* **370** (2003) 789.
[24] M. Sneep, W. Ubachs, *J. Quant. Spectrosc. Radiat. Transfer* **78** (2003) 171.
[25] A.E. Gekhman, N.I. Moiseeva, V.V. Minin, G.M. Larin, Yu.V. Kovalev, Yu.A. Ustynyuk, V.A. Rosniatovskii, E.N. Timokhina, K.V. Bozhenko, I.I. Moiseev, *Kinet. Catal.* **44** (2003) 40.
[26] V.E. Klymenko, V.M. Rozenbaum, *Mol. Phys.* **101** (2003) 2855.
[27] A.J.C. Varandas, *J. Phys. Chem. A* **108** (2004) 758.
[28] F. Dayou, J. Campos-Martinez, M.I. Hernandez, R. Hernández-Lamoneda, *J. Chem. Phys.* **120** (2004) 10355;
F. Dayou, M. Bartolomei, J. Campos-Martinez, M.I. Hernandez, R. Hernández-Lamoneda, *Mol. Phys.* **102** (2004) 2323.
[29] Y.A. Freimann, H.J. Jodl, *Phys. Rep.* **401** (2004) 1.
[30] J.C.F. de Paula, A.C. Pavao, C.A. Taft, *THEOCHEM* **713** (2005) 33.
[31] R. Hernández-Lamoneda, M.I. Hernandez, J. Campos-Martinez, *Chem. Phys. Lett.* **414** (2005) 11;
R. Hernández-Lamoneda, M. Bartolomei, M.I. Hernandez, J. Campos-Martinez, F. Dayou, *J. Phys. Chem. A* **109** (2005) 11587.
[32] J. Liu, K. Morokuma, *J. Chem. Phys.* **123** (2005) 204319.
[33] B. Bussery, P.E.S. Wormer, *J. Chem. Phys.* **99** (1993) 1230;
B. Bussery, S.Ya. Umanskii, M. Aubert-Frécon, O. Bouty, *J. Chem. Phys.* **101** (1994) 416;
B. Bussery, *Chem. Phys.* **184** (1994) 29;
B. Bussery-Honvault, V. Veyret, *J. Chem. Phys.* **108** (1998) 3243.
[34] Z. Slanina, H. Uhlik, S.L. Lee, S. Nagase, *J. Quant. Spectrosc. Radiat. Transfer* **97** (2006) 415.
[35] M. Sneep, D. Ityaksov, I. Aben, H. Linnartz, W. Ubachs, *J. Quant. Spectrosc. Radiat. Transfer* **98** (2006) 405.
[36] M. Abbaspour, E.K. Goharshadi, J.S. Emampour, *Chem. Phys.* **326** (2006) 620.
[37] R. Steudel, M.W. Wong, *Angew. Chem. Int. Ed.* **46** (2007) 1768.
[38] M.T. Nguyen, T.L. Nguyen, A.M. Mebel, S.H. Lin, M.T. Nguyen, *J. Phys. Chem. A* **107** (2003) 5452.
[39] M.T. Nguyen, *Coord. Chem. Rev.* **244** (2003) 93.
[40] E.E. Rennie, P.M. Mayer, *J. Chem. Phys.* **120** (2004) 10561.
[41] M.H.K. Jafari, A. Maghari, S. Shahbazian, *Chem. Phys.* **314** (2005) 249.
[42] P.C. Samartzis, A.M. Wodtke, *Int. Rev. Phys. Chem.* **25** (2006) 527.
[43] C.A. Long, G. Henderson, G.E. Ewing, *Chem. Phys.* **2** (1973) 485.
[44] D. Cappelletti, F. Vecchiocattivi, F. Pirani, E.L. Heck, S. Dickinson, *Mol. Phys.* **93** (1998) 485.
[45] J. Tennyson, A. van der Avoird, *J. Chem. Phys.* **77** (1982) 5664;
A. van der Avoird, P.E.S. Wormer, A.P.J. Jansen, *J. Chem. Phys.* **84** (1986) 1629.

[46] M.C. van Hemert, P.E.S. Wormer, A. van der Avoird, *Phys. Rev. Lett.* **51** (1993) 1167.
[47] O. Couronne, Y. Ellinger, *Chem. Phys. Lett.* **306** (1999) 71.
[48] A. Wada, H. Kanamori, S. Iwata, *J. Chem. Phys.* **109** (1998) 9434.
[49] A.H. Hamdani, A. Shen, Y. Dong, H. Gao, Z. Ma, *Chem. Phys. Lett.* **325** (2000) 610.
[50] F. Uhlik, Z. Slanina, A. Hinchliffe, *THEOCHEM* **282** (1993) 271.
[51] J.R. Stallcop, H. Partridge, *Chem. Phys. Lett.* **281** (1997) 212.
[52] M.H. Karimi Jafari, A. Maghari, S. Shahbazian, *Chem. Phys.* **314** (2005) 249.
[53] Z. Slanina, F. Uhlik, W.B. De Almeida, *Thermochim. Acta* **231** (1994) 55.
[54] A.J. Brewer, J.W. Vaughn, *J. Chem. Phys.* **50** (1969) 2960;
J.H. Dymond, E.B. Smith, *The Virial Coefficient of Pure Gases and Mixtures: A Critical Compilation*, Clarendon Press, Oxford, 1980.
[55] G.E. Ewing, J.P.M. Trusler, *Physica A* **184** (1992) 415.
[56] K.R. Hall, G.A. Iglesias-Silva, *J. Chem. Eng. Data* **39** (1994) 873.
[57] V. Aquilanti, D. Ascenzi, D. Cappelletti, F. Pirani, *Nature* **371** (1994) 399.
[58] S. Green, *J. Chem. Phys.* **62** (1975) 2271.
[59] R. Cambi, D. Cappelletti, G. Liuti, F. Pirani, *J. Chem. Phys.* **95** (1991) 1852.
[60] J.M. Hutson, BOUND, computer code, version 5 (1993), distributed by collaborative computational project no. 6 of the science and engineering, Research Council, UK.
[61] J.M. Hutson, *Comput. Phys. Commun.* **84** (1994) 1.
[62] B. Bussery-Honvault, V. Veyret, *Phys. Chem. Chem. Phys.* **1** (1999) 3387;
V. Veyret, B. Bussery-Honvault, S.Y. Umanskii, *Phys. Chem. Chem. Phys.* **1** (1999) 3395.
[63] A.E. Thornley, J.M. Hutson, *J. Chem. Phys.* **101** (1994) 5578.
[64] J.B.L. Martins, J.R.D. Politi, A.D. Braga, R. Gargano, *Chem. Phys. Lett.* **431** (2006) 51;
A.F.A. Vilela, P.R.P. Barreto, R. Gargano, C.R.M. Cunha, *Chem. Phys. Lett.* **427** (2006) 29;
R. Bukowski, K. Szalewicz, G.C. Groenenboom, A. Van der Avoird, *Science* **315** (2007) 1249.
[65] J.S. Daniel, S. Solomon, H.G. Kjaergaard, D.P. Schofield, *Geophys. Res. Lett.* **31** (2004) L06118;
I.V. Ptashnik, K.M. Smith, K.P. Shine, D.A. Newnham, *Quarterly J. R. Meteorol. Soc.* **130** (2004) 2391.
[66] D. Cappelletti, V. Aquilanti, E. Cornicchi, M.M. Teixidor, F. Pirani, *J. Chem. Phys.* **123** (2005) 024302.
[67] M.P. Hodges, R.J. Wheatley, A.H. Harvey, *J. Chem. Phys.* **116** (2002) 1397.
[68] V. Aquilanti, E. Cornicchi, M.M. Teixidor, N. Saendig, F. Pirani, D. Cappelletti, *Angew. Chem. Int. Ed.* **44** (2005) 2356.
[69] Y. Tatamitani, B. Liu, J. Shimada, T. Ogata, P. Ottaviani, A. Maris, W. Caminati, J.L. Alonso, *J. Am. Chem. Soc.* **124** (2002) 2739.
[70] I.V. Alabugin, M. Manoharan, S. Peabody, F. Weihold, *J. Am. Chem. Soc.* **125** (2003) 5973.
[71] J.L. Alonso, S. Antolinez, S. Bianco, A. Lesarri, J.C. Lopez, W. Caminati, *J. Am. Chem. Soc.* **126** (2004) 3244.
[72] K. Pfeilsticker, A. Lotter, C. Peters, H. Bösch, *Science* **300** (2003) 2078.
[73] A. Vaida, J.S. Daniel, H.G. Kjaergaard, L.M. Goss, *A.F., Quarterly J. Royal Meteorol. Soc.* **127** (2001) 1627;
H.G. Kjaergaard, T.W. Robinson, D.L. Howard, J.S. Daniel, J.E. Headrick, V. Vaida, *J. Phys. Chem.* **107** (2003) 10680.
[74] R.W. Bickes Jr, G. Duquette, C.J.N. Van der Meijdenberg, A.M. Rulis, G. Scoles, K.M. Smith, *J. Phys. B* **8** (1975) 3034.
[75] T. Slankas, M. Keil, A. Kuppermann, *J. Chem. Phys.* **70** (1979) 1482.
[76] J. Brudermann, C. Steinbach, U. Buck, K. Patkiwski, R. Moszynski, *J. Chem. Phys.* **117** (2002) 11166.
[77] A. Palma, S. Green, D.J. DeFrees, A.D. McLean, *J. Chem. Phys.* **89** (1988) 1401.
[78] S. Maluendas, A.D. McLean, S. Green, *J. Chem. Phys.* **96** (1992) 8150.
[79] B. Kukawska-Tarnawska, G. Chalasinski, M.M. Szczesniak, *J. Mol. Struct.: THEOCHEM* **297** (1993) 313.
[80] F.-M. Tao, Z. Li, Y.-K. Pan, *Chem. Phys. Lett.* **255** (1996) 179.
[81] M.P. Hodges, R.J. Wheatley, A.H. Harvey, *J. Chem. Phys.* **117** (2002) 7169.
[82] K. Patkowski, T. Korona, R. Moszynski, B. Jeziorski, K. Szalewicz, *J. Mol. Struct.: THEOCHEM* **591** (2002) 231.
[83] G. Calderoni, F. Cargnoni, M. Raimondi, *Chem. Phys. Lett.* **370** (2003) 233.

[84] F. Pirani, G.S. Maciel, D. Cappelletti, V. Aquilanti, *Int. Rev. Phys. Chem.* **25** (2006) 165.
[85] F. Pirani, D. Cappelletti, G. Liuti, *Chem. Phys. Lett.* **350** (2001) 286.
[86] V. Aquilanti, D. Cappelletti, F. Pirani, *Chem. Phys.* **209** (1996) 299.
[87] R.C. Cohen, R.J. Saykally, *J. Chem. Phys.* **98** (1993) 6007.
[88] M.J. Frisch, G.W. Trucks, H.B. Schlegel, G.E. Scuseria, M.A. Robb, J.R. Cheeseman, J.A. Montgomery Jr., T. Vreven, K.N. Kudin, J.C. Burant, J.M. Millam, S.S. Iyengar, J. Tomasi, V. Barone, B. Mennucci, M. Cossi, G. Scalmani, N. Rega, G.A. Petersson, H. Nakatsuji, M. Hada, M. Ehara, K. Toyota, R. Fukuda, J. Hasegawa, M. Ishida, T. Nakajima, Y. Honda, O. Kitao, H. Nakai, M. Klene, X. Li, J.E. Knox, H.P. Hratchian, J.B. Cross, V. Bakken, C. Adamo, J. Jaramillo, R. Gomperts, R.E. Stratmann, O. Yazyev, A.J. Austin, R. Cammi, C. Pomelli, J.W. Ochterski, P.Y. Ayala, K. Morokuma, G.A. Voth, P. Salvador, J.J. Dannenberg, V.G. Zakrzewski, S. Dapprich, A.D. Daniels, M.C. Strain, O. Farkas, D.K. Malick, A.D. Rabuck, K. Raghavachari, J.B. Foresman, J.V. Ortiz, Q. Cui, A.G. Baboul, S. Clifford, J. Cioslowski, B.B. Stefanov, G. Liu, A. Liashenko, P. Piskorz, I. Komaromi, R.L. Martin, D.J. Fox, T. Keith, M.A. Al-Laham, C.Y. Peng, A. Nanayakkara, M. Challacombe, P.M.W. Gill, B. Johnson, W. Chen, M.W. Wong, C. Gonzalez, J.A. Pople, Gaussian 03, Revision C. 02, Gaussian, Inc., Wallingford CT, 2004.
[89] A.E. Reed, L.A. Curtiss, F. Weinhold, *Chem. Rev.* **88** (1988) 899.
[90] V. Aquilanti, D. Ascenzi, D. Cappelletti, S. Franceschini, F. Pirani, *Phys. Rev. Lett.* **74** (1995) 2929.
[91] F. Pirani, M. Bartolomei, V. Aquilanti, M. Scotoni, M. Vescovi, D. Ascenzi, D. Bassi, D. Cappelletti, *Phys. Rev. Lett.* **86** (2001) 5035.
[92] F. Pirani, M. Bartolomei, V. Aquilanti, M. Scotoni, M. Vescovi, D. Ascenzi, D. Bassi, D. Cappelletti, *J. Chem. Phys.* **119** (2003) 265.
[93] V. Aquilanti, M. Bartolomei, F. Pirani, D. Cappelletti, F. Vecchiocattivi, Y. Shimizu, T. Kasai, *Phys. Chem. Chem. Phys.* **7** (2005) 291.
[94] D. Cappelletti, A.F.A. Vilela, P.R.P. Barreto, R. Gargano, F. Pirani, V. Aquilanti, *J. Chem. Phys.* **125** (2006) 133111.
[95] R. Viswanathan, T.R. Dyke, *J. Chem. Phys.* **82** (1985) 1674.
[96] G. de Oliveira, C.E. Dykstra, *J. Chem. Phys.* **106** (1997) 5616.
[97] G. de Oliveira, C.E. Dykstra, *J. Chem. Phys.* **110** (1999) 289.
[98] S. François, I. Sowka, A. Monod, B. Temime-Roussel, J.M. Laugier, H. Wortham, *Atmos. Res.* **74** (2005) 525.
[99] M. Lee, B.G. Heikes, D.W. O'Sullivan, *Atmos. Environ.* **34** (2000) 3475.
[100] V. Vaida, *Int. J. Photoenergy* **7** (2005) 61.
[101] T.H. Lay, J.W. Bozzelli, *J. Phys. Chem. A* **1001** (1997) 9505.
[102] J. Matthews, A. Sinha, J.S. Francisco, *J. Chem. Phys.* **122** (2005) 221101.
[103] S.S. Bychov, B.A. Grishanic, V.N. Zadkov, H. Takahashi, *J. Raman Spectrosc.* **33** (2002) 962.
[104] K. Ruud, T. Helgaker, *Chem. Phys. Lett.* **352** (2002) 533.
[105] M. Elango, R. Parthasarathi, V. Subramanian, C.N. Ramachandran, N. Sathyamurthy, *J. Phys. Chem. A* **110** (2006) 6294.
[106] S.A. Kulkarni, L.J. Bartolotti, R.K. Pathak, *Chem. Phys. Lett.* **372** (2003) 620.
[107] I. Alkorta, K. Zborowsky, J. Elguero, *Chem. Phys. Lett.* **427** (2006) 289.
[108] M.C. Daza, J.A. Dobado, J.M. Molina, P. Salvador, M. Duran, J.L. Villaveces, *J. Chem. Phys.* **110** (1999) 11806.
[109] M.C. Daza, J.A. Dobado, J.M. Molina, J.L. Villaveces, *Phys. Chem. Chem. Phys.* **2** (2000) 4094.
[110] J.M. Molina, J.A. Dobado, M.C. Daza, J.L. Villaveces, *THEOCHEM* **117** (2002) 580.
[111] V. Aquilanti, G.S. Maciel, *Origins Life Evol. Biosphere* **36** (2006) 435.
[112] G.S. Maciel, A.C.P. Bitencourt, M. Ragni, V. Aquilanti, *Chem. Phys. Lett.* **432** (2006) 383.
[113] G.S. Maciel, A.C.P. Bitencourt, M. Ragni, V. Aquilanti, *Int. J. Quant. Chem.* **107** (2007) 2697.
[114] R.R. Moreno, A.M. Grana, R.A. Mosquera, *Struc. Chem.* **11** (2000) 9.
[115] D.R. Lide, *Handbook of Chemistry and Physics*, 80th ed., CRC Press, Boca Raton, 2000–2002.
[116] S.C. Homitsky, S.M. Dragulin, L.M. Haynes, S. Hsieh, *J. Chem. Phys. A* **108** (2004) 9492.
[117] J.D. Watts, J.S. Francisco, *J. Chem. Phys.* **125** (2006) 104301.
[118] L.M. Haynes, K.M. Vogelhuber, J.L. Pippen, S. Hsieh, *J. Chem. Phys.* **124** (2005) 234306.
[119] S. Tonmunphean, V. Parasuk, A. Karpfen, *J. Phys. Chem. A* **106** (2002) 438.
[120] E.A. Cohen, H.P. Pickett, *J. Mol. Spectrosc.* **87** (1981) 582.

[121] M. Gotttselig, M. Luckhaus, M. Quack, J. Stohner, M. Willeke, *Helvetica Chim. Acta* **84** (2001) 1846.
[122] P.R.P. Barreto, A.F.A. Vilela, A. Lombardi, G.S. Maciel, F. Palazzetti, V. Aquilanti, *J. Phys. Chem. A* **111** (2007) 12754.
[123] G.S. Maciel, A.C.P. Bitencourt, M. Ragni, V. Aquilanti, *J. Phys. Chem. A* **111** (2007) 12604.

CHAPTER **16**

The Study of Dynamically Averaged Vibrational Spectroscopy of Atmospherically Relevant Clusters Using *Ab Initio* Molecular Dynamics in Conjunction with Quantum Wavepackets

Srinivasan S. Iyengar[1*], **Xiaohu Li**[*] and **Isaiah Sumner**[*]

Contents

1. Introduction 334
2. Computational Methods for Soft Vibrational Mode Clusters 335
   2.1 Extended Lagrangian *ab initio* molecular dynamics 335
   2.2 Including nuclear quantum effects in *ab initio* molecular dynamics: the quantum wavepacket *ab initio* molecular dynamics (QWAIMD) method 339
3. Cluster Dynamics Simulations using ADMP and QWAIMD 342
   3.1 Vibrational properties 342
   3.2 The "amphiphilicity" of the hydrated proton 347
4. Conclusions 348
Acknowledgements 348
References 348

Abstract

The *ab initio* atom-centered density matrix propagation (ADMP) and the quantum wavepacket *ab initio* molecular dynamics (QWAIMD) computational methods are briefly described. Studies on vibrational and electronic properties obtained utilizing these methods are highlighted.

[*] Department of Chemistry and Department of Physics, Indiana University, 800 E. Kirkwood Ave, Bloomington, IN 47405, USA
[1] Corresponding author. E-mail: iyengar@indiana.edu

Advances in Quantum Chemistry, Vol. 55
ISSN 0065-3276, DOI: 10.1016/S0065-3276(07)00216-X

1. INTRODUCTION

Water clusters provide sites for heterogeneous chemical reactions in the earth's atmosphere [1–12]. Numerous molecules are known to solvate in such water clusters. This has led to several studies analyzing the stability of HCl [3,13,14], H_2SO_4 [15], N_2O_5, $ClONO_2$ [15], HO_2 [5,6,16–20] and ions such as H_3O^+ [21–50], OH^- [51–54], halides [55–58] and metal ions [59,60] in water clusters. Medium sized clusters have been used to model cloud droplets [16] and aerosols [61]. Understanding the stability, dynamics, spectroscopy and mechanism of incorporation of ionic species and free-radicals into water clusters constitutes an important challenge to both theory and experiment with wide implications. Furthermore, the impact of hydrogen bonded systems that are intricately present in water cluster systems extends beyond fundamental atmospheric chemistry [12–14,24,26,27,54,62], and well into the areas of materials [63–67], biological sciences [68–70], condensed [25,50–52] and gas phase cluster chemistry [21,24,26,27,54,71,72].

Recently, gas phase single photon [21–23,73] and multiple-photon [74] vibrational action spectroscopy experiments and condensed phase multi-dimensional infra-red experiments [75,76] have become critical in deciphering the precise vibrational signatures that contribute to dynamics in soft-mode hydrogen bonded systems. These measurements provide a "direct probe" of the vibrational dynamics in medium sized clusters. The challenge to theory in modeling these experiments arises from the fact that weak intermolecular interactions, polarizability of water [71,77] and the rapid change in topography of the potential energy surface at finite temperature, dominate important structural and spectroscopic properties in water-clusters [24,26,27,54,62]. These features render the problem attractive to *ab initio* quantum chemical treatment. In addition, dynamical aspects become significant at finite temperatures [27,54,62] and nuclear quantum effects may be critical through hydrogen tunneling and zero-point effects [71,72]. In this contribution, we review a few computational methods that attempt to understand the dynamical, electronic and spectroscopic properties of ions and radicals embedded in soft-mode hydrogen bonded systems such as water clusters. The areas that are covered in this article include the field of *ab initio* molecular dynamics and extensions thereof that include contributions from nuclear quantum effects. In addition we also present the applications of these methodologies to solve the vibrational spectroscopic problems in fundamental systems such as those highlighted above.

This article is organized as follows. In Section 2 *ab initio* molecular dynamics methods are described. Specifically, in Section 2.1 we discuss the extended Lagrangian atom-centered density matrix (ADMP) technique for simultaneous dynamics of electrons and nuclei in large clusters, and in Section 2.2 we discuss the quantum wavepacket *ab initio* molecular dynamics (QWAIMD) method. Simulations conducted and new insights obtained from using these approaches are discussed in Section 3 and the concluding remarks are given in Section 4.

2. COMPUTATIONAL METHODS FOR SOFT VIBRATIONAL MODE CLUSTERS

There exists a number of computational methods that attempt to solve the vibrational spectroscopic problems in fundamental chemical systems. One of the most direct approaches is through harmonic analyses of optimized nuclear configurations available in standard electronic structure packages. This approach, however, is not adequate for soft vibrational mode systems [26,27,54,62,72]. The effect of nuclear dynamics must be considered through quantum [78–114], semi-classical [80,115–130] or classical treatments where the electronic structure is accurately computed [115,131–138]. For systems with more than a few degrees of freedom, a classical or semi-classical approximation of nuclei is generally desireable, if only to keep the computational expense tractable. In this regard, the availability of "on-the-fly" approaches to electron–nuclear dynamics (or *ab initio* molecular dynamics (AIMD), as they are commonly referred to in the literature), has been rather critical in recent years. It should be noted that "on-the-fly" approaches are nearly as old as quantum mechanics itself (see *for example* Ref. [139,140] for a description of the Dirac–Frenkel time-dependent variational principle, which constitutes a formally exact "on-the-fly" dynamics scheme). More recently, many novel approaches to "on-the-fly" AIMD have been developed. Here, an approximation to the electronic wavefunction is propagated along with the nuclear degrees of freedom to simulate dynamics on the Born–Oppenheimer surface. If the nuclei are treated classically [133–137,141] then the forces on the nuclei are determined from the electronic structure. All of these approaches can be broadly categorized into: (a) Born–Oppenheimer dynamics (BOMD) approaches, where the electronic structure is converged self-consistently, or (b) extended Lagrangian approaches [142,143], where an approximation to the electronic structure is propagated through an adjustment of the relative time-scales of electrons and nuclei. The Car–Parrinello method [133,136,141] is a well-known example of the latter approach. The AIMD approaches when combined with full quantum or semi-classical dynamics schemes, has the potential to treat large problems accurately with the complete machinery of quantum dynamics. Several steps have been taken in this direction [115,144–150].

2.1 Extended Lagrangian *ab initio* molecular dynamics

The method of *ab initio* molecular dynamics (AIMD) relies on a calculation of the electronic potential energy surface traversed by the nuclei "on-the-fly" during the dynamics procedure. Both Born–Oppenheimer (BO) molecular dynamics (MD) [115,131–138], as well as Car–Parrinello (CP) dynamics [133,134,136,141,151] are part of this category. The CP scheme differs from the BO dynamics approach in that the wavefunctions are propagated together with the classical nuclear degrees of freedom using an extended Lagrangian. This, in turn, relies on an adjustment of the relative nuclear and electronic time-scales, which facilitates the adiabatic propagation of the electronic wavefunction in response to the nuclear motion with suitably large time-steps. This adjustment of time-scales through the use of a fictitious electronic wavefunction kinetic energy and mass, enables the CP approach to predict effectively similar nuclear dynamics on the BO surface at significantly reduced cost. In this respect, CP differs from methods which rigorously treat the detailed dynamics (rather than structure) of the electrons. (See Ref. [115] and references therein.) The CP method is essentially an extended Lagrangian [142,143] dynamics scheme in which the electronic degrees of freedom are not iterated to convergence at each step, but are instead treated as fictitious dynamical variables and propagated along with the nuclear degrees of freedom by a simple adjustment of time scales. The resultant energy surface remains close to a converged adiabatic electronic surface. Numerous important examples of applications with density functional theory and the CP method are now well documented in the literature (see, e.g., Ref. [136,141]). In the original CP approach, the KohnSham molecular orbitals, expanded in a plane-wave basis, were chosen as dynamical variables to represent the electronic degrees of freedom [133]. However, this is not the only possible choice. An alternative approach is to propagate the individual elements of the reduced one-particle density matrix, **P**.

In Atom-centered Density Matrix Propagation (ADMP) [135,152–157], atom-centered Gaussian basis sets are employed to represent the single-particle electronic density matrix within an extended Lagrangian formalism. Here the basis functions follow the nuclei. The ADMP method has several attractive features. Systems can be simulated by accurately treating all electrons or by using pseudopotentials. Through the use of smaller values for the tensorial fictitious mass, relatively large time-steps can be employed and lighter atoms such as hydrogens are routinely used. A wide variety of exchange-correlation functionals can be utilized, including hybrid density functionals such as B3LYP. Atom centered functions can be used with the appropriate physical boundary conditions for molecules, polymers, surfaces and solids, without the need to treat replicated images to impose 3d periodicity. This is particularly relevant to atmospheric clusters that are described here. QM/MM generalization has been demonstrated [157]. ADMP has been demonstrated through the treatment of several interesting problems including [24,26,27,62,153,155,157–159]: (a) a recent demonstration that dynamical effects are critical in obtaining good vibrational spectroscopic properties of flexible systems [27,54,62], (b) the prediction of the "amphiphilic" nature of the hydrated proton [24–26] which has now been confirmed by many experimental and theoretical studies.

The ADMP equations of motion for the nuclei and density matrix are

$$\mathbf{M}\frac{d^2\mathbf{R}}{dt^2} = -\left.\frac{\partial E(\mathbf{R},\mathbf{P})}{\partial \mathbf{R}}\right|_{\mathbf{P}}, \tag{1}$$

$$\underline{\mu}^{1/2}\frac{d^2\mathbf{P}}{dt^2}\underline{\mu}^{1/2} = -\left[\left.\frac{\partial E(\mathbf{R},\mathbf{P})}{\partial \mathbf{P}}\right|_{\mathbf{R}} + \mathbf{\Lambda}\mathbf{P} + \mathbf{P}\mathbf{\Lambda} - \mathbf{\Lambda}\right], \tag{2}$$

where **R**, **V** and **M** are the nuclear positions, velocities and masses, and **P**, **W** and $\underline{\mu}$ are the density matrix, the density matrix velocity and the fictitious mass tensor for the electronic degrees of freedom. $\mathbf{\Lambda}$ is a Lagrangian multiplier matrix used to impose N-representability of the single particle density matrix. The energy, $E(\mathbf{R},\mathbf{P})$, is calculated using McWeeny purification, $\tilde{\mathbf{P}} = 3\mathbf{P}^2 - 2\mathbf{P}^3$,

$$E = Tr\left[\mathbf{h}'\tilde{\mathbf{P}}' + \frac{1}{2}\mathbf{G}'(\tilde{\mathbf{P}}')\tilde{\mathbf{P}}'\right] + E_{xc} + V_{NN}$$
$$= Tr\left[\mathbf{h}\tilde{\mathbf{P}} + \frac{1}{2}\mathbf{G}(\tilde{\mathbf{P}})\tilde{\mathbf{P}}\right] + E_{xc} + V_{NN}. \quad (3)$$

Here, $\mathbf{h}'$ is the one electron matrix in the non-orthogonal Gaussian basis and $\mathbf{G}'(\tilde{\mathbf{P}}')$ is the two electron matrix for Hartree–Fock calculations, but for DFT it represents the Coulomb potential. The term E_{xc} is the DFT exchange-correlation functional (for Hartree–Fock $E_{xc} = 0$), while V_{NN} represents the nuclear repulsion energy. In the orthonormal basis, these matrices are $\mathbf{h} = \mathbf{U}^{-\mathbf{T}}\mathbf{h}'\mathbf{U}^{-\mathbf{1}}$, etc., where the overlap matrix for the non-orthogonal Gaussian basis, $\mathbf{S}'$, is factorized to yield $\mathbf{S}' = \mathbf{U}^{\mathbf{T}}\mathbf{U}$. There are a number of choices for the transformation matrix $\mathbf{U}$, e.g., $\mathbf{U}$ can be obtained from Cholesky decomposition [160] of $\mathbf{S}'$ or $\mathbf{U} = \mathbf{S}'^{1/2}$ for Löwdin symmetric orthogonalization. The matrix $\mathbf{U}$ can also include an additional transformation so that overall rotation of the system is factored out of the propagation of the density. The density matrix in the orthonormal basis, $\mathbf{P}$, is related to the density matrix in the non-orthogonal Gaussian basis, $\mathbf{P}'$, by $\mathbf{P} \equiv \mathbf{U}\mathbf{P}'\mathbf{U}^{\mathbf{T}}$. The gradient terms involved in the equations of motion are

$$\left.\frac{\partial E(\mathbf{R},\mathbf{P})}{\partial \mathbf{P}}\right|_{\mathbf{R}} = 3\mathbf{FP} + 3\mathbf{PF} - 2\mathbf{FP}^2 - 2\mathbf{PFP} - 2\mathbf{P}^2\mathbf{F}, \quad (4)$$

where $\mathbf{F}$ is the Fock matrix and in the non-orthogonal basis:

$$\mathbf{F}'_{\nu,\sigma} \equiv \mathbf{h}'_{\nu,\sigma} + \mathbf{G}'(\tilde{\mathbf{P}}')_{\nu,\sigma} + \frac{\partial E_{xc}}{\partial \mathbf{P}'}, \quad (5)$$

while the orthogonal basis Fock matrix is $\mathbf{F} = \mathbf{U}^{-\mathbf{T}}\mathbf{F}'\mathbf{U}^{-1}$. The nuclear gradients are

$$\left.\frac{\partial E}{\partial \mathbf{R}}\right|_{\mathbf{P}} = \left\{Tr\left[\frac{d\mathbf{h}'}{d\mathbf{R}}\tilde{\mathbf{P}}' + \frac{1}{2}\left.\frac{\partial \mathbf{G}'(\mathbf{P}')}{\partial \mathbf{R}}\right|_{\mathbf{P}'}\tilde{\mathbf{P}}'\right]\right.$$
$$\left. - Tr\left[\mathbf{F}'\tilde{\mathbf{P}}'\frac{d\mathbf{S}'}{d\mathbf{R}}\tilde{\mathbf{P}}'\right] + \left.\frac{\partial E_{xc}}{\partial \mathbf{R}}\right|_{\mathbf{P}} + \frac{\partial V_{NN}}{\partial \mathbf{R}}\right\}$$
$$+ Tr\left[[\tilde{\mathbf{P}},\mathbf{F}]\left(\tilde{\mathbf{Q}}\frac{d\mathbf{U}}{d\mathbf{R}}\mathbf{U}^{-1} - \tilde{\mathbf{P}}\mathbf{U}^{-\mathbf{T}}\frac{d\mathbf{U}^{\mathbf{T}}}{d\mathbf{R}}\right)\right], \quad (6)$$

where $\tilde{\mathbf{Q}} \equiv \mathbf{I} - \tilde{\mathbf{P}}$. Note that as the commutator $[\tilde{\mathbf{P}},\mathbf{F}] \to 0$, the nuclear forces tend to those used in the standard Born–Oppenheimer MD [135,154,161]. However, in ADMP, the magnitude of the commutator $[\tilde{\mathbf{P}},\mathbf{F}]$ is non-negligible and hence the general expression for the nuclear gradients [135,154] in Eq. (6) is used.

Like CP, ADMP represents fictitious dynamics where the density matrix is propagated instead of being converged. The accuracy and efficiency is governed by the choice of the fictitious mass tensor, $\underline{\mu}$; hence one must be aware of the limits on this quantity. We have derived two independent criteria [152,154] that place bounds on the choice of the fictitious mass. Firstly, the choice of the fictitious mass determines the magnitude of the commutator $[\tilde{\mathbf{P}}, \mathbf{F}]$ thus determining the extent of deviation from the Born–Oppenheimer surface [154]:

$$\left\| [\mathbf{F}, \mathbf{P}_{\mathrm{approx}}] \right\|_{\mathrm{F}} \geqslant \frac{1}{\|[\mathbf{P}_{\mathrm{approx}}, \mathbf{W}]\|_{\mathrm{F}}} \left| Tr\left[\mathbf{W} \underline{\boldsymbol{\mu}}^{1/2} \frac{d\mathbf{W}}{dt} \underline{\boldsymbol{\mu}}^{1/2} \right] \right|, \tag{7}$$

where $\|[\ldots]\|_{\mathrm{F}}$ is the Frobenius norm [160,162] of the commutator and is defined as $\|A\|_{\mathrm{F}} = \sqrt{\sum_{i,j} A_{i,j}{}^2}$. Secondly, the rate of change of the fictitious kinetic energy,

$$\begin{aligned} \frac{d\mathcal{H}_{\mathrm{fict}}}{dt} &= Tr\left[\mathbf{W} \underline{\boldsymbol{\mu}}^{1/2} \frac{d^2\mathbf{P}}{dt^2} \underline{\boldsymbol{\mu}}^{1/2} \right] \\ &= -Tr\left[\mathbf{W} \left(\left. \frac{\partial E(\mathbf{R}, \mathbf{P})}{\partial \mathbf{P}} \right|_{\mathbf{R}} + \boldsymbol{\Lambda}\mathbf{P} + \mathbf{P}\boldsymbol{\Lambda} - \boldsymbol{\Lambda} \right) \right], \end{aligned} \tag{8}$$

is to be bounded and oscillatory and this again is determined by the choice of fictitious mass tensor. We have shown that ADMP gives results that are in good agreement with BOMD and is computationally superior to BOMD [153]. However, one must monitor the quantities in Eqs. (7) and (8) to ascertain that the ADMP dynamics is physically consistent. In all applications studied to date [24,26,27,62,135, 152,153,157] these conditions are satisfied thus yielding a computationally efficient and accurate approach to model dynamics on the Born–Oppenheimer surface.

Current implementation of the ADMP approach has been found to be computationally superior to Born–Oppenheimer dynamics [153]. This important result can be conceptualized based on the following: In Born–Oppenheimer dynamics, the density matrix is to be converged at every dynamics step. Assuming that the largest possible time-step is used during dynamics, SCF convergence requires approximately 8–12 SCF steps. (This depends on the convergence threshold and difficult cases such as transition metal complexes may require more SCF steps.) In ADMP, on the contrary, only the equivalent of 1 SCF step is required per dynamics step; this 1 SCF step is necessary to calculate the Fock matrix required for propagating the density matrix. (A brief review of ADMP is presented in Section 2.1.) Both BOMD and ADMP evaluate the gradient of energy with respect to nuclear coordinates and this calculation requires approximately the same amount of time in both methods. Note that the gradients used in ADMP are more general than those used in BOMD [135,154] on account of the non-negligible magnitude of the commutator of the Fock and density matrix. (See Section 2.1 and Ref. [154] for details.) However, the additional terms require no significant computation over the standard BOMD gradient calculation. The calculation of nuclear force requires approximately 3 times as much computation time as a single SCF cycle. This makes ADMP faster than BOMD by over a factor of 4 per dynamics step. However, the requirement that the ADMP energies oscillate about the BO values with small amplitudes [152] implies that ADMP step sizes cannot be as large at those in BO dynamics. But good energy conservation, which applies to both methods, limits the BO steps to at most twice those of ADMP [153]. (ADMP already uses reasonably large time-steps on account of smaller values for the fictitious mass and an innovative tensorial fictitious mass scheme [152].) This allows ADMP to be over a factor of 2 superior to BOMD, but this estimate is only for cases where the SCF convergence in BOMD is not difficult [153]. The hard to converge cases would require more SCF steps (or a better SCF convergence algorithm) thus making ADMP more efficient as compared to BOMD for these cases. Furthermore, computational improvements that speed up the gradient evaluation will tilt this comparison further towards ADMP.

2.2 Including nuclear quantum effects in *ab initio* molecular dynamics: the quantum wavepacket *ab initio* molecular dynamics (QWAIMD) method

To include quantum nuclear effects in *ab initio* molecular dynamics, we have recently developed an efficient, parallel quantum dynamical methodology for simultaneous treatment of electrons and nuclei [147–149] that can be applied to large systems. Starting from the full time-dependent Schrödinger equation and invoking the time-dependent self consistent field approximation, a system is partitioned into three sections based on chemical complexity [147–149]. Subsystem A comprises particles that display *critical quantum dynamical effects*. These are studied using a time-dependent Schrödinger equation. (While for polyatomic chemical applications, subsystem A may comprise protons, hydrogen atoms or hydride ions; the choice of particles in this subsystem is based on the physics of the problem.) Subsystems B and C comprise the surrounding "bath" particles that dynamically influence the properties of subsystem A. We include the bulk of the nuclear degrees of freedom and electrons within subsystems B and C; the nuclei in subsystem B are treated classically. AIMD [115,131–135,137] is used to treat the dynamics of subsystems B and C; both extended Lagrangian [133,135,136,141,152–155] and Born–Oppenheimer treatment [134,137,138] options are available. We have derived and tested a scheme [147–149] that allows simultaneous dynamics of all three subsystems coupled through a time-dependent procedure. Salient features include: (a) The quantum dynamical free-propagation is formally exact and computationally efficient. This is achieved using a banded Toeplitz[1] representation of the distributed approximating functional (DAF) propagator [147–149,163,164]. (b) One of the principal bottlenecks in the approach is computation of the interaction potential of the quantum wavepacket, on a grid. Improvements in efficiency are achieved by performing electronic structure calculations only at grid points chosen based on the time-dependent gradients of the potential, its magnitude and wavepacket intensity [149]. We call this procedure time-dependent deterministic sampling (TDDS) [149] which is used for efficient "on-the-fly" quantum propagation in multiple dimensions and results in a method that currently allows the simultaneous dynamical treatment of medium sized systems. (Applications to biological enzymes are in progress [165].) (c) The overall computational scheme for conducting quantum wavepacket-*ab initio* molecular dynamics [147–149] displays

[1] The (i, j)th element of a Toeplitz matrix depends only on $|i - j|$. This property of the DAF propagator yields an efficient scheme for quantum dynamics where only the first (banded) row of a matrix needs to be stored.

high efficiency and is implemented in parallel. Time-scales of the order of picoseconds are accessible. QM/MM Generalizations are in progress.

If ADMP is used to study the dynamics of subsystems B and C, the equations of motion are:

$$\imath\hbar\frac{\partial}{\partial t}\psi(R_{QM};t) = \left[-\frac{\hbar^2}{2M_{QM}}\nabla^2_{R_{QM}} + E(\{\mathbf{R_C},\mathbf{P_C}\},R_{QM})\right]\psi(R_{QM};t), \tag{9}$$

$$\mathbf{M}\frac{d^2\mathbf{R_C}}{dt^2} = -\left\langle\psi\left|\left.\frac{\partial E(\{\mathbf{R_C},\mathbf{P_C}\},R_{QM})}{\partial \mathbf{R_C}}\right|_{\mathbf{P_C}}\right|\psi\right\rangle, \tag{10}$$

$$\underline{\boldsymbol{\mu}}^{1/2}\frac{d^2\mathbf{P_C}}{dt^2}\underline{\boldsymbol{\mu}}^{1/2} = -\left\langle\psi\left|\left.\frac{\partial E(\{\mathbf{R_C},\mathbf{P_C}\},R_{QM})}{\partial \mathbf{P_C}}\right|_{\mathbf{R_C}}\right|\psi\right\rangle - [\boldsymbol{\Lambda}\mathbf{P_C} + \mathbf{P_C}\boldsymbol{\Lambda} - \boldsymbol{\Lambda}]. \tag{11}$$

Here $\psi(R_{QM};t)$ represents the quantum wavepacket value at the spatial point R_{QM} at time t, $\mathbf{R_C}$ represents the classical nuclear coordinates and $\mathbf{P_C}$ represents the single-particle electronic density matrix. It is important to note that Eqs. (10) and (11) are wavepacket averaged forms of Eqs. (1) and (2). The system of equations (9), (10) and (11) are coupled, and solved simultaneously as a single initial value problem using multiple time-scales [166]. If Born–Oppenheimer (BO) dynamics is used (instead of ADMP) to represent subsystems B and C, Eq. (11) is substituted by self consistent field (SCF) convergence of $\mathbf{P_C}$.

The quantum propagation is performed using a symmetric split operator methodology where the free propagation is achieved using the DAF propagator [147–149,163,164]:

$$\tilde{K}(R^i_{QM}, R^j_{QM}; \Delta t) = \frac{(2\pi)^{-1/2}}{\sigma(0)}\left[e^{-\frac{(R_{QM}-R'_{QM})^2}{2\sigma(\Delta t)^2}}\right] \times \sum_{n=0}^{M/2}\frac{(-1/4)^n}{n!}\left(\frac{\sigma(0)}{\sigma(\Delta t)}\right)^{2n+1} H_{2n}\left(\frac{R_{QM}-R'_{QM}}{\sqrt{2}\sigma(\Delta t)}\right), \tag{12}$$

where $\{\sigma(\Delta t)\}^2 = \sigma(0)^2 + \imath\Delta t\hbar/M_{QM}$ and $H_{2n}(x)$ are even order Hermite polynomials. Equation (12) is obtained from the well-known analytical expression for free evolution of a Gaussian function [167] and the fact that Hermite functions are generated from Gaussians as

$$H_n(x)\exp\left[-\frac{x^2}{2\sigma^2}\right] = (-1)^n\frac{d^n}{dx^n}\exp\left[-\frac{x^2}{2\sigma^2}\right].$$

(See Ref. [148] for a physically appealing derivation.) Equation (12) yields an efficient, accurate and formally exact scheme to perform quantum dynamics. *It involves the action of a banded, sparse and Toeplitz matrix on a vector.* The propagation scheme has been shown to accurately represent [147] all quantum dynamical features including zero-point effects, tunneling as well as over-barrier reflections and in this sense differs from standard semi-classical treatments. The approach also substantially differs from other formalisms such as centroid dynamics [97,98,168–171], where the Feynman path centroid is propagated in a classical-like manner using forces obtained by averaging over trajectories spanned by a set of Feynman path discretizations (beads) [147].

2.2.1 Time dependent deterministic sampling

Unlike standard approaches to quantum dynamics [78], QWAIMD precludes the need to obtain a fitted potential energy surface. The "on-the-fly" evaluation of the electronic structure energy and gradients, however, scales with the number of real space points needed for wavepacket propagation, and comprises an important computational bottleneck. Consequently, the number of real space points where the electronic structure energy and gradients are evaluated is optimized *to bring the overall scaling of the algorithm down*. We have derived an efficient deterministic, dynamically determined sampling procedure [149,150] for this purpose. *The sampling function is defined to be inversely proportional to the electronic energy, and directly proportional to the gradients of potential energy and wavepacket amplitude.* Only when the sampling function has large values, electronic structure calculations are performed. The potential in all other regions are obtained through Hermite curve interpolation [149,150]. The physical reasons behind the chosen form of the sampling function are as follows. Classically allowed regions should be well represented through accurate potential evaluations and hence the sampling function is inversely proportional to the potential energy. Similarly, a classical turning point region often signifies the advent of tunneling and generally has a large value for the potential gradient. Hence the sampling function is directly proportional to the potential gradient. Finally it is important to accurately represent regions of the potential where the wavepacket is currently present and hence the sampling function has the chosen form

$$\omega(\mathbf{x}) \propto \frac{[\tilde{\rho} + 1/I_\chi] \times [\tilde{E}' + 1/I_{E'}]}{\tilde{E} + 1/I_E}. \tag{13}$$

The parameters I_E, $I_{E'}$ and I_χ are chosen such that *the distribution of points where the wavepacket is sampled is similar in the classically allowed and classically forbidden (tunneling) regions* which allows us to accurately describe time-dependent quantum mechanical phenomena such as tunneling.

The TDDS procedure is implemented through the utilization of sophisticated techniques from computational mathematics. A wavelet scheme is used to expand the TDDS function and forces on the classical nuclei, obtained from the compressed real space grid, are computed through a low-pass filtering scheme introduced to attenuate Lagrange interpolation [150]. We find that this procedure greatly reduces the computational cost (by approximately four orders of magnitude) while maintaining accuracy. Furthermore, the approach allows a parallel implementation over a large number of processors. Physical reasons behind such an extraordinary compression are closely examined in Ref. [149,150] and it is shown that the sampling function provides a greater sample of grid points in regions where the standard semi-classical approximations generally fail. Thus, important progress has been made to convert the current procedure into a robust and practical method for quantum-classical dynamics in large systems.

3. CLUSTER DYNAMICS SIMULATIONS USING ADMP AND QWAIMD

In this section we discuss new chemical insights gleaned from the utilization of ADMP and QWAIMD. All simulations are conducted using a development version of the Gaussian series of electronic structure programs [172]. The ADMP methodology is available as part of the Gaussian-03 series of electronic structure programs [173]. The QWAIMD methodology is still under development and will be available in future.

3.1 Vibrational properties

Vibrational analysis inclusive of anharmonic, nuclear dynamical and temperature effects is vital towards the understanding of chemistry. While much of vibrational analysis in quantum chemistry is still conducted within the harmonic approximation, it has been shown that for medium sized clusters dynamical effects can dominate the observed spectroscopic behavior [26,27,54,62,150,158].

In recent publications [26,27], the structure, dynamics and ro-vibrational spectrum of the hydrated proton in the "magic" 21 water cluster [21–23,174] ($H^+(H_2O)_{21}$) was studied using ADMP. It was found that dynamical effects were critical in determining the vibrational properties of such clusters. (Please see Figure 16.1.) Furthermore, these results are in good agreement with previous experiment [21,23] through the reproduction of the sharp free OH stretch peak at $\approx$3700 cm^{-1} and the almost complete lack of intensity in the 2000–3000 cm^{-1} region. The latter was perhaps most puzzling since the harmonic spectra do reveal sharp peaks with large intensities in this region (Figure 4 in Refs. [21,26]) resulting from the OH stretch of the protonated (Eigen-like) species. Based on the agreement of our dynamical results with experiment [21] we concluded the importance of dynamical effects in vibrational spectroscopy of hydrogen bonded clusters.

Similar thermal effects have also been noted in hydroxide water clusters [54]. While proton transfer is widely accepted to follow the "Grotthuss" mechanism [28], the "hole hopping" hydroxide transfer process [175] has been challenged [176,177] and the existence and role of a four-coordinated (pentavalent) hydroxide oxygen as an intermediate in hydroxide transport in aqueous phase is debated [176–178]. In Ref. [54], we studied the energetic stability, dynamical evolution and spectroscopic properties of the hydroxide water cluster, $OH^-(H_2O)_6$, in gas phase that (a) could support a stable hydroxide ion with a four-coordinated hydroxide oxygen [179], and (b) display hydroxide ion migration.

Our results from finite temperature AIMD simulations indicate the presence of both pentavalent and tetra-hedral coordinated hydroxide ions. For example, the structural evolution of the system inclusive of the various isomers (or inherent structures) sampled during the dynamics at 220 K are shown in Figure 16.2. Starting from the pentavalent structure (A), the system evolves through accessible intermediates (B and C) into five-membered ring structures D and E *that facilitate a proton hop and a resultant hydroxide migration* eventually leading into another four-coordinated central hydroxide (H). The five-membered ring that supports the

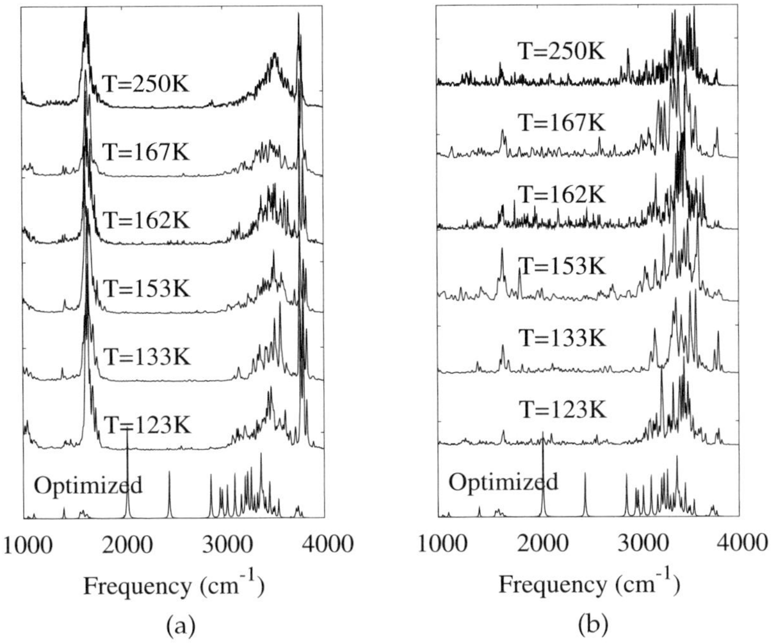

FIGURE 16.1 Fourier transform of the nuclear velocity–velocity autocorrelation function (a) and dipole–dipole autocorrelation function (b) obtained using ADMP simulations for a range of temperatures. Bottom panels in both figures (a) and (b) represent harmonic frequencies computed at the B3LYP/6-31+G** level starting from a snapshot obtained from low temperature ADMP simulation. This shows intensity in the 2000–3000 cm^{-1} region, which gets averaged out as a result of dynamical sampling of the an-harmonic potential surface in the FT-VAC and FT-DAC. A scaling factor of 0.962 is used for the geometry optimized spectrum while no scaling factor is necessary for the ADMP spectrum since it already contains contributions from anharmonicity. See Refs. [26,27] for details.

hydroxide through three water molecules donating hydrogen bonds seems central to the hydroxide transfer process occurring in this cluster. Furthermore, the spectroscopic features of these isomers are very different and the contribution of each isomer is modulated by the internal temperature of the cluster. For example, we present the dynamically averaged vibrational spectra obtained from our simulations in Figure 16.3. The reason for this temperature dependence in the spectrum is because the energy domains and regions of the potential surface sampled are different at different temperatures. Our studies indicate that the lower temperature dynamics, on average, samples structures that are 5–10 kcal/mol above structure A while the 220 K dynamics samples structures that are on average about 15 kcal/mol above the energy of structure A. This difference in thermal sampling in conjunction with the polarizability of the hydroxide ion is responsible for the temperature dependent effect found here.

To consider if these simulations have an experimental connection, we consider here two different kinds of experiments that are presently being employed by

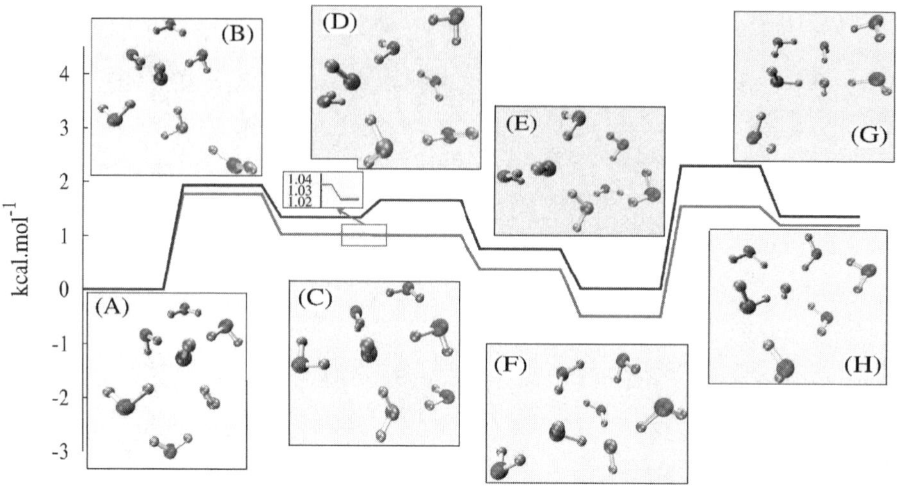

FIGURE 16.2 Structural evolution with energies obtained from geometry optimization of structures picked at intervals of ≈550 ps from AIMD (red lines: B3LYP, blue lines: MP2). Water oxygens are red, hydrogens are grey. The initial hydroxide oxygen is brown. The atom shown in aquamarine starts as a water oxygen but loses a proton during dynamics to become a hydroxide (structures F–H). (For interpretation of the references to color in this figure legend the reader is referred to the web version of this book.)

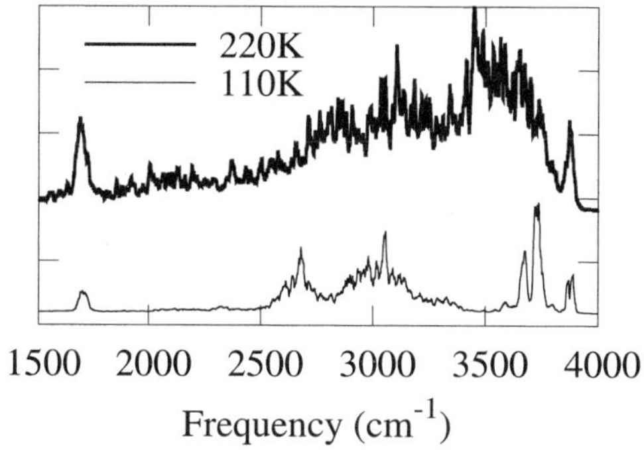

FIGURE 16.3 Fourier transform of dipole auto-correlation (FT-DAC) with quantum nuclear corrections. The lower temperature spectrum shows critical features (see text) which get "averaged out" due to dynamics at higher temperature. See Ref. [54] for details.

many groups to understand these kinds of clusters. In one set of experiments [180] an argon tagged cluster is isolated using mass-spectrometry and then irradiated using a single IR photon of a given frequency. The photon energy, if absorbed, induces the cluster to undergo intra-cluster vibrational redistribution which may result in a suitable amount of energy being transferred into the argon–cluster weak interaction and the subsequent ejection of argon. This loss of argon could then be detected using a second mass-spectrometric measurement. Clearly, in this situation the time scale of argon ejection and the intra-cluster vibrational redistribution is determined by (a) the amount of energy in the incident photon, and (b) the structure of the cluster at the time when the photon is absorbed. These correspond to the initial structure in our simulations and the initial kinetic energy imparted to the molecules. It is important to note that the sampling of the potential energy surface (and hence the resultant spectrum) is completely governed in this case by the two factors described above. In other words, the incident photon energy and the initial geometrical factors completely determine the dynamics both in the argon-tagged experimental action spectrum as well as the AIMD results presented here. Furthermore, these two factors also limit the region of conformational space that can be sampled, purely from (constant) energetic considerations.

In a second experimental situation [74,181], which is also used by many groups to describe cluster vibrations, the internal vibrational energy of the system is elevated above the dissociation threshold by sequential absorption of *many* photons incident in a non-coherent fashion. This spectroscopic technique is called infra-red multi-photon dissociation (IRMPD). The higher temperature AIMD simulations may be expected to simulate the pre-dissociation dynamics in these experiments [182] and there again the initial structure and initial kinetic energy dominate the sampling of the potential surface. As a result we expect that, within the practical limitations of finite time AIMD, our lower internal temperature simulations may be closer to the argon tagged single-photon action spectrum experiments [180], while our higher internal temperature AIMD simulations may be closer to multi-photon dissociation experiments [74]. This last statement has been independently confirmed through simulations on the proton-bound dimethyl ether system which has been studied using both argon-tagged vibrational action spectroscopy and infra-red multi-photon dissociation. Our higher temperature AIMD results conform with the IRMPD measurements while low temperature AIMD results yield spectra similar to argon-tagged vibrational action spectroscopy. It has also been found to be the case by Bowman and coworkers in Ref. [183].

We have also studied the behavior of gas-phase radicals, such as the hydroperoxyl radical (HO_2) [62], in water clusters which is important in atmospheric science (Figure 16.4). The hydroperoxyl radical is a major species in the HO_X chemical family [2] that affects the budgeting of many chemical systems in the atmosphere. The HO_X system plays a central role (along with the OH radical) in oxidative chemistry in the troposphere and ultimately controls the production rate of tropospheric ozone [7,16]. It is hence considered significant in atmospheric [2,5] and combustion chemistry [184]. Recent theoretical studies [16,17] have indicated the HO_2 radical to possess stable interactions with water clusters. Such stability provides an important sink for HO_X compounds [16,17,61,185] in the troposphere. As a result, the structural and dynamical features of water clusters play a vital role on HO_2 related chemistry.

There are two significant results that we have obtained as a result of finite temperature AIMD studies on HO_2 water clusters [62]. Firstly, the solvated HO_2 system displays an isolated, low-lying LUMO on account of being a free-radical. Furthermore, the dynamically averaged HOMO–LUMO gap monotonically de-

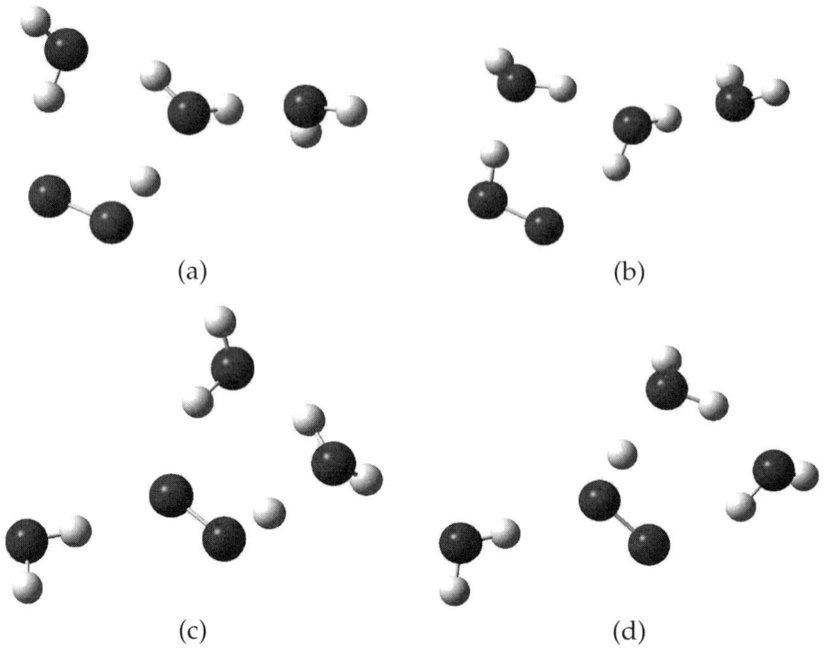

FIGURE 16.4 The figures depict the dynamical nature of the HO_2–water interaction. The structures in (a) and (b) are related by one single concerted (cyclic) proton hop. This is also the case for the structures in (c) and (d). See Ref. [62] for details.

creases as the number of water molecules is increased in an HO_2 water cluster. For a cluster of 20-water molecules solvating a single HO_2, the gap is already in the wide band gap semi-conductor region. This aspect is interesting because a low-lying isolated LUMO presents a trap for an excess electron to promote solvated electron chemistry. In aqueous aerosol chemistry, it is well-known [60,61] that HO_2 solvated in water clusters reacts with divalent cations such as Fe^{2+} and monovalent cations such as Cu^+ and this comprises an important HO_2 sink in the troposphere. Such reactions could be assisted by the lower HOMO–LUMO gap and isolated low-lying LUMO in solvated HO_2. Another important conclusion that arises from our AIMD study [62] is that the dynamical averaging of the potential energy surface greatly affects the excitation as well as infra-red spectra of such flexible water cluster systems. We find that the dynamics could *spread* the excitation energy spectrum by as much as 1 eV at temperatures that are relevant in the Earth's atmosphere. This further illustrates that single point excitation spectrum calculations on flexible systems will not, in general, lead to reliable results. We also find similar effects for vibrational properties; dynamical effects can give rise to marked deviations from the corresponding harmonic spectra as noted for the other cases discussed here.

To obtain molecular vibrational signatures using QWAIMD, we have recently introduced [149,150] a scheme using the Fourier transform of a unified velocity/flux autocorrelation function:

$$C(\omega) = \int_{-\infty}^{+\infty} e^{[-\imath\omega t]} \{ \langle v(t)v(0) \rangle_C + \langle \mathbf{J}(t)\mathbf{J}(0) \rangle_Q \}, \tag{14}$$

where the average flux, $\mathbf{J}(t)$, of the quantum wavepacket is

$$\mathbf{J}(t) = \mathcal{R}\left[\left\langle \psi(t) \left| \frac{-\imath\hbar}{m} \nabla \right| \psi(t) \right\rangle \right],$$

that is the time-dependent expectation value of flux. (Since the quantum wavepacket is complex, $\mathbf{J}(t)$ is also the expectation value of the semi-classical velocity operator.) Here $\mathcal{R}[\cdots]$ represents the real part of the complex number. The symbols $\langle \cdots \rangle_C$ and $\langle \cdots \rangle_Q$ represent classical and quantum ensemble average. The various components in these equations, namely, $|\psi(t)\rangle$ and the classical nuclear velocities are obtained from QWAIMD.

To demonstrate the cumulative velocity/flux correlation function formalism, in Ref. [150], we performed a comparative study of the vibrational density of states and infra-red spectra obtained in $[\text{Cl–H–Cl}]^-$ using electronic structure Harmonic frequencies, classical *ab initio* molecular dynamics, computation of nuclear quantum-mechanical eigenstates and by employing quantum wavepacket *ab initio* dynamics to compute the velocity/flux correlation function. It was found that the harmonic frequencies obtained from even high level electronic structure methods (including MP2, CCSD, CCSD(T)) are in poor agreement with experiment in this case on account of the highly anharmonic nature of the potential surface. Classical *ab initio* molecular dynamics improves considerably on the harmonic electronic structure frequencies but the velocity/flux correlation function of Eq. (14), obtained from QWAIMD, provides results in good agreement with experiment [150].

3.2 The "amphiphilicity" of the hydrated proton

An important contribution from the ADMP treatment of water clusters has been the so-called amphiphilic nature, or directional hydrophobicity, of the hydrated proton. Through ADMP treatment of protonated water clusters [24], and through the study of large water-vacuum interface systems [25] using the computationally efficient second generation Multi-State Empirical Valence Bond (MS-EVB2) approach [186,187] it was demonstrated that a protonated species tends to reside on the surface of a water vacuum interface with its lone pairs directed away from the neighboring water molecules. These results have subsequently been confirmed by many experimental [188–191] and theoretical studies [192,193].

The proposed reason for this effect as discussed in Ref. [25] is as follows. A single water molecule has on average four hydrogen bonds: two of these hydrogen bonds are donated from the oxygen atom to neighboring water molecules and one each is donated to the hydrogens from neighboring waters. When a water molecule gains an excess proton and becomes a hydronium, it gains a net positive charge. On average, the center of positive charge resides on the oxygen atom. Hence, while the hydrogen atoms in the hydronium retain their hydrogen bonds to the neighboring water molecules, the oxygen, on being the center of positive charge in a hydronium, is unable to support its hydrogen bond with any neighboring water molecule. Due to this reason, the solvation shell of a hydronium complex only comprises three water molecules on average. This reduction creates a lowering of the water density around the hydronium in the region directly in front of the lone pairs on the oxygen atom. Such a reduced density is entropically unfavorable to support inside the cluster, seemingly reminiscent of the hydrophobic effect. But if the hydronium were to be situated on the surface, with the lone pairs on the oxygen atom pointing "outwards", this reduction in density becomes a non-factor. This is an important result that may have critical bearing on many heterogeneous reactions occurring on cloud droplets.

4. CONCLUSIONS

We have discussed two *ab initio* molecular dynamics techniques that are useful in studying hydrogen bonded clusters of significance in atmospheric chemistry. Atom-centered density matrix propagation (ADMP) is an extended Lagrangian formalism that allows the simultaneous dynamics of electronic configurations along with classical nuclear dynamics. Critical quantum nuclear effects can be added to *ab initio* molecular dynamics through the quantum wavepacket *ab initio* molecular dynamics (QWAIMD) formalism.

Both approaches have been demonstrated for a variety of flexible hydrogen bonded systems. In this article we have utilized the techniques to discuss the role of (classical and quantum) dynamics on vibrational properties and electronic excitation and band gap properties. In addition we have also shown how these techniques can be used to obtain new and rather intriguing physical results such as the amphiphilic nature of the hydrated proton and have also shown how AIMD simulations can be utilized as a tool to understand differences between different experimental spectroscopic measurements.

ACKNOWLEDGEMENTS

The research is supported by the American Chemical Society Petroleum Research Fund, the Camille and Henry Dreyfus foundation, the Arnold and Mabel Beckman Foundation and the IU-STARS (Indiana University Science, Technology And Research Scholars) program.

REFERENCES

[1] S. Solomon, R.R. Garcia, F.S. Rowland, D.J. Wuebbles, *Nature* **321** (1986) 755–758.
[2] B.J. Finlayson-Pitts, J.N. Pitts Jr., *Chemistry of the Upper and Lower Atmosphere*, Academic Press, San Diego, 2000.
[3] R.P. Turco, O.B. Toon, P. Hamil, *J. Geophys. Res.* **94** (1989) 16493–16510.
[4] S. Solomon, *Rev. Geophys.* **37** (1999) 275–316.

[5] D.E. Heard, M.J. Pilling, *Chem. Rev.* **103** (2003) 5163–5198.
[6] J. Troe, *Proc. Combust. Inst.* **28** (2000) 1463–1469.
[7] J.A. Logan, M.J. Prather, S.C. Wofsy, M.B. McElroy, *J. Geophys. Res.* **86** (1981) 7210.
[8] M.J. McEwan, L.F. Phillips, *Acc. Chem. Res.* **3** (1970) 9–17.
[9] A.W. Castleman, R.G. Keesee, *Chem. Rev.* **86** (1986) 589–618.
[10] M.J. McEwan, L.F. Phillips, *Chemistry of the Atmosphere*, Eward Arnold, London, 1975.
[11] R.P. Wayne, *Chemistry of the Atmosphere*, Clarendon Press, Oxford, 1994.
[12] S. Aloisio, J.S. Francisco, *Acc. Chem. Res.* **33** (2000) 825–830.
[13] B.J. Gertner, J.T. Hynes, *Science* **271** (1996) 1563–1566.
[14] J.P. Devlin, N. Uras, J. Sadlej, V. Buch, *Nature* **417** (2002) 269–271.
[15] D.R. Hanson, A.R. Ravishankara, S. Solomon, *J. Geophys. Res.-Atmos.* **99** (1994) 3615–3629.
[16] Q. Shi, S.D. Belair, J.S. Francisco, S. Kais, *Proc. Natl. Acad. Sci. USA* **100** (17) (2003) 9686–9690.
[17] S.D. Belair, H. Hernandez, J.S. Francisco, *J. Am. Chem. Soc.* **126** (2004) 3024–3025.
[18] S. Aloisio, Y. Li, J.S. Francisco, *J. Chem. Phys.* **110** (1999) 9017–9019.
[19] S. Aloisio, J.S. Francisco, *J. Phys. Chem. A* **102** (1998) 1899–1902.
[20] R. Vacha, P. Slavicek, M. Mucha, B. Finlayson-Pitts, P. Jungwirth, *J. Phys. Chem. A* **108** (2004) 11573–11579.
[21] J.-W. Shin, N.I. Hammer, E.G. Diken, M.A. Johnson, R.S. Walters, T.D. Jaeger, M.A. Duncan, R.A. Christie, K.D. Jordan, *Science* **304** (2004) 1137–1140.
[22] T.S. Zwier, *Science* **304** (2004) 1119–1120.
[23] M. Miyazaki, A. Fujii, T. Ebata, N. Mikami, *Science* **304** (2004) 1134–1137.
[24] S.S. Iyengar, T.J.F. Day, G.A. Voth, *Int. J. Mass Spectrom.* **241** (2005) 197–204.
[25] M.K. Petersen, S.S. Iyengar, T.J.F. Day, G.A. Voth, *J. Phys. Chem. B* **108** (2004) 14804–14806.
[26] S.S. Iyengar, M.K. Petersen, T.J.F. Day, C.J. Burnham, V.E. Teige, G.A. Voth, *J. Chem. Phys.* **123** (2005) 084309.
[27] S.S. Iyengar, *J. Chem. Phys.* **126** (2007) 216101.
[28] N. Agmon, *Chem. Phys. Lett.* **244** (1995) 456–462.
[29] S. Wei, Z. Shi, A.W. Castleman Jr., *J. Chem. Phys.* **94** (1991) 3268–3270.
[30] M. Eigen, L.D. Maeyer, *Proc. R. Soc. London, Ser. A* **247** (1958) 505.
[31] G. Zundel, in: P. Schuster, G. Zundel, C. Sandorfy (Eds.), *The Hydrogen Bond*, vol. 2, North-Holland, Amsterdam, 1976, pp. 683–766.
[32] R.P. Bell, *The Proton in Chemistry*, Cornell University Press, Ithaca, NY, 1973.
[33] R.M. Lynden-Bell, J.C. Rasaiah, *J. Chem. Phys.* **105** (1996) 9266.
[34] R. Pomes, B. Roux, *Biophys. J.* **71** (1996) 19.
[35] H. Decornez, K. Drukker, S. Hammes-Schiffer, *J. Phys. Chem. A* **103** (1999) 2891.
[36] M.L. Brewer, U.W. Schmitt, G.A. Voth, *Biophys. J.* **80** (2001) 1691.
[37] C.J. Tsai, K.D. Jordan, *Chem. Phys. Lett.* **213** (1993) 181.
[38] S.S. Xantheas, *J. Chem. Phys.* **102** (1995) 4505.
[39] S. Sadhukhan, D. Munoz, C. Adamo, G.E. Scuseria, *Chem. Phys. Lett.* **306** (1999) 83.
[40] Y. Xie, R.B. Remington, H.F. Schaefer III, *J. Chem. Phys.* **101** (1994) 4878–4884.
[41] S.J. Singer, S. McDonald, L. Ojamae, *J. Chem. Phys.* **112** (2) (2000) 710.
[42] R.R. Sadeghi, H.-P. Cheng, *J. Chem. Phys.* **111** (1999) 2086–2094.
[43] D. Wei, D.R. Salahub, *J. Chem. Phys.* **106** (1997) 6086–6094.
[44] M.E. Tuckerman, K. Laasonen, M. Sprik, M. Parrinello, *J. Phys. Chem.* **99** (1995) 5749–5752.
[45] A. Banerjee, A. Quigley, R.F. Frey, D. Johnson, J. Simons, *J. Am. Chem. Soc.* **109** (1987) 1038.
[46] X.-D. Xiao, V. Vogel, Y.R. Shen, *Chem. Phys. Lett.* **163** (1989) 555.
[47] C. Radüge, V. Pflumio, Y.R. Shen, *Chem. Phys. Lett.* **274** (1997) 140.
[48] M. Okumura, L.I. Yeh, J.D. Myers, Y.T. Lee, *J. Phys. Chem.* **94** (1990) 3416.
[49] T.F. Magnera, D.E. David, J. Michl, *Chem. Phys. Lett.* **182** (1991) 363.
[50] D. Marx, M.E. Tuckerman, J. Hutter, M. Parrinello, *Nature* **397** (1999) 601.
[51] D. Asthagiri, L.R. Pratt, J.D. Kress, M.A. Gomez, *Proc. Natl. Acad. Sci.* **101** (2004) 7229–7233.
[52] M.E. Tuckerman, D. Marx, M. Parrinello, *Nature* **417** (2002) 925–929.
[53] N. Agmon, *Chem. Phys. Lett.* **319** (2000) 247–252.
[54] X. Li, V.E. Teige, S.S. Iyengar, *J. Phys. Chem. A* **111** (2007) 4815–4820.
[55] L. Perera, M.L. Berkowitz, *J. Chem. Phys.* **95** (1991) 1954–1963.

[56] L. Perera, M.L. Berkowitz, *J. Chem. Phys.* **96** (1992) 8288–8294.
[57] E.M. Knipping, M.J. Lakin, K.L. Foster, P. Jungwirth, D.J. Tobias, R.B. Gerber, D. Dabdub, B.J. Finlayson-Pitts, *Science* **288** (2000) 301–306.
[58] U. Achatz, B.S. Fox, M.K. Beyer, V.E. Bondybey, *J. Am. Chem. Soc.* **123** (2001) 6151.
[59] E.D. Sloan Jr., *Clathrate Hydrates of Natural Gases*, Marcel Dekker, Inc., New York, 1998.
[60] D.J. Jacob, *Atmos. Environ.* **34** (2000) 2131–2159.
[61] J. Thornton, J.P.D. Abbat, *J. Geophys. Res.* **110** (2005) D08309.
[62] S.S. Iyengar, *J. Chem. Phys.* **123** (2005) 084310.
[63] S.M. Haile, D.A. Boysen, C.R.I. Chisholm, R.B. Merle, *Nature* **410** (2001) 910.
[64] M.F. Schuster, W.H. Meyer, *Annu. Rev. Mater. Res.* **33** (2003) 233–261.
[65] K.D. Kreuer, A. Fuchs, M. Ise, M. Spaeth, J. Maier, *Electrochim. Acta* **43** (1998) 1281–1288.
[66] M. Iannuzzi, M. Parrinello, *Phys. Rev. Lett.* **93** (2004) 025901.
[67] B.S. Hudson, N. Verdal, *Phys. B Condens. Matter* **385** (2006) 212–215.
[68] Z. Nagel, J. Klinman, *Chem. Rev.* **106** (8) (2006) 3095–3118.
[69] W.W. Cleland, M.M. Kreevoy, *Science* **264** (1994) 1887.
[70] A. Warshel, A. Papazyan, P.A. Kollman, *Science* **269** (1995) 102.
[71] N.I. Hammer, E.G. Diken, J.R. Roscioli, M.A. Johnson, E.M. Myshakin, K.D. Jordan, A.B. McCoy, X. Huang, J.M. Bowman, S. Carter, *J. Chem. Phys.* **122** (24) (2005) 244301.
[72] E.G. Diken, J.M. Headrick, J.R. Roscioli, J.C. Bopp, M.A. Johnson, A.B. McCoy, *J. Phys. Chem. A* **109** (2005) 1487–1490.
[73] J.R. Roscioli, L.R. McCunn, M.A. Johnson, *Science* **316** (5822) (2007) 249–254.
[74] D.T. Moore, J. Oomens, L. van der Meer, G. von Helden, G. Meijer, J. Valle, A.G. Marshall, J.R. Eyler, *ChemPhysChem* **5** (2004) 740–743.
[75] I.V. Rubtsov, R.M. Hochstrasser, *J. Phys. Chem. B* **106** (2002) 9165.
[76] K.D. Rector, D. Zimdars, M.D. Fayer, *J. Chem. Phys.* **109** (1998) 5455.
[77] D. Eisenberg, W. Kauzman, *The Structure and Properties of Water*, Oxford University Press, Oxford, 1969.
[78] R.E. Wyatt, J.Z.H. Zhang (Eds.), *Dynamics of Molecules and Chemical Reactions*, Marcel Dekker, Inc., New York, 1996.
[79] J.M. Bowman, *Acc. Chem. Res.* **19** (1986) 202–208.
[80] A.B. McCoy, R.B. Gerber, M.A. Ratner, *J. Chem. Phys.* **101** (1994) 1975.
[81] A.B. McCoy, X. Huang, S. Carter, J.M. Bowman, *J. Chem. Phys.* **123** (6) (2005) 064317.
[82] S.S. Iyengar, G.A. Parker, D.J. Kouri, D.K. Hoffman, *J. Chem. Phys.* **110** (1999) 10283–10298.
[83] H.-D. Meyer, U. Manthe, L.S. Cederbaum, *Chem. Phys. Lett.* **165** (1990) 73.
[84] C.L. Lopreore, R.E. Wyatt, *Phys. Rev. Lett.* **82** (1999) 5190.
[85] G.C. Schatz, A. Kupperman, *J. Chem. Phys.* **65** (1976) 4642.
[86] J.B. Delos, *Rev. Mod. Phys.* **53** (1981) 287.
[87] M.D. Feit, J.A. Fleck, *J. Chem. Phys.* **78** (1982) 301.
[88] R. Kosloff, *Annu. Rev. Phys. Chem.* **45** (1994) 145.
[89] C. Leforestier, R.H. Bisseling, C. Cerjan, M.D. Feit, R. Freisner, A. Guldberg, A. Hammerich, D. Jolicard, W. Karrlein, H.D. Meyer, N. Lipkin, O. Roncero, R. Kosloff, *J. Comput. Phys.* **94** (1991) 59.
[90] P. DeVries, in: A.D. Bandrauk (Ed.), *Physics*, in: *NATO ASI Series B*, vol. 171, Plenum Press, New York, 1988, p. 481.
[91] H.W. Jang, J.C. Light, *J. Chem. Phys.* **102** (1995) 3262–3268.
[92] S.C. Althorpe, D.C. Clary, *Annu. Rev. Phys. Chem.* **54** (2003) 493–529.
[93] S.C. Althorpe, F. Fernandez-Alonso, B.D. Bean, J.D. Ayers, A.E. Pomerantz, R.N. Zare, E. Wrede, *Nature* **416** (2002) 67–70.
[94] Y. Huang, S.S. Iyengar, D.J. Kouri, D.K. Hoffman, *J. Chem. Phys.* **105** (1996) 927.
[95] W.H. Miller, S.D. Schwartz, J.W. Tromp, *J. Chem. Phys.* **79** (1983) 4889.
[96] N. Makri, *Comput. Phys. Commun.* **63** (1991) 389–414.
[97] J. Cao, G.A. Voth, *J. Chem. Phys.* **100** (1994) 5106.
[98] S. Jang, G.A. Voth, *J. Chem. Phys.* **111** (1999) 2357.
[99] M.D. Feit, J.A. Fleck, *J. Chem. Phys.* **79** (1983) 301.
[100] M.D. Feit, J.A. Fleck, *J. Chem. Phys.* **80** (1984) 2578.

[101] D. Kosloff, R. Kosloff, *J. Comput. Phys.* **52** (1983) 35.
[102] D. Kosloff, R. Kosloff, *J. Chem. Phys.* **79** (1983) 1823.
[103] H. Tal-Ezer, R. Kosloff, *J. Chem. Phys.* **81** (1984) 3967.
[104] B. Hartke, R. Kosloff, S. Ruhman, *Chem. Phys. Lett.* **158** (1986) 223.
[105] S.S. Iyengar, D.J. Kouri, D.K. Hoffman, *Theor. Chem. Acc.* **104** (2000) 471.
[106] J.V. Lill, G.A. Parker, J.C. Light, *Chem. Phys. Lett.* **89** (1982) 483.
[107] J.C. Light, I.P. Hamilton, J.V. Lill, *J. Chem. Phys.* **82** (1985) 1400.
[108] D.T. Colbert, W.H. Miller, *J. Chem. Phys.* **96** (1992) 1982–1991.
[109] Y. Huang, D.J. Kouri, M. Arnold, I. Thomas, L. Marchioro, D.K. Hoffman, *Comput. Phys. Commun.* **80** (1994) 1.
[110] T. Peng, J.Z.H. Zhang, *J. Chem. Phys.* **105** (1996) 6072–6074.
[111] S.C. Althorpe, *J. Chem. Phys.* **114** (2001) 1601–1616.
[112] R.B. Gerber, M.A. Ratner, *J. Chem. Phys.* **70** (1988) 97–132.
[113] J.O. Jung, R.B. Gerber, *J. Chem. Phys.* **105** (1996) 10332.
[114] N. Matsunaga, G.M. Chaban, R.B. Gerber, *J. Chem. Phys.* **117** (2002) 3541.
[115] E. Deumens, A. Diz, R. Longo, Y. Öhrn, *Rev. Mod. Phys.* **66** (1994) 917.
[116] M.D. Hack, D.G. Truhlar, *J. Phys. Chem. A* **104** (2000) 7917–7926.
[117] A.W. Jasper, C. Zhu, S. Nangia, D.G. Truhlar, *Faraday Discuss.* **127** (2004) 1–22.
[118] W.H. Miller, *J. Phys. Chem. A* **105** (2001) 2942–2955.
[119] E.J. Heller, *J. Chem. Phys.* **62** (1975) 1544–1555.
[120] G.A. Fiete, E.J. Heller, *Phys. Rev. A* **68** (2003) 022112.
[121] S. Hammes-Schiffer, J. Tully, *J. Chem. Phys.* **101** (1994) 4657–4667.
[122] M. Ben-Nun, J. Quenneville, T.J. Martnez, *J. Phys. Chem. A* **104** (2000) 5161.
[123] J.D. Coe, T.J. Martinez, *J. Am. Chem. Soc.* **127** (2005) 4560.
[124] T.J. Martinez, M. Ben-Nun, G. Ashkenazi, *J. Chem. Phys.* **104** (1996) 2847.
[125] T.J. Martinez, R.D. Levine, *J. Chem. Phys.* **105** (1996) 6334.
[126] D.A. Micha, *J. Phys. Chem. A* **103** (1999) 7562–7574.
[127] R.B. Gerber, V. Buch, M.A. Ratner, *J. Chem. Phys.* **77** (1982) 3022.
[128] R.H. Bisseling, R. Kosloff, R.B. Gerber, M.A. Ratner, L. Gibson, C. Cerjan, *J. Chem. Phys.* **87** (1987) 2760–2765.
[129] N. Makri, W.H. Miller, *J. Chem. Phys.* **87** (1987) 5781–5787.
[130] J.H. Skone, M.V. Pak, S. Hammes-Schiffer, *J. Chem. Phys.* **123** (2005) 134108.
[131] I.S.Y. Wang, M. Karplus, *J. Am. Chem. Soc.* **95** (1973) 8160.
[132] C. Leforestier, *J. Chem. Phys.* **68** (1978) 4406.
[133] R. Car, M. Parrinello, *Phys. Rev. Lett.* **55** (1985) 2471.
[134] M.C. Payne, M.P. Teter, D.C. Allan, T.A. Arias, J.D. Joannopoulos, *Rev. Mod. Phys.* **64** (1992) 1045.
[135] H.B. Schlegel, J.M. Millam, S.S. Iyengar, G.A. Voth, A.D. Daniels, G.E. Scuseria, M.J. Frisch, *J. Chem. Phys.* **114** (2001) 9758.
[136] D. Marx, J. Hutter, in: J. Grotendorst (Ed.), *Modern Methods and Algorithms of Quantum Chemistry*, vol. 1, John vonNeumann Institute for Computing, Julich, 2000, pp. 301–449.
[137] K. Bolton, W.L. Hase, G.H. Peslherbe, in: D.L. Thompson (Ed.), *Modern Methods for Multidimensional Dynamics*, World Scientific, Singapore, 1998, p. 143.
[138] H.B. Schlegel, *J. Comput. Chem.* **24** (2003) 1514–1527.
[139] P.A.M. Dirac, in: *The Principles of Quantum Mechanics*, fourth ed., in: *The Int. Ser. Monogr. Phys.*, vol. 27, Oxford University Press, New York, 1958.
[140] J. Frenkel, *Wave Mechanics*, Oxford University Press, Oxford, 1934.
[141] D.K. Remler, P.A. Madden, *Mol. Phys.* **70** (1990) 921.
[142] H.C. Andersen, *J. Chem. Phys.* **72** (1980) 2384–2393.
[143] M. Parrinello, A. Rahman, *Phys. Rev. Lett.* **45** (1980) 1196–1199.
[144] M. Pavese, D.R. Berard, G.A. Voth, *Chem. Phys. Lett.* **300** (1999) 93–98.
[145] M.E. Tuckerman, D. Marx, *Phys. Rev. Lett.* **86** (2001) 4946–4949.
[146] B. Chen, I. Ivanov, M.L. Klein, M. Parrinello, *Phys. Rev. Lett.* **91** (2003) 215503.
[147] S.S. Iyengar, J. Jakowski, *J. Chem. Phys.* **122** (2005) 114105.
[148] S.S. Iyengar, *Theor. Chem. Acc.* **116** (2006) 326.
[149] J. Jakowski, I. Sumner, S.S. Iyengar, *J. Chem. Theory Comput.* **2** (2006) 1203–1219.

[150] I. Sumner, S.S. Iyengar, *J. Phys. Chem. A* **111** (2007) 10313.
[151] M.E. Tuckerman, M. Parrinello, *J. Chem. Phys.* **101** (1994) 1302.
[152] S.S. Iyengar, H.B. Schlegel, J.M. Millam, G.A. Voth, G.E. Scuseria, M.J. Frisch, *J. Chem. Phys.* **115** (2001) 10291.
[153] H.B. Schlegel, S.S. Iyengar, X. Li, J.M. Millam, G.A. Voth, G.E. Scuseria, M.J. Frisch, *J. Chem. Phys.* **117** (2002) 8694.
[154] S.S. Iyengar, H.B. Schlegel, G.A. Voth, J.M. Millam, G.E. Scuseria, M.J. Frisch, *Isr. J. Chem.* **42** (2002) 191–202.
[155] S.S. Iyengar, M.J. Frisch, *J. Chem. Phys.* **121** (2004) 5061.
[156] S.S. Iyengar, H.B. Schlegel, G.A. Voth, *J. Phys. Chem. A* **107** (2003) 7269–7277.
[157] N. Rega, S.S. Iyengar, G.A. Voth, H.B. Schlegel, T. Vreven, M.J. Frisch, *J. Phys. Chem. B* **108** (2004) 4210–4220.
[158] X. Li, D.T. Moore, S.S. Iyengar, *J. Chem. Phys.* (2008), in press.
[159] J.S. Newhouse, T. Newhouse, J. Saito, S.S. Iyengar, M.-A. Gilles-Gonzalez, E.I. Newhouse, J. Perry, M. Alam, *Biophys. J.* (2007), submitted for publication.
[160] G.H. Golub, C.F. van Loan, *Matrix Computations*, The Johns Hopkins University Press, Baltimore, 1996.
[161] P. Pulay, *Mol. Phys.* **17** (1969) 197.
[162] F. Riesz, B.-Sz. Nagy, *Functional Analysis*, Dover Publications, Inc., New York, 1990.
[163] D.J. Kouri, Y. Huang, D.K. Hoffman, *Phys. Rev. Lett.* **75** (1995) 49–52.
[164] D.K. Hoffman, N. Nayar, O.A. Sharafeddin, D.J. Kouri, *J. Phys. Chem.* **95** (1991) 8299.
[165] S.S. Iyengar, J. Jakowski, I. Sumner, *J. Phys. Chem. B* (2007), submitted.
[166] M.E. Tuckerman, B.J. Berne, A. Rossi, *J. Chem. Phys.* **94** (1991) 1465.
[167] R.P. Feynman, A.R. Hibbs, *Quantum Mechanics and Path Integrals*, McGraw–Hill Book Company, New York, 1965.
[168] J. Cao, G.A. Voth, *J. Chem. Phys.* **100** (1994) 5093.
[169] J. Cao, G.A. Voth, *J. Chem. Phys.* **101** (1994) 6157.
[170] J. Cao, G.A. Voth, *J. Chem. Phys.* **101** (1994) 6168.
[171] S. Jang, G.A. Voth, *J. Chem. Phys.* **111** (1999) 2371.
[172] M.J. Frisch, G.W. Trucks, H.B. Schlegel, G.E. Scuseria, M.A. Robb, J.R. Cheeseman, J.A. Montgomery Jr., T. Vreven, K.N. Kudin, J.C. Burant, J.M. Millam, S.S. Iyengar, J. Tomasi, V. Barone, B. Mennucci, M. Cossi, G. Scalmani, N. Rega, G.A. Petersson, H. Nakatsuji, M. Hada, M. Ehara, K. Toyota, R. Fukuda, J. Hasegawa, M. Ishida, T. Nakajima, Y. Honda, O. Kitao, H. Nakai, M. Klene, X. Li, J.E. Knox, H.P. Hratchian, J.B. Cross, C. Adamo, J. Jaramillo, R. Gomperts, R.E. Stratmann, O. Yazyev, A.J. Austin, R. Cammi, C. Pomelli, J.W. Ochterski, P.Y. Ayala, K. Morokuma, G.A. Voth, P. Salvador, J.J. Dannenberg, V.G. Zakrzewski, S. Dapprich, A.D. Daniels, M.C. Strain, O. Farkas, D.K. Malick, A.D. Rabuck, K. Raghavachari, J.B. Foresman, J.V. Ortiz, Q. Cui, A.G. Baboul, S. Clifford, J. Cioslowski, B.B. Stefanov, G. Liu, A. Liashenko, P. Piskorz, I. Komaromi, R.L. Martin, D.J. Fox, T. Keith, M.A. Al-Laham, C.Y. Peng, A. Nanayakkara, M. Challacombe, P.M.W. Gill, B. Johnson, W. Chen, M.W. Wong, C. Gonzalez, J.A. Pople, Gaussian Development Version, revision b.01; Gaussian, Inc., Pittsburgh PA.
[173] M.J. Frisch, G.W. Trucks, H.B. Schlegel, G.E. Scuseria, M.A. Robb, J.R. Cheeseman, J.A. Montgomery Jr., T. Vreven, K.N. Kudin, J.C. Burant, J.M. Millam, S.S. Iyengar, J. Tomasi, V. Barone, B. Mennucci, M. Cossi, G. Scalmani, N. Rega, G.A. Petersson, H. Nakatsuji, M. Hada, M. Ehara, K. Toyota, R. Fukuda, J. Hasegawa, M. Ishida, T. Nakajima, Y. Honda, O. Kitao, H. Nakai, M. Klene, X. Li, J.E. Knox, H.P. Hratchian, J.B. Cross, C. Adamo, J. Jaramillo, R. Gomperts, R.E. Stratmann, O. Yazyev, A.J. Austin, R. Cammi, C. Pomelli, J.W. Ochterski, P.Y. Ayala, K. Morokuma, G.A. Voth, P. Salvador, J.J. Dannenberg, V.G. Zakrzewski, S. Dapprich, A.D. Daniels, M.C. Strain, O. Farkas, D.K. Malick, A.D. Rabuck, K. Raghavachari, J.B. Foresman, J.V. Ortiz, Q. Cui, A.G. Baboul, S. Clifford, J. Cioslowski, B.B. Stefanov, G. Liu, A. Liashenko, P. Piskorz, I. Komaromi, R.L. Martin, D.J. Fox, T. Keith, M.A. Al-Laham, C.Y. Peng, A. Nanayakkara, M. Challacombe, P.M.W. Gill, B. Johnson, W. Chen, M.W. Wong, C. Gonzalez, J.A. Pople, Gaussian 03, revision b.02, Gaussian, Inc., Pittsburgh PA, 2003.
[174] C.-C. Wu, C.-K. Lin, H.-C. Chang, J.-C. Jiang, J.-L. Kuo, M.L. Klein, *J. Chem. Phys.* **122** (2005) 074315.

[175] J.D. Bernal, R.H. Fowler, *J. Chem. Phys.* **1** (1933) 515–548.
[176] N. Agmon, *Chem. Phys. Lett.* **319** (2000) 247–252.
[177] M.E. Tuckerman, D. Marx, M. Parrinello, *Nature* **417** (2002) 925–929.
[178] D. Asthagiri, L.R. Pratt, J.D. Kress, M.A. Gomez, *Proc. Natl. Acad. Sci.* **101** (2004) 7229–7233.
[179] H.M. Lee, P. Tarkeshwar, K.S. Kim, *J. Chem. Phys.* **121** (2004) 4657.
[180] E.G. Diken, J.M. Headrick, J.R. Roscioli, J.C. Bopp, M.A. Johnson, A.B. McCoy, *J. Phys. Chem. A* **109** (2005) 1487–1490.
[181] G. Groenewold, A. Gianotto, K. Cossel, M. VanStipdonk, D. Moore, N. Polfer, J. Oomens, W. de-Jong, L. Visscher, *J. Am. Chem. Soc.* **128** (14) (2006) 4802–4813.
[182] D.T. Moore, X. Li, S.S. Iyengar, manuscript in preparation.
[183] M. Kaledin, A.L. Kaledin, J.M. Bowman, *J. Phys. Chem. A* **110** (2006) 2922–2939.
[184] J.A. Miller, R.J. Kee, C.K. Westbrook, *Annu. Rev. Phys. Chem.* **41** (1990) 345.
[185] R.V. Martin, D.J. Jacob, R.M. Yantosca, M. Chin, P. Ginoux, *J. Geophys. Res.* **108** (2003) 4097.
[186] U.W. Schmitt, G.A. Voth, *J. Chem. Phys.* **111** (1999) 9361.
[187] T.J.F. Day, A.V. Soudachov, M. Cuma, U.W. Schmitt, G.A. Voth, *J. Chem. Phys.* **117** (2002) 5839.
[188] P.B. Petersen, R.J. Saykally, *J. Phys. Chem. B* **109** (2005) 7976.
[189] Y.R. Shen, V. Ostroverkhov, *Chem. Rev.* **106** (2006) 1140.
[190] L.M. Pegram, M.T. Record, *Proc. Natl. Acad. Sci. USA* **103** (2006) 14278.
[191] V. Buch, A. Milet, R. Vacha, P. Jungwirth, J.P. Devlin, *Proc. Natl. Acad. Sci. USA* **104** (2007) 7342.
[192] M. Mucha, T. Frigato, L.M. Levering, H.C. Allen, D.J. Tobias, L.X. Dang, P. Jungwirth, *J. Phys. Chem. B* **109** (2005) 7617.
[193] J.A. Morrone, K.E. Hasllinger, M.E. Tuckerman, *J. Phys. Chem. B* **110** (2006) 3712.

CHAPTER 17

From Molecules to Droplets

Allan Gross[*], **Ole John Nielsen**[**] and **Kurt V. Mikkelsen**[**]

Contents

1. Introduction 356
2. Energy Functional and Hamiltonians 358
   2.1 A structured environment model based on the heterogeneous dielectric media method 359
   2.2 A structured environment model based on the quantum mechanical–classical system method 362
3. The Introduction of the Multiconfigurational Self-Consistent Wave Function 366
4. Derivation of Response Equations for Quantum–Classical Systems 369
   4.1 A structured environment response approach: the multiconfigurational self-consistent field response theory for the heterogeneous dielectric media model 373
   4.2 A structured environment model given by the multiconfigurational self-consistent field response theory for the quantum mechanical–classical system model 376
5. Brief Overview of Results 381
6. Conclusion 382
Acknowledgement 383
References 383

Abstract

We consider methods for investigating the interactions between aerosol particles and molecules and how to calculate properties of molecules interacting with aerosol particles. The basic models include a heterogeneous dielectric media approach and a quantum mechanical–classical mechanical approach. Both models describe the electronic structure of the molecule at the level of correlated electronic approaches or density functional theory approximations.

[*] Research Department, Danish Meteorological Institute, Lyngbyvej 100, DK-2100 Copenhagen Ø, Denmark
[**] Department of Chemistry, University of Copenhagen, Universitetsparken 5, DK-2100 Copenhagen Ø, Denmark

Advances in Quantum Chemistry, Vol. 55
ISSN 0065-3276, DOI: 10.1016/S0065-3276(07)00217-1

1. INTRODUCTION

The interaction between molecules in the gas phase and the surface of particles and droplets in the atmosphere is central to important phenomena such as ozone depletion in polar areas, hygroscopic growth of particles, aging, acid rain, particle growth and cloud droplet formation.

1150 Tg C [1] of natural volatile organic compounds (VOCs) are emitted to the atmosphere yearly and 110 Tg C VOCs [1] from human activities including fossil fuel burning. Upon release to the atmosphere VOCs are photo-oxidized to form polyfunctional molecules which are water soluble, reactive and with low vapor pressures allowing for partitioning to the particle phase. It is important to understand the fate of biogenic as well as anthropogenic emissions. A number of molecules in the atmosphere are only produced by human activity. An obvious example is the chloro–fluoro–carbons (CFCs) and halons as well as their first and second generation alternatives the hydro–fluoro–carbons (HFCs), and SF_6 and CF_3SF_5. These compounds have been very useful for a large number of applications e.g., air-condition, degreasing of electronics, solvents, fire extinguishing equipment. These compounds are not oxidized in the troposphere but transported to the stratosphere where photolysis by UV radiation leads to the release of halogen atoms which in turn act as catalysts for destruction of the ozone layer [1]. Special circumstances apply over Antarctica where the role of polar stratospheric clouds in the formation of the Antarctic ozone hole highlights the importance of including heterogeneous chemistry in the description of the atmosphere.

Following the Montreal protocol and its amendments the use of CFCs and halons has been phased out as has the use of some of the first generation alternatives. New alternatives with shorter atmospheric lifetimes are being investigated in order to comply with stricter regulations on the use of greenhouse gases (GHG).

Greenhouse gases are by definition molecules that absorb infrared radiation in the so-called atmospheric window and have fairly long lifetimes on the order of years. The most important green house gases are H_2O, CO_2, N_2O, CH_4 and man-made halogenated compounds. Concern about the consequences of rising CO_2 levels in the atmosphere was expressed as early as 1896 and the recent release of the 4th assessment report by the Intergovernmental Panel of Climate Change (IPCC) emphasizes their warming effect on climate. Atmospheric particles have an effect on climate that could be just as large as the effect of greenhouse gases, however, much remains to be learned and further research is urgently needed. Examples of issues that should be investigated include the effects of chemical composition on nucleation and particle growth.

We consider methods for describing how molecules interact with aerosol particles and how to obtain molecular properties and rate constants of relevance when studying the molecular level mechanisms for the formation of aerosol particles and how these provide the basis for heterogeneous chemistry. For understanding mass and heat transfer to and from aerosol particles we need to focus on the processes related to a gas molecule as it approaches the surface of an aerosol particle. A macroscopic property related to these processes is the sticking probabilities/mass accommodation coefficients that are used when modelling evaporation, condensation and heterogeneous chemistry in relation to aerosol particles. For modelling at the molecular level heterogeneous chemistry/aerosol dynamics we need to know:

(1) absorption rate constants,
(2) desorption rate constants,
(3) sticking coefficients, and
(4) surface reaction coefficients (if the gas-phase molecule is able to react on the surface of the aerosol).

Presently, we will review microscopic models that provide molecular insight to the interactions between gas molecules and aerosol particles.

We outline methods that are used to calculate interaction energies and linear and nonlinear molecular properties. Hereby, we provide a fundamental understanding of how molecules stick to aerosol particles and the basis for calculating macroscopic properties such as sticking-coefficients, diffusion coefficients along with nonlinear optical signals. The perturbations of homogeneous and heterogeneous environments on molecular systems can in some cases result in very different observations and this can be illustrated when considering molecules with an inversion symmetry center solvated in solution or in a structured environment. The solvated molecule, due to symmetry rules, does not exhibit nonlinear optical properties whereas for the molecule interacting with a heterogeneous/structured environment, such as an interface, one is able to induce nonlinear optical molecular properties within molecules possessing inversion symmetry.

We cover the theoretical background for two basic models for interactions between molecules and aerosol particles and outline procedures for determining interaction energies and linear and nonlinear optical properties of molecules interacting with an aerosol particle. The models are a heterogeneous solvation model and a quantum mechanical–classical mechanical model. For both we outline the procedures for calculating at the level of correlated electronic structure both interaction energies and molecular properties for molecules interacting with aerosol particles.

Coupling of quantum mechanical molecular subsystems with larger classically treated subsystems has traditionally involved electronic structure models describing molecules embedded in a dielectric medium and this is a research area that has expanded tremendously over the last three decades [2–36]. Most of this work has involved electronic structure methods that have been based on uncorrelated electronic structure methods [2–12,15–19]. Accurate description of the electronic structure of molecular systems requires that the correlated electronic motion in the molecule is incorporated and therefore a number of correlated electronic structure methods have been developed such as the second order Møller–Plesset (MP2) [28,30,90,91], the multiconfigurational self-consistent reaction field (MCSCRF) [13,20] and the coupled-cluster self-consistent reaction field (CCSCRF) method [36].

For molecules situated in/at and interacting with a structured environment, it is important to determine molecular properties of the molecules when they are exposed to an external time-dependent electromagnetic field. Methods for doing this are clearly related to the work of bringing electron correlation descriptions of modern response theory for molecular systems in the gas phase [37–44], to the situation where the molecules are solvated [21,24,45–47].

We will focus on methods that are able to perform investigations of electromagnetic properties of molecules exposed to structured environments such as a molecule interacting with an aerosol particle and the computation of energy terms along with linear and nonlinear electromagnetic properties of a molecular system exposed to a structured environment. Presently, we focus on correlated electronic structure methods for general electronic structure systems in conjunction with two different methods for investigating molecular properties of molecules in structured environments: (i) a heterogeneous dielectric media method by Jørgensen *et al.* [48–50] and (ii) a quantum mechanical-classical system method by Poulsen *et al.* [51–53]. For closed shell electronic systems, coupled cluster and density functional theory electronic structure methods have been developed for structured environments [54–67,92–95].

We consider correlated electronic structure methods for molecular systems interacting with a structured environment and the utilization of response theory makes it possible to calculate molecular properties of the molecular subsystem coupled to an aerosol particle and the molecular properties could be

- frequency-dependent polarizabilities (α),
- frequency-dependent first hyperpolarizability tensors (β),
- frequency-dependent second hyperpolarizabilities (γ),
- frequency-dependent polarizabilities of excited states,
- excitation and deexcitation energies along with their corresponding transition moments,
- two-photon matrix elements,
- three-photon absorptions,
- two-photon absorption between excited states.

We start out with a section on the energy functionals and Hamiltonians that are relevant for molecular systems interacting with a structured environment. We continue with a section that briefly describes the correlated electron structure method, the multiconfigurational self-consistent field (MCSCF) electronic structure method. In the following section we cover the procedure for obtaining the correlated MCSCF response equations for the two different models describing molecules in structured environments. The final sections provide a brief overview of the results obtained using the two methods and a conclusion.

2. ENERGY FUNCTIONAL AND HAMILTONIANS

In order to investigate the energies and molecular properties of molecules interacting with aerosol particles, it is crucial to establish the Hamiltonians and the energy functionals for the two structural environment methods. The basic principle for both structural environment methods is the same and it is one that has been utilized successfully within quantum chemistry [2–33] and molecular reaction dynamics [19,68–71,96], we divide a large system into two subsystems. The focus is on the smaller of the two subsystems and the smaller subsystem is described by quantum mechanics, whereas the other and larger subsystem is treated by a much coarser method, typically by using classical mechanics.

For both methods, we describe the interactions between the quantum subsystem and the classical subsystem as interactions between charges and/or induced charges/dipoles and a van der Waals term [2–18]. The coupling between the quantum subsystem and the classical subsystem is introduced into the quantum mechanical Hamiltonian by finding effective interaction operators for the interactions between the two subsystems. This provides an effective Schrodinger equation for determining the MCSCF electronic wave function of the molecular system exposed to a classical environment, a structured environment, such as an aerosol particle.

2.1 A structured environment model based on the heterogeneous dielectric media method

For this approach, the molecule subsystem is encapsulated in a cavity denoted C and the cavity is determined by the surfaces Σ_m and Σ_l (see Figure 17.1). The cavity is surrounded by a structured environment that is approximated by two parts denoted S_m and S_l. The two linear, homogeneous and isotropic dielectric media, S_m and S_l, are in contact with the cavity through the surfaces Σ_m and Σ_l. The dielectric media, S_m and S_l, are characterized by the dielectric constants ϵ_m and ϵ_l, respectively, and they have the following spatial positions:

$$V_{S_m} = \left\{ \vec{\mathbf{r}} = (r,\theta,\phi) \mid \theta \notin \left[-\frac{\pi}{2}; \frac{\pi}{2} \right] \right\},$$
$$V_{S_l} = \left\{ \vec{\mathbf{r}} = (r,\theta,\phi) \mid r \geqslant R, \theta \in \left[-\frac{\pi}{2}; \frac{\pi}{2} \right] \right\}. \quad (1)$$

Thereby, we have defined a heterogeneous dielectric model with a hemisperical cavity, C, having a radius of R and positioned such that the cavity C is on the surface of S_m and embedded in S_l. The volume of the cavity C is

$$V_C = \left\{ \vec{\mathbf{r}} = (r,\theta,\phi) \mid r \leqslant R, \theta \in \left[-\frac{\pi}{2}; \frac{\pi}{2} \right] \right\}. \quad (2)$$

We describe the electronic structure of the molecular system (denoted M) by a MCSCF wave function (denoted Ψ), and the charge distribution of the molecular system is given by ρ_M.

Polarization charges in the structured environment are induced due to the charge distribution of the molecular system within the cavity. The interactions between the induced polarization charges and the charge distribution of the molecular system are described by a polarization potential, U_{pol} and the polarization energy is determined using the following expression:

$$E_{\mathrm{pol}} = \frac{1}{2} \int d\vec{\mathbf{r}}\, \rho_M(\vec{\mathbf{r}}) U_{\mathrm{pol}}(\vec{\mathbf{r}}), \quad (3)$$

where $\vec{\mathbf{r}}$ denotes position vectors within the cavity. For fixed nuclear configurations, we solve the nonlinear Schrödinger equation where the Hamiltonian

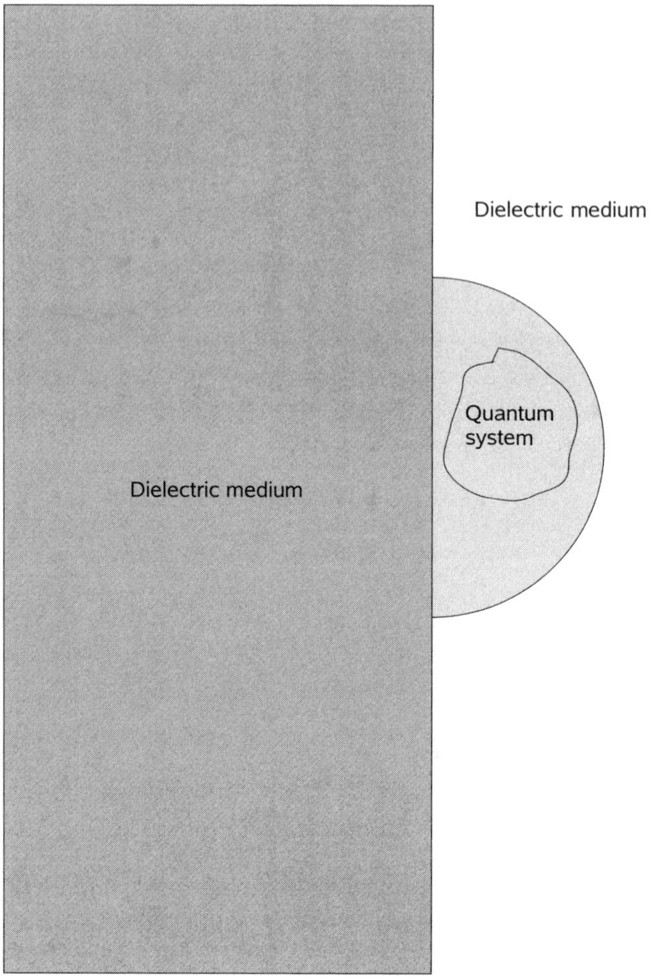

FIGURE 17.1 Illustration of the model presented in Section 2.1.

includes the couplings to the structured environment. The solution gives the energy and wave function of the molecular system interacting with the structured environment.

We denote the coordinates of the N_{el} electrons as

- $\mathbf{q} \equiv \mathbf{q}_1, \ldots, \mathbf{q}_{N_{\mathrm{el}}}$

and we denote the coordinates of the M nuclei as

- $\mathbf{Q} \equiv \mathbf{Q}_1, \ldots, \mathbf{Q}_M$.

Based on the above definitions and approximations, we have the following representation of the Hamiltonian for the molecular system coupled to the two dielectric media

$$H_M(\mathbf{q};\mathbf{Q})\Psi(\mathbf{q};\mathbf{Q}) = E(\mathbf{Q})\Psi(\mathbf{q};\mathbf{Q}), \tag{4}$$

where

$$H_M(\mathbf{q};\mathbf{Q}) = H_M^0(\mathbf{q};\mathbf{Q}) + W_{\text{pol}} \tag{5}$$

and the Hamiltonian for the molecular system is given by

- the Hamiltonian, H_M^0, for the isolated molecular system in vacuum, and
- the interaction operator, W_{pol}.

The interaction operator, W_{pol}, that describes the interactions between the quantum mechanical and classical subsystems is given by

- the induced potential, U_{pol}, in the two dielectric media and
- the molecular charge distribution $\rho_m(\vec{\mathbf{r}})$

$$\rho_m(\vec{\mathbf{r}}) = \sum_{i=1}^{M} Q_i \delta(\vec{\mathbf{r}} - \vec{\mathbf{R}}_i), \tag{6}$$

where the partial charge on the ith nucleus (having the position vector $\vec{\mathbf{R}}_i$) is given by Q_i. A reliable method for calculating the partial charges is the method by Cioslowski [72,73] for which it has been shown that this approach follows the normal convergency rules with respect to one and many-electron basis sets [74].

The quantum mechanical subsystem is exposed to a total electrostatic potential, Φ, which is given by

- the sum of a potential arising from the molecular charge distribution, U_{ρ_M},
- and a potential arising from the induced charges in the two dielectric media, U_{pol},

and therefore we have that

$$\Phi(\vec{\mathbf{r}}) = U_{\rho_M}(\vec{\mathbf{r}}) + U_{\text{pol}}(\vec{\mathbf{r}}), \quad \vec{\mathbf{r}} \in V_C. \tag{7}$$

The electrostatic potential Φ within the cavity is determined by solving the Poisson and Laplace equations with the appropriate boundary conditions

$$\nabla^2 \Phi(\vec{\mathbf{r}}) = \begin{cases} -4\pi\rho_M(\vec{\mathbf{r}}), & \vec{\mathbf{r}} \in V_C, \\ 0, & \vec{\mathbf{r}} \in V_{S_1}, \\ 0, & \vec{\mathbf{r}} \in V_{S_2}, \end{cases} \tag{8}$$

with the additional condition that the charge distribution of the quantum mechanical subsystem is nonzero within the cavity, C, i.e. $\rho(\vec{\mathbf{r}}) = \rho_M(\vec{\mathbf{r}})$ for $\vec{\mathbf{r}} \in V_C$ and zero outside the cavity.

We apply the method of image charges [75,76] for determining the induced potential and for a charge q_i located at the position $\vec{\mathbf{r}}_i = (r_i, \theta_i, \phi_i)$ within the cavity we determine the following contributions to the induced potential:

- one charge $q_{i,1} = -q_i$ at $\vec{\mathbf{r}}_{i,1} = (r_i, \pi - \theta_i, \phi_i)$ in the metal surface, S_2, due to the perfect conductor/vacuum interface.

- A second charge $q_{i,2} = -q_i a(\epsilon - 1)/(\epsilon + 1)r_i$ at $\vec{\mathbf{r}}_{i,2} = (a^2/r_i, \theta_i, \phi_i)$ due to the hemispherical environment between the vacuum and the dielectric medium.
- A third $q_{i,3} = -q_{i,2}$ at $\vec{\mathbf{r}}_{i,3} = (a^2/r_i, \pi - \theta_i, \phi_i)$ due to the perfect conductor/dielectric medium.

The determination of the image charges/induced charges makes it possible to obtain the induced potential

$$U_{\text{pol}}^{(i)}(\vec{\mathbf{r}}) = \sum_{j}^{3} \frac{q_{i,j}}{|\vec{\mathbf{r}} - \vec{\mathbf{r}}_{i,j}|}. \tag{9}$$

As the molecular system contains a certain number of partial charges (denoted N) we evaluate the total induced potential to be

$$U_{\text{pol}}(\vec{\mathbf{r}}) = \sum_{i=1}^{N} U_{\text{pol}}^{(i)}(\vec{\mathbf{r}}). \tag{10}$$

At this point by introducing the wave function for the quantum mechanical subsystem we are able to determine the interaction operator between the quantum mechanical and the classical subsystems

$$W_{\text{pol}} = \frac{1}{2} \sum_{j} \sum_{k} U_{\text{pol}}^{(k)}(\mathbf{r}_j) \sum_{pq} \langle \phi_p | q_j | \phi_q \rangle E_{pq}, \tag{11}$$

where ϕ_p and ϕ_q represent the molecular orbitals p and q. We define the excitation operator, E_{pq}, as:

$$E_{pq} = \sum_{\sigma} a_{p\sigma}^{\dagger} a_{q\sigma}, \tag{12}$$

where the operators $a_{p\sigma}^{\dagger}$ and $a_{p\sigma}$ are the creation and annihilation operators for an electron in the spin orbital $\phi_{p\sigma}$.

Concluding this section, we note that the interaction operator describing the interactions between the molecular compound (the quantum mechanical subsystem) and the structured environment (the classical subsystem) is effectively written as a product of two one-electron operators.

2.2 A structured environment model based on the quantum mechanical–classical system method

In order to establish the procedure for calculating MCSCF wave functions for this model (see Figure 17.2), we define the Hamiltonian for the total system as:

$$\hat{H} = \hat{H}_{\text{QM}} + \hat{H}_{\text{QM/CM}} + \hat{H}_{\text{CM}}, \tag{13}$$

where we have the following definition for the three terms

- the Hamiltonian of the quantum mechanical system in vacuum is denoted by $\hat{\mathbf{H}}_{\text{QM}}$;

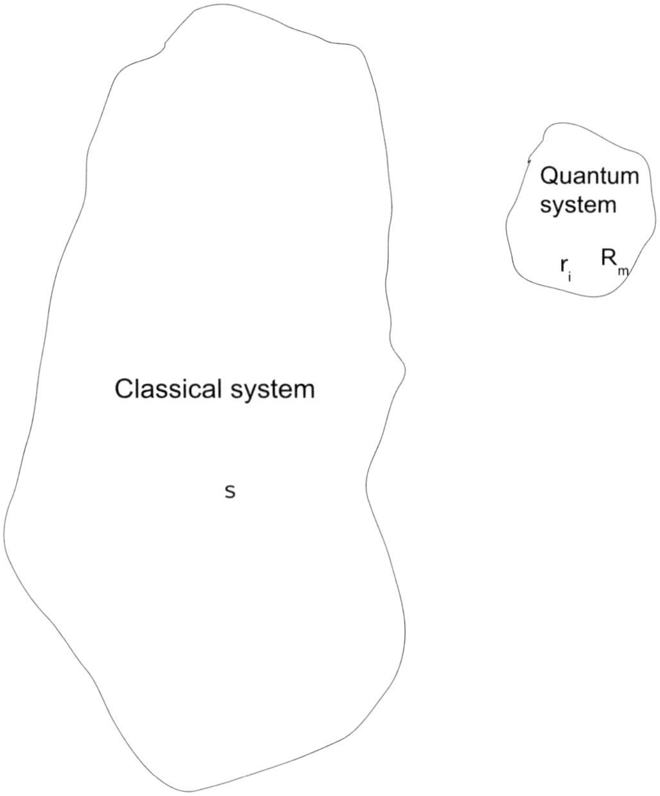

FIGURE 17.2 Illustration of the model presented in Section 2.2.

- the Hamiltonian for the classical system is denoted by $\hat{\mathbf{H}}_{\mathrm{CM}}$ and this term is represented as a force field;
- and the term that describes the interactions between the quantum mechanical and the classical system is denoted by $\hat{\mathbf{H}}_{\mathrm{QM/CM}}$.

The interaction operator between the quantum mechanical and classical mechanical subsystems is written as

$$\hat{H}_{\mathrm{QM/CM}} = \hat{H}^{\mathrm{el}} + \hat{H}^{\mathrm{vdw}} + \hat{H}^{\mathrm{pol}} \tag{14}$$

and the interactions are composed of three terms related to

- the electrostatic interactions denoted by $\hat{H}^{\mathrm{el}}$,
- the polarization interactions given by $\hat{H}^{\mathrm{pol}}$,
- and the van der Waals interactions represented by $\hat{H}^{\mathrm{vdw}}$.

We make use of the following notation for the quantum and classical subsystems (i) i and $\bar{r}_i$, are the index and coordinates for the electrons, respectively, (ii) m and $\bar{R}_m$, denote, respectively, the index and coordinates for the nuclei in the quantum subsystem, and (iii) the index and coordinates for the classical subsystem are given by s and $\bar{R}_s$, respectively.

We define the electrostatic interaction operator for the electrostatic interactions between the charges in the quantum subsystem and the charges in the classical subsystem as

$$\hat{H}^{\text{el}} = -\sum_{s=1}^{S}\sum_{i=1}^{N}\frac{q_s}{|\bar{r}_i - \bar{R}_s|} + \sum_{s=1}^{S}\sum_{m=1}^{M}\frac{q_s Z_m}{|\bar{R}_s - \bar{R}_m|}. \tag{15}$$

The interactions between the electrons in the quantum subsystem and the classical charges in the classical subsystem are described by the first term in the equation. The electrostatic interactions between the classical charges and the charges of the nuclei are given by the second term.

For the van der Waals interactions, $\hat{H}^{\text{vdw}}$, we represent these as

$$\begin{aligned}\hat{H}^{\text{vdw}} &= \sum_{s=1}^{A}\sum_{m:\text{center}} 4\epsilon_{ms}\left[\left(\frac{\sigma_{ms}}{|\bar{R}_m - \bar{R}_s|}\right)^{12} - \left(\frac{\sigma_{ms}}{|\bar{R}_m - \bar{R}_s|}\right)^{6}\right] \\ &= \sum_{s=1}^{A}\sum_{m:\text{center}}\left[\frac{A_{ms}}{|\bar{R}_m - \bar{R}_s|^{12}} - \frac{B_{ms}}{|\bar{R}_m - \bar{R}_s|^{6}}\right],\end{aligned} \tag{16}$$

where we sum over the van der Waals interactions between the nuclei (m) and the classical sites (s). The two terms A_{ms} and B_{ms} are interaction terms.

We represent the polarization interactions where the index a runs over all polarization sites

$$\hat{H}^{\text{pol}} = \frac{1}{2}\sum_{i=1}^{N}\sum_{a=1}^{A}\frac{\bar{\mu}_a^{\text{ind}}\cdot(\bar{R}_a - \bar{r}_i)}{|\bar{R}_a - \bar{r}_i|^3} - \frac{1}{2}\sum_{m=1}^{M}\sum_{a=1}^{A}\frac{Z_m\bar{\mu}_a^{\text{ind}}\cdot(\bar{R}_a - \bar{R}_m)}{|\bar{R}_a - \bar{R}_m|^3}. \tag{17}$$

For each polarization site we have an induced dipole moment, $\bar{\mu}_a^{\text{ind}}$, which is given as

$$\bar{\mu}_a^{\text{ind}} = \alpha\big(\vec{E}^e(\bar{R}_a) + \vec{E}^m(\bar{R}_a) + \vec{E}^s(\bar{R}_a) + \vec{E}^{\text{ind}}(\bar{R}_a)\big). \tag{18}$$

The electric fields appear due to (i) the nuclei belonging to the quantum mechanical subsystem ($\vec{E}^m$), (ii) the electrons belonging to the quantum mechanical subsystem ($\vec{E}^e$), (iii) for the classical subsystem the induced dipole moments ($\vec{E}^{\text{ind}}$), and (iv) the charges in the classical subsystem ($\vec{E}^s$).

At this point we write the interaction energy ($E_{\text{QM/CM}}$), between the quantum and classical subsystems as

$$\begin{aligned}E_{\text{QM/CM}} &= E^{\text{el}} + E^{\text{pol}} + E^{\text{vdw}} \\ &= \tilde{O}_{mm'}^{As'} - \sum_{s=1}^{S}\langle N_s\rangle - \frac{1}{2}\alpha\sum_{a=1}^{A}\langle Rr_a\rangle\{\langle Rr_a\rangle + \bar{O}_{mm'}^{aS'}\} \\ &\quad + \underbrace{E^{\text{vdw}} + E_{S,M}^{\text{el,nuc}}}_{\text{independent of electrons}}.\end{aligned} \tag{19}$$

In this expression we have defined the following terms:

- $$\bar{O}^{aS'}_{mm'} = 2\sum_{m=1}^{M} \frac{Z_m(\bar{R}_a - \bar{R}_m)}{|\bar{R}_a - \bar{R}_m|^3} + \sum_{s' \notin a} \frac{q_{s'}(\bar{R}_a - \bar{R}_{s'})}{|\bar{R}_a - \bar{R}_{s'}|^3} + \sum_{a' \neq a} \left\{ \frac{3(\bar{\mu}^{\text{ind}}_{a'} \cdot (\bar{R}_a - \bar{R}_{a'}))(\bar{R}_a - \bar{R}_{a'})}{|\bar{R}_a - \bar{R}_{a'}|^5} - \frac{\bar{\mu}^{\text{ind}}_{a'}}{|\bar{R}_a - \bar{R}_{a'}|^3} \right\}, \quad (20)$$

- $$\tilde{O}^{As'}_{mm'} = \frac{1}{2}\alpha \sum_{m=1}^{M} \sum_{a=1}^{A} -Z_m \frac{(\bar{R}_a - \bar{R}_m)}{|\bar{R}_a - \bar{R}_m|^3} \times \left[\sum_{m'=1}^{M} Z_{m'} \frac{(\bar{R}_a - \bar{R}_{m'})}{|\bar{R}_a - \bar{R}_{m'}|^3} + \sum_{s' \notin a} q_{s'} \frac{(\bar{R}_a - \bar{R}_{s'})}{|\bar{R}_a - \bar{R}_{s'}|^3} + \sum_{a' \neq a} \left\{ \frac{3(\bar{\mu}^{\text{ind}}_{a'} \cdot (\bar{R}_a - \bar{R}_{a'}))(\bar{R}_a - \bar{R}_{a'})}{|\bar{R}_a - \bar{R}_{a'}|^5} - \frac{\bar{\mu}^{\text{ind}}_{a'}}{|\bar{R}_a - \bar{R}_{a'}|^3} \right\} \right], \quad (21)$$

- when calculating the contribution of the polarization interactions we are concerned with terms related to Rr_a

  $$Rr_a = \sum_{pq} t^a_{pq} \boldsymbol{E}_{pq} \quad (22)$$

  and we determine the expectation value as

  $$\langle Rr_a \rangle = \frac{\langle 0|Rr_a|0\rangle}{\langle 0|0\rangle} = \sum_{pq} D_{pq} t^a_{pq}, \quad (23)$$

  where

  $$t^a_{pq} = \langle \phi_p | \frac{\bar{r}_i - \bar{R}_a}{|\bar{r}_i - \bar{R}_a|^3} | \phi_q \rangle, \quad (24)$$

- and for the electrostatic interactions we have that N_s is given as

  $$N_s = \sum_{pq} n^s_{pq} \boldsymbol{E}_{pq}, \quad (25)$$

  and we calculate the expectation value as:

  $$\langle N_s \rangle = \sum_{pq} D_{pq} n^s_{pq}, \quad (26)$$

  where the one-electron integral is written as

  $$n^s_{pq} = \langle \phi_p | \frac{q_s}{|\bar{R}_s - \bar{r}_i|} | \phi_q \rangle, \quad (27)$$

- additionally, we have the term for the electrostatic interaction involving the nuclear part of the molecular subsystem

$$E_{S,M}^{\text{el,nuc}} = \sum_{s=1}^{S} \sum_{m=1}^{M} \frac{q_s Z_m}{|\bar{R}_m - \bar{R}_s|}. \tag{28}$$

Therefore, we have established an expression for the total energy and the Hamiltonian of the system.

3. THE INTRODUCTION OF THE MULTICONFIGURATIONAL SELF-CONSISTENT WAVE FUNCTION

We will in this section introduce the representation of the electronic wave function of the molecular system and describe how we determine the wave function of the quantum mechanical subsystem including interactions with the structured environment. We consider the situation where the MCSCF electronic wave function of the quantum mechanical subsystem is optimized while interacting with a classical system represented by charges, polarization sites and van der Waals sites. We start out by expressing the total electronic free energy for the QM/MM-system as

$$\mathcal{F}_{\text{QM/CM}}(\lambda) = \mathcal{E}_{\text{vac}}(\lambda) + \mathcal{E}_{\text{QM/CM}}(\lambda), \tag{29}$$

and this energy expression is given as a functional of the electronic structure parameters, denoted by λ, and for a MCSCF electronic wave function, where the function $|\Theta_i\rangle$ denotes the set of configuration state functions (CSFs), we have

$$|0\rangle = \exp\left[\sum_{r>s} \kappa_{sr}(\boldsymbol{E}_{sr} - \boldsymbol{E}_{rs})\right] \sum_i c_i |\Theta_i\rangle, \tag{30}$$

where our parameter set, λ, is divided into two groups: (i) orbital parameters denoted $\{\kappa\}$ and (ii) configurational parameters denoted $\{c_i\}$.

The determination of the optimal electronic structure parameters requires that we establish a procedure for optimizing the energy functional and this is achieved by expanding the energy function to second order with respect to the orbital and configurational parameters. The non-redundant electronic parameters are $\lambda, \lambda^{(k)}$ and k provides the number of the current iteration and we have the following energy difference to second order between the current iteration and the next optimization step

$$\begin{aligned} \Delta\mathcal{E}^{(2)}(\lambda - \lambda^{(k)}; \lambda^{(k)}) &= \mathcal{E}^{(2)}(\lambda - \lambda^{(k)}; \lambda^{(k)}) - \mathcal{E}(\lambda^{(k)}) \\ &= \boldsymbol{g}^T(\lambda - \lambda^{(k)}) + \frac{1}{2}(\lambda - \lambda^{(k)})^T \boldsymbol{H}(\lambda - \lambda^{(k)}). \end{aligned} \tag{31}$$

We need to understand how the interactions between the quantum and classical subsystems can be incorporated into the optimization procedures of the MCSCF electronic wave function. We do this by deriving the gradient and Hessian contributions of the quantum-classical interactions to the expansion of the energy functional and we find the following gradient terms

$$\frac{\partial E_{\mathrm{QM/CM}}}{\partial \lambda_i} = -\sum_{s=1}^{S} \frac{\partial \langle N_s \rangle}{\partial \lambda_i} - \alpha \sum_{a=1}^{A} \frac{\partial \langle Rr_a \rangle}{\partial \lambda_i} \left\{ \langle Rr_a \rangle + \frac{1}{2} \bar{O}^{aS'}_{mm'} \right\}, \tag{32}$$

and we define an effective one-electron operator

$$T^g = \sum_{s=1}^{S} (-N_s) - \alpha \sum_{a=1}^{A} \left\{ \langle Rr_a \rangle + \frac{1}{2} \bar{O}^{aS'}_{mm'} \right\} Rr_a, \tag{33}$$

and this definition provides rather compact expressions for the gradient of the energy functional. For the configurational part of the gradient due to the quantum-classical interactions we have

$$\begin{aligned} \frac{\partial E_{\mathrm{QM/CM}}}{\partial c_\mu} &= \sum_{s=1}^{S} \left(-\frac{\partial \langle N_s \rangle}{\partial c_\mu} \right) - \alpha \sum_{a=1}^{A} \frac{\partial \langle Rr_a \rangle}{\partial c_\mu} \left\{ \langle Rr_a \rangle + \frac{1}{2} \bar{O}^{aS'}_{mm'} \right\} \\ &= 2\big[\langle \mu | T^g | 0 \rangle - \langle 0 | T^g | 0 \rangle c_\mu \big], \end{aligned} \tag{34}$$

and for the orbital part of the gradient we find the following contribution

$$\begin{aligned} \frac{\partial E_{\mathrm{QM/CM}}}{\partial \kappa_{pq}} &= \sum_{s=1}^{S} \left(-\frac{\partial \langle N_s \rangle}{\partial \kappa_{pq}} \right) - \alpha \sum_{a=1}^{A} \frac{\partial \langle Rr_a \rangle}{\partial \kappa_{pq}} \left\{ \langle Rr_a \rangle + \frac{1}{2} \bar{O}^{aS'}_{mm'} \right\} \\ &= 2 \langle 0 | \big[\boldsymbol{E}_{pq}, T^g \big] | 0 \rangle. \end{aligned} \tag{35}$$

Clearly, it is an advantage to define an effective one-electron operator T^g when constructing the gradient contributions to the energy functional due to the interactions between the two subsystems.

Our next step for achieving a second order optimization procedure of the energy functional is to obtain the Hessian contribution, denoted by $\sigma_i^{(k)}$, due to the interactions between the quantum and classical subsystems. This is effectively done by performing linear transformations using configuration state function trial vectors and orbital trial vectors. The trial vectors are denoted $\boldsymbol{b}^{(k)}$ and we obtain the following expressions

$$\sigma_i^{(k)} = \sum_j \frac{\partial^2 E}{\partial \lambda_j \partial \lambda_i} \boldsymbol{b}_i^{(k)}. \tag{36}$$

When we utilize a configuration state function trial vector we find

$$\begin{aligned} \sigma_j^{c,\mathrm{QM/CM}} &= \sum_\nu \frac{\partial^2 E_{\mathrm{QM/CM}}}{\partial c_\nu \partial \lambda_j} \boldsymbol{b}^{(c)} \\ &= -\sum_{s=1}^{S} \sum_\nu \frac{\partial^2 \langle N_s \rangle}{\partial c_\nu \partial \lambda_j} \boldsymbol{b}_\nu^{(c)} - \alpha \sum_{a=1}^{A} \left\{ \langle Rr_a \rangle + \frac{1}{2} \bar{O}^{aS'}_{mm'} \right\} \\ &\quad \times \sum_\nu \frac{\partial^2 \langle Rr_a \rangle}{\partial c_\nu \partial \lambda_j} \boldsymbol{b}_\nu^{(c)} - \alpha \sum_{a=1}^{A} \frac{\partial \langle Rr_a \rangle}{\partial \lambda_j} \sum_\nu \frac{\partial \langle Rr_a \rangle}{\partial c_\nu} \boldsymbol{b}_\nu^{(c)}. \end{aligned} \tag{37}$$

On the other hand we have the following for an orbital trial vector,

$$\sigma_j^{o,\mathrm{QM/CM}} = \sum_{pq} \frac{\partial^2 E_{\mathrm{QM/CM}}}{\partial\lambda_j \partial\kappa_{pq}} \boldsymbol{b}_{pq}^{(o)}$$
$$= -\sum_{s=1}^{S} \sum_{pq} \frac{\partial^2 \langle N_s \rangle}{\partial\lambda_j \partial\kappa_{pq}} \boldsymbol{b}_{pq}^{(o)} - \alpha \sum_{a=1}^{A} \left\{ \langle Rr_a \rangle + \frac{1}{2} \bar{O}_{mm'}^{aS'} \right\}$$
$$\times \sum_{pq} \frac{\partial^2 \langle Rr_a \rangle}{\partial\lambda_j \partial\kappa_{pq}} \boldsymbol{b}_{pq}^{(o)} - \alpha \sum_{a=1}^{A} \frac{\partial \langle Rr_a \rangle}{\partial\lambda_j} \sum_{pq} \frac{\partial \langle Rr_a \rangle}{\partial\kappa_{pq}} \boldsymbol{b}_{pq}^{(o)}. \quad (38)$$

As for the case of the gradient terms to the energy functional, we define effective one-electron operators and thereby we are able to express these terms as:

- for a configuration state function trial vector,

$$\sigma_\mu^{c,\mathrm{QM/CM}} = 2\big[\langle\mu|T^g|B\rangle - \langle 0|T^g|0\rangle \boldsymbol{b}_\mu^{(c)}\big] + 2\big[\langle\mu|T^{xc}|0\rangle - \langle 0|T^{xc}|0\rangle c_\mu\big], \quad (39)$$

and

$$\sigma_{pq}^{c,\mathrm{QM/CM}} = 2\big[\langle 0|[\boldsymbol{E}_{pq}, T^g]|B\rangle + \langle 0|[\boldsymbol{E}_{pq}, T^{xc}]|B\rangle\big], \quad (40)$$

- and for an orbital trial vector

$$\sigma_j^{o,\mathrm{QM/CM}} = 2\langle\mu|T^{yo}|0\rangle + 2\big[\langle\mu|T^{xo}|0\rangle - \langle 0|T^{xo}|0\rangle c_\mu\big], \quad (41)$$

and

$$\sigma_{pq}^{o,\mathrm{QM/CM}} = 2\langle 0|[\boldsymbol{E}_{pq}, T^{yo}]|0\rangle + 2\langle 0|[\boldsymbol{E}_{pq}, T^{xo}]|0\rangle$$
$$+ \sum_t \big(\langle 0|[\boldsymbol{E}_{tq}, T^g]|0\rangle \boldsymbol{b}_{pt} - \langle 0|[\boldsymbol{E}_{tp}, T^g]|0\rangle \boldsymbol{b}_{qt}\big). \quad (42)$$

The definitions of the effective one-electron operators are:

-

$$T^{xc} = -2\alpha \sum_{a=1}^{A} \langle 0|Rr_a|B\rangle Rr_a, \quad (43)$$

with

$$|B\rangle = \sum_\mu \boldsymbol{b}_\mu^{(c)} |\Theta_\mu\rangle. \quad (44)$$

-

$$T^{yo} = \sum_{s=1}^{S} (-V^s) - \alpha \sum_{a=1}^{A} \left\{ \langle Rr_a \rangle + \frac{1}{2} \bar{O}_{mm'}^{aS'} \right\} Q^a \quad (45)$$

-

$$T^{xo} = -\alpha \sum_{a=1}^{A} \langle 0|Q^a|0\rangle Rr_a, \tag{46}$$

- – the term for V^s is represented as

$$V^s = \sum_{pq} V^s_{pq} \boldsymbol{E}_{pq} \tag{47}$$

with

$$V^s_{pq} = \sum_r \left[\kappa_{pr} n^s_{rq} - n^s_{pr}\kappa_{rq}\right], \tag{48}$$

–

$$Q^a = \sum_{pq} Q^a_{pq} \boldsymbol{E}_{pq}, \tag{49}$$

$$Q^a_{pq} = \sum_r \left[\kappa_{pr} t^a_{rq} - t^a_{pr}\kappa_{rq}\right]. \tag{50}$$

We note that the above optimization procedure for an energy functional concerning an MCSCF electronic wave function describing a quantum mechanical subsystem interacting with a classical subsystem is very similar to that seen for vacuum [77,78] and reaction field [20,79] methods utilizing MCSCF wave functions. We observe that the mathematical structure for the implementation is the same but we have to provide different effective one-electron operators.

4. DERIVATION OF RESPONSE EQUATIONS FOR QUANTUM–CLASSICAL SYSTEMS

We will in this section consider the mathematical structure for computational procedures when calculating molecular properties of a quantum mechanical subsystem coupled to a classical subsystem. Molecular properties of the quantum subsystem are obtained when considering the interactions between the externally applied time-dependent electromagnetic field and the molecular subsystem in contact with a structured environment such as an aerosol particle. Therefore, we need to study the time evolution of the expectation value of an operator A and we express that as

$$\frac{d\langle A\rangle}{dt} = \left\langle \frac{\partial A}{\partial t} \right\rangle - i\big\langle [A, H]\big\rangle, \tag{51}$$

where we write the total Hamiltonian as

$$H = H_0 + W_{se} + V(t) \tag{52}$$

and we have the following terms: (i) the time independent Hamiltonian of the quantum subsystem is $(H_0 + W_{se})$ where H_0 is the Hamiltonian of the isolated quantum subsystem and the operator W_{se} gives the interactions between the molecular subsystem and the structured environment, (ii) the $V(t)$ operator takes care of the interactions between the externally applied time-dependent electromagnetic field and the quantum subsystem within the structured environment. We determine the time-dependent expectation values using the time dependent wave function, $|0^t\rangle$, at a given time t

$$\langle \cdots \rangle = \langle {}^t 0 | \cdots | 0^t \rangle. \tag{53}$$

We define the time-dependent wave function $|0^t\rangle$ at time t as

$$|0^t\rangle = \exp[i\kappa(t)] \exp[iS(t)] |0\rangle \tag{54}$$

and we have that the electronic wave function $|0\rangle$ is an optimized multiconfigurational self-consistent reference wave function that has been optimized in the presence of the structured environment

$$(H_0 + W_{se})|0\rangle = E_0|0\rangle, \tag{55}$$

and furthermore, we require that the electronic wave function for both the orbital and configurational variation parameters, λ, satisfies the generalized Brillouin condition

$$\langle 0|[\lambda, H_0 + W_{se}]|\,\rangle. \tag{56}$$

Our starting point is the following, we have a wave function and an energy functional that has been optimized in the presence of the structured environment and therefore, we have the physically correct state of the quantum mechanical subsystem coupled to the structured environment.

Our wish to study how the quantum mechanical subsystem evolves in time requires that we select appropriate operators, that will provide time propagation of the wave functions. These unitary transformation operators are

- for the orbital part $\exp[i\kappa(t)]$ where

$$\kappa(t) = \sum_k \left[\kappa_k(t) c_k^\dagger + \kappa_k^*(t) c_k\right], \tag{57}$$

  with the excitation operator, c_k,

$$c_k = E_{pq}, \quad p > q, \tag{58}$$

- and for the configuration part $\exp[iS(t)]$ with

$$S(t) = \sum_n \left[S_n(t) R_n^\dagger + S_n^*(t) R_n\right] \tag{59}$$

  and the states given by $|n\rangle$ belong to the orthogonal complement space to $|0\rangle$. Therefore we write the state transfer operator as

$$R_m^\dagger = |n\rangle\, \langle 0|. \tag{60}$$

Furthermore, we define a set of operators, $\mathbf{T}$, that will enable us to follow the evolution of the quantum subsystem:

$$\mathbf{T} = \left(\mathbf{c}^{\dagger}, \mathbf{R}^{\dagger}, \mathbf{c}, \mathbf{R}\right) \tag{61}$$

and the individual operators are

$$c_k^{\dagger,t} = \exp\left[i\kappa(t)\right]c_k^{\dagger}\exp\left[-i\kappa(t)\right], \tag{62}$$
$$c_k^{t} = \exp\left[i\kappa(t)\right]c_k\exp\left[-i\kappa(t)\right], \tag{63}$$
$$R_n^{\dagger,t} = \exp\left[i\kappa(t)\right]\exp\left[iS(t)\right]R_n^{\dagger}\exp\left[-iS(t)\right]\exp\left[-i\kappa(t)\right], \tag{64}$$
$$R_n^{t} = \exp\left[i\kappa(t)\right]\exp\left[iS(t)\right]R_n\exp\left[-iS(t)\right]\exp\left[-i\kappa(t)\right]. \tag{65}$$

We determine how a general operator evolves in time by determining the time evolution of the expectation value of the above time transformed operators. For doing this we utilize the Ehrenfest equation $\mathbf{T}^{t,\dagger}$

$$\frac{d}{dt}\langle \mathbf{T}^{\dagger,t}\rangle = \left\langle \frac{\partial \mathbf{T}^{\dagger,t}}{\partial t}\right\rangle - i\langle[\mathbf{T}^{\dagger,t}, H_0]\rangle - i\langle[\mathbf{T}^{\dagger,t}, V(t)]\rangle - i\langle[\mathbf{T}^{\dagger,t}, W_{se}]\rangle \tag{66}$$

and we calculate the expectation values from the time dependent wave function, $|O^t\rangle$.

We solve the Ehrenfest equation for each order of the perturbation and we collect for each order the necessary terms for calculating response functions at the given order. The only part within the Ehrenfest equation that is not covered by the ordinary response theory of quantum mechanical systems in vacuum is represented by the last term on the right hand side of Eq. (66) [80–83]. The contributions to the response functions arising from the last term are related to the presence of the structured environment coupled to the quantum mechanical subsystem. It is the main contributor to changes in molecular properties when transferring the quantum subsystem from vacuum to the structured environment.

The perturbation arising from the externally applied electromagnetic field is written as

$$V(t) = \int_{-\infty}^{\infty} d\omega V^{\omega}\exp\left[(-i\omega + \epsilon)t\right], \tag{67}$$

where we let the term ϵ be a positive infinitesimal number ensuring that the perturbation is zero at $t = -\infty$, the term V^{ω} denotes the Fourier transform of the perturbation operator, $V(t)$, and as the perturbation is Hermitian we have that $(V^{\omega})^{+} = V^{-\omega}$.

In the case of an arbitrary time-independent operator A, we have

$$\langle 0^t|A|0^t\rangle = \langle 0|A|0\rangle + \int_{-\infty}^{\infty} d\omega_1 \exp\left[(-i\omega_1 + \epsilon)t\right]\langle\langle A; V^{\omega_1}\rangle\rangle_{\omega_1}$$
$$+ \frac{1}{2}\int_{-\infty}^{\infty} d\omega_1 \int_{-\infty}^{\infty} d\omega_2 \exp\left[(-i(\omega_1 + \omega_2) + 2\epsilon)t\right]\langle\langle A; V^{\omega_1}, V^{\omega_2}\rangle\rangle_{\omega_1,\omega_2} + \cdots \tag{68}$$

and this enables us to define the following response functions [80]:

- the linear response function: $\langle\langle A; V^{\omega_1}\rangle\rangle_{\omega_1}$, and
- the quadratic response functions: $\langle\langle A; V^{\omega_1}, V^{\omega_2}\rangle\rangle_{\omega_1,\omega_2}$.

Next we will utilize the sum over states or spectral approach when we assume that it is possible to determine an exact reference state $|0\rangle$ together with eigenfunctions, $|n\rangle$, that are solutions to a Hamiltonian given by $H_0 + W_{se}$. Using this basis of exact states and that the energy difference between the state given by $|n\rangle$ and the reference state $|0\rangle$ is given as $\omega_n = E_n - E_0$, we are able to write the linear response function as

$$\langle\langle A; V^{\omega}\rangle\rangle_{\omega} = \lim_{\epsilon\to 0}\sum_{n\neq 0}\frac{\langle 0|A|n\rangle\langle n|V^{\omega}|0\rangle}{\omega-\omega_n+i\epsilon} - \lim_{\epsilon\to 0}\sum_{n\neq 0}\frac{\langle 0|V^{\omega}|n\rangle\langle n|A|0\rangle}{\omega+\omega_n+i\epsilon}, \tag{69}$$

where the frequency of the external electric field is ω. Based on this expression of the linear response function we observe that this function has poles at frequency $\omega = \pm\omega_n$ which are the excitation and deexcitation energies of the quantum subsystem and residues giving the corresponding transition moments.

For the quadratic response function, we have that the energy difference between the excited state $|q\rangle$ and the ground state $|0\rangle$ is $\omega_{q0} = E_q - E_0$ and for the external electric field we have the frequencies ω_m and ω_n. We are able to write the spectral representation of the quadratic response function as

$$\begin{aligned}
\langle\langle B; V^{\omega_m}, V^{\omega_n}\rangle\rangle_{\omega_m,\omega_n} = \sum_{p,q>0}\Bigg\{ & \frac{\langle 0|B|p\rangle[\langle p|V^{\omega_m}|q\rangle - \delta_{pq}\langle 0|V^{\omega_m}|0\rangle]\langle q|V^{\omega_n}|0\rangle}{(\omega_m+\omega_n-\omega_{p0})(\omega_n-\omega_{q0})} \\
&+ \frac{\langle 0|V^{\omega_n}|q\rangle[\langle q|V^{\omega_m}p\rangle - \delta_{pq}\langle 0|V^{\omega_m}|0\rangle]\langle p|B|0\rangle}{(\omega_m+\omega_n+\omega_{p0})(\omega_n+\omega_{q0})} \\
&- \frac{\langle 0|V^{\omega_m}|p\rangle[\langle p|B|q\rangle - \delta_{pq}\langle 0|B|0\rangle]\langle q|V^{\omega_n}|0\rangle}{(\omega_m+\omega_{p0})(\omega_n-\omega_{q0})} \\
&+ \frac{\langle 0|B|p\rangle[\langle p|V^{\omega_n}|q\rangle - \delta_{pq}\langle 0|V^{\omega_n}|0\rangle]\langle q|V^{\omega_p}|0\rangle}{(\omega_m+\omega_n-\omega_{p0})(\omega_m-\omega_{q0})} \\
&+ \frac{\langle 0|V^{\omega_m}|q\rangle[\langle q|V^{\omega_n}|p\rangle - \delta_{pq}\langle 0|V^{\omega_n}|0\rangle]\langle p|B|0\rangle}{(\omega_m+\omega_n+\omega_{p0})(\omega_m+\omega_{q0})} \\
&- \frac{\langle 0|V^{\omega_n}|p\rangle[\langle p|B|q\rangle - \delta_{pq}\langle 0|B|0\rangle]\langle q|V^{\omega_m}|0\rangle}{(\omega_n+\omega_{p0})(\omega_m-\omega_{q0})}\Bigg\}.
\end{aligned} \tag{70}$$

We observe that we obtain poles when either ω_n or ω_m equals an excitation or deexcitation energy of one-photon transitions and when the sum of two frequencies, $\omega_n + \omega_m$, is equal to an excitation or deexcitation energy of two-photon transitions. The residues provide information about one- and two-photon transition matrix elements.

In the case of the cubic response function, we have the following

$$\langle\langle\langle B; V^{\omega_m}, V^{\omega_n}, V^{\omega_l}\rangle\rangle\rangle_{\omega_m,\omega_n,\omega_l}$$
$$= \sum P(m,n,l)$$
$$\times \sum_{p,q,r>0} \frac{\langle 0|B|p\rangle[\langle p|V^{\omega_m}|q\rangle - \delta_{pq}\langle 0|V^{\omega_m}|0\rangle][\langle q|V^{\omega_n}|r\rangle - \delta_{qr}\langle 0|V^{\omega_n}|0\rangle]\langle r|V^{\omega_l}|0\rangle}{(\omega_m + \omega_n + \omega_l - \omega_{p0})(\omega_n + \omega_l - \omega_{q0})(\omega_r - \omega_{l0})} \tag{71}$$

and

- the energy difference between the state, $|p\rangle$ and reference state is $\omega_{p0} = E_p - E_0$,
- the permutation operator is the operator P, and
- the frequencies of the external electric field are given by ω_m, ω_n and ω_l.

Based on this representation we find that the cubic response function has poles at frequencies

- ω_n, ω_m or ω_l equal to an excitation or deexcitation energy for a one photon transition,
- $\omega_n + \omega_m$, $\omega_n + \omega_l$, $\omega_m + \omega_l$ equal to an excitation or deexcitation energy for a two-photon transition, and
- $\omega_n + \omega_m + \omega_l$ equal to an excitation or deexcitation energy for a three photon transition.

and correspondingly that the cubic response function has the residues

- three-photon transition matrix elements, and
- transition moments between non-reference states (excited states).

The sum-over-states expressions that we have presented in Eqs. (69), (70), and (71) are only true for exact wave functions and they are rather cumbersome methods for calculating time-dependent electromagnetic properties of a quantum mechanical subsystem within a structured environment. The advantage of the sum-over-states expressions is that they illustrate the type of information that is obtainable from response functions. We have utilized modern versions of response theory where the summation over states is eliminated when performing actual calculations, that involve approximative wave functions [21,24,45–47,80–83].

4.1 A structured environment response approach: the multiconfigurational self-consistent field response theory for the heterogeneous dielectric media model

Our next step concerns the mathematical background for calculating time-dependent electromagnetic properties of quantum mechanical subsystems that are interacting with a structured environment. In this section we let the structured environment, the aerosol particle, be represented as a heterogeneous dielectric media and we consider the modifications of the last term in Eq. (66). Based on these modifications we are able to determine the changes to the response function that arise due to the interactions between the molecular subsystem and the heterogeneous dielectric media. The interactions are given by Eq. (11) and we rewrite the last term of Eq. (66) as

$$-i\langle[\mathbf{T}^{\dagger,t}, W_{\text{pol}}]\rangle = -i\sum_j \sum_k U_{\text{pol}}^{(k)}(\vec{\mathbf{r}}_j)\sum_{pq}\langle\phi_p|q_j|\phi_q\rangle\langle 0^t|[\mathbf{T}^{\dagger,t}, E_{pq}]|0^t\rangle. \quad (72)$$

We determine the contributions to the linear, quadratic and cubic response equations by expanding the time-dependent wave function, $|O^t\rangle$, and the time-dependent operator, $\mathbf{T}^{\dagger,t}$, to the appropriate perturbation order. Finally, we collect the terms related to a given order of the perturbation and we have that the terms relevant for

- linear response equations are those that provide first order contributions,
- quadratic response equations are those that provide second order contributions, and
- cubic response equations are those that provide third order contributions.

Generally, we have

$$-i\langle[\mathbf{T}^{\dagger,t}, W_{\text{pol}}]\rangle = G^{(0)}(\mathbf{T}^{\dagger,t}) + G^{(1)}(\mathbf{T}^{\dagger,t}) + G^{(2)}(\mathbf{T}^{\dagger,t}) + G^{(3)}(\mathbf{T}^{\dagger,t}) + \cdots. \quad (73)$$

For the linear response function, we determine the modifications of the MCSCF equations due to the interactions between a quantum mechanical subsystem and a structured environment and we focus on the linear terms in $\kappa(t)$ and $S(t)$. For the orbital operators, c_k^t and $c_k^{\dagger,t}$, we determine the modifications as:

-

$$G^{(1)}(c_k^t) = -\sum_j \sum_k U_{\text{pol}}^{(k)}(\vec{\mathbf{r}}_j)\sum_{pq}\langle\phi_p|q_j|\phi_q\rangle \times \left(\langle 0|[S(t),[c_k, E_{pq}]]|0\rangle + \langle 0|[c_k,[\kappa(t), E_{pq}]]|0\rangle\right) \quad (74)$$

- and

$$G^{(1)}(c_k^{\dagger,t}) = -\sum_j \sum_k U_{\text{pol}}^{(k)}(\vec{\mathbf{r}}_j)\sum_{pq}\langle\phi_p|q_j|\phi_q\rangle \times \left(\langle 0|[S(t),[c_k^\dagger, E_{pq}]]|0\rangle + \langle 0|[c_k^\dagger,[\kappa(t), E_{pq}]]|0\rangle\right). \quad (75)$$

Next we consider the set of configurational operators R_n^t and $R_n^\dagger$ and we obtain:

-

$$G^{(1)}(R_n^t) = -\sum_j \sum_k U_{\text{pol}}^{(k)}(\vec{\mathbf{r}}_j)\sum_{pq}\langle\phi_p|q_j|\phi_q\rangle \times \left(\langle 0|[R_n,[S(t), E_{pq}]]|0\rangle + \langle 0|[R_n,[\kappa(t), E_{pq}]]|0\rangle\right), \quad (76)$$

- $$G^{(1)}(R_n^{\dagger,t}) = -\sum_j \sum_k U_{\text{pol}}^{(k)}(\vec{\mathbf{r}}_j) \sum_{pq} \langle \phi_p | q_j | \phi_q \rangle \times \left(\langle 0 | [R_n^\dagger, [S(t), E_{pq}]] | 0 \rangle + \langle 0 | [R_n^\dagger, [\kappa(t), E_{pq}]] | 0 \rangle \right). \quad (77)$$

We determine the contributions to the quadratic response equations by collecting terms that involve the quadratic terms in $\kappa(t)$ and $S(t)$ and as before we consider the orbital and the configurational parts separately and we have

- for the orbital operator c_k

$$G^{(2)}(c_k^t) = \frac{i}{2!} \sum_j \sum_k U_{\text{pol}}^{(k)}(\mathbf{r}_j) \sum_{pq} \langle \phi_p | q_j | \phi_q \rangle \big(\langle 0 | [S(t), [S(t), [c_k, E_{pq}]]] | 0 \rangle + \langle 0 | [c_k, [\kappa(t), [\kappa(t), E_{pq}]]] | 0 \rangle + 2 \langle 0 | [S(t), [c_k, [\kappa(t), E_{pq}]]] | 0 \rangle \big). \quad (78)$$

- For R_n^t

$$G^{(2)}(R_n^t) = \frac{i}{2!} \sum_j \sum_k U_{\text{pol}}^{(k)}(\mathbf{r}_j) \sum_{pq} \langle \phi_p | q_j | \phi_q \rangle \big(\langle 0 | [R_n, [S(t), [S(t), E_{pq}]]] | 0 \rangle + \langle 0 | [R_n, [\kappa(t), [\kappa(t), E_{pq}]]] | 0 \rangle + 2 \langle 0 | [R_n, [S(t), [\kappa(t), E_{pq}]]] | 0 \rangle \big). \quad (79)$$

We determine the contributions for the orbital operator $c_k^\dagger$ and the configurational operator $R_n^{\dagger,t}$ similarly and in the above equations we can replace c_k with $c_k^\dagger$ and R_n with $R_n^\dagger$, respectively.

For the cubic response function we collect the cubic terms in the expansion of $\kappa(t)$ and $S(t)$ and we have for the two operators c_k and R_n (the corresponding equations for $c_k^\dagger$ and $R_n^\dagger$ are given by substituting c_k with $c_k^\dagger$ and R_n with $R_n^\dagger$, respectively):

- for c_k

$$\begin{aligned} G^{(3)}(c_l^t) = \frac{1}{3!} \sum_j \sum_k U_{\text{pol}}^{(k)}(\mathbf{r}_j) \sum_{pq} \langle \phi_p | q_j | \phi_q \rangle \big\{ & \langle 0 | [S(t), [S(t), [S(t), [c_l, E_{pq}]]]] | 0 \rangle \\ & + \langle 0 | [c_l, [\kappa(t), [\kappa(t), [\kappa(t), E_{pq}]]]] | 0 \rangle \\ & + 2 \langle 0 | [S(t), [S(t), [c_l, [\kappa(t), E_{pq}]]]] | 0 \rangle \\ & + 2 \langle 0 | [S(t), [c_l, [\kappa(t), [\kappa(t), E_{pq}]]]] | 0 \rangle \big\}, \end{aligned} \quad (80)$$

- for R_n^t

$$\begin{aligned} G^{(3)}(R_n^t) = \frac{1}{3!} \sum_j \sum_k U_{\text{pol}}^{(k)}(\mathbf{r}_j) \sum_{pq} \langle \phi_p | q_j | \phi_q \rangle \big\{ & \langle 0 | [R_n, [S(t), [S(t), [S(t), E_{pq}]]]] | 0 \rangle \\ & + \langle 0 | [R_n, [\kappa(t), [\kappa(t), [\kappa(t), E_{pq}]]]] | 0 \rangle \\ & + 2 \langle 0 | [R_n, [S(t), [S(t), [\kappa(t), E_{pq}]]]] | 0 \rangle \\ & + 2 \langle 0 | [R_n, [\kappa(t), [\kappa(t), [S(t), E_{pq}]]]] | 0 \rangle \big\}. \end{aligned} \quad (81)$$

We observe that these contributions to the response equations have the same structure as the response equations for the molecule in vacuum [83]. We note that the implementation of these modifications to the vacuum MCSCF response equations gives us the possibility of investigating linear and nonlinear molecular properties of molecular systems interacting with heterogeneous dielectric media.

4.2 A structured environment model given by the multiconfigurational self-consistent field response theory for the quantum mechanical–classical system model

Here, we consider the time evolution of a reference state ($|0\rangle$) using the following definition of the Hamiltonian for the quantum mechanical subsystem

$$H = H_0 + W_{\mathrm{QM/CM}} + V(t) \tag{82}$$

and the three terms in the Hamiltonian describe the following

- the perturbation operator, $V(t)$, related to an external field takes care of the interactions between the quantum subsystem and the externally applied time-dependent electromagnetic field.
- We describe the isolated quantum mechanical subsystem by the Hamiltonian H_0.
- The operator $W_{\mathrm{QM/CM}}$ represents the interactions between the quantum mechanical and the classical mechanical subsystems.

From the determination of the electronic wave function for the time-independent Hamiltonian, $H_0 + W_{\mathrm{QM/CM}}$, we obtain the reference state of the quantum mechanical subsystem, $|0\rangle$,

$$(H_0 + W_{\mathrm{QM/CM}})|0\rangle = E_0|0\rangle, \tag{83}$$

where we have used the following expression for the interaction operator [21]

$$W_{\mathrm{QM/CM}} = E^{\mathrm{vdw}} + E^{\mathrm{el,nuc}}_{S,M} + \tilde{O}^{As'}_{mm'} + T^{g} \tag{84}$$

and defined the effective one-electron operator T^g as

$$T^g = \sum_{s=1}^{S}(-N_s) - \alpha \sum_{a=1}^{A}\left\{\langle Rr_a\rangle + \frac{1}{2}\bar{O}^{aS'}_{mm'}\right\} Rr_a. \tag{85}$$

Our next step is to determine the modifications of the response functions due to the interactions between the molecular subsystem and the structured environment, such as an aerosol particle. In order to do so, we consider the time-evolution of expectation values of the time-transformed operators $\boldsymbol{T}^{\dagger,t}$ from Eq. (65) and we do this using Ehrenfest's equation

$$\frac{d}{dt}\langle \boldsymbol{T}^{\dagger,t}\rangle = \left\langle \frac{\partial \boldsymbol{T}^{\dagger,t}}{\partial t}\right\rangle - i\langle[\boldsymbol{T}^{\dagger,t}, H_0]\rangle - i\langle[\boldsymbol{T}^{\dagger,t}, V(t)]\rangle - i\langle[\boldsymbol{T}^{\dagger,t}, W_{\mathrm{QM/CM}}]\rangle. \tag{86}$$

We utilize the representation of the unitary operators in orbital and configuration space given in Eq. (54) to describe the time evolution of the electronic wave function. We focus on the contributions to the response functions that involve $W_{\mathrm{QM/CM}}$ and we have

$$-i\langle[\boldsymbol{T}^{\dagger,t}, W_{\mathrm{QM/CM}}]\rangle = -i\langle[\boldsymbol{T}^{\dagger,t}, T^{g}]\rangle = G^{(a)}_{\mathrm{QM/CM}} + G^{(b)}_{\mathrm{QM/CM}} + G^{(c)}_{\mathrm{QM/CM}}, \quad (87)$$

where these three terms represent the following contributions

- the term $G^{(a)}_{\mathrm{QM/CM}}$ describes polarization interactions between the two subsystems,

$$G^{(a)}_{\mathrm{QM/CM}} = -i(-\alpha) \sum_{a,pq,p'q'} t^{a}_{pq} \langle {}^{t}0|[\boldsymbol{T}^{\dagger,t}, \boldsymbol{E}_{pq}]|0^{t}\rangle t^{a}_{p'q'} \langle {}^{t}0|\boldsymbol{E}_{p'q'}|0^{t}\rangle, \quad (88)$$

- and similar for the term $G^{(b)}_{\mathrm{QM/CM}}$

$$G^{(b)}_{\mathrm{QM/CM}} = -i\left(-\frac{1}{2}\alpha\right) \sum_{a,pq} \bar{O}^{aS'}_{mm'} \langle {}^{t}0|[\boldsymbol{T}^{\dagger,t}, \boldsymbol{E}_{pq}]|0^{t}\rangle t^{a}_{pq}. \quad (89)$$

- The last term $G^{(c)}_{\mathrm{QM/CM}}$ concerns the electrostatic interactions between the two subsystems

$$G^{(c)}_{\mathrm{QM/CM}} = -i\left(-\sum_{s,pq} \langle {}^{t}0|[\boldsymbol{T}^{\dagger,t}, \boldsymbol{E}_{pq}]|0^{t}\rangle n^{s}_{pq}\right). \quad (90)$$

Based on these contributions we are able to identify the modifications for solving Eq. (66) and we are able, based on the response functions, to determine frequency dependent electromagnetic properties of a quantum mechanical subsystem in contact with a structured environment.

4.2.1 Linear response equations

In order to determine the modifications to the linear response functions, we consider the first order contributions of the perturbation related to the three terms $G^{(a)}_{\mathrm{QM/CM}}$, $G^{(b)}_{\mathrm{QM/CM}}$, and $G^{(c)}_{\mathrm{QM/CM}}$. Presently, we focus on the changes due to the term $G^{(a)}_{\mathrm{QM/CM}}$ and the operator

$$c^{t}_{k} = \exp[i\kappa(t)]c_{k} \exp[-i\kappa(t)] \quad (91)$$

whereby we find after introducing the time-dependent wave function

$$\begin{aligned} G^{(a)}_{\mathrm{QM/CM}} = -i(-\alpha) \sum_{a,pq,p'q'} & t^{a}_{pq} t^{a}_{p'q'} \langle 0| \exp[-iS(t)] \exp[-i\kappa(t)] \\ & \times \left[\exp[i\kappa(t)]c_{k} \exp[-i\kappa(t)], \boldsymbol{E}_{pq}\right] \exp[i\kappa(t)] \exp[iS(t)]|0\rangle \\ & \times \langle 0| \exp[-i\kappa(t)] \exp[-iS(t)]\boldsymbol{E}_{p'q'} \exp[i\kappa(t)] \exp[iS(t)]|0\rangle. \end{aligned} \quad (92)$$

We are able to write this expression compactly by using the following states

$$|0^R\rangle = -\sum_n S_n R_n^\dagger|0\rangle = -\sum_n S_n|n\rangle,$$
$$\langle 0^L| = \sum_n \langle 0|(S_n' R_n^\dagger) = \sum_n S_n'\langle n|, \tag{93}$$

and we determine the contributions for the term $G^{(a)}_{\mathrm{QM/CM}}$ to be

$$G^{(a)}_{\mathrm{QM/CM}} = -\alpha \sum_{a,pq,p'q'} t^a_{p'q'}\{(\langle 0|[c_k, \boldsymbol{E}_{pq}]|0^R\rangle + \langle 0^L|[c_k, \boldsymbol{E}_{pq}]|0\rangle)t^a_{pq}$$
$$+ Q^a_{pq}\langle 0|[c_k, \boldsymbol{E}_{pq}]|0\rangle\}\langle 0|\boldsymbol{E}_{p'q'}|0\rangle - \alpha \sum_{a,pq,p'q'} t^a_{pq}\{(\langle 0|\boldsymbol{E}_{p'q'}|0^R\rangle$$
$$+ \langle 0^L|\boldsymbol{E}_{p'q'}|0\rangle)t^a_{p'q'} + Q^a_{p'q'}\langle 0|\boldsymbol{E}_{p'q'}|0\rangle\}\langle 0|[c_k, \boldsymbol{E}_{pq}]|0\rangle. \tag{94}$$

Furthermore, we have defined the one-index transformed integrals as

$$Q^a_{pq} = \sum_r [\kappa_{pr} t^a_{rq} - t^a_{pr}\kappa_{rq}], \tag{95}$$

$$Q^a = \sum_{pq} Q^a_{pq}\boldsymbol{E}_{pq}, \tag{96}$$

where Q^a_{pq} (Q^a) is an index transformed integral.

Furthermore, we need to determine the modification arising from the other two terms $G^{(b)}_{\mathrm{QM/CM}}$ and $G^{(c)}_{\mathrm{QM/CM}}$ and as a start we consider the time propagation of the operator c_k^t and we have

$$G^{(b)}_{\mathrm{QM/CM}} = -\frac{1}{2}\alpha \sum_{a,pq} \bar{O}^{aS'}_{mm'}\{(\langle 0|[c_k, \boldsymbol{E}_{pq}]|0^R\rangle$$
$$+ \langle 0^L|[c_k, \boldsymbol{E}_{pq}]|0\rangle)t^a_{pq} + Q^a_{pq}\langle 0|[c_k, \boldsymbol{E}_{pq}]|0\rangle\} \tag{97}$$

and

$$G^{(c)}_{\mathrm{QM/CM}} = -\sum_{s,pq}\{(\langle 0|[c_k, \boldsymbol{E}_{pq}]|0^R\rangle$$
$$+ \langle 0^L|[c_k, \boldsymbol{E}_{pq}]|0\rangle)n^s_{pq} + V^s_{pq}\langle 0|[c_k, \boldsymbol{E}_{pq}]|0\rangle\}. \tag{98}$$

Our aim is to obtain compact and useful expressions that are easy to implement and that is achieved through the construction of the following effective operators, T^g (given in Eq. (85)), T^{xc}, T^{xo}, and T^{yo}. We have defined the latter three operators as

- the T^{xc} operator

$$T^{xc} = -\alpha \sum_{a=1}^{A}\{\langle 0|Rr_a|0^R\rangle + \langle 0^L|Rr_a|0\rangle\}Rr_a, \tag{99}$$

- the T^{xo} operator

$$T^{xo} = -\alpha \sum_{a=1}^{A} \langle 0|Q^a|0\rangle Rr_a, \tag{100}$$

- the T^{yo} operator

$$T^{yo} = \sum_{s=1}^{S}(-V^s) - \alpha \sum_{a=1}^{A} \left\{ \langle Rr_a\rangle + \frac{1}{2}\bar{O}^{aS'}_{mm'} \right\} Q^a, \tag{101}$$

Based on these four effective operators we are able to write the modifications of the linear response equations due to the presence of the structured environment as

$$\begin{aligned} W^{[2]}(c_k) &= -i\langle [c_k, W_{\mathrm{QM/CM}}]\rangle \\ &= -\langle 0^L|\big[c_k, T^g\big]|0\rangle - \langle 0|\big[c_k, T^g\big]|0^R\rangle \\ &\quad - \langle 0|\big[c_k, T^{xc}\big]|0\rangle - \langle 0|\big[c_k, T^{yo} + T^{xo}\big]|0\rangle. \end{aligned} \tag{102}$$

Furthermore, we find for the operator $c_k^{\dagger,t}$ a result that is the same as the one above but $c_k^\dagger$ has replaced c_k. We find the following expression for the state transfer operators R_n^t and $R_n^{\dagger,t}$:

- $$\begin{aligned} W^{[2]}(R_n) &= -i\langle [R_n, W_{\mathrm{QM/CM}}]\rangle \\ &= -\langle n|T^g|0^R\rangle - \langle 0|T^g|0\rangle S_n(t) - \langle n|T^{xc}|0\rangle - \langle n|T^{yo} + T^{xo}|0\rangle. \end{aligned} \tag{103}$$

- $$\begin{aligned} W^{[2]}(R_n) &= -i\langle \big[R_n^\dagger, W_{\mathrm{QM/CM}}\big]\rangle \\ &= -\langle 0^L|T^g|n\rangle - \langle 0|T^g|0\rangle S_n^*(t) - \langle 0|T^{xc}|n\rangle - \langle 0||T^{yo} + T^{xo}|n\rangle. \end{aligned} \tag{104}$$

Finally, we can write the contributions, due to the interactions between the molecular and classical subsystems, quantum and classical subsystems, to the linear response function as

$$\begin{aligned} W^{[2]}_{j(k)}N^1 = &-\begin{pmatrix} \langle 0|[c_j, W]|0^{1R}\rangle + \langle 0^{1L}|[c_j, W]|0\rangle \\ \langle j|W|0^{1R}\rangle \\ \langle 0|[c_j^\dagger, W]|0^{1R}\rangle + \langle 0^{1L}|[c_j^\dagger, W]|0\rangle \\ \langle 0^{1L}|W|j\rangle \end{pmatrix} \\ &- \begin{pmatrix} \langle 0|[c_j, W(^1\kappa) + 2A^1]|0\rangle \\ \langle j|W(^1\kappa) + 2A^1|0\rangle \\ \langle 0|[c_j^\dagger, W(^1\kappa) + 2A^1]|0\rangle \\ -\langle 0|W(^1\kappa) + 2A^1|j\rangle \end{pmatrix} - \langle 0|W|0\rangle \begin{pmatrix} 0 \\ ^1S_j \\ 0 \\ ^1S_j' \end{pmatrix}, \end{aligned} \tag{105}$$

and we do that effectively by defining the following operators

$$T = -\alpha \sum_{a=1}^{A} \langle 0|Rr_a|0\rangle Rr_a, \tag{106}$$

$$W = T - \frac{1}{2}\alpha \sum_{a=1}^{A} \bar{O}_{mm'}^{aS'} - \sum_{s=1}^{S} N_s = T^g, \tag{107}$$

$$2A^1 = -\alpha \sum_{a=1}^{A} \{\langle 0|Rr_a(^1\kappa)|0\rangle + \langle 0^{1L}|Rr_a|0\rangle + \langle 0|Rr_a|0^{1R}\rangle\} Rr_a. \tag{108}$$

4.2.2 Quadratic response equations

When we perform investigations of nonlinear time-dependent properties of quantum mechanical systems coupled to a structured environment given by a classical system we need to consider the modifications to the response equations. Therefore, we will focus on the modifications related to the term $\langle [\boldsymbol{T}^{\dagger,t}, W_{\mathrm{QM/CM}}]\rangle$.

Once more we expand $|O^t\rangle$ and $\mathbf{T}^{\dagger,t}$ and then collect to second order the appropriate terms for the quadratic response equations. It is convenient for us to define the following effective operators

- $$T = -\alpha \sum_{a=1}^{A} \langle 0|Rr_a|0\rangle Rr_a, \tag{109}$$

- $$W = T - \frac{1}{2}\alpha \sum_{a=1}^{A} \bar{O}_{mm'}^{aS'} - \sum_{s=1}^{S} N_s, \tag{110}$$

- $$2A^1 = -\alpha \sum_{a=1}^{A} \{\langle 0|Rr_a(^1\kappa)|0\rangle + \langle 0^{1L}|Rr_a|0\rangle + \langle 0|Rr_a|0^{1R}\rangle\} Rr_a, \tag{111}$$

- $$\begin{aligned} 2A^{12} = -\alpha \sum_{a=1}^{A} \{ & \langle 0|Rr_a(^1\kappa, {^2\kappa})|0\rangle + 2(\langle 0^{1L}|Rr_a(^2\kappa)|0\rangle + \langle 0|Rr_a(^2\kappa)|0^{1R}\rangle) \\ & + \langle 0^{1L}|Rr_a|0^{2R}\rangle + \langle 0^{2L}|Rr_a|0^{1R}\rangle\} Rr_a. \end{aligned} \tag{112}$$

We use these effective one-electron operators for expressing the QM/CM contributions to the quadratic response equations

$$
\begin{aligned}
&\left(W^{[3]}_{jl_1l_2}+W^{[3]}_{jl_2l_1}\right){}^1N_{l_1}\,{}^2N_{l_2}\\
&=\frac{1}{2}P(1,2)\begin{pmatrix}\langle 0|[q_j,W({}^1\kappa,{}^2\kappa)+2A^{12}+4A^1({}^2\kappa)]|0\rangle\\ \langle j|W({}^1\kappa,{}^2\kappa)+2A^{12}+4A^1({}^2\kappa)|0\rangle\\ \langle 0|[q_j^\dagger,W({}^1\kappa,{}^2\kappa)+2A^{12}+4A^1({}^2\kappa)]|0\rangle\\ -\langle 0|W({}^1\kappa,{}^2\kappa)+2A^{12}+4A^1({}^2\kappa)|j\rangle\end{pmatrix}\\
&+P(1,2)\begin{pmatrix}\langle 0|[q_j,W({}^2\kappa)+2A^2]|0^{1R}\rangle+\langle 0^{1L}|[q_j,W({}^2\kappa)+2A^2]|0\rangle\\ \langle j|W({}^2\kappa)+2A^2|0^{1R}\rangle\\ \langle 0|[q_j^\dagger,W({}^2\kappa)+2A^2]|0^{1R}\rangle+\langle 0^{1L}|[q_j^\dagger,W({}^2\kappa)+2A^2]|0\rangle\\ -\langle 0^{1L}|W({}^2\kappa)+2A^2|j\rangle\end{pmatrix}\\
&+\frac{1}{2}P(1,2)\begin{pmatrix}\langle 0^{1L}|[q_j,W]|0^{2R}\rangle+\langle 0^{2L}|[q_j,W]|0^{1R}\rangle\\ 0\\ \langle 0^{1L}|[q_j^\dagger,W]|0^{2R}\rangle+\langle 0^{2L}|[q_j^\dagger,W]|0^{1R}\rangle\\ 0\end{pmatrix}\\
&+2P(1,2)\langle 0|A^2|0\rangle\begin{pmatrix}0\\ {}^1S_j\\ 0\\ {}^1S'_j\end{pmatrix}+P(1,2)\,{}^1S_n\,{}^2S'_n\begin{pmatrix}\langle 0|[q_j,T]|0\rangle\\ \langle j|T|0\rangle\\ \langle 0|[q_j^\dagger,T]|0\rangle\\ -\langle 0|T|j\rangle\end{pmatrix}. \qquad (113)
\end{aligned}
$$

5. BRIEF OVERVIEW OF RESULTS

In this section we present a brief overview of some of the results related to atmospheric processes obtained using the two methods.

We have studied how interactions between molecular dipoles and molecular charges affect the uptake of sulfuric acid molecules [84]. We have illustrated the effect of the structure and dipole moments of the gas-phase sulfuric acid and its hydrates on the uptake of the sulfuric acid molecules by the charged clusters/ultrafine particles. From different structures of the complexes having large variations in the total dipole moment of hydrated complex we have obtained large enhancements of the uptake coefficients.

The dielectric media model has been used to describe a molecule interacting with an aerosol particle [85]. The aerosol particle was given by a dielectric medium and the macroscopic dielectric constant depends on the average composition and temperature. The model was used for investigating the interactions between an aqueous particle and a succinic acid molecule. From the investigations, it was clearly observed that interaction energies and linear response properties such as excitation energies, transition moments and polarizabilities depend on the orientation of the organic dicarboxylic acid relative to the particle surface and on the dielectric properties of the particle.

The quantum mechanical/molecular mechanical aerosol model was developed to describe the interaction between gas phase molecules and atmospheric particles. This method has been utilized for the calculation of interaction energies and time-dependent molecular properties. Particularly, the model has been used for investigating how SO_2, phenol, succinic acid molecules, and dicarboxylic acids interact with an aqueous particle [86–89]. We have determined how the interaction energies and linear response properties (excitation energies, transition moments, and polarizabilities) depend on the actual configuration of the molecule and the aerosol particle. We have observed large effects of the molecular properties with respect to the distance between aerosol particle and molecule and on their relative orientations. The aerosol particles were constructed from molecular dynamics simulations of clusters of 128 to 512 water molecules treated classically. The interaction energies calculated using a quantum mechanics/molecular mechanics approach were compared to the interaction energies obtained from a purely classical approach. It was clearly shown that for specific interactions between the molecules and the aerosol particle were treated correctly using the quantum mechanics/molecular mechanics approach and that the purely classical approach was not able to capture correctly the interactions when the molecule and the aerosol particle were interacting strongly. Based on the energy surfaces we determined the reaction pathways with the largest interaction between the molecule and the particle. The utilization of the quantum mechanics/molecular mechanics (QM/MM) for calculating potential energy curves for each reaction path enabled us to determine mass accommodation coefficients. We showed how the mass accommodation coefficient depended on the relative orientations of the molecules with respect to the aerosol particle.

6. CONCLUSION

Based on the mathematical structure of methods describing the situation where a quantum mechanical system is coupled to a homogeneous dielectric medium, heterogeneous dielectric media or a classical subsystem, we observe that the changes of the energy functionals and response equations have the same structure. The three models differ with respect to how we have constructed the effective operators. The similarity between the three different approaches is also an advantage when implementing the three approaches in modern electronic structure programs. Furthermore, we mentioned that the modifications to the response equations have the mathematical structure that was also found in the response equations for the molecule in vacuum [83].

We have presented response methods that provide procedures for calculating frequency-dependent molecular properties for a molecular subsystem coupled to a structured environment. We have shown that the molecular subsystem is treated on a quantum mechanical level and the structured environment as a classical subsystem. We have presented the structured environment, classical subsystem, as a heterogeneous dielectric media or a molecular mechanics force field. We have demonstrated that the interactions between the quantum mechanical and classical subsystems are part of the energy functional used for optimizing the MCSCF electronic wave function.

The advantage of having access to an electronic structure program containing the implementation of the above models for calculation of linear and nonlinear response functions is clearly seen by the multitude of molecular properties that one is able to obtain for molecules interacting with a structured environment, e.g. an aerosol particle. Based on the structured environment methods utilizing MCSCF electronic wave functions, we are able to investigate molecules in the electronic ground state, ionized and excited states. For these states we are able to calculate frequency-dependent linear and nonlinear polarizabilities, transition moments, and vertical excitation energies.

ACKNOWLEDGEMENT

K.V.M. thanks Statens Naturvidenskabelige Forskningsråd, Statens Tekniske Videnskabelige Forskningsråd, the Danish Center for Scientific Computing and the EU-network NANOQUANT for support. We thank M. Bilde and M.S. Johnson for helpful conversations. O.J.N. thanks Statens Naturvidenskabelige Forskningsråd and the Villum Kann Rasmussen fund.

REFERENCES

[1] J.H. Seinfeld, S.N. Pandis, *Atmospheric Chemistry and Physics*, Wiley, New York, 2006.
[2] M.D. Newton, *J. Phys. Chem.* **79** (1975) 2795.
[3] J.O. Noell, K. Morokuma, *Chem. Phys. Lett.* **36** (1975) 465.
[4] D.L. Beveridge, G.W. Schnuelle, *J. Phys. Chem.* **79** (1975) 2562.
[5] J. Hylton, R.E. Christoffersen, G.G. Hall, *Chem. Phys. Lett.* **24** (1974) 501.
[6] R. Contreras, A. Aizman, *Int. J. Quantum Chem.* **27** (1985) 193.
[7] H. Hoshi, M. Sakurai, Y. Inone, R. Chujo, *J. Chem. Phys.* **87** (1987) 1107.
[8] O. Tapia, in: H. Ratajczak, W.J. Orville-Thomas (Eds.), *Molecular Interactions*, Wiley, New York, 1980.
[9] A. Warshel, *Chem. Phys. Lett.* **55** (1978) 454.
[10] E. Sanchez-Marcos, B. Terryn, J.L. Rivail, *J. Phys. Chem.* **87** (1985) 4695.
[11] D. Rinaldi, *Comput. Chem.* **6** (1982) 155.
[12] O. Tapia, in: R. Daudel, A. Pullman, L. Salem, A. Veillard (Eds.), *Quantum Theory of Chemical Reactions*, vol. 3, Wiley, Dordrecht, 1980, p. 25.
[13] G. Karlström, *J. Phys. Chem.* **93** (1989) 4952.
[14] M. Karelson, M. Zerner, *J. Am. Chem. Soc.* **112** (1990) 9405.
[15] P.N. Day, J.H. Jensen, M.S. Gordon, S.P. Webb, W.J. Stevens, M. Krauss, D. Garmer, H. Basch, D. Cohen, *J. Chem. Phys.* **105** (1996) 1968.
[16] W. Chen, M.S. Gordon, *J. Chem. Phys.* **105** (1996) 11081.
[17] C.J. Cramer, D.G. Truhlar, *J. Am. Chem. Soc.* **113** (1991) 8305.
[18] C.J. Cramer, D.G. Truhlar, *Science* **256** (1992) 213.
[19] K.V. Mikkelsen, E. Dalgaard, P. Svanstøm, *J. Phys. Chem.* **91** (1987) 3081.
[20] K.V. Mikkelsen, H. Ågren, H.J.A. Jensen, T. Helgaker, *J. Chem. Phys.* **89** (1988) 3086.
[21] K.V. Mikkelsen, P. Jørgensen, H.J.A. Jensen, *J. Chem. Phys.* **100** (1994) 6597–6607.
[22] K.V. Mikkelsen, Y. Luo, H. Ågren, P. Jørgensen, *J. Chem. Phys.* **100** (1994) 8240.
[23] K.V. Mikkelsen, Y. Luo, H. Ågren, P. Jørgensen, *J. Chem. Phys.* **102** (1995) 9362.
[24] K.V. Mikkelsen, K.O. Sylvester-Hvid, *J. Phys. Chem.* **100** (1996) 9116.
[25] S. Di Bella, T.J. Marks, M.A. Ratner, *J. Am. Chem. Soc.* **116** (1994) 4440.
[26] J. Yu, M.C. Zerner, *J. Chem. Phys.* **100** (1994) 7487.

[27] R. Cammi, M. Cossi, B. Mennucci, J. Tomasi, *J. Chem. Phys.* **105** (1996) 10556.
[28] A. Willetts, J.E. Rice, *J. Chem. Phys.* **99** (1993) 426.
[29] R. Cammi, M. Cossi, J. Tomasi, *J. Chem. Phys.* **104** (1996) 4611.
[30] M.W. Wong, M.J. Frisch, K.B. Wiberg, *J. Am. Chem. Soc.* **113** (1991) 4776.
[31] J.G. Angyan, *Chem. Phys. Lett.* **241** (1995) 51.
[32] F.J. Olivares del Valle, J. Tomasi, *Chem. Phys.* **150** (1991) 139.
[33] C. Chipot, D. Rinaldi, J.L. Rivail, *Chem. Phys. Lett.* **191** (1991) 287.
[34] C.J. Cramer, D.G. Truhlar, *Chem. Rev.* **99** (1999) 2161.
[35] J. Tomasi, R. Cammi, B. Mennucci, *Int. J. Quantum Chem.* **75** (1999) 783.
[36] O. Christiansen, K.V. Mikkelsen, *J. Chem. Phys.* **110** (1999) 1365.
[37] D.M. Bishop, *Adv. Quantum Chem.* **25** (1994) 1.
[38] D.P. Shelton, J.E. Rice, *Chem. Rev.* **94** (1994) 195.
[39] D.R. Kanis, M.A. Ratner, T.J. Marks, *Chem. Rev.* **94** (1994) 195.
[40] B.L. Hammond, J.E. Rice, *J. Chem. Phys.* **97** (1992) 1138.
[41] S.P. Karna, G.B. Talapatra, W.M.K.P. Wijekoon, P.N. Prasad, *Phys. Rev. A* **45** (1992) 2763.
[42] H. Sekino, R.J. Bartlett, *J. Chem. Phys.* **98** (1993) 3022.
[43] D. Jonsson, P. Norman, Y. Luo, H. Ågren, *J. Chem. Phys.* **105** (1996) 581.
[44] D. Jonsson, P. Norman, H. Ågren, *J. Chem. Phys.* **105** (1996) 6401.
[45] O. Christiansen, K.V. Mikkelsen, *J. Chem. Phys.* **110** (1999) 8348.
[46] D. Jonsson, P. Norman, H. Ågren, Y. Luo, K.O. Sylvester-Hvid, K.V. Mikkelsen, *J. Chem. Phys.* **109** (1998) 6351.
[47] K.O. Sylvester-Hvid, K.V. Mikkelsen, D. Jonsson, P. Norman, H. Ågren, *J. Chem. Phys.* **109** (1998) 5576.
[48] S. Jørgensen, M.A. Ratner, K.V. Mikkelsen, *J. Chem. Phys.* **115** (2001) 3792.
[49] S. Jørgensen, M.A. Ratner, K.V. Mikkelsen, *J. Chem. Phys.* **115** (2001) 8185.
[50] S. Jørgensen, M.A. Ratner, K.V. Mikkelsen, *J. Chem. Phys.* **116** (2002) 10902.
[51] T.D. Poulsen, J. Kongsted, A. Osted, P.R. Ogilby, K.V. Mikkelsen, *J. Chem. Phys.* **115** (2001) 2393.
[52] T.D. Poulsen, P.R. Ogilby, K.V. Mikkelsen, *J. Chem. Phys.* **116** (2002) 3730.
[53] T.D. Poulsen, P.R. Ogilby, K.V. Mikkelsen, *J. Chem. Phys.* **115** (2001) 7843.
[54] J. Kongsted, A. Osted, K.V. Mikkelsen, P.-O. Åstrand, O. Christiansen, *J. Chem. Phys.* **121** (2004) 8435.
[55] K. Aidas, J. Kongsted, A. Osted, K.V. Mikkelsen, O. Christiansen, *J. Phys. Chem. A* **35** (2005) 8001.
[56] J. Kongsted, A. Osted, K.V. Mikkelsen, O. Christiansen, *J. Phys. Chem. A* **107** (2003) 2578.
[57] J. Kongsted, A. Osted, K.V. Mikkelsen, O. Christiansen, *J. Chem. Phys.* **118** (2003) 1620.
[58] J. Kongsted, A. Osted, K.V. Mikkelsen, O. Christiansen, *J. Chem. Phys.* **119** (2003) 10519.
[59] J. Kongsted, A. Osted, K.V. Mikkelsen, O. Christiansen, *J. Chem. Phys.* **120** (2004) 3787.
[60] J. Kongsted, A. Osted, K.V. Mikkelsen, O. Christiansen, *Mol. Phys.* **100** (2002) 1813.
[61] A. Osted, J. Kongsted, K.V. Mikkelsen, O. Christiansen, *J. Phys. Chem. A* **108** (2004) 8646.
[62] A. Osted, J. Kongsted, K.V. Mikkelsen, O. Christiansen, *Mol. Phys.* **101** (2003) 2055.
[63] L. Jensen, P.Th. van Duijnen, J.G. Snijders, *J. Chem. Phys.* **118** (2003) 514.
[64] L. Jensen, P.Th. van Duijnen, J.G. Snijders, *J. Chem. Phys.* **119** (2003) 3800.
[65] L. Jensen, M. Swart, P.Th. van Duijnen, *J. Chem. Phys.* **122** (2005) 034103.
[66] L. Jensen, P.Th. van Duijnen, *Int. J. Quantum Chem.* **102** (2005) 612.
[67] P.Th. van Duijnen, A.H. de Vries, M. Swart, F. Grozema, *J. Chem. Phys.* **117** (2002) 8442.
[68] D. Borgis, S. Lee, J.T. Hynes, *Chem. Phys. Lett.* **162** (1989) 19.
[69] D. Borgis, J.T. Hynes, *J. Chem. Phys.* **94** (1991) 3619.
[70] D. Borgis, J.T. Hynes, *Chem. Phys.* **170** (1993) 315.
[71] K.V. Mikkelsen, *Z. Phys. Chem.* **170** (1991) 129–142.
[72] J. Cioslowski, *Phys. Rev. Lett.* **62** (1989) 1469.
[73] J. Cioslowski, *J. Am. Chem. Soc.* **111** (1989) 8333.
[74] P.-O. Åstrand, K. Ruud, K.V. Mikkelsen, T. Helgaker, *J. Phys. Chem. A* **102** (1998) 7686.
[75] W.R. Smythe, *Static and Dynamic Electricity*, 2nd ed., McGraw–Hill, New York, 1950.
[76] W. Greiner, *Classical Electrodynamics*, Springer, New York, 1996.
[77] H.J.A. Jensen, H. Ågren, *Chem. Phys. Lett.* **110** (1984) 140.
[78] H.J.A. Jensen, H. Ågren, *Chem. Phys.* **104** (1986) 229.

[79] K.V. Mikkelsen, A. Cesar, H. Ågren, H.J.A. Jensen, *J. Chem. Phys.* **103** (1995) 9010–9023.
[80] J. Olsen, P. Jørgensen, *J. Chem. Phys.* **82** (1985) 3235.
[81] P. Jørgensen, H.J.A. Jensen, J. Olsen, *J. Chem. Phys.* **89** (1988) 3654.
[82] J. Olsen, H.J.A. Jensen, P. Jørgensen, *J. Comput. Phys.* **74** (1988) 265.
[83] H. Hettema, H.J.A. Jensen, J. Olsen, P. Jørgensen, *J. Chem. Phys.* **97** (1992) 1174; K.O. Sylvester-Hvid, K.V. Mikkelsen, D. Jonsson, P. Norman, H. Ågren, *J. Chem. Phys.* **109** (1998) 5576–5584.
[84] M. Sloth, S. Jorgensen, M. Bilde, K.V. Mikkelsen, *J. Phys. Chem. A* **107** (2003) 8623–8629.
[85] A.B. Nadykto, A. Al Natsheh, F. Yu, K.V. Mikkelsen, J. Ruuskanen, *Aerosol Sci. Technol.* **38** (2004) 349–353.
[86] M. Sloth, M. Bilde, K.V. Mikkelsen, *J. Chem. Phys.* **118** (2003) 10085–10092.
[87] M. Sloth, M. Bilde, K.V. Mikkelsen, *Mol. Phys.* **102** (2004) 2361–2368.
[88] A. Gross, I. Barnes, R.M. Sorensen, J. Kongsted, K.V. Mikkelsen, *J. Phys. Chem. A* **108** (2004) 8659–8671.
[89] H. Falsig, A. Gross, J. Kongsted, A. Osted, M. Sloth, K.V. Mikkelsen, O. Christiansen, *J. Phys. Chem. A* **110** (2006) 660–670.
[90] C.B. Nielsen, K.V. Mikkelsen, S.P.A. Sauer, *J. Chem. Phys.* **114** (2001) 7753–7760.
[91] C.B. Nielsen, S.P.A. Sauer, K.V. Mikkelsen, *J. Chem. Phys.* **119** (2003) 3849–3870.
[92] C.B. Nielsen, J. Kongsted, O. Christiansen, K.V. Mikkelsen, *J. Chem. Phys.* **126** (2007), Art. No. 154112.
[93] K. Aidas, J. Kongsted, C.B. Nielsen, K.V. Mikkelsen, O. Christiansen, K. Ruud, *Chem. Phys. Lett.* **442** (2007) 322–328.
[94] K. Aidas, A. Møgelhøj, H. Kjær, C.B. Nielsen, K.V. Mikkelsen, K. Ruud, O. Christiansen, J. Kongsted, *J. Phys. Chem. A* **111** (2007) 4199–4210.
[95] J. Kongsted, C.B. Nielsen, K.V. Mikkelsen, O. Christiansen, K. Ruud, *J. Chem. Phys.* **126** (2007) 034510.
[96] S. Jakobsen, K.V. Mikkelsen, S.U. Pedersen, *J. Phys. Chem.* **100** (1996) 7411.

CHAPTER 18

Theoretical Studies of the Dissociation of Sulfuric Acid and Nitric Acid at Model Aqueous Surfaces

Roberto Bianco[1*] **Shuzhi Wang**[*] and **James T. Hynes**[2*,**]

Contents

1. Introduction 388
2. Methodology 389
   2.1 Electronic structure calculations, acid molecule placement and reaction paths 389
   2.2 Further electronic structure issues 390
   2.3 Reaction free energetics 391
3. First Acid Dissociation of Sulfuric Acid 393
   3.1 Atmospheric relevance 393
   3.2 H_2SO_4 acid dissociation results [23,24] 396
4. Acid Dissociation of Nitric Acid 397
   4.1 Atmospheric relevance 397
   4.2 HNO_3 acid dissociation results [25] 400
5. Concluding Remarks 401

Acknowledgements 402

References 402

Abstract

A brief review is given of recent electronic structure calculations addressed to the acid dissociation at aqueous surfaces of sulfuric acid H_2SO_4 and nitric acid HNO_3, proton transfer reactions which are important in various atmospheric chemistry contexts. Two of many examples of their atmospheric

[*] Department of Chemistry and Biochemistry, University of Colorado, Boulder, CO 80309-0215, USA
[**] Département de Chimie, CNRS UMR 8640 PASTEUR, Ecole Normale Supérieure, 24 rue Lhomond, Paris 75231, France
[1] Corresponding author. E-mail: roberto.bianco@colorado.edu
[2] Corresponding author. E-mails: hynes@spot.colorado.edu; hynes@chimie.ens.fr

Advances in Quantum Chemistry, Vol. 55
ISSN 0065-3276, DOI: 10.1016/S0065-3276(07)00218-3

relevance are sulfate aerosol surfaces acting as heterogeneous reaction sites for reactions related to ozone depletion in the mid-latitude stratosphere, and the uptake of gas phase HNO_3 by ice aerosols in the upper troposphere to provide surfaces for heterogeneous reactions, again related to ozone depletion. Despite the fact that these acids are usually regarded as strong and readily dissociate in the more familiar room temperature, bulk water solution context, it is found that both—particularly HNO_3—can remain molecular at an aqueous surface under a wide range of atmospherically relevant conditions.

1. INTRODUCTION

Acid dissociations at the surfaces of aerosols play a key role in atmospheric chemistry. The realization of this fact may be said to have commenced with the seminal suggestion [1] of heterogeneous reactions involving hydrochloric acid, HCl occurring on aerosol surfaces (such as the ice aerosols in polar stratospheric clouds) functioning as critical components in the dramatic depletion of ozone in the "Ozone Hole" in the Antarctic stratosphere [2,3]. This drew intense attention to the importance of the chemical behavior of this acid—long familiar in bulk aqueous solution circumstances—in an aqueous surface context. In addition to another hydrohalic acid, HBr, which is important for similar reasons in e.g. the Arctic tropospheric boundary layer [4–9], two other well-known "strong" acids—sulfuric acid (H_2SO_4) and nitric acid (HNO_3)—are of wide significance in the atmosphere. Here we cite just two instances of this relevance. First, sulfate aerosols—supercooled concentrated aqueous solutions of H_2SO_4—are the primary heterogeneous reaction sites for chemistry related to ozone depletion in the mid-latitude stratosphere, i.e. away from polar regions [10–12]. Second, HNO_3 is an end product of many of the atmospheric heterogeneous reactions, in effect removing gas phase NO_x species which are key in regulating the atmospheric ozone levels [1–3].

In all these (and other) circumstances, a key question which arises is the chemical state of the acid in the surface region, i.e. is it molecular or is it acid-dissociated to the appropriate anion and a hydronium ion, H_3O^+? Thus for example, for HCl, this question arises for this acid's reaction with chlorine nitrate ($ClONO_2$) to produce molecular chlorine and HNO_3 on an ice surface; this reaction is believed to be the most important heterogeneous reaction in the Antarctic stratosphere connected to the Ozone Hole [1]. As an example for H_2SO_4, the issue of whether its first acid dissociation occurs in a surface layer of a sulfate aerosol is obviously relevant for the surface chemical composition of these aerosols and thus for the mechanisms of the heterogeneous reactions upon them; one such reaction is the hydrolysis of dinitrogen pentoxide N_2O_5, again related to ozone depletion [10–12]. A similar issue arises for HNO_3 in connection with this acid's uptake from the gas phase by ice aerosols in the upper troposphere [13–16].

Among the acids mentioned above, most attention has been paid in the past to HCl [7,8,17–22]. In this chapter, we refer to HCl only in passing, and instead focus on the question of the acid dissociations of H_2SO_4 and HNO_3 at an aqueous surface, as these acids have been the most recent foci in the Hynes group [23–26]. In particular, we focus exclusively here on the study, via electronic structure methods, of the mechanisms and free energetics of the following two proton transfer reactions at an aqueous surface: the first acid dissociation of sulfuric acid

$$H_2SO_4 + H_2O \longrightarrow HSO_4^- + H_3O^+ \quad (1)$$

to produce the bisulfate ion (HSO_4^-) and the hydronium ion, and the acid dissociation of nitric acid

$$HNO_3 + H_2O \longrightarrow NO_3^- + H_3O^+ \quad (2)$$

to produce the nitrate ion (NO_3^-) and the hydronium ion. In each case, the proton transfer reaction is followed to the contact ion pair stage, rather than to the separated ions. No pretense is made to a complete discussion and only the key features are discussed within; for further details, discussion and references, the interested reader may consult the original references [23–25].

The outline of the remainder of this chapter is as follows. In Section 2, we give a brief outline of the methodology employed to study reactions Eqs. (1) and (2). Section 3 then gives a more detailed exposition of the atmospheric relevance of H_2SO_4 and a discussion of its first acid ionization. Section 4, focused on HNO_3, has an analogous structure. Some concluding remarks are given in Section 5.

2. METHODOLOGY

Our general methodology, which has been described in detail in Ref. [23], and implemented in Ref. [24] for H_2SO_4 and Ref. [25] for HNO_3, consists of several key steps, now briefly discussed in turn. For ease of presentation in much of this section, we designate the acid molecule as HA, with A = HSO_4 or NO_3, so that the common strategy for the acids is emphasized.

2.1 Electronic structure calculations, acid molecule placement and reaction paths

In general, for each acid HA, the $HA{\cdot}(H_2O)_n{\cdot}W_m$ model reaction system (MRS) comprises a $HA{\cdot}(H_2O)_n$ core reaction system (CRS), described quantum chemically, embedded in a cluster of W_m classical, polarizable waters of fixed internal structure (effective fragment potentials, EFPs) [27]. The CRS is treated at the Hartree–Fock (HF) level of theory, with the SBK [28] effective core potential basis set complemented by appropriate polarization and diffused functions. The W-waters not only provide solvation at a low computational cost; they also prevent the unwanted collapse of the CRS towards structures typical of small gas phase clusters by enforcing natural constraints representative of the H-bonded network of a surface environment. In particular, the structure of the W_m cluster equilibrates to the CRS structure along the whole reaction path, without any constraints on its shape other than those resulting from the fixed internal structure of the W-waters.

GAMESS [29] was used for the calculations. The HF/SBK+* optimized structure of each isolated acid molecule compares well with both the experimental one and that obtained at a higher level of theory. Since the W-waters were parameterized to interact with a CRS described at the HF/DH(d,p) level of theory [27], we tested the SBK+* basis set vs. DH(d,p) for certain HA–water complexes involving hydrogen (H)-bonds typical of the larger clusters used for the main calculations. The structures were in each case found to have a large overlap, so that the HF/SBK+* method could be considered to be a good substitute for HF/DH(d,p) when using water-effective fragment potentials (Ws) [24,25].

The next part of the methodology concerned the placement of the acid molecule, the selection of the MRS and the selection of the CRS for the various cases studied (three for H_2SO_4 and four for HNO_3). Here we outline the procedure for the more complex case of H_2SO_4, with its two acidic protons; the procedure was quite similar for the simpler case of HNO_3.

Beginning with a classical lattice, in each of the three different placements studied, the H_2SO_4 was initially placed atop or embedded within the surface layer of a W_{50} cluster with the structure of hexagonal ice [30], and the MRS structure was partially relaxed via geometry optimization. This allowed identification of the two Ws H-bonded to the two H_2SO_4 protons. Following this step, the MRS underwent several cycles of visual inspections and reductions in the number of Ws, further partial optimizations, and enlargements of the CRS to its final size. This MRS, which was then used for the reaction path calculations, had both protons on H_2SO_4 coordinated to quantum waters, with the H_2O accepting the dissociating proton being further H-bonded to two other quantum waters, able to stabilize the ensuing H_3O^+ ion. (The H_2O EFP is not suitable for coordination to the H_3O^+ ion: the H-bonded proton from H_3O^+ promptly transfers to the H_2O EFP forming a bond of 0.5 Å.)

The initial search for the transition state (TS) region for the proton transfer (PT) was performed by mapping the MRS potential energy surface (PES) at fixed (and decreasing) $HO_3SOH\cdots OH_2$ H-bond distances, with all other MRS internal coordinates being optimized. When the region of the maximum for this one-dimensional path was identified, the PES mapping was expanded in that region as a function of the two $HO_3SO\text{–}H\cdots OH_2$ distances, and the saddle point region was located with greater accuracy. Finally, the O–H constraints were released, the TS was fully optimized, and its Hessian calculated. With the fully characterized TS in hand, the intrinsic reaction coordinate (IRC) path [31] was obtained, and its end points were further optimized to the reactant complex (RC)—involving a molecular pair—and product complex (PC)—involving a contact ion pair—and characterized by a frequency calculation.

2.2 Further electronic structure issues

A second aspect of electronic structure issues was the estimation of electron correlation effects. As discussed further within, the 0 K HF reaction free energy calculations for several of the cases examined, when augmented by zero point energy and thermal effects, give reaction free energies of the acid ionization that are small in magnitude, ~0–6 kcal/mol either favorable or unfavorable. Such small magnitudes implied the need for an estimate of electron correlation effects in the 0 K reaction energies.

These correlation effects for the 0 K reaction energies for proton transfer from HA to a water molecule were estimated in two steps. The first step compared the electronic contributions (PA_e) to the gas-phase proton affinities (PA) of H_2O and the appropriate anion A^- at the HF, MP2 [32] and CBS levels [33]. The MP2 value for the difference $PA_e(A^-) - PA_e(H_2O)$, which is the reaction energy $\Delta E_{rxn}(0\text{ K})$ for the $HA + H_2O \rightarrow A^- + H_3O^+$ gas phase reaction, was found to be less than the high level CBS level value by several kcal/mol. This indicated a bias, here labelled B, in the MP2 calculations *in favor* of the proton transfer from HA to H_2O compared to the CBS result. Thus, to correct the "intrinsic" deficiencies of the MP2 estimates of electron correlation effects in the aqueous surface calculations, we need to apply the correction +B kcal/mol to the 0 K reaction energy. The second step assesses electron correlation effects for stable reactant and product complexes found in the HF calculations for the several aqueous surface cases examined (see Sections 3 and 4). These calculations, which were performed on a significant portion of the MRS for the various cases, indicate non-negligible electron correlation effects on the 0 K HF reaction energies. These effects were negative contributions (here labelled N), in the range of several kcal/mol in magnitude, favoring the acid dissociation of H_2SO_4 or HNO_3 by this amount more than the HF values. However, consideration of the first portion of this paragraph indicated that these needed to be corrected by +B kcal/mol. The net effect was that the 0 K HF reaction energies for the H_2SO_4 acid dissociation were corrected by +0.96 to −0.72 kcal/mol in the various surface placement cases, while those for the HNO_3 acid dissociation were corrected by −1.97 to −0.70 kcal/mol. (For the cases reported in the present contribution, these corrections can be seen in Tables 18.1 and 18.2 as the difference between the 0 K reaction energies ΔE(HF) at the HF level and their corrected values ΔE.)

2.3 Reaction free energetics

The reaction free energies associated with the production of the $NO_3^-{\cdot}H_3O^+$ or $HSO_4^-{\cdot}H_3O^+$ CIP were calculated as

$$\Delta G = (E_{PC} - E_{RC}) + (ZPE_{PC} - ZPE_{RC}) + \left(H_{PC}^{therm} - H_{RC}^{therm}\right) - T(S_{PC} - S_{RC}), \tag{3}$$

where E is the energy at 0 K, ZPE is the zero point energy, H^{therm} is the thermal contribution from translational, rotational, and vibrational motions, and S is the entropy contribution from these motions. The ZPEs and the vibrational contributions to the thermodynamic quantities from intermolecular modes were calculated using frequencies scaled by the average factors 1.0962 and 0.9173, respectively for frequencies below and above 1000 cm^{-1}, obtained via comparison between MP2/aug-cc-pVDZ [34] and HF frequencies for the cyclic water tetramer as detailed in Ref. [24]. The 0.9173 factor was also used to scale the intramolecular modes for H_3O^+ in the $HSO_4^-{\cdot}H_3O^+$ CIP, whereas both H_2SO_4 and HSO_4^-

TABLE 18.1 Energetics and thermodynamics of H_2SO_4 dissociation at model water layers[a]

	T									
	30	60	90	120	150	180	210	240	270	300
A $H_2SO_4{\cdot}(H_2O)_6{\cdot}W_{27} \rightarrow HSO_4^-{\cdot}H_3O^+{\cdot}(H_2O)_5{\cdot}W_{27}$										
ΔG	3.44	3.42	3.43	3.43	3.44	3.44	3.48	3.49	3.53	3.56
ΔH	3.44	3.43	3.43	3.41	3.38	3.34	3.32	3.28	3.26	3.23
ΔS	0.08	0.09	−0.02	−0.20	−0.40	−0.58	−0.75	−0.89	−1.01	−1.10
	$\Delta E(\mathrm{HF}) = 2.76$; $\Delta E = 2.58$; $\Delta E + \Delta \mathrm{ZPE} = 3.44$									
B $H_2SO_4{\cdot}(H_2O)_4{\cdot}W_{26} \rightarrow HSO_4^-{\cdot}H_3O^+{\cdot}(H_2O)_3{\cdot}W_{26}$										
ΔG	−1.47	−1.25	−0.99	−0.69	−0.39	−0.07	0.29	0.63	1.01	1.39
ΔH	−1.63	−1.74	−1.83	−1.90	−1.98	−2.07	−2.15	−2.26	−2.36	−2.47
ΔS	−5.33	−8.17	−9.34	−10.04	−10.60	−11.10	−11.58	−12.03	−12.45	−12.84
	$\Delta E(\mathrm{HF}) = -5.97$; $\Delta E = -5.07$; $\Delta E + \Delta \mathrm{ZPE} = -1.53$									

a T in K. Energy ΔE, free energy ΔG, enthalpy ΔH, and ZPEs in kcal/mol. Entropy ΔS in cal/(mol K). See Ref. [24]. Thermal contributions calculated using frequencies scaled as described in Section 2.

TABLE 18.2 Energetics and thermodynamics of HNO_3 dissociation at model water layers[a]

	T					
	0	60	120	180	240	300
A $HNO_3{\cdot}(H_2O)_3{\cdot}W_{33} \rightarrow NO_3^-{\cdot}H_3O^+{\cdot}(H_2O)_2{\cdot}W_{33}$						
ΔG	4.62	4.68	4.84	5.05	5.29	5.57
ΔH	4.62	4.56	4.47	4.36	4.23	4.12
ΔS	0.00	−1.98	−3.09	−3.83	−4.41	−4.85
	$\Delta E_{\mathrm{HF}} = 5.57$; $\Delta E = 4.87$; $\Delta E + \Delta \mathrm{ZPE} = 4.62$					
B $HNO_3{\cdot}(H_2O)_3{\cdot}W_{50} \rightarrow NO_3^-{\cdot}H_3O^+{\cdot}(H_2O)_2{\cdot}W_{50}$						
ΔG	2.08	2.05	2.10	2.22	2.43	2.67
ΔH	2.08	2.04	1.91	1.74	1.55	1.33
ΔS	0.00	−0.15	−1.55	−2.69	−3.66	−4.46
	$\Delta E_{\mathrm{HF}} = 2.06$; $\Delta E = 1.01$; $\Delta E + \Delta \mathrm{ZPE} = 2.08$					

a T in K. Reaction energy ΔE, free energy ΔG, and enthalpy ΔH, and zero point energy difference ΔZPE in kcal/mol. Reaction entropy ΔS in cal/(mol K). Thermal contributions calculated using frequencies scaled as described in Section 2.

intramolecular frequencies were assigned by visual inspection via Molden [35] and scaled by an average scaling factor 0.882. For the nitric acid problem [25], the intramolecular modes of HNO_3, H_2O, NO_3^-, and H_3O^+ in the CRS of each RC and PC were identified via an automated analysis of the atomic motion components and confirmed visually via Molden [35], and each mode frequency was scaled according to its mode-specific scaling factor.

3. FIRST ACID DISSOCIATION OF SULFURIC ACID

3.1 Atmospheric relevance

Sulfuric acid (H_2SO_4) and the two anions bisulfate (HSO_4^-) and sulfate ($SO_4^=$) resulting from its successive acid dissociations are present in various atmospheric locations in the form of sulfate aerosols (SAs), liquid supercooled aqueous solutions of sulfuric acid. The bulk composition of SAs is strongly dependent on the temperature and the water vapor pressure [12,36,37], and thus varies in the polar stratosphere [1,2,38–41], the midlatitude stratosphere (away from the poles) [10,37] and the Arctic boundary layer [6,42–44]. SAs can catalyze heterogeneous reactions important for ozone depletion [2] such as the hydrolysis of N_2O_5, $N_2O_5 + H_2O \rightarrow 2HNO_3$, in the midlatitude stratosphere [37] and the reactions of $HOX + HBr \rightarrow BrX + H_2O$ ($X = Br, Cl$) in the Arctic boundary layer [6,44]. They can also influence the adsorption of strong acids like HCl and HBr [5,44,45], and oxidize organic species like acetaldehyde (CH_3CHO) [46]. There is even the possibility that SAs provide the key heterogeneous reaction site in the Antarctic stratosphere [1], as opposed to the standard view that the dominant site is provided by ice aerosols.

The catalytic properties of SAs for heterogeneous reactions are believed to derive from their high acidity and the presumed ability to assist reacting species by proton transfer (see e.g. Refs. [5,10,47]) as well as by simple solvation. The comprehension of such reactions would of course greatly benefit from a molecular level description of the surface region of SAs. However, while the bulk ionic composition of SAs is considered to be well described by thermodynamic models [12,36], the ionic composition of the surface region is uncertain, despite a number of experimental investigations addressing this issue [48,49]. For example, under conditions where the SA bulk ionic composition is characterized by the complete first acid ionization of H_2SO_4 Eq. (1) to produce the bisulfate and hydronium ions without any significant production of $SO_4^=$ ions via the second acid dissociation

$$HSO_4^- + H_2O \longrightarrow SO_4^= + H_3O^+ \tag{4}$$

it is not clear that the surface ionic composition is identical to that of the bulk. Thus, it is conceivable that molecular H_2SO_4 exists at the surface of such an SA due to the occurrence of the reverse of reaction Eq. (1) due to possibly reduced polarity conditions at the surface.

In view of the widespread perception that H_2SO_4 is a "strong" acid (in its first dissociation), as well as the implied ease of its dissociation in small water clusters [50,51], one might have difficulty entertaining the notion that H_2SO_4 might not be dissociated in an aqueous environment. Beyond the issue of reduced polarity conditions at an interface noted above, it is relevant to note in this connection that the free energy of the first acid dissociation for H_2SO_4 in the bulk is estimated to be not very negative (experimental estimates for the 'infinite dilution' value of ΔG_a are in the range -2.73 to -4.89 kcal/mol [52], and both enthalpic and entropic components are likely to be strongly affected by the number of H-bonds formed by water molecules with, for example, the sulfate moiety of the acid, as well as by the temperature.

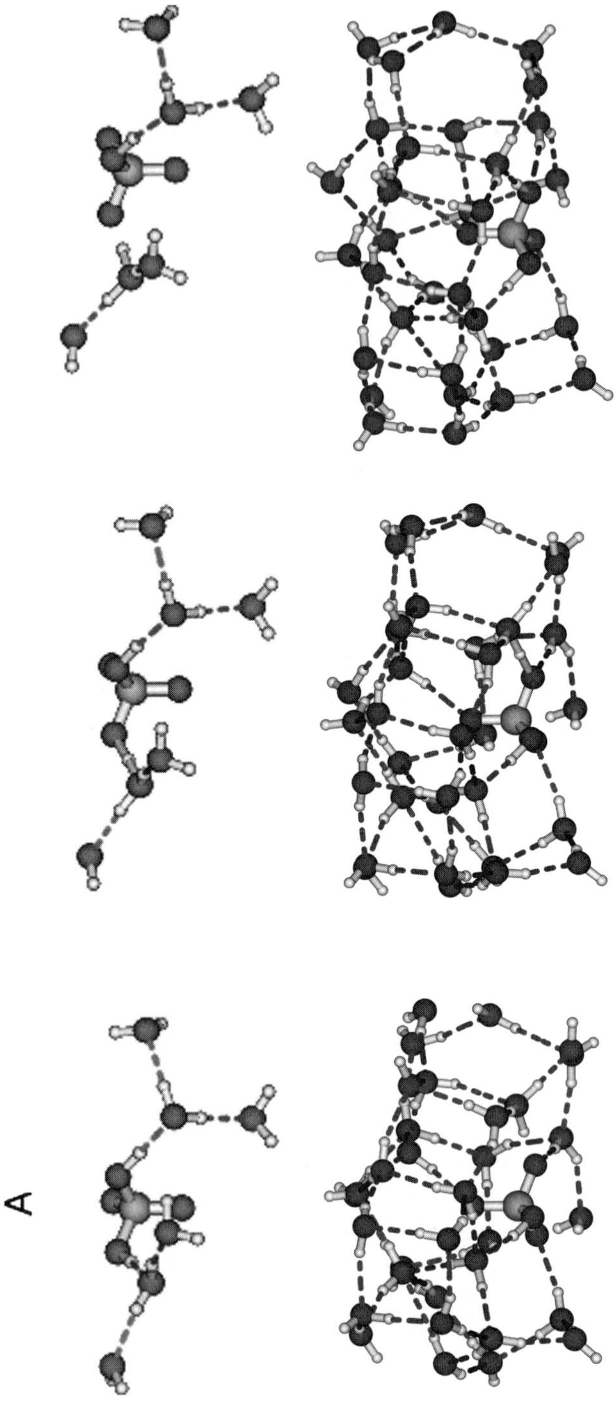

FIGURE 18.1 Sulfuric acid case A. The reactant complex is shown for the dissociation of one of the surface protons of H_2SO_4 for a semi-embedded case. Only the reactant complex is shown. CRS (top) and MRS (bottom).

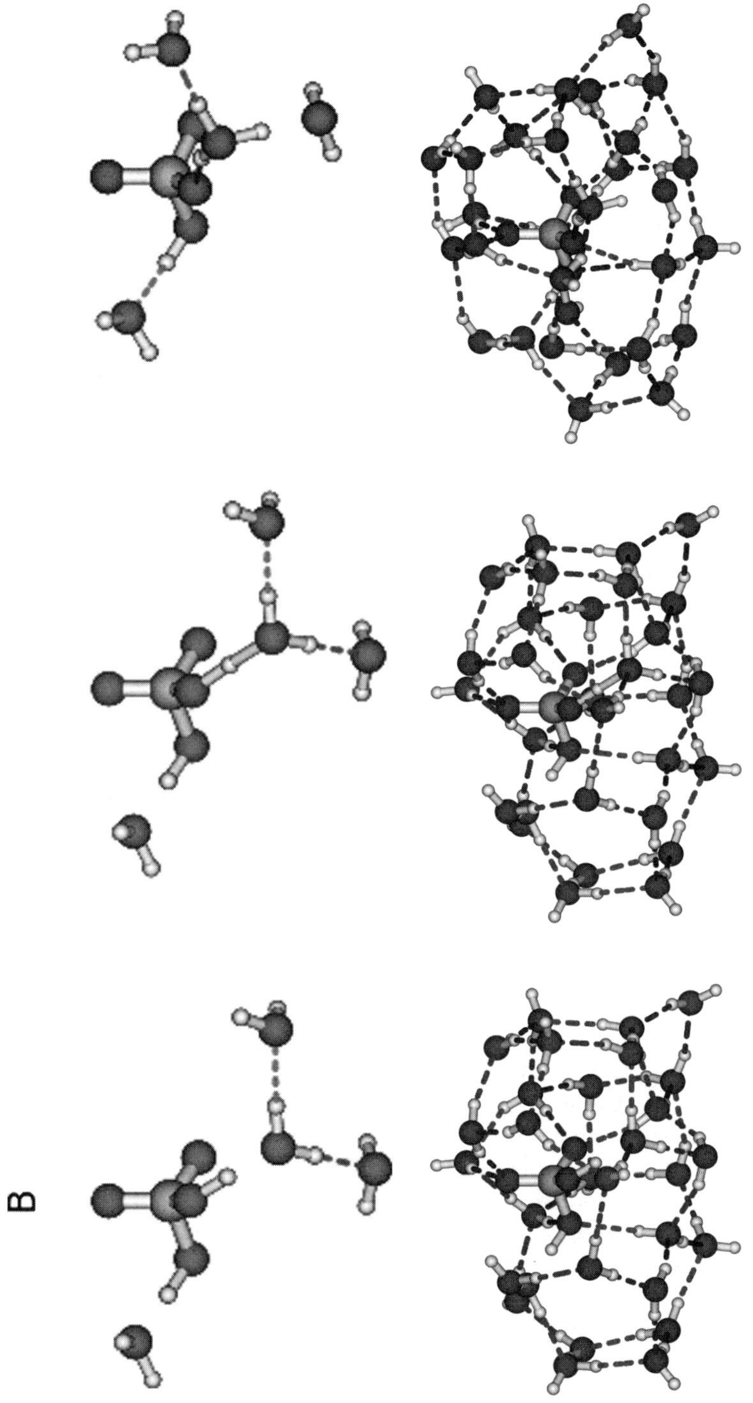

FIGURE 18.2 Sulfuric acid case B. The reactant complex is shown for the dissociation of the bulk proton of H_2SO_4. CRS (top) and MRS (bottom).

3.2 H_2SO_4 acid dissociation results [23,24]

Using the methodology described in Section 2, the first H_2SO_4 acid dissociation reaction Eq. (1) to produce the $H_3O^+ \cdots HSO_4^-$ contact ion pair was studied [23,24] by placing an H_2SO_4 molecule at four different locations at an aqueous interface, with differing solvation of the sulfate moiety of the acid and of the proton-accepting water molecule. The first of these positionings—with the least solvation—placed the H_2SO_4 atop the surface with its two potential proton donating OH groups hydrogen-bonded to surface waters. The acid dissociation was strongly endothermic ($\Delta E(0\ \mathrm{K}) \simeq 8$ kcal/mol) and was not considered further. The remaining three cases varied in their degree of embedding in the water surface. The proton transfer reaction free energetics for these three cases, as well the reaction enthalpies and entropies, were calculated from 0 to 300 K. Here we focus the discussion exclusively on the cases of those two where the embedding of the H_2SO_4 in the surface is least (case A) and where it is maximum (case B), aspects of which are shown in Figures 18.1 and 18.2, respectively. The energetic and thermodynamic data for cases A and B are collected in Table 18.1. In case A, which we have termed a "semi-embedded, surface proton dissociation" situation (an H_2SO_4 proton is donated to a surface layer water, see Figure 18.1), the story is remarkably the same over the entire temperature range: H_2SO_4 acid dissociation is unfavorable. The dissociation is endothermic at 0 K, $\Delta E + \Delta\mathrm{ZPE} = 3.44$ kcal/mol, with the ΔZPE contributing only ~0.9 kcal/mol. The reaction free energy ΔG in Table 18.1 remains close to this value over the entire temperature range reported, differing by only ~0.1 kcal/mol; both the entropy contribution $-T\Delta S$ and the thermal contribution to ΔH are small, and tend to compensate each other, preserving the 0 K result both qualitatively and quantitatively.

A more varied H_2SO_4 acid dissociation behavior is exhibited by case B, where compared to case A, the effect of further direct solvation on the sulfate group oxygens has been modeled by embedding the H_2SO_4 molecule deeper into the aqueous layer. Further, in this case, we considered the situation with the acidic OH pointing toward the bulk, and we have labelled case B as "embedded, bulk proton dissociation". Table 18.1 shows that, in contrast to case A, the 0 K energetics are now favorable for the dissociation: $\Delta E + \Delta\mathrm{ZPE} = -1.53$ kcal/mol, a value reduced in magnitude from the more favorable bare energetics ΔE by ~3.5 kcal/mol by ZPE effects. The enhanced ZPE of the product complex compared to that for the reactant complex should be related to entropic effects for the reaction, discussed below. Turning to the thermodynamics in Table 18.1, the dissociation is thermodynamically favored, with $\Delta G < 0$, up to ~190 K, although ΔG is only of the order of -1 kcal/mol or less above ~90 K. In the temperature range 190–250 K (the sulfate aerosol range), the acid dissociation is not favored, although ΔG is less than 1 kcal/mol in magnitude. Even at 300 K, ΔG is only ~1 kcal/mol. The reaction is exothermic at all temperatures examined, with the ΔH values approximately equal to -2 kcal/mol above 120 K, but the unfavorable entropy effects ($-T\Delta S$) significantly reduce the magnitude of ΔG compared to this, resulting in the slightly positive ΔG value above ~200 K. These negative ΔS values are expected as a consequence of, e.g., orientational restrictions of water molecules by the ions in the ion pair product. Although not discussed in any detail here, we have considered [24] a different embedded case where the acidic OH is H-bonded to a surface water (rather than pointing towards the bulk as in case B), and the unfavorable entropy effects result in ΔG being negative only at temperatures below 70 K. In the sulfate aerosol temperature range 190–250 K, ΔG is $\sim$1 kcal/mol, and at 300 K, ΔG is $\sim$2 kcal/mol.

The calculated reaction free energies in Table 18.1 (and in Ref. [24] for another case) for the first acid dissociation of H_2SO_4 on/at model aqueous surfaces with varying degrees of solvation indicate that the feasibility of the dissociation is a quite sensitive function of the solvation and the temperature. While these calculations involving a single H_2SO_4 molecule are not directly applicable to the surface of atmospheric concentrated sulfate aerosols, they do suggest that molecular H_2SO_4 could well be found at those surfaces, or more cautiously stated, its presence should not be ruled out *a priori*. Molecular H_2SO_4's presence would mean that, if a heterogeneous reaction such as N_2O_5's hydrolysis occurring at the surface were to be acid-catalyzed [37], then molecular H_2SO_4 would have to be the proton source, rather than H_3O^+. (This conclusion would not apply if the hydrolysis occurred well below the surface where H_2SO_4 is presumably acid dissociated.)

4. ACID DISSOCIATION OF NITRIC ACID

4.1 Atmospheric relevance

Nitric acid HNO_3 is an important chemical species in a broad range of heterogeneous reactions important in various atmospheric contexts. A non-exhaustive listing would include the following illustrations. In the Antarctic stratosphere, nitrate aerosols (NAT, nitric acid trihydrate) form one of the several important sites for heterogeneous chlorine chemistry key for ozone depletion (the other being ice aerosols) [1]. In addition, nitric acid is itself a product of these chlorine reactions, e.g. that of chlorine nitrate with hydrochloric acid [1,18]. At lower altitudes, the troposphere provides numerous instances of heterogeneous nitric acid chemistry. One example is the reaction of nitric oxide NO with (molecular) nitric acid on aqueous surfaces, which has been shown to generate reactive nitrogen species (renoxification) significant in tropospheric pollution [53], while in the marine troposphere, nitric acid uptake by sea-salt aerosols is thought to be a significant NO_x removal process [54]. A further tropospheric role for HNO_3 involves its uptake by water–ice aerosol particles in a special focus of this chapter, the upper troposphere (UT), where cirrus clouds are formed, as are airplane condensation trails (contrails) [13–16,55–57]. This uptake is important for ozone concentrations for several reasons. First, HNO_3 can generate NO_2 which ultimately produces ozone; the uptake of HNO_3 by water–ice particles removes ("sequesters") HNO_3 from this activity, ultimately affecting ozone production levels [56]. Second, heterogeneous reactions occurring on the UT particles surface such as the hydrolysis of chlorine nitrate, $ClONO_2$, and the reaction of $ClONO_2$ with hydrochloric acid, HCl—both related to ozone depletion—can occur on ice particles, but are much less efficient on HNO_3-coated ice. Such chlorine activation reactions occurring on cirrus clouds in the tropopause region may have an important impact on ozone [57].

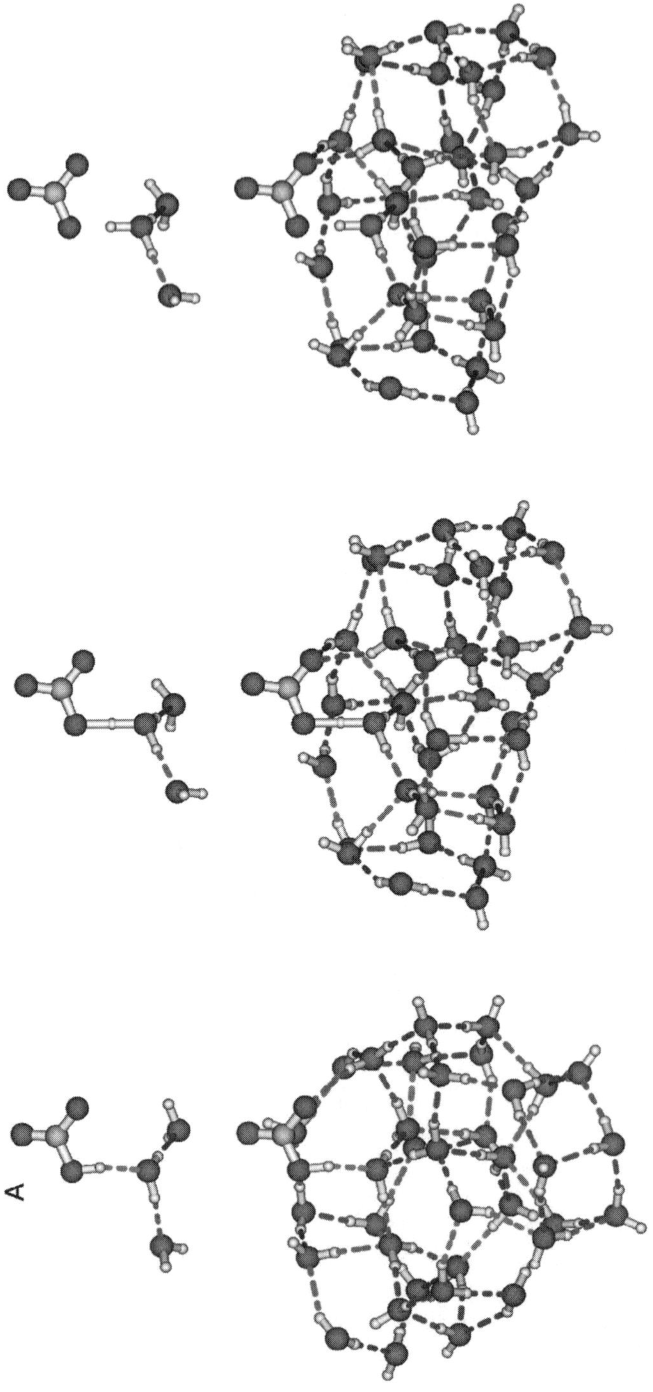

FIGURE 18.3 Nitric acid case A, $HNO_3 \cdot (H_2O)_3 \cdot W_{33} \rightarrow NO_3^- \cdot H_3O^+ \cdot (H_2O)_2 \cdot W_{33}$. The reactant complex is shown. Top row: CRS. Bottom row: MRS.

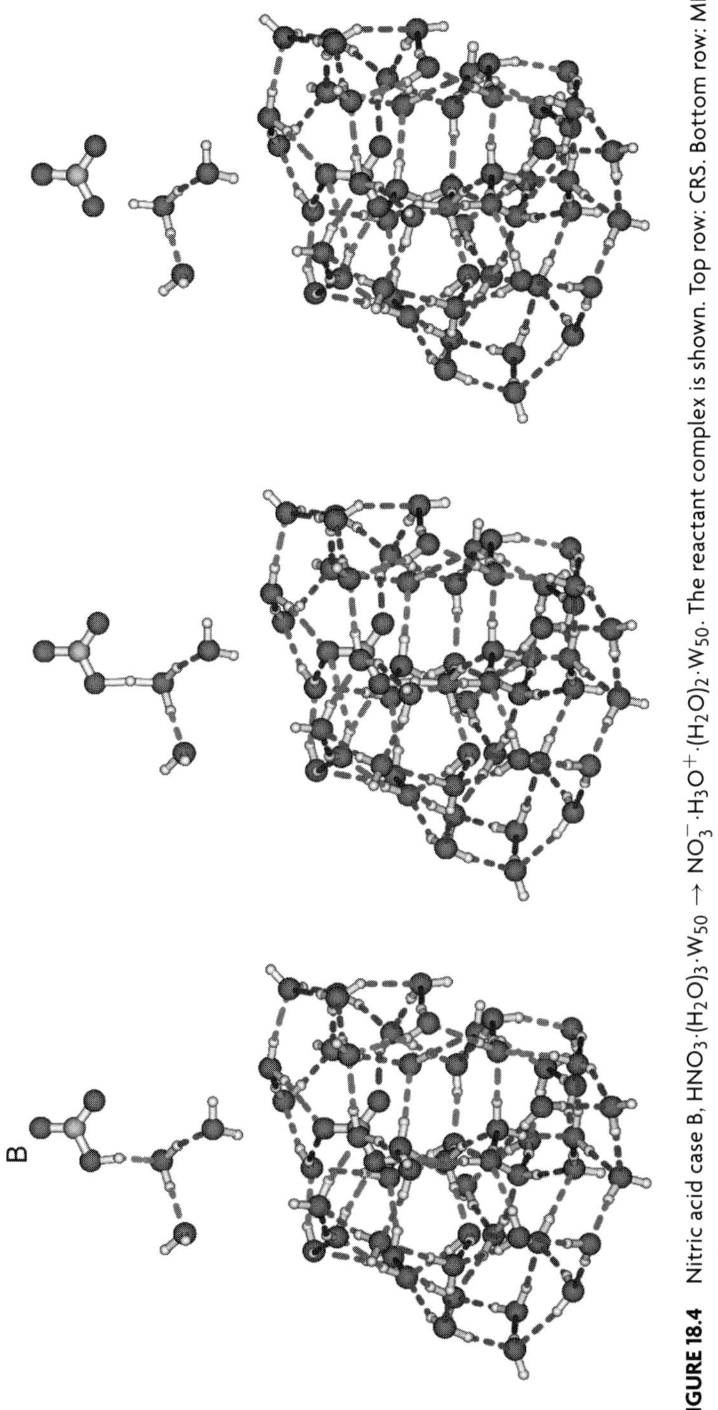

FIGURE 18.4 Nitric acid case B, $HNO_3 \cdot (H_2O)_3 \cdot W_{50} \rightarrow NO_3^- \cdot H_3O^+ \cdot (H_2O)_2 \cdot W_{50}$. The reactant complex is shown. Top row: CRS. Bottom row: MRS.

We have examined theoretically [25] and discuss in this chapter the possibility of acid dissociation of nitric acid at an aqueous surface (Eq. (2)), a proton transfer reaction of the acid with a water molecule to produce the hydronium ion H_3O^+ and the nitrate ion NO_3^-. This acid ionization has been reasonably suggested by Abbatt [13] to be a chemisorptive mechanism involved in the significant HNO_3 uptake in the UT [13–16]. It (or rather its lack of occurrence) has also been argued to be important in the uptake of HNO_3 by sea-salt aerosols [54] and is significant for the renoxification process, which involves molecular HNO_3 rather than the acid ionization product NO_3^- [53].

It should be remarked at the outset that since HNO_3 is a much weaker acid (in bulk solution) than is sulfuric acid (H_2SO_4) [58], which, as discussed in Section 3, was predicted to acid ionize at an aqueous surface only under certain lower temperature conditions [24], it cannot be taken for granted *a priori* that HNO_3 will acid ionize in the surface layer. While laboratory experiments [59–64] suggest that molecular nitric acid is dominant at the surfaces of highly concentrated nitric acid solutions (for which molecular nitric acid is also dominant in the bulk [65,66], so that the presence of molecular HNO_3 cannot be regarded as surprising), the situation at lower concentrations—relevant for the low concentration conditions of interest in the present chapter—has been less clear. Several surface-sensitive Sum Frequency Generation studies from the Schultz [60,61,63] and Richmond [64] groups bear upon this issue, and are discussed within in connection with our results. At low coverage of HNO_3 on UT ice aerosols up to about 0.1 monolayer [13, 14] only a single HNO_3 molecule need be considered, which is the case addressed in Ref. [25] and recounted here. This case was addressed via electronic structure methods similar to those employed previously [8] for the acids HCl [7], HBr [7] and especially the other focus acid of this chapter, H_2SO_4 [23,24]. We focussed on several different degrees of solvation of the HNO_3 at an aqueous surface, two of which are discussed here. The feasibility of the acid ionization Eq. (2) was assessed [25] by reaction free energy calculations, over the wide temperature range 0–300 K. This range includes that relevant for the UT 206–264 K (5–15 km) [67] and at its upper end makes contact with recent spectroscopic experiments [64].

4.2 HNO_3 acid dissociation results [25]

Using the methodology described in Section 2, the HNO_3 acid dissociation reaction Eq. (2) to produce the contact ion pair was studied [25] by placing an HNO_3 molecule at four different locations at an aqueous interface, with increasing solvation of the NO_3 moiety of the acid. The PT reaction free energetics, as well the reaction enthalpies and entropies, were calculated from 0 to 300 K. Here we focus the discussion exclusively on the cases where the NO_3 moiety is least solvated (case A) and best solvated (case B), aspects of which are shown in Figures 18.3 and 18.4.

The basic story is simple: HNO_3 remains molecular over the entire temperature range considered. Some results from Ref. [25] are collected in Table 18.2. In both cases shown (and in the other two considered [25]) the 0 K reaction is endoergic. At finite temperatures, the unfavorable reaction free energy ΔG has a negative entropy change ΔS contribution—a feature intuitively expected upon ion pair formation from the molecular pair, with increased H-bonding from surface region waters. This negative ΔS is increasingly important as T increases.

If the calculated ΔG value for the least unfavorable case, case B, is taken at a typical temperature of the upper troposphere (UT), 205–265 K [67], then the equilibrium constant for the HNO_3 acid dissociation to the ion pair is $\sim 10^{-2}$, and it will be even smaller at higher temperatures relevant in other locations, e.g. the marine boundary layer. Thus the quite reasonable suggestion [13] that the efficient uptake of HNO_3 by ice aerosols in the UT is due to the acid dissociation Eq. (2) is not supported by these results. However, one could imagine that the originally suggested acid ionization mechanism [13] could become involved either by diffusion of molecular HNO_3 deeper into the aerosol [20] where it could dissociate or by a "burying" of the molecular HNO_3 via the accretion of water molecules at a dynamic ice surface [17,68], in each case benefiting from increased solvation of the HNO_3, and especially of its NO_3 moiety.

These results can also be placed in perspective with respect to laboratory experiments. A number of experimental spectroscopic and other experiments [59–61, 63,64] have suggested that there is molecular HNO_3 at aqueous surfaces; however, it needs to be emphasized that most of these experiments examine the surfaces of high concentrations of HNO_3 in water, and it is already known [65,66] that the bulk of such solutions contains molecular nitric acid, increasingly so as the concentration increases. A very recent surface sensitive infrared spectroscopic study [64] for solutions more closely approaching the low concentration of the present study (although still much more concentrated, ~5.2 M HNO_3, such that ~25% molecular HNO_3 remains in the bulk [64,66]) has identified molecular HNO_3 at the surface, and our results are consistent with that study. The authors of Ref. [64] have made certain spectroscopic assignments—associated with strong and weak solvation of the molecular HNO_3—in support of the presence of molecular HNO_3. Examination [25] of these vibrations for the two present solvation limits of cases A and B gave vibrational frequencies in good agreement with those assignments, thus supporting both the experimental conclusions [64] and the present calculations. Further, lack of HNO_3 acid dissociation at a 298 K water surface has also been recently found via *ab initio* molecular dynamics [69].

5. CONCLUDING REMARKS

We have recounted here some of the key ingredients and highlights of our recent electronic structure work on the acid dissociations of sulfuric and nitric acids at aqueous surfaces, and have tried to place these reactions and the results presented in an atmospheric chemistry context. Perhaps the most striking aspect of these results is that both acids, which are conventionally regarded as "strong" acids from the homogeneous aqueous solution perspective, are decidedly not "strong" at aqueous interfaces, this statement applying to H_2SO_4 in the higher temperature range 210–300 K where ΔG is positive, and to HNO_3 over the entire temperature range up to 300 K. A key role in this diminution of the effective acid strength is played by the increasing importance of the unfavorable reaction entropy with increasing temperature, with this unfavorable character primarily associated with increased hydrogen bonding of the surrounding waters with the contact ion pair dissociation product compared to the neutral molecular reactant pair. The much reduced effective strength of these acids at aqueous surfaces should prove relevant in the various atmospheric situations mentioned within in Sections 3.1 and 4.1.

We have not discussed here the limitations of the present calculations due e.g. to the finite size of the systems considered or to the harmonic approximations for the frequencies used in the zero point energy and thermodynamic contributions to the free energies, enthalpies and entropies. These are discussed at some length in Refs. [24,25], and are not expected to be important. For these issues and indeed many further details concerning the H_2SO_4 and HNO_3 acid dissociation calculations, the interested reader is referred to those references. We also note that all of our calculations concerned the acid dissociation to the contact ion pair stage. Generally, there would be a positive cost in free energy to further separate the ions from this stage [70,71].

As noted in Sections 3.1 and 4.1, the atmospheric relevance of the two acids often extends beyond the issue of their surface dissociation in the conditions that we have considered here. Thus, for example, the widely occurring sulfate aerosols are concentrated sulfuric acid solutions and nitric acid trihydrate (NAT) is also clearly fairly concentrated. Elucidation of the acid ionization state of such aerosol surfaces—and thus the theoretical characterization of the mechanisms and rates of chemical reaction occurring on them—requires other tools to deal with the multiple proton transfer possibilities present in these systems. One such tool, a generalization of available reaction Monte Carlo methodologies [72] to treat such multiple proton transfers, is under development [73].

ACKNOWLEDGEMENTS

The work on H_2SO_4 was supported in part by NSF grant ATM-0000542. That research was performed in part using the Molecular Science Computing Facility (MSCF) in the William R. Wiley Environmental Molecular Sciences Laboratory, a national scientific user facility sponsored by the U.S. Department of Energy's Office of Biological and Environmental Research and located at the Pacific Northwest National Laboratory. Pacific Northwest is operated for the Department of Energy by Battelle. The work on HNO_3 was supported in part by NSF grant CHE-0417570. JTH also acknowledges support via an ANR grant (NT05-4-43154) and ECOS grant A01U03.

REFERENCES

[1] S. Solomon, *Rev. Geophys.* **37** (1999) 275.
[2] J.H. Seinfeld, S.N. Pandis, *Atmospheric Chemistry and Physics: From Air Pollution to Climate Change*, Wiley–Interscience, New York, 1998.

[3] (a) C.E. Kolb, D.R. Worsnop, M.S. Zahniser, P. Davidovits, L.F. Keyser, M.-T. Leu, M.J. Molina, D.R. Hanson, A.R. Ravishankara, L.R. Williams, M.A. Tolbert, Laboratory Studies of Atmospheric Heterogeneous Chemistry, in: J.R. Barker (Ed.), *Progress and Problems in Atmospheric Chemistry*, World Scientific Publishing, Co., River Edge, NJ, 1995;
(b) T. Peter, *Ann. Rev. Phys. Chem.* **48** (1997) 785;
(c) M.J. Molina, L.T. Molina, C.E. Kolb, *Annu. Rev. Phys. Chem.* **47** (1996) 327;
(d) J.P.D. Abbatt, M.J. Molina, *Annu. Rev. Energy Environ.* **18** (1993) 1;
(e) B.J. Finlayson-Pitts, J.N. Pitts, *Chemistry of the Upper and Lower Atmosphere: Theory, Experiments, and Applications*, Academic Press, New York, 1999.
[4] (a) D.R. Hanson, A.R. Ravishankara, *J. Phys. Chem.* **96** (1992) 9441;
(b) H. Niki, K.H. Becker (Eds.), *The Tropospheric Chemistry of Ozone in the Polar Regions, NATO ASI Series, Series I, Global Environmental Change*, vol. 7, Springer-Verlag, New York, 1993;
(c) L.A. Barrie, J.W. Bottenheim, R.C. Schnell, P.J. Crutzen, R.A. Rasmussen, *Nature* **334** (1988) 138;
(d) L.A. Barrie, *Features of Polar Regions Relevant to Tropospheric Ozone, NATO ASI Series, Series I, Global Environmental Change*, vol. 7, Springer-Verlag, New York, 1993, pp. 3–24;
(e) J.C. McConnell, G.S. Henderson, L. Barrie, J. Bottenheim, H. Niki, C.H. Langford, E.M.J. Templeton, *Nature* **355** (1992) 150.
[5] G.C.G. Waschewsky, J.P.D. Abbatt, *J. Phys. Chem. A* **103** (1999) 5312.
[6] S.-M. Fan, D.J. Jacob, *Nature* **359** (1992) 522.
[7] A. Al-Halabi, R. Bianco, J.T. Hynes, *J. Phys. Chem. A* **106** (2002) 7639. For HCl, see also Ref. [17].
[8] R. Bianco, J.T. Hynes, *Acc. Chem. Res.* **39** (2006) 159.
[9] (a) B.J. Gertner, G.H. Peslherbe, J.T. Hynes, *Isr. J. Chem.* **39** (1999) 273;
(b) C. Conley, F.M. Tao, *Chem. Phys. Lett.* **301** (1999) 29;
(c) S.H. Robertson, D.C. Clary, *Faraday Discuss.* **100** (1995) 309.
[10] G.N. Robinson, D.R. Worsnop, J.T. Jayne, C.E. Kolb, P. Davidovits, *J. Geophys. Res. Atmos.* **102** (1997) 3583.
[11] D.R. Hanson, A.R. Ravishankara, *Geophys. Res. Lett.* **22** (1995) 385.
[12] K.S. Carslaw, T. Peter, S.L. Clegg, *Rev. Geophys.* **35** (1997) 125.
[13] J.P.D. Abbatt, *Geophys. Res. Lett.* **24** (1997) 1479.
[14] M.A. Zondlo, S.B. Barone, M.A. Tolbert, *Geophys. Res. Lett.* **24** (1997) 1391.
[15] M.A. Zondlo, P.K. Hudson, A.P. Prenni, M.A. Tolbert, *Ann. Rev. Phys. Chem.* **51** (2000) 473.
[16] P.K. Hudson, J.E. Shilling, M.A. Tolbert, O.B. Toon, *J. Phys. Chem. A* **106** (2002) 9874.
[17] B.J. Gertner, J.T. Hynes, *Science* **271** (1996) 1563;
B.J. Gertner, J.T. Hynes, *Faraday Discuss.* **110** (1998) 301.
[18] R. Bianco, J.T. Hynes, *J. Phys. Chem. A* **103** (1999) 3797.
[19] (a) M. Svanberg, J.B.C. Pettersson, K. Bolton, *J. Phys. Chem. A* **104** (2000) 5787;
(b) J.P. Devlin, N. Uras, J. Sadlej, V. Buch, *Nature* **417** (2002) 269;
(c) K. Bolton, *Int. J. Quantum Chem.* **96** (2004) 607;
(d) Y.A. Mantz, F.M. Geiger, L.T. Molina, M.J. Molina, B.L. Trout, *Chem. Phys. Lett.* **348** (2001) 285;
(e) For ionization of HCl at, rather than atop, an ice surface, see Ref. [17].
[20] C. Toubin, S. Picaud, P.N.M. Hoang, C. Girardet, R.M. Lynden-Bell, J.T. Hynes, *J. Chem. Phys.* **118** (2003) 9814.
[21] (a) G.J. Kroes, *Adv. At. Mol. Phys.* **34** (1999) 259;
(b) D.C. Clary, L.C. Wang, *J. Chem. Soc. Faraday Trans.* **93** (1997) 2763;
(c) G.J. Kroes, D.C. Clary, *J. Phys. Chem.* **96** (1992) 7079.
[22] R. Bianco, B.J. Gertner, J.T. Hynes, *Ber. Bunsenges. Phys. Chem.* **102** (1998) 518.
[23] R. Bianco, J.T. Hynes, *Theor. Chem. Acc.* **111** (2004) 182.
[24] R. Bianco, S. Wang, J.T. Hynes, *J. Phys. Chem. B* **109** (2005) 21313.
[25] R. Bianco, S. Wang, J.T. Hynes, Theoretical study of the dissociation of nitric acid HNO_3 at a model aqueous surface, *J. Phys. Chem. A* **111** (2007) 11033.
[26] R. Bianco, S. Wang, J.T. Hynes, Infrared signatures of HNO_3 and NO_3^- at a model aqueous surface. A theoretical study, submitted for publication.
[27] (a) P.N. Day, J.H. Jensen, M.S. Gordon, S.P. Webb, W.J. Stevens, M. Krauss, D. Garmer, H. Basch, D. Cohen, *J. Chem. Phys.* **105** (1996) 1968;
(b) W. Chen, M.S. Gordon, *J. Chem. Phys.* **105** (1996) 11081.

[28] W.J. Stevens, H. Basch, M. Krauss, *J. Chem. Phys.* **81** (1984) 6026.
[29] M.W. Schmidt, K.K. Baldridge, J.A. Boatz, S.T. Elbert, M.S. Gordon, J.H. Jensen, S. Koseki, N. Matsunaga, K.A. Nguyen, S.J. Su, T.L. Windus, M. Dupuis, J.A. Montgomery, *J. Comput. Chem.* **14** (1993) 1347.
[30] J.A. Hayward, J.R. Reimers, *J. Chem. Phys.* **106** (1997) 1518.
[31] K. Ishida, K. Morokuma, A. Komornicki, *J. Chem. Phys.* **66** (1977) 2153, and references therein.
[32] C. Møller, M.S. Plesset, *Phys. Rev.* **46** (1934) 618.
[33] K.A. Peterson, S.S. Xantheas, D.A. Dixon, T.H. Dunning, *J. Phys. Chem. A* **102** (1998) 2449.
[34] S.S. Xantheas, T.H. Dunning Jr., *J. Chem. Phys.* **99** (1993) 8774.
[35] G. Schaftenaar, J.H. Noordik, *J. Comput. Aided Mol. Des.* **14** (2000) 123. Molden is available free of charge for academic use at http://www.cmbi.ru.nl/molden/molden.html.
[36] A. Tabazadeh, O.B. Toon, S.L. Clegg, P. Hamill, *Geophys. Res. Lett.* **24** (1997) 1931.
[37] M.J. Molina, L.T. Molina, D.M. Golden, *J. Phys. Chem.* **100** (1996) 12888.
[38] E.L. Fleming, S. Chandra, J.J. Barnett, M. Corney, *Adv. Space Res.* **10** (1990) 1211.
[39] A.R. Ravishankara, D.R. Hanson, *J. Geophys. Res. Atmos.* **101** (1996) 3885.
[40] S. Borrmann, S. Solomon, J.E. Dye, D. Baumgardner, K.K. Kelly, K.R. Chan, *J. Geophys. Res. Atmos.* **102** (1997) 3639.
[41] M.A. Tolbert, *Science* **264** (1994) 527;
M.A. Tolbert, *Science* **272** (1996) 1597.
[42] W.R. Leaitch, R.M. Hoff, J.I. MacPherson, *J. Atmos. Chem.* **9** (1989) 187.
[43] B.A. Michalowski, J.S. Francisco, S.-M. Li, L.A. Barrie, J.W. Bottenheim, P.B. Shepson, *J. Geophys. Res. Atmos.* **105** (2000) 15131.
[44] J.P.D. Abbatt, J.B. Nowak, *J. Phys. Chem. A* **101** (1997) 2131.
[45] J.R. Morris, P. Behr, M.D. Antman, B.R. Ringeisen, J. Splan, G.M. Nathanson, *J. Phys. Chem. A* **104** (2000) 6738.
[46] J.L. Duncan, L.R. Schindler, J.T. Roberts, *J. Phys. Chem. B* **103** (1999) 7247.
[47] D.J. Donaldson, A.R. Ravishankara, D.R. Hanson, *J. Phys. Chem. A* **101** (1997) 4717.
[48] D.H. Fairbrother, H. Johnston, G. Somorjai, *J. Phys. Chem.* **100** (1996) 13696.
[49] (a) S. Baldelli, C. Schntzer, M.J. Shultz, D.J. Campbell, *J. Phys. Chem. B* **101** (1997) 10435;
(b) C. Schnitzer, S. Baldelli, M.J. Shultz, *Chem. Phys. Lett.* **313** (1999) 416;
(c) C. Radüge, V. Pflumio, Y.R. Shen, *Chem. Phys. Lett.* **274** (1997) 140.
[50] (a) L.J. Larson, M. Kuno, F.M. Tao, *J. Chem. Phys.* **112** (2000) 8830;
(b) S. Re, Y. Osamura, K. Morokuma, *J. Phys. Chem. A* **103** (1999) 3535;
(c) H. Arstila, K. Laasonen, A. Laaksonen, *J. Chem. Phys.* **108** (1998) 1031;
(d) E.V. Akhmatskaya, C.J. Apps, I.H. Hillier, A.J. Masters, N.E. Watt, J.C. Whitehead, *Chem. Commun.* 7 (1997) 707.
[51] R. Hofmannsievert, A.W. Castleman, *J. Phys. Chem.* **88** (1984) 3329.
[52] D.M. Petkovic, *J. Chem. Soc. Dalton Trans.* **12** (1982) 2425.
[53] (a) A.M. Rivera-Figueroa, A.L. Sumner, B.J. Finlayson-Pitts, *Environ. Sci. Technol.* **37** (2003) 548;
(b) M. Mochida, B.J. Finlayson-Pitts, *J. Phys. Chem. A* **104** (2000) 9705;
(c) N.A. Saliba, H. Yang, B.J. Finlayson-Pitts, *J. Phys. Chem. A* **105** (2001) 10339;
(d) K.A. Ramazan, L.M. Wingen, Y. Miller, G.M. Chaban, R.B. Gerber, S.S. Xantheas, B.J.J. Finlayson-Pitts, *J. Phys. Chem. A* **110** (2006) 6886.
[54] (a) C. Guimbaud, F. Arens, L. Gutzwiller, H.W. Gäggeler, M. Ammann, *Atmos. Chem. Phys. Discuss.* **2** (2002) 739;
(b) J.A. Davies, R.A. Cox, *J. Phys. Chem. A* **102** (1998) 7631;
(c) P. Beichert, B.J. Finlayson-Pitts, *J. Phys. Chem. A* **100** (1996) 15218;
(d) S. Ghosal, J.C. Hemminger, *J. Phys. Chem. B* **108** (2004) 14102.
[55] S.K. Meilinger, B. Kärcher, Th. Peter, *Atmos. Chem. Phys. Discuss.* **4** (2004) 4455.
[56] M.G. Lawrence, P.J. Crutzen, *Tellus* **50B** (1998) 263.
[57] (a) S. Solomon, S. Borrmann, R.R. Garcia, R. Portmann, L. Thomason, L.R. Poole, D. Winker, M.P. McCormick, *J. Geophys. Res.* **102** (1997) 21411;
(b) S. Borrmann, S. Solomon, J.E. Dye, B.P. Luo, *Geophys. Res. Lett.* **23** (1996) 2133;
(c) A. Meier, J. Hendricks, *J. Geophys. Res.* **107** (2002) D23, Art. no. 4696.
[58] D.W. Oxtoby, N.H. Nachtrieb, *Principles of Modern Chemistry*, second ed., Saunders College Publishing, Orlando, FL, 1990, p. 206.

[59] (a) D.J. Donaldson, D. Anderson, *Geophys. Res. Lett.* **26** (1999) 3625;
(b) D. Clifford, T. Bartels-Rauschwa, D.J. Donaldson, *Phys. Chem. Chem. Phys.* **9** (2007) 1362.
[60] C. Schnitzer, S. Baldelli, M.J. Shultz, *Chem. Phys. Lett.* **313** (1999) 416.
[61] M.J. Shultz, C. Schnitzer, D. Simonelli, S. Baldelli, *Int. Rev. Phys. Chem.* **18** (2000) 123.
[62] H.S. Yang, B.J. Finlayson-Pitts, *J. Phys. Chem. A* **105** (2001) 1890.
[63] M.J. Shultz, S. Baldelli, C. Schnitzer, D. Simonelli, *J. Phys. Chem. B* **106** (2002) 5313.
[64] M.C. Kido Soule, P.G. Blower, G.L. Richmond, *J. Phys. Chem. A* **111** (2007) 3349.
[65] T.F. Young, L.F. Maranville, H.M. Smith, in: W.J. Hamer (Ed.), *The Structure of Electrolyte Solutions*, Wiley, New York, 1959, chap. 4.
[66] N. Minogue, E. Riordan, J.R. Sodeau, *J. Phys. Chem. A* **107** (2003) 4436.
[67] K.D. Evans, S.H. Melfi, R.A. Ferrare, D.N. Whiteman, *Appl. Opt.* **36** (1997) 2594.
[68] D.R. Haynes, N.J. Tro, S.M. George, *J. Phys. Chem.* **96** (1992) 8502.
[69] E.S. Shamay, V. Buch, M. Parrinello, G.L. Richmond, *J. Am. Chem. Soc.* **129** (2007) 12910.
[70] K. Ando, J.T. Hynes, *J. Phys. Chem.* **101** (1997) 10464.
[71] K. Ando, J.T. Hynes, *J. Phys. Chem. A* **103** (1997) 10398.
[72] J.K. Johnson, A.Z. Panagiotopoulos, K.E. Gubbins, *Mol. Phys.* **81** (1994) 717.
[73] R. Bianco, S. Wang, J.T. Hynes, in progress.

CHAPTER 19

Investigating Atmospheric Sulfuric Acid–Water–Ammonia Particle Formation Using Quantum Chemistry

Theo Kurtén* and **Hanna Vehkamäki** *

Contents

1. Introduction 407
2. Theoretical Methods for Free Energy Calculations 411
3. Applications of Quantum Chemistry to Atmospheric Nucleation Phenomena 415
   - 3.1 Construction of classical potentials 415
   - 3.2 Investigating the fundamental chemistry of nucleating systems 416
   - 3.3 Computing formation free energies for nucleating clusters 422
4. Challenges 423
5. Conclusions 424

Acknowledgements 425

References 425

1. INTRODUCTION

The scope of this paper is to provide an overview of methods used to study properties of electrically neutral molecular clusters initiating particle formation in the troposphere, with focus on quantum chemistry. The review of results is intended to be complete with regard to water–sulfuric acid–ammonia clusters. Concerning studies on clusters including other molecular species, we review representative examples and newest publications. Ionic clusters and clusters involving iodine, related to coastal nucleation, are mentioned in passing.

Atmospheric aerosols are liquid or solid particles floating in the air. Their size ranges from nanometer-scale molecular clusters to, for example, road dust with a

* Department of Physical Sciences, P.O. Box 64, 00014 University of Helsinki, Finland
E-mail: theo.kurten@helsinki.fi
E-mail: hanna.vehkamaki@helsinki.fi

Advances in Quantum Chemistry, Vol. 55
ISSN 0065-3276, DOI: 10.1016/S0065-3276(07)00219-5

diameter of a few hundred micrometers. Cloud droplets are also aerosols with diameters above one micrometer, and they always contain a non-aqueous aerosol particle as a condensation nucleus. Aerosol particles can be emitted to the atmosphere by both natural and anthropogenic processes, for example desert dust and soot from combustion, respectively. Gaseous condensable vapors can condense, or heterogeneously nucleate, on the surfaces of existing particles, changing their size and compositions, but vapors can also form completely new particles in the atmosphere by homogeneous nucleation. The smallest ultrafine atmospheric particles are products of gas-to-particle nucleation, and over continents it has been estimated that 30% of the aerosol particles are formed this way [1].

Aerosols affect climate, visibility and human health. The fourth assessment report of the Intergovernmental Panel on Climate Change [2] concludes that anthropogenic contributions to aerosols together produce a cooling effect, and aerosols remain the dominant uncertainty in predicting radiative forcing. Aerosols affect our planets radiation budget by scattering and absorbing incoming solar radiation as well as outgoing long-wave radiation. Since cloud formation depends on the aerosol present, they also have an indirect effect on climate by influencing the occurrence, properties and lifetime of clouds and precipitation.

Although the ultrafine aerosol particles constitute a negligible fraction of the total aerosol mass, they make up a significant fraction of the number concentration. In polluted urban conditions, traffic-related nucleated particles can even dominate the aerosol number distribution. The smaller the particles, the deeper they penetrate into the respiratory passages, and there is evidence that nano-scale particles can find their way from the lungs to the circulations system, causing cardiovascular diseases and premature deaths [3]. Despite the evidence of the importance of high concentrations of ultrafine particles in aerosol related phenomena, present legislative constraints do not control the number concentration, only the mass of aerosol particles. In most countries, current legislation restricts the total mass of particles below 10 micrometers in aerodynamic diameter (PM10); legal limitations on the total mass of particles below 2.5 micrometer in diameter (PM2.5) are in effect only in the United States of America. WHO has guidelines for the upper limit of PM2.5, and binding restrictions on PM2.5 are currently in preparation in the European Union. Before limitations on the number concentrations can be prepared, scientist need to achieve a significantly improved understanding of the role of both natural and anthropogenic ultrafine particles, and their measurements techniques; the road to this objective starts from understanding the formation processes and properties of atmospheric ultrafine particles.

Particle formation events from gaseous precursors are observed frequently almost everywhere in the troposphere, both in polluted cities and remote clean areas [4]. It is likely that different nucleation mechanisms are at work in different conditions, but no formation mechanism has been identified so far. It is, however, clear that particles are formed by nucleation of a multicomponent vapor mixture. Water vapor is the most abundant condensable gas in the atmosphere, but it can not form particles on its own: homogeneous nucleation requires such a high supersaturation, that heterogeneous nucleation on omnipresent pre-existing particles always starts first and consumes the vapor. However, vapor that is undersaturated with respect to pure liquids, for example sulfuric acid and water, can be highly supersaturated with respect to the mixture of these liquids. In coastal areas [5] iodine-containing species are believed to be responsible for particle formation, and in several cases the gas-phase concentration of sulfuric acid correlates well with the appearance of ultrafine particles. In many cases photochemistry and/or mixing induced by sunrise seems to trigger particle formation, although at some sites also night-time nucleation has been observed. Nucleation around atmospheric ions is one promising pathway to explain at least some of the observed formation events [6–8]. Newly nucleated particles contain less than one hundred molecules, possibly only around ten or less. Electrically neutral particles can not even be detected before they have grow to a diameter above 2.5 nm, when they already contain thousands of molecules. The composition of the particles can be uncovered by mass spectrometry only when they contain around one million molecules. At this stage, the substances added during the growth process completely overshadow the composition of the original critical cluster produced by nucleation. Theoretical tools are thus vital for understanding the first stages of particle formation.

The nucleation process is often thought to be simple cluster kinetics: clusters grow when they collide with vapor molecules (monomers) and each other, and break up by evaporating vapor molecules or splitting into two smaller clusters. However, for some organic nucleation mechanisms it is possible that the kinetics involved is significantly more complicated (see below). In most cases it is enough to take into account only monomer–cluster collisions and the evaporation of single molecules. Collisions with inert air molecules keep the clusters close to the ambient temperature despite the latent heat involved in condensation and evaporation of molecules. The critical cluster is the smallest cluster for which growth is more likely than decay. Growth probabilities can for many cluster types be justifiably calculated using simple kinetic gas theory, and the evaporation probability of a cluster depends on the internal stability of the cluster, which is characterized by its formation free energy. The formation free energy is the difference between the free energy of the cluster and the free energies the constituent molecules would have when unbound in the vapor phase, and it depends on the vapor densities as well as the temperature. The formation free energy can be split into an attractive volume term (corresponding to the attractive interactions between cluster molecules) and a repulsive surface term (corresponding to the work required to form the surface of the cluster). Since the volume term depends on the third power and the surface term on the second power of the number of molecules in the cluster, the free energy curve plotted against the number of molecules will exhibit a maximum. The location and height of the maximum depends on the temperature and the concentration of the nucleating vapor. From a macroscopic thermodynamic point of view, the formation free energy can be split into enthalpy and entropy contributions. It should be noted that even though the enthalpy term may, for many substances, always be favourable for cluster formation, it is overweighed by the unfavourable entropy term for small cluster sizes.

When the formation free energy is plotted against the number of molecules, in a one-component vapor the critical cluster size is the location of the maximum of

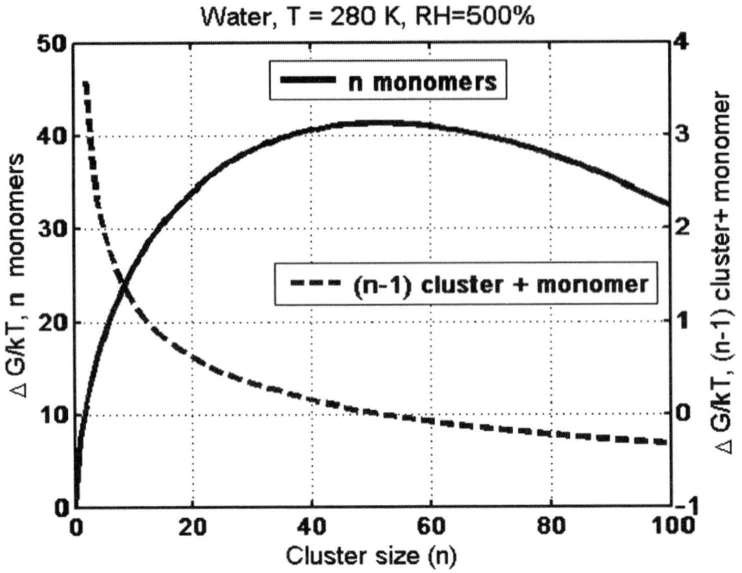

FIGURE 19.1 Free energy surface for cluster formation in the one-component case, here for pure water (see McDonald [9] for original data). Both free energy related to the formation of a cluster from single molecules (monomers) and the free energy of adding one monomer to a cluster are plotted against the number of molecules in the cluster. The free energy of monomer addition is positive for small clusters, but turns negative at the critical size.

the free energy curve. See Figure 19.1 where the free energy curve is plotted using water as an example (see Abraham [10] and Kashchiev [11] for details). In multi-component systems, one axis per component is needed to represent the number of molecules, and the critical cluster size is the saddle point of the free energy surface: it is a maximum on one direction, but minimum in all other directions. When the cluster has overcome the critical size, it can freely grow by going downhill along the free energy surface in one direction.

It should be noted that the thermodynamic free energy surface seen in Figure 19.1 is somewhat different from the potential energy surfaces commonly encountered in physical chemistry or molecular physics. The latter are plotted with respect to the positions of atomic nuclei (usually expressed in terms of one or more reaction coordinates; see Figure 19.2 for a schematic example). Equilibrium structures are represented as minima on this surface, and the differences between the minima are the thermodynamic barriers. The free energy surfaces encountered in nucleation studies (such as Figure 19.1) can be viewed as a small subset of these potential energy surfaces, where the arrangement of atoms in the clusters is not accounted for explicitly, but is instead implicitly included via the method by which the free energies are calculated. When nucleation rates are calculated from free energy surfaces such as that given in Figure 19.1, it is implicitly assumed that there are no reaction barriers (also called kinetic barriers) for cluster formation processes. (On the atomic potential energy surface, reaction barriers are repre-

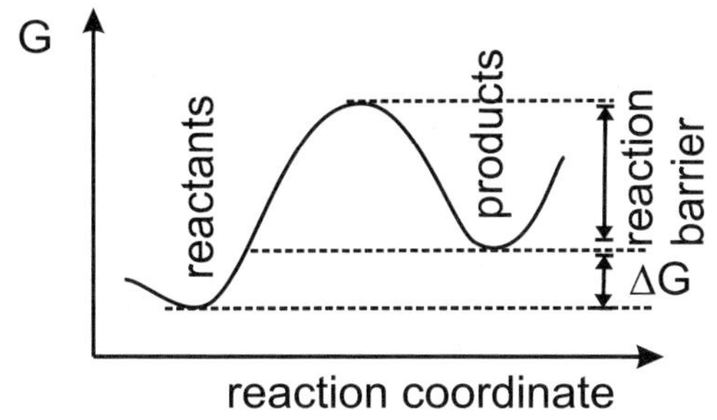

FIGURE 19.2 Schematic free energy surface for a chemical reaction leading from "reactants" to "products". The formation free energies of Figure 19.1 are related to the differences between the minima. (Note that free energies are rigorously defined only for stationary points on the atomic potential energy surface, but are used here loosely to facilitate comparison with the formation free energies of Figure 19.1.)

sented by saddle points, or maxima in the one-dimensional case, for example in Figure 19.2.) As long as the studied systems are weakly bound molecular clusters, this assumption is not problematic. However, as the focus of studies shifts toward cluster formation involving more chemically complex species and bond-breaking reactions, the distinction between thermodynamics (related to minima on the atomic potential energy surface) and kinetics (related to saddle points on the atomic potential energy surface) should be emphasized.

The nucleation rate (number of particles formed per volume and time unit) is inversely proportional to the exponential of the critical cluster formation free energy. Steady-state conditions with time-independent concentrations of vapor and clusters are most often assumed to calculate the nucleation rate, but in reality the process in the atmosphere is highly dynamic: cluster formation consumes the vapor, sources and/or gas-phase chemistry produce it, while condensation on pre-existing particles and chemistry can act as vapor sinks. Fully dynamic studies of actual cluster formation processes in atmospheric varying conditions are desirable in the future, if or when computer resources permit them, but presently theoretical efforts are focused on predicting the formation free energy of the critical cluster. Besides the interaction energy of the molecules, also the entropic contributions arising from vibrations and rotations at atmospheric temperature far above 0 K must be determined. The size- and composition-dependent properties of real atmospheric clusters are averages over countless cluster configurations populated at ambient conditions, not properties of the most stable minimum free energy structures.

2. THEORETICAL METHODS FOR FREE ENERGY CALCULATIONS

The most widely used method for calculating the formation free energy is the classical nucleation theory (CNT) [12–16] based on thermodynamics. The molecular clusters are treated as spherical droplets of bulk liquid having sharp boundaries associated with surface tension. The dominant role of CNT is understandable since it is applicable as soon as a few key properties of the liquid forming the droplet are known: liquid density, saturation vapor pressures of all the components above the mixture, and the surface tension. It should be noted that even this basic information is not available for many of the atmospherically interesting molecular mixtures, for example highly concentrated aqueous solutions of ammonia and sulfuric acid, let alone mixtures of, for example, water, sulfuric acid and various organic molecules. The fact that bulk liquid density is used to model a small molecular cluster can be corrected for within CNT, but the use of size-independent flat surface tension causes inaccuracies and, in the case of surface active mixtures where certain components accumulate in the surface layer, even a complete breakdown of the theory. For mixtures that are not surface active, CNT is valid down to surprisingly small clusters, for example 8–10 molecules in the case of pure water [17].

Classical density functional theory (DFT) [18,19] treats the cluster formation free energy as a functional of the average density distributions of atoms or molecules. The required input information is an intermolecular potential describing the substances at hand. The boundary between the cluster and the surrounding vapor is not anymore considered sharp, and surface active systems can be studied adequately. DFT discussed here is not to be confused with the quantum mechanical density functional theory (discussed below), where the equivalent of the Schrödinger equation is expressed in terms of the electron density. Classical DFT has been used successfully to uncover why and how CNT fails for surface active systems using simple model molecules [20], but it is not practically applicable to real atmospheric clusters: if the molecules are not chain-like, the numerical solution of the problem gets too burdensome, unless the whole molecule is treated in terms of an effective potential.

Molecular dynamics (MD), also knows as molecular mechanics (MM), simulations [21] solve Newton's equations of motion for the chosen set of molecules, and Monte Carlo (MC) simulations [22–24] study statistical averages of cluster properties at certain conditions, for example, constant vapor concentration and temperature. Both single clusters, clusters surrounded by vapor phase, and whole cluster distributions immersed into vapor can be studied. The larger the number of molecules, the higher the computational cost. The input information required is again the potential energy related to molecular interactions, including both intermolecular and intramolecular contributions in the case of non-rigid molecules. In principle these interactions can be calculated using quantum chemistry, but enormous computational costs of such a method force us mostly to resort to classical force fields to describe the interactions. MD simulations capture the true dynamic nature of cluster formation. MD can answer questions about timescales of various processes, for example, the rate of evaporations/condensation, relaxation time or the lifetime of a cluster [25]. A truly dynamic method is the only way to calculate the actual nucleation rate, the time required to form a critical cluster when starting with supersaturated vapor. The problems of MD are related to moderately high computational costs (compared to CNT or MC, though not to quantum chemical methods), incomplete sampling of cluster configurations, and temperature control. Established thermostatting methods may not work well for small clusters [26,27], and the most realistic way of temperature control with an excess of inert gas molecules requires a lot of computing time. MC methods used in atmospheric nucleation studies are computationally less expensive, and can thus be applied to larger systems, and/or to gather more statistics. Temperature control is trivial, but we lose information on the realistic temperature fluctuations in the growing clusters. MC methods are well suited to calculations of average formation free energies and critical cluster sizes, but they yield no information about the absolute timescales of the cluster formation process. MC is an effective tool for sampling the configuration space.

Classical interaction potentials for molecular systems can be pairwise or take manybody interactions into account, molecules can be rigid or flexible, and both stretching and torsion motions can be considered [28]. Polarizability can be introduced, but for example for water this does not seem to make the potentials mimic reality unambiguously better. The potentials, also called force fields, are fitted to reproduce certain sets of properties, and can fail to reproduce other characteristics. The potentials suffer from problems with transferability: for example a water–sulfuric acid potential developed for the two-component mixture does not necessarily describe the interaction between water and sulfuric acid in a three-component mixture of water, sulfuric acid and ammonia. Also, potentials created for bulk liquid do not always work for surface layers or small clusters.

Quantum chemistry, also called electronic structure methods, represents the most chemically accurate, but also computationally expensive, simulation method. Whereas classical potentials treat electronic interactions only implicitly, quantum chemical methods are based on the numerical solution of the Schrödinger equation (or the related Kohn–Sham equations in case of DFT) for systems of nuclei and electrons, subject to a number of approximations with varying degrees of severity. For all methods applicable to the (relatively large) molecular clusters of interest to atmospheric nucleation studies, these approximations include, for example, treating the motion of atomic nuclei using classical mechanics (with the forces computed from the electronic wavefunction), and expressing the electronic wavefunction in terms of single-electron wavefunctions composed of basis functions selected from some basis set. In studies of atmospheric molecular clusters, this basis set usually consists of atom-centered gaussian functions, though slater-type functions, numerical basis functions and plane waves have also been used. Electron–electron correlation has to be accounted for if chemical accuracy is desired. This is done either through an exchange-correlation functional (in quantum density functional theory) or via the use of various excitation operators (in correlated *ab initio* methods). Due to their prohibitively high computational costs, only the lowest-order correlated *ab initio* methods such as MP2 (2nd order Møller–Plesset perturbation theory) can generally be applied to systems of atmospheric interest, though single-point energy corrections can sometimes be computed at higher levels such as MP4 or CCSD(T) (see *e.g.* Jensen [29] for definitions and explanations of the various methods). Hartee–Fock calculations, which ignore electronic correlation, have also been performed for some atmospherically relevant molecular clusters, but as their computational cost is roughly similar to

DFT methods while their accuracy is significantly lower, they are seldom used anymore.

The advantages and disadvantages of density functional theory in describing hydrogen-bonded and other weakly bound systems has been extensively debated (see *e.g.* Refs. [30,31]), but there is no clear consensus as to whether or not DFT methods are capable of describing hydrogen bonds as well as for example MP2. For some sets of reference molecules, heavily parametrized density functionals (see *e.g.* Ref. [32]) often replicate experimental binding energies and/or molecular geometries very well, but analogously to classical potentials, their transferability to different types of systems is questionable. MP2 and related methods may have greater mean errors in comparison to some reference datasets, but they are more systematically reliable. Unlike density functionals, the results can also be systematically improved upon by using higher-order corrections, though these are often computationally too expensive to be of practical use.

The greatest advantage of quantum chemistry in molecular simulations is that there are no system-specific parameters to be determined, and the methods can thus—with the possible exception of the heavily parametrized functionals mentioned above—be applied to any chemical system. In principle, quantum chemistry methods can do everything that classical methods can, but the computational time required for one energy or gradient evaluation is typically several orders of magnitude higher, which usually restricts the dynamical sampling to simple energy minimizations, with thermal contributions to enthalpies and entropies being computed using very simple rigid rotor and harmonic oscillator models. Exceptions to this rule exist, for example Choe *et al.* [33] very recently used quantum chemistry methods (PBE functional with periodic boundary conditions, pseudopotentials and an adaptive-finite element basis) to study the dynamics of proton transfer reactions of aqueous sulfuric acid solutions. Also, recently developed [34–36] "black box"—applications have made anharmonic vibrational frequency evaluations possible for at least the smallest cluster structures. For example, anharmonic vibrational frequencies have been computed for the $H_2SO_4{\cdot}H_2O$ cluster by Miller *et al.* [37], and for the $H_2SO_4{\cdot}(H_2O)_2$ and $HSO_4^-{\cdot}(H_2O)_2$ clusters by Kurtén *et al.* [38].

One application for which quantum chemical methods are inherently superior is the study of nucleation processes involving proper chemical reactions (*i.e.* the breaking and formation of covalent bonds). Classical force fields are usually by construction unable to treat bond breaking and formation, though some potential schemes exist which enable modeling chemical reactions: for covalent molecular systems the most widely used ones are those based on the Brenner and ReaxFF potentials [39,40]. While sulfuric acid–water–ammonia cluster formation probably does not involve other reactions than proton transfer (discussed below), it is possible that nucleation mechanisms involving organic molecules or some unstable precursors of sulfuric acid may play significant roles in the atmosphere. These mechanisms can probably not be understood without accounting for complicated chemical reactions that can not be modeled by any classical interaction potentials.

Atmospherically relevant molecular clusters often contain strongly acidic or basic molecules (*e.g.* sulfuric acid and ammonia), which gives rise to various different proton transfer reactions. Accurate modeling of proton transfer and its role in cluster thermodynamics and/or kinetics is one of the key challenges for all simulation methods. Classical nucleation theory implicitly assumes that the extent of proton transfer in small cluster is equal to that in bulk liquid. Though classical potentials generally have difficulties in describing bond breaking and formation, there are various tricks that can be used to model proton transfer, as described in Section 3.1. Quantum chemical methods are able to model bond breaking and formation without trouble, and can accurately predict the relative energetics of clusters with different proton transfer states. Dynamical features of proton transfer processes have also been studied using quantum chemical methods [33]. As already mentioned, even quantum chemical methods usually model the motion of nuclei using classical mechanics. Proton transfer reactions are known [41] to involve a large degree of quantum-mechanical tunneling. While tunneling processes can be modeled by applying advanced kinetics models to the computed potential energy surfaces (see *e.g.* Ref. [42], where this is done for the atmospherically relevant SO_3+H_2O reaction), these methods are computationally too demanding to be routinely used in simulations of clusters containing multiple molecules. Therefore, though quantum chemical studies of molecular clusters capture the energetics of proton transfer states correctly, it is uncertain whether they correctly describe the dynamics of the process even when dynamical simulations are attempted.

3. APPLICATIONS OF QUANTUM CHEMISTRY TO ATMOSPHERIC NUCLEATION PHENOMENA

3.1 Construction of classical potentials

One of the most straightforward applications of quantum chemistry to nucleation phenomena is the construction of pair potentials for molecular dynamics studies. Geometry optimizations and binding energies computed using quantum chemical methods can be used to determine all of the potential parameters, as done for example by Ding *et al.* [43] in their H_2SO_4–HSO_4^-–H_2O–H_3O^+ potential. Ding *et al.* fitted the parameters of their intermolecular potential consisting of a Coulomb term and a Lennard-Jones 6-12 term to reproduce the energies and geometries obtained by Re *et al.* [44]. The intramolecular degrees of freedom were represented by a harmonic potential with the equilibrium distances matching those of Re *et al.* Another alternative is to determine only part of the parameters, or the relations between the parameters, using quantum chemistry, and fit the rest to experimental results. This was done by Kusaka *et al.* [45] for the H_2SO_4–HSO_4^-–H_2O–H_3O^+ system and Kathmann and Hale [46] in their potential for $SO_4^{2\delta-}$–H_2O–$H^{\delta+}$, where δ is a partial, empirical charge less than 1 *e*. Like Ding *et al.*, both groups used Coulomb and Lennard-Jones 6–12 potential terms, but Kusaka *et al.* used experimental geometries and dipole moments together with the quantum chemical results of Kurdi and Kochanski [47] as a basis for their potential, while Kathmann and Hale based their potential on experimental dipole moments, liquid solution surface tension and partial equilibrium vapor pressures, and their own quantum chemical results (at the HF/DZV+3P level).

As mentioned above, the modeling of proton transfer reactions with classical potentials is problematic. There are two main pathways to describe the different protolysis states of sulfuric acid with pair potentials. One is to develop separate potentials for H_2SO_4 and HSO_4^-, as done by Ding *et al.* [43]. Energetic parameters can then in principle be computed by simulating all the different possible proton transfer states and combining the results using some suitable statistical analysis. For example, for a system with two acid molecules, there are three different proton transfer states if only the first proton transfer reaction is considered, and six if also the second, less favorable, reaction is included. Conceivably, this could also be done "on the fly" by using Monte Carlo sampling at certain intervals to determine whether or not a proton transfer step would be energetically favorable at the given molecular configuration. However, the computational demands of such an approach might well outweigh the benefits, and to our knowledge no such model has been published for any atmospherically relevant molecular cluster system.

Another, in principle more chemically justified, method for modeling proton transfer is to treat the proton as a separate molecule, with its own potential parameters [46]. However, in order to produce reasonable results the charge assigned to the free proton has to be considerably lower than $+1\ e$; for example Kathmann *et al.* used the value $+0.1627\ e$. This counterintuitive adjustment decreases the chemical justification for the model, and raises the question of whether the dynamics predicted for the "partially charged protons" have any physical significance. However, as mentioned above, real proton transfer reactions are known to be influenced by quantum mechanical tunneling, which is replicated neither by molecular dynamics nor conventional quantum chemical methods, both of which assume atomic nuclei to move according to classical physics. Thus, a "partially charged" proton is not necessarily that much worse for modeling proton transfer than any other model based on classical nuclear motion.

3.2 Investigating the fundamental chemistry of nucleating systems

Another application of quantum chemical methods is the investigation of the fundamental chemical behaviour of molecular systems potentially relevant to nucleation. Within the field of tropospheric nucleation mechanisms, two questions which have merited considerable study under the last decade are the modeling of the hydration of sulfuric acid, and the role of ammonia in sulfuric acid–water nucleation.

3.2.1 Sulfuric acid hydrates

The concentration of water molecules in the atmosphere is around $10^{15}/\text{cm}^3$, while that of sulfuric acid molecules rarely exceeds $10^8/\text{cm}^3$. Thus, water molecules frequently collide with acid molecules. Chemical intuition and experiments [48,49] show that sulfuric acid molecules gather water around them in the atmosphere: these small stable clusters containing sulfuric acid and water are called hydrates, and although stable, they have not nucleated. Hydrates correspond to a local formation free energy minimum, and to grow further they face an uphill of the

TABLE 19.1 Gibbs free energy (at 298 K and 1 atm reference pressure) for the addition of water molecules to sulfuric acid, from various calculations and experiments. For the quantum chemical results, the use of harmonic or anharmonic vibrational frequencies has been indicated

	$H_2SO_4 + H_2O \Rightarrow H_2SO_4{\cdot}H_2O$	$H_2SO_4{\cdot}H_2O + H_2O \Rightarrow H_2SO_4{\cdot}(H_2O)_2$
B3LYP/6-311++G(2d,2p), harmonic [53]	−0.6 kcal/mol	−0.1 kcal/mol
B3LYP/D95++G(d,p), harmonic [44]	−2.4 kcal/mol	−2.0 kcal/mol
PW91/DNP, harmonic [54]	−2.1 kcal/mol	−1.4 kcal/mol
PW91/ATZ2P, harmonic [55]	−2.6 kcal/mol	−2.6 kcal/mol
Extrapolated MP4/ aug-cc-pV(T + d)Z, anharmonic [38]	−3.4 kcal/mol	−2.3 kcal/mol
CNT [59]	−4.1 kcal/mol	−2.5 kcal/mol
Experimental [48]	-3.6 ± 1 kcal/mol	-2.3 ± 0.3 kcal/mol

free energy surface. Formation of the critical cluster from hydrates has a higher free energy barrier than formation from free sulfuric acid molecules, and taking into account the effect of hydrates lowers the nucleation rate [50,51]. Relevant hydrates contain only 1–5 water molecules and one sulfuric acid molecule, and thus they are small enough to be treated with quantum chemical methods. Several groups have studied the structure and energetics of different protolysis states of the small hydrates [38,44,47,52–56]. Although hydrates with two sulfuric acid molecules contain a negligible fraction of the total gas-phase sulfuric acid, they are important for the nucleation process, as the critical clusters (at least in pure sulfuric acid–water nucleation) probably contain more than one sulfuric acid molecule, and they have also been the focus of several studies [56–58].

Table 19.1 shows experimental results [48] and the results of Re *et al.* [44], Bandy and Ianni [53], Ding and Laasonen [54], Al Natsheh *et al.* [55] and Kurtén *et al.* [38] for the Gibbs free energies (at 298 K and a reference pressure of 1 atm for all reactants) for the first two water addition reactions to a sulfuric acid molecule. Figure 19.3 shows the corresponding relative concentrations of free sulfuric acid molecules, monohydrates and dihydrates at 298 K and RH 50%. Only clusters with one sulfuric acid molecule are considered in this comparison. The prediction of CNT [59] (with the Clegg activity model [60]) is also included in the table and figure, and it matches the experiments rather well. The comparison between the results of Kurtén *et al.* (calculated using a computationally very expensive combination of MP2/aug-cc-pV(D + d)Z geometries and anharmonic vibrational frequencies, MP2/aug-cc-pV(T + d)Z and MP4/aug-cc-pV(D + d)Z electronic energies), those of Re *et al.* (computed at the B3LYP/D95++(d,p) level), Bandy and Ianni (computed at the B3LYP/6-311++G(2d,2p level), Ding *et al.* (computed at the PW91/DNP level) and Al Natsheh *et al.* (computed at the PW91/ATZ2P level) show that both high-level electronic energies and advanced thermochemical approaches are needed before experimental results can be qualitatively replicated using systematically reliable theories. Heavily parameterized methods can give

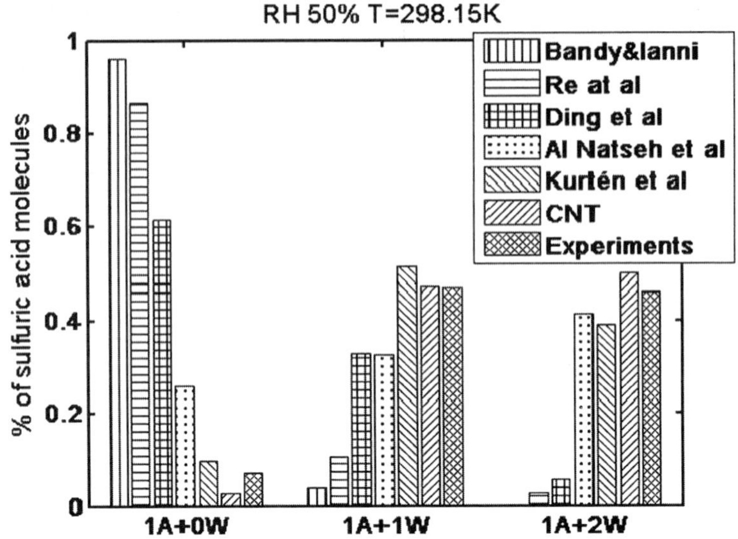

FIGURE 19.3 Distribution of sulfuric acid molecules to free acid molecules (1A + 0W) and clusters containing one sulfuric acid molecule and one water molecules (monohydrate, 1A + 1W) or two water molecule (dihydrate, 1A + 2W) according to experimental data [48], classical nucleation theory [59] and quantum chemical calculations of Re *et al.* [44], Bandy and Ianni [53], Ding and Laasonen [54], Al Natsheh *et al.* [55] and Kurtén *et al.* [38].

results compatible to experiments at the few existing experimental points by accident, but this is no guarantee that the results are realistic for other clusters.

3.2.2 The role of ammonia in water–sulfuric acid nucleation

Experimental studies have indicated that the presence of ammonia has a clear nucleation-enhancing effect [61] in the sulfuric acid–water system. However, different theoretical methods have yielded very different predictions for the strength of this effect. CNT predicts the enhancing effect to be stronger than what is experimentally observed. In terms of the average NH_3:H_2SO_4 mole ratio of nucleating clusters in atmospheric conditions, state-of-the art thermodynamics and updated models based on classical nucleation theory [60,62,63] have predicted ratios around or even above 1:1. In contrast, density functional studies on H_2SO_4–H_2O–NH_3 clusters containing one sulfuric acid molecule by Ianni and Bandy [64] at the B3LYP/6-311++G(2d,2p) level predicted that ammonia does not enhance particle formation, corresponding to a mole ratio of close to 0. Larson *et al.* [65] studied the same cluster stoichiometries at the MP2/6-311++G(d,p) level, which yielded somewhat larger sulfuric acid–ammonia binding energies, and led them to conclude that ammonium bisulfate forms in the atmosphere. However, using their formation energetics (and computing free energies from their reported rotational constants and vibrational frequencies) together with atmospherically realistic temperatures and partial pressures for ammonia still leads to very low NH_3:H_2SO_4 mole ratios for the one-acid clusters.

An error analysis study by Kurtén *et al.* [66] found that the B3LYP density functional underestimates the binding energy of sulfuric acid–water and sulfuric acid–ammonia clusters compared to high-level coupled-cluster methods. The underestimation of B3LYP binding energies for sulfuric acid–water was also noted by Al Natsheh *et al.* [55] and Nadykto and Yu [67]. The neglect of vibrational anharmonicity also leads to underestimated binding energies. However, neither of these errors were large enough to explain the difference between experimental and computational predictions on the role of ammonia. More importantly, recent studies by Kurtén *et al.* [68], Nadykto and Yu [67] and Torpo *et al.* [69] showed that the main drawback of the earlier computational studies was not so much the quantum chemical method (though it plays a role, too) than the limitation of the cluster dataset to only one-acid clusters. Computations on larger clusters containing two or three sulfuric acid molecules demonstrate the nucleation-enhancing effect of ammonia. Specifically, the presence of ammonia at atmospherically realistic partial pressures was shown to significantly assist the growth of two-acid clusters to three-acid clusters [69]. This implies a lower limit of 1:3 for the NH_3:H_2SO_4 mole ratio in atmospheric conditions, and qualitatively eliminates the contradiction between experimental and quantum chemical results.

In another recent study by Kurtén *et al.* [70], an upper limit to the NH_3:H_2SO_4 mole ratio of atmospheric clusters was estimated by studying only the $(H_2SO_4)_n\cdot(NH_3)_m$ "core clusters" (with $n = 2$ and $m = 0,\ldots,4$) without including any water molecules. This treatment was justified by the observation made in previous studies [64,67,68] that the number of water molecules in the cluster affects the acid–ammonia binding relatively weakly, and also systematically: the addition of water molecules tends to increase the free energies of ammonia addition reactions (in other words make them less favorable thermodynamically). Thus, NH_3:H_2SO_4 mole ratios computed for the "core clusters" represent an upper limit to the real mole ratios in atmospheric conditions. The omission of water molecules decreased the computational effort significantly, and the authors were able to use the quite high-level method RI-MP2/aug-cc-pV(T + d)Z//RI-MP2/aug-cc-pV(D + d)Z. The effect of possible systematic errors due to, for example, basis-set effects, higher-level correlation or vibrational anharmonicity was assessed via a sensitivity analysis approach. The results indicate that NH_3:H_2SO_4 mole ratios of atmospheric clusters are unlikely to exceed 1:1 in any atmospheric conditions, and are likely to be around 1:2 in most conditions. These mole ratios are somewhat lower that those typically measured for large (>10 nm in diameter) clusters, which indicates that the composition of nucleating particles may differ significantly from larger particles. The precise reasons for the differences are still unknown, and worthy of further study.

3.2.3 Studies on clusters containing other molecular species

Quantum chemistry has also been applied to investigating whether other sulfur-containing molecules than sulfuric acid itself might participate in nucleation reactions. Some recent experimental evidence [71,72] indicates that the mixture of sulfuric acid and some intermediate products of the SO_2 oxidation chain may promote nucleation even more effectively than sulfuric acid on its own. Similar speculations have also been presented in the past [73]. To our knowledge, no comprehensive and generally accepted explanation for the effect exists yet, though several studies have addressed issues related to the subject. For example, $SO_3 \cdot H_2O$ complexes have been studied by several groups, most recently by Morokuma and Muguruma [74], Meijer and Sprik [75], Larson *et al.* [76,77], Loerting and Liedl [42, 78], Ida *et al.* [79], Standard *et al.* [80] and Fliegl *et al.* [81]. (Note that most of these studies do not explicitly focus on atmospheric nucleation mechanisms.) One of the key results of these investigations has been that the presence of additional water molecules significantly lowers the activation barrier of the $SO_3 + H_2O \rightarrow H_2SO_4$ reaction. The studies also show that quantitatively accurate modeling of the formation and reactions of the $SO_3 \cdot H_2O$ complex require both high-level electronic energies (for example, the explicitly correlated RI-MP2-R12 calculations by Fliegl *et al.* [81] or the QCISD calculations by Standard *et al.* [80]) and, at least at lower temperatures, multidimensional tunneling corrections [42]. Especially the activation energies are sensitive to basis-set effects, as shown by Standard *et al.* [80].

The $HOSO_2$ radical (formed in the addition of OH to SO_2, the first step in the SO_2 oxidation chain) and its complexes and reactions have also been the object of several studies, for example those by Majumdar *et al.* [82], Wierzejewska and Olbert-Majkut [83] and Aaltonen and Francisco [84]. Earlier studies [73] have speculated on the possible role of molecular species containing two sulfur atoms such as $H_2S_2O_6$ and $H_2S_2O_8$ in atmospheric nucleation processes. Such clusters have to our knowledge not been studied with quantum chemistry in the context of tropospheric nucleation, though Rosén *et al.* [85] used HF/6-31+G(d) calculations to qualitatively augment experimental results on the $H_2S_2O_7^- \cdot (H_2SO_4)_x$ ion cluster system, which might form from $HSO_4^- \cdot (H_2SO_4)_x$ and SO_3 in the stratosphere. Despite the numerous studies, there is yet no consensus on the mechanism of nucleation in the SO_2–H_2O–O_2–ultraviolet light system, and the topic will remain the focus of intensive studies. Quantum chemistry is likely to be a key tool in these investigations, as the relevant systems involve both radicals and chemical reactions, which are difficult to study using other methods such as classical potentials.

A few studies have also considered the effect of other inorganic species together with sulfuric acid or its precursors. Larson and Tao [77] and Pawlowski *et al.* [86] investigated $SO_3 \cdot NH_3 \cdot H_2O$ clusters, and concluded that in very high ammonia concentrations, $NH_3 \cdot SO_3$ clusters rather than H_2SO_4-based clusters might act as nucleating agents. Ida *et al.* [79] also studied clusters containing SO_3 along with both H_2O, OH, NH_3 and several other species (such as methylated ammonia), but did not speculate on the atmospheric implications of their results. Bienko and Latajka [87] studied $H_2SO_4 \cdot CO$ complexes, but focused mainly on their IR spectra, without much speculation of their potential role in the atmosphere.

DMS (dimethylsulfate) is produced by algae in the sea water, and is a likely nucleation precursor over the oceans. Its oxidation reactions yields SO_2 and finally sulfuric acid. Other products of DMS oxidation than SO_2, SO_3 or H_2SO_4 have also been investigated using quantum chemistry, for example in the recent studies by Li *et al.* [88] and Wang [89], which both predicted that methanesulfonic acid (MSA), like sulfuric acid, is strongly hydrated in atmospheric conditions, and suggested that hydrated clusters of MSA might participate in nucleation processes.

Some recent quantum chemical studies have addressed the participation of organic molecules in nucleation. Recent experimental evidence [90–92] indicates that various organic molecules, possibly together with sulfuric acid [93,94] may be involved in some atmospheric nucleation events. There are innumerable quantum chemical studies of gas-phase organic reactions potentially relevant to aerosol processes, but few of these have focused specifically on nucleation. Exceptions among these are the PW91 study of Sloth *et al.* [95] on interaction energies between aerosol precursors formed in the photo-oxidation of α-pinene, and the X3LYP study by Tong *et al.* [96] on secondary organic aerosol formation from reactions of aldehydes. Sulfuric acid–organic acid complexes have recently been investigated by Nadykto and Yu [67], who studied complexes of sulfuric acid with formic acid or acetic acid at the PW91/6-311++G(3df,3pd) level and by Zhang *et al.* [93], who considered complexes of sulfuric acid with aromatic acids formed in combustion processes, using the B3LYP functional with CCSD(T) energy corrections. Both of these studies found that sulfuric acid is quite strongly bound to organic acids, indicating that such clusters may play some role in nucleation. However, the experimental evidence [90–92,94] indicates that the sulfuric acid–organic nucleation mechanisms may involve real chemical reactions as opposed to simple clustering (and associated proton transfer reactions). An example of such a reaction was recently studied by Kurtén *et al.* [97], who investigated the reaction of sulfuric acid with stabilized Criegee Intermediates formed in the ozonolysis of biogenic terpenes using the B3LYP and RI-CC2 methods. The reaction was found to be essentially barrierless, but its atmospheric relevance depends strongly on the formation rates and lifetime of the biogenic stabilized Criegee Intermediates, for which no definite and reliable values are yet available.

Nucleation involving iodine oxide species has recently been investigated by several groups, some of which have supported their experimental work with quantum chemical computations. For example, Saunders and Plane [98] used a combination of electron microscopy and density functional theory (B3LYP with RRKM kinetics modeling) computations, and predicted that the ultra-fine iodine oxide particles formed in the coastal marine boundary layer consist of I_2O_5, while Begović *et al.* [99] computed energetics (using the G96PW91 functional) for HOIO formation reactions in order to explain observed (laboratory) mass spectra of iodine-containing aerosols.

Ion-induced nucleation has also been studied using quantum chemical methods. Low-level quantum chemistry calculations have been used to interpret and complement experimental results in studies by Froyd and Lovejoy [100] and Rosén *et al.* [85] on HSO_4^- and $H_2S_2O_7^-$-based ion clusters, respectively. Recently, Nadykto *et al.* [101] showed that the sign preference in ion-induced nucleation (*e.g.* water tends to nucleate more efficiently on negatively charged ions) can be quite easily explained and predicted by quantum chemical calculations on small charged clusters. Kurtén *et al.* [38] recently studied the hydration of HSO_4^-, and also tentatively concluded that the role of ammonia in ion-induced sulfuric acid–water nucleation is likely to be significantly smaller than in neutral sulfuric acid–water nucleation.

Obtaining qualitatively useful data for nucleation studies does not necessarily require the computation of quantitatively accurate free energies. For example, the relative importance of different nucleation pathways is primarily related to the differences in formation free energies between different types of clusters rather than to the absolute free energies. Even computational methods with relatively large systematic errors can thus yield useful and reliable data, as long as the errors are constant for the set of systems compared. For example, a test comparison [66] of 12 computational methods (3 theories and 4 basis sets, with and without counterpoise corrections) yielded differences as large as 5 kcal/mol for the absolute value of the binding energy of the $H_2SO_4{\cdot}NH_3$ cluster, but all predicted values within about 2 kcal/mol from each other for the difference in binding energies of the $H_2SO_4{\cdot}NH_3$ and $H_2SO_4{\cdot}H_2O$ clusters. The latter value, as discussed in [38] and [68] is much more important for assessing the role of ammonia in sulfuric acid–water nucleation. As another example, the studies of Froyd and Lovejoy [100] and Rosén *et al.* [85] show that even very modest HF-level calculations can provide qualitatively useful data that assists the interpretation of experimental results.

3.3 Computing formation free energies for nucleating clusters

The most ambitious application of quantum chemistry to nucleation is the calculation of the formation free energy of a critical cluster. In order to verify that a cluster is critical at the given conditions, the minimum requirement is that the free energy surface as a function of the molecular composition contains a maximum. Due to computational constraints, this has to our knowledge not been done for any system relevant to tropospheric nucleation mechanisms. For pure water clusters such data may exist, but as mentioned in Section 1, homogeneous nucleation of water does not occur anywhere in the atmosphere. Partial free energy surfaces restricted to the smallest cluster sizes have been computed by Kurdi and Kochanski [47], Re *et al.* [44], Arstila *et al.* [52], Ianni and Bandy [53,57], Al Natsheh *et al.* [55], Ding and Laasonen [54,58], Arrouvel *et al.* [56] and Kurtén *et al.* [38] for H_2SO_4–H_2O clusters, by Ianni and Bandy [64], Larson *et al.* [65], Kurtén *et al.* [68,70], Nadykto and Yu [67] and Torpo *et al.* [69] for H_2SO_4–NH_3–H_2O clusters, and by Larson and Tao [77] and Pawlowski *et al.* [86] for SO_3–NH_3–H_2O clusters.

For the calculations to be useful for computing formation free energies and further nucleation rates, both the electronic energy and the thermal contributions need to be computed using quite an advanced method to yield quantitatively reliable results. A recent high-level study [38] on $HSO_4^-{\cdot}(H_2O)_{1,\ldots,4}$ and $H_2SO_4{\cdot}(H_2O)_{1,\ldots,4}$ clusters demonstrates that a combination of MP2/aug-cc-pV(D + d)Z geometries, high-level correlation (MP4) and large basis-set (aug-cc-pV(T + d)Z) corrections, perturbative anharmonic vibrational frequency calculations and internal rotor analysis can replicate experimental formation enthalpies and entropies for very small clusters (see Figure 19.3). However, the computational effort involved is very large; the method could not be directly applied to clusters larger than $H_2SO_4{\cdot}(H_2O)_2$ or $HSO_4^-{\cdot}(H_2O)_2$, and scaling-factor based extensions failed severely for the $HSO_4^-{\cdot}(H_2O)_4$ cluster, presumably due to unidentified internal rotations.

In the future, the extent of accurate free energy surfaces will probably increase gradually. This is due both to the increases in computing power, and the development and testing of methods that allow a greater degree of accuracy for a certain amount of computing power. These methods include the resolution-of-identity approximations, which have recently been used in atmospheric studies for example by Fliegl *et al.* [81] and Kurtén *et al.* [70], and the density fitting approximation, which has been successfully applied by Nadykto *et al.* [102] to the vibrational spectra of sulfuric acid monohydrate and formic acid dimer clusters. Intriguing possibilities are also offered by various QM/MM (Quantum Mechanics/Molecular Mechanics) methods and fragmentation or embedding models, see for example Heyden *et al.* [103] and Dahlke and Truhlar [104] for recent applications to water clusters, and Falsig *et al.* [105] for an atmospherically relevant study of phenol–water cluster reactions.

4. CHALLENGES

One of the main difficulties of quantum chemical clusters studies is the generation of initial input structure for the geometry optimizations. The number of cluster conformers (structural isomers) increases combinatorially with the number of participating molecules, and a complete sampling of all conformers with quantum chemical methods becomes impossible already for relatively small clusters (around 6–8 molecules). Unfortunately, most cluster studies do not explicitly address the issue of input structure generation, which makes it difficult or impossible to reproduce their results, and decreases their usefulness as a starting point for further research. Presumably, they have usually employed a combination of chemical intuition and comparisons to earlier studies on similar systems. Classical potentials or semi-empirical methods together with various annealing simulations can be used to generate multiple conformers from a smaller number of initial structures, and lower-level quantum chemical computations are often used to narrow down the number of structures selected for the higher-level optimization. The development of generally available systematic Monte Carlo-based conformer sampling packages (for example within the Spartan 06 [106] program suite) will hopefully help decrease the randomness of the input structure generation process. In the future, we predict that input structures will increasingly be generated by a combination of system-specific interaction potentials with Monte Carlo-based sampling algorithms.

However, as the size of the studied cluster structures grows, it should be remembered that modeling the cluster as a single, harmonically vibrating and rigidly rotating minimum energy structure becomes an increasingly inadequate approach. In the context of atmospheric clusters, especially water molecules are quite weakly bound to the clusters, and the minimum energy geometry, harmonic oscillator and rigid rotor approximations are not well justified for extensively hydrated clusters. In a recent molecular mechanics study by Kathmann *et al.* [107], the effect of anharmonicity on the thermochemistry of hydrated ions was found to be notable for $i > 4$ for $Na^+ \cdot (H_2O)_i$ clusters and $i > 1$ for $Cl^- \cdot (H_2O)_i$ clusters. It should be noted that Kathmann *et al.* use the term "anharmonicity" to cover two essentially separate issues: the anharmonicity of vibrational frequencies themselves, and the contribution of other conformers than the minimum energy (or minimum free energy) structure to the thermochemical parameters of the cluster. While the former can be accounted for by anharmonic vibrational calculations for smaller clusters, and various scaling factor approaches for larger clusters, the latter can not, according to Kathmann *et al.*, be fully addressed by anything else than a complete sampling of the configuration space—an impossible task for even the computationally cheapest quantum chemical methods. As described in Section 3.3, we have recently [38] shown that a combination of anharmonic frequency calculations and internal rotation analysis can replicate the experimental thermochemical parameters for the very smallest sulfuric acid–water clusters. While anharmonic frequency calculations are computationally prohibitively expensive, internal rotor analysis is somewhat less demanding, and could potentially help decrease some of the worst errors induced by the harmonic approximation for medium-sized (on the order of 10 molecules) clusters. For qualitative studies, sensitivity analysis approaches such as that employed in our recent study [70] can also give order-of-magnitude estimates of the possible role of anharmonicity.

The apparent controversy between classical and quantum chemical modeling is in any case somewhat misleading. Real molecular clusters are both "quantum mechanical" and "anharmonic"; neither classical anharmonic nor quantum mechanical harmonic approaches can be expected to yield perfect results. The approaches should be seen as complementary rather than competing, with each method helping to complement the drawbacks of the other. For classical simulations, this could mean for example zero-point energy corrections (when they are not already implicitly included by fitting to experimental results), or correcting the computed potential energy surfaces using a limited set of quantum chemical single-point energy calculations. For quantum chemical simulations, it could imply more extensive conformational sampling using classical potentials, or QM/MM simulations with, for example, water molecules in outer hydration shells being treated using molecular dynamics. The ultimate objective would be to combine the chemical accuracy and universal nature of quantum chemistry with the extensive configurational sampling of Monte Carlo methods and the dynamic character of molecular dynamic studies.

5. CONCLUSIONS

Quantum chemical methods are valuable tools for studying atmospheric nucleation phenomena. Molecular geometries and binding energies computed using electronic structure methods can be used to determine potential parameters for classical molecular dynamic simulations, which in turn provide information on the dynamics and qualitative energetics of nucleation processes. Quantum chemistry calculations can also be used to obtain accurate and reliable information on the fundamental chemical and physical properties of molecular systems relevant to nucleation. Successful atmospheric applications include investigations on the hydration of sulfuric acid and the role of ammonia, sulfur trioxide and/or ions in sulfuric acid–water nucleation. Recently, atmospheric nucleation mechanisms involving short-lived precursors of H_2SO_4 or various organic molecules have also been investigated using quantum chemical methods. As these mechanisms often involve complicated chemical reactions, they can not reliably be studied using classical potentials, and future studies on these nucleation processes are likely to rely heavily on quantum chemical tools.

Unfortunately, quantitatively reliable quantum chemical calculations of nucleation rates for atmospherically relevant systems would require the application of both high-level electronic structure methods and complicated anharmonic thermochemical analysis to large cluster structures. Such computations are therefore computationally too expensive for currently available computer systems, and will likely remain so for the foreseeable future. Instead, a synthesis of different approaches will probably be necessary. In the future, successful nucleation studies are likely to contain combinations of the best features of both classical (Monte Carlo and molecular dynamics) and quantum chemical methods, with the ultimate objective being a chemically accurate, complete configurational sampling.

ACKNOWLEDGEMENTS

The authors thank the Academy of Finland for financial support, and Kai Nordlund, Ismo Napari, Kari E. Laasonen, Adam Foster and Martta Salonen for assistance.

REFERENCES

[1] D.V. Spracklen, K.S. Carslaw, M. Kulmala, V.-M. Kerminen, G.W. Mann, S.-L. Sihto, *Atmos. Chem. Phys.* **6** (2006) 5631.
[2] The full report will be published by Cambridge Univ. Press, New York. Online version available at: http://ipcc-wg1.ucar.edu/wg1/wg1-report.html.
[3] S. Von Klot, A. Peters, P. Aalto, T. Bellander, N. Berglind, D. D'Ippoliti, R. Elosua, A. Hörmann, M. Kulmala, T. Lanki, H. Löwel, J. Pekkanen, S. Picciotto, J. Sunyer, F. Forastriere, the {HEAPSS} study group, *Circulation* **112** (2005) 3073.
[4] M. Kulmala, H. Vehkamäki, T. Petäjä, M. Dal Maso, A. Lauri, V.-M. Kerminen, W. Birmili, P.H. McMurry, *J. Aerosol Sci.* **35** (2004) 143.
[5] C.D. O'Dowd, J.L. Jimenez, R. Bahreini, R.C. Flagan, J.H. Seinfeld, K. Hämeri, L. Pirjola, M. Kulmala, S.G. Jennings, T. Hoffmann, *Nature* **417** (2002) 632.
[6] M. Vana, E. Tamm, U. Horrak, A. Mirme, H. Tammet, L. Laakso, P.P. Aalto, M. Kulmala, *Atmos. Res.* **82** (2006) 536.
[7] F.L. Eisele, E.R. Lovejoy, E. Kosciuch, K.F. Moore, R.L. Mauldin III, J.N. Smith, P.H. McMurry, K. Iida, *J. Geophys. Res. (D)* **111** (2006) D04305.
[8] V. Kanawade, S.N. Tripathi, *J. Geophys. Res. (D)* **111** (2006) D02209.
[9] J.E. McDonald, *Am. J. Phys.* **31** (1963) 31.
[10] F.F. Abraham, *Homogeneous Nucleation Theory*, Academic Press, New York, 1974.
[11] D.D. Kashchiev, *Nucleation—Basic Theory with Applications*, Butterworth–Heinemann, Oxford, 2000.
[12] H. Reiss, *J. Chem. Phys.* **18** (1950) 840.
[13] D. Stauffer, *J. Aerosol Sci.* **7** (1976) 319.
[14] J.B. Zeldovich, *Acta Physicochim. U.R.S.S.* **XVIII** (1943) 1/22/01.

[15] J.B. Zeldovich, *Zh. Eksp. Theor. Fiz.* **12** (1942) 525.
[16] H. Vehkamäki, *Classical Nucleation Theory in Multicomponent Systems*, Springer, Berlin–Heidelberg, 2006.
[17] J. Merikanto, E. Zapadinsky, A. Lauri, H. Vehkamäki, *Phys. Rev. Lett.* **98** (2007) 145702.
[18] R. Evans, *Adv. Phys.* **28** (1979) 143.
[19] D. Henderson, *Fundamentals of Inhomogeneous Fluids*, Marcel Dekker, New York, 1992.
[20] A. Laaksonen, I. Napari, *J. Phys. Chem. B* **105** (2001) 11678.
[21] K. Binder, J. Horbach, W. Kob, W. Paul, F. Varnik, *J. Phys. Condens. Matter* **16** (2004) S429.
[22] N. Metropolis, A.W. Rosenbluth, M.N. Rosenbluth, A.H. Teller, E. Teller, *J. Chem. Phys.* **21** (1953) 1087.
[23] J. Yao, R.A. Greenkorn, C. Chao, *Mol. Phys.* **46** (1982) 587.
[24] J. Merikanto, H. Vehkamäki, E. Zapadinsky, *J. Chem. Phys.* **121** (2004) 914.
[25] I. Napari, H. Vehkamäki, *J. Chem. Phys.* **125** (2006) 094313.
[26] S.A. Harris, I.J. Ford, *J. Chem. Phys.* **118** (2003) 9216.
[27] E. Kelly, M. Seth, T. Ziegler, *J. Phys. Chem A* **108** (2004) 2167.
[28] D.C. Rapaport, *The Art of Molecular Dynamics Simulation*, second ed., Cambridge University Press, Cambridge, UK, 2004.
[29] F. Jensen, *Introduction to Computational Chemistry*, second ed., Wiley, West Sussex, UK, 2007.
[30] S. Tsusuki, H.P. Lüthi, *J. Chem. Phys.* **114** (2001) 3949.
[31] T. van Mourik, R.J. Gdanitz, *J. Chem. Phys.* **116** (2002) 9620.
[32] Y. Zhao, N.E. Schultz, D.G. Truhlar, *J. Chem. Theory Comput.* **2** (2006) 364.
[33] Y.-K. Choe, E. Tsuchida, T. Ikeshoji, *J. Chem. Phys.* **126** (2007) 154510.
[34] G.M. Chaban, J.O. Jung, R.B. Gerber, *J. Phys. Chem. A* **104** (2000) 2772.
[35] L. Pele, B. Brauer, R.B. Gerber, *Theor. Chem. Acc.* **117** (2007) 69.
[36] V. Barone, *J. Chem. Phys.* **122** (2005) 014108.
[37] Y. Miller, G.M. Chaban, R.B. Gerber, *J. Phys. Chem. A* **109** (2005) 6565.
[38] T. Kurtén, M. Noppel, H. Vehkamäki, M. Salonen, M. Kulmala, *Boreal Environ. Res.* **12** (2007) 431.
[39] D.W. Brenner, *Phys. Rev. B* **42** (1990) 9458.
[40] A.C.T. van Duin, S. Dasgupta, F. Lorant, W.A. Goddard III, *J. Phys. Chem. A* **105** (2001) 9396.
[41] R.J. McMahon, *Science* **299** (2003) 833.
[42] T. Loerting, K.R. Liedl, *J. Phys. Chem. A* **105** (2001) 5137.
[43] C.-G. Ding, T. Taskila, K. Laasonen, A. Laaksonen, *Chem. Phys.* **287** (2003) 7.
[44] S. Re, Y. Osamura, K. Morokuma, *J. Phys. Chem. A* **103** (1999) 3535.
[45] I. Kusaka, Z.-G. Wang, J. Seinfeld, *J. Chem. Phys.* **108** (1998) 6829.
[46] S.M. Kathmann, B.N. Hale, *J. Phys. Chem. B* **105** (2001) 11719.
[47] L. Kurdi, E. Kochanski, *Chem. Phys. Lett.* **158** (1989) 111.
[48] D.R. Hanson, F. Eisele, *J. Phys. Chem. A* **104** (2000) 1715.
[49] J.J. Marti, A. Jefferson, X.P. Cai, C. Richert, P.H. McMurry, F. Eisele, *J. Geophys. Res.* **102** (1997) 3725.
[50] R.H. Heist, H. Reiss, *J. Chem. Phys.* **61** (1974) 573.
[51] A. Jaecker-Voirol, P. Mirabel, H. Reiss, *J. Chem. Phys.* **87** (1987) 4849.
[52] H. Arstila, K. Laasonen, A. Laaksonen, *J. Chem. Phys.* **108** (1998) 1031.
[53] A.R. Bandy, J.C. Ianni, *J. Phys. Chem. A* **102** (1998) 6533.
[54] C.-G. Ding, K. Laasonen, *Chem. Phys. Lett.* **390** (2004) 307.
[55] A. Al Natsheh, A.B. Nadykto, K.V. Mikkelsen, F. Yu, J. Ruuskanen, *J. Phys. Chem. A* **108** (2004) 8914;
Correction published in *J. Phys. Chem. A* **110** (2006) 7982.
[56] C. Arrouvel, V. Viossat, C. Minot, *J. Mol. Struct. THEOCHEM* **718** (2005) 71.
[57] J.C. Ianni, A.R. Bandy, *J. Mol. Struct. THEOCHEM* **497** (2000) 19.
[58] C.-G. Ding, K. Laasonen, A. Laaksonen, *J. Phys. Chem. A* **107** (2003) 8648.
[59] M. Noppel, H. Vehkamäki, M. Kulmala, *J. Chem. Phys.* **116** (2002) 218.
[60] S.L. Clegg, P. Brimblecombe, A.S. Wexler, *J. Phys. Chem. A* **102** (1998) 2137.
[61] S.M. Ball, D.R. Hanson, F.L. Eisele, P.H. McMurry, *J. Geophys. Res. D* **104** (1999) 23709.
[62] H. Vehkamäki, I. Napari, M. Kulmala, M. Noppel, *Phys. Rev. Lett.* **93** (2004) 148501.
[63] T. Anttila, H. Vehkamäki, I. Napari, M. Kulmala, *Boreal Environ. Res.* **10** (2005) 511.

[64] J.C. Ianni, A.R. Bandy, *J. Phys. Chem. A* **103** (1999) 2801.
[65] L.J. Larson, A. Largent, F.-M. Tao, *J. Phys. Chem. A* **103** (1999) 6786.
[66] T. Kurtén, M.R. Sundberg, H. Vehkamäki, M. Noppel, J. Blomqvist, M. Kulmala, *J. Phys. Chem. A* **110** (2006) 7178.
[67] A.B. Nadykto, F. Yu, *Chem. Phys. Lett.* **435** (2007) 14.
[68] T. Kurtén, L. Torpo, C.-G. Ding, H. Vehkamäki, M.R. Sundberg, K. Laasonen, M. Kulmala, *J. Geophys. Res.* **112** (2007) D04210.
[69] L. Torpo, T. Kurtén, H. Vehkamäki, M.R. Sundberg, K. Laasonen, M. Kulmala, *J. Am. Chem. Soc. A* **111** (2007) 10671.
[70] T. Kurtén, L. Torpo, M.R. Sundberg, V.-M. Kerminen, H. Vehkamäki, M. Kulmala, *Atmos. Chem. Phys. Discuss.* **7** (2007) 2937.
[71] T. Berndt, O. Böge, F. Stratmann, J. Heintzenberg, M. Kulmala, *Science* **307** (2005) 698.
[72] T. Berndt, O. Böge, F. Stratmann, *Geophys. Res. Lett.* **33** (2006) L15817.
[73] J.P. Friend, R.A. Burnes, R.M. Vasta, *J. Phys. Chem.* **84** (1980) 2423.
[74] K. Morokuma, C. Muguruma, *J. Am. Chem. Soc.* **116** (1994) 10316.
[75] E.J. Meijer, M. Sprik, *J. Phys. Chem. A* **102** (1998) 2893.
[76] L.J. Larson, M. Kuno, F.-M. Tao, *J. Chem. Phys.* **112** (2000) 8830.
[77] L.J. Larson, F.-M. Tao, *J. Phys. Chem. A* **105** (2001) 4344.
[78] T. Loerting, K.R. Liedl, *Proc. Natl. Acad. Sci.* **97** (2000) 8874.
[79] B.N. Ida, P.S. Fudacz, D.H. Pulsifer, J.M. Standard, *J. Phys. Chem. A* **110** (2006) 5831.
[80] J.M. Standard, I.S. Buckner, D.H. Pulsifer, *J. Mol. Struct. THEOCHEM* **673** (2004) 1.
[81] H. Fliegl, A. Glöß, O. Welz, M. Olzmanna, W. Klopper, *J. Chem. Phys.* **125** (2006) 054312.
[82] D. Majumdar, G.-S. Kim, J. Kim, K.S. Oh, J.Y. Lee, K.S. Kima, W.Y. Choi, S.-H. Lee, M.-H. Kang, B.J. Mhin, *J. Chem. Phys.* **112** (2000) 723.
[83] M. Wierzejewska, A. Olbert-Majkut, *J. Phys. Chem. A* **111** (2007) 2790.
[84] E.T. Aaltonen, J.S. Francisco, *J. Phys. Chem. A* **107** (2003) 1216.
[85] S. Rosén, K.D. Froyd, J. Curtius, E.R. Lovejoy, *Int. J. Mass Spectrom.* **232** (2004) 9.
[86] P.M. Pawlowski, S.R. Okimoto, F.-M. Tao, *J. Phys. Chem. A* **107** (2003) 5327.
[87] A.J. Bienko, Z. Latajka, *Chem. Phys.* **282** (2002) 207.
[88] S. Li, W. Qian, F.-M. Tao, *J. Phys. Chem. A* **111** (2007) 190.
[89] L. Wang, *J. Phys. Chem. A* **111** (2007) 3642.
[90] B. Bonn, G. Schuster, G.K. Moortgat, *J. Phys. Chem. A* **106** (2002) 2869.
[91] B. Bonn, G.K. Moortgat, *Geophys. Res. Lett.* **30** (2003) 1585.
[92] N.L. Ng, J.H. Kroll, M.D. Keywood, R. Bahreini, V. Varutbangkul, R.C. Flagan, J.H. Seinfeld, A. Lee, A.H. Goldstein, *Environ. Sci. Technol.* **40** (2006) 2283–2297.
[93] R. Zhang, I. Suh, J. Zhao, D. Zhang, E.C. Fortner, X. Tie, L.T. Molina, M.J. Molina, *Science* **2004** (2004) 1487.
[94] J.D. Surratt, J.H. Kroll, T.E. Kleindienst, E.O. Edney, M. Claeys, A. Sorooshian, N.L. Ng, J.H. Offenberg, M. Lewandowski, M. Jaoui, R.C. Flagan, J.H. Seinfeld, *Environ. Sci. Technol.* **41** (2007) 517.
[95] M. Sloth, M. Bilde, K.V. Mikkelsen, *Mol. Phys.* **102** (2004) 2361.
[96] C. Tong, M. Blanco, W.A. Goddard III, J.H. Seinfeld, *Environ. Sci. Technol.* **40** (2006) 2333.
[97] T. Kurtén, B. Bonn, H. Vehkamäki, M. Kulmala, *J. Phys. Chem. A* **111** (2007) 3391.
[98] R.W. Saunders, J.M.C. Plane, *Environ. Chem.* **2** (2005) 299.
[99] N. Begović, Z. Marković, S. Anić, L. Kolar-Anić, *Environ. Chem. Lett.* **2** (2004) 65.
[100] K.D. Froyd, E.R. Lovejoy, *J. Phys. Chem. A* **107** (2003) 9812.
[101] A.B. Nadykto, A. Al Natsheh, F. Yu, K.V. Mikkelsen, J. Ruuskanen, *Phys. Rev. Lett.* **96** (2006) 125701.
[102] A.B. Nadykto, H. Du, F. Yu, *Vib. Spectrosc.* **44** (2007) 286.
[103] A. Heyden, H. Lin, D.G. Truhlar, *J. Phys. Chem. B* **111** (2007) 2231.
[104] E.E. Dahlke, D.G. Truhlar, *J. Chem. Theory Comput.* **3** (2007) 46.
[105] H. Falsig, A. Gross, J. Kongsted, A. Osted, M. Sloth, K.V. Mikkelsen, O. Christiansen, *J. Phys. Chem. A* **110** (2006) 660.
[106] Wavefunction, Inc., Spartan '06, Wavefunction, Inc. Irvine, CA (2006). See also: www.wavefun.com.
[107] S. Kathmann, G. Schenter, B. Garrett, *J. Phys. Chem. C* **111** (2007) 4977.

CHAPTER 20

The Impact of Molecular Interactions on Atmospheric Aerosol Radiative Forcing

Shawn M. Kathmann*, **Gregory K. Schenter*** and **Bruce C. Garrett***

Contents

1. Introduction 429
2. Atmospheric Physics 434
3. Background on Nucleation Theories 438
4. Dynamical Nucleation Theory 440
5. Why Accurate Chemical Physics is Important to Nucleation 444
6. Summary and Future Directions 445
Acknowledgements 446
References 446

Abstract

We review the chemical physics of nucleation and its connection to atmospheric aerosol radiative forcing. The scientific community has demanded a comprehensive investigation of aerosol radiative forcing of climate. Particular emphasis has been placed on gaining the fundamental knowledge necessary to accurately predict aerosol formation and growth and their subsequent radiative affects on climate. The chemical physics of the molecular clusters underlying nucleation will be outlined and future developments towards modeling multi-component nucleation in the atmosphere discussed.

1. INTRODUCTION

Aerosol radiative forcing is nonlinearly dependent on surface albedo, aerosol composition and size distribution, relative humidity, and solar angle [1]. Albedo is simply the ratio of reflected to incident electromagnetic radiation (e.g., fresh clean

* Chemical & Materials Sciences Division, Pacific Northwest National Laboratory, Richland, WA 99352, USA

Advances in Quantum Chemistry, Vol. 55
ISSN 0065-3276, DOI: 10.1016/S0065-3276(07)00220-1

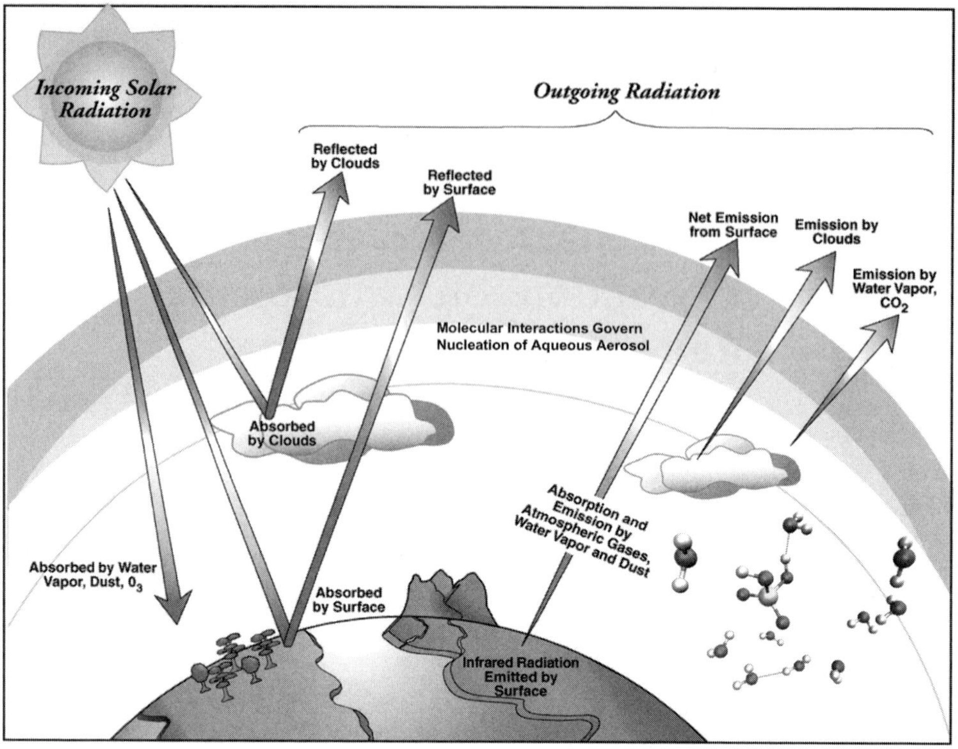

FIGURE 20.1 The balance of the Earth's energy requires the incoming energy flux from the Sun to be balanced by the outgoing energy flux. This illustration highlights the role molecular interactions and chemical physics of nucleation play in aerosol radiative forcing.

snow = 0.85, fields and woods = 0.02, Earth mean = 0.4) [2]. The concentration of atmospheric water vapor and aqueous aerosols are highly variable and governed by evaporation and condensation processes. Water is the most important greenhouse gas as its availability governs hydrological feedback cycles (e.g., oceans, clouds, ice, and precipitation) that influence the distribution of the Sun's energy on Earth. The Earth's climate is governed by the balance between absorption of incoming solar radiation and emission of thermal radiation—see Figure 20.1. Of the many agents that affect climate, water vapor is known to play a major role and aerosols are the major source of uncertainty in understanding the impact of radiative forcing on climate change—see IPCC 2007 report [3]. Aerosols can directly affect Earth's radiative balance by either reflecting or absorbing radiation and indirectly affect the radiative balance by modifying cloud properties. Atmospheric aerosols are produced from direct emission from sources on Earth to form primary aerosols (e.g., dust and sea salt particles) or through gas-to-particle nucleation to form secondary aerosols (e.g., sulfate and secondary organic aerosols). Both direct and indirect effects contribute to cooling i.e., the sign of the climate forcing is negative. The indirect effect ($\approx -2\,\mathrm{W/m^2}$) is about twice that of the direct effect, however, the uncertainty of the indirect effect is twice as large as the direct effect. To put aerosol radiative forcing in perspective, consider that the radiative forcing due to clouds is about 50 W/m^2—thus, non-cloud direct and indirect aerosol effects represent a small but important fraction of the total (i.e., cloud and non-cloud) aerosol forcing.

The thermodynamics and kinetics of the nucleating clusters leading to aerosol formation are dictated by molecular interactions. *The reason why molecular interactions and chemical physics are relevant to aerosol radiative forcing is simply because nucleation creates new particles and the rate of nucleation, which affects particle number and size, is extremely sensitive to molecular interactions.* Since nucleation of tiny aerosols are the end result of tens to hundreds of clustering reactions, the overall nucleation rate can change by many orders of magnitude as a consequence of very slight changes in the underlying clustering thermo-kinetics [4–8]. Thus, reliable modeling of atmospheric aerosol formation requires a fundamental understanding of the chemical physics of nucleation.

The impact of secondary aerosols on indirect radiative forcing is the most variable and is the least understood [3]. The reasons why the indirect effect of secondary aerosols is so difficult to describe is that it depends upon [1] (1) a series of molecular-microphysical processes that connect aerosol nucleation to cloud condensation nuclei to cloud drops and then ultimately to cloud albedo and (2) complex cloud-scale dynamics on scales of 100–1000 km involve a consistent matching of multiple spatial and time scales and are extremely difficult to parameterize and incorporate in climate models. Nucleation changes aerosol particle concentrations that cause changes in cloud droplet concentrations, which in turn, alter cloud albedo. Thus, macro-scale cloud properties that influence indirect forcing result from both micro-scale and large-scale dynamics. To date, the micro-scale chemical physics has not received the appropriate attention.

The competition between primary and secondary aerosol formation in the atmosphere is not well understood. For example, organic precursors favorable to nucleation are products of chemical or photo-oxidative reactions involving volatile compounds. It has been widely assumed that nucleation of new particles is only important when the surface area concentration of existing aerosol particles is low. Direct field measurements provide evidence that nucleation occurs in the presence of pre-existing particles. In order to predict the impact on radiative processes due to aerosols with any confidence, accurate models for predicting the formation and distribution of important aerosols are needed. Figure 20.2 illustrates the different sizes of atmospheric particles, from molecular clusters through cloud drops, which are important in the evolution of aerosol-particle populations and their resulting physical, chemical, and optical properties. Currently, the chemical content of clusters smaller than 3 nm cannot be measured using the best technology. Since critical clusters are typically 1 nm or smaller, our ability to predict which nucleation pathway occurs is severely hindered. We simply don't know the essential chemical species and mechanisms that exist at these scales. Knowing the details of the chemical species and mechanisms is required to develop robust nucleation/aerosol/climate models. A fundamental chemical physics approach is a powerful guide to provide some of the necessary information.

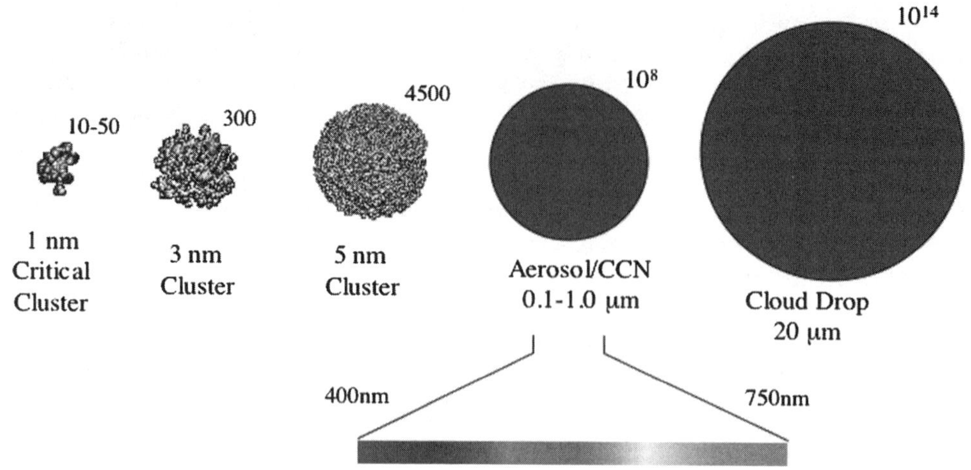

FIGURE 20.2 Illustration of the size (below particle) and number of molecules (upper right) in atmospheric particles ranging from critical clusters up to cloud drops. The visible spectrum is shown to emphasize the size range (0.1 to 1.0 mm) relevant to aerosol radiative forcing. Nucleation creates new particles that evolve into larger particles relevant to aerosol radiative forcing.

A major limitation currently in predicting aerosol number concentration and size distributions is the simplistic treatment of the nucleation process. The sizes of the critical clusters that represent the limiting stages of the nucleation process are extremely small (ca. 10 to 50 molecules). The nucleation free energy barrier that determines the rate of these processes depends delicately upon the molecular interactions between the cluster species. Currently, when nucleation processes are included in global climate models, continuum descriptions [9,10] are typically employed (e.g., Classical Nucleation Theory), which treat the molecular clusters as tiny spherical liquid droplets that are characterized using bulk-liquid surface tension and density. These assumptions certainly limit our ability to predict the properties and behavior of molecular clusters [11]. Nonetheless, Classical Nucleation Theory (CNT) [9,10] has provided useful qualitative insights into nucleation phenomena and its use, even with its limitations, is widespread throughout the scientific community. Recent work by Kulmala *et al.* [12], employing CNT on ternary nucleation of sulfuric acid (sulfuric acid concentrations ranging from 10^4 to 10^{10} cm^{-3}), ammonia, and water (RH $= 90\%$), found that 25 ppt of ammonia increases the nucleation rate by nearly 20 orders of magnitude! Moreover, they found that the largest influence on the nucleation rate (the first 16 of the 20 orders of magnitude) occurs for ammonia concentrations below 5 ppt; 4 ppt is currently the detection limit for ammonia measurements. The sensitivity of CNT can be understood in terms of the variation in the surface tension when ammonia is added. Given the dubious validity of using bulk surface tension to treat molecular clusters containing tens of molecules, the sensitivity that Kulmala and coworkers found provides further impetus to address the problem at the molecular chemical physics level (e.g., the critical cluster they inferred from their calculations is comprised of 8 H_2SO_4, 7 H_2O, and 5 NH_3 molecules). CNT cannot provide the essential chemical physics required to mechanistically and quantitatively understand nucleation because the thermo-kinetics of the nucleation process depend on dynamical barriers determined by small clusters and hence have nothing to do with bulk properties such as surface tension [4]. A fundamental chemical physics model that incorporates cluster thermochemistry and kinetics is necessary because nucleation is extremely sensitive to these properties.

In 2004 Kulmala *et al.* [13] published a review summarizing the findings of more than 100 field observations (ground, ship, and aircraft) of nucleation (aerosol size $\leqslant$20 nm) and growth (aerosol size >20 nm) spanning a broad range of locations and conditions all over the Earth. An important point to keep in mind is that instrumental limitations preclude the observation of the smallest aerosol particles (<3 nm) so that nucleation processes cannot be measured directly—the exact chemical content is unknown. The Kulmala *et al.* review reports several key findings: (1) several studies using multiple measuring stations have shown that regional nucleation events can occur uniformly over hundreds of kilometers, (2) nucleation in the outflow of convective clouds and storms can be globally important due to the large volumes of air, (3) during most of the observed nucleation bursts the aerosol distributions are peaked at sizes below 15 nm, (4) nucleation events took place *almost exclusively* during breakup of the morning inversion layer, (5) nucleation rates ranged from 0.01 to 10^5 $cm^{-3}\,s^{-1}$ and growth rates ranged from 0.1 to 200 nm h^{-1}, (6) it is not yet possible, based upon the available field data, to decide what are the most relevant nucleation pathways (ions, ammonia, nitric acid, organics, etc.) in the atmosphere although there is general support for the involvement of sulfuric acid, (7) aerosol formation and subsequent growth seem to occur everywhere, (8) depending upon the location, nucleation can increase the concentration of cloud condensation nuclei by more than a factor of two over one day, (9) measurements of multi-component nucleation at various concentrations would significantly increase our understanding, and (10) new-particle production is an important process that must be understood and included when developing climate models. *In summary, no additional field studies are required to strengthen the conclusion that nucleation is an essential process that must be accurately described in atmospheric models.*

The extreme sensitivity of nucleation to physical conditions (supersaturation and temperature) as well as chemical content points to the need to better understand the influence of multi-component nucleation as well as other aerosol processes (coagulation, condensation, evaporation, sources, and sinks) on the number concentration in the 0.1 to 1.0 μm size range. The reason for focusing on the 0.1 to 1.0 μm size range (corresponding to the visible or optical wavelengths) is that these aerosols (often referred to as the "accumulation mode" since aerosol removal mechanisms are least efficient) have the longest atmospheric lifetimes. Also, by coincidence, the solar blackbody spectrum is peaked in the 0.1 to 1.0 μm size range, further underscoring their importance to aerosol radiative forcing. Developing accurate computational models for the determination of clustering mechanisms and the quantification of rates is essential for a deeper understanding of aerosol lifecycle and evaluating their radiative impact. Furthermore, if climate predictions are to be useful in formulating energy policy, then these uncertainties in aerosol radiative forcing must be reduced. Computational models based on state-of-the-art chemical physics (i.e., requiring accurate energies and statistical mechanical simulation methods), benchmarked against careful laboratory measurements, are needed to predict: aerosol concentration, size distribution, chemical content, optical properties, and other physicochemical properties.

2. ATMOSPHERIC PHYSICS

An understanding of the influence that aerosols have on climate has become increasingly important over the last several decades [3]. Primary and secondary aerosols can affect the Earth's radiative balance by scattering and absorbing light directly and can act indirectly as cloud condensation nuclei and therefore influence the distribution, duration, precipitation processes, and radiative properties of clouds. Thus, developing the ability to understand, model, and predict aerosol formation with confidence is essential to determine the impact of aerosol radiative forcing in climate models.

The influence of aerosols on radiative forcing depends on the balance between three essential quantities: single-scattering albedo, upscatter fraction, and optical depth—all are sensitive functions of the aerosol chemical composition and size distribution in the 0.1 to 1.0 μm range [1,14]. Aerosol climate forcing, ΔF, depends on geophysical and aerosol parameters and can be expressed as

$$\begin{aligned}
&\Delta F = F_0(1 - A_{\text{cloud}})T_{\text{a}}^2 \Delta R,\\
&\Delta R = \left(r + \frac{t^2 R_{\text{s}}}{1 - rR_{\text{s}}}\right) - R_{\text{s}} = \text{change in reflectance due to aerosols},\\
&r = \left(1 - e^{-\tau}\right)\omega\beta = \text{fraction reflected upward},\\
&t = e^{-\tau} + \omega(1 - \beta)\left(1 - e^{-\tau}\right) = \text{total downward transmitted fraction},
\end{aligned}$$

where F_0 = incident solar flux, A_{cloud} = fraction of surface covered by clouds, T_{a} = fractional transmittance of the atmospheric due to gases, R_{s} = albedo of Earth's surface, ω = single scattering albedo of aerosol, β = upscatter fraction of aerosol, and τ = aerosol optical depth. The change in reflectance, ΔR, is due to the presence of aerosols. Climate forcing concerns the change in incoming radiative flux, so that the net change in forcing is $-\Delta F$. For example: if $\Delta R > 0$, then $-\Delta F < 0$ and we get a net cooling effect, if $\Delta R < 0$, then $-\Delta F > 0$ and we get a net warming effect. The single-scattering albedo, ω, and optical depth, τ, are determined from results of the Mie [1,14–16] solution of Maxwell's equations for the interaction of electromagnetic waves with a spherical aerosol particle with index of refraction n convoluted with the aerosol size distribution. The upscatter fraction, β, is determined from radiative transfer theory and depends on the aerosol size distribution and the solar zenith angle. Molecular interactions, rate constants, and free energies determine nucleation rates, J, as a function of temperature and vapor concentrations, and the nucleation rates, in turn, influence the aerosol size distribution. The turning point for the fundamental predictability of aerosol radiative forcing models hinges on the ability to accurately quantify the relevant mechanisms including the chemical physics of multi-component aerosol formation—see Figure 20.3.

Estimating the "indirect effect" is difficult because the mechanisms underlying the chain of microphysical processes connecting emissions with cloud albedo are uncertain—twice as uncertain as the direct effect as mentioned previously. Clouds form when an air parcel is cooled through vertical lifting and the vapor becomes

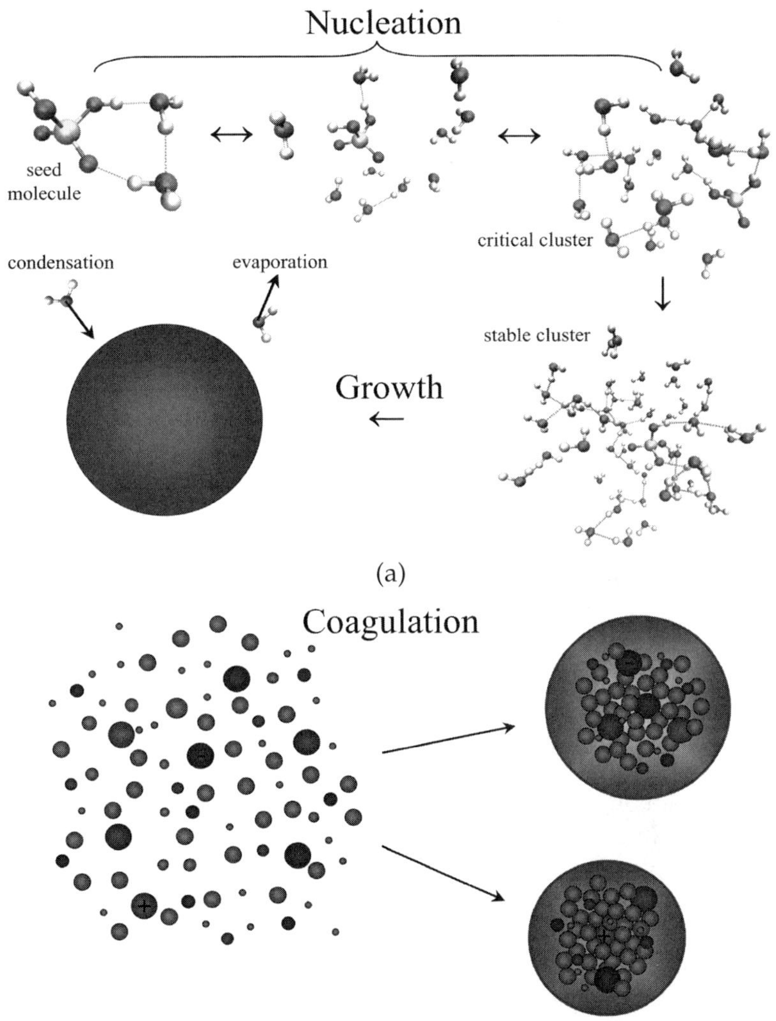

FIGURE 20.3 Relevant processes: (a) nucleation and growth, (b) coagulation, and (c) aerosol–light interactions.

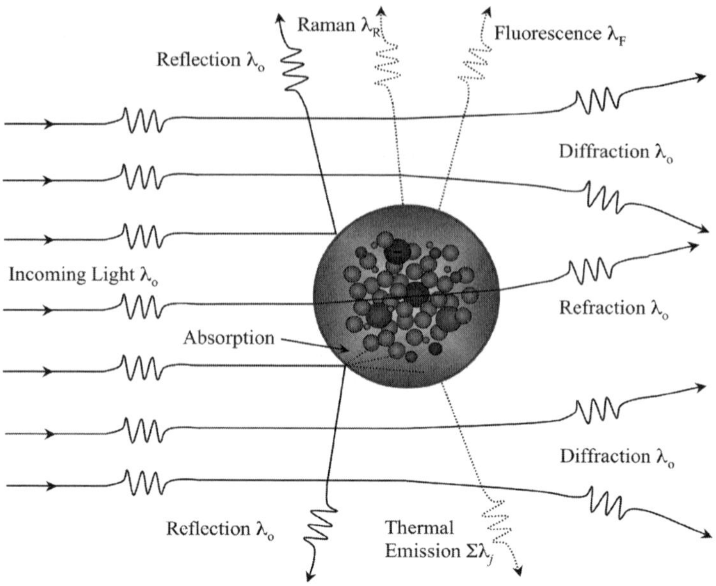

(c)

FIGURE 20.3 (*Continued.*)

supersaturated. This process, in turn, activates cloud condensation nuclei to become cloud droplets. Aerosol mass is created by gas-to-particle conversion, which can occur by nucleation of fresh aerosol or by the growth of existing aerosols. The resulting mechanism depends on the local aerosol number concentrations and vapor concentrations. Nucleation leads to a large population of small aerosol particles; growth leads to a small population of large aerosol particles. These competing routes affect the lifetimes and spatial distributions of clouds; for example, aerosol particles in the range of 0.1 to 1.0 μm have long lifetimes whereas smaller particles (via growth, evaporation/condensation, and coagulation) and larger particles (via precipitation) aerosol have short lifetimes. Atmospheric aerosols are created by nucleation or through direct emission. Aerosol lifecycle is modeled using the General Dynamic Equation (GDE) [1,14]. The GDE treats several aerosol processes simultaneously: coagulation, condensation, evaporation, nucleation, direct sources, and sinks. Nucleation creates a fresh burst—see Figure 20.4—of very tiny aerosols that grow into long-lived particles in the 0.1 to 1.0 μm size range where they can alter scattering and absorption of light. It is this transfer of aerosol particles between these two size modes that is relevant to aerosol radiative forcing. Since nucleation can increase aerosol number concentration by many orders of magnitude, it seems straightforward to ask the question "What is the sensitivity of the 0.1 to 1.0 μm aerosol number concentration and chemical composition to various mechanisms underlying multi-component nucleation?"

After a nucleation burst occurs, the aerosol population (fresh clusters and pre-existing particles) evolves through coagulation, condensation, evaporation, emis-

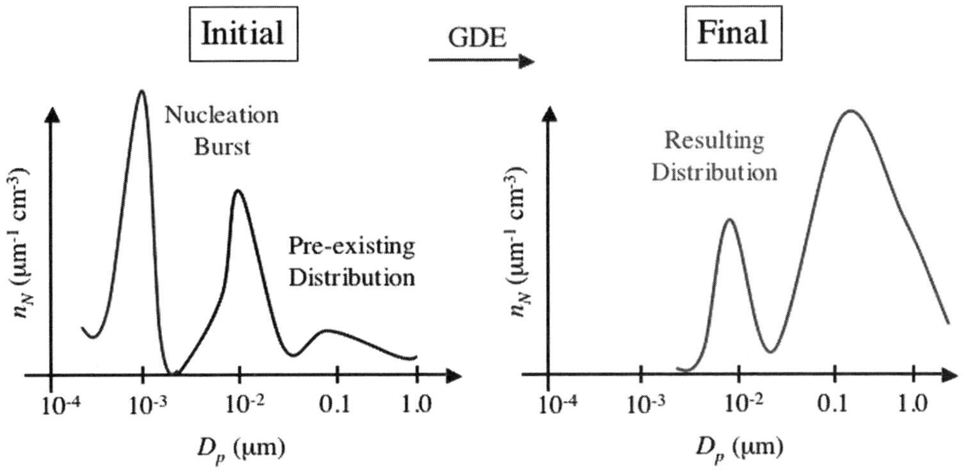

FIGURE 20.4 Schematic representation of the potential influence of a nucleation burst (blue) of clusters peaked at 10 Å with a pre-existing aerosol distribution (black) resulting in a distribution (green) with an increased aerosol number concentration n_N for aerosol having diameters D_p in the 0.1 to 1.0 μm size range. The temporal evolution of the initial aerosol distribution is modeled using the General Dynamic Equation (GDE). (For interpretation of the references to colour in this figure legend, the reader is referred to the web version of this book.)

sion of fresh particles, and removal—see Figure 20.3. Progress in understanding the molecular nature of aerosol radiative forcing can be made by answering the question "What is the individual and collective sensitivity of the aerosol population with respect to uncertainties in the underlying nucleation processes?"

The GDE is expressed as [1,14]

$$\frac{\partial n(v,t)}{\partial t} = \frac{1}{2}\int_0^v K(v-q,q)n(v-q)n(q,t)\,dq - n(v,t)\int_0^v K(q,v)n(q,t)\,dq$$
$$-\frac{\partial}{\partial v}\big[I(v)n(v,t)\big] + J(v)\delta(v-v_0) + S(v) - R(v),$$

where the 1st and 2nd terms on the right hand side (involving the kernel, K) describe aerosol coagulation, the 3rd term is due to evaporation and condensation, the 4th term is due to nucleation rate J of fresh clusters of volume v_0 (the volume v_0 corresponding to the volume of the critical clusters), and the 5th and 6th terms are for sources and sinks, respectively. By the time the aerosols grow into the 0.1 to 1.0 μm size range, the aerosol chemical content is represented by an index of refraction. Thus, the totality of the chemical speciation of aerosols, and their associated absorption and scattering properties, is subsumed into a complex index of refraction (i.e., $n = n_{\mathrm{Re}} + in_{\mathrm{Im}}$) with the imaginary component n_{Im} commonly referred to as the extinction coefficient.

The following schematic summarizes how nucleation theories such as Dynamical Nucleation Theory (DNT), the GDE, and radiative forcing from Mie calculations can be coupled:

$$\begin{pmatrix} \alpha \\ A \end{pmatrix} \underset{\text{DNT}}{\longrightarrow} J([H_2O],[X_j]) \underset{\text{GDE}}{\longrightarrow} n(v,t) \underset{\text{Mie}}{\longrightarrow} \Delta F,$$

where α is the evaporation rate constant at temperature T, A is the cluster Helmholtz free energy, J is the nucleation rate of critical clusters, $[H_2O]$ is the ambient water vapor concentration, $[X_j]$ denotes the vapor concentration of the other chemical species, $n(v,t)$ is the time-dependent aerosol size distribution (t denotes the time dependence and v is the volume of aerosol particle) determined from numerical solution of the GDE, and ΔF is the forcing due to the presence of aerosols. It is important to note that there are potentially a great number of chemical species relevant to atmospheric aerosol formation. Given the large amount of water vapor in the atmosphere it should be of no surprise that of the various chemical species that may play a role in aerosol formation the great majority involve water. Furthermore, atmospheric conditions never achieve the supersaturations necessary to nucleate water homogeneously (i.e., pure water) due to the existence of favorable interactions between water molecules and the ubiquitous trace chemical species (acids, organics, ions, minerals, etc.). However, our hope is that by investigating simpler systems using theory, molecular simulation, and careful laboratory studies, we can understand both the limitations and strengths of computational chemical physics to further predict aerosol formation with confidence.

3. BACKGROUND ON NUCLEATION THEORIES

Theoretical approaches to nucleation go back almost 80 years to the development of Classical Nucleation Theory (CNT) by Volmer and Weber, Becker and Döring and Zeldovich [9,10,17–20]. CNT is an approximate nucleation model based on continuum thermodynamics, which views nucleation embryos as tiny liquid drops of molecular dimension. In CNT, the steady-state nucleation rate J, can be written in the form $J \propto \beta_{i*} N_{i*}^{\mathrm{EQ}}$, where β_{i*} is the monomer condensation rate constant for a cluster containing i molecules (an i-cluster) and N_{i*}^{EQ} is the concentration of i^*-clusters (the * denotes quantities pertaining to the critical cluster). CNT uses condensation rate constants approximated by gas-kinetic collision rates whose geometric cross-sections are determined by the bulk liquid phase density. The critical cluster concentrations are given by $N_{i*}^{\mathrm{EQ}} \propto \exp[\sigma^3]$ where σ is the bulk surface tension. Hence, a small uncertainty in the surface tension can affect J by many orders of magnitude. For example, a change in critical cluster surface tension of water by 15% would change the nucleation rate by 10^{14}. This is not to say that the bulk surface tension is uncertain by 15%, but that the surface energy for a molecular cluster of critical size ($\sim$1 nm) cannot adequately be quantified by using the bulk surface tension—one would expect it to be different from the bulk surface tension. The difficulty of using CNT is further compounded by the fact that atmospheric nucleation often occurs under conditions where physico-chemical data (trace concentration, surface tension, vapor pressure, and liquid number density) are unavailable. Thus, while CNT has proven useful for prediction of the nucleation threshold at some very limited conditions, it is inadequate for a truly molecular understanding of nucleation. Generally, only the slopes of the experimental nucleation rates versus the saturation ratio are in agreement with CNT. Careful analysis shows that CNT cannot predict the correct temperature dependence of the nucleation rate [21,22]. Many extensions and modifications for unary and multicomponent systems have been made to CNT (see reviews by Oxtoby *et al.* [23] and Kulmala *et al.* [13,24]), yet none of these changes have addressed the fundamental deficiencies in the basic assumptions (i.e., continuum surface tension, condensation rate constants approximated by gas kinetic collision rates, cluster definition, etc.). Thus, these changes have not provided a general theory able to quantitatively predict nucleation rates nor the underlying mechanisms and atmospheric chemical physics.

The relatively small size of critical clusters (tens to hundreds of molecules) provides a compelling argument for explicitly treating the molecular interactions at the molecular level in the nucleation process. Promising alternatives to CNT employ molecular simulations [25,26] in order to calculate cluster properties [4,6–8,27–55], which can then be used to determine nucleation rates. Many of these molecular approaches take a similar view as CNT, approaching nucleation from a liquid-phase perspective, assuming condensation rate constants to be approximated by gas-phase collision rates with liquid droplets. The cluster distribution functions are obtained using molecular simulations to compute the relevant partition functions or Helmholtz free energies for the clusters. There are typically two approaches employed to calculate a cluster's Helmholtz free energy: (1) using quantum mechanical electronic structure calculations to develop a harmonic model of molecular interactions and use the rigid-rotor harmonic oscillator approximation to determine free energies, and (2) using analytic empirical or semi-empirical interaction potentials and calculating the full anharmonic free energy through statistical mechanical sampling. The first approach, using *ab initio* electronic structure calculations, has the general benefit that bonds can be made or broken and that the accuracy of the energetics can, in principle, be systematically improved. However, these benefits come at a high computational cost and thus one's ability to perform statistical mechanical sampling becomes intractable. Consequently, only a single configuration of nuclei is considered in the estimation of free energies (i.e., the rigid-rotor harmonic oscillator approximation) that severely limits the relevance of the resulting free energies toward atmospheric conditions. Furthermore, even faster electronic structure methods like Density Functional Theory (DFT) are not fast enough to adequately sample the relevant nuclear configuration space. It is precisely this reason why researchers in chemical physics turn to analytic interaction potentials (either empirical or semi-empirical)—you can go beyond the rigid-rotor harmonic oscillator approximation so that one can actually do the appropriate statistical mechanical sampling. This issue will be discussed in more detail in the following sections.

Before a molecular simulation can be performed the "*cluster*" relevant to nucleation must be defined [11,27–29,56,57]. The *definition of the cluster* is important because there must be some set of criteria that differentiates it from the rest of the system—some way of partitioning space to include the essential molecular configurations. Some molecular theories have defined a cluster to be a collection of i molecules lying within a sphere of predetermined radius whose center is at the center-of-mass of the cluster, the so-called physically consistent cluster [58]. Others have utilized the Stillinger cluster [59] definition where a molecule only belongs to a cluster if it within an assumed distance from at least one other molecule in the cluster. In many studies using a spherical constraint, the Helmholtz free energy of the i-cluster was assumed to be independent of the constraining radius or alternatively the constraining volume. The "correct" cluster volume and corresponding Helmholtz free energy to use in the cluster distribution function has been a source of confusion and controversy for almost 40 years [11,33,60,61]. Alternatively, thermodynamic density functional theory has been used to calculate the free energy barrier to nucleation [23,62]. Although great promise has been shown by these approaches, until recently they have been applied primarily to model systems (e.g., Lennard-Jones systems) and have focused on cluster properties rather than on obtaining nucleation rates for practical systems (e.g., water or multi-component systems). Furthermore, attention has not been focused on the sensitivity of the cluster properties to the underlying interaction energies or the role of harmonic versus anharmonic or classical versus quantum statistical mechanics.

4. DYNAMICAL NUCLEATION THEORY

Nucleation has long been recognized to be extremely sensitive to chemical and physical conditions. Coulier [63], Aitken [64], and Wilson [65,66] realized the importance of dust, ions, and trace contaminants in condensation processes. Nucleation requires surmounting an activation barrier. Trace species can decrease the activation barrier through favorable interactions leading to enhanced stability of the embryonic clusters, or stated another way, by decreasing the range that the system can extend into thermodynamic metastability. All experimental measurements of nucleation rates are characterized by some degree of impurity [67,68]. The extreme sensitivity of nucleation to trace impurities demands that the underlying chemical physics be measured and modeled as carefully and accurately as possible. Experimentally, this requires ultra-high purity reagents, impeccably clean equipment, and mass spectrometers characterizing the system before, during and after the nucleation process. Computationally, this requires using sufficiently accurate estimates of the molecular interaction energies and converged statistical sampling of ensemble averages. But how accurate and sufficiently converged do we need to be? Since nucleation is a multi-step process, typically requiring 10's or 100's of individual reaction steps, each elementary step must be accurately computed in order to predict the overall nucleation rate. This represents a formidable challenge to computational chemistry even with ready access to some of the world's fastest supercomputers. The errors that pile up exponentially over these many reactions can lead to very large errors in the nucleation rate. To accurately predict nucleation rates requires: (1) accurate representations of the molecular interactions, (2) a theoretical formalism connecting interaction energies to rate constants, and (3) appropriate statistical mechanical sampling to obtain accurate free energies or, equivalently, equilibrium constants. Recently we developed a new approach that provides a molecular understanding of the nucleation process—Dynamical Nucleation Theory (DNT) [4–8,27–29,57]. DNT is a chemical physics approach to study homogeneous vapor-phase nucleation. DNT has provided fundamental insights into the underlying mechanisms, kinetics, and sensitivity of homogeneous nucleation. Our theoretical studies have shown that slight variations in the interactions between cluster molecules can have enormous consequences on the overall nucleation rate. For example, a free energy change of $-0.5\ \mathrm{kcal\,mol^{-1}}$ in each H_2O monomer addition reaction can increase the nucleation rate of water by 10 orders of magnitude! This sensitivity has dramatic consequences on the level of molecular detail required to understand and predict clustering mechanisms, thermodynamics, and kinetics of nucleation. To illustrate the chemical physics influences that can give rise to an energy difference of $-0.5\ \mathrm{kcal\,mol^{-1}}$, consider that this is equivalent to the difference in the zero-point-energy between $(H_2O)_2$ and $(D_2O)_2$—the consequence of a simple isotopic substitution. Moreover, the difference in binding energy between the sulfuric acid monohydrate, $H_2O{\cdot}H_2SO_4$ (BE $= 12.5\ \mathrm{kcal\,mol^{-1}}$) and $(H_2O)_2$ (BE $= 5.0\ \mathrm{kcal\,mol^{-1}}$) is $-7.5\ \mathrm{kcal\,mol^{-1}}$ underscoring why H_2SO_4 is an energetically favorable nucleation seed. Figure 20.5 shows the basic scheme for the reaction processes involved in binary homogeneous nucleation of sulfuric acid and water assuming monomer addition and loss. From the outset it should be clear that even though atmospheric aerosol measurements agree on the general involvement of sulfuric acid, the nucleation of fresh aerosols likely involves additional chemical species that will ultimately alter the simplified mechanism, thermodynamics, kinetics, and nucleation rates implied in Figure 20.5.

DNT utilizes a gas-phase reaction kinetics perspective and thus provides a natural setting in which the *i*-cluster Helmholtz free energies and evaporation and condensation rate constants, necessary for the construction of a consistent nucleation theory, can be obtained and used to obtain nucleation rates. As a molecular theory, DNT does not require bulk thermodynamic properties such as the liquid density or surface tension. Using the mathematical framework of DNT, a systematically improvable nucleation model for multi-component systems (e.g., $XYZ{\cdot}(H_2O)_i$ where XYZ = ammonia, inorganic/organic acids, or ions), has been constructed employing a fundamentally sound description of the thermodynamic and kinetic properties of molecular clusters. Although, we have not applied DNT to systems like sulfuric acid [30], water and ammonia, this is not a limitation of DNT, but a result of not having appropriate analytic interaction potentials that allow for acid dissociation. DNT provides a framework in which increasingly more accurate calculations on clusters can be readily incorporated, as there is no restriction on whether the system interaction energies used in DNT come from analytic potentials or determined directly from high-level *ab initio* or DFT electronic structure calculations. Again, the problem with electronic structure methods is that they are too computationally expensive to do the statistical mechanical sampling.

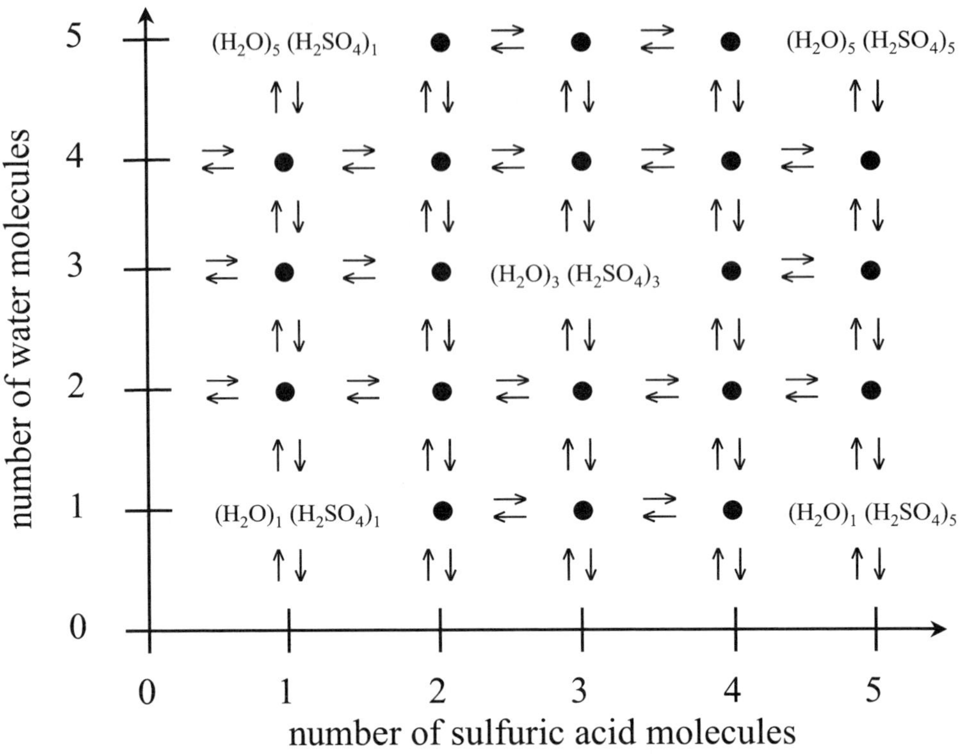

FIGURE 20.5 Illustration of the all the evaporation and condensation rate constants the are required for multi-component nucleation of sulfuric acid (H_2SO_4) and water (H_2O) including the relevant products of sulfuric acid dissociation: bisulfate (HSO_4^-), sulfate (SO_4^{2-}), and hydronium (H_3O^+).

Significant progress has been made in developing DNT over the last few years and these advances provide the basis for a true molecular-level understanding of nucleation. In DNT, application of Variational Transition State Theory (VTST) removes the ambiguity associated with determining the size of the relevant regions of configuration space by explicitly locating the bottleneck of individual dynamical processes. The dynamical bottleneck may be characterized by a spherical dividing surface (with constraining radius r_{cut}) separating reactant (i-cluster) and product ($i-1$ cluster plus a monomer at infinity) regions in configuration space. As a variational theory, any dividing surface will yield a rate constant greater than the exact rate calculated from explicit dynamics. Improving the estimate of the rate constant is achieved by varying the dividing surface to minimize the reactive flux. Application of VTST assumes a one-way flux through the dividing surface and thus precludes trajectories that return from the product region of phase space and subsequently pass through the dividing surface in the opposite direction toward reactants—this effect is called re-crossing. Explicit molecular dynamical simulation and computation of the transmission coefficient correcting for re-crossing effects can further improve the DNT estimates of the rate constants [57].

We have applied DNT to the nucleation of pure water using the Dang–Chang [69] polarizable water model at a temperature of 243 K. Our results show good agreement with experiment and thus demonstrate the fidelity of DNT, however, future studies will explore the temperature dependence of the thermodynamics and kinetics of small water clusters [7]. Our treatment of multi-component nucleation is similar to that of our treatment of unary nucleation and can be reviewed in our previous publications [5]. It is important to note that the results of our sensitivity analysis are quite general and thus do not apply simply to the test cases considered in our previous publications. The majority of atmospheric aerosols (both cloud and non-cloud aerosols) involve water to varying degrees. Reaction channels involving addition and loss of multi-component clusters of various sizes (as opposed to monomer addition and loss) can be included by constructing increasingly more general dividing surfaces. The steady-state nucleation rate, J, is expressed as [4,9,10]

$$J = \left[\sum_{i=1}^{\infty} \left(\beta_i N_i^{\mathrm{EQ}}\right)^{-1} \right]^{-1} = \left[\sum_{i=1}^{\infty} \left(\beta \alpha_{i+1} N_{i+1}^{\mathrm{EQ}}\right)^{-1} \right]^{-1},$$

where β_i and α_{i+1} are the condensation rate constants onto an i-cluster and evaporation rate constants from an $i+1$ cluster, respectively. N_i^{EQ} is the i-cluster distribution function. Once the DNT evaporation rate constants, α_{i+1}, and i-cluster Helmholtz free energies, A_i, are computed, the cluster distribution functions, N_i^{EQ}, can be used in the detailed balance condition, $K_{i,i+1}^{\mathrm{EQ}} = N_{i+1}^{\mathrm{EQ}}/N_i^{\mathrm{EQ}} = \beta_i/\alpha_{i+1}$, to calculate the equilibrium constants, $K_{i,i+1}^{\mathrm{EQ}}$, and the condensation rate constants β_i. In the application of DNT we have employed Monte Carlo methods to compute the necessary ensemble averages, however, a variety of molecular simulation techniques can also be employed [70].

Sensitivity analysis can help to identify those kinetic parameters upon which the nucleation rate depends most. We have applied sensitivity analysis [4,5] to both unary and multi-component steady-state nucleation rates by considering the variation of the nucleation rate with respect to independent fluctuations in the kinetic parameters. It was found that the nucleation rate is most sensitive to all *precritical* cluster rate constants when the evaporation and condensation rate constants are treated as the independent variables. Alternatively, the nucleation rate is most sensitive to those clusters around the critical size when the evaporation rate constants (or condensation rate constants) and equilibrium cluster populations are treated as the independent variables. Additionally, we have demonstrated that the sensitivity of the nucleation rate to the interaction energetics is governed by the cluster Helmholtz free energies. In particular, we tested two popular water models [4] (polarizable Dang–Chang and non-polarizable TIP4P) and found that the rate constants can differ up to an order of magnitude. The two water models also gave discrepancies in the Helmholtz free energies ranging from 1 to 9 kcal mol^{-1} for the dimer up to the decamer, respectively.

5. WHY ACCURATE CHEMICAL PHYSICS IS IMPORTANT TO NUCLEATION

In order to appreciate the fidelity of computational chemical physics in the context of nucleation rate prediction, we must have (1) a consistent theoretical formalism for obtaining the rate constants, (2) accurate interaction energetics, and (3) the appropriate statistical mechanical sampling to obtain the relevant properties of molecular clusters. In response to point (1), we developed DNT, which is the first consistent molecular theory of nucleation based on cluster properties. The goal is to calculate cluster thermodynamic and kinetic properties (e.g., free energies and rate constants) in terms of molecular properties or quantum states (i.e., molecular interaction energies including bond making/breaking, rotations, vibrations, and translations) using the partition function (PF). In order to evaluate the PF—point (3)—we must have a representation of the potential energy surface (PES), $U(\mathbf{r}^i)$, for all the degrees of freedom needed to sample the microstates appropriate for molecules undergoing the clustering reactions relevant to nucleation—point (2). PESs based on *ab initio*, DFT, or analytic PESs do an acceptable job of predicting or reproducing some properties, however, both can fail for other properties for a variety of reasons. By construction, empirically derived analytic potentials implicitly include some quantum mechanical effects on nuclear motion (e.g., zero-point-motion) since they are parameterized so that classical simulations reproduce a particular set of observables that embody quantum nuclear degrees of freedom. In contrast, *ab initio* or DFT calculations (within the Born–Oppenheimer approximation) require explicit consideration of the quantum nuclear degrees of freedom. Furthermore, given the expense of *ab initio* energetics, it becomes computationally intractable to obtain individual energetics for the millions or billions of relevant configurations underlying the nucleating clusters. As a consequence, *ab initio* calculations must employ the rigid-rotor harmonic oscillator approximation (RRHOA) to exclude the troublesome issue of statistical mechanical sampling. For the clusters underlying the nucleation process, the statistical mechanical sampling should account for *anharmonicity*. There are two essential types of anharmonicity: (1) local anharmonicity of the intramolecular vibrations of the molecules within a cluster, and (2) global anharmonicity due to sampling between the exceedingly large number of inherent configurations within the relevant volume of configuration space. This is why analytic interaction potentials are so useful in real-world statistical mechanics [71]—the configuration space can be more thoroughly sampled without appeal to the RRHOA. Furthermore, "anharmonic" corrections (e.g., local anharmonicity, rotational–vibrational coupling, and centrifugal distortion) do indeed extend the range of RRHOA's validity, however, generally not enough to overcome the neglected global anharmonic contributions.

Our calculations have shown that the global anharmonicities in the Helmholtz free energies for both pure water clusters [4] and aqueous ionic clusters [6,8,71] fundamental to nucleation are essential to include in the prediction of accurate equilibrium constants and free energies. For example, the size-dependent anharmonic chemical potentials of water clusters are wildly different from the RRHOA results underscoring the importance of configurations far removed from minima. Additionally, our theoretical results using empirical potentials to describe the free energetics of aqueous ionic clusters agree extremely well with experimental data [71]. Although *ab initio* calculations can provide accurate energetics for individual cluster configurations, the RRHOA can invalidate the resulting thermodynamics and hence the predictability of even the highest level electronic structure or DFT data.

The sensitivity of nucleation to conditions and the underlying interaction potentials demands an understanding of how this sensitivity propagates to aerosol concentrations and size, and ultimately the direct and indirect effects of aerosols on radiative forcing. Conditions in the atmosphere are heterogeneous and dynamic. Fluctuations in precursor concentrations can lead to nucleation that will not be captured in models that treat nucleation as a source term that depends on averaged concentrations of precursors. As indicated above, CNT displays sensitivity to conditions, but based only upon how the bulk liquid properties change as a function of conditions. Our work on DNT has shown that it is cluster properties that control nucleation rates, not bulk properties, and therefore an understanding of nucleation requires treatment of the chemical physics of molecular clusters. The potential chemical species involved in atmospheric nucleation are many and the underlying clustering reactions should embrace this chemical complexity and avoid the unqualified use of simplified models or CNT. Therefore, an explicit molecular-level/cluster-based treatment is mandated. Using a molecular-level approach, a nucleation module can be constructed based on parameters for molecular clusters rather than the bulk liquid surface tension and density used in CNT. The parameters in a cluster-based model could be obtained from a combination of field, laboratory, theory/computations, or simpler chemical descriptors (e.g., group additivity rules). Such a cluster-based model for nucleation has the advantage of being systematically improved as more accurate energetics and more complete statistical mechanical sampling becomes available.

6. SUMMARY AND FUTURE DIRECTIONS

Although new particle formation by multi-component vapor-to-liquid nucleation has been identified as an important process affecting aerosol concentrations, a fundamental understanding of the mechanisms, kinetics, thermodynamics, and rates of nucleation does not exist. The current paper describes a pathway for the use of DNT to determine the mechanisms and rates of nucleation, and to use this new knowledge to develop improved chemical physics models of aerosol formation in the atmosphere. DNT determines nucleation rates from knowledge of the molecular interactions that govern the thermodynamic and kinetic properties of molecular clusters leading to aerosol formation. Using DNT, nucleation rates can enable atmospheric modelers to address the importance of multi-component nucleation to aerosol radiative forcing with confidence. Given the large number of potential species (ions, organics, inorganics, acids, metals, etc.) that interact favorably with water and hence affect aerosol formation it should be evident that fundamental theoretical developments at the molecular level can be used to better model aerosol formation. Furthermore, the current treatment of nucleation using CNT cannot begin to deal with trace chemical species taking part in the clustering reactions. Theoretical and computational methods have advanced to the point where calculations of important thermodynamic and kinetic parameters for cluster formation are now possible. The statistical thermodynamics and kinetics of molecular clusters underlying nucleation has been outlined and future developments towards modeling aerosol radiative forcing will greatly benefit from a better description of the essential chemical physics.

ACKNOWLEDGEMENTS

This work was supported by the Division of Chemical Sciences, Office of Basic Energy Sciences of the U.S. Department of Energy (DOE) and calculations were performed in part using the Molecular Science Computing Facility in the William R. Wiley Environmental Molecular Sciences Laboratory (EMSL) at the Pacific Northwest National Laboratory. S.M.K. would like to acknowledge helpful discussions with Nels S. Laulainen, Richard C. Easter, and Steven J. Ghan from the Atmospheric Sciences Division at PNNL. The EMSL is funded by the DOE Office of Biological and Environmental Research. Battelle operates Pacific Northwest National Laboratory for DOE.

REFERENCES

[1] J.H. Seinfeld, S.N. Pandis, *Atmospheric Chemistry and Physics: From Air Pollution to Climate Change*, Wiley, New York, 1998.
[2] H.J. Gray, A. Isaacs, in: *A New Dictionary of Physics*, Longman Group Limited, London, 1975.
[3] S. Solomon, D. Qin, M. Manning, Z. Chen, M. Marquis, K.B. Averyt, M. Tignor, H.L. Miller (Eds.), *IPCC, 2007: The Physical Science Basis. Contribution of Working Group I to the Fourth Assessment Report of Intergovernmental Panel on Climate Charge*, Cambridge University Press, Cambridge, 2007.
[4] S.M. Kathmann, G.K. Schenter, B.C. Garrett, *J. Chem. Phys.* **116** (2002) 5046.
[5] S.M. Kathmann, G.K. Schenter, B.C. Garrett, *J. Chem. Phys.* **120** (2004) 9133.
[6] S.M. Kathmann, G.K. Schenter, B.C. Garrett, *Phys. Rev. Lett.* **94** (2005) 116104.
[7] S.M. Kathmann, *Theor. Chem. Acc.* **116** (2006) 169.
[8] S.M. Kathmann, G.K. Schenter, B.C. Garrett, *Phys. Rev. Lett.* **98** (2007) 109603.
[9] F.F. Abraham, *Homogeneous Nucleation Theory*, Academic Press, New York, 1974.
[10] A.C. Zettlemoyer, *Nucleation*, Marcel Dekker, New York, 1969.
[11] H. Reiss, in: B.N. Hale, M. Kulmala (Eds.), *15th International Conference on Nucleation and Atmospheric Aerosols, AIP, Conf. Proc., Rolla, Missouri, USA*, vol. 534, American Institute of Physics, Melville, NY, 2000, p. 181.
[12] M. Kulmala, P. Korhonen, I. Napari, *et al.*, *J. Geophys. Res.* **107** (2002) 8111.
[13] M. Kulmala, H. Vehkamaki, T. Petaja, *et al.*, *J. Aerosol Sci.* **35** (2004) 143.
[14] S.K. Friedlander, *Smoke, Dust, and Haze: Fundamentals of Aerosol Dynamics*, Oxford University Press, New York, 2000.
[15] C.F. Bohren, D.R. Huffman, *Absorption and Scattering of Light by Small Particles*, Wiley, New York, 1998.
[16] G. Mie, *Ann. Phys.* **330** (1908) 377.
[17] M. Volmer, *Kinetik der Phasenbildung*, Theodor Steinkopff Verlag, Dresden, 1939.
[18] M. Volmer, A. Weber, *Z. Phys. Chem.* **119** (1926) 277.
[19] R. Becker, W. Doering, *Ann. Phys.* **24** (1935) 719.

[20] J. Zeldovich, *J. Exp. Theor. Phys.* **12** (1942) 525.
[21] C.L. Weakliem, H. Reiss, *J. Chem. Phys.* **99** (1993) 5374.
[22] C.L. Weakliem, H. Reiss, *J. Chem. Phys.* **101** (1994) 2398.
[23] D.W. Oxtoby, *J. Phys.: Condens. Matter* **4** (1992) 7627.
[24] M. Kulmala, *Science* **302** (2003) 1000.
[25] M.P. Allen, D.J. Tildesley, *Computer Simulation of Liquids*, Oxford University Press, New York, 1987.
[26] D. Frenkel, B. Smit, *Understanding Molecular Simulation: From Algorithms to Applications*, Academic Press, San Diego, CA, USA, 1996.
[27] G.K. Schenter, S.M. Kathmann, B.C. Garrett, *Phys. Rev. Lett.* **82** (1999) 3484.
[28] G.K. Schenter, S.M. Kathmann, B.C. Garrett, *J. Chem. Phys.* **110** (1999) 7951.
[29] S.M. Kathmann, G.K. Schenter, B.C. Garrett, *J. Chem. Phys.* **111** (1999) 4688.
[30] S.M. Kathmann, B.N. Hale, *J. Phys. Chem. B* **105** (2001) 11719.
[31] B.N. Hale, *Aust. J. Phys.* **49** (1996) 425.
[32] I. Kusaka, *J. Chem. Phys.* **111** (1999) 3769.
[33] I. Kusaka, D. Oxtoby, *J. Chem. Phys.* **110** (1999) 5249.
[34] I. Kusaka, Z. Wang, J. Seinfeld, *J. Chem. Phys.* **103** (1995) 8993.
[35] I. Kusaka, Z. Wang, J. Seinfeld, *J. Chem. Phys.* **102** (1995) 913.
[36] I. Kusaka, Z.-G. Wang, J.H. Seinfeld, *J. Chem. Phys.* **108** (1998) 3416.
[37] I. Kusaka, Z.-G. Wang, J.H. Seinfeld, *J. Chem. Phys.* **108** (1998) 6829.
[38] K. Oh, X. Zeng, H. Reiss, *J. Chem. Phys.* **107** (1997) 1242.
[39] K. Oh, X. Zeng, *J. Chem. Phys.* **110** (1999) 4471.
[40] K.J. Oh, G.T. Gao, X.C. Zeng, *Phys. Rev. Lett.* **86** (2001) 5080.
[41] P. Schaaf, B. Senger, H. Reiss, *J. Phys. Chem.* **101** (1997) 8740.
[42] P. Schaaf, B. Senger, J.-C. Voegel, *et al.*, *Phys. Rev. E* **60** (1999) 771.
[43] P. Schaaf, B. Senger, J.C. Voegel, *et al.*, *J. Chem. Phys.* **114** (2001) 8091.
[44] K. Suzuki, in: M. Kulmala, P.E. Wagner (Eds.), *Nucleation and Atmospheric Aerosols 1996*, Pergamon, New York, 1996.
[45] B. Senger, P. Schaaf, D. Corti, *et al.*, *J. Chem. Phys.* **110** (1999) 6438.
[46] B. Senger, P. Schaaf, D. Corti, *et al.*, *J. Chem. Phys.* **110** (1999) 6421.
[47] H. Vehkamaki, I. Ford, *J. Chem. Phys.* **112** (2000) 4193.
[48] A. Bandy, J. Ianni, *J. Phys. Chem. A* **102** (1998) 6533.
[49] J. Ianni, A.R. Bandy, *J. Phys. Chem.* **103** (1999) 2801.
[50] J. Ianni, A.R. Bandy, *J. Mol. Struct.* **497** (2000) 19.
[51] T. Kurten, M.R. Sundberg, H. Vehkamaki, *et al.*, *J. Phys. Chem. A* **110** (2006) 7178.
[52] J. Merikanto, E. Zapadinsky, A. Lauri, *et al.*, *Phys. Rev. Lett.* **98** (2007) 145702.
[53] J. Merikanto, H. Vehkamaki, E. Zapadinsky, *J. Chem. Phys.* **121** (2004) 914.
[54] A.B. Nadykto, A. Al Natsheh, K.V. Mikkelsen, *et al.*, *Phys. Rev. Lett.* **96** (2006) 125701.
[55] A.B. Nadykto, F. Yu, *Phys. Rev. Lett.* **93** (2004) 016101.
[56] G.K. Schenter, *J. Chem. Phys.* **108** (1998) 6222.
[57] G.K. Schenter, S.M. Kathmann, B.C. Garrett, *J. Chem. Phys.* **116** (2002) 4275.
[58] J.K. Lee, J.A. Barker, F.F. Abraham, *J. Chem. Phys.* **58** (1973) 3166.
[59] F.H. Stillinger, *J. Chem. Phys.* **38** (1963) 1486.
[60] P. Debenedetti, H. Reiss, *J. Chem. Phys.* **108** (1998) 5498.
[61] P. Debenedetti, H. Reiss, *J. Chem. Phys.* **111** (1999) 3771.
[62] A. Laaksonen, V. Talanquer, D.W. Oxtoby, *Annu. Rev. Phys. Chem.* **46** (1995) 489.
[63] P.J. Coulier, *J. Pharm. Chem.* **22** (1875) 165.
[64] J. Aitken, *Proc. R. Soc. Edinb.* **11** (1880) 14.
[65] C.T.R. Wilson, *Trans. R. Soc. (London) A* **189** (1897) 265.
[66] C.T.R. Wilson, *Trans. R. Soc. (London) A* **193** (1900) 289.
[67] V.B. Mikheev, P.M. Irving, N.S. Laulainen, *et al.*, *J. Chem. Phys.* **116** (2002) 10772.
[68] J.L. Schmitt, K.V. Brunt, G.J. Doster, in: B.N. Hale, M. Kulmala (Eds.), *15th International Conference on Nucleation and Atmospheric Aerosols, AIP, Conf. Proc., Rolla, Missouri, USA*, vol. 534, American Institute of Physics, Melville, NY, 2000, p. 51.
[69] L.X. Dang, T.M. Chang, *J. Chem. Phys.* **106** (1997) 8149.
[70] B.J. Palmer, S.M. Kathmann, M. Krishnan, *et al.*, *J. Chem. Theory Comput.* **3** (2007) 583.
[71] S.M. Kathmann, G.K. Schenter, B.C. Garrett, *J. Phys. Chem. C* **111** (2007) 4977.

CHAPTER 21

Computational Quantum Chemistry: A New Approach to Atmospheric Nucleation

Alexey B. Nadykto[1*], **Anas Al Natsheh**[**], **Fangqun Yu**[*], **Kurt V. Mikkelsen**[***] and **Jason Herb**[*]

Contents

1. Introduction 450
   1.1 Binary homogeneous nucleation (BHN) of H_2SO_4 and H_2O 451
   1.2 Ternary homogeneous nucleation (THN) of H_2SO_4–H_2O–NH_3 451
   1.3 Ion-mediated nucleation (IMN) of H_2SO_4–H_2O–ion 452
   1.4 Organics-enhanced nucleation H_2SO_4–H_2O–organics 453
2. Nucleation Theory 454
3. Why Should We Apply the Quantum Theory to Atmospheric Problems? 455
4. Quantum Methods 456
5. Application of Quantum Methods to Atmospheric Species 457
   5.1 Neutral binary H_2SO_4–H_2O clusters 458
   5.2 Neutral ternary H_2SO_4–H_2O–NH_3 clusters 463
   5.3 Stabilization of H_2SO_4–H_2O complexes by low-molecular organic acids 468
   5.4 Neutral gas-phase hydrates of HNO_3 in the stratosphere 468
   5.5 Relevance of dipolar properties of neutral monomers and pre-nucleation clusters to nucleation of ions 470
   5.6 Ionic $H_3O^+(H_2O)_n$ clusters 471
   5.7 Ionic $HSO_4^-(H_2O)_n$ clusters 472
   5.8 From CNT to "first principle" nucleation theory 473
6. Concluding Remarks 475

[*] Atmospheric Sciences Research Center, State University of New York at Albany, 251 Fuller Rd., Albany 12203, NY, USA
[**] Kajaani University of Applied Sciences, Kuntokatu 5, 87101 KAJAANI, Finland
[***] Department of Chemistry, University of Copenhagen, Universitetparken 5, DK-2100 Copenhagen Ø, Denmark
[1] Corresponding author. E-mail: alexn@asrc.cestm.albany.edu

Advances in Quantum Chemistry, Vol. 55
ISSN 0065-3276, DOI: 10.1016/S0065-3276(07)00221-3

Acknowledgements 475
References 475

Abstract

A clear understanding of the nucleation and particle formation mechanisms is critically important for assessing the lifecycle of atmospheric particles and predicting the climate changes associated with aerosol radiative forcing. The main assumption of the liquid droplet approach used in the classical nucleation theory is that the properties of nucleating and pre-nucleation clusters are the same as those of the bulk liquid. However, molecular clusters formed during the first few nucleation steps are quantum-mechanical systems, and cannot be adequately treated using the capillarity approximation and/or empirical potentials parameterized to the bulk liquid. This Review discusses the recent developments of theoretical methods to solve problems in the atmospheric nucleation theory, and is primarily focused on the quantum chemistry of common atmospheric nucleation precursors. The importance of the quantum-mechanical treatment of molecular clusters underlying nucleation will be outlined and future developments towards the "first principles" nucleation theory will be discussed.

1. INTRODUCTION

Atmospheric particles influence the Earth climate indirectly by affecting cloud properties and precipitation [1,2]. The indirect effect of aerosols on climate is currently a major source of uncertainties in the assessment of climate changes. New particle formation is an important source of atmospheric aerosols [3]. While the contribution of secondary particles to total mass of the particulate matter is insignificant, they usually dominate the particle number concentration of atmospheric aerosols and cloud condensation nuclei (CCN) [4]. Another important detail is that high concentrations of ultrafine particles associated with traffic observed on and near roadways [5–7] lead, according to a number of recent medical studies [8–11] to adverse health effects.

The critical importance of clear understanding and insight of the new particle formation mechanisms for quantitatively assessing the climate-related, health and environmental impacts of atmospheric aerosols is commonly accepted [12]. Although nucleation phenomena have been intensively studied in the past, there are still major uncertainties concerning nucleation mechanisms in the atmosphere. The dominant constituent of nucleating vapours in the atmosphere, water is incapable of self-nucleation under typical atmospheric conditions due to the very low supersaturation with respect to pure water. Although pure water never nucleates in the atmosphere, nucleating water vapour is a perfect model system, on which different nucleation theories can be validated against experimental data from a number of recent state-of-art laboratory studies [13–15]. Another important detail is that the classical theory of homogeneous nucleation of water vapour is a theoretical foundation for various multi-component nucleation models widely used in the atmospheric studies. Nucleation in the atmosphere is essentially multicomponent and involves a number of different species including e.g. sulfuric acid, ammonia, ions and organics. At the present time, there are several candidate mechanisms that could possibly be responsible for atmospheric nucleation:

1.1 Binary homogeneous nucleation (BHN) of H_2SO_4 and H_2O

Nucleation rates in the classical BHN are sensitive to the hydration thermodynamics, which is not well understood. Vehkamaki *et al.* [16] modified the widely used parameterization of Kulmala *et al.* [17] by correcting some inconsistencies in kinetics and using the updated hydration model of Noppel *et al.* [18]. The improved parameterization [16] predicts the nucleation rates by a factor of 10^1–10^4 higher than model of Kulmala *et al.* [17]. Nucleation rates obtained using the recently developed kinetic BHN model by Yu [19,20] are ~2–4 orders of magnitude lower than those predicted by Vehkamaki *et al.* [16]. Although the kinetic BHN model has been further recently improved by using multiple experimental data and quantum-mechanical calculations, considerable uncertainties still exist in the BHN calculations. The reduction of uncertainties in the BHN theory is critically important because the theoretical formalism of BHN is widely used in other multicomponent nucleation theories. At the present time, a thorough evaluation of existing BHN theories is quite difficult due to large uncertainties in the laboratory experiments caused by difficulties in precise determining H_2SO_4 concentrations in the nucleation zone. Moreover, laboratory studies [21,22] show that in the case of in situ produced H_2SO_4 (via the reaction of OH radicals with SO_2), the formation of new particles begins at H_2SO_4 concentrations of $\sim 10^7\ cm^{-3}$, which is about 3 orders of magnitude lower than the corresponding H_2SO_4 concentration in the nucleation experiments with liquid sulfuric acid. However, a more recent independent laboratory study [23] indicates that H_2SO_4 concentration of $\sim 10^{10}\ cm^{-3}$ is needed to achieve considerable BHN rates, even if H_2SO_4 is produced in situ via $SO_2 + OH$ reaction. This finding disagree with the experiments [21,22]. Further research is needed to advance the understanding of BHN processes.

1.2 Ternary homogeneous nucleation (THN) of H_2SO_4–H_2O–NH_3

BHN was long considered as the dominant nucleation mechanism in the atmosphere. However, the BHN was not able to explain all the recently observed nucleation events. In response, the THN theory, with ammonia as the exclusive principle stabilizer of H_2SO_4–H_2O clusters, has been developed. While NH_3 is known as the efficient neutralizer of sulfuric acid solutions, its efficiency of in neutralizing and stabilizing small acid clusters remains unclear [24]. Classical model of THN [25] predicts that NH_3 at ppt level can enhance the H_2SO_4–H_2O nucleation by up to ~30 orders of magnitude. However, all the laboratory studies [26–30] indicate that the presence of NH_3 at much higher ppb–ppm levels enhances the H_2SO_4–H_2O nucleation by up to only ~2 orders of magnitude [24]. Moreover, Nadykto and Yu [31] have pointed out that ammonia is certainly not an exclusive stabilizer of H_2SO_4–H_2O clusters in the atmosphere. The kinetic simulations carried out using the THN model, which was constrained by experiment data, indicate that the contribution of THN to the formation of new particles in the boundary layer is likely very small [24]. In contrast, Jung *et al.* [32] reported good agreement between the model simulations based on the original THN parameterization of Napari *et al.* [25] and nucleation measurements in Pittsburgh region. However, the widely used classical THN model [25] used by Jung *et al.* [32] has been recently corrected and revised [33]. The corrected THN model by Anttila *et al.* [33] predicts much smaller THN rates that are negligible in the lower troposphere [33]. Jung *et al.* [32] may have to reconsider their conclusions, if the more accurate THN model [33] was used in their study. The effect of ammonia on atmospheric nucleation rates remains unclear and further studies are needed to the role of ammonia in the atmospheric nucleation [24,31].

1.3 Ion-mediated nucleation (IMN) of H_2SO_4–H_2O–ion

The presence of ions, which essentially work as a catalyst, helps to overcome the nucleation barrier and leads to significant enhancement in nucleation rates. Strong interactions between ionic clusters and dipolar atmospheric precursors, common pollutants, toxic and chemically-active substances promote the formation of ultrafine aerosol particles that are associated directly with the adverse public health effects [34–39]. It has been shown [40,41] that IMN could be an important particle formation mechanism. Most recently, a new second generation IMN model [42], which is built upon on the substantial progress in understanding the molecular nature and thermodynamics of small charged and neutral clusters achieved in the last few years [20,34–39,43] has been developed. The model predictions suggest that IMN can lead to significant nucleation in the lower atmosphere [42,44].

The state-of-art measurements of evolving air–ion mobility spectra and size-resolved charged fraction (CF) of freshly formed particles became available recently [45–48]. While ion charge and mobility measurements recorded mainly in Hyytiälä (Finland) indicate that ions are involved in more than 90% of the particle formation events [47,49], the relative contribution of ion-mediated and homogeneous nucleation processes to measured nucleation rates remains unclear [47–49]. Laakso *et al.* [48] claim, based on a simplified analysis of the experimental data, that their measurements indicate a relatively small contribution of ion nucleation. Iida *et al.* [46] made similar conclusion based on their analysis of the charged fractions of 3–5.5 nm particles measured at NCAR's Marshall Field Site in Boulder, Colorado. However, both the simplified analysis of Yu [49] and detailed kinetic simulations of the evolution of charged and neutral cluster size spectra of Yu and Turco [50] showed that observations analyzed by Laakso *et al.* [48] may indicate the dominance of IMN. The main reason for the disagreement in the interpretation of the experimental data between studies [49,50] and [46,48] is the assumption about the source of neutral clusters/particles smaller than 3 nm. Iida *et al.* [46] and Laakso *et al.* [48] assume that most of sub-3 nm neutral clusters/particles appear due to a homogeneous nucleation of undetermined nature. In contrast, Yu and Turco [50] argue that most of the sub-3 nm neutral clusters/particles are a product of the ion–ion recombination or neutralization of charged clusters formed on ions, and that the aforementioned clusters should thus be considered as formed via IMN. More recently, Kulmala *et al.* [51] reported measurements of continuous pool of neutral clusters in the sub-3 nm size range and concluded that homogeneous nucleation dominates over the ion-induced nucleation in the boreal forest conditions. Further research, both experimental and theoretical, is needed to quantify the relative contributions of ion-mediated and homogeneous nucleation to observed new particle formation in the atmosphere.

1.4 Organics-enhanced nucleation H_2SO_4–H_2O–organics

The pioneering experimental work of Zhang *et al.* [52] has shown that the presence of organic acids can enhance the nucleation of H_2SO_4 and H_2O. More recently, it has been pointed out [31] that ammonia is certainly not an exclusive stabilizer of H_2SO_4–H_2O clusters in the atmosphere. Low-molecular formic and acetic acids, which are among the most abundant organic acids in the atmosphere, appeared to be as efficient in stabilizing H_2SO_4–H_2O clusters as NH_3 [31]. Further work is needed in order to determine the organic species that can enhance the atmospheric nucleation and to quantify their effect on atmospheric nucleation rates.

While the sulfuric acid is key nucleation precursor in the low troposphere, its contribution to the polar stratospheric chemistry is a lot more modest. Another strong acid–nitric-plays a major role as the dominant reservoir for ozone destroying odd nitrogen radicals (NO_x) in the lower and middle polar stratosphere. Nitric acid is an extremely detrimental component in the polar stratosphere clouds (PSCs), where nitric acid and water are the main constituents, whose presence significantly increases the rate of the ozone depletion by halogen radicals. Gas-phase hydrates of the nitric acid that condense and crystallize in the stratosphere play an important role in the physics and chemistry of polar stratospheric clouds (PSCs) related directly to the ozone depletion in Arctic and Antarctic.

It is well known that nucleation and particle formation rates are very sensitive to the thermodynamics of initial steps of the cluster formation, where reliable thermodynamic data are often lacking. It is clear that a theoretical tool that could predict, with sufficient degree of confidence, the thermodynamic properties of nucleating particles/clusters is urgently needed. At the present time, most of the nucleation theories are based on so-called liquid droplet or capillarity approximation. The theoretical foundation of the commonly used bulk liquid model did not experience significant changes since being developed back in 19th century by Lord Kelvin (W. Thomson). The applicability of such an approach for the description of initial steps of the cluster formation is generally limited. Therefore, an accurate theoretical treatment of a matter in an indeterminate, neither gaseous nor liquid, state of molecular cluster requires a more rigorous method. Another problem of theoretical foundation of the Classical Nucleation Theory (CNT) is the application of the steady-state approximation. In the case, when multicomponent water-trace impurities mixtures are nucleating, the nucleation rates may be limited kinetically. Recently developed kinetic nucleation models are able to correct some kinetic inconsistencies in the CNT models and incorporate the thermochemical data for small clusters from quantum-mechnical calculations and laboratory studies [20,24,42,53]. Molecular-based models are more advanced theoretically than the CNT and they compute nucleation kinetically. However, they often use spherically symmetric approximation and are very sensitive to empirical interaction potentials that are typically parameterized in such a way as to reproduce the bulk liquid.

Quantum methods have been progressing continuously since Schrödinger's original work and their development has reached the stage when "chemical properties can often be calculated with wave function-based methods as well or better than they can be measured" [54]. Unlike the existing nucleation models (see e.g. review [55] and references therein), which are sensitive to poorly defined input parameters and empirical interaction models, the quantum theory can treat a system of an arbitrary chemical composition. Thermochemical data obtained from "first principles" can be used in explicit kinetic simulations of nucleation rates, with the forward/condensation rates controlled by collision kinetics and backward/escape rates governed by stepwise Gibbs free energy changes.

2. NUCLEATION THEORY

The classical theory of homogeneous and ion-induced nucleation developed back in 1930s by Volmer and Weber [56], Becker and Döring [57], and Volmer [58]. The classical bulk liquid approach, which has been criticized in a number of studies for the application of the bulk surface tension and density to molecular clusters and sensitivity to poorly defined input parameters, is still used almost exclusively in the interpretation of nucleation experiments. Simple and completely analytical expression for nucleation rates and reasonable predictivity of the CNT in the case of water explain the longstanding success of the CNT. However, classical nucleation rates are very sensitive to the key input parameters that are not well defined for a number of liquids. The cluster thermochemistry in the classical theory is based on the liquid droplet approach, with the surface tension and bulk density as the key parameters controlling the Gibbs free energy of the nucleating cluster. The strong exponential dependency of nucleation rates of *i*-mer on the Gibbs free energy explains the excessive sensitivity of nucleation rates to the input parameters. A large number of modifications and extensions of the CNT of homogeneous nucleation of single-component vapours have been published since the original work of Volmer and Weber [56] and Becker and Döring [57]. At the present time, the theoretical approaches used to correct the classical thermodynamics of unary clusters can be divided into two distinct categories—(i) scaling theories; (ii) molecular models based on empirical interaction potentials [59,60]. Scaling [61–63] is a legitimate theoretical approach and an efficient tool that allows achieving some degree of agreement between theory and experiments, however it is unable to provide any new information about the molecular nature of nucleation phenomena. Molecular-based methods that are often imposed as a viable alternative to the classical nucleation theory do not explicitly use the properties of bulk liquid to compute nucleation rates. However, interactions between vapour monomers and *i*-mers in the existing molecular studies are usually approximated using empirical (e.g. TIP4P, TIP5P, Dang–Chang) potentials parameterized to the bulk liquid properties. The aforementioned empirical potentials "were never intended to treat small clusters" [64] and their applicability to the pre-existing and nucleating clusters is questionable. In our view, the existing molecular-based nucleation approaches should be classified as semi-classical because they implicitly treat molecular clusters as liquid-like droplets. It is important to note that the above-mentioned conclusion does not imply that molecular-based methods using empirical potentials are invalid or defective [43,65]. However, it is necessary to bear in mind that nucleation rates are very sensitive to thermodynamics of initial cluster growth steps. This means that empirical potentials used in the molecular-based nucleation studies must provide an adequate description of thermochemical properties of molecular clusters [65].

The extension of the CNT to homogeneous nucleation in atmospheric, essentially multicomponent, systems have faced significant problems due to difficulties in determining the activity coefficients, surface tension and density of binary and ternary solutions. The BHN and THN theories have been experiences a number of modifications and updates. At the present time, the updated quasi-steady state BHN model [16] and kinetic quasi-unary nucleation theory [24,66], and classical THN theory [25,33] and kinetic THN model constrained by the experimental data [42] are alternative approaches used in the atmospheric studies. While the aforementioned purely classical models [16,25,33] are based on the conventional liquid droplet approach, the theoretical foundation of the kinetic models [42,66] has been advanced by employing experimental data and quantum-mechanical calculations for the thermochemistry of molecular pre-nucleation clusters. There exist only a few molecular-based studies for binary H_2SO_4–H_2O mixtures [67,68] and none for the ternary H_2SO_4–H_2O–NH_3 systems.

Unlike the CNT for homogeneous nucleation, whose theoretical foundation has not been advanced since late 30s, the theoretical formalism of the ion-induced nucleation theory has been recently improved and extended. The critical importance of the dipole moment of condensing monomers and pre-nucleation clusters have been pointed out in the series of recent publications of Nadykto with co-authors [34–39] and, more recently, Leopold with co-authors [69]. The classical ion-induced nucleation theory has been advanced through the incorporation of the effect of the polar host vapour molecule-charged cluster interactions and some of the serious shortcomings of the original model have been successfully corrected [36].

3. WHY SHOULD WE APPLY THE QUANTUM THEORY TO ATMOSPHERIC PROBLEMS?

Nucleation in the atmosphere is essentially multicomponent process. However, a commonly used classical approach incapable of the quantitative treatment of multicomponent systems due to (a) excessive sensitivity to poorly defined activity coefficients, density and surface tension of multicomponent solutions; (b) strong dependence of nucleation rates on thermochemistry of initial growth steps where the liquid droplet model is invalid; (c) impossibility of the application of the scaling approach due to the absence of reliable experimental data. Application of more sophisticated molecular-based methods could possibly advance the multicomponent nucleation theory; however, they need reliable empirical potentials for multcomponent systems that are not well defined. Unlike the existing nucleation models, the quantum theory treats a system of an arbitrary chemical composition based on its wave function and needs neither bulk liquid properties nor spherically symmetric approximation.

4. QUANTUM METHODS

The most common computational quantum methods [70,71] can be divided into two distinct categories: (a) *ab initio*; (b) density functional theory (DFT). In principle, all knowledge about a system can be obtained from the wave function, which is obtained by solving the Schrödinger equation for a complete multi-electron system. The term *ab initio* literally means "from first principles"; however, it does not imply that the Schrödinger equation is solved exactly. The main idea of *ab initio* is to select a method that can lead to a reasonable approximation to the solution of the Schrödinger equation and then to choose a basis set that will implement this method properly. The original Hartree–Fock method is incapable of giving a correct solution to the Schrödinger equation even if a very large and flexible basis set is employed (Hartree–Fock limit) because in the reality electrons are not paired up in the way that the original Hartree–Fock theory suggests. The difference in energy between the exact result and Hartree–Fock limit is called the "correlation energy". There are two classes of methods dealing with the "correlation problem"—variational methods (e.g. CID, CISD, full CI) and perturbation methods (e.g. MP*n*, QCISD). At the present time, the Møller and Plesset second order method (MP2) is the most commonly used *ab initio* method. In addition to pure *ab initio* methods, there exist compound *ab initio* methods (e.g. CBS-APNO, G2, G3, G2MP2, G3MP2) involving a variety of basis sets and advanced theoretical corrections to provide good energy estimates. The main idea of the DFT is to describe an interacting electronic system via its density—a one-body quantity—instead of its many-body wave function, providing in this way accurate results comparable in quality with *ab initio* calculations at lower computational costs. DFT is neither a Hartree–Fock method nor a post-Hartree–Fock method because it is constructed in a different way and the resulting orbitals are often referred to as "Kohn-Sham" orbitals. The most common density functionals have been derived by careful comparison with experiments. At the present time, numbers of atoms in a complex/cluster that can be treated by *ab initio* (MP2) and DFT are ~40 and ~150. In order to extend the size range, hybrid (e.g. quantum mechanics–molecular mechanics (QM/MM)) methods [72–78] can be applied. Interactions of gas-phase monomers with an aerosol particle and the corresponding absorption, desorption and surface reaction rate constants can be simulated using quantum–statistical (QM-ST) methods [74–78]. The QM-ST method is based on quantum statistical mechanics and phase-space theory. A process involving a molecule and an aerosol particle is divided into elementary steps such as (i) diffusion of the molecule to the aerosol particle; (ii) adsorption of the molecule at the aerosol particle surface; (iii) chemical reactions on the aerosol particle; (iv) desorption of products from the aerosol particle, and (v) diffusion of products away from the aerosol particle. Usually, it is assumed that the first and last steps are fast and that the calculated rates are largely controlled by steps two, three and four. When a molecule and an aerosol particle interact, the molecular system can be attached either through physisorption or chemisorption. The calculations of the corresponding interaction energies can be done by utilizing QM/MM Molecular Dynamics methods, where the statistical molecular dynamics simulations are performed for selecting relevant structures of the molecule-aerosol particle system. The methods [74–78] provide procedures for calculating at the level of correlated electronic structure both interaction energies and molecular properties for molecules interacting with aerosol particles that can be utilized for QM-ST simulations of rate constants, reaction probabilities or sticking coefficients. Although hybrid methods have been successfully applied to calculate interactions between aerosol particles and gas molecules [74–78] and to model the sulfuric acid solution [73], their capability to predict the free energies of molecular nucleation and pre-nucleation clusters is yet to be studied, and, thus, further research is needed in order to determine the applicability of hybrid methods to atmospheric nucleation.

5. APPLICATION OF QUANTUM METHODS TO ATMOSPHERIC SPECIES

At the present time, quantum computations are quite demanding, and, thus, the size of a system that can be treated using computational quantum methods is limited. A perfectly logical question, "Why the application of quantum methods can systematically improve the nucleation theory?" leads to very simple answers: (a) nucleation rates are very sensitive to free energies of the formation of initial (or pre-critical) clusters; (b) the pre-nucleation clusters are essentially quantum systems, and must be treated using a quantum approach; (c) the number of H_2SO_4 molecules in critical clusters in typical atmospheric nucleation events are not excessively large and are generally $\sim$10; (d) recently developed kinetic nucleation models can effectively use the clustering energy changes obtained from quantum calculations.

The importance of rigorous theoretical treatment of initial steps of the cluster growth has been pointed out in Nadykto *et al.* [43], where a classical problem of the sign preference in ion-induced nucleation has been solved. Wilson's 1897 pioneering cloud chamber experiments [79] demonstrated that ions of opposite sign exhibited significantly different nucleation rates. Later nucleation experiments have confirmed that in a number of systems nucleation rates of positive and negative ions are different [80–82]. This phenomenon is known as the sign preference and has remained a mystery for more than a century.

The theoretical formalism of the commonly used classical ion-induced nucleation theory has been recently improved and extended [34–36]; however, a theoretical description of the effect of the chemical identity of the core ion on the gas-to-particle conversion was still lacking. Continuously growing interest in the ion-induced nucleation during the last decade has resulted in a number of molecular-based theories dedicated to the sign preference. Most of molecular-based theories [83–87] conclude that the sign preference is governed by the ion sign alone. In contrast, a novel Dynamical Nucleation Theory (DNT) [60,64,88–90] suggests [60] that taking into account the ion sign alone is insufficient and that core ion properties can be adequately described using ion size and ion sign as the key parameters.

We have investigated [43] the sign preference using a quantum approach and showed that this puzzling phenomenon is essentially quantum in nature. It is shown that the effect of the chemical identity of the core ion is controlled by the electronic structure of the core ion through the influence on the intermolecular bonding energies during the initial steps of cluster formation. Nadykto *et al.* [43] have shown that in order to answer a practical question—"Which ion is a better nucleator?", one need not to perform multiple complicate and expensive nucleation experiments. The answer can now be obtained by carrying out a relatively simple quantum-mechanical study of pre-critical clusters over the limited size range, or by looking at the data on the thermochemistry of cluster ions available for a number of substances Our results show, in agreement with the experimental data on ion clustering and nucleation study of Rabeony and Mirabel [82], that sign preference for water is weak (slightly positive). DNT [60], in contrast, gives a much stronger negative sign preference and predicts that the addition of 40 water molecules to a 1 nm core ion, which is already more than 5 times large than a water molecule, is still not enough to reach the limit where the effects of electrical fields of positively and negatively charged core ions become undistinguishable. These predictions clearly disagree with the experimental data and short-ranged nature of electrical interactions. In our view, the failure of the DNT and other molecular-based theories is caused by the usage of improper empirical potentials that were not parameterized is such a way as to reproduce the properties of *i*-mers at low *i*.

5.1 Neutral binary H_2SO_4–H_2O clusters

The modern era of quantum studies of atmospheric nucleation precursors [91–117] began with the pioneering systematic investigation of sulfuric acid hydrates by Bandy and Ianni [92]. This is not surprising because the hydration of the monomeric sulfuric acid in the atmosphere is a fundamental phenomenon whose description is included in the most of up-to-date nucleation models. Molecular hydrate complexes have been detected in sulfate aerosols and their stability has been corroborated in experiments. Unfortunately, the first systematic study [92] performed using the B3LYP method was unable to predict the hydration free energies in agreement with latter experiments [117]. Nadykto and Yu [31] and, independently, Kurten *et al.* [99] pointed out that the disagreement with experimental data was caused by problems of B3LYP in the description of the hydrogen bonding. At the present time, PW91PW91 is the most common density functional used in the quantum-mechanical studies of atmospheric species. Although it would be an exaggeration to present the PW91PW91 as an ultimate development in the quantum chemistry of hydrogen bonded systems, it provides good geometries, excellent vibrational frequencies and quite accurate cluster free energies [31,95,98,99,104–106, 113–115].

The computational quantum methods express molecule/complex/cluster energies using the optimized geometry and calculated vibrational frequencies. Figure 21.1 presents the equilibrium geometries of most stable isomers of H_2SO_4 (H_2O) ($n = 1$–5).

Adequate description of both the equilibrium geometry and vibrational spectra is critically important for the free energy calculations. Table 21.1 presents a comparison of equilibrium geometries of sulfuric acid monohydrate obtained at different levels of theory with experimental data of Leopold with co-authors [118]. As may be seen from Table 21.1, both *ab initio* and DFT methods reproduce the hydrate structure with sufficient accuracy. The PW91PW91 and MP2 in combination 6-311++G(3df,3pd) basis set provide the best overall agreement with experimental data.

The calculations of the vibrational spectra are an essential part of the free energy calculations. The vibrational spectra can be computed using either the most common Rigid Rotor Harmonic Approximation (RRHOA), or different an-

TABLE 21.1 A comparison of $(H_2SO_4)(H_2O)$ geometry obtained using *ab initio* and DFT methods with experiments. Abbreviations MP2/6-311++G(3df,3pd) and PW91/6-311++G(3df,3pd), B3LYP/D95++, B3LYP/6-311++G(2d,2p), PW91/DNP, PW91/TZP and Exp. refer to the Nadykto and Yu [31], Re *et al.* [102], Bandy and Ianni [92], Ding *et al.* [106], Al Natsheh *et al.* [95] and Fiacco *et al.* [118], respectively

	MP2/ (3df,3pd)	B3LYP/ D95++	B3LYP/ (2d,2p)	PW91/ DNP	PW91/ TZP	PW91/ (3df,3pd)	Exp.
R(H1–O2)	0.967	0.975		0.978	0.98	0.976	0.95
R(O2–S)	1.579	1.636	1.611	1.634	1.63	1.614	1.578
R(O3–S)	1.557	1.603		1.604	1.59	1.576	1.567
R(O4–S)	1.43	1.466	1.439	1.463	1.46	1.446	1.464
R(O5–S)	1.42	1.458	1.43	1.453	1.44	1.435	1.41
R(H6–O3)	0.997	1.009	0.999	1.023	1.03	1.022	1.04
R(H6–O8)	1.654	1.651	1.682	1.621	1.61	1.61	1.645
R(H9–O8)	0.976					0.98	0.98
R(H10–O8)	0.96					0.97	0.98
R(H9–O4)	2.15	2.23	2.207	2.141		2.05	2.04
A(H1–O2–S)	108.23					107.82	108.5
A(O2–S–O4)	104.37					104.15	104.71
A(O3–S–O4)	109.18					109.18	106.71
A(O3–S–O5)	107.33					107.8	106.7
A(H6–O3–S)	108.23					108.15	108.6
A(O4–S–O5)	122.7					122.37	123.3
A(O2–S–O3)	102.93					103.2	101.8
A(H9–O8–H10)	105.74					105.73	107

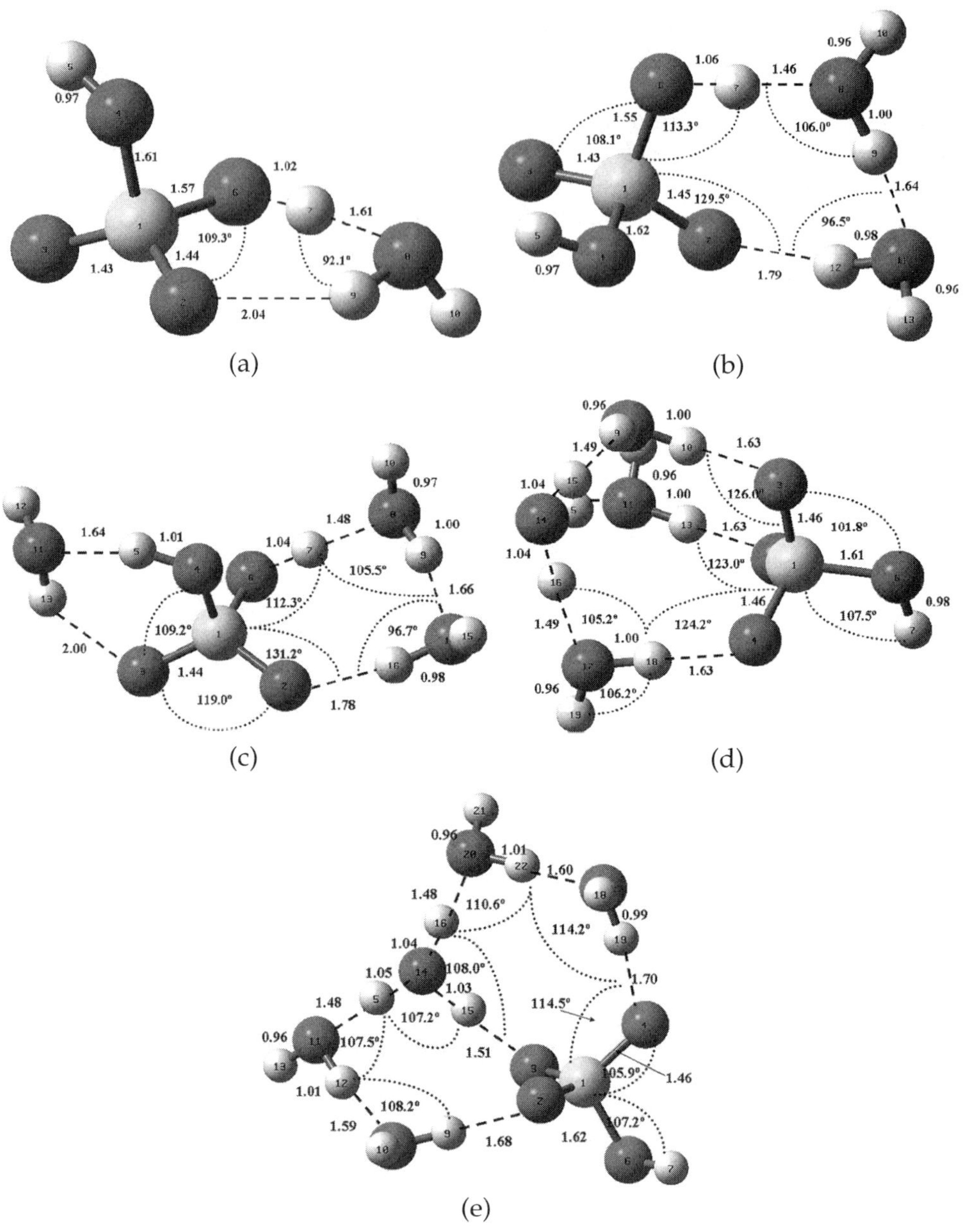

FIGURE 21.1 Most stable isomers of (a) $H_2SO_4(H_2O)$; (b) $H_2SO_4(H_2O)_2$; (c) $H_2SO_4(H_2O)_3$; (d) $H_2SO_4(H_2O)_4$; (e) $H_2SO_4(H_2O)_5$ calculated at PW91PW91/6-311++G(3df,3pd) level of theory.

harmonic approximations. Rigorous theoretical treatment of anharmonic effects is important indeed. However, drawing a meaningful conclusion about the fidelity of anharmonic approximations is impossible without referring to a spe-

TABLE 21.2 Experimental and theoretical frequencies of $(H_2O)_2$ (cm^{-1})

	VPT2/MP2/ a-vtz[a]	MP2(full)/ a-vqz[b]	VSCF MP2/TZP[c]	CC-VSCF/ MP2/TZP[c]	Exp.[c]	Exp.[b]
1	3753	3776	3763	3724	3745	3745
2	3722	3769	3733	3745	3730	3735
3	3615	3671	3768	3647	3600	3660
4	3554	3597	3560	3565	3530	3601
5	1592	1600	1612	1605	1601	1616
6	1580	1590	1565	1564	1599	1599
7	502	517	769	732		523
8	**310**	**308**	**550**	**521**		**311**
9	**138**	**144**	**451**	**409**		**143**
10	**114**	**108**	**259**	**147**	**150**	**108**
11	**113**	**113**	**414**	**309**		**103**
12	**60**	**65**	**545**	**419**		**>88**

a VPT2 calculations at MP2/aug-cc-pvtz level [108].

b Anharmonic calculations at MP2/aug-cc-pvqz and compilation of experimental data [107].

c VSCF and CC-VSCF anharmonic calculations at MP2/TZP level and compilation of experimental data [109].

cific electronic structure method and comparing results produced by this method with reliable spectroscopic experiments. Due to excessively large computational requirements imposed by anharmonic calculations, very little relevant computational information is available at the present time. Another important detail is that the deviation in model predictions obtained using different anharmonic corrections is excessively large. In the case of water dimer, which is the simplest hydrogen-bonded complex, the deviation in low-laying vibrational frequencies predicted using different anharmonic approximations exceeds [65] 400 cm^{-1} or 500% (see also Table 21.2). In addition, there is strong method and considerable basis set dependences of the *ab initio* anharmonic results whose costs are prohibitively high, when large basis sets are applied [107]. Considerable uncertainties in the few available experimental low-laying frequencies responsible for the vibrational contribution to the Gibbs free energies of aqueous clusters, sensitivity of the experimental spectra to the matrix and absence of a complete list of gas-phase frequencies for simplest hydrogen bonded systems such as e.g. water dimer further complicate this issue [107].

The accuracy of the commonly used RRHOA or, in plain words, Harmonic Approximation (HA) has been studied in detail in the past. Scaling factors for both fundamentals (weighted to high and low ends of the spectra) and thermodynamic properties (enthalpy, entropy, zero-point energies (ZPE)), which allow achieving good agreement between theory and experiments, have been proposed by Scott and Radom [119]. The analysis of the sensitivity of the stepwise Gibbs free energy changes to the scaling factor shows that in order to reach a mean average uncertainty of 1 kcal mol^{-1}, which is well below the typical experimental uncertainty,

TABLE 21.3 Experimental and theoretical frequencies of (H_2SO_4) (cm^{-1})

	CC-VSCF MP2/TZP[a]	PW91/ TZP[b]	PW91/ a-vtz[c]	PW91/ (3df,3pd)[c]	Exp. 1[a]	Exp. 2[b]
1	3590	3632	3646	3663	3609	3567
2	3500	3627	3642	3659		3563
3	1434	1426	1387	1423	1465	1452
4	1182	1160	1141	1176	1220	1216
5	1128	1153	1125	1126	1157	1157
6	1118	1138	1114	1110	1138	1136
7	821	801	793	816	891	882
8	779	745	737	767	834	831
9	525	513	500	521	568	558
10	519	499	490	508	550	548
11	480	463	450	467		506
12	489	413	401	415	422	422
13	379	341	338	351	281	379
14	459	321	300	320		288
15	380	221	221	238	215	224

a Anharmonic CC-VSCF MP2/TZP calculations and compilation of experimental data by Miller *et al.* [111].

b Harmonic calculations and compilation of experimental data by Al Natsheh *et al.* [95].

c Nadykto *et al.* [104].

the value of the scaling factor should be either <~0.65 or >~1.5. However, none of the existing quantum chemical methods in RRHOA is capable of giving such gross predictions of the vibrational spectra [119]. It is important to note that harmonic DFT frequencies are typically closer to experiments than harmonic *ab initio* frequencies. The scaling factors for some density functionals (e.g. BLYP, PW91PW91) in the harmonic approximation are very close to unity [104,105,110,119] that can be used as a valuable argument in favor of the application of low-cost harmonic approximation for DFT calculations of free energies.

The comparison of the theoretical frequencies obtained using the DFT in the Rigid Rotor Harmonic Approximation (RRHOA) with the best anharmonic *ab initio* predictions for (H_2SO_4) (Table 21.3) and $(H_2SO_4)(H_2O)$ (Table 21.4) shows that PW91PW91 functional provides an adequate description of the vibrational spectra in good agreement with both experiments and higher-level theory. The more detailed benchmarks for these and other molecules/complexes/clusters composed of atmospheric precursors can be found in studies [104,105,110].

Kathmann *et al.* [89,90] claim based on classical MC simulations with TIP4P potential never intended to treat small clusters [64] that the effect of the anharmonicity on free energies of small clusters is large. This claim contradicts the results of quantum-mechanical DFT and *ab initio* studies [65,98,108,115].

Tables 21.5, 21.6 and 21.7 summarize the thermochemical data for (H_2SO_4)-$(H_2O)_n$ and $(H_2SO_4)_2(H_2O)_n$ clusters. As may be seen from Table 21.5, the affinity

TABLE 21.4 Experimental and theoretical frequencies of $(H_2SO_4)(H_2O)$ (cm^{-1})

	CC-VSCF MP2/TZP[a]	MP2/ TZP[a]	PW91/ TZP[b]	PW91/ a-vtz[c]	PW91/ (3d,3pd)[c]	Exp.[a]
1	3738	3986	3730	3757	3770	3745
2	3593	3841	3638	3650	3668	3640
3	3628	3822	3508	3562	3576	3573
5	1573	1619	1607	1593	1594	1600
6	1436	1468	1464	1437	1444	1500
8	1172	1187	1140	1128	1155	1205
10	868	884	917	882	893	887
11	834	820	856	844	871	834
19	529	349	349	337	349	554

a Anharmonic CC-VSCF MP2/TZP calculations, harmonic MP2/TZP calculations and compilation of experimental data Miller *et al.* [111].
b Harmonic calculations Al Natsheh *et al.* [95].
c Nadykto *et al.* [104].

of water to $(H_2SO_4)(H_2O)_n$ cluster given by DFT at PW91PW91/6-311++G(3df, 3pd) level varies from -1.2 to -3.4 kcal mol^{-1}. While both PW91PW91 studies are in a reasonable agreement with diffusion experiments [117], B3LYP6-311++G(2d, 2p) deviates from the experimental data by 2–3 kcal mol^{-1}. As seen from Table 21.6, B3LYP/6-311++G(2d,2p) [94] consistently underestimates the stability of affinity of water to $(H_2SO_4)_2(H_2O)_n$ by ~2–4 kcal mol^{-1} per step compared to PW91PW91/DNP. The comparison of hydration free energies for sulfuric acid monomer and dimer shows that the conversion of sulfuric acid monomer into a dimer yields a moderate enhancement in the hydration free energies. In contrast to the relatively weak dependence of H_2O affinity on the number of water molecules in the cluster, the Gibbs free energy associated with the attachment of H_2SO_4 (see Table 21.7) depends strongly on the water content.

5.2 Neutral ternary H_2SO_4–H_2O–NH_3 clusters

The structure and thermochemical properties of H_2SO_4–H_2O–NH_3 clusters have been studied theoretically and reported in number of publications [31,93,99]. However, neither structural nor spectroscopic experimental data are available at the present time. Optimized geometries of the most stable isomers of $(NH_3)(H_2SO_4)$-$(H_2O)_n$ (n = 1–5) are presented in Figure 21.2. The thermochemical data for $(NH_3)(H_2SO_4)(H_2O)_n$ and $(NH_3)(H_2SO_4)_2(H_2O)_n$ are presented in Figures 21.3 and 21.4.

Nadykto and Yu [31,113] and, independently, Kurten *et al.* [99] have reached the following conclusions about the thermochemistry of ternary clusters:

TABLE 21.5 Enthalpies, entropies, and Gibbs free energy of hydration calculated at $T = 298.15$ K and $P = 101.3$ kPa. Abbreviations PW91/(3df,3pd), PW91/DNP, B3LYP/(2d,2p), and Exp. refer to Nadykto and Yu [31], Ding *et al.* [106], Bandy and Ianni [92], and Hanson and Eisile [117], respectively

Reaction	PW91/(3df,3pd)			PW91/DNP	B3LYP/(2d,2p)	Exp.
	ΔH	ΔS	ΔG	ΔG	ΔG	ΔG
$H_2SO_4 + H_2O \Leftrightarrow (H_2SO_4)(H_2O)$	−11.76	−31.80	−2.28	−2.5	−0.6	−3.6 ± 1
$(H_2SO_4)(H_2O) + H_2O \Leftrightarrow (H_2SO_4)(H_2O)_2$	−12.57	−32.08	−3.00	−1.8	−0.1	2.6 ± 0.3
$(H_2SO_4)(H_2O)_2 + H_2O \Leftrightarrow (H_2SO_4)(H_2O)_3$	−11.34	−31.71	−1.89	−2.2	−0.5	
$(H_2SO_4)(H_2O)_3 + H_2O \Leftrightarrow (H_2SO_4)(H_2O)_4$	−14.49	−37.15	−3.42	−3.7		
$(H_2SO_4)(H_2O)_4 + H_2O \Leftrightarrow (H_2SO_4)(H_2O)_5$	−9.99	−29.42	−1.22	−2.4		

TABLE 21.6 Enthalpies, entropies, and Gibbs free energy of hydration for sulfuric acid dimer calculated at $T = 298.15$ K and $P = 101.3$ kPa. Abbreviations PW91/(3df,3pd), PW91/DNP, B3LYP refer to the present study, Ding *et al.* [106] and Ianni and Bandy [94], respectively

Reaction	PW91/6-311++G(3df,3pd)			PW91/DNP	B3LYP
	ΔH	ΔS	ΔG	ΔG	ΔG
$(H_2SO_4)_2 + H_2O \Leftrightarrow (H_2SO_4)_2(H_2O)$	−14.44	−36.99	−3.41	−2.6	0.2
$(H_2SO_4)_2(H_2O) + H_2O \Leftrightarrow (H_2SO_4)_2(H_2O)_2$	−12.85	−33.12	−2.98	−2.6	−0.3
$(H_2SO_4)_2(H_2O)_2 + H_2O \Leftrightarrow (H_2SO_4)_2(H_2O)_3$	−14.56	−37.59	−3.36	−4.2	−0.2
$(H_2SO_4)_2(H_2O)_3 + H_2O \Leftrightarrow (H_2SO_4)_2(H_2O)_4$	−14.81	−36.95	−3.80	−5.1	−0.5
$(H_2SO_4)_2(H_2O)_4 + H_2O \Leftrightarrow (H_2SO_4)_2(H_2O)_5$	−11.95	−32.00	−2.41	−4.3	−0.3

TABLE 21.7 Enthalpies, entropies, and Gibbs free energy changes of associated with formation of $(H_2SO_4)_2(H_2O)_n$ by attachment of sulfuric acid. $T = 298.15$ K and $P = 101.3$ kPa. Abbreviations PW91/(3df,3pd), PW91/DNP and B3LYP refer to the present study, Ding *et al.* [106] and Ianni and Bandy [94], respectively

Reaction	PW91/(3df,3pd)			PW91/DNP	B3LYP
	ΔH	ΔS	ΔG	ΔG	ΔG
$H_2SO_4 + H_2SO_4 \Leftrightarrow (H_2SO_4)_2$	−16.16	−35.46	−5.59	−3.1	−2.5
$(H_2SO_4)(H_2O) + (H_2SO_4) \Leftrightarrow (H_2SO_4)_2(H_2O)$	−18.84	−40.65	−6.72	−3.3	−1.7
$(H_2SO_4)(H_2O)_2 + (H_2SO_4) \Leftrightarrow (H_2SO_4)_2(H_2O)_2$	−19.12	−41.69	−6.69	−4.0	−1.9
$(H_2SO_4)(H_2O)_3 + (H_2SO_4) \Leftrightarrow (H_2SO_4)_2(H_2O)_3$	−22.34	−47.57	−8.16	−6.6	−1.2
$(H_2SO_4)(H_2O)_4 + (H_2SO_4) \Leftrightarrow (H_2SO_4)_2(H_2O)_4$	−22.66	−47.37	−8.54	−8.0	−2.0
$(H_2SO_4)(H_2O)_5 + (H_2SO_4) \Leftrightarrow (H_2SO_4)_2(H_2O)_5$	−24.62	−49.96	−9.72	−9.9	−5.1

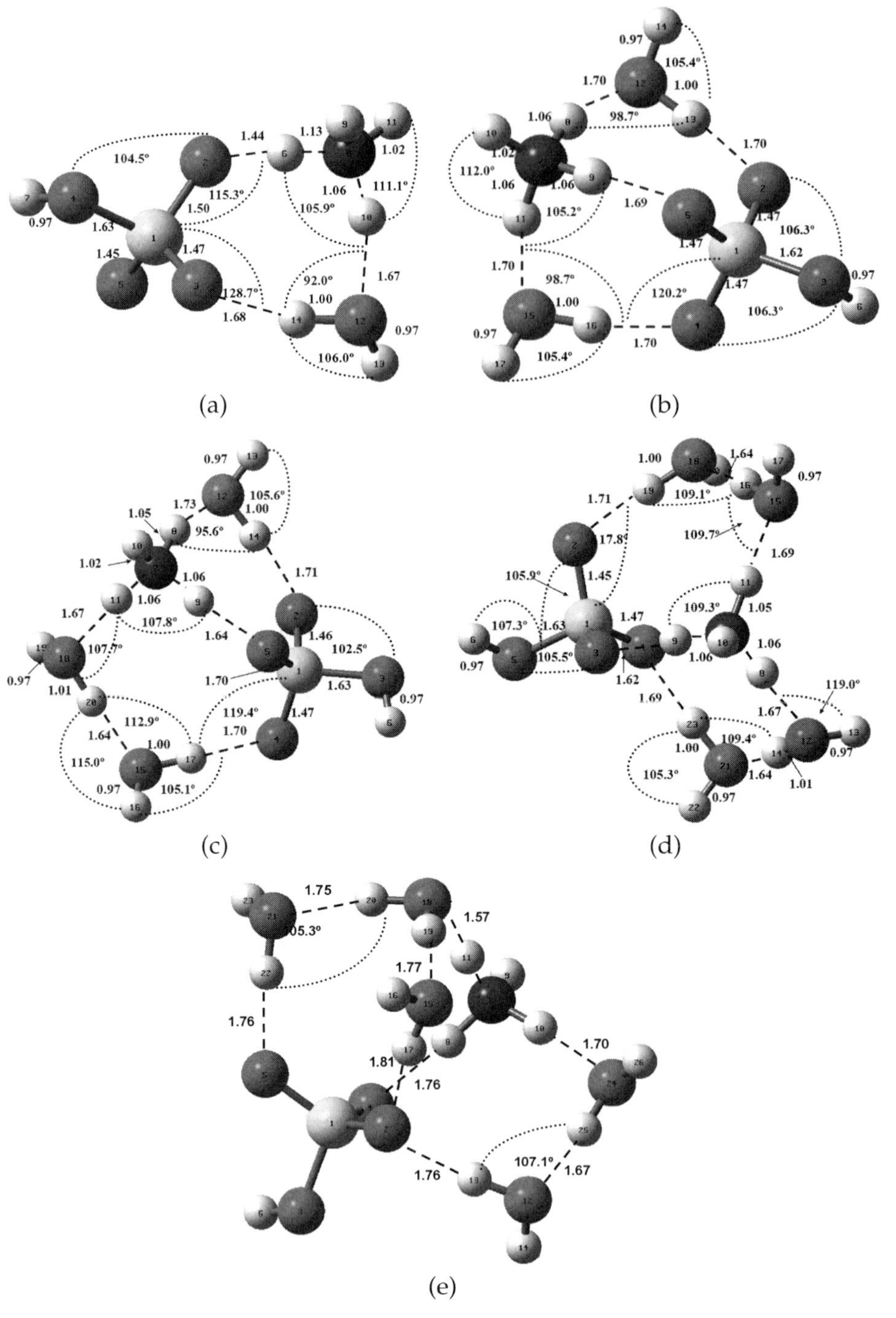

FIGURE 21.2 Most stable isomers of (a) $(NH_3)(H_2SO_4)(H_2O)$; (b) $(NH_3)(H_2SO_4)(H_2O)_2$; (c) $(NH_3)(H_2SO_4)(H_2O)_3$; (d) $(NH_3)(H_2SO_4)(H_2O)_4$; (e) $(NH_3)(H_2SO_4)(H_2O)_5$ calculated at PW91PW91/6-311++G(3df,3pd) level of theory.

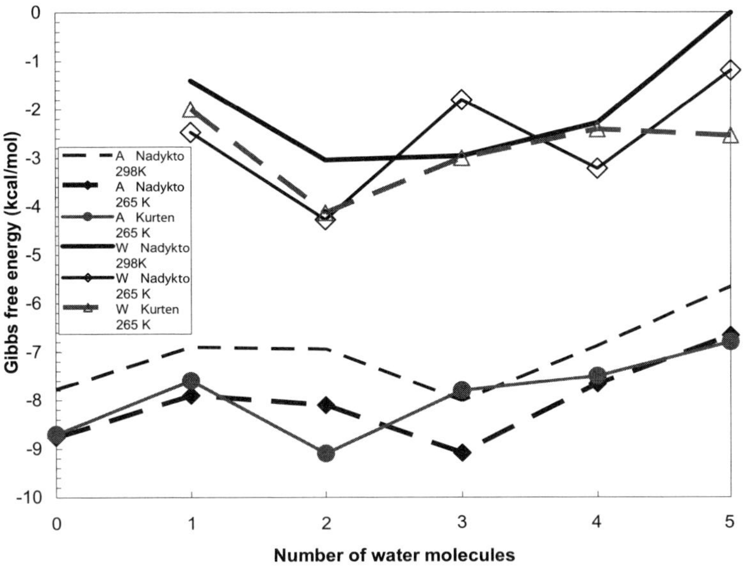

FIGURE 21.3 Gibbs free energy changes associated with the formation of $(H_2SO_4)_2(H_2O)_n$-(NH_3) by attachment of H_2O (W) and ammonia (A). Abbreviations Nadykto and Kurten refer to PW91/6-311++G(3df,3pd) and PW91/DNP studies by Nadykto and Yu [31,113] and Kurten *et al.* [99], respectively. Gibbs free energy changes were calculated at $P = 101.3$ kPa.

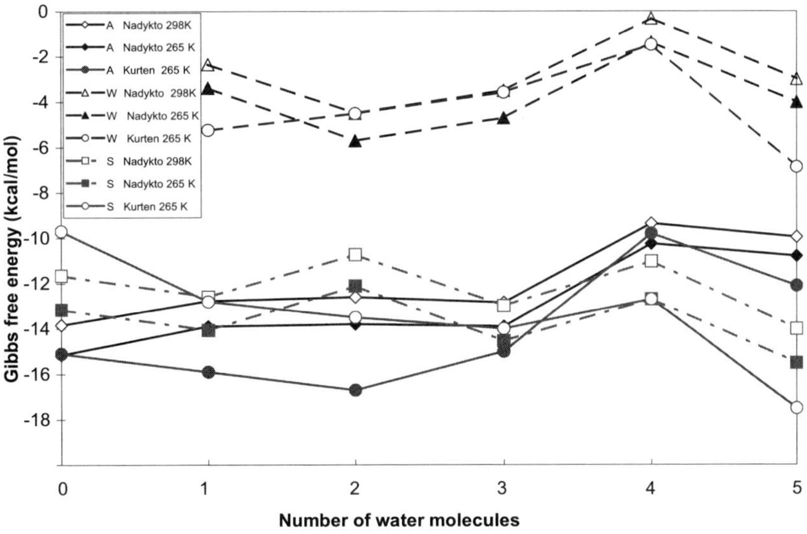

FIGURE 21.4 Gibbs free energy changes associated with the formation of $(H_2SO_4)_2(NH_3)$-$(H_2O)_n$ by addition of water (W), sulfuric acid (SA) and ammonia (A). Abbreviations Nadykto and Kurten refer to PW91/6-311++G(3df,3pd) and PW91/DNP studies by Nadykto and Yu [31,113] and Kurten *et al.* [99], respectively. Gibbs free energy changes were calculated at $P = 101.3$ kPa. $T = 298.15$ K unless specified.

(a) The stabilizing effect of ammonia on the formation of small H_2SO_4–H_2O clusters is likely to increase with the number of sulfuric acid molecules in the cluster.
(b) Thermochemistry of H_2SO_4 and NH_3 in small H_2SO_4–H_2O–NH_3 clusters is either virtually independent of or depends weakly on the water content. This suggests that the stabilizing effect of ammonia at initial steps of the cluster growth is associated mainly with the sulfuric acid.

5.3 Stabilization of H_2SO_4–H_2O complexes by low-molecular organic acids

Ammonia has been considered as an exclusive principle stabilizer of H_2SO_4–H_2O clusters in the atmosphere since 1990s. However, recent study of Nadykto and Yu [31] have shown that common low molecular organic acid such as formic and acetic acids can also efficiently stabilize H_2SO_4–H_2O clusters. Interaction of formic acid and acetic acids, whose concentrations in the atmosphere are as high or higher than ammonia concentrations, with sulfuric acid and water leads to the formation of stable hydrogen-bonded complexes, whose stability is close to the stability of complexes of sulfuric acid with ammonia. In addition to the formation of strong hydrogen-bonded complexes with sulfuric acid and water, formic and acetic acids form quite thermodynamically stable complexes with ammonia. This indicates that ammonia is certainly not an exclusive stabilizer of H_2SO_4–H_2O clusters in the atmosphere and that the involvement of common organic species, alongside with or without ammonia, in clustering and subsequent nucleation of sulfuric acid and water should be studied in detail further.

5.4 Neutral gas-phase hydrates of HNO_3 in the stratosphere

Theoretical studies of the coexistence of gas-phase hydrates of HNO_3 under the stratospheric conditions have been carried out by Al Natsheh with co-authors [96,97]. The family of nitric acid trihydrates (NAT), which dominate the composition of the Ia PSCs, are considered to be thermodynamically stable. However,

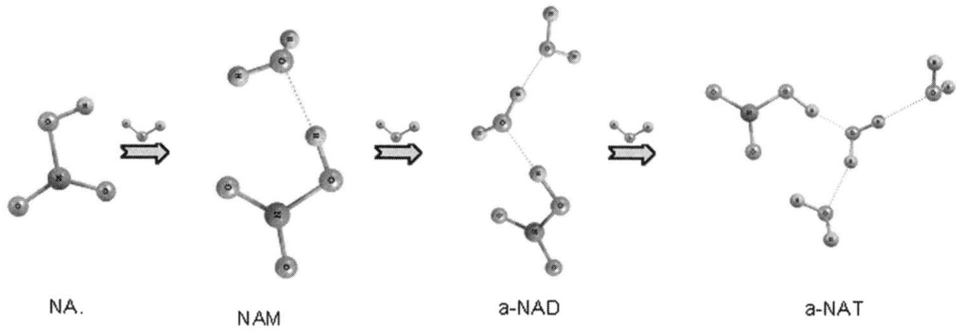

FIGURE 21.5 HNO_3–NAM–NAD–NAT transformations.

their formation mechanisms and thermodynamic properties were poorly understood, and the formation pathways for different conformer of NAD and NAT, as well as the NAD–NAT transformations, remained puzzling (see Figure 21.5). Arnold [120] suggests that physical chemistry of stratospheric clouds may involve a formation of metastable nitric acid hydrates. According to Arnold, such hydrates are less stable than the NAT and they differ from NAT in both catalytic and optical properties and HCl solubility. A clear understanding of the role of

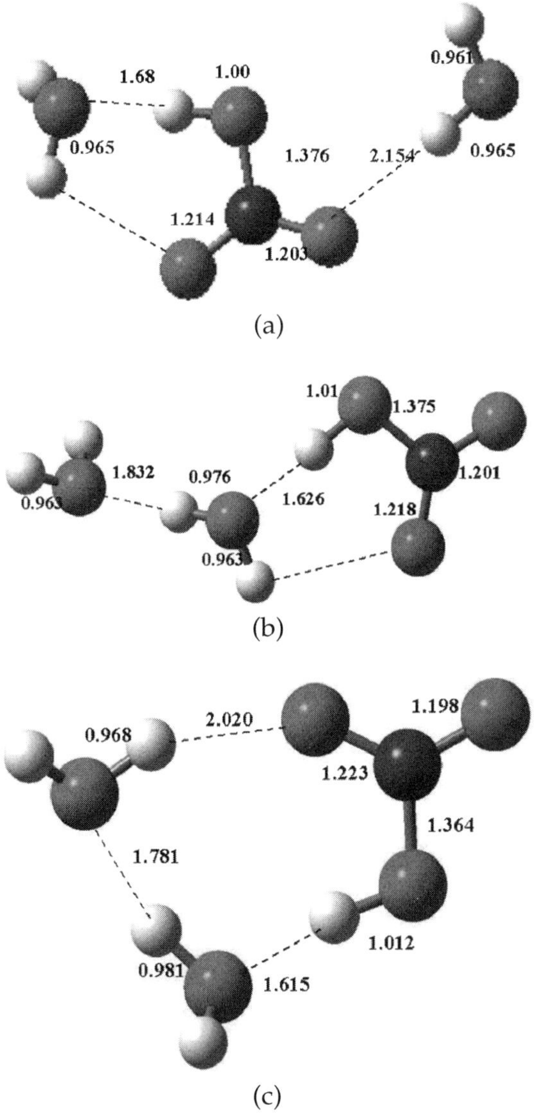

FIGURE 21.6 Equilibrium conformations of NAD: (a) NAD-a (β), (b) NAD-b (α), (c) NAD-c (γ).

the metastable NAD in the stratosphere is important because NAD particles grow much faster than NAT crystals. Another important detail is that difference in the chemical and physical properties of NAD and NAT can affect heterogeneous surface chemistry of PSCs, especially the transformation of chlorine reservoir species, such as $ClONO_2$ and HCl, into active chlorine forms that readily photolyze in sunlight.

NAD has been investigated in a number of experiments. In the past, it was assumed that there exists one isomer of NAD only. However, recent spectroscopic studies have hinted at the presence of a second NAD conformation. In the recent study quantum-mechnical study [96], three equilibrium structures of the gas-phase NAD, including two new structures not found in the earlier theoretical earlier, have been identified (see Figure 21.6).

Al Natsheh *et al.* [96,97] found that although the formation of the gas-phase β-NAD is thermodynamically favorable under stratospheric conditions, the difference in the formation free energy between β-NAD and α-NAD is small. Therefore, both forms are expected to be abundant. The comparison of the thermodynamic data for α- and β-NAD and α- and β-NAT suggests that the formation of NAT from NAD is thermodynamically favorable only if the α-NAD $\Rightarrow$ α-NAT transformation occurs. The formation of NAT from the nitric acid and NAM is more favorable thermodynamically than the NAD $\Rightarrow$ NAT transformation; however, the direct nitric acid $\Rightarrow$ NAD, nitric acid $\Rightarrow$ NAT, and NAM $\Rightarrow$ NAT processes under stratospheric conditions are likely kinetically limited. The small differences in the formation free energies of NAM, α-NAD, β-NAD, α-NAT and β-NAT suggest that hydrated nitric acid in the stratosphere is a mixture composed mainly of NAM, α-NAD, β-NAD, α-NAT and β-NAT [96]. On the other hand, the availability of water and strongly hygroscopic nature of nitric acid may drive the acquisition of more water molecules into clusters that reduces the NAM fraction. NAM grows into α-NAD and β-NAD, which transforms into α-NAT and β-NAT, respectively. These considerations suggest that nitric acid crystals under stratospheric conditions are formed mainly from the gas-phase β-NAD and β-NAT. Further research is needed to determine the extend to which the structure of gas-phase hydrates of nitric acid changes during the crystals formation [121,122].

5.5 Relevance of dipolar properties of neutral monomers and pre-nucleation clusters to nucleation of ions

One of the major nucleation mechanisms in the Earth's atmosphere, nucleation on ions is controlled by the dipole moment of condensable monomers and pre-existing clusters [36] whose involvement in strong short-ranged dipole–charge interaction with airborne ions increases nucleation rates by many orders of magnitude. The high polarity of H_2SO_4 hydrates was predicted in our quantum-mechanical study [95]. Two years later, this finding has received the experimental conformation in the work of Leopold with co-authors [69], who have reported dipole moment of 3.05 Debyes for the most stable isomer of hydrogen-bonded $(H_2SO_4)(H_2O)$ at $T = 3$ K. A comparison of theoretical and experimental dipole moments of the sulfuric acid and sulfuric acid hydrates is presented in Table 21.8.

TABLE 21.8 Dipole moment (Debyes) of most stable isomers of $(H_2SO_4)(H_2O)_n$ as a function of the hydration number n

n	PW91/6-311++G(3df,3pd)	PW91/TZP [95]	MP2/a-vtz [69]	Exp. [69]
0	3.09	2.817	3.42	2.73
1	2.51	2.14	2.97	3.02
2	3.35	2.42		
3	2.86	3.84		
4	2.59			
5	3.09			

As may be seen from Table 21.8, the sulfuric acid hydrates are highly polar that implies [34–38] their active involvement in the strong dipole–charge interaction with atmospheric ions and substantial enhancement in nucleation rates due to the dipole–charge interaction. Further work is needed in order to quantify the effect of other atmospheric molecules and molecular clusters on ion nucleation rates.

5.6 Ionic $H_3O^+(H_2O)_n$ clusters

The importance of the ambient ionization on the cluster formation is well established [34–38,123–127]. The thermochemistry of the hydration of simple and common atmospheric ion H_3O^+ has been studied using the computational quantum methods in the past. Table 21.9 presents the comparison of calculated hydration

TABLE 21.9 Comparison of hydration enthalpies for $H_3O^+(H_2O)_{n-1} + H_2O \rightarrow H_3O^+(H_2O)_n$ reaction calculated using DFT PW91PW91/6-311++G(3df,3pd) in Nadykto *et al.* [43] with available *ab initio* studies (MP2/cc-pVTZ (MP2), MCCM-UT-CCSD//MP2/cc-pVTZ (MCCM)) [128] and experimental data of Froyd and Lovejoy [125]. The entropies obtained using the DFT are also compared with experimental values (entropies were not reported in the *ab initio* study [128]). The data are given at ambient temperature of 298.15 K and ambient pressure of 101.3 kPa

n	ΔH				ΔS	
	DFT	MP2	MCCM	Exp.	DFT	Exp.
1	−37.87	−37.21	−35.3		−30.05	
2	−22.75	−22.98	−21.72	−20.9	−27.35	−24.1
3	−18.12	−19.35	−18.24	−18	−26.05	−27.9
4	−13.65	−13.71	−12.66	−12.8	−27.22	−23.8
5	−12.07	−12.4	−11.42	−12.2	−26.53	−26.9
6	−11.16			−10.6	−25.72	−25.1
7	−9.78			−10	−24.58	−25.8

enthalpies and entropies for $H_3O^+(H_2O)_{n-1} + H_2O \rightarrow H_3O^+(H_2O)_n$ reaction with the recent experiments [125].

As may be seen from Table 21.9, the computational quantum methods (both *ab initio* and DFT) agree well with each other and recent experiments of Froyd and Lovejoy [125]. Moreover, they were able to predict the abrupt drop in the enthalpy change at $n = 4$ related to the formation of the second hydration/solvation shell in good agreement with recent experiments [125].

5.7 Ionic $HSO_4^-(H_2O)_n$ clusters

The hydration of hydrogensulfate ion HSO_4^- has been studied using MP2/aug-cc-pv(T + d)z [100] and DFT PW91PW91/6-311++G(3df,3pd) [114] methods. Figure 21.7 presents the comparison of the theoretical Gibbs free energies with the experimental data [126]. The comparison shows that both methods agree well with the experimental data and each other. Although the individual stepwise Gibbs free energies predicted by MP2/aug-cc-pv(T+d)z and presented in work [100] without the counterpoise correction, and DFT PW91PW91/6-311++G(3df,3pd) vary, the cumulative Gibbs free energy change associated with the 4-mer formation differ by <0.1 kcal mol^{-1}. The difference in the cumulative Gibbs free energy change associated with the pentamer formation between DFT PW91PW91/6-311++G(3df,3pd) and experimental data is <0.7 kcal mol^{-1} that is a clear indication of the reasonable performance of the aforementioned DFT method.

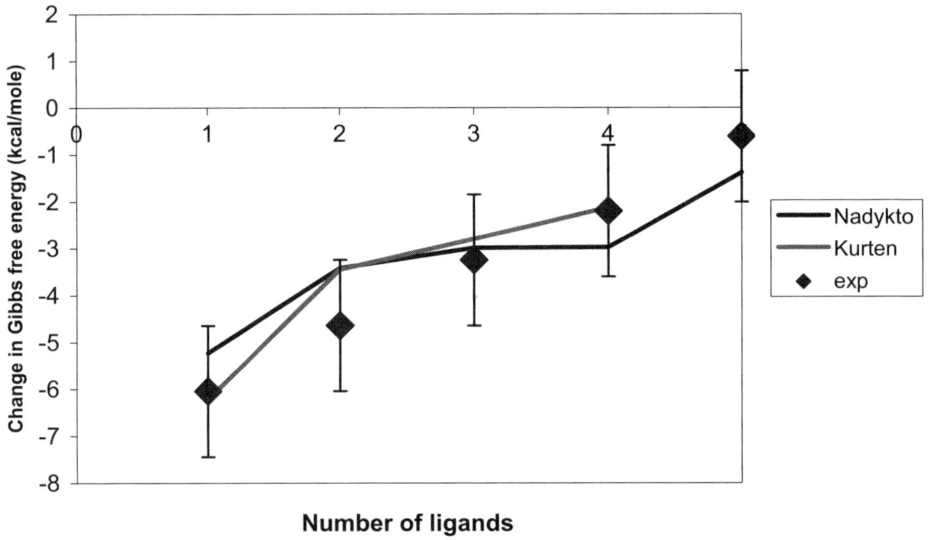

FIGURE 21.7 The hydration free energies for hydrogensulfate ion HSO_4^- as a function of the hydration number.

5.8 From CNT to "first principle" nucleation theory

Nadykto *et al.* [43,65] and more recently Merikanto *et al.* [129] have suggested that the application of quantum correction to the thermodynamics of aqueous systems could be a viable alternative to existing molecular models, and that a physically sound correction could be obtained by studying only the small clusters ($i < \sim 10$). This suggestion is based on a quite fast convergence of the stepwise Gibbs free energy changes associated with the hydration to the bulk value [43] and experimental observations showing that the CNT begin to fail at $i < \sim 10$. In our recent work [53] we applied a quantum correction to the classical thermochemistry for small water clusters ($(H_2O)_i$ $i = 1$–10) of the CNT. Recently published G3 data for $i = 2$–6 and more affordable, yet accurate G3MP2, for the larger i up to 10, have been used. The test results show that G3 and G3MP2 predictions for $i = 2$–6 agree within ~ 0.3 kcal mol^{-1}. Over 30 equilibrium conformers of 6, 7, 8, 9 and 10-mers have been identified.

The comparison of curves in Figures 21.8 and 21.9 shows that the kinetically consistent classical nucleation theory corrected using quantum data successfully predicts nucleation rates in nearly all the cases studied here. The accuracy of its model predictions is higher than that of the (self-consistency corrected) SCC CNT and the best molecular-based nucleation studies, and stays in line with the best up-to-date empirical scaling theories. As seen from the comparison of curves rep-

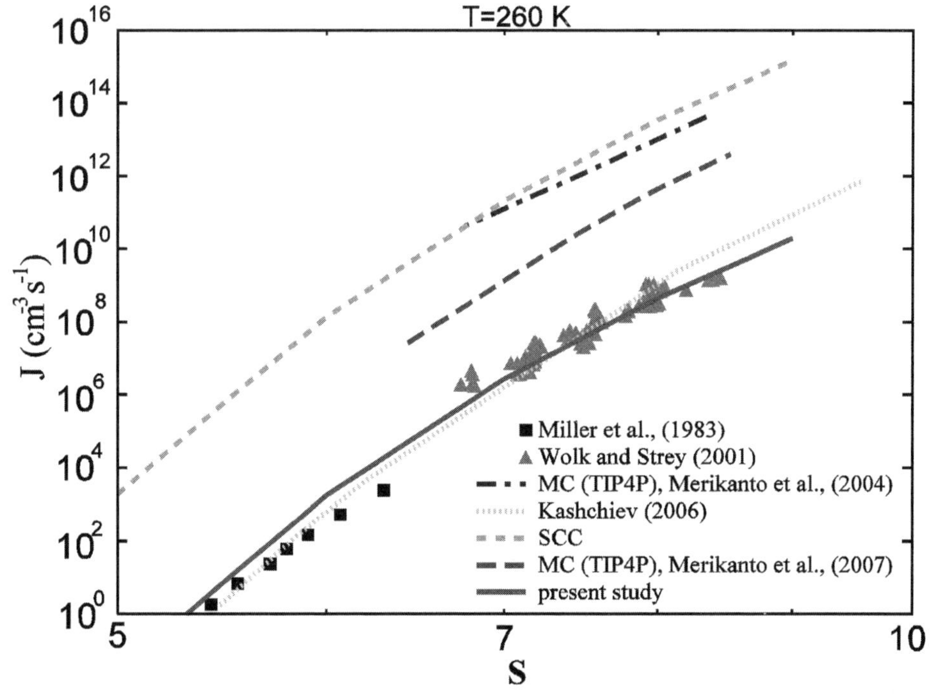

FIGURE 21.8 Comparison of nucleation rates obtained using the CNT constrained by quantum calculations with experiments and other theoretical studies (from Du *et al.* [53]).

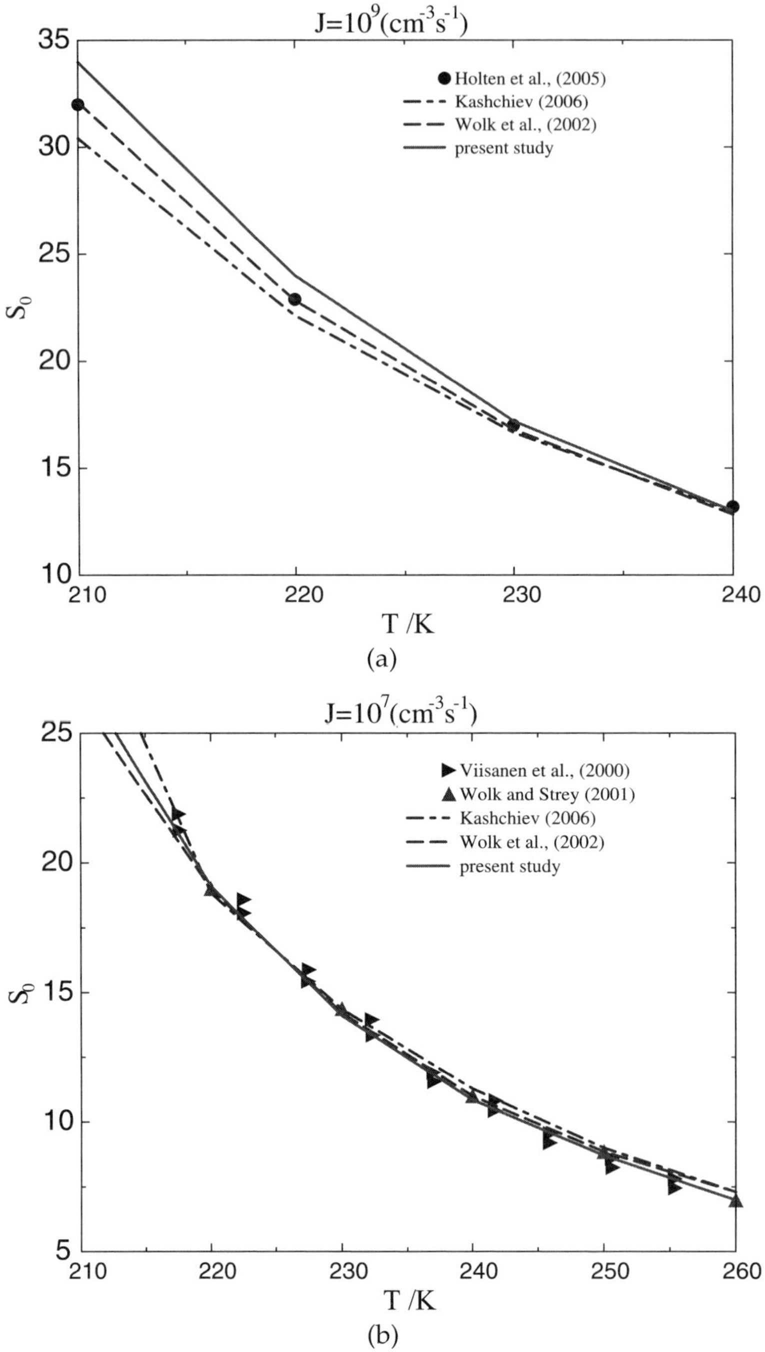

FIGURE 21.9 Comparison of predicted onset saturation ratios with experiments and the best up-to-date empirical scaling theories for (a) $J = 10^9$ cm^{-3} s^{-1}; (b) $J = 10^7$ cm^{-3} s^{-1} (from Du *et al.* [53]).

resenting data of Merikanto *et al.* [129,130], Merikanto with co-workers [129] have undertaken an impressive progress in their Monte-Carlo TIP4P simulations. However, a scaling factor of 10^2–10^3 is still needed in order to achieve the agreement between their model and experiments. The possible reason for this deviation is an insufficiently accurate description of the initial cluster growth steps due to the application of the TIP4P model parameterized to the bulk liquid.

6. CONCLUDING REMARKS

Although the theory of atmospheric nucleation has been developing for several decades, a clear and insight understanding of the molecular nature of nucleation phenomena and mechanisms controlling the burst of new particles in the atmosphere is yet to be achieved. The further progress in the atmospheric nucleation theory and reduction of uncertainties in nucleation rates is impossible without a rigorous theoretical treatment of nucleating and pre-nucleation clusters. Those clusters are quantum-mechanical systems and must be treated using quantum methods. Although a claim that a completely quantum theory of atmospheric nucleation can be developed within couple of years would be an exaggeration, the quick growth of the computational power and impressive progress achieved in quantum theory since the Schrödinger's original work leave us a hope that it may happen soon. Numerous benefits of the application of the quantum theory to the multicomponent atmospheric systems are obvious. Quantum theory is applicable to clusters of any chemical composition and treats them explicitly using their chemical identities without referring to properties of assumable/hypothetical bulk liquid. The quantum calculations can readily be utilized directly to develop new hydration/clustering models, to constrain existing kinetic nucleation models and to investigate the effect of trace species on nucleation processes. Thermochemical data obtained using computational quantum methods can be conveniently incorporated in the framework of kinetic nucleation models that can explicitly simulate the observed nucleation events.

ACKNOWLEDGEMENTS

The support of this study by NSF under grant by 0618124 and the NOAA/DOC under grant NA05OAR4310103 is gratefully acknowledged. KVM thanks The Danish Research Councils and The Danish Center for Scientific Computing for the support.

REFERENCES

[1] S. Twomey, *J. Atmos. Sci.* **34** (1977) 1149.
[2] R.J. Charlson, *et al.*, *Science* **255** (1992) 423.
[3] M. Kulmala, *et al.*, *J. Aerosol Sci.* **35** (2004) 143.
[4] J.H. Seinfeld, S.N. Pandis, *Atmospheric Chemistry and Physics: From Air Pollution to Climate Change*, Wiley–Interscience, New York, 1998.

[5] D.B. Kittelson, W.F. Watts, J.P. Johnson, *Atmos. Environ.* **38** (2004) 9.
[6] Y. Zhu, W.C. Hinds, *Atmos. Environ.* **39** (2005) 1557.
[7] D. Westerdahl, *et al.*, *Atmos. Environ.* **39** (2005) 3597.
[8] A. Peters, *et al.*, *Am. J. Respir. Crit. Care Med.* **155** (1997) 1376.
[9] P. Penttinen, *Eur. Respir. J.* **17** (2001) 428.
[10] G. Oberdorster, M. Utell, *Environ. Health Perspect.* **110** (2002) 440.
[11] D. Saxon, D. Diaz-Sanchez, *Nature Immunol.* **6** (3) (2005) 223.
[12] M. Kulmala, *Science* **302** (2003) 1000.
[13] Y.J. Kim, *et al.*, *J. Phys. Chem. A* **108** (2004) 4365.
[14] V. Holten, D.G. Labetski, M.E.H. van Dongen, *J. Chem. Phys.* **123** (2005) 104505.
[15] Y. Viisanen, R. Strey, H. Reiss, *J. Chem. Phys.* **112** (2000) 8205.
[16] H. Vehkamäki, *et al.*, *J. Geophys. Res.* **107** (D22) (2002) 4622, doi:10.1029/2002JD002184.
[17] M. Kulmala, A. Laaksonen, L. Pirjola, *J. Geophys. Res.* **103** (1998) 8301.
[18] M. Noppel, H. Vehkamaki, M. Kulmala, *J. Chem. Phys.* **116** (2002) 218.
[19] F. Yu, *J. Chem. Phys.* **122** (2005) 074501.
[20] F. Yu, *J. Geophys. Res.* **111** (2006) D04201, doi:10.1029/2005JD006358.
[21] T. Berndt, *et al.*, *Science* **307** (2005) 698.
[22] T. Berndt, O. Böge, F. Stratmann, *Geophys. Res. Lett.* **33** (2006) L15817.
[23] D. Benson, L.-H. Young, S.-H. Lee, *Geophys. Res. Lett.* (2008), in press.
[24] F. Yu, *J. Geophys. Res.* **111** (2006) D01204, doi:10.1029/2005JD005968.
[25] I. Napari, M. Noppel, H. Vehkamäki, M. Kulmala, *J. Geophys. Res.* **107** (D19) (2002) 4381, doi:10.1029/2002JD002132.
[26] S.M. Ball, D.R. Hanson, F.L. Eisele, P.H. McMurry, *J. Geophys. Res.* **104** (D19) (1999) 23709, doi:10.1029/1999JD900411.
[27] T.O. Kim, T. Ishida, M. Adachi, K. Okuyama, J.H. Seinfeld, *Aerosol Sci. Technol.* **29** (1998) 112.
[28] P.S. Christensen, S. Wedel, H. Livbjerg, *Chem. Eng. Sci.* **49** (1994) 4605.
[29] G.L. Diamond, J.V. Iribarne, D.J. Corr, *J. Aerosol Sci.* **16** (1985) 43.
[30] P.M. Nolan, Condensation nuclei formation by photooxidation of sulfur dioxide with OH radical in the presence of water vapor and ammonia, Thesis, Drexel University (1987).
[31] A.B. Nadykto, F. Yu, *Chem. Phys. Lett.* **435** (2007) 14.
[32] J.G. Jung, P.J. Adams, S.N. Pandis, *Atmos. Environ.* **40** (2006) 2248.
[33] T. Anttila, H. Vehkamaki, I. Napari, M. Kulmala, *Boreal Environ. Res.* **10** (2005) 511.
[34] A.B. Nadykto, *et al.*, *Chem. Phys. Lett.* **382** (1–2) (2003) 6.
[35] A.B. Nadykto, F. Yu, *J. Geophys. Res.* **108** (D23) (2003) 4717, doi:10.1029/2003JD003664.
[36] A.B. Nadykto, F. Yu, *Phys. Rev. Lett.* **93** (2004) 016101.
[37] A.B. Nadykto, F. Yu, *Atmos. Chem. Phys.* **4** (2004) 385.
[38] A.B. Nadykto, *et al.*, *Aerosol Sci. Technol.* **38** (4) (2004) 349.
[39] F. Yu, *J. Chem. Phys.* **122** (2005) 084503.
[40] F. Yu, R.P. Turco, *Geophys. Res. Lett.* **27** (2000) 883.
[41] F. Yu, R.P. Turco, *J. Geophys. Res.* **106** (2001) 4797.
[42] F. Yu, *Atmos. Chem. Phys.* **6** (12) (2006) 5193.
[43] A.B. Nadykto, *et al.*, *Phys. Rev. Lett.* (2006) 125701.
[44] F. Yu, Z. Wang, G. Luo, R.P. Turco, *Atmos. Chem. Phys. Discuss.* **7** (2007) 13597.
[45] M. Vana, *et al.*, *Atmos. Res.* **82** (2006) 536.
[46] K. Iida, *et al.*, *J. Geophys. Res.* **111** (2006) D23201, doi:10.1029/2006JD007167.
[47] A. Hirsikko, *et al.*, *Atmos. Chem. Phys.* **7** (2007) 201.
[48] L. Laakso, *et al.*, *Atmos. Chem. Phys.* **7** (2007) 1333.
[49] F. Yu, *Atmos. Chem. Phys. Discuss.* **6** (2006) S4727.
[50] F. Yu, R.P. Turco, in: *Proc. 17th Int. Conf. Nucleation & Atmospheric Aerosols*, Galway, Ireland, 13–17 August 2007 (2007).
[51] M. Kulmala, *et al.*, Science Express Report, published online 30 August 2007, DOI:10.1126/science.1144124.
[52] R. Zhang, *et al.*, *Science* **304** (2004) 1487.
[53] H. Du, A.B. Nadykto, F. Yu, *Phys. Rev. Lett.* (2007), under review.
[54] B.C. Garrett, *et al.*, *Chem. Rev.* **105** (1) (2005) 355.

[55] D.W. Oxtoby, *J. Phys. Condens. Matter* **4** (1992) 7627.
[56] M. Volmer, A. Weber, *Z. Phys. Chem.* **119** (1926) 277.
[57] R. Becker, W. Doering, *Ann. Phys.* **24** (1935) 719.
[58] M. Volmer, *Kinetik der Phasenbildung*, Theodor Steinkopff Verlag, Dresden, 1939.
[59] H. Oh, G.T. Gao, X.C. Zeng, *Phys. Rev. Lett.* **86** (2001) 5080.
[60] S.M. Kathmann, G.K. Schenter, B.C. Garrett, *Phys. Rev. Lett.* **94** (2005) 116104.
[61] B.N. Hale, *J. Chem. Phys.* **122** (2005) 204509.
[62] J. Wolk, R. Strey, C.H. Heath, B.E. Wyslouzil, *J. Chem. Phys.* **117** (2002) 4954.
[63] D. Kashchiev, *J. Chem. Phys.* **125** (2006) 044505.
[64] S.M. Kathmann, G.K. Schenter, B.C. Garrett, *J. Chem. Phys.* **116** (2002) 5046.
[65] A.B. Nadykto, *et al.*, *Phys. Rev. Lett.* **98** (10) (2007) 109604.
[66] F. Yu, *J. Chem. Phys.* **127** (5) (2007), art. no. 054301.
[67] S.M. Kathmann, B.N. Hale, *J. Phys. Chem. B* **105** (47) (2001) 11719.
[68] C.-G. Ding, T. Taskila, K. Laasonen, A. Laaksonen, *Chem. Phys.* **287** (1–2) (2003) 7.
[69] C.S. Brauer, G. Sedo, K. Leopold, *Geophys. Res. Lett.* **33** (23) (2006) L23805.
[70] T.U. Helgaker, P. Jorgensen, J. Olsen, *Molecular Electronic—Structure Theory*, Wiley, New York, 2000.
[71] I.N. Levine, *Quantum Chemistry*, Prentice Hall, Englewood Cliffs, NJ, 1999.
[72] T. Kerdcharoen, U. Birkenheuer, S. Krüger, A. Woiterski, N. Rösch, *Theor. Chem. Acc.* **109** (6) (2003) 285.
[73] Y.-K. Choe, E. Tsuchida, T. Ikeshoji, *J. Chem. Phys.* **126** (15) (2007), art. no. 154510.
[74] M. Sloth, *et al.*, *J. Phys. Chem. A* (2008), in press.
[75] K.V. Mikkelsen, *Ann. Rev. Phys. Chem.* **57** (2006) 365.
[76] A. Gross, K.V. Mikkelsen, *Adv. Quantum Chem.* **50** (2005) 125.
[77] M. Sloth, M. Bilde, K.V. Mikkelsen, *Mol. Phys.* **102** (2004) 2361.
[78] M. Sloth, M. Bilde, K.V. Mikkelsen, *J. Chem. Phys.* **118** (2003) 10085.
[79] C.T.R. Wilson, *Philos. Trans. R. Soc. London A* **189** (1897) 265.
[80] L.B. Loeb, A.F. Kip, A.W. Einarsson, *J. Chem. Phys.* **6** (1937) 264.
[81] A.W. Castleman, I.N. Tang, *J. Chem. Phys.* **57** (1972) 3629.
[82] H. Rabeony, P. Mirabel, *J. Phys. Chem.* **91** (7) (1987) 1815.
[83] I. Kusaka, Z. Wang, J.H. Seinfeld, *J. Chem. Phys.* **102** (1995) 913.
[84] I. Kusaka, Z. Wang, J.H. Seinfeld, *J. Chem. Phys.* **103** (1995) 8993.
[85] E.P. Brodskaya, A.P. Lyubartsev, A. Laaksonen, *J. Phys. Chem. B* **106** (2002) 6479.
[86] K.J. Oh, G.T. Gao, X.C. Zeng, *Phys. Rev. Lett.* **86** (2001) 5080.
[87] V.B. Warshavsky, X.C. Zeng, *Phys. Rev. Lett.* **89** (2002) 246104.
[88] G.K. Schenter, S.M. Kathmann, B.C. Garrett, *Phys. Rev. Lett.* **82** (17) (1999) 3484.
[89] S.M. Kathmann, G.K. Schenter, B.C. Garrett, *Phys. Rev. Lett.* **98** (2007) 109603.
[90] S.M. Kathmann, G.K. Schenter, B.C. Garrett, *J. Phys. Chem. C* **111** (13) (2007) 4977.
[91] L. Kurdi, E. Kochanski, *Chem. Phys. Lett.* **158** (1989) 111.
[92] A.R. Bandy, J.C. Ianni, *J. Phys. Chem. A* **102** (32) (1998) 6533.
[93] J.C. Ianni, A.R. Bandy, *J. Phys. Chem. A* **103** (1999) 2801–2811.
[94] J.C. Ianni, A.R. Bandy, *J. Mol. Struct. (Theochem)* **497** (2000) 19.
[95] A. Al Natsheh, *et al.*, *J. Phys. Chem. A* **108** (41) (2004) 8914;
A. Al Natsheh, *et al.*, *J. Phys. Chem. A* **110** (25) (2006) 7982.
[96] A. Al Natsheh, *et al.*, *Chem. Phys.* **324** (2006) 210.
[97] A. Al Natsheh, *et al.*, *Chem. Phys. Lett.* **426** (2006) 20.
[98] T. Kurten, *et al.*, *J. Phys. Chem. A* **110** (2006) 7178.
[99] T. Kurten, *et al.*, *J. Geophys. Res.* **112** (D4) (2006) D04210, doi:10.1029/2006JD007391, (2007a).
[100] T. Kurten, *et al.*, *Boreal Environ. Res.* **12** (3) (2007) 431.
[101] T. Kurten, *et al.*, *Atmos. Chem. Phys.* **7** (10) (2007) 2765.
[102] S. Re, Y. Osamura, K. Morokuma, *J. Phys. Chem. A* **103** (1999) 3535.
[103] C. Arrouvel, V. Viossat, C. Minot, *J. Mol. Struct. (THEOCHEM)* **718** (2005) 71.
[104] A.B. Nadykto, H. Du , F. Yu, *Vib. Spectrosc.* **44** (2) (2007) 286.
[105] H. Lewandowski, E. Koglin, R.J. Meier, *Vib. Spectrosc.* **39** (2005) 15.
[106] C.-G. Ding, K. Laasonen, A. Laaksonen, *J. Phys. Chem. A* **107** (41) (2003) 8648.

[107] M. Dunn, *et al.*, *J. Phys. Chem. A* **110** (2006) 303.
[108] K. Diri, E.M. Myshakin, K.D. Jordan, *J. Phys. Chem. A* **109** (2005) 4005.
[109] G.M. Chaban, J.O. Jung, R.B. Gerber, *J. Phys. Chem. A* **104** (2000) 4952.
[110] A.B. Nadykto, F. Yu, A. Al Natsheh, Vib. Spectrosc. (2008), under review.
[111] Y. Miller, G.M. Chaban, R.B. Gerber, *J. Phys. Chem. A* **109** (29) (2005) 6565.
[112] L.J. Larson, A. Largent, F.-M. Tao, *J. Phys. Chem. A* **103** (1999) 6786.
[113] A.B. Nadykto, F. Yu, in: K. O'Dowd, P.E. Wagner (Eds.), *Nucleation and Atmospheric Aerosols*, Springer, 2007, p. 297.
[114] A.B. Nadykto, F. Yu, J. Herb, *Chem. Phys.* (2008), under review.
[115] A.B. Nadykto, F. Yu, A. Al Natsheh, *J. Mol. Struct. (THEOCHEM)* (2008), revised.
[116] R. Bianco, S. Wang, J.T. Hynes, *J. Phys. Chem. B* **109** (45) (2005) 21313.
[117] D.R. Hanson, F.L. Eisele, *J. Phys. Chem. A* **104** (2000) 1715.
[118] D.L. Fiacco, S.W. Hunt, K.R. Leopold, *J. Am. Chem. Soc.* **124** (16) (2002) 4504.
[119] A. Scott, L. Radom, *J. Phys. Chem.* **100** (41) (1996) 16502.
[120] F. Arnold, *Ber. Bunsen-Ges. Phys. Chem. Chem. Phys.* **96** (1992) 339.
[121] V.J. Herrero, *et al.*, *Chem. Phys.* **331** (2006) 186.
[122] M.A. Zondlo, P.K. Hudson, A.J. Prenni, M.A. Tolbert, *Annu. Rev. Phys. Chem.* **51** (2000) 473.
[123] F. Arnold, A.A. Viggiano, H. Schlager, *Nature* **297** (1982) 371.
[124] A. Sorokin, F. Arnold, D. Wiedner, *Atmos. Environ.* **40** (11) (2006) 2030.
[125] K.D. Froyd, E.R. Lovejoy, *J. Phys. Chem. A* **107** (2003) 9800.
[126] K.D. Froyd, E.R. Lovejoy, *J. Phys. Chem. A* **107** (2003) 9812.
[127] K.D. Froyd, Ion induced nucleation in the atmosphere: Studies of NH_3, H_2SO_4, and H_2O cluster ions, Ph.D. thesis, Univ. Colorado, Boulder, (2002).
[128] Ya. Kim, Yo. Kim, *Chem. Phys. Lett.* **362** (2002) 419.
[129] H. Merikanto, E. Zapadinsky, A. Lauri, H. Vehkamaki, *Phys. Rev. Lett.* **98** (2007) 145702.
[130] J. Merikanto, H. Vehkamaki, E. Zapadinsky, *J. Chem. Phys.* **121** (2004) 914.

SUBJECT INDEX

1-bromopropane 215
1-butanol 253
1-propanol 252
2-pentanone 261
2-propanol 252

A
α abstraction 218
α-dicarbonyl compounds 266
ab initio 179, 334
ab initio energetics 444
ab initio molecular dynamics 333, 335, 339, 345, 347
ab initio spectrum 63
absorption cross section 75–97, 99, 112, 119, 124, 127
absorption rate constants 357
absorption spectrum 60, 110, 116
abstraction 215
acetaldehyde 257
acetic acid 264
activation barriers 220
activation energy 208
addition to carbon 247
adiabatic 105
aerosol
  aqueous 430
  atmospheric 407
  climate forcing 429, 434
  direct effect 430
  formation 438
  indirect effect 430, 435
    – light interaction 435
  particle(s) 355–359, 369, 373, 381–383
  secondary 430
  size distribution 429
  sulfate aerosols 388
albedo 429
alcohols 252
aldehydes 257
ammonia 424, 441
ammonia in sulfuric acid–water nucleation 416, 418, 421, 422
"amphiphilicity" of the hydrated proton 347
anharmonic coupling 60
anharmonicity 162, 342, 444
anisotropy 314, 315
aqueous surfaces 387
aromatic–OH adduct 299, 305
Arrhenius equation 249
Arrhenius plots 261
asymmetric isotopomers and energy redistribution 11
atom-centered density matrix propagation (ADMP) 336–348, 340, 342, 347, 348
autocorrelation function 347
avoided crossings 105

B
β abstraction 218
B//A approach 251
basis set extrapolation 106
bending mode progression 59
benzene 275–283, 285–290, 292, 294
benzene–chlorine 292
BHandHLYP 251
bicyclic radical formation 306
bicyclic radical reaction 307
Billing, Gert Due 1
binary homogeneous nucleation 451
binary solution 455, 458
biogenic hydrocarbons 177
Birge–Sponer 143, 146
bisulfate 389
Born–Oppenheimer approximation 105
Born–Oppenheimer dynamics (BOMD) 335, 338
BOUND 317, 318
bound complexes 257
BrO 161
bromine 219
bromopropyl alkyl radicals 221
$BrONO_2$ 170
BSSE 251

C
calcium–aluminum-rich inclusions (CAIs) 6, 9
Canonical Variational Theory 249
Car–Parrinello 335
CC2 150, 151
CC3 150, 151
centrifugal barrier 171
centrifugal maxima 171, 172
chaperon and energy transfer mechanism 12, 15
Chebychev polynomial 114
chemical potential 444
chirality 324
Cl radical reaction 230
Cl_2 77, 83–90, 92, 99, 275, 276, 279–286, 289, 290, 292, 294
Classical Nucleation Theory (CNT) 432, 438
classical subsystem 382
climate 430
ClO 162
$ClONO_2$ 165
cloud condensation nuclei 431
cloud drops 432
cluster 312, 389, 455, 457, 458, 461–463, 468
  critical 431
  quantum 457
cluster definition 439
cluster distribution function 443
cluster dynamics 342
cluster free energy 438, 441
clustering reactions 431
CO 16, 17, 58
CO_2 76, 77, 89, 90, 93, 95, 96, 99
coagulation 435
collisional complexes 313
complete active space self consistent field (CASSCF) 104
complete basis limit 162
complex mechanism 259, 262, 265
complex path 261
condensation 435
configuration space 444
configuration state functions 366
configurational parameters 366
contact ion pair 389
continuum absorption 319
core reaction system (CRS) 389
Coriolis coupling 317
correlation energy 106
counterpoise 251
coupled cluster including singles, doubles and perturbative triples (CCSD(T)) 144, 151, 152
coupled cluster singles and doubles (CCSD) 145, 148, 150
coupled-cluster self-consistent reaction field 357
coupled-cluster theory 161
Criegee intermediate 192
CS_2 59
cubic response function 373, 374, 375
curvature 77, 83, 88, 89, 90, 93, 96, 97

D
Davidson correction 104
Density Functional Theory (DFT) 439, 462
derivative couplings 105
desorption rate constants 357
diabatization 105
dielectric medium 357
differential cross section 28
diffusion coefficients 357
dimer 312–314
direct abstraction 261, 262
direct photodissociation 58
discrete variable representation 107
dispersion forces 313
dissociation 387
distributed approximating functional (DAF) 339, 340
dividing surface 442
dynamical bottleneck 442
dynamical effects 342
Dynamical Nucleation Theory (DNT) 440

E
effective core potential (ECP) 105
effective fragment potential (EFP) 389
Ehrenfest equation 371, 376
electron correlation 390
electron–nuclear dynamics 335
electron-transfer process 264
electronic structure calculations 103, 335, 389, 439
electronic transitions 149
electrostatic interaction 363, 365, 366
electrostatic interaction operator 364
energies and molecular properties 358
energy 357, 360
energy functional 358, 369, 382
energy terms 358
enrichment factor 121
ensemble averages 443
enthalpies of reaction 220
entropy change 248
equilibrium constants 443
ethanol 252
evaporation 435

excitation and de-excitation energies 358, 372, 373
excitation energies 381, 382
experimental 44–53, 275, 277–279, 282, 283, 285, 288, 289, 294
extended Lagrangian *ab initio* molecular dynamics 333
extinction coefficient 437

F
first derivative 78, 79
fluorescence 64
formaldehyde 125, 257
formic acid 264
fractionation factor 124
Franck–Condon 112, 119
free energy barrier 432
frequency-dependent first hyperpolarizability tensors 358
frequency-dependent linear and nonlinear polarizabilities 383
frequency-dependent molecular properties 382
frequency-dependent polarizabilities 358
frequency-dependent polarizabilities of excited states 358
frequency-dependent second hyperpolarizabilities 358

G
γ abstraction 218
G3 theory 219
GAMESS 390
gas-to-particle conversion 436
gas-to-particle nucleation 408
General Dynamic Equation (GDE) 437
geometries 208
glory 315, 321
glycolaldehyde 266
growth 435

H
H_2S 59
H_2SO_4, see sulfuric acid
H_2SO_4 photolysis, see sulfuric acid photolysis
H_3O^+ 389
halogens 44, 48, 50, 53
Hamiltonian 108, 358–363, 366, 369, 372, 376
harmonic analyses 335
harmonic approximation 342, 444
harmonic frequencies 347
HCl 115
heat of formation ($\Delta H_{0,f}$) 221
Helmholtz free energy 441, 443
Hermite 114
Hessian 77, 90, 93
heterogeneous dielectric media 355, 358, 373, 374, 376, 382
heterogeneous environments 357
heterogeneous reactions 388
heterogeneous solvation model 357
HNO_3 387, 401, 468
HO_2 345, 346
HOI 164, 167
homogeneous 357
homogeneous dielectric medium 382
HSO_4^- ion 389
hydrated proton 347
hydration of sulfuric acid 416, 424
hydrogen abstraction 220, 237, 247, 258, 264, 267
hydrogen bond 319, 321, 334, 348
hydrogen peroxide 324
hydrogen persulfide 324
hydronium ion 389
hydroperoxyl radical 345
hydroxide water clusters 342
hydroxyacetone 266
hydroxyalkoxy radicals 184
hydroxyalkyl peroxy radicals 184
hydroxyalkyl radicals 184
hydroxyl radical 178
hyperspherical coordinate 109

I
indirect photodissociation 58
industrial solvent 241
infra-red multi-photon dissociation (IRMPD) 345
interaction energies 357, 364, 370, 381, 382
interaction operator 361, 363
interaction, quantum-classical 363
intermediate organic radicals 178
intermolecular interactions 312, 314
interpolation 107
interpolation moving least squares 107
intramolecular bond 266
intramolecular hydrogen bond 229, 267
intramolecular interactions 253, 254, 265
IO 161, 162
ionic 452, 471, 472
$IONO_2$ 165, 168
isodesmic 161
isoprene 177
isoprene nitrate 189
isotope effects 101, 126
isotopes 102
isotopic branching 31
isotopic fractionation 2
isotopologues 102, 110, 119, 120, 124, 126

isotopomers 102, 110, 325
isotropic 316

J
J-value 139, 140, 153
Jacobi coordinates 108

K
ketones 258, 259, 263
kinetic 276–278, 289
kinetic energy operator 108
kinetics 44, 45, 48, 50, 53, 278

L
Lagrange interpolation 341
Lanczos 114
large isotope-specific effects 13
linear 357
linear, homogeneous and isotropic dielectric media 359
linear and nonlinear electromagnetic properties 358
linear and nonlinear molecular properties 376
linear and nonlinear optical properties 357
linear and nonlinear response functions 383
linear response function 372, 374, 377
linear response properties 381, 382
local mode frequency 144, 145
local mode model 141, 142
Lyman-α 150, 153, 155

M
mass accommodation coefficients 356
mass-anomalous fractionation 16
mass-dependent fractionation 58
mass-independent fractionation 6–8, 12, 58
mechanisms 178, 270
mercury 44, 45, 47–50, 52, 53
methanol 252
methylglyoxal 266
Mie scattering 434
model reaction system (MRS) 389
molecular beam scattering 312, 313
molecular clusters 407, 411, 413–415, 424
molecular interactions 431
molecular properties 356–358, 369, 383
molecular simulations 439
molecule-centered primitive basis functions 150
monoterpenes 177
Monte Carlo 443
multiconfigurational self-consistent field (MCSCF) 104, 357–359, 362, 366, 369, 383
multi-reference configuration interaction (MRCI) 104

N
n-bromopropane 215
N_2–N_2 313–315
N_2–O_2 314, 315
N_2O 2, 119
negative activation energy 259
neural network 107, 108
neutral 452, 453, 458, 463, 468
neutralizer 451
new corrections to SAR 255
nitrate ion 389
nitrate radical 178
nitric acid, see HNO_3
non-methane hydrocarbons 177
nonlinear optical molecular properties 357
nonlinear optical properties 357
nonlinear optical signals 357
nuclear quantum effects 334, 339
nuclear spin electron spin coupling 16
nucleation 407, 413, 420, 421, 424, 425, 429, 432, 435
  burst 437
  ion-induced 441
  mechanisms 420
  rate prediction 432
  rates 433, 443
  sensitivity 433
Nucleation Theory 438

O
O_2–O_2 313–315
O_3 2, 22, 58, 76, 77, 80, 89–92, 96, 99
OClO 161
OCS 123
OH 224, 246, 263, 268, 269, 270
one- and two-photon transition matrix elements 372
one-particle density matrix 336
opacity 35, 64
optimized nuclear configurations 335
orbital 366
organic acids 441
oscillator strength 142
overtone induced photolysis 140
oxidation 177
oxygen atom 22, 23
oxygenated VOCs 246, 268, 269, 270
ozone depletion 388
ozone depletion potentials (ODP) 217

P
partial cross section 113

particles 381
partition function 444
photodissociation 101, 115
photodissociation quantum yield 64
photodissociation rate coefficients 65
physics 434
pinenes 177
Podolsky 108
polarizabilities 381, 382
polarization energy 359
polarization interactions 363–365
poles 372
post-reactive complexes 215, 227
potential energy surface 23, 25, 104, 444
pre-reactive complexes 236
precritical cluster 443
predissociation lifetime 60
propagation 113
propanone 259, 261
proton transfer 389

Q
quadratic response 371, 374, 375, 380
quantum 77, 80, 98
  chemical calculations 193
  chemistry 1
  correction 80
  dynamical calculations
    – time dependent 26, 27
  effect 80, 82, 83, 88, 89, 93, 98
  mechanical–classical mechanical 355, 357
  nuclear effects 339
  states 444
  wavepacket *ab initio* molecular dynamics (QWAIMD) 333, 339, 341, 342, 346–348
quasiclassical trajectory 26, 27

R
radiative forcing 430
rate coefficient 248, 258, 263, 269
rate constant 37, 253, 356, 440, 444
reactant complex 247, 248, 254, 257, 260, 262, 265
reaction coordinate 249
reaction cross section 35, 36
reaction energy 208
reaction free energy 391, 392
reaction paths 389
reactive flux 442
recrossing effects 260, 267
reflection principle 76–80, 82, 90, 109, 110, 115
relativistic pseudopotential (effective core potential) 161
reproducing kernel Hilbert space 107
response equations 376, 380, 382
response functions 377
response theory 358, 373
RRKM 10, 171, 173, 181

S
S 451
SAR 255
saddle point 390
scalar relativistic effects 161
second derivative 77, 79
second order 79
second order Møller–Plesset 357
second order Møller–Plesset perturbation theory (CASPT2) 104
second virial coefficient 314, 315
self-shielding 58
sensitivity analysis 443
sesquiterpenes 177
Shepard interpolation 107
single photon action spectroscopy 344, 345
slope 76, 77, 96, 99
Small-Curvature Tunneling 250
SO_2 59, 68, 76, 77, 89, 90, 93–95, 99, 154, 155
SO_2 hot bands 71
SO_3 59
$SO_4^{=}$ 393
solar flux 102
spherical harmonic 315
split operator propagation 113
statistical mechanics 444
sticking coefficient 357
sticking probabilities 356
stratospheric bromine chemistry 217
structured environment 357–360, 370, 371, 373, 380, 382, 383
structured environment methods 358, 383
sulfate ion 393
sulfur cycle 138
sulfuric acid 387, 441, 451–453, 455, 457–460, 462, 463, 468
sulfuric acid hydrates 416
sulfuric acid photolysis 137–139
sulfuric acid–water–ammonia systems 455
sulfuric acid–water nucleation 425
surface environment 389
surface reaction coefficients 357
surface tension 438
symmetry restrictions on intramolecular energy redistribution 10

T
ternary 455, 463
ternary clusters 463
ternary homogeneous nucleation 451

theoretical 44, 45, 49–51, 53, 177, 275, 277, 279, 289, 292, 294
three-photon absorptions 358
three-photon transition matrix elements, and 373
time-dependent deterministic sampling 339, 341
time-dependent methods 111
time-dependent molecular properties 382
time-dependent Schrödinger equation (TDSE) 103, 111, 339
time-independent methods 109
toluene 275–279, 283–286, 289, 291, 292, 294
torsion 324
total electronic free energy 366
total energy 366
trace impurity 440
transition dipole moment 106
transition moments 381–383
transition moments between non-reference states 373
transition state 231, 253, 265
transition state theory 179, 249
transition state (TS), search for 390
transition structures 254
Troe's unimolecular formalism 171
tunneling 249, 260, 262, 264, 268
two-photon absorption between excited states 358
two-photon matrix elements 358

V
van der Waals complexes 317
van der Waals interaction 316, 363, 364
Variational Transition State Theory (VTST) 442
vertical excitation energies 383
vibrational properties 333, 342
vibrational redistribution 344
vibrational zero-point energy 162
vibronic structure 59

W
water 23, 441
water cluster 319, 334, 342, 345, 347
wavelet 341
wavepacket 111, 333, 339, 347
wavepacket propagation 2, 116

Z
zero point energy (ZPE) 391, 392, 444
zero point energy model 109, 110, 115